Praise for *Millimeter Wave Wireless Communications*

"This is a great book on mmWave systems that covers many aspects of the technology targeted for beginners all the way to the advanced users. The authors are some of the most credible scholars I know of who are well respected by the industry. I highly recommend studying this book in detail."

—Ali Sadri, PhD, Sr. Director, Intel Corporation, MCG mmWave Standards and Advanced Technologies

"The most comprehensive book covering all aspects of 60 GHz/mm-Wave communication, from digital bits and signal processing all the way to devices, circuits, and electromagnetic waves. A great reference for engineers and students of mm-Wave communication."

—Ali Niknejad, Berkeley Wireless Research Center (BWRC)

"Due to the huge availability of spectrum in 30-100 GHz bands, millimeter wave communication will be the next frontier in wireless technology. This book is the first in-depth coverage addressing essential aspects of millimeter wave communication including channel characteristics and measurements at millimeter wave bands, antenna technology, circuits, and physical layer and medium access control design. It also has an interesting chapter on 60 GHz unlicensed band wireless standards. I found the book extremely useful and recommend it to researchers and practicing engineers who are keen on shaping the future of wireless communication. Thank you Rappaport, Heath, Daniels, and Murdock for giving us *Millimeter Wave Wireless Communications*."

—Amitabha (Amitava) Ghosh, Head, North America Radio Systems, Nokia

"I highly recommend *Millimeter Wave Wireless Communications* to anyone looking to broaden their knowledge in mmWave communication technology. The authors have introduced the key technologies relevant to the rapidly evolving world of wireless access communications while providing an excellent bibliography for anyone seeking to learn about specific topics in greater depth."

—Bob Cutler, Principal Solutions Architect, Agilent Technologies Inc.

"This timely, ambitious, and well-written book is the first to cover all aspects of millimeter wave wireless communications. The authors' interdisciplinary approach illustrates how the unique characteristics of millimeter wave hardware and signal propagation affect and can be mitigated or exploited in the physical, multiple access, and network layers of the overall system design. The authors are renowned wireless communication experts uniquely qualified to write a comprehensive book on this emerging field, which strikes the perfect balance of breadth and depth. This book is likely to become an immediate classic, as well as required reading for students, researchers, and practitioners."

—Andrea Goldsmith, Stephen Harris Chair Professor, Department of Electrical Engineering, Stanford University

"Mm-wave communications systems promise to alleviate the spectrum crunch and be a major part of future WLAN as well as cellular systems. The authors, leading experts in the field, have admirably succeeded in illuminating all the diverse aspects — ranging from semiconductor technology to wave propagation to MAC layer and standards — that impact the design and deployment. The book is a must-read for anybody working on this important emerging class of systems."

—Professor Andy Molisch, University of Southern California, FIEEE, FAAAS, FIET, MAuAcSc

"This is the first book that addresses the technologies of millimeter wave design needed to implement multi-gigabit communication links. It provides in one place the communication theory background as well as the unique characteristics of millimeter wave communication systems."

—Bob Brodersen, Berkeley Wireless Research Center, Department of Electrical Engineering and Computer Science, University of California, Berkeley

"With the advent of broadly addressing the millimeter wave spectrum from 30 GHz-300 GHz, new groundbreaking advances in communications are to be expected. This book provides a fantastic overview as well as in-depth background material for millimeter wave communications. It is a must-buy to be in the hands of any wireless communications engineer active in advancing technology beyond its current boundaries."

—Gerhard P. Fettweis, cfAED Coordinator, HAEC Coordinator, Vodafone Chair Professor, Technische Universität, Dresden

"This timely monograph is expected to play an influential role in the definition of future generations of wireless systems by formulating a future-proof road-map...."

—Professor Lajos Hanzo, FREng, FIEEE, DSc, Head of Communications, Signal Processing and Control, University of Southampton

Millimeter Wave Wireless Communications

Prentice Hall Communications Engineering and Emerging Technologies Series
Theodore S. Rappaport, Series Editor

Visit **informit.com/communicationengineering**
for a complete list of available publications

Prentice Hall Professional's Communications Engineering and Emerging Technologies Series provides leading-edge learning and information about wireless and other innovative technologies that are revolutionizing communications around the world. This series reveals to readers the minds of leading experts in these fields, often as the technologies are being developed and implemented.

Make sure to connect with us!
informit.com/socialconnect

ALWAYS LEARNING   PEARSON

Millimeter Wave Wireless Communications

Theodore S. Rappaport
Robert W. Heath Jr.
Robert C. Daniels
James N. Murdock

PRENTICE HALL

Upper Saddle River, NJ • Boston • Indianapolis • San Francisco
New York • Toronto • Montreal • London • Munich • Paris • Madrid
Capetown • Sydney • Tokyo • Singapore • Mexico City

This book is not for sale or distribution in the U.S.A. or Canada

Many of the designations used by manufacturers and sellers to distinguish their products are claimed as trademarks. Where those designations appear in this book, and the publisher was aware of a trademark claim, the designations have been printed with initial capital letters or in all capitals.

The authors and publisher have taken care in the preparation of this book, but make no expressed or implied warranty of any kind and assume no responsibility for errors or omissions. No liability is assumed for incidental or consequential damages in connection with or arising out of the use of the information or programs contained herein.

For information about buying this title in bulk quantities, or for special sales opportunities (which may include electronic versions; custom cover designs; and content particular to your business, training goals, marketing focus, or branding interests), please contact our corporate sales department at corpsales@pearsoned.com or (800) 382-3419.

For government sales inquiries, please contact governmentsales@pearsoned.com.

For questions about sales outside the United States, please contact international@pearsoned.com.

Visit us on the Web: informit.com/ph

Library of Congress Cataloging-in-Publication Data
Rappaport, Theodore S., 1960- author.
  Millimeter wave wireless communications / Theodore S. Rappaport, Robert W. Heath, Robert C. Daniels, James N. Murdock.
       pages   cm
  Includes bibliographical references and index.
  ISBN 978-0-13-217228-8 (hardcover : alk. paper)
 1. Millimeter wave communication systems. 2. Wireless communication systems. I. Heath, Robert W. (Robert William), 1931- II. Daniels, Robert C. (Robert Clark), 1980- III. Murdock, James N. (James Nelson), 1986- IV. Title.
  TK5103.4835.R37 2014
  621.384—dc23
                                                                                    2014014143

Copyright © 2015 Pearson Education, Inc.

Cover illustration by Jennifer Rappaport, www.jennyrap.com.

All rights reserved. Printed in the United States of America. This publication is protected by copyright, and permission must be obtained from the publisher prior to any prohibited reproduction, storage in a retrieval system, or transmission in any form or by any means, electronic, mechanical, photocopying, recording, or likewise. To obtain permission to use material from this work, please submit a written request to Pearson Education, Inc., Permissions Department, One Lake Street, Upper Saddle River, New Jersey 07458, or you may fax your request to (201) 236-3290.

ISBN-13: 978-0-13-217228-8
ISBN-10:   0-13-217228-3
Text printed in the United States on recycled paper at Courier in Westford, Massachusetts.
First printing, September 2014

To my wife, Brenda, and our children, Matthew, Natalie, and Jennifer. Their love is a gift from God that I am thankful for every day.

—TSR

To my family, Garima, Pia, and Rohan, for their love and support; to my parents, Bob and Judy Heath, for their encouragement; and to Dr. Mary Bosworth for her passion in the pursuit of higher education.

—RWH

To my parents, Richard and Lynn Daniels, for their steadfast support.

—RCD

To my family, Peter, Bonnie, and Thomas Murdock.

—JNM

Contents

Preface — xvii
Acknowledgments — xxi
About the Authors — xxiii

Part I Prerequisites — 1

1 Introduction — 3
  1.1 The Frontier: Millimeter Wave Wireless — 3
  1.2 A Preview of MmWave Implementation Challenges — 17
  1.3 Emerging Applications of MmWave Communications — 19
    1.3.1 Data Centers — 19
    1.3.2 Replacing Wired Interconnects on Chips — 21
    1.3.3 Information Showers — 22
    1.3.4 The Home and Office of the Future — 23
    1.3.5 Vehicular Applications — 24
    1.3.6 Cellular and Personal Mobile Communications — 25
    1.3.7 Aerospace Applications — 26
  1.4 Contributions of This Textbook — 27
  1.5 Outline of This Textbook — 28
    1.5.1 Illustrations for this Textbook — 30
  1.6 Symbols and Common Definitions — 31
  1.7 Chapter Summary — 32

2 Wireless Communication Background — 33
  2.1 Introduction — 33
  2.2 Complex Baseband Representation — 34

2.3	Digital Modulation		39
	2.3.1	Symbols	40
	2.3.2	Symbol Detection	41
	2.3.3	Binary Phase Shift Keying and Variants	43
	2.3.4	Amplitude Shift Keying and Variants	45
	2.3.5	Quadrature Phase Shift Keying and Variants	45
	2.3.6	Phase Shift Keying	47
	2.3.7	Quadrature Amplitude Modulation	48
2.4	Equalization in the Time Domain		49
	2.4.1	Linear Equalization	51
	2.4.2	Decision Feedback Equalization	53
	2.4.3	Maximum Likelihood Sequence Estimation	55
2.5	Equalization in the Frequency Domain		56
	2.5.1	Single Carrier Frequency Domain Equalization	57
	2.5.2	OFDM Modulation	60
2.6	Error Control Coding		62
	2.6.1	Block Codes for Error Detection	63
	2.6.2	Reed-Solomon Code	64
	2.6.3	Low Density Parity Check Codes	65
	2.6.4	Convolutional Codes	67
	2.6.5	Trellis Coded Modulation	69
	2.6.6	Time Domain Spreading	70
	2.6.7	Unequal Error Protection	71
2.7	Estimation and Synchronization		72
	2.7.1	Structure to Facilitate Communication	73
	2.7.2	Frequency Offset Synchronization	75
	2.7.3	Frame Synchronization	78
	2.7.4	Channel Estimation	79
2.8	Multiple Input Multiple Output (MIMO) Communication		81
	2.8.1	Spatial Multiplexing	82
	2.8.2	Spatial Diversity	84
	2.8.3	Beamforming in MIMO Systems	85
	2.8.4	Hybrid Precoding	87
2.9	Hardware Architectures		88
2.10	System Architecture		91
2.11	Chapter Summary		95

Part II Fundamentals — 97

3 Radio Wave Propagation for MmWave — 99
- 3.1 Introduction — 99
- 3.2 Large-Scale Propagation Channel Effects — 101
  - 3.2.1 Log-Distance Path Loss Models — 105
  - 3.2.2 Atmospheric Effects — 106
  - 3.2.3 Weather Effects on MmWave Propagation — 108
  - 3.2.4 Diffraction — 111
  - 3.2.5 Reflection and Penetration — 111
  - 3.2.6 Scattering and Radar Cross Section Modeling — 117
  - 3.2.7 Influence of Surrounding Objects, Humans, and Foliage — 119
  - 3.2.8 Ray Tracing and Site-Specific Propagation Prediction — 122
- 3.3 Small-Scale Channel Effects — 126
  - 3.3.1 Delay Spread Characteristics — 129
  - 3.3.2 Doppler Effects — 130
- 3.4 Spatial Characterization of Multipath and Beam Combining — 132
  - 3.4.1 Beam-Combining Procedure — 132
  - 3.4.2 Beam-Combining Results — 133
- 3.5 Angle Spread and Multipath Angle of Arrival — 135
- 3.6 Antenna Polarization — 138
- 3.7 Outdoor Channel Models — 139
  - 3.7.1 3GPP-Style Outdoor Propagation Models — 149
  - 3.7.2 Vehicle-to-Vehicle Models — 165
- 3.8 Indoor Channel Models — 166
  - 3.8.1 Ray-Tracing Models for Indoor Channels — 169
  - 3.8.2 Rayleigh, Rician, and Multiwave Fading Models — 169
  - 3.8.3 IEEE 802.15.3c and IEEE 802.11ad Channel Models — 171
  - 3.8.4 IEEE 802.15.3c — 177
  - 3.8.5 IEEE 802.11ad — 178
- 3.9 Chapter Summary — 184

4 Antennas and Arrays for MmWave Applications — 187
- 4.1 Introduction — 187
- 4.2 Fundamentals of On-Chip and In-Package MmWave Antennas — 189
  - 4.2.1 Antenna Fundamentals — 191
  - 4.2.2 Fundamentals of Antenna Arrays — 194

	4.3	The On-Chip Antenna Environment	198
		4.3.1 Complementary Metal Oxide Semiconductor Technology (CMOS)	200
	4.4	In-Package Antennas	209
	4.5	Antenna Topologies for MmWave Communications	211
	4.6	Techniques to Improve Gain of On-Chip Antennas	225
		4.6.1 Integrated Lens Antennas	231
	4.7	Adaptive Antenna Arrays — Implementations for MmWave Communications	235
		4.7.1 Beam Steering for MmWave Adaptive Antenna Arrays	237
		4.7.2 Antenna Array Beamforming Algorithms	242
		4.7.3 Specific Beamforming Algorithms — ESPRIT and MUSIC	249
		4.7.4 Case Studies of Adaptive Arrays for MmWave Communications	251
	4.8	Characterization of On-Chip Antenna Performance	252
		4.8.1 Case Studies of MmWave On-Chip Antenna Characterization	253
		4.8.2 Improving Probe Station Characterizations of On-Chip or In-Package Antennas	255
	4.9	Chapter Summary	257
5	**MmWave RF and Analog Devices and Circuits**		**259**
	5.1	Introduction	259
	5.2	Basic Concepts for MmWave Transistors and Devices	260
	5.3	S-Parameters, Z-Parameters, Y-Parameters, and ABCD-Parameters	263
	5.4	Simulation, Layout, and CMOS Production of MmWave Circuits	267
	5.5	Transistors and Transistor Models	273
	5.6	More Advanced Models for MmWave Transistors	279
		5.6.1 BSIM Model	286
		5.6.2 MmWave Transistor Model Evolution — EKV Model	287
	5.7	Introduction to Transmission Lines and Passives	288
		5.7.1 Transmission Lines	292
		5.7.2 Differential versus Single-Ended Transmission Lines	298
		5.7.3 Inductors	298
		5.7.4 Parasitic Inductances from Bond Wire Packaging	304
		5.7.5 Transformers	304
		5.7.6 Interconnects	308
	5.8	Basic Transistor Configurations	308
		5.8.1 Conjugate Matching	309
		5.8.2 Miller Capacitance	310

	5.8.3 Poles and Feedback	311
	5.8.4 Frequency Tuning	313
5.9	Sensitivity and Link Budget Analysis for MmWave Radios	314
5.10	Important Metrics for Analog MmWave Devices	317
	5.10.1 Non-Linear Intercept Points	317
	5.10.2 Noise Figure and Noise Factor	322
5.11	Analog MmWave Components	323
	5.11.1 Power Amplifiers	323
	5.11.2 Low Noise Amplifiers	334
	5.11.3 Mixers	342
	5.11.4 Voltage-Controlled Oscillators (VCOs)	350
	5.11.5 Phase-Locked Loops	364
5.12	Consumption Factor Theory	370
	5.12.1 Numerical Example of Power-Efficiency Factor	374
	5.12.2 Consumption Factor Definition	376
5.13	Chapter Summary	382

6 Multi-Gbps Digital Baseband Circuits — 383

6.1	Introduction	383
6.2	Review of Sampling and Conversion for ADCs and DACs	384
6.3	Device Mismatches: An Inhibitor to ADCs and DACs	393
6.4	Basic Analog-to-Digital Conversion Circuitry: Comparators	394
	6.4.1 Basic ADC Components: Track-and-Hold Amplifiers	397
6.5	Goals and Challenges in ADC Design	403
	6.5.1 Integral and Differential Non-Linearity	406
6.6	Encoders	407
6.7	Trends and Architectures for MmWave Wireless ADCs	409
	6.7.1 Pipeline ADC	409
	6.7.2 Successive Approximation ADCs	410
	6.7.3 Time-Interleaved ADC	411
	6.7.4 Flash and Folding-Flash ADC	413
	6.7.5 ADC Case Studies	420
6.8	Digital-to-Analog Converters (DACs)	421
	6.8.1 Basic Digital-to-Analog Converter Circuitry: The Current DAC	422
	6.8.2 Case Studies of DAC Circuit Designs	429
6.9	Chapter Summary	431

Part III MmWave Design and Applications — 433

7 MmWave Physical Layer Design and Algorithms — 435
- 7.1 Introduction — 435
- 7.2 Practical Transceivers — 436
  - 7.2.1 Signal Clipping and Quantization — 436
  - 7.2.2 Power Amplifier Non-linearity — 439
  - 7.2.3 Phase Noise — 441
- 7.3 High-Throughput PHYs — 444
  - 7.3.1 Modulation, Coding, and Equalization — 445
  - 7.3.2 A Practical Comparison of OFDM and SC-FDE — 447
  - 7.3.3 Synchronization and Channel Estimation — 459
- 7.4 PHYs for Low Complexity, High Efficiency — 461
  - 7.4.1 Frequency Shift Keying (FSK) — 462
  - 7.4.2 On-Off, Amplitude Shift Keying (OOK, ASK) — 463
  - 7.4.3 Continuous Phase Modulation — 463
- 7.5 Future PHY Considerations — 464
  - 7.5.1 Ultra-Low ADC Resolution — 464
  - 7.5.2 Spatial Multiplexing — 466
- 7.6 Chapter Summary — 469

8 Higher Layer Design Considerations for MmWave — 471
- 8.1 Introduction — 471
- 8.2 Challenges when Networking MmWave Devices — 472
  - 8.2.1 Directional Antennas at the PHY — 472
  - 8.2.2 Device Discovery — 475
  - 8.2.3 Collision Detection and Collision Avoidance — 476
  - 8.2.4 Channel Reliability Due to Human Blockage — 478
  - 8.2.5 Channel Utilization and Spatial Reuse — 480
- 8.3 Beam Adaptation Protocols — 481
  - 8.3.1 Beam Adaptation in IEEE 802.15.3c — 482
  - 8.3.2 Beam Adaptation in IEEE 802.11ad — 483
  - 8.3.3 Beam Adaptation for Backhaul — 484
  - 8.3.4 Beam Adaptation through Channel Estimation — 484
- 8.4 Relaying for Coverage Extension — 487
- 8.5 Support for Multimedia Transmission — 493
- 8.6 Multiband Considerations — 497
- 8.7 Performance of Cellular Networks — 500
- 8.8 Chapter Summary — 504

Contents

9	**MmWave Standardization**		**507**
9.1	Introduction		507
9.2	60 GHz Spectrum Regulation		509
	9.2.1	International Recommendations	509
	9.2.2	Regulations in North America	509
	9.2.3	Regulations in Europe	510
	9.2.4	Regulations in Japan	510
	9.2.5	Regulations in Korea	511
	9.2.6	Regulations in Australia	511
	9.2.7	Regulations in China	511
	9.2.8	Comments	511
9.3	IEEE 802.15.3c		512
	9.3.1	IEEE 802.15.3 MAC	512
	9.3.2	IEEE 802.15.3c MmWave PHY	520
9.4	WirelessHD		550
	9.4.1	Application Focus	550
	9.4.2	WirelessHD Technical Specification	551
	9.4.3	The Next Generation of WirelessHD	554
9.5	ECMA-387		555
	9.5.1	Device Classes in ECMA-387	555
	9.5.2	Channelization in ECMA-387	556
	9.5.3	MAC and PHY Overview for ECMA-387	557
	9.5.4	Type A PHY in ECMA-387	559
	9.5.5	Type B PHY in ECMA-387	560
	9.5.6	Type C PHY in ECMA-387	561
	9.5.7	The Second Edition of ECMA-387	561
9.6	IEEE 802.11ad		562
	9.6.1	IEEE 802.11 Background	562
	9.6.2	Important IEEE 802.11ad MAC Features	564
	9.6.3	Directional Multi-Gigabit PHY Overview for IEEE 802.11ad	572
9.7	WiGig		582
9.8	Chapter Summary		583
Bibliography			**585**
List of Abbreviations			**653**
Index			**657**

Preface

When the cellular telephone revolution began in the 1970s, it was hard to imagine how wireless communication would become such a fundamental part of today's world. Indeed, the Internet had not yet been invented, personal computers did not exist, and long-distance data communication was carried out over landline phones using analog audio modems with data rates no greater than 300 bits per second. The launch of the commercial cellular telephone industry gave birth to unprecedented freedom and functionality, the wireless age was born, and tetherless communications captured the hearts and minds of a new generation of engineers and technologists, and most importantly, the public. As the computer and Internet revolutions sprang forward in the 1990s and into the 21st century, the wireless industry followed. Despite its remarkable growth, however, wireless has failed to reach its full potential.

Wireless technologies are pervasive, with over 5 billion cellphones on planet Earth. Today's fourth-generation (4G) Long Term Evolution (LTE) cellular technology and the IEEE 802.11n wireless Internet standards provide enormous data rates — transfer rates of hundreds of megabits per second between users. The wireless industry is estimated to be a 1 trillion USD global business, approaching the size of the global construction industry. It is remarkable to consider, however, that in one important dimension, wireless is still in its infancy with enormous room to expand.

Consider this astonishing fact: Since the first cellphone call was made 40 years ago, computer clock speeds have increased from less than 1 MHz to today's clock rates of 5 GHz, more than three orders of magnitude. The memory and storage sizes of computers have exploded from a few kilobytes to today's terabyte-sized hard drives, an expansion of seven orders of magnitude. Yet, during their 40-year lifetime, the mobile communications and personal area network industries have been range-bound in operating carrier frequency, stuck between approximately 500 MHz (used by the early analog mobile phone systems and the recently allocated 700 MHz cellular spectrum) and 5.8 GHz (used by modern WiFi enterprise systems on the IEEE 802.11a standard). Remarkably, in 40 years, the wireless industry has seen little movement in its operating frequency. While all other technological evolutions have exploited Moore's law to gain an increase in scale of many orders of magnitude, the operating carrier frequencies of practically all mobile and portable wireless communication systems have barely budged.

Why would wireless be so delinquent in moving up in the frequency bands? Why would the wireless industry wait until now to exploit the vast frontier of spectrum that, to date, has seen little use, yet offers such vast potential for greater capacity? After all, as we show in Chapter 1 of this book, England's Ofcom and the US FCC considered millimeter wave (mmWave) for mobile communications in the 1980s. And more than 100 years ago, Jagadis Chandra Bose and Pyotr Lebedev (also spelled "Lebedew") reportedly conducted the first 60 GHz wireless transmissions. To address these questions, and to offer the fundamental technical details needed by the next generation of engineers who will be fortunate to conquer this new frontier of mmWave wireless, we have written this textbook.

The answers to why consumer wireless networks have not exploited mmWave frequencies have to do with many factors, such as the slow standards process and even slower global spectrum regulatory process; myths and lack of fundamental understanding of radio propagation, antennas, circuits, and networks; the amount of entrenched investments and the competition between traditional (and emerging) business models to provide broadband wireless coverage and capacity; the competing standards that dilute capital needed to bring about a new cellular or WiFi technology; the existing and expanding infrastructure needed to carry such large bandwidths; and, most of all, the cost of designing, fabricating, and deploying widespread consumer technologies that deliver an unmet need, such that they justify the massive investments needed to bring them to the marketplace.

Today, semiconductor technologies are able to make reliable radio frequency circuits with gate lengths at or below 30 nm. The ability to fabricate low-cost radio frequency circuits and on-chip antennas systems at frequencies much higher than 5 GHz is now firmly in place. Optical fiber backbones are now being deployed throughout the world, and mmWave wireless systems will enable a much greater proliferation of backhaul to support data rates that meet or exceed those of 4G cellular systems. The processing power of smartphones and tablets, and the insatiable public demand for content offered through these devices, is also now proven. New wireless local area network (WLAN) products based on IEEE 802.11ad are offering multi-gigabit per second data rates. Recent work has shown that outdoor radio propagation is viable at mmWave frequencies when directional, steerable phased-array antennas are used. Thus, all of the key components are set for wireless to expand from its current low microwave carrier frequency to a new frequency regime that promises several orders of magnitude more capacity for future cellphone and local area network users.

This textbook has evolved from our research programs in many technical areas that must come together to make mmWave mobile communications a reality. We have sought to explore the literature and we have used our own personal experiences as researchers, entrepreneurs, inventors, and consultants to build a textbook that empowers engineers at all levels to work on the exciting future of mmWave wireless. Fundamental principles in many important areas of mmWave communications, circuits, antennas, and propagation are treated in this book. We also provide a wealth of references to assist the reader in exploring specific areas of interest where specific challenges or advances must still be made.

The material in this book is designed to provide a solid foundation in mmWave fundamentals, including communication theory, channel propagation, circuits, and antennas. Chapter 1 provides an introduction and illustrates the vast capabilities and new architectures that will evolve as wireless moves from the UHF and microwave bands to the mmWave spectrum. Conventional applications of mmWave, including WLANs, wireless personal area networks (WPANs), and cellular networks are described, along with new applications of mmWave for applications in the office of the future, data centers, personal interconnects, and the automotive and aerospace industries. Chapter 2 provides an introduction to the fundamentals of digital communication. Important topics including baseband signal and channel models, modulation, equalization, and error control coding are discussed with an emphasis on the techniques that are already found in early mmWave systems. Multiple input multiple output (MIMO) wireless communication principles, which leverage large-antenna arrays, are also reviewed. Background is provided on hardware architectures for upconversion and downconversion and the fundamentals of the network stack.

The treatment of the fundamentals of mmWave starts in Chapter 3, where elemental principles on radio wave propagation for both indoor and outdoor applications of mmWave communication are taught. Radio propagation characteristics for mmWave are reviewed in detail, with special attention paid to the fundamental issues that are pertinent to building mmWave wireless networks. Chapter 4 delves into antennas and antenna arrays for mmWave communication systems. Because mmWave antennas will be very small and integrated, important background is provided on the fundamentals of on-chip and in-package antennas, describing the challenges associated with fabricating efficient mmWave antennas as part of the chip fabrication or packaging. Chapter 5 provides in-depth treatment of analog circuit design and provides fundamental treatment of key design challenges and operating considerations when building RF circuitry. Background is provided on several topics, including analog millimeter wave transistors, their fabrication, and important circuit design approaches for the basic building blocks within a transceiver. The treatment of analog circuits has intentionally included details of complementary metal oxide semiconductor (CMOS) and metal oxide semiconductor field-effect transistors (MOSFET) semiconductor theory in sufficient detail to allow communications engineers to appreciate and understand the fundamentals and challenges of mmWave analog circuits, and to understand the capabilities and approaches used to create circuits that will enable the mmWave revolution. Key parameters used to characterize active and passive analog components as well as key qualities of merit are also reviewed with great detail. Design approaches for transmission lines, amplifiers (both power amplifiers and low-noise amplifiers) frequency synthesizers, voltage-controlled oscillators (VCOs), and frequency dividers are also reviewed. Chapter 6 delves into baseband circuit design. It presents a wide range of technical design issues and references to help the reader understand the fundamentals of designing multi-gigabit per second high-fidelity digital-to-analog (DAC) and analog-to-digital (ADC) converters, and the challenges with reaching such high bandwidths.

The book concludes with a detailed treatment of mmWave design and applications. Chapter 7 presents physical layer aspects of the design of a mmWave communication system. The emphasis is on the physical layer algorithmic choices and design considerations for 60 GHz communication systems. Important concepts include practical impairments such as clipping, quantization, nonlinearity, phase noise, and emerging physical layer design concepts such as spatial multiplexing. The choice of modulation and equalization is discussed in terms of tradeoffs that can be made between complexity and throughput. Chapter 8 reviews higher-layer (above the physical layer) design issues for mmWave systems, with a particular emphasis on techniques relevant to 60 GHz systems, but also speculates on the impact in mmWave cellular and backhaul systems. Chapter 8 provides background on the challenge associated with networking mmWave devices. Select topics are treated, including beam adaptation protocols, relaying, multimedia transmission, and multiband considerations. Finally, Chapter 9 concludes the technical content with a review of design elements from the standardization efforts for 60 GHz wireless communication systems. Several standards are reviewed, including IEEE 802.15.3c for WPAN, Wireless HD, ECMA-387, IEEE 802.11ad, and the Wireless Gigabit Alliance (WiGig) standard. The key features of each standard are presented along with important details about their physical layer and medium access control design choices.

Just as the cellphone has morphed from a voice-only analog communications device into today's impressive smartphone, the frontier of mmWave wireless communications is sure to usher in even more astounding capabilities and is certain to spawn new businesses, new consumer use cases, and complete transformations of how we live and work. The vast mmWave spectrum, and the new technologies that will conquer it, will bring wireless into its renaissance where it pervades all aspects of our lives. Our hope is that this textbook offers some assistance for the engineering explorers who are called to create this exciting future.

Acknowledgments

A book of this nature is not a small undertaking, and we have been fortunate to benefit from many colleagues who have helped provide insights, materials, and suggestions in the preparation of this book over the past several years. Leaders from industry who have provided us with valuable and detailed suggestions and who have helped inspire us to write this book include Wonil Roh and Jerry Z. Pi of Samsung, Amitava Ghosh of Nokia, Ali Sadri of Intel Corp., and Bob Cutler of Agilent Technologies. This text has benefited greatly from their thorough reviews, sponsorship, and inspiration, and we are grateful for their involvement. Graduate and undergraduate students at New York University (NYU) and The University of Texas at Austin (UT), as well as many industry practitioners, leading academics from many areas, and our corporate sponsors, have contributed immeasurably to this textbook, and we wish to express our sincere gratitude to them here. Colleagues Bob Brodersen and Ali Niknejad (University of California, Berkeley), Lajos Hanzo (University of Southampton), Andrea Goldsmith (Stanford University), Gerhard Fettweis (TU Dresden), and Andreas Molisch (University of Southern California) all have offered important suggestions that have made this book better. NYU graduate students Krupa Vijay Panchal, Nikhil Suresh Patne, George Robert MacCartney, Shuai Nie, Shu Sun, Abhi Shah (who helped collect inputs from the NYU students), Mathew Samimi, Sijia Ding, and Junhong Zhang offered critical suggestions to make the material accessible and better organized. NYU undergraduate students George Wong, Yaniv Azar, Jocelyn Shulz, Kevin Wang, and Hang (Celine) Zhao also provided proofreading and editing and were involved in the pioneering measurements covered in Chapter 3. UT graduate students Andrew Thornburg, Tianyang Bai, and Ahmed Alkhateeb provided proofreading, editing, and help with the figures. Vutha Vu provided special help with proofreading as well as tracking and implementing many critical edits. Professors Hao Ling, Andrea Alu, Ranjit Gharpurey, and Eric Swanson should also be mentioned for providing critical insights into antenna and circuit design.

Once the manuscript was complete, the Prentice Hall team worked tirelessly to turn this manuscript into a book. Our editor, Bernard Goodwin, commissioned the work and ensured that a seamless and professional typesetting and editing team would be available through production. Julie Nahil, Pearson's Full-Service Production Manager, brought in the stellar production team—copy editor Stephanie Geels and project manager Vicki Rowland. Stephanie's expert editing and insightful questions pushed us to clarify and refine the manuscript, while Vicki deftly kept track of myriad loose ends and seemed to be available at whatever odd hour if a question or issue popped up. Additionally, proofreaders Linda Begley and Archie Brodsky's eagle eyes caught numerous little glitches that somehow slipped by everyone else. We are most grateful to this superb and dedicated Pearson/Prentice Hall team for their commitment and dedication to this project.

About the Authors

Theodore (Ted) S. Rappaport is the David Lee/Ernst Weber Professor of Electrical and Computer Engineering at New York University's Polytechnic School of Engineering. He also holds professorships in NYU's Courant Institute and Medical School. During his career, he founded three major academic wireless research centers at Virginia Tech, The University of Texas at Austin, and New York University, and he founded and sold two wireless technology companies to publicly traded firms. His technical expertise encompasses antennas, propagation, systems, and circuits for wireless communications, and he has over 100 US or international patents issued or pending. His other textbooks include *Wireless Communications: Principles and Practice* (2002), which has been translated into seven other languages, *Smart Antennas for Wireless Communications* (1999), and *Principles of Communication Systems Simulation with Wireless Applications* (2004). When he is not working Ted enjoys jogging and amateur (ham) radio (his call sign is N9NB).

Robert W. Heath Jr., is a Cullen Trust Endowed Professor in the Department of Electrical and Computer Engineering at The University of Texas at Austin. He is founder of MIMO Wireless Inc. and cofounder of Kuma Signals, LLC. His research includes many aspects of multiple antenna communication: limited feedback techniques, multiuser, multihop, multicell networking, and millimeter wave, as well as other areas of signal processing, including adaptive video transmission, manifold signal processing, and applications of machine learning to wireless communication. He has published more than 300 refereed conference or journal publications and has nearly 50 US patents issued. His other textbooks include *Digital Wireless Communication: Physical Layer Exploration Lab Using the NI USRP* (2012) and several others forthcoming. Robert also enjoys cycling, windsurfing, scuba diving, and sailing. His amateur radio call sign is KE5NCG.

Robert (Bob) C. Daniels is the cofounder and CTO at Kuma Signals, LLC, in Austin, Texas. At Kuma, he serves as the principal investigator on wireless communications research and development contracts, and he guides the technical direction of internal development. Bob is an expert in the theory and practical application of digital wireless communications, especially for the design of multiuser, large bandwidth, multiple-antenna wireless communication networks. He has published dozens of peer-reviewed IEEE publications and has also been the sole architect of several RF-proven physical layer implementations through general purpose and special purpose (i.e., field-programmable graphic array) hardware for various standards, including IEEE 802.11n and the Wideband Networking Waveform (WNW). He holds B.S. degrees in Electrical Engineering and Mathematics from The Pennsylvania State University, and M.S.E. and Ph.D. degrees in Electrical and Computer Engineering from The University of Texas at Austin.

James N. Murdock is an RF and analog engineer at Texas Instruments (TI), where he focuses on low-power and mmWave frequency circuits. Prior to joining TI, James studied for his B.S. and M.S. degrees in Electrical Engineering at The University of Texas at Austin under the mentorship of Dr. Rappaport. James's master's thesis covered low-power techniques for RF systems and applications of these techniques to large-scale communication systems, in addition to mmWave channel modeling. James has cowritten more than ten conference papers in topics ranging from low-power timing circuits (ISSCC 2014) to on-chip mmWave antennas (MTT-S 2011) to mmWave channel modeling (WCNC 2012). James has cowritten three journal papers.

Part I

Prerequisites

Chapter 1

Introduction

1.1 The Frontier: Millimeter Wave Wireless

Emerging millimeter wave (mmWave) wireless communication systems represent more than a century of evolution in modern communications. Since the early 1900s, when Guglielmo Marconi developed and commercialized the first wireless telegraph communication systems, the wireless industry has expanded from point-to-point technologies, to radio broadcast systems, and finally to wireless networks. As the technology has advanced, wireless communication has become pervasive in our world. Modern society finds itself immersed in wireless networking, as most of us routinely use cellular networks, wireless local area networks, and personal area networks, all which have been developed extensively over the past twenty years. The remarkable popularity of these technologies causes device makers, infrastructure developers, and manufacturers to continually seek greater radio spectrum for more advanced product offerings.

Wireless communication is a transformative medium that allows our work, education, and entertainment to be transported without any physical connection. The capabilities of wireless communications continue to drive human productivity and innovation in many areas. Communication at mmWave operating frequencies represents the most recent game-changing development for wireless systems. Interest in mmWave is in its infancy and will be driven by consumers who continue to desire higher data rates for the consumption of media while demanding lower delays and constant connectivity on wireless devices. At mmWaves, available spectrum is unparalleled compared to cellular and wireless local area network (WLAN) microwave systems that operate at frequencies below 10 GHz. In particular, the unlicensed spectrum at 60 GHz offers 10× to 100× more spectrum than is available for conventional unlicensed wireless local area networks in the Industrial, Scientific, and Medical (ISM) bands (e.g., at 900 MHz, 2.4 GHz, 5 GHz) or for users of WiFi and 4G (or older) cellular systems that operate at carrier frequencies below 6 GHz. To reinforce this perspective, Fig. 1.1 shows the magnitude of spectrum resources at 28 GHz (Local Multipoint Distribution Service [LMDS]) and 60 GHz in comparison to other modern wireless systems. Over 20 GHz of spectrum is waiting to be used for cellular or WLAN traffic in the 28, 38, and 72 GHz bands alone, and hundreds of gigahertz more spectrum could be used at frequencies above 100 GHz. This is a staggering amount of

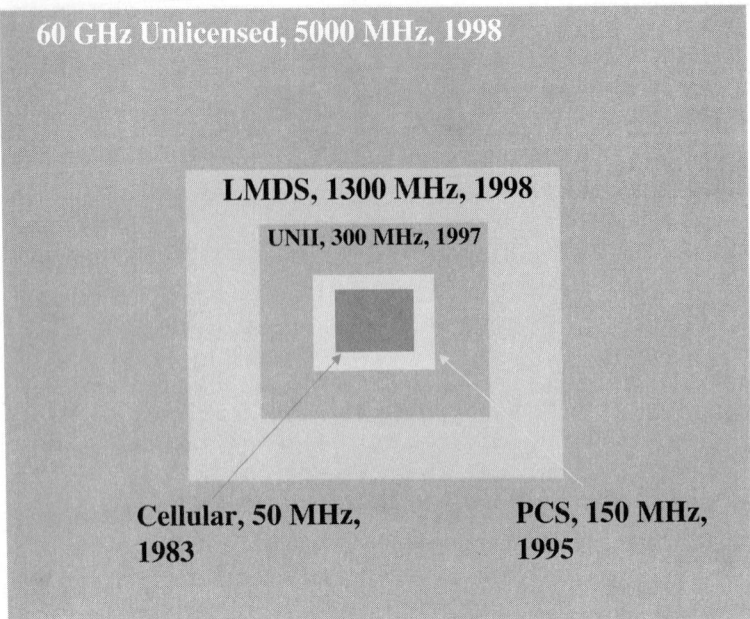

Figure 1.1 Areas of the squares illustrate the available licensed and unlicensed spectrum bandwidths in popular UHF, microwave, 28 GHz LMDS, and 60 GHz mmWave bands in the USA. Other countries around the world have similar spectrum allocations [from [Rap02]].

available new spectrum, especially when one considers that all of the world's cellphones currently operate in less than 1 GHz of allocated spectrum. More spectrum makes it possible to achieve higher data rates for comparable modulation techniques while also providing more resources to be shared among multiple users.

Research in mmWave has a rich and exciting history. According to [Mil],

> In 1895, Jagadish Chandra Bose first demonstrated in Presidency College, Calcutta, India, transmission and reception of electromagnetic waves at 60 GHz, over 23 meters distance, through two intervening walls by remotely ringing a bell and detonating some gunpowder. For his communication system, Bose pioneered the development of entire millimeter-wave components like: spark transmitter, coherer, dielectric lens, polarizer, horn antenna and cylindrical diffraction grating. This is the first millimeter wave communication system in the world, developed more than 100 years ago.

A pioneering Russian physicist, Pyotr N. Lebedew, also studied transmission and propagation of 4 to 6 mm wavelength radio waves in 1895 [Leb95].

Today's radio spectrum has become congested due to the widespread use of smartphones and tablets. Fig. 1.1 shows the relative bandwidth allocations of different spectrum bands in the USA, and Fig. 1.2 shows the spectrum allocations from 30 kHz to 300 GHz according to the Federal Communications Commission (FCC). Note that although Figs. 1.1 and 1.2 represent a particular country (i.e., the USA), other countries around the world have remarkably similar spectrum allocations stemming from the global allocation of spectrum by the World Radiocommunication Conference (WRC) under the auspices of the International Telecommunication Union (ITU). Today's cellular and

1.1 The Frontier: Millimeter Wave Wireless

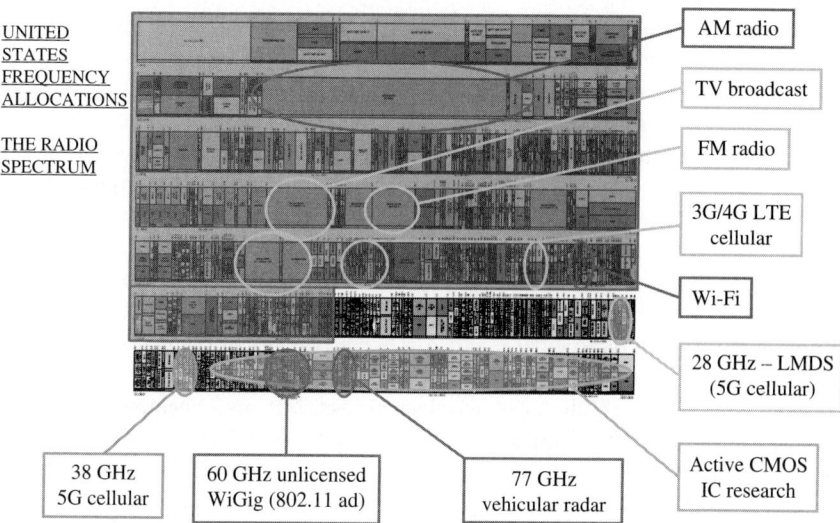

Figure 1.2 Wireless spectrum used by commercial systems in the USA. Each row represents a decade in frequency. For example, today's 3G and 4G cellular and WiFi carrier frequencies are mostly in between 300 MHz and 3000 MHz, located on the fifth row. Other countries around the world have similar spectrum allocations. Note how the bandwidth of all modern wireless systems (through the first 6 rows) easily fits into the unlicensed 60 GHz band on the bottom row [from [Rap12b] U.S. Dept. of Commerce, NTIA Office of Spectrum Management]. See page C1 (immediately following page 8) for a color version of this figure.

personal communication systems (PCS) mostly operate in the UHF ranges from 300 MHz to 3 GHz, and today's global unlicensed WLAN and wireless personal area network (WPAN) products use the Unlicensed National Information Infrastructure (U-NII) bands of 900 MHz, 2.4 MHz and 5.8 MHz in the low microwave bands. The wireless spectrum right now is already allocated for many different uses and very congested at frequencies below 3 GHz (e.g., UHF and below). AM Radio broadcasting, international shortwave broadcasting, military and ship-to-shore communications, and amateur (ham) radio are just some of the services that use the lower end of the spectrum, from the hundreds of kilohertz to the tens of megahertz (e.g., medium-wave and shortwave bands). Television broadcasting is done from the tens of megahertz to the hundreds of megahertz (e.g., VHF and UHF bands). Current cellphones and wireless devices such as tablets and laptops works at carrier frequencies between 700 MHz and 6 GHz, with channel bandwidths of 5 to 100 MHz. The mmWave spectrum, ranging between 30 and 300 GHz, is occupied by military, radar, and backhaul, but has much lower utilization. In fact, most countries have not even begun to regulate or allocate the spectrum above 100 GHz, as wireless technology at these frequencies has not been commercially viable at reasonable cost points. This is all about to change. Given the large amount of spectrum available, mmWave presents a new opportunity for future mobile communications to use channel bandwidths of 1 GHz or more. Spectrum at 28 GHz, 38 GHz, and 70-80 GHz looks especially promising for next-generation cellular systems. It is amazing to note from Fig. 1.2 that the unlicensed band at 60 GHz contains more spectrum than has been used by every satellite, cellular, WiFi, AM Radio, FM Radio, and television station in the world! This illustrates the massive bandwidths available at mmWave frequencies.

MmWave wireless communication is an enabling technology that has myriad applications to existing and emerging wireless networking deployments. As of the writing of this book, mmWave based on the 60 GHz unlicensed band is seeing active commercial deployment in consumer devices through IEEE 802.11ad [IEE12]. The cellular industry is just beginning to realize the potential of much greater bandwidths for mobile users in the mmWave bands [Gro13][RSM+13]. Many of the design examples in this book draw from the experience in 60 GHz systems and the authors' early works on mmWave cellular and peer-to-peer studies for the 28 GHz, 38 GHz, 60 GHz, and 72 GHz bands. But 60 GHz WLAN, WPAN, backhaul, and mmWave cellular are only the beginning — these are early versions of the next generation of mmWave and terahertz systems that will support even higher bandwidths and further advances in connectivity.

New 60 GHz wireless products are exciting, not only because of their ability to satisfy consumer demand for high-speed wireless access, but also because 60 GHz products may be deployed worldwide, thanks to harmonious global spectrum regulations. Harmonized global spectrum allocations allow manufacturers to develop worldwide markets, as demonstrated through the widespread adoption and commercial success of IEEE 802.11b WLANs in 1999, and more recent innovations such as IEEE 802.11a, IEEE 802.11g, IEEE 802.11n, and IEEE 802.11ac WLANs that all operate in the same globally allocated spectrum. WLAN succeeded because there was universal international agreement for the use of the 2.4 GHz ISM and 5 GHz Unlicensed National Information Infrastructure bands, which allowed major manufacturers to devote significant resources to create products that could be sold and used globally. Without international spectral agreements, new wireless technologies will founder for lack of a global market. This was demonstrated by early incarnations of Ultra-wide band (UWB) at the turn of the century, whose initial hype dramatically waned in the face of nonuniform worldwide spectral interference regulations. Fortunately, the governments of the USA, Europe, Korea, Japan, and Australia have largely followed the recommendations of the ITU, which designate frequencies between 57 and 66 GHz for unlicensed communications applications [ITU]. In the USA, the FCC has designated bands from 57 to 64 GHz for unlicensed use [Fed06]. In Europe, the European CEPT has allocated bands from 59 to 66 GHz for some form of mobile application [Tan06]. Korea and Japan have designated bands from 57 to 66 GHz and 59 to 66 GHz, respectively [DMRH10]. Australia has dedicated a smaller band from 59.3 to 62.9 GHz. Consequently, there is roughly 7 GHz of spectrum available worldwide for 60 GHz devices.

At the time of this writing, the cellular industry is just beginning to explore similar spectrum harmonization for the use of mobile cellular networks in frequency bands that are in the mmWave spectrum.[1] Dubbed "Beyond 4G" or "5G" by the industry, new cellular network concepts that use orders of magnitude more channel bandwidth, for simultaneous mobility coverage as well as wireless backhaul, are just now being introduced to governments and the ITU to create new global spectrum bands at carrier frequencies that are at least an order of magnitude greater than today's fourth-generation (4G) Long

---

1. Although the mmWave band is formally defined as the spectrum between 30 and 300 GHz, the industry has loosely considered the term *mmWave* to denote all frequencies between 10 and 100 GHz. The term "sub-terahertz" has been used loosely to define frequencies above 100 GHz but lower than 300 GHz.

Term Evolution (LTE) and WiMax mobile networks. Thus, just as the WLAN unlicensed products have moved from the carrier frequencies of 1 to 5 GHz in their early generations, now to 60 GHz, the 1 trillion USD cellular industry is about to follow this trend: moving to mmWave frequency bands where massive data rates and new capabilities will be supported by an immense increase in spectrum.

Unlicensed spectrum at 60 GHz is readily available throughout the world, although this was not always the case. The FCC initiated the first major regulation of 60 GHz spectrum for commercial consumers through an unlicensed use proposal in 1995 [Mar10a], yet the same idea was considered a decade earlier by England's Office of Communications (OfCom) [RMGJ11]. At that time, the FCC considered the mmWave band to be "desert property" due to its perceived unfavorable propagation characteristics and lack of low-cost commercial circuitry. However, the allocation of new spectrum has ignited and will continue to ignite the inventiveness and creativity of engineers to create new consumer products at higher frequencies and greater data rates. This perception of poor propagation due to low distance coverage is heavily influenced by the O_2 absorption effect where a 60 GHz carrier wave interacts strongly with atmospheric oxygen during propagation, as illustrated in Fig. 1.3 [RMGJ11][Wel09]. This effect is compounded by other perceived unfavorable qualities of mmWave communication links: increased free space path loss, decreased signal penetration through obstacles, directional communication due to high-gain antenna requirements, and substantial intersymbol interference (ISI, i.e., frequency

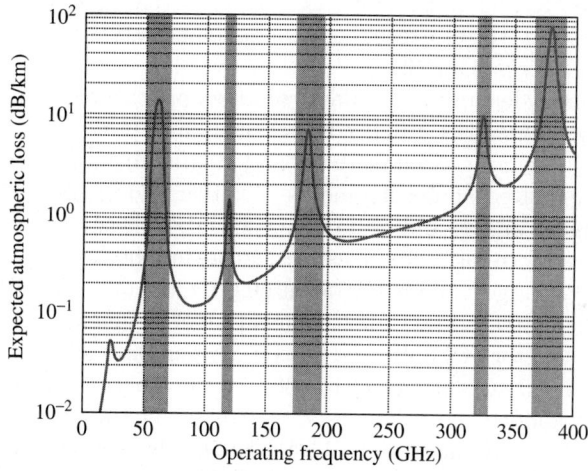

Figure 1.3 Expected atmospheric path loss as a function of frequency under normal atmospheric conditions (101 kPa total air pressure, 22° Celsius air temperature, 10% relative humidity, and 0 g/m^3 suspended water droplet concentration) [Lie89]. Note that atmospheric oxygen interacts strongly with electromagnetic waves at 60 GHz. Other carrier frequencies, in dark shading, exhibit strong attenuation peaks due to atmospheric interactions, making them suitable for future short-range applications or "whisper radio" applications where transmissions die out quickly with distance. These bands may service applications similar to 60 GHz with even higher bandwidth, illustrating the future of short-range wireless technologies. It is worth noting, however, that other frequency bands, such as the 20-50 GHz, 70-90 GHz, and 120-160 GHz bands, have very little attenuation, well below 1 dB/km, making them suitable for longer-distance mobile or backhaul communications.

selectivity) due to many reflective paths over massive operating bandwidths. Furthermore, 60 GHz circuitry and devices have traditionally been very expensive to build, and only in the past few years have circuit solutions become viable in low-cost silicon.

In the early days of 60 GHz wireless communication, many viewed fixed wireless broadband (e.g., fiber backhaul replacement) as the most suitable 60 GHz application, due to requirements for highly directional antennas to achieve acceptable link budgets. Today, however, the propagation characteristics that were once seen as limitations are now either surmountable or seen as advantages. For example, 60 GHz oxygen absorption loss of up to 20 dB/km is almost negligible for networks that operate within 100 meters. The shift away from long-range communications actually benefits close-range communications because it permits aggressive frequency reuse with simultaneously operating networks that do not interfere with each other. Further, the highly directional antennas required for path loss mitigation can actually work to promote security as long as network protocols enable antenna directions to be flexibly steered. Thus, many networks are now finding a home at 60 GHz for communication at distances less than 100 m. Also, the 20 dB/km oxygen attenuation at 60 GHz disappears at other mmWave bands, such as 28, 38, or 72 GHz, making them nearly as good as today's cellular bands for longer-range outdoor mobile communications. Recent work has found that urban environments provide rich multipath, especially reflected and scattered energy at or above 28 GHz — when smart antennas, beamforming, and spatial processing are used, this rich multipath can be exploited to increase received signal power in non-line of sight (NLOS) propagation environments. Recent results by Samsung show that over 1 Gbps can be carried over mmWave cellular at ranges exceeding 2 km, demonstrating that mmWave bands are useful for cellular networks [Gro13].

Although consumer demand and transformative applications fuel the need for more bandwidth in wireless networks, rapid advancements and price reductions in integrated mmWave (>10 GHz) analog circuits, baseband digital memory, and processors have enabled this progress. Recent developments of integrated mmWave transmitters and receivers with advanced analog and radio frequency (RF) circuitry (see Fig. 1.4) and new phased array and beamforming techniques are also paving the way for the mmWave future (such as the product in Fig. 1.5). Operation at 60 GHz and other mmWave frequencies at reasonable costs is largely the result of a continuation of advancements in complementary metal oxide semiconductor (CMOS) and silicon germanium (SiGe technologies). Signal generation into terahertz frequencies (1 to 430 THz) has been possible since at least the 1960s through photodiodes and other discrete components not amenable to small-scale integration and/or mass production [BS66]. Packaging the analog components needed to generate mmWave RF signals along with the digital hardware necessary to process massive bandwidths, however, has only been possible in the last decade. Moore's Law, which has accurately predicted that integrated circuit (IC) transistor populations and computations per unit energy will double at regular intervals every two years [NH08, Chapter 1], explains the dramatic advancements that now allow 60 GHz and other mmWave devices to be made inexpensively. Today, transistors made with CMOS and SiGe are fast enough to operate into the range of hundreds of gigahertz [YCP+09], as shown in Fig. 1.6. Further, due to the immense number of transistors required for modern digital circuits (on the order of billions) each transistor is extremely cheap. Inexpensive circuit production processes will make system-on-chip (SoC) mmWave radios — a complete integration of

1.1 The Frontier: Millimeter Wave Wireless

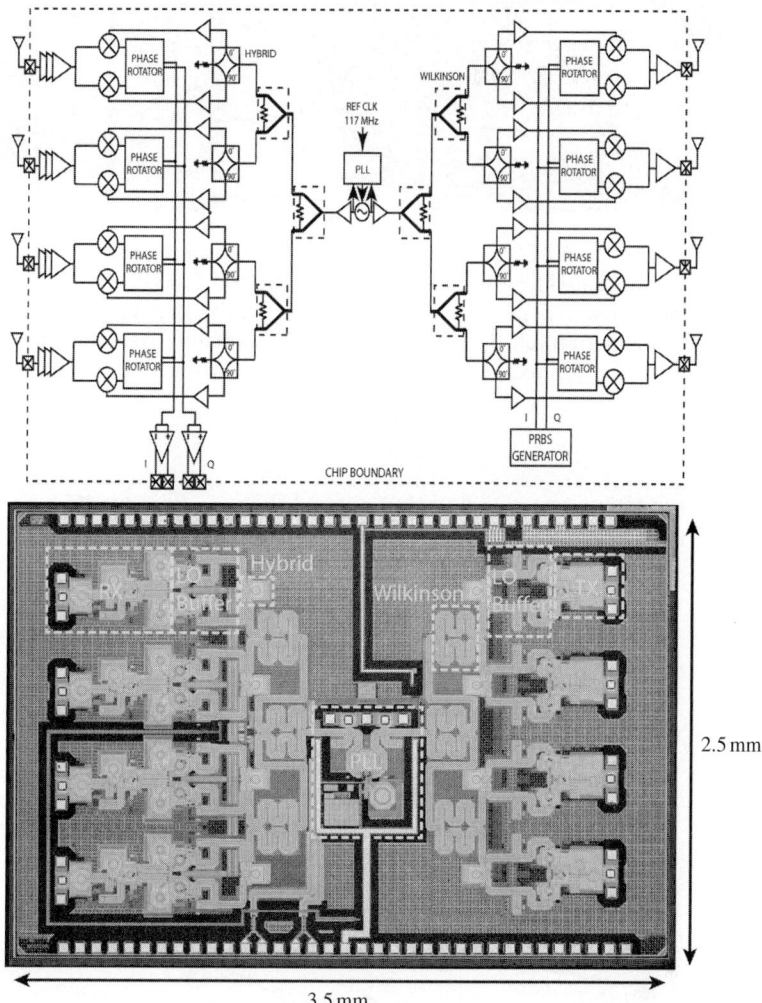

Figure 1.4 Block diagram (top) and die photo (bottom) of an integrated circuit with four transmit and receive channels, including the voltage-controlled oscillator, phase-locked loop, and local oscillator distribution network. Beamforming is performed in analog at baseband. Each receiver channel contains a low noise amplifier, inphase/quadrature mixer, and baseband phase rotator. The transmit channel also contains a baseband phase rotator, up-conversion mixers, and power amplifiers. Figure from [TCM+11], courtesy of Prof. Niknejad and Prof. Alon of the Berkeley Wireless Research Center [© IEEE].

all analog and digital radio components onto a single chip — possible. For mmWave communication, the semiconductor industry is finally ready to produce cost-effective, mass-market products.

Wireless personal area networks (WPANs) provided the first mass-market commercial applications of short-range mmWave using the 60 GHz band. The three dominant 60 GHz WPAN specifications are WirelessHD, IEEE 802.11ad (WiGig), and IEEE 802.15.3c.

Figure 1.5 Third-generation 60 GHz WirelessHD chipset by Silicon Image, including the Sil6320 HRTX Network Processor, Sil6321 HRRX Network Processor, and Sil6310 HRTR RF Transceiver. These chipsets are used in real-time, low-latency applications such as gaming and video, and provide 3.8 Gbps data rates using a steerable 32 element phased array antenna system (courtesy of Silicon Image) [EWA+11] [© IEEE]. See page C2 (immediately preceding page 9) for a color version of this figure.

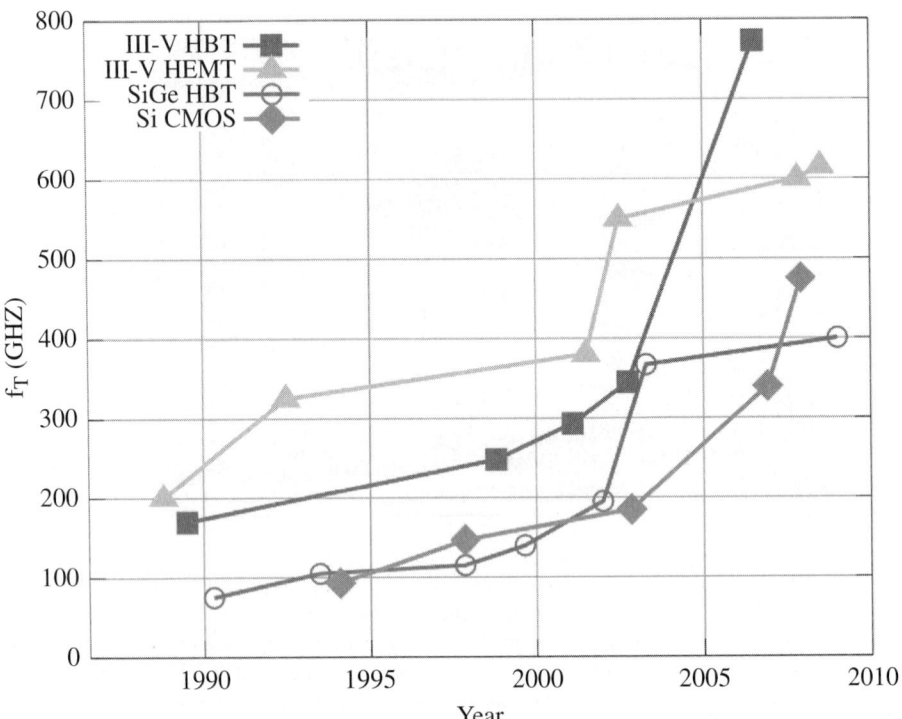

Figure 1.6 Achievable transit frequency (f_T) of transistors over time for several semiconductor technologies, including silicon CMOS transistors, silicon germanium heterojunction bipolar transistor (SiGe HBT), and certain other III-V high electron mobility transistors (HEMT) and III-V HBTs. Over the last decade CMOS (the current technology of choice for cutting edge digital and analog circuits) has become competitive with III-V technologies for RF and mmWave applications [figure reproduced from data in [RK09] © IEEE].

1.1 The Frontier: Millimeter Wave Wireless

WPANs support connectivity for mobile and peripheral devices; a typical WPAN realization is demonstrated in Fig. 1.7, where products such as those shown in Fig. 1.5 may be used. Currently, the most popular application of WPAN is to provide high-bandwidth connections for cable replacement using the high-definition multimedia interface (HDMI), now proliferating in consumer households. The increasing integration of 60 GHz silicon devices allows implementation on small physical platforms while the massive spectrum allocations at 60 GHz allow media streaming to avoid data compression limitations, which are common at lower frequencies with reduced bandwidth resources. Easing compression requirements is attractive because it reduces signal processing and coding circuitry requirements, thereby reducing the digital complexity of a device. This may lead to lower cost and longer battery life in a smaller form factor. Due to major technical and marketing efforts by the Wireless Gigabit Alliance (WiGig), the IEEE 802.11ad standard has been designed to incorporate both WPAN and WLAN capabilities, and WiGig-compliant devices are just starting to ship in laptops, tablets, and smartphones around the world, whereas WirelessHD-compliant devices have been shipping since 2008. The success of today's USB standard in consumer electronics has demonstrated how harmonious interfaces lead to a proliferation of compatible devices. 60 GHz is poised to fill this role for high-definition multimedia systems, as illustrated in Fig. 1.8.

WLANs, which extend the communication range beyond WPAN, also employ mmWave technology in the 60 GHz band. WLANs are used to network computers through a wireless access point, as illustrated in Fig. 1.9, and may connect with other wired networks or to the Internet. WLANs are a popular application of unlicensed spectrum that is being incorporated more broadly into smartphones, tablets, consumer devices, and cars. Currently, most WLAN devices operate under the IEEE 802.11n standard and have

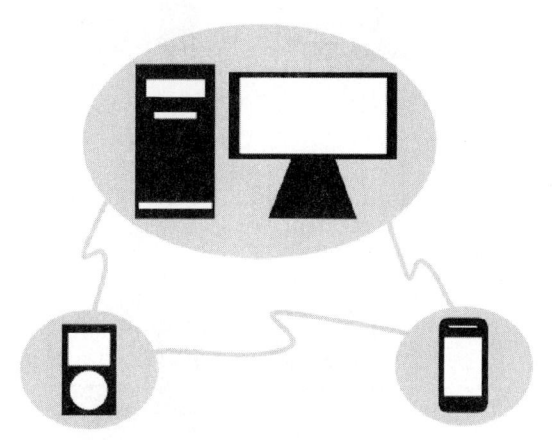

Figure 1.7 Wireless personal area networking. WPANs often connect mobile devices such as mobile phones and multimedia players to each other as well as desktop computers. Increasing the data-rate beyond current WPANs such as Bluetooth and early UWB was the first driving force for 60 GHz solutions. The IEEE 802.15.3c international standard, the WiGig standard (IEEE 802.11ad), and the earlier WirelessHD standard, released in the 2008–2009 time frame, provide a design for short-range data networks (≈ 10 m). All standards, in their first release, guaranteed to provide (under favorable propagation scenarios) multi-Gbps wireless data transfers to support cable replacement of USB, IEEE 1394, and gigabit Ethernet.

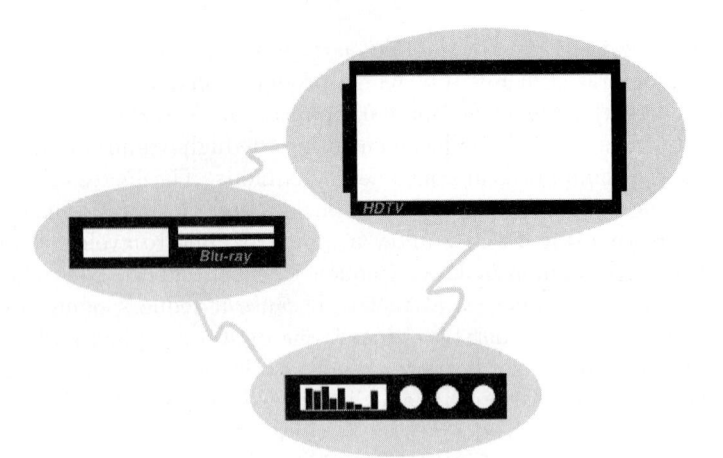

Figure 1.8 Multimedia high-definition (HD) streaming. 60 GHz provides enough spectrum resources to remove HDMI cables without sophisticated joint channel/source coding strategies (e.g., compression), such as in the wireless home digital interface (WHDI) standard that operates at 5 GHz frequencies. Currently, 60 GHz is the only spectrum with sufficient bandwidth to provide a wireless HDMI solution that scales with future HD television technology advancement.

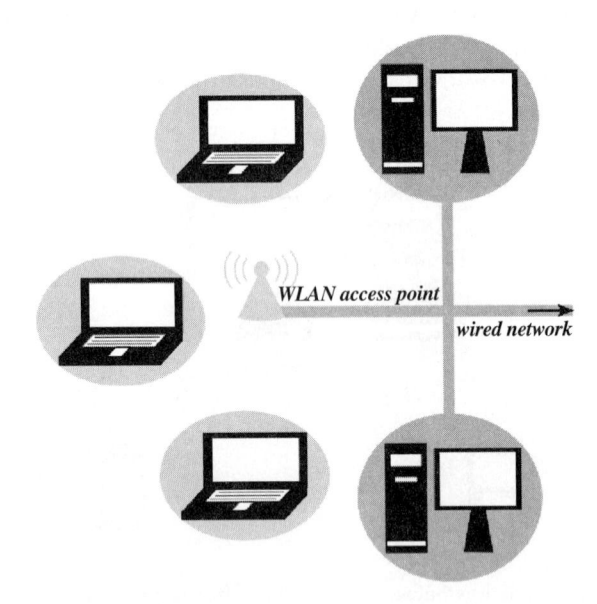

Figure 1.9 Wireless local area networking. WLANs, which typically carry Internet traffic, are a popular application of unlicensed spectrum. WLANs that employ 60 GHz and other mmWave technology provide data rates that are commensurate with gigabit Ethernet. The IEEE 802.11ad and WiGig standards also offer hybrid microwave/mmWave WLAN solutions that use microwave frequencies for normal operation and mmWave frequencies when the 60 GHz path is favorable. Repeaters/relays will be used to provide range and connectivity to additional devices.

1.1 The Frontier: Millimeter Wave Wireless

the ability to communicate at hundreds of megabits per second. IEEE 802.11n leverages multiple transmit and receive antennas using multiple input multiple output (MIMO) communication methods. These devices carry up to four antennas and operate in the 2.4 GHz or 5.2 GHz unlicensed bands. Until IEEE 802.11n, standard advancements (in terms of data rate capabilities) have been largely linear, that is, a single new standard improves on the previous standard for the next generation of devices. The next generation of WLAN, however, has two standards for gigabit communication: IEEE 802.11ac and IEEE 802.11ad. IEEE 802.11ac is a direct upgrade to IEEE 802.11n through higher-order constellations, more available antennas (up to 8) per device, and up to 4 times more bandwidth at microwave frequencies (5 GHz carrier). IEEE 802.11ad takes a revolutionary approach by exploiting 50 times more bandwidth at mmWave frequencies (60 GHz). It is supported by device manufacturers that recognize the role of mmWave spectrum in the continued bandwidth scaling for next-generation applications. IEEE 802.11ad and mmWave technology will be critical for supporting wireless traffic with speeds competitive not only with gigabit Ethernet, but also 10 gigabit Ethernet and beyond. The largest challenges presented to 60 GHz and mmWave WLAN are the development of power-efficient RF and phased-array antennas and circuitry, and the high attenuation experienced by mmWaves when propagating through certain materials. Many strategies will be employed to overcome these obstacles, including 60 GHz repeaters/relays, adaptive beam steering, and hybrid wired/microwave/mmWave WLAN devices that use copper or fiber cabling or low microwave frequencies for normal operation, and mmWave frequencies when the 60 GHz path loss is favorable. Although the WPAN and WLAN network architectures provide different communication capabilities, several wireless device companies, including Panasonic, Silicon Image, Wilocity, MediaTek, Intel, and Samsung, are aggressively investing in both technologies.

MmWave technology also finds applications in cellular systems. One of the earliest applications of mmWave wireless communication was backhaul of gigabit data along a line-of-sight (LOS) path, as illustrated in Fig. 1.10. Transmission ranges on the order of 1 km are possible if very high-gain antennas are deployed. Until recently, however, 60 GHz and mmWave backhaul has largely been viewed as a niche market and has not

Figure 1.10 Wireless backhaul and relays may be used to connect multiple cell sites and subscribers together, replacing or augmenting copper or fiber backhaul solutions.

drawn significant interest. 60 GHz backhaul physical layer (PHY) design traditionally assumed expensive components to provide high reliability and to maximize range, resulting in bulky equipment and reducing the cost advantage over wired backhaul; however, a new application for wireless backhaul is emerging. Cellular systems are increasing in density (resulting in 1 km or less distances between base stations). Concurrently, cellular base stations require higher-capacity backhaul connections to provide mobile high-speed video and to implement advanced multicell cooperation strategies. If wireless backhaul devices are able to leverage recent mmWave hardware cost reductions, they may be able to service this growing need at a lower cost with more infrastructure flexibility. Further, backhaul systems are investigating LOS MIMO strategies to scale throughput into fiber capabilities [SST+09]. As operators continue to move to smaller cell sizes to exploit spatial reuse, the cost per base station will drop as they become more plentiful and more densely distributed in urban areas. Thus, wireless backhaul will be essential for network flexibility, quick deployment, and reduced ongoing operating costs. Consequently, wireless backhaul is likely to reemerge as an important application of 60 GHz and mmWave wireless communications. In fact, we envisage future cellular and WLAN infrastructure to be able to simultaneously handle backhaul, *fronthaul*, and position location connections using mmWave spectrum.

We foresee that mmWave will play a leading role in fifth-generation (5G) cellular networks. In the past generations of cellular technology, various PHY technologies have been successful in achieving ultra-high levels of spectral efficiency (bits/sec/Hz), including orthogonal frequency division multiplexing, multiple antennas, and efficient channel coding [GRM+10][STB09][LLL+10][SKM+10][CAG08][GMR+12]. Heterogeneous networks, coordinated multipoint transmission, relays, and the massive deployment of small cells or distributed antennas promise to further increase area spectral efficiency (bits/s/Hz/km^2) [DMW+11][YHXM09][PPTH09][HPWZ13][CAG08][GMR+12]. The focus on area spectral efficiency is a result of extremely limited bandwidths available in the UHF and microwave frequency bands where cellular systems are deployed, as illustrated in Fig. 1.11. MmWave cellular will change the current operating paradigm using the untapped mmWave spectrum.

Cellular systems may use mmWave frequencies to augment the currently saturated 700 MHz to 2.6 GHz radio spectrum bands for wireless communications [KP11a]. The combination of cost-effective CMOS technology that can now operate well into the mmWave frequency bands, and high-gain, steerable antennas at the mobile and base station, strengthens the viability of mmWave wireless communications [RSM+13]. MmWave spectrum would allow service providers to offer higher channel bandwidths well beyond the 20 MHz typically available to 4G LTE users. By increasing the RF channel bandwidth for mobile radio channels, the data capacity is greatly increased, while the latency for digital traffic is greatly decreased, thus supporting much better Internet-based access and applications that require minimal latency. Given this significant jump in bandwidth and new capabilities offered by mmWave, the base station-to-device links, as well as backhaul links between base stations, will be able to handle much greater capacity than today's cellular networks in highly populated areas.

Cellular systems that use mmWave frequencies are likely to be deployed in licensed spectrum at frequencies such as 28 GHz or 38 GHz or at 72 GHz, because licensed spectrum better guarantees the quality of service. The 28 GHz and 38-39 GHz bands

1.1 The Frontier: Millimeter Wave Wireless

Band	Uplink (MHz)	Downlink (MHz)	Carrier bandwidth (MHz)
700 MHz	746–763	776–793	1.25 5 10 15 20
AWS	1710–1755	2110–2155	1.25 5 10 15 20
IMT extension	2500–2570	2620–2690	1.25 5 10 15 20
GSM 900	880–915	925–960	1.25 5 10 15 20
UMTS core	1920–1980	2110–2170	1.25 5 10 15 20
GSM 1800	1710–1785	1805–1880	1.25 5 10 15 20
PCS 1900	1850–1910	1930–1990	1.25 5 10 15 20
Cellular 850	824–849	869–894	1.25 5 10 15 20
Digital dividend	470–854		1.25 5 10 15 20

Figure 1.11 United States spectrum and bandwidth allocations for 2G, 3G, and 4G LTE-A (long-term evolution advanced). The global spectrum bandwidth allocation for all cellular technologies does not exceed 780 MHz. Currently, allotted spectrum for operators is dissected into disjoint frequency bands, each of which possesses different radio networks with different propagation characteristics and building penetration losses. Each major wireless provider in each country has, at most, approximately 200 MHz of spectrum across all of the different cellular bands available to them [from [RSM+13] © IEEE].

are currently available with spectrum allocations of over 1 GHz of bandwidths, and the E-Band above 70 GHz has over 14 GHz available [Gho14]. Originally intended for LMDS use in the late 1990s, the 28 GHz and 38 GHz licenses could be used for mobile cellular as well as backhaul [SA95][RSM+13].

MmWave cellular is a growing topic of research interest [RSM+13]. The use of mmWave for broadband access has been pioneered by Samsung [KP11a][KP11b][PK11][PKZ10][PLK12], where data rates were reported in the range of 400 Mbps to 2.77 Gbps for a 1 GHz bandwidth at 1 km distance. Nokia has recently demonstrated that 73 GHz could be used to provide peak data rates of over 15 Gbps [Gho14]. Propagation characteristics of promising mmWave bands have been evaluated in [RQT+12], [MBDQ+12], [RSM+13], and [MSR14], and show path loss is slightly larger in NLOS conditions compared with today's UHF and microwave bands due to the higher carrier frequency. The scattering effects also become important at mmWave frequencies, causing weak signals to become an important source of diversity, and NLOS paths are weaker, making blockage and coverage holes more pronounced. To allow high-quality links, directional beamforming will be needed at both the base station and at the handset where propagation can be improved [GAPR09][RRE14]. Hybrid architectures for beamforming appear especially attractive as they allow both directional beamforming and more complex forms of precoding while using limited hardware [EAHAS+12a][AELH13]. Applications to picocellular networks are also promising [ALRE13], indicating 15-fold improvements in data rates compared with current 3GPP LTE 4G cellular deployments. Work in [RRE14] shows over 20-fold improvement in end-user data rates over the most advanced 4G LTE networks in New York City. Results in [BAH14] show 12-fold improvements compared with other competing microwave technologies, and results in [ALS+14], [RRE14], and [Gho14] predict 20 times or more capacity improvements using mmWave technologies. As 5G is developed and implemented, we believe the main differences compared to 4G will be the use of much greater spectrum allocations at untapped mmWave frequency bands, highly directional beamforming antennas at both the mobile device and base station, longer battery life, lower outage probability, much higher bit rates in larger portions of the coverage area, cheaper infrastructure costs, and higher aggregate capacity for many simultaneous users in both licensed and unlicensed spectrum, in effect creating a user experience in which massive data-rate cellular and WiFi services are merged.

The architecture of mmWave cellular networks is likely to be much different than in microwave systems, as illustrated in Fig. 1.12. Directional beamforming will result in high gain links between base station and handset, which has the added benefit of reducing out-of-cell interference. This means that aggressive spatial reuse may be possible. Backhaul links, for example, may share the same mmWave spectrum, allowing rapid deployment and mesh-like connectivity with cooperation between base stations. MmWave cellular may also make use of microwave frequencies using, for example, the phantom cell concept [KBNI13] where control information is sent on microwave frequencies and data is sent (when possible) on mmWave frequencies.

A number of universities have research programs in mmWave wireless communication. The University of Surrey, England, has set up a research hub for 5G mobile technology

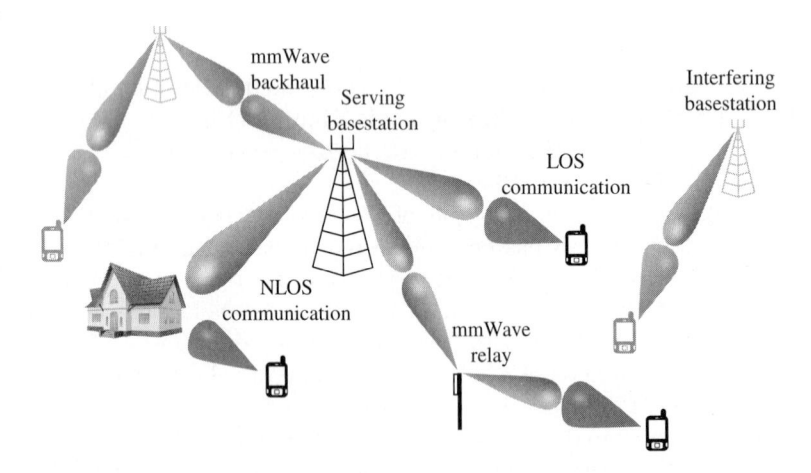

Figure 1.12 Illustration of a mmWave cellular network. Base stations communicate to users (and interfere with other cell users) via LOS, and NLOS communication, either directly or via heterogeneous infrastructure such as mmWave UWB relays.

with a goal to expand UK telecommunication research and innovation [Surrey]. New York University (NYU) recently established the NYU WIRELESS research center to create new technologies and fundamental knowledge for future mmWave wireless devices and networks [NYU12]. Aalborg University has an active mmWave research effort. The Wireless Networking and Communications Group (WNCG) at The University of Texas at Austin has a vibrant research program on 5G cellular technologies including mmWave [Wi14]. Aalto University has an active mmWave research effort. The University of Southern California, the University of California at Santa Barbara, the University of California at Berkeley, the California Institute of Technology, the University of Bristol, and the Korea Advanced Institute of Science and Technology (KAIST) are just some of the many universities that have substantial research efforts on mmWave for future wireless networks.

WPANs, WLANs, and cellular communication mark the beginning of mass consumer applications of mmWave technologies, where we evolve to a world where data is transported to and from the cloud and to each other in quantities we cannot fathom today. We believe that mmWave is the "tip of the iceberg" for dramatic new products and changes in our way of life, and will usher in a new generation of engineers and technologists with new capabilities and expertise. This exciting future will bring about revolutionary changes in the way content is distributed, and will completely change the form factor of many electronic devices, motivating the use of larger bandwidths found in the mmWave spectrum for many other types of networks, far beyond 60 GHz [RMGJ11][Rap12a]. For this to happen, however, many challenges must be overcome. Although we predict that future inexpensive UWB wireless cellular and personal area networks will be enabled through a move to mmWave frequencies and continued advancements in highly integrated digital and analog circuitry, we do not predict that all future advancements will be carried on the shoulders of solid-state process engineers, alone. Future wireless engineers will need to understand not only communications engineering and wireless system design principles, but also circuit design, antenna and propagation models, and mmWave electromagnetic theory to successfully codevelop their designs of future wireless solutions.

1.2 A Preview of MmWave Implementation Challenges

Implementation challenges for mmWave communication involve many layers of the communications stack. At the hardware level of the PHY, antennas are a major challenge. To minimize costs, mmWave chipset vendors may prefer to exploit the short carrier wavelength by incorporating antennas or antenna arrays directly on-chip or in-package. For the simplest and lowest-cost solutions, high-gain single chip solutions are attractive [RGAA09]. Single-antenna solutions, however, must overcome the challenges of low on-chip efficiencies whereas in-package antennas must overcome lossy package interconnects. MmWave systems may also employ many closely spaced antennas in packages or on circuit boards that are much smaller than a centimeter when using high-permittivity materials. Adaptive or switched-beam antenna arrays can provide transmit and receive antenna gain, but require protocol modifications at the signal processing level of the PHY and the data link layer to direct the beams.

A cornerstone of low-cost mmWave circuits is the use of CMOS or SiGe technology. Silicon on Insulator (SOI) CMOS processes are also attractive for high-end applications as they provide impressive quality (Q) factors due to reduced values for parasitic capacitances and inductances. SOI processes, however, suffer from increased costs compared with standard CMOS with a joined device channel and substrate. Because CMOS processes have now reached transit frequencies of hundreds of gigahertz, single-chip mmWave systems are feasible, complete with a digital baseband and mmWave analog front end. On-chip integration will also facilitate techniques like mixed-signal equalization [TKJ+12] [HRA10] that may improve the performance of complete systems versus multichip solutions. Unfortunately, foundries do not yet report relative permittivities or loss-tangents for process materials at mmWave frequencies in process design kits (PDKs), forcing early developers to measure these critical parameters until they are provided.

Communications signal processing at mmWave is also met with new challenges. Although mmWave wireless links can be modeled using conventional linear complex baseband system theory, the characteristics of mmWave wireless propagation combined with mmWave hardware design requirements produce unique design decisions at the PHY. Modulation and equalization algorithm selection must take into account the derived tradeoff between beam steering complexity and equalization complexity. For example, a mmWave system that uses omni directional antennas can suffer from severe ISI due to the multipath channels that cause successive symbols arriving at a receiver to overlap and interfere with each other [Rap02]. Fig. 1.13 illustrates the temporal and spatial variations of a typical omnidirectional 60 GHz impulse response and shows how multipath components may induce tens or even hundreds of nanoseconds of delay. The channel shown in

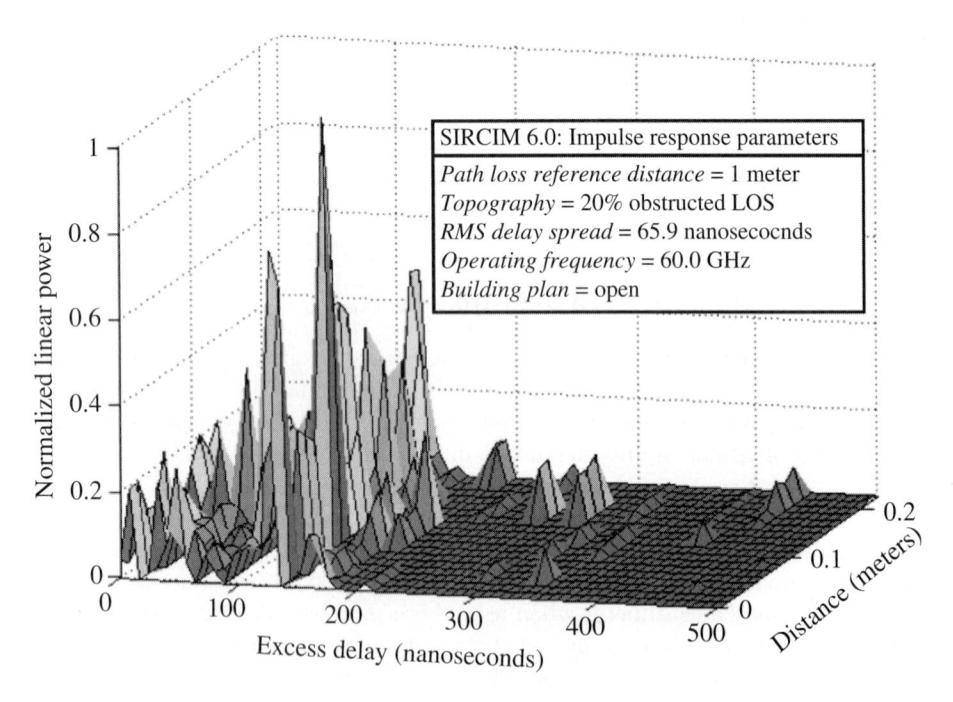

Figure 1.13 Long delay spreads characterize wideband 60 GHz channels and may result in severe inter-symbol interference, unless directional beamforming is employed. Plot generated with Simulation of Indoor Radio Channel Impulse Response Models with Impulse Noise (SIRCIM) 6.0 [from [DMRH10] © IEEE].

Fig. 1.13 has a delay spread of 65.9 ns, which could potentially spread a 60 GHz signal over tens to hundreds of symbol periods (e.g., this much spread would smear a signal over 120 symbols in the Single Carrier-Physical Layer of the IEEE 802.15.3c standard) [DMRH10]. A device operating in this environment would either need nonlinear equalization algorithms in the PHY and/or very long equalization periods, both of which would increase the complexity of the device (possibly erasing digital complexity benefits of 60 GHz relative to lower frequency systems). Directional beam steering antennas such as antenna arrays may be used to reduce the RMS delay spread seen by the device, but beam steering also results in an additional computational burden.

Above the PHY, the medium access control (MAC) of mmWave devices must also consider unique design factors. Most of the computational burden of beam steering would fall in the MAC layer. In addition to reducing complexity by optimal co-design of beam-steering and modulation algorithms, beam steering presents problems related to neighbor discovery and hidden and exposed nodes in a network. Neighbor discovery, which refers to the link protocol that manages link activation and maintenance, is especially difficult with beam steering and mobile devices. The hidden node problem, in which a coordinating device is unable to prevent an interfering device from transmitting, is challenging enough in microwave systems with omnidirectional antennas. The addition of very directional mmWave antennas (to combat mmWave path loss) only compounds this problem. Exposed nodes, which are prevented from communicating due to interference, are more likely to occur with conventional MAC protocols at mmWave because of the directionality of "all clear" messages with mmWave antennas [DMRH10].

1.3 Emerging Applications of MmWave Communications

60 GHz WPAN and WLAN are only the first step in a mmWave communications revolution. In addition to providing the first mass-market mmWave devices and enhancing cross-disciplinary communications design, 60 GHz communications will also have a substantial impact on other network technologies. Data centers may cut costs by employing mmWave communication links to interconnect the computers for high bandwidth, flexibility, and low power. Further, computational platforms may replace lossy, wired interconnects with high-speed wireless interconnects. Together, data center and computational platform improvements extend the reach of cloud computing through new non-traditional wireless applications. Cellular systems may incorporate mmWave to provide higher bandwidths to solve the spectrum crunch by providing mobile networks, peer-to-peer data transfers, and backhaul in the same bands. Not all emerging applications of 60 GHz and mmWave wireless devices, however, are unprecedented. Backhaul wireless links, broadband cellular communication, intra-vehicular communication, inter-vehicular communication, and aerospace communication have all been the subject of research and some market developments. Several technology breakthroughs at mmWave, however, hope to bring these applications to larger markets with vast capabilities.

1.3.1 Data Centers

To accommodate continued growth in the Internet and cloud-based applications, Internet service providers and major Web portals are building thousands of data centers each year. Data centers are used by all major Internet companies, including Google,

Microsoft, Yahoo, and Amazon, to distribute processing, memory storage, and caching throughout the global Internet. As multimedia content, for example, high-definition movies, increasingly streams over the Internet, data center buildout continues to accelerate. The buildout of data centers is comparable to the rapid buildout of towers in the early years of the cellular telephone industry.

Individual data centers often provide thousands of co-located computer servers [BAHT11]. Each data center can consume up to 30 megawatts of power, equivalent to the power drain of a small city, and must be built near a large water source (such as a lake or river) to accommodate cooling requirements. Remarkably, over 30% of the power dissipation in a typical data center is for cooling systems, for switching bottlenecks, and for broadband communication connections/circuitry between servers. Broadband circuitry is likely to become problematic as the Internet continues to expand over both wired and wireless connections [Kat09].

There are three types of communication in data centers: chip-to-chip, shelf-to-shelf, and rack-to-rack (less than 100 m). At present, data centers employ wired connections for all three types of data communication. Shelf-to-shelf and rack-to-rack communication is implemented using electrical copper connections and is the biggest bottleneck at present. Table 1.1 compares different copper solutions in terms of their power per port, reach, and link costs.

The broadband wired connections within data centers will not be able to accommodate future bandwidth requirements due to the increase in metal wire signal loss with increasing frequency. Data centers are expected to make a transition to other technologies.

Table 1.1 A representative sample of technology choices for computer interconnections within a data center. This table makes the case for a different interconnect technology [Hor08].

Solution	Power per Port (W)	Port Type	Reach	Interconnect	Link Cost
CX4	up to 1.6W	Dedicated copper SAS SFF8470	upto 15m	4 lanes of 3.125G copper in heavy-gauge casing	$ 250
10GBASE-T	~4W	Dedicated copper RJ45	30 m (or 100 m)	CAT5/CAT6 copper cable	$ 500
Active Twin-ax	1W	Hot pluggable SFP + or XFP	up to 30 m	Thin-gauge twin-ax copper Cable	$ 150
10GBASE-SR	1W	up to 300 m	Hot pluggable SFP + or XFP	Optical glass fiber	$ 500

Solution	Power per Port (W)	California Elec $/kWh	Cost per Year	CO_2 per Year per 1600 Ports (ton)	OPEX Cost per Year per DataCenter Cluster ($K)
CX4	up to 1.6W	20.72	$ 291	17	465
10GBASE-T	~4W	20.72	$ 727	42	1162
Active Twin-ax	1W	20.72	$ 182	11	291
10GBASE-SR	1W	20.72	$ 182	11	291

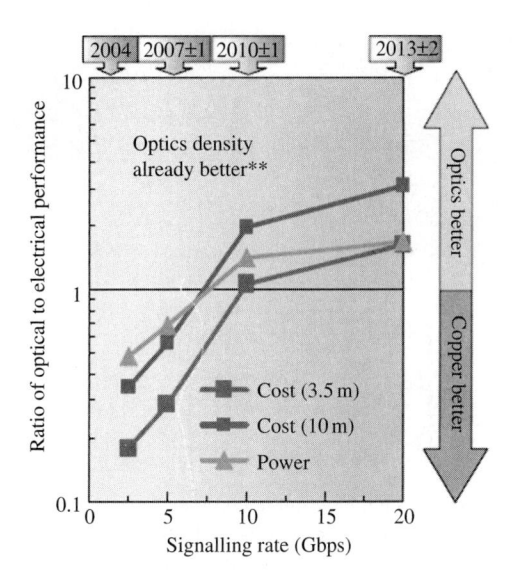

1.3 Emerging Applications of MmWave Communications

Figure 1.14 Comparison between optical and electrical performance in terms of cost and power for short cabled interconnects. The results show that optical connections are preferred to electrical copper connections for higher data rates, assuming wires are used [adapted from [PDK$^+$07] © IEEE].

For example, Fig. 1.14 shows how optical interconnects have cost and power advantages over copper interconnects for longer ranges and/or higher power. Both cable technologies, though, have disadvantages. For example, electrical connections typically have lower bandwidth and higher dielectric losses in FR4 (a common material used for the construction of printed circuit boards), whereas optical connections typically are not standardized and installation may be costly.

MmWave wireless communication using 60 GHz is an alternative to wired connections in data centers that could offer lower cost, lower power consumption, and greater flexibility. For example, a 10 m wireless 60 GHz link has a power budget in which 200 mW is dissipated before the power amplifier (e.g., by mixers or a voltage-controlled oscillator), 200 mW dissipated by the transmitter/antenna power amplifiers, and 600 mW of power dissipated in the channel/antennas giving a total of 1 W [Rap12b] [Rap09], which is comparable to the solutions in Table 1.1. A wireless solution allows flexible design of the data center, for example, placement of the servers, and permits easy reconfiguration. More flexible designs and a reduction in the numbers of cables and conduits allow for better placement of heat sources, and in turn, results in less stringent cooling and power requirements.

1.3.2 Replacing Wired Interconnects on Chips

The integrated antennas used to link individual 60 GHz devices may serve as the precursor of antennas used to link different components on a single chip or within a package, or within a close proximity as illustrated in Fig. 1.15. These links may be used for power combining, or more critically, signal delivery. On-chip antenna connections for power combining were evaluated as early as the mid 1980s [Reb92], but the market for high-frequency systems was limited and the technology was ahead of its time. Many

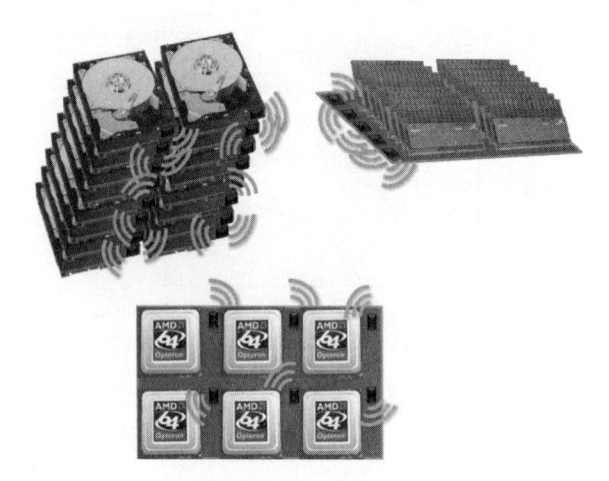

Figure 1.15 MmWave wireless will enable drastic changes to the form factors of today's computing and entertainment products. Multi-Gbps data links will allow memory devices and displays to be completely tetherless. Future computer hard drives may morph into personal memory cards and may become embedded in clothing [Rap12a][Rap09][RMGJ11].

researchers have experimented with on-chip or in-package wireless signal delivery (i.e., wireless interconnects) using highly integrated antennas [OKF+05]. This research demonstrates several challenges facing digital circuit design, including clock skew [FHO02] and interconnect delay [ITR09].[2,3] Of these challenges, interconnect delay may be the most important to consider. The International Roadmap for Semiconductors (ITRS) identified interconnect delay as the most critical phenomenon affecting high-performance products [ITR09].

The bandwidth of copper interconnects used on-chip is also an important issue. When clock frequencies increase, the passband bandwidth is decreased due to the increased resistance exhibited by a metal wire as frequency increases. This is exhibited by the square-root dependence of metal surface resistance on frequency in addition to the skin and proximity effects that also increase resistance. An on-chip or in-package antenna may mitigate these challenges because it would reduce the total length of wire seen by a signal. Therefore, the antennas developed for 60 GHz systems may provide value across many future applications requiring very high data rates within a chip or package.

1.3.3 Information Showers

With massive mmWave spectrum and low-cost electronics now available for the first time ever, the transfer of information will become truly ubiquitous and virtually unlimited. By replacing copper wiring with massive bandwidth radio links that are located at building entrances, hallways, roadway on-ramps, and lampposts, it will soon be possible to beam

---

2. Clock skew limits the size of a digital chip due to reduced component synchronization as the chip becomes larger.

3. Interconnect delay is the time required for a signal to propagate across a connection between different components on a chip.

entire libraries of information to people as they walk or drive. Consider today's student, who carries a heavy backpack full of books between classes. By using a concept known as the *information shower*, enormous amounts of content may be transferred in seconds, with or without the student's knowledge, as illustrated in Fig. 1.16.

Memory storage and content delivery will be revolutionized using the information shower, making real-time updates and access to the latest versions of books, media, and Web content appear seamless and automatic. The student of the future will merely need a handheld communicator to obtain all of the content for her entire educational lifetime, downloaded in a matter of seconds, and updated through continual access to information showers. Furthermore, peer-to-peer networking will enable very close range wireless communications between different users, so that massive downloads by an individual user may be shared to augment content of another nearby user. Information showers will exploit both cellular and personal area networks so that future consumers of content may use low-power and lightweight devices that will replace today's bulky and power-hungry televisions, personal computers, and printed matter.

1.3.4 The Home and Office of the Future

As mmWave devices and products evolve over the next couple of decades, the way in which our homes and offices are wired will radically change. As content from Web servers moves closer to the edge of the network, the bandwidth carried around our homes and enterprises will skyrocket by orders of magnitude. Also, the number of wireless devices that we rely upon will increase dramatically [Rap11][RMGJ11]. Today's Internet cables will likely be replaced with massive-bandwidth mmWave radio networks, obviating the need for wired ports for Internet and telephone service, as shown in Fig. 1.17. Many low-power wireless memory devices will replace books and hard drives that are bulky and inefficient. Untethered access to information within a room and between rooms will become the norm, as humans adapt to the renaissance of wireless communications, in which our personal devices are linked by massive-bandwidth data links that carry tens of gigabits of data per second. Even today's building wiring (e.g., Cat6 Ethernet cables) will be replaced by low-cost, high-bandwidth, rapidly deployable wireless systems that

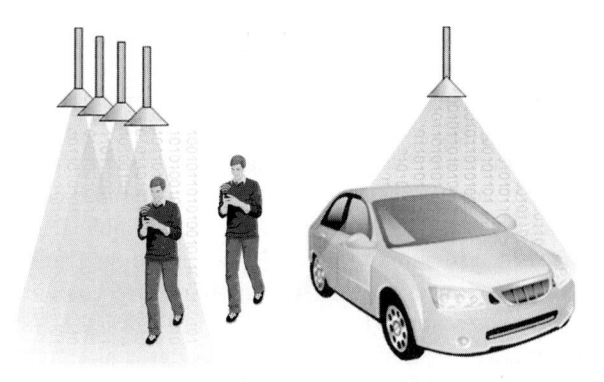

Figure 1.16 Future users of wireless devices will greatly benefit from the pervasive availability of massive bandwidths at mmWave frequencies. Multi-Gbps data transfers will enable a lifetime of content to be downloaded on-the-fly as users walk or drive in their daily lives [Rap12a][Rap09][RMGJ11].

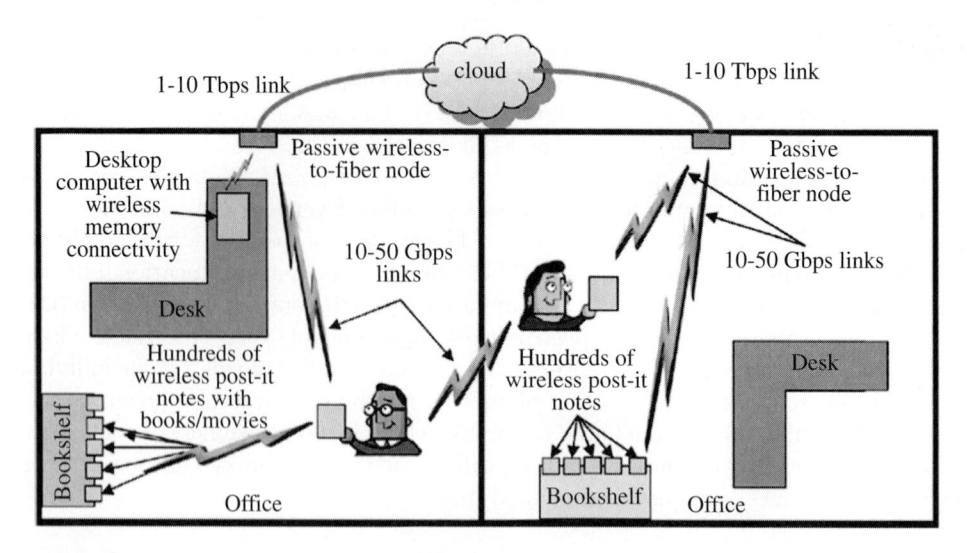

Figure 1.17 The office of the future will replace wiring and wired ports with optical-to-RF interconnections, both within a room and between rooms of a building. UWB relays and new distributed wireless memory devices will begin to replace books and computers. Hundreds of devices will be interconnected with wide-bandwidth connections through mmWave radio connections using adaptive antennas that can quickly switch their beams [Rap11] [from [RMGJ11] © IEEE].

have switchable beams to adapt coverage and capacity for any building floor plan. Later chapters of this text provide the technical details needed to engineer such systems.

1.3.5 Vehicular Applications

There are many applications of mmWave in the context of vehicles. Broadband communication within an automobile is being pursued to remove wired connections of vehicular devices (e.g., wires between dashboard DVD player and backseat displays) as well as to provide multimedia connectivity of portable devices inside the vehicle (e.g., MP3 players, cellphones, tablets, laptops). MmWave is especially attractive for intravehicle communications due to its inability to easily penetrate and interfere with other vehicular networks (due to high vehicle penetration losses). There are other applications outside the vehicle, as illustrated in Fig. 1.18. Vehicle-to-vehicle (V2V) communication may be used for collision avoidance or to exchange traffic information. Vehicle-to-infrastructure (V2I) links may also be used to communicate traffic information or to provide range and coverage extension of mobile broadband networks. Realization of intervehicle communication at mmWave is challenging due to the high Doppler and variable PHY and MAC conditions, which increase overhead for maintaining links, and lower transmitter height above ground, which limits the distance between automobiles of a connected network. While the current vehicle-to-vehicle standard IEEE 802.11p uses a 5.9 GHz band allocated for intelligent transportation systems, mmWave transmission is already employed at 24 and 77 GHz for automotive radar and cruise control. This makes it foreseeable that mmWave will find its way into other vehicular applications in the coming years.

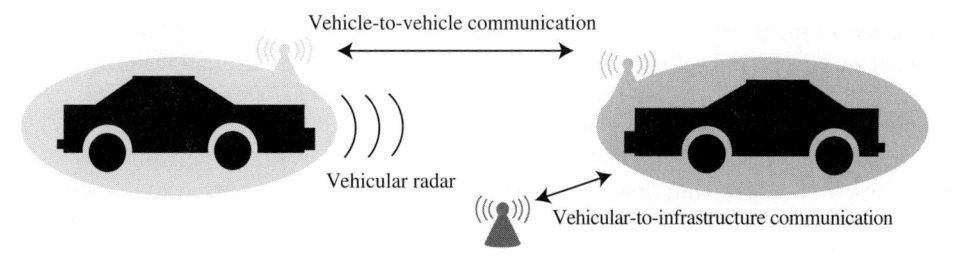

Figure 1.18 Different applications of mmWave in vehicular applications, including radar, vehicle-to-vehicle communication, and vehicle-to-infrastructure communication.

1.3.6 Cellular and Personal Mobile Communications

Today's cellular networks throughout the world use frequencies in the UHF and low microwave spectrum bands, between 400 MHz and 4.0 GHz. The use of these relatively low frequency spectrum bands has not changed in the 40 years of the cellular radio industry [RSM+13]. Even today, tiny slivers of spectrum (e.g., tens of MHz) within these bands continue to be allocated by governments around the world for the deployment of the fourth generation (i.e., 4G) of cellular technologies based on the LTE standard.

Demand for cellular data, however, has been growing at a staggering pace, and capacity projections are clear — cellular networks will require much greater spectrum allocations than have ever been available before. Conservative estimates of per-user data consumption growth range from 50% to 70% per year. Some wireless carriers, such as China Mobile, are already reporting even greater data consumption increases (e.g., 77% per year increase in data consumption per user from 2011 to 2012), and operators continue to experience incredible increases in video and live streaming traffic on their networks. This trend will only accelerate with time, especially as new social networking and machine-to-machine applications evolve, and as the Internet of things becomes a reality [CIS13].

The wireless community is steadily beginning to realize that the radio propagation at mmWave frequencies (dubbed "Beyond 4G," and called "5G" by some early researchers) may not only be viable, but may actually have greater benefits than today's cellular networks, when one considers the ability to use miniature, high-gain directionally steerable antennas, spatial multiplexing, new low-power electronics, advanced signal processing, and dormant or lightly used spectrum bands that have many tens of gigahertz of bandwidth available to them. The key technological components are about to become mature to enable multi-Gpbs mobile data rates for future mmWave wireless networks using cellular radio architectures.

Recent capacity results show that future mmWave cellular networks may use 1 or 2 GHz channels, instead of LTE's 40 MHz RF channel bandwidths, and by using Time Division Duplexing (TDD) in a relatively small cell (200 m radius) scenario, end-user data rates will easily be increased by a factor of 20 over most LTE networks, enabling multi-Gbps mobile links for cellphone users [RRE14].

As shown in the remaining chapters of this textbook, particularly in Chapters 3-8, the frequencies above 10 GHz are a new frontier for the cellular communications field, as many orders of magnitude greater bandwidth are available for immediate use. The smaller wavelength of mmWave cellular will enable great capacity gains by exploiting spatial and temporal multipath in the channel, in a far greater manner than today's 4G wireless networks. When additional capacity gains from beamforming and spatial multiplexing are combined with the vastly larger channel bandwidths available at mmWave carrier frequencies, it is clear that low-cost, UWB mobile communication systems with data rates and system capacities that are orders of magnitudes greater than today's wireless networks will evolve.

Such advances in capacity are not only required as today's cellular users demand more video and cloud-based applications, they are also logical when one considers that fact that advances in Moore's law have brought similar order-of-magnitude increases to computer clock speeds and memory sizes over the past four decades. Wireless communications, and cellular and WiFi networks in particular, are about to realize massive increases in data rates through the use of much more bandwidth than ever available before, and with this massive bandwidth will come new architectures, capabilities, and use cases for cellphone subscribers [PK11]. Such advances will usher in the renaissance of wireless communications [Rap12a].

1.3.7 Aerospace Applications

Because of the significant absorption of signals in oxygen, the 60 GHz spectrum is ideal for aerospace communication where terrestrial eavesdropping must be avoided [Sno02].[4] Consequently, many spectrum regulations, including FCC regulations in the USA [ML87], have allocated 60 GHz for intersatellite communication. Intersatellite communication links are LOS, and special design considerations for satellite systems result in few technology translations to consumer applications. One emerging 60 GHz aerospace application is multimedia distribution in aircraft, as illustrated in Fig. 1.19, to reduce the cabin wiring [GKT+09]. The localization of 60 GHz signals and the massive bandwidth resources make 60 GHz attractive versus microwave frequencies [BHVF08]. Unfortunately, to protect intersatellite communication from wireless in-aircraft applications, regulations currently disallow 60 GHz wireless communication in aircraft.

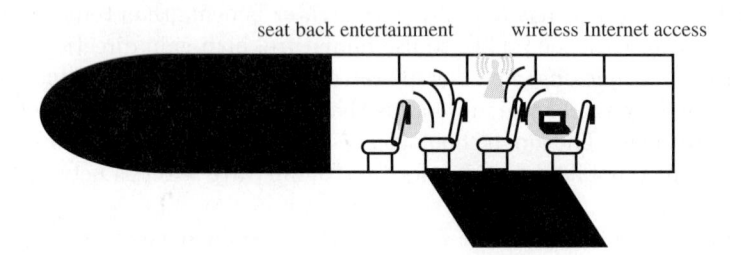

Figure 1.19 Different applications of mmWave in aircraft including providing wireless connections for seat-back entertainment systems and for wireless cellular and local area networking. Smart repeaters and access points will enable backhaul, coverage, and selective traffic control.

---

4. Figure 1.3 shows that the 180, 325, and 380 GHz bands are also well suited for "whisper radios" that are hard to eavesdrop on.

Regulations, however, are likely to change in the future with enough industry pressure and demonstration of the feasibility of network coexistence. Also, as mmWave wireless becomes more mature, additional high-attenuation bands, such as 183 and 380 GHz, will find use in aerospace applications.

1.4 Contributions of This Textbook

Today, active mmWave wireless device and product research programs exist at several major companies, such as Samsung [EAHAS+12] [KP11a] [PK11], Intel [CRR08], L3, Qualcomm, Huawei, Ericsson, Broadcom, and Nokia, and universities such as Georgia Tech [DSS+09], New York University [RSM+13][RRE14][PR07][RBDMQ12] [RQT+12] [AWW+13] [SWA+13] [RGAA09] [AAS+14] [ALS+14] [SR14] [Gho14] [SR14a] [MSR14], UC Berkeley [SB08][SB09a], UCLA [Raz08], UC San Diego [AJA+14] [BBKH06][DHG+13], UC Santa Barbara [RVM12][TSMR09], University of Florida [OKF+05], USC [BGK+06a], The University of Texas at Austin [GAPR09][GJRM10] [RGAA09][DH07][PR07][PHR09][DMRH10][BH13b][BH13c][BH14][EAHAS+12][AEAH12], and The University of Texas at Dallas [CO06][SMS+09]. Further, multiple textbooks and research books on the subject of mmWave devices and communications are available [Yng91][NH08][HW11]. Despite this research and progress, we have endeavored to create the first comprehensive text that brings communications and network-centric viewpoints to the intersection of antennas, propagation, semiconductors, and circuit design and fabrication for future wireless systems. Some existing texts on mmWave evolved from the circuits or packaging area but lack fundamental communications and network expertise. This book is distinct, since the ability to create future wireless communication systems requires a deep and fundamental understanding of multiple-user communications, antennas and propagation, and network theory, in addition to fundamental circuit design and microelectronics knowledge. It is rare that communications and network researchers work with circuit designers or semiconductor scientists in a university setting; this gap portends a huge void in the world's wireless research capabilities. Innovation and leadership are enhanced through interdisciplinary approaches to the creation and fabrication of broadband mmWave wireless devices and networks. This book endeavors to guide engineering practitioners, researchers, and students to find new, interdisciplinary ways to create mmWave broadband wireless devices and the networks they will form.

In this text, we demonstrate the state of the art in the areas of antennas, propagation, semiconductors, analog and digital circuits, communications, networks, and standards, while identifying critical interdependencies that will impact the future of communications at the edge of the network. By combining previously separated research fields of semiconductor devices and circuit design with the fundamentals of antennas and propagation, communications, and networking research, this book illustrates the problems and provides the knowledge to create the next generation of devices that will operate at the spectral frontier of mmWave frequencies.

For communications engineers, this book provides key insights into circuits challenges and fundamental semiconductor physics, along with antenna and propagation fundamentals. This is important because the formation of design tradeoffs requires insights into multiple fields. For example, instead of using higher-order signaling constellations, the analog-to-digital converter can be simplified or even eliminated by using simple binary modulation, even something as simple as on-off keying or differential phase shift keying. Essentially, this becomes a tradeoff between low-cost communication efficiency and mixed-signal power efficiency. As semiconductor devices continue to scale toward higher frequencies, even into the terahertz (300 GHz and above) range by 2020 [SMS+09], communications researchers will be unable to produce working sensors, channel measurement systems, and other critical research tools that help to provide fundamental knowledge [RGAA09] unless they have core circuit design knowledge at the mmWave regime and above.

For analog, mixed-signal, and RF circuit designers, this book provides foundations in the operations of the higher layers, including radio channel aspects, digital signal processing, and network protocols. This will facilitate better technical interactions with communication engineers. While the wide majority of today's wireless devices still use the standard superheterodyne and homodyne (direct conversion) architectures developed by Major Edwin Armstrong over a century ago, completely new receiver architectures, which fuse the detection and memory capabilities for pipelining received data, must be developed to handle such massive transmission bandwidths with low power consumption. New concepts in organizing memory cells at the chip level need to be integrated with communications coding techniques in order to implement the power efficient devices of the future, especially when considering that high-gain advanced antenna techniques, such as MIMO and phased arrays, will be implemented in such tiny physical sizes.

Coverage in the book is intentionally broad, but is also fundamental in nature, transferring key knowledge in mmWave communication, propagation, antennas, circuits, algorithms, design approaches, and standards. Such knowledge is crucial for understanding and balancing the demands for power, capacity, and delay in the era of wireless networks with unprecedented bandwidths.

1.5 Outline of This Textbook

This book is organized to allow the engineering practitioner, researcher, or student to rapidly find useful information on specific topics that are central to the infant world of mmWave wireless communications, including the nascent but commercially viable world of 60 GHz communication. Each chapter begins with an introduction that previews the material in each section and is completed with a summary that reviews salient points of each topic discussed. Chapter 1 serves as an introduction to the entire book, and motivates the study of mmWave communication.

Chapter 2 provides background material for wireless communication system design. This chapter begins with an introduction to the complex baseband signal representation and its relationship to the wireless medium that provides the physical channel for communication. Then, using the complex baseband model, the design of discrete-time wireless communication systems to send and receive information through the transmission of data symbols is discussed. This includes a summary of equalization concepts to deal with channel distortion effects and error-correcting codes to deal with degradations due to impairments in the channel and communication hardware. A special section is included on Orthogonal Frequency Division Multiplexing (OFDM) modulation, which is popular

1.5 Outline of This Textbook

in many commercial standards such as 4G LTE and IEEE 802.11n. Finally, Chapter 2 concludes with implementation topics including the estimation and detection of signals at the receiver, the architecture used for RF/analog/digital circuits in a communication system, and the layering of a communication system.

Chapter 3 transitions into the fundamentals of mmWave propagation and summarizes the physical characteristics of the wireless channel at operating frequencies around 60 GHz and other mmWave frequencies. This chapter consists of several different aspects of the wireless channel, each of which builds a complete picture of a mmWave wireless channel model. New results for the 28, 38, and 73 GHz outdoor urban cellular environments are given in this chapter, and they demonstrate the improvements that adaptive antennas can make in both link budget and reduction of multipath delay spread. First, measurement results that characterize the large-scale path loss are summarized. Then the penetration/reflection ability of mmWave signals is reviewed, which will be important to determine the feasibility of NLOS communication. A special section is devoted to the loss experienced by mmWave signals due to atmospheric effects such as energy absorption of oxygen and water molecules. Ray tracing is also described, as this approach will be critical for accurate site selection and deployment of future mmWave systems, where both indoor and outdoor channel conditions are considered. Finally, the indoor and outdoor mmWave channels are summarized in terms of their temporal, spectral, and spatial characteristics with respect to realistic mobility scenarios.

Chapter 4 provides background on antenna theory with an emphasis on techniques that are relevant for mmWave communication: in-package and on-chip antennas. The high cable losses at mmWave frequencies motivate pushing the antennas as close to the signal processing as possible. An in-package integrated antenna is one that is manufactured as part of the packaging process whereas an on-chip antenna is one that is built as part of the semiconductor process. Cost savings can potentially be realized with on-chip antennas if research can provide designs of high efficiency. Potential antenna topologies for mmWave are reviewed including planar, lens, aperture, and array antennas. Although many classic textbooks have dealt with the important area of antennas, we focus on the key concepts that are vital for on-chip and in-package antennas that will be used in mmWave consumer electronic products in the future. Also, array theory and fundamental semiconductor properties are treated, so readers can understand the challenges and approaches for implementing on-chip antennas. Although these approaches are nascent, and far from perfected at the time of this writing, future integrated wireless devices operating in the 30-300 GHz range will likely rely on tight integration not used at conventional UHF microwave bands. The chapter concludes with a survey of classical results on array processing, which are relevant for mmWave using adaptive antenna arrays.

Chapter 5 describes semiconductor device basics and enumerates the hardware design challenges at mmWave carrier frequencies. This includes a discussion of the RF hardware design issues including antenna design and amplifier design in the front end. Amplifier design is summarized by first presenting the challenges associated with characterizing and measuring mmWave signals. To address these challenges, S-parameters and Y-parameters are defined, and the design/cost issues that surface with different technologies including GaAs, InP, SiGe, and CMOS are interpreted. Circuit design at traditional frequencies (<10 GHz) takes advantage of lumped element assumptions because circuit dimensions are much smaller than the wavelength of the carrier frequency. Unfortunately, with mmWave frequencies, these assumptions cannot be made. This problem is discussed in detail via transmission line modeling followed by a summary of the design of passive and active elements in mmWave circuits. The key analog circuit components of mmWave transceivers are covered in detail in Chapter 5, and the chapter concludes with a novel and powerful figure of merit, the consumption factor, for determining and comparing power efficiencies for any mmWave circuit or system.

Chapter 6 discusses digital baseband issues. Much of the discussion is devoted to analog-to-digital conversion (ADC) and digital-to-analog conversion (DAC), as this consumes a substantial amount of power in mmWave circuit implementations. The impact of device fabrication mismatch, design architectures, fundamentals of DAC and ADC circuit design, and promising techniques for achieving multi-Gbps sampling and signal reproduction are given in this chapter.

Chapter 7 presents the design and applications of mmWave systems through a summary of 60 GHz PHY algorithms. The design of 60 GHz baseband algorithms is intrinsically linked to the wireless channel and hardware constraints discussed in Chapters 3 through 6. This relationship between the constraints and the PHY design is presented in the beginning of this chapter. Following this discussion, PHY design rules within these constraints are offered through sections on modulation, coding, and channel equalization. This chapter ends with a section that analyzes the impact of future/emerging hardware technology and its ability to relax certain design constraints for mmWave PHYs.

Chapter 8 reviews higher layer (above the PHY) design issues for mmWave systems with a particular emphasis on techniques relevant to 60 GHz and emerging cellular and backhaul systems. The use of directional beam steering, the limited coverage of mmWave signal propagation, and sensitivity to effects like human blockage of dominant signal paths present challenges that must be addressed at higher layers. This chapter reviews the key problems from a higher layer-perspective then expands on select topics in more detail. First, the incorporation of beam steering into a MAC protocol is described in more detail. Then, multihop operation using relays is reviewed as a way to achieve better coverage and to provide resilience to human blockages. Next, because multimedia is an important application for indoor systems, the cross-layer incorporation of video using unequal error protection is described in more detail. Finally, multiband strategies are discussed in which low frequency control signals are used to make network establishment and management easier.

Chapter 9 concludes the technical content of this text with a review of design elements from the standardization efforts for 60 GHz wireless communication systems. Three different WPAN standards are presented including IEEE 802.15.3c for WPAN, Wireless HD for uncompressed high-definition video streaming, and ECMA-387. Each of these WPAN standards has a distinct approach to the physical and MAC layer of the wireless communication system design, and these differences will be highlighted in this chapter. Two different WLAN standards are also presented including IEEE 802.11ad and WiGig (from which IEEE 802.11ad was based), which stretch WLAN into gigabit capabilities through 60 GHz spectrum.

1.5.1 Illustrations for this Textbook

You can find the color versions of the illustrations in this book at informit.com/title/9780132172288.

1.6 Symbols and Common Definitions

We use the notation in Table 1.2 and assign specific definitions to the variables in Table 1.3 throughout this text.

Table 1.2 Generic notation used in this text.

$\triangleq$	by definition
$\star$	convolution operator
$\mathbf{a}$	bold lowercase is used to denote column vectors
$\mathbf{A}$	bold uppercase is used to denote matrices
a, A	non-bold letters are used to denote scalar values
$|a|$	magnitude of scalar a
$\|\mathbf{a}\|$	vector 2-norm of $\mathbf{a}$
$\|\mathbf{A}\|_F$	Frobenius norm of $\mathbf{A}$
$\mathcal{A}$	calligraphic letters denote sets
$|\mathcal{A}|$	cardinality of set $\mathcal{A}$
$\mathbf{A}^T$	matrix transpose
$\mathbf{A}^*$	conjugate transpose
$\mathbf{A}^c$	conjugate
$\mathbf{A}^{1/2}$	matrix square root
$\mathbf{A}^{-1}$	matrix inverse
$\mathbf{A}^\dagger$	Moore-Penrose pseudo inverse
$\mathbf{a}_k$	k^{th} entry of vector $\mathbf{a}$
$[\mathbf{A}]_{k,l}$	scalar entry of $\mathbf{A}$ in k^{th} row l^{th} column
$[\mathbf{A}]_{:,k}$	k^{th} column of matrix $\mathbf{A}$
$[\mathbf{A}]_{:,k:m}$	column consisting of rows $k, k+1, \ldots, m$ of matrix $\mathbf{A}$
$(\cdot)$	used to index a continuous signal
$a(t)$	continuous scalar signal and value at t
$\mathbf{a}(t)$	continuous vector signal and value at t
$\mathbf{A}(t)$	continuous matrix signal and value at t
$[\cdot]$	used to index a discrete-time signal
$a[n]$	denotes discrete-time scalar signal and value at n
$\mathbf{a}[n]$	discrete-time vector signal and value at n
$\mathbf{A}[n]$	discrete-time matrix signal and value at n
$\mathbf{a}[n]$	denotes discrete-time vector signal in frequency domain at subcarrier n
$\mathbf{A}[n]$	discrete-time matrix signal in frequency domain at subcarrier n
log	denotes $\log_2$ unless otherwise mentioned

Table 1.3 Common definitions used in this text.

E_s	signal energy
N_o	noise energy
L	channel order
$\{h[\ell]\}_{\ell=0}^{L}$	discrete-time ISI channel impulse response with $(L+1)$ taps
$\mathsf{H}[k] = \sum_{\ell=0}^{L} h[\ell] e^{-j2\pi k\ell/N}$	frequency domain channel transfer function
$y[n]$	symbol sampled received signal
$x[n]$	symbol sampled transmitted signal
$s[n]$	symbol sampled transmitted signal before precoding
$\mathbf{I}_N$	$N \times N$ identity matrix
$\mathbf{0}_{N,M}$	$N \times M$ all zeros matrix
j	imaginary number $j = \sqrt{-1}$
$\mathbb{E}$	expectation operator
$x \sim \mathcal{N}(m, \sigma^2)$	means that x is a Gaussian random variable with mean m and variance σ^2
$x \sim \mathcal{N}_c(m, \sigma^2)$	means that x is a circularly symmetric complex Gaussian random variable with complex mean m, total variance σ^2, the real and imaginary parts of x are independent, and the variance of the real and imaginary parts are each $\sigma^2/2$
A_{eff}	effective aperture of an antenna (square meters)
A_{max}	maximum effective aperture of an antenna (square meters)
d	transmitter-receiver separation distance (meters)
EIRP	effective isotropic radiated power
λ	wavelength (meters)
c	speed of light in free space $= 3 \times 10^8$ m/s

1.7 Chapter Summary

Communications and network researchers, circuit designers, and antenna engineers seldom interact at universities or within the industry, leading to the potential for a huge void in mmWave wireless research capabilities. The innovative skills needed to ensure global leadership in the next revolution of wireless communications at the edge of the network, and in future mobile cellular systems, are not presently supported in an interdisciplinary manner by the government or the industrial research complex. We must teach researchers new, interdisciplinary strategies to create the ever-evolving broadband wireless devices and systems at mmWave, and beyond. To this end, it is our hope that you find this text to be a useful guide, enabling the creation of myriad new devices and applications that will soon be using the mmWave spectrum.

Chapter 2

Wireless Communication Background

2.1 Introduction

This chapter provides a background for wireless digital communication, since virtually all modern wireless networks (including those provided by 60 GHz standards and emerging cellular mmWave systems) communicate using binary digital data. We begin by introducing the complex baseband signal representation and its relationship to the wireless medium in Section 2.2. The complex baseband representation provides a convenient, universal framework for representing transmitted waveforms, the propagation channel, and the received signal, without dependence on the carrier frequency. By leveraging the Nyquist sampling theorem, the complex baseband communication model also leads to a convenient discrete-time representation [RF91][Rap02].

Next, we review communication transmitter and receiver (i.e., transceiver) signal processing components, including modulation and detection in Section 2.3, time domain equalization in Section 2.4, and frequency domain equalization in Section 2.5. Here, we highlight two subsections of special relevance to mmWave transceivers: single carrier modulation with frequency domain equalization (SC-FDE) in Section 2.5.1 and orthogonal frequency division multiplexing (OFDM) in Section 2.5.2. The principles of error control coding, which enhance the robustness of binary data to wireless channel impairments, are summarized in Section 2.6. Section 2.7 describes synchronization in digital wireless systems. Because of the importance of multiple antennas, Section 2.8 includes a review of key principles from multiple input multiple output (MIMO) communication. We conclude the chapter with a discussion of the architecture choices in the RF/analog/digital circuits in Section 2.9 and the layering of a communication system in Section 2.10.

Examples are provided from commercial systems, in particular IEEE 802.15.3c [802.15.3-03], ECMA 387 [ECMA10], and IEEE 802.11ad [802.11-12], since their published standards (discussed in Chapter 9) are publicly available. Note that each of the aforementioned standards has multiple physical layer (PHY) options, creating multiple potential transceiver signal processing embodiments. The techniques described are relevant for emerging mmWave cellular applications where multiple base stations share spectrum to provide seamless coverage over a wide geographic area [AELH14][ALS+14][BAH14][BDH14][Gho14][SNS+14][RSM+13][RRE14].

2.2 Complex Baseband Representation

Modern wireless communication systems are complicated, with many stages of data formatting, routing, authentication, verification, etc. To keep things simple, this chapter will primarily focus on the fundamental model of communication, that is, the transference of bits from one device (the transmitter) to another device (the receiver) through a wireless medium (the channel). Let $x_c(t)$ denote the continuous-time transmitted waveform, $h_c(t)$ the linear time-invariant multipath channel impulse response, and $y_c(t)$ the continuous-time received signal. Note that $x_c(t)$, $y_c(t)$, and $h_c(t)$ are real functions, that is, their domain and support are both real, since both physical quantities being modeled (time and voltage) are scalars.

Assume that binary data (bits) are mapped onto a superposition of electromagnetic waves with variable amplitude, which we call a waveform. This waveform is equivalently characterized by $x_c(t)$, a continuous-time voltage signal. The waveform is emitted by antennas at the transmitter and distorted by the wireless channel, that is, the wireless medium, before it produces current on the antenna at the receiver, rendering a corresponding superposition of variable amplitude electromagnetic waves, which we call a received signal $y_c(t)$. The function $h_c(t)$, the impulse response of the medium, characterizes the mapping between $x_c(t)$ and $y_c(t)$.

Wireless communication systems transmit and receive *passband signals* because of how spectrum is allocated to disparate systems to keep them from interfering. If we consider the equivalent relationship between time and frequency in linear systems, this implies, as illustrated in Fig. 2.1, that the energy in $x_c(t)$ is concentrated in frequency components around the carrier frequency f_c. The carrier frequency is also known as the operating or center frequency. At mmWave frequencies, the carrier frequency might be in the range of 10 to 300 GHz. The spectral width, W, of frequencies with nonzero energy centered around f_c (i.e., the RF bandwidth) is assumed to be much smaller

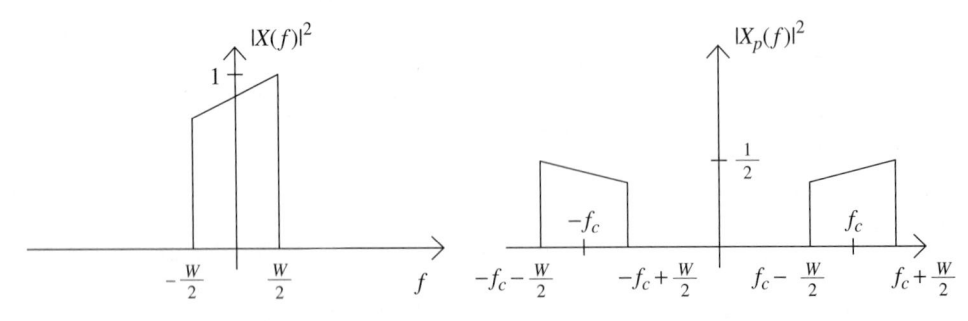

Figure 2.1 Illustration of the concept of a baseband signal (left) and a passband signal (right). Note that $W/2$ denotes the baseband bandwidth of the signal, following a signal processing convention where W represents the Nyquist sampling rate of the baseband signal.

2.2 Complex Baseband Representation

than f_c itself.[1] We refer to this as narrowband communication (not to be confused with narrowband channels, as will be discussed later). As an example, in the IEEE 802.15.3c and IEEE 802.11ad standards, the carrier frequency f_c is between 59 and 64 GHz, and the RF bandwidth W is approximately 2 GHz.

Passband signals have a convenient mathematical representation through the *complex envelope*. A passband signal can be written as

$$x_c(t) = A(t)\cos(2\pi f_c t + \phi(t))$$

where $A(t)$ is an amplitude function and $\phi(t)$ is a phase function. In amplitude modulation (AM), information (binary data or otherwise) is encoded entirely in $A(t)$, whereas in phase and frequency modulation (PM and FM) information is encoded in $\phi(t)$ (specifically, FM encodes information into $\phi'(t)$, the phase derivative). Applying trigonometric identities

$$\begin{aligned}
x_c(t) &= A(t)\cos(\phi(t))\cos(2\pi f_c t) - A(t)\sin(\phi(t))\sin(2\pi f_c t) \\
&= x_i(t)\cos(2\pi f_c t) - x_q(t)\sin(2\pi f_c t) \quad &(2.1) \\
&= \text{Re}\left\{x(t)e^{j2\pi f_c t}\right\} \quad &(2.2)
\end{aligned}$$

where the final step is obtained from (2.1) by employing Euler's formula $e^{j\theta} = \cos\theta + j\sin\theta$, and $x_i(t)$ and $x_q(t)$ are known as the *in-phase* and *quadrature* components. The signal $x(t) = x_i(t) + jx_q(t)$ is the *complex envelope* or *baseband equivalent* of the real signal $x_c(t)$. Note that the complex envelope is bandlimited because both the in-phase and quadrature components are bandlimited.[2] Also note that if it is possible to produce $x_c(t)$ from $x(t)$ and vice versa for any f_c, it is possible to design communication waveforms and signal decoding processes at baseband, without carrier frequency specification [RF91][RHF93].

Baseband communication system design, which produces $x(t)$ at the transmitter and its corresponding estimated received signal $y(t)$ at the receiver is enabled by *upconversion* and *downconversion*. Upconversion is the process of creating $x_c(t)$ from $x_i(t)$ and $x_q(t)$ through multiplication by a sinusoid. This may involve multiple stages of upconversion for implementation efficiency (see Chapter 5). Conversely, downconversion takes a passband signal and creates the complex envelope. If $y_c(t)$ is the passband signal (centered at f_c) observed through the antennas at the receiver, there exists an equivalent baseband signal $y(t)$ such that

$$y_c(t) = \text{Re}\left\{y(t)e^{j2\pi f_c t}\right\}.$$

Exploiting trigonometric identities, it can be shown that

$$y_c(t)\cos(2\pi f_c t) = \frac{1}{2}y_i(t) + \frac{1}{2}y_i(t)\cos(4\pi f_c t) - \frac{1}{2}y_q(t)\sin(4\pi f_c t) \quad (2.3)$$

$$y_c(t)\sin(2\pi f_c t) = -\frac{1}{2}y_q(t) + \frac{1}{2}y_q(t)\cos(4\pi f_c t) + \frac{1}{2}y_i(t)\sin(4\pi f_c t) \quad (2.4)$$

---

1. Although the illustration shows an ideal bandlimited signal, all practical signals must have finite temporal duration. This implies that all practical signals have infinite bandwidth (W is not finite). Nevertheless, practical signals can be *effectively* bandlimited whereby the majority of signal energy is confined to $f_c \pm W/2$. In general, W is a measure of the RF bandwidth of the signal [Sle76].
2. Different authors might use different normalization factor (e.g., $1/\sqrt{2}$) in this representation.

for $y(t) = y_i(t) + jy_q(t)$. The cosine and sine components in (2.3) and (2.4) do not spectrally overlap with the in-phase and quadrature baseband components (assuming that $W < f_c$). Consequently, a low pass filter with an ideal cutoff at $W/2$ and minimum Nyquist sampling rate W will reproduce the baseband components (up to scaling factors). This filter is not unique as any low pass filter with cutoff at least $W/2$ and less than f_c will work. A block diagram for this system is illustrated in Fig. 2.2. Note that an initial bandpass filter is included to filter out unwanted frequencies coming into the RX antenna. Expressed mathematically, if $f_{1,W/2}(t)$ is the impulse response of an ideal low pass filter with cutoff $W/2$, then

$$y(t) = 2f_{1,W/2}(t) \star y_c(t)e^{-j2\pi f_c t} \tag{2.5}$$

where $\star$ is the convolution operation in a linear system[3] and the factor of 2 accounts for the $\frac{1}{2}$ in (2.3) and (2.4). Practical architectures for upconversion and downconversion are discussed in Section 2.9.

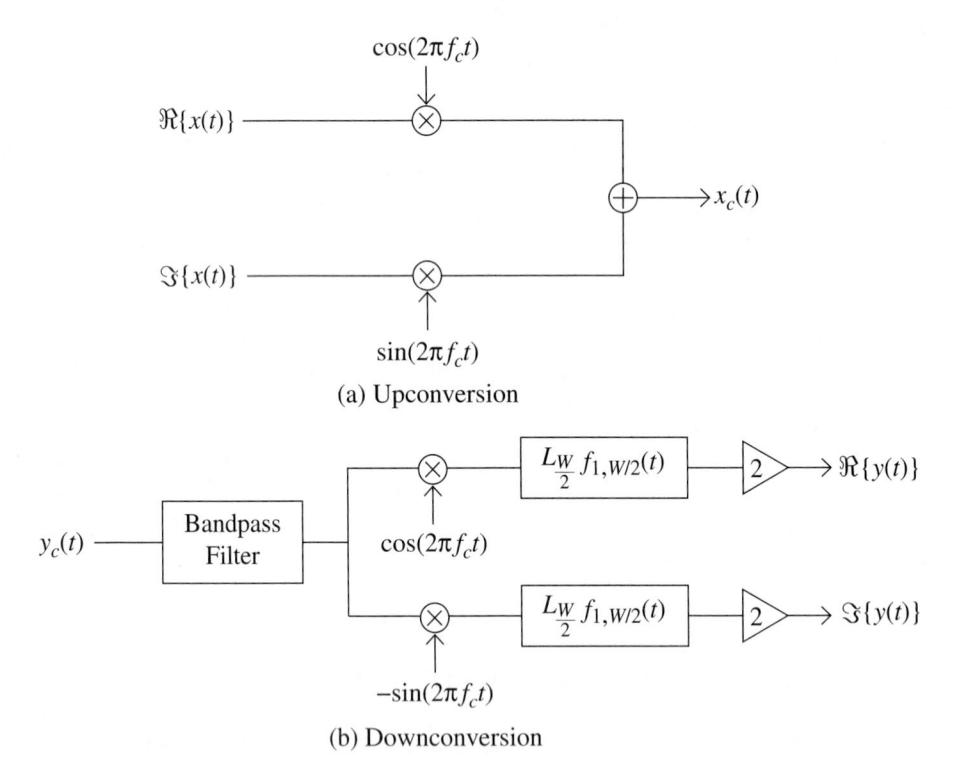

Figure 2.2 Illustration of upconversion in (a) and downconversion in (b) to obtain quadrature representations of a real signal. The bandpass filter is sometimes referred to as an anti-aliasing filter. These are ideal architectures; more practical approaches are discussed in Section 2.9.

---

3. In a linear system if $a(x)$ is the input signal and $b(x)$ is the impulse response of the linear system, then $c(x) = a(x) \star b(x) = \int_{-\infty}^{\infty} b(x-y)a(y)dy$ is the output of the system. The convolution is used to represent the filtering operation.

2.2 Complex Baseband Representation

Ideally, the wireless channel (which includes the wireless medium and the analog circuit components) does not distort the transmitted waveform, resulting in upconversion and downconversion being inverse processes. In practice, the wireless channel attenuates and disperses the transmitted waveform. Hence, we need to accurately represent $y(t)$ as a function of $x(t)$ and some (yet undetermined) model of the wireless channel, which will depend on f_c but can be represented by some baseband equivalent channel. Several assumptions are required to develop a relationship between $x(t)$ and $y(t)$ that involves a complex baseband equivalent channel response.

Assuming ideal RF components (nonlinear distortion is neglected) and neglecting noise (for now), signal propagation can be modeled as a linear time-invariant system $h_c(t)$. The channel distortion effect itself is time varying; however, communication typically occurs in bursts such that time invariance assumptions are valid.[4] Under the time invariance (also known as *block fading*) assumption, the received passband signal is related to the transmitted passband signal through a convolution integral

$$y_c(t) = \int_{-\infty}^{\infty} h_c(t-\tau)x_c(\tau)d\tau. \tag{2.6}$$

Alternatively, at baseband, the received signal can be written as

$$y(t) = \int_{-\infty}^{\infty} h(t-\tau)x(\tau)d\tau \tag{2.7}$$

where $h(t)$ is the complex baseband equivalent channel. The complex baseband equivalent channel generally includes multipath propagation between the transmitter and receiver *and* the various filtering operations at the transmitter and receiver.

To find an expression for $h(t)$, recall that $x_c(t)$ is a passband signal, so it is only necessary to model the channel effects over a bandwidth W, centered at f_c as illustrated

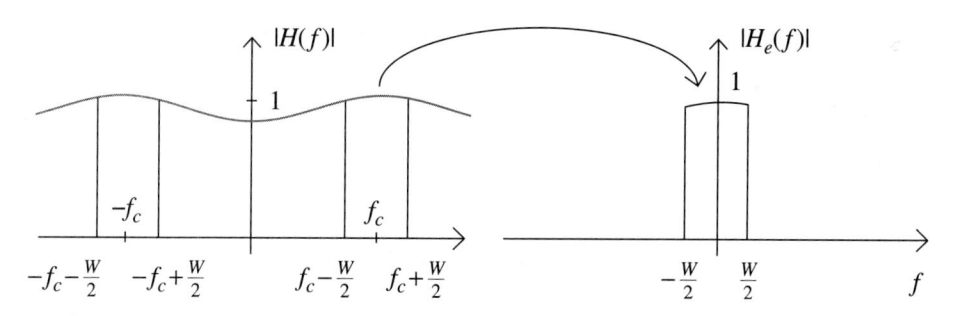

Figure 2.3 The propagation channel itself is not bandlimited. From a signal processing perspective, the receiver only cares about the part of the channel that distorts the transmitted signal. The baseband equivalent channel results from converting the channel of interest down to baseband and rescaling.

---

4. The propagation time scale is typically much smaller than the observation time leading to transmission of bursts of information. For details on how to determine whether a channel is time invariant or time varying, the reader is encouraged to reference coherence time discussions in [Rap02].

in Fig. 2.3. Hence, bandpass filtering $h_c(t)$, scaling, and downconverting provides the formula for $h(t)$. Let

$$f_{\text{p},W}(t) = 2W \frac{\sin(\pi W t)}{\pi W t} \cos(2\pi f_c t) \qquad (2.8)$$

denote the impulse response of an ideal bandpass filter with bandwidth W centered around f_c. Then, the complex baseband channel (or baseband equivalent channel) can be written

$$h_c(t) = f_{1,W/2}(t) \star (h_c(t) \star f_{p,W}(t)) \, e^{-j2\pi f_c t}. \qquad (2.9)$$

A factor of 2 is not necessary here since the effective channel results from filtering a non-bandlimited signal, rather than starting with a passband signal. It is also useful to consider an alternative representation of the channel impulse response that provides a more intuitive model (in the physical propagation sense).

Suppose that the wireless channel impulse response, or channel for short, is the superposition of L delayed and attenuated components, or impulses, known as *multipath components*,[5]

$$h_c(t) = \sum_{\ell=1}^{L} \alpha_\ell \delta(t - \tau_\ell) \qquad (2.10)$$

where α_ℓ is the amplitude of the fading coefficient of the ℓth multipath component having a delay of τ_ℓ [Rap02]. Inserting (2.10) into (2.6) yields

$$y_c(t) = \sum_{\ell=1}^{L} \alpha_\ell x_c(t - \tau_\ell) \qquad (2.11)$$

$$= \sum_{\ell=1}^{L} \text{Re}\{\alpha_\ell e^{-j2\pi f_c \tau_\ell} x(t - \tau_\ell) e^{j2\pi f_c t}\}. \qquad (2.12)$$

Using the complex baseband received signal formulation in (2.7)

$$y(t) = \sum_{\ell=1}^{L} \alpha_\ell e^{-j2\pi f_c \tau_\ell} x(t - \tau_\ell) \qquad (2.13)$$

$$= \underbrace{\left(\sum_{\ell=1}^{L} \alpha_\ell e^{-j2\pi f_c \tau_\ell} \delta(t - \tau_\ell) \right)}_{h_e(t)} \star x(t). \qquad (2.14)$$

We refer to $h_e(t)$ as the pseudo-complex baseband equivalent channel. This provides a relationship between $y(t)$ and $x(t)$ but, unlike $h(t)$, it is not bandlimited. The baseband equivalent channel for this case is therefore

$$h(t) = f_{1,W/2}(t) \star h_e(t). \qquad (2.15)$$

---

5. A multipath component may result from many physical electromagnetic propagation mechanisms including reflection, scattering, refraction, and/or diffraction. All of these mechanisms produce attenuation and delay.

Consequently, $h(t)$ can be derived from $h_e(t)$ simply by filtering with an ideal low pass filter.

For accurate digital circuit implementation of complex baseband signal processing, it will often be desirable to analyze complex baseband communication in discrete time. At the transmitter, the complex envelope, $x(t)$, is created by passing a digital (sampled) representation to the digital-to-analog converter (DAC). At the receiver, $y(t)$ is sampled by the analog-to-digital converter (ADC). Since (2.9) is bandlimited, $x(t)$ can be fully represented from its samples $\{x[n]\}_n$, where $x[n] = x(nT)$ and $T < 1/W$ such that the sampling frequency is greater than twice the bandwidth of the signal.[6] In practice, the ADC implements the sampling operation (and also a quantization, although not specified here). The bandlimited signal can be reconstructed from its samples using

$$x(t) = \sum_n x[n] \frac{\sin(\pi(t-nT)/T)}{\pi(t-nT)/T}, \quad (2.16)$$

which is approximated in hardware through a DAC (e.g., by employing a sample-and-hold followed by a reconstruction filter instead of the ideal interpolating filter). Because $x(t)$ and $y(t)$ can be represented in terms of their samples, we would expect that an input-output relationship also exists in terms of the sampled version of $h(t)$. From [OS09, Chapter 4] it is possible to show that if T is chosen to satisfy Nyquist's theorem, then

$$y[n] = \sum_{k=-\infty}^{\infty} h[k]x[n-k] \quad (2.17)$$

where $h[n] = T\,h(nT)$ since $h(t)$ is already bandlimited. In subsequent sections, this connection can be used to develop channel estimation and linear equalization algorithms.

2.3 Digital Modulation

While analog data can be communicated over wireless channels, the efficiency, security, integrity, and universal representation of digital data has led to its dominance in information and computing systems. Digital communication systems convey digital information through binary sequences that are encoded onto transmitted waveforms, represented by $x(t)$ in the complex baseband model. Digital modulation is the process that completes the mapping from bit sequences into encoded waveforms at the transmitter. Conversely, digital demodulation is the process by which information bit sequences are extracted from received signals, that is, from $y(t)$ in the complex baseband model at the receiver. This section surveys digital modulation techniques that may be useful in mmWave systems, with an emphasis on those already specified for use in 60 GHz. The methods here may be used with channel data to obtain realistic error rate performance [RF91]. First, some background on symbol mapping and detection is given, followed by a description of various modulation formats.

---

6. Since the bandwidth of $x(t)$ and $y(t)$ is $W/2$, this means that $1/T > W$ or $T < 1/W$. The fundamental connection between bandlimited signals and sampled signals is established by the Nyquist sampling theorem. Essentially, a (baseband) bandlimited signal with bandwidth $W/2$ can be fully represented from its samples if $T \leq 1/W$, where equality holds except for some special signals. More background on the connection between discrete- and continuous-time channel representations is available in [OS09].

2.3.1 Symbols

In most general references on digital communication [PS07], digital modulation is described from an abstract vector space representation. Such generality is not needed here as the modulation processes in this chapter can be characterized through two operations: symbol mapping and waveform synthesis.

- *Symbol mapping* is the process of encoding the sequence of bits into a sequence of symbols from a finite *constellation* (with real and/or complex symbols). If $\mathcal{C} = \{s_0, ..., s_{M-2}, s_{M-1}\} \subset \mathbb{C}^M$ is the constellation, then the number of possible symbols or cardinality of the set $\mathcal{C}$ is $|\mathcal{C}| = M = 2^b$ for $b \in \mathbb{N}$. Essentially, to encode a sequence of bits onto digital symbols, b bits of the sequence are mapped at a time onto symbols in $\mathcal{C}$. Constellations available in 60 GHz standards exhibit cardinalities from $M = 2$ to $M = 64$, which means that each symbol represents between 1 and 6 bits. The process of mapping bits to constellation points must be fixed and also typically observes Gray coding principles (minimizes adjacent bit differences) to reduce the number of flipped bits in the case of incorrect symbol estimation at the receiver. The exact labeling of bits to points in the constellation is not necessarily universal and may vary between standards. In this book, we consider energy-normalized constellations where $\sum_m |s_m|^2 = M$, such that the constellation choice does not change average transmit power (we call this a *unit-norm constellation*).

- *Waveform synthesis* is the process of mapping symbols into transmitted waveforms. Many waveform synthesis embodiments employ complex pulse amplitude modulation or *quadrature amplitude modulation* and may be represented at baseband through

$$x(t) = \sqrt{E_s} \sum_{n=-\infty}^{\infty} s[n] g_{\text{tx}}(t - nT_s) \quad (2.18)$$

where E_s is the *symbol energy*, T_s is the *symbol period* (not to be confused with sample period T), $s[n] \in \mathcal{C}$ is the constellation symbol, and $g_{\text{tx}}(t)$ is the *pulse shaping filter*.[7] Waveform synthesis in (2.18) sends information on successive pulses; hence, the spectral properties of $x(t)$ are determined by $g_{\text{tx}}(t)$. Fig. 2.4 illustrates how digital symbols are synthesized as time waveforms for both rectangular pulse shapes and raised cosine rolloff pulse shapes. Assuming that the symbols in the sequence are independent and identically distributed (i.i.d.), and that the constellation is normalized to unity, then the power spectral density of $x(t)$ equals

$$P_x(f) = \frac{E_s}{T_s} |G_{\text{tx}}(f)|^2 \quad (2.19)$$

where $G_{\text{tx}}(f)$ is the Fourier transform of $g_{\text{tx}}(t)$. The pulse shape is assumed to be normalized such that $\int_f |G_{\text{tx}}(f)|^2 df = 1$ (implying that the average power in

---

7. The root raised cosine pulse shaping filter is commonly implemented, characterized by β, the excess bandwidth factor (e.g., $\beta = 0.25$ is common) [Cou07].

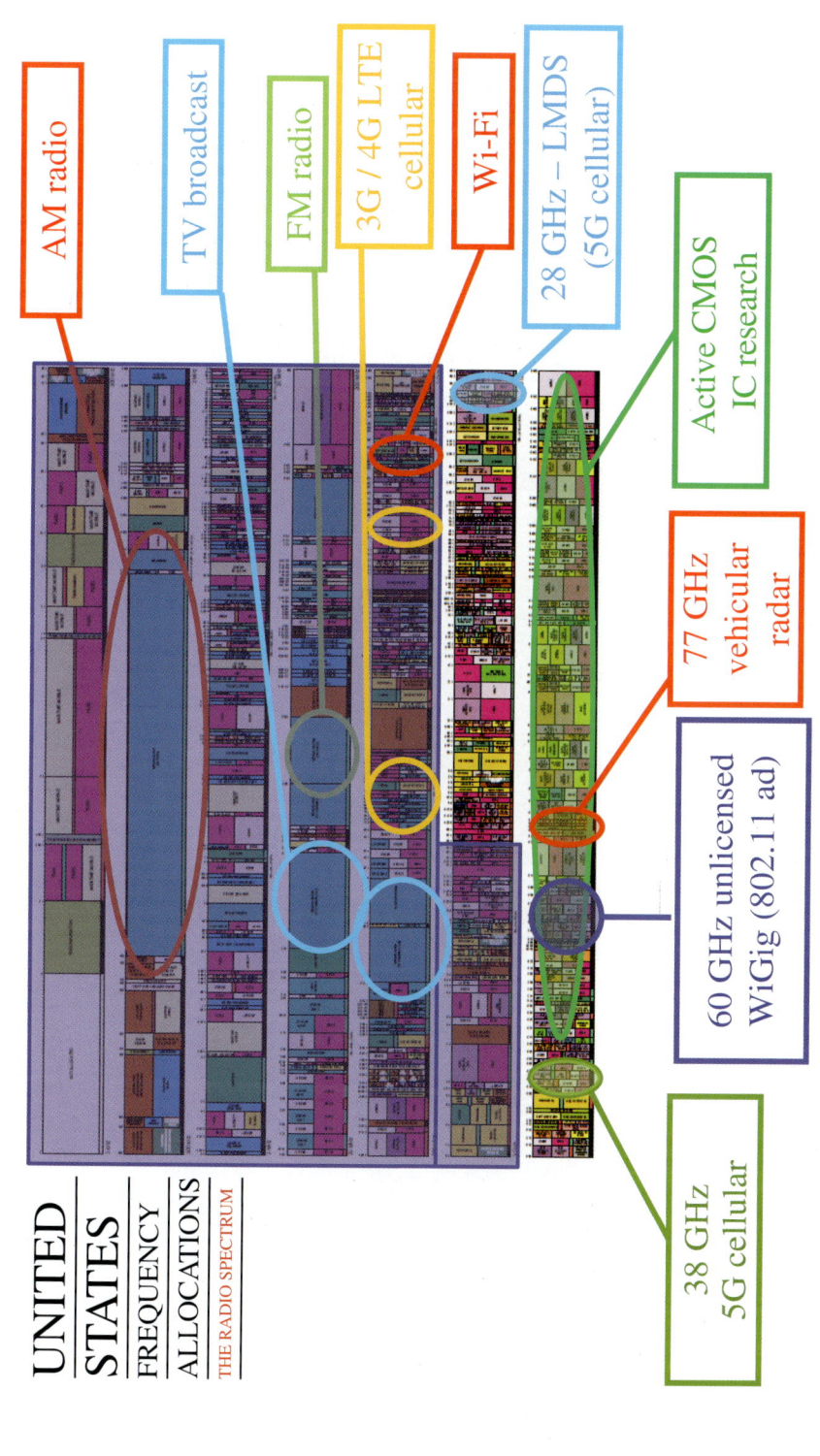

Figure C.1 Wireless spectrum used by commercial systems in the USA. Each row represents a decade in frequency. For example, today's 3G and 4G cellular and WiFi carrier frequencies are mostly in between 300 MHz and 3000 MHz, located on the fifth row. Other countries around the world have similar spectrum allocations. Note how the bandwidth of all modern wireless systems (through the first six rows) easily fits into the unlicensed 60 GHz band on the bottom row [from [Rap12b] U.S. Dept. of Commerce, NTIA Office of Spectrum Management].

Figure C.2 Third-generation 60 GHz WirelessHD chipset by Silicon Image, including the Sil6320 HRTX Network Processor, Sil6321 HRRX Network Processor, and Sil6310 HRTR RF Transceiver. These chipsets are used in real-time, low-latency applications such as gaming and video, and provide 3.8 Gbps data rates using a steerable 32 element phased array antenna system (courtesy of Silicon Image) [EWA+11] [©IEEE].

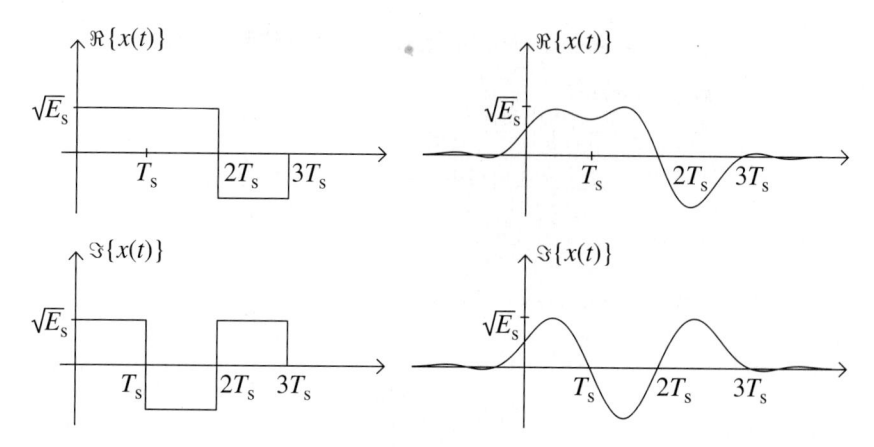

Figure 2.4 Illustration of a baseband waveform with a QPSK signal. The left (top and bottom) shows the output with a square pulse shape and the right (top and bottom) with a raised cosine pulse shape and 25% excess bandwidth for the symbols $1+j, 1-j, -1+j$.

$x(t)$ equals $\int_f P_x(f)df = E_s/T_s$). Practically speaking, E_s is rendered through power amplifiers (PAs) in the transmit processing.

2.3.2 Symbol Detection

In the discussion of the complex baseband model, the only communication impairment that was considered was the wireless channel, which is usually assumed to be constant over short time periods and thus can be measured and equalized. Communication must also deal with random or stochastic impairments — usually called noise. The most common form of noise in digital communication systems is additive noise.[8] The fundamental additive noise phenomenon is thermal noise, whose amplitude distribution is characterized by a spectrally white normal (Gaussian) distribution. The complex baseband communication model with additive noise becomes

$$y(t) = x(t) \star h(t) + v(t) \tag{2.20}$$

through the inclusion of additive white Gaussian noise (AWGN), where the continuous time noise signal $v(t)$ is an i.i.d. random process. For real constellations, that is, $\mathcal{C} \subset \mathbb{R}$, $v(t)$ is a zero mean Gaussian random variable with variance $N_o W/2 = kT_e W/2$, where $k = 1.38 \times 10^{-23} J/K$ is Boltzmann's constant, T_e is the effective noise temperature of the device in Kelvin, and W is the RF bandwidth of the passband signal translating to a single-sided baseband bandwidth of $W/2$. For complex constellations, $v(t)$ is a zero-mean circularly symmetric complex Gaussian random variable with variance of $N_o W/2$ per dimension and total variance of $\sigma_v^2 = N_o W$ in units of hertz. When the order is low ($M = 2$), most constellations can be realized in a single dimension, that is, real

---

8. Non-additive noise is also important but is not considered to be fundamental as it arises from non-ideal hardware. The most common kind of non-additive noise — phase noise — is discussed in more detail in Chapters 5 and 7.

constellations. A single dimensional representation is desirable to, for example, reduce transmitter and receiver complexity. Real constellations, however, are substantially outperformed by two-dimensional complex constellations in high orders ($M > 2$), because the quadrature dimension can be leveraged to increase the minimum distance between constellation points (assuming a fixed average transmit power). Complex constellations, however, require more complexity in the transceiver design.

The goal of symbol detection is to predict each transmitted symbol at the receiver from the received signal (after distortion by channel and noise impairments) in some optimal way. In general, the detection process explicitly depends on the type of modulation, the channel impairments, and the optimality criterion. Detection may be further complicated if error control coding is employed or if there is a propagation channel $h(t)$.

Here, the general procedure for maximum likelihood symbol detection (the most common optimality criterion) for complex pulse amplitude modulations, as described by (2.18), is discussed with $h(t) = \delta(t)$. Note that the non-ideal channel $h(t)$ is not considered; techniques for dealing with the channel are deferred to Sections 2.4 and 2.5.

Classic communication theory has shown that the maximum likelihood (ML) receiver must first process $y(t)$ through a *matched filter* $g_{\text{rx}}(t) = g_{\text{tx}}^*(-t)$, followed by sampling at the symbol rate (assuming perfect synchronization) to produce received samples $y[n]$. Given that the pulse shape is normalized, the received signal model with an ideal channel after sampling becomes

$$y[n] = \sqrt{E_s}s[n] + v[n] \qquad (2.21)$$

where $v[n]$ is the sampled filtered Gaussian noise sequence. If $g_{\text{tx}}(t) \star g_{\text{rx}}(t)$ is zero for all periods of T_s except at $t = 0$ (a Nyquist pulse shape), then adjacent symbols do not interfere if correctly sampled and the additive noise maintains its i.i.d. statistical properties [PS07]. The ML receiver optimally selects

$$\widehat{s}[n] = \arg\min_{s \in \mathcal{C}} |y[n] - \sqrt{E_s}s|^2 \qquad (2.22)$$

as the predicted symbol at each sample, which turns out to be the closest constellation point in $\mathcal{C}$ to the observed sample $y[n]$ in the Euclidean distance metric [PS07]. Consequently, this practice is also known as *minimum distance decoding*. Note that the structure of the constellation can be used to simplify the calculation in (2.22) to avoid a brute-force search.

The performance of a detector is measured by the probability of error. In AWGN channels, the performance is a function of the ratio of the average symbol energy E_s and the noise variance, or the *signal-to-noise ratio* (SNR), defined

$$\text{SNR} = \frac{E_s}{\sigma_v^2}. \qquad (2.23)$$

Later, when channels are not ideal, SNR must also account for energy lost in the wireless channel. The exact calculation of the probability of symbol error depends on the constellation. A useful upper bound on the probability of symbol error is

$$P(\text{SNR}) \leq (M-1)Q\left(\sqrt{\text{SNR}\frac{d_{\min}^2}{2}}\right) \qquad (2.24)$$

2.3 Digital Modulation

where $Q(x) = (\sqrt{2\pi})^{-1} \int_x^\infty \exp(-z^2/2) dz$ is the tail probability of a Gaussian distribution with zero mean and unit variance and $d_{\min}$ is the minimum distance between two different points in the constellation. It is also useful to know that $Q(x) \leq 0.5 \exp(-x^2/2)$. Based on (2.24), it can be observed that the probability of error of the ML detector goes down if the SNR is high and goes up if the constellation is tightly packed (i.e., the constellation points are close together).

Because of the noise in the communication channel, forward error correction via error control codes is used at the receiver to correct errors in the received signal. Most error control codes provide improved performance when the detector offers auxiliary information or *soft information* that relates to its confidence about the decision. If the detector has already computed or been given the noise statistics (e.g., the noise variance) then it is possible to numerically compute bit decision reliability metrics. The most common reliability metric is the log-likelihood ratio (LLR). Suppose that b bits are mapped to each constellation symbol and let $s_m^{(k)}$ denote the kth bit ($k \in \{1, 2, \ldots, b\}$) in the mth constellation symbol ($m \in \{0, 1, \ldots, M-1\}$). Let the set $\mathcal{C}_0^{(k)}$ account for all constellation points (symbols) that have 0 at the kth bit (i.e., $\mathcal{C}_0^{(k)} = \{s_m : s_m^{(k)} = 0\}$) and $\mathcal{C}_1^{(k)} = \{s_m : s_m^{(k)} = 1\}$ is defined similarly. Let $\Pr\{\cdot\}$ denote the probability function and let $\Pr\{a|b\}$ denote the probability of a given b. The LLR of the kth bit for the nth received sample, $y[n]$, with a known noise variance σ_v^2 is

$$L\left(k|y[n], \sigma_v^2\right) = \log \left(\frac{\sum_{s_m \in \mathcal{C}_0^{(k)}} \Pr\{y[n]|x[n] = s_m\}}{\sum_{s_{m'} \in \mathcal{C}_1^{(k)}} \Pr\{y[n]|x[n] = s_{m'}\}} \right). \quad (2.25)$$

For AWGN, as described by (2.21), the max-log approximation yields

$$L\left(k|y[n], \sigma_v^2\right) \approx \min_{s \in \mathcal{C}_k^{(1)}} \frac{1}{\sigma_v^2} |y[n] - \sqrt{E_s}s|^2 - \min_{s \in \mathcal{C}_k^{(0)}} \frac{1}{\sigma_v^2} |y[n] - \sqrt{E_s}s|^2 \quad (2.26)$$

where a strongly positive value implies that the kth bit is most likely a 0 whereas a strongly negative value implies that kth bit is most likely 1. An ambiguous bit decision means that the LLR is close to zero. Detectors with error control coding can use this information to determine which bits are more likely to be decoded with low fidelity and can incorporate this into their decoding algorithms.

2.3.3 Binary Phase Shift Keying and Variants

Thus far, constellations have been abstractly defined as a real or complex set of points. For the remainder of Section 2.3, we will consider specific constellation realizations, symbol mapping procedures onto these constellations, and the resultant consequences in communication system performance. Binary phase shift keying (BPSK) is a simple constellation and symbol mapping procedure used to communicate one bit per symbol. BPSK and variations are illustrated in Fig. 2.5 with common bit labeling (mappings of bits to symbols). BPSK can be expressed using the transmit waveform in (2.18) with standard constellation $\mathcal{C} = \mathcal{C}_{\text{BPSK}} = \{1, -1\}$. Constellation mapping generally assumes

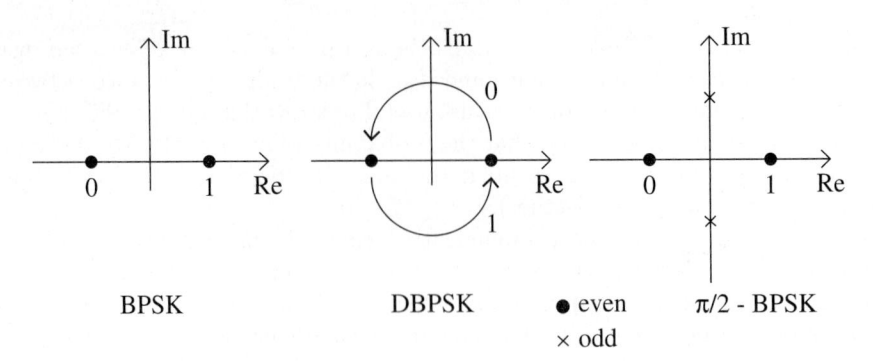

Figure 2.5 BPSK with bit labeling from IEEE 802.15.3c is shown in the left. The same labeling is used in IEEE 802.11ad. With DBPSK (middle), the information is encoded into the successive phase differences. For $\pi/2$–BPSK (right), the • denotes even symbols and × for odd symbols.

one of two demodulation architectures: (1) coherent demodulation in which the phase is precisely known through synchronization algorithms and (2) differentially coherent demodulation in which the receiver is not aware of the initial waveform phase.

Standard BPSK generally assumes coherent demodulation and if $\{b[n]\}$ is the sequence of transmitted bits, the bit-to-symbol mapping provides $s[n] = (-1)^{b[n]}$ (thus the bit labeling $b[n] = 0 \Rightarrow s[n] = 1$ and $b[n] = 1 \Rightarrow s[n] = -1$). Alternatively, differential BPSK (DBPSK) is compatible with a certain kind of noncoherent demodulation in which the initial phase of the received waveform is unknown. In DBPSK, the bit sequence $\{b[n]\}$ is differentially encoded to produce $s[n] = (-1)^{(b[n]-b[n-1])}s[n-1]$ with the first element in $\{s[n]\}$ arbitrarily initialized to 1 or -1. Hence, if the phase of adjacent samples at the receiver changes by $\pm\pi$ radians, this indicates a 1 bit and no phase change indicates a 0 bit. This is compatible with noncoherent demodulation since $y^*[n]y[n-1]$ does not depend on knowledge of the initial received phase.

$\pi/2$-BPSK is a special variation of BPSK that is commonly implemented for 60 GHz radios. One drawback of standard BPSK and DBPSK is that the phase of adjacent symbols can change up to π radians at a time. Large phase transitions lead to spectral outgrowth (inefficient use of bandwidth) which, in turn, leads to complex envelope variations after bandlimiting filters. Complex envelope variation is undesirable for the PA in the transmitter since, in general, low complex envelope variation allows for cheap and efficient PA implementation.[9] $\pi/2$-BPSK reduces complex envelope variation by reducing phase transitions from π radians to $\pi/2$ radians at the cost of turning BPSK into a complex constellation. Essentially, $\pi/2$-BPSK maps adjacent samples to two different constellations, either $\{1,-1\}$ or $\{j,-j\}$, each $\pi/2$ rotations of the other. In IEEE 802.15.3c, a successive rotation feature is added to provide compatibility with Gaussian Minimum Shift Keying (GMSK)[10]. Mathematically, if $\{s[n]\}$ is a BPSK-mapped sequence, then $\{j^n s[n]\}$ is the $\pi/2$-BPSK-mapped sequence in IEEE 802.15.3c. This is discussed further in Chapter 9.

---

9. Complex envelope variation is characterized through the peak-to-average-power ratio (PAPR) of transmitted signals and is discussed more in Chapters 5 and 7.
10. GMSK is treated in [Rap02].

2.3 Digital Modulation

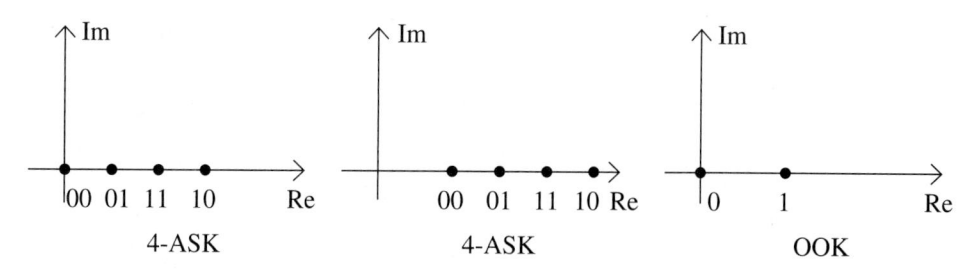

Figure 2.6 Left and middle, two variations of 4-ASK with a gray coded bit labeling. Right, On-Off Keying (OOK) with a conventional bit labeling.

2.3.4 Amplitude Shift Keying and Variants

M-Amplitude Shift Keying (M-ASK) is a type of digital pulse amplitude modulation that encodes information onto a discrete set of amplitude levels. ASK and some variations are illustrated in Fig. 2.6. ASK is different from BPSK and M-ary quadrature amplitude modulation (M-QAM) in that it does not use the phase to convey information. In M-ASK, the amplitude levels are generally equally spaced. The constellation might take the form $\{0, 1, \ldots, M-1\}$, which includes a zero value, or might instead use $\{1, 2, \ldots, M\}$ when it is desirable not to include the zero. The pulse shaping filter for ASK is usually assumed to be a square pulse $g_{\mathrm{tx}}(t) = 1$ for $t \in [0, T_\mathrm{s})$ and zero otherwise. Though inefficient from a bandwidth perspective (the square pulse shape creates a wide spectrum in frequency), ASK modulation has some practical advantages: (1) it is not especially sensitive to phase noise or frequency offset and (2) the peak-to-average power ratio is not too large, making ASK more friendly for implementation. 60 GHz systems have considered primarily 4-ASK and 2-ASK, in lower-cost and shorter-range applications.

On-Off Keying (OOK) is a special case of 2-ASK in which the constellation takes the form $\{0, 1\}$. OOK is known to provide low power consumption [SUR09] and has been the subject of a number of early 60 GHz integrated transceivers [KSM+98]. Essentially, with OOK, a sine wave is sent during a 1 and nothing is sent during a 0. Consequently, modulation and demodulation (if there are no channel impairments) can be very simple. For example, at the transmitter, it is also possible to directly modulate the RF carrier. At the receiver, envelope demodulation can be used at the receiver that does not require using a local oscillator or clock recovery circuit [DC07], though this can be tricky in practice since noise thresholds need to be computed.

2.3.5 Quadrature Phase Shift Keying and Variants

Quadrature phase shift keying (QPSK) is a complex modulation scheme that sends two bits of information per symbol period. It can be represented using (2.18) with complex symbols and conceptually is equivalent to using BPSK modulation for each of the in-phase and quadrature components. The constellation for QPSK is $\mathcal{C}_{\mathrm{QPSK}} = \frac{1}{\sqrt{2}}\{1+j, -1+j, -1-j, 1-j\}$. QPSK and related constellations and their bit-to-symbol mappings are illustrated in Fig. 2.7. Differential QPSK (DQPSK) can be employed in a way similar to DBPSK. If $\{q[n]\}$ is a sequence of QPSK modulated symbols then the symbols are differentially encoded to provide $s[n] = q[n]s[n-1]$. Noncoherent demodulation can be

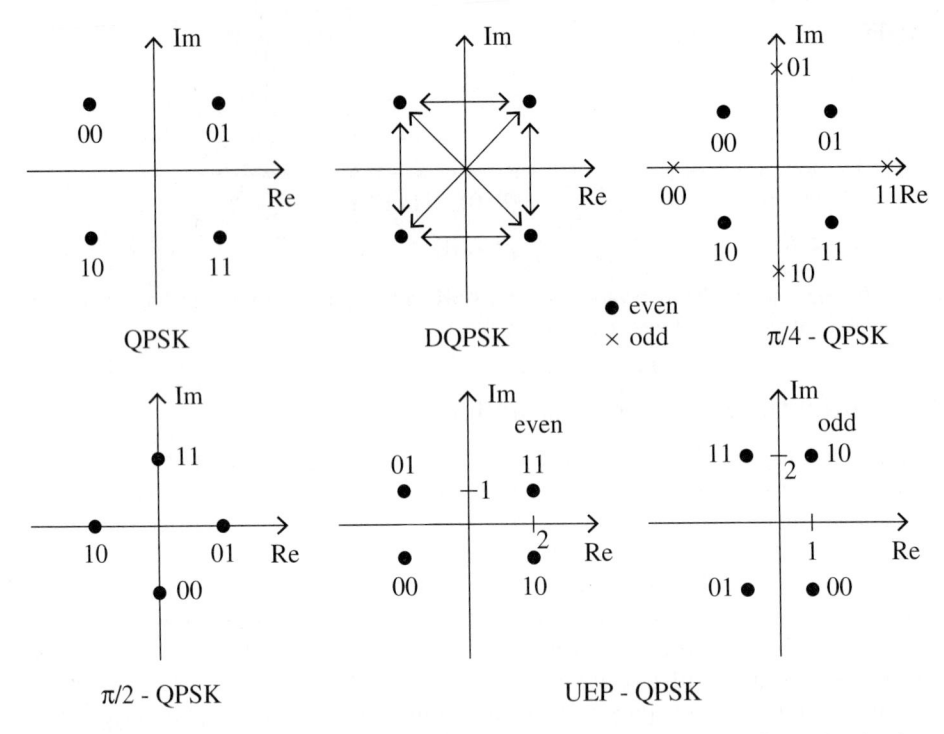

Figure 2.7 Top, left, QPSK with common bit labeling. DQPSK encodes information in the transitions between bits (encoding not shown). Top, right, $\pi/4$–QPSK uses two sets of constellations: one for the even and one for the odd symbols. Bottom, left, $\pi/2$–QPSK with bit labeling from IEEE 802.15.3c has a non-quadrature constellation. Bottom, middle, and right, Unequal error protection (UEP) QPSK has its information encoded using the same rotation operation as for $\pi/2$–BPSK.

performed through operations on $y[n]y^*[n-1]$, making DQPSK robust to phase offsets at the expense of additional noise (the optimal detector also becomes more complex due to noise products).

While the standard implementation of QPSK also suffers from large potential phase transitions (up to π radians) between adjacent symbols, leading to undesirable spectral outgrowth and complex envelope variation, envelope variations are reduced through $\pi/4$–QPSK [JMW72] by limiting the maximum phase transition between adjacent samples to $3\pi/4$ radians by alternating between the QPSK constellation and the $\pi/4$-rotated QPSK constellation. For example, even symbols will be mapped into the constellation $\{e^{j\pi/2}, e^{j3\pi/2}, e^{j5\pi/2}, e^{j7\pi/2}\}$, whereas odd symbols will be mapped into the $\pi/4$ phase-shifted constellation $\{e^{j\pi/2+j\pi/4}, e^{j3\pi/2+j\pi/4}, e^{j5\pi/2+j\pi/4}, e^{j7\pi/2+j\pi/4}\}$.

Offset QPSK (OQPSK) is another method for reducing envelope fluctuations in QPSK (also known as *staggered QPSK*) [Rho74]. OQPSK offsets the quadrature component by $T_s/2$ in the transmitted waveform such that

$$x(t) = \sqrt{E_s} \sum_{n=-\infty}^{\infty} s_I[n]g_{\text{tx}}(t-nT_s) + js_Q[n]g_{\text{tx}}(t-nT_s-T_s/2) \qquad (2.27)$$

2.3 Digital Modulation

where $s_I[n]$ is the real (in-phase) part of $s[n]$ and $s_Q[n]$ is the imaginary (quadrature) part of $s[n]$. Because the in-phase and quadrature components change at different time instants, the received signal fluctuations are less drastic, instantaneously making it more suitable for lower-performance PAs at the cost of increased transmitter and receiver complexity.

Unequal error protection QPSK (UEP-QPSK) is an asymmetric constellation, or nonuniform constellation, in which one bit is given more power than the other bit. For example, an unnormalized UEP-QPSK constellation might look like $\{\alpha + j, -\alpha + j, \alpha - j, -\alpha - j\}$ where $\alpha > 1$. The motivation for using unequal error protection is for multimedia transmission (audio or video) where sensitive information may require more protection while less sensitive (or optional) information is given less protection. UEP-QPSK provides a convenient way of protecting two different classes of data and with lower complexity than a pure coding solution [SKLG98]. The type of UEP-QPSK used in 60 GHz systems uses a different constellation for even and odd symbols, as illustrated in Fig. 2.7. In this way the in-phase and quadrature signals maintain the same average power, which is advantageous for transceiver design.

2.3.6 Phase Shift Keying

M-Phase Shift Keying (M-PSK) is a constellation constructed by taking equally spaced points on the complex unit circle. Two examples of M-PSK are illustrated in Fig. 2.8. It also uses the same complex pulse amplitude waveform represented using (2.18). The unnormalized M-PSK constellation is given by $\left\{e^{\frac{j2\pi k}{M}}\right\}_{k=0}^{M-1}$. The normalization factor is simply $1/\sqrt{M}$. Although M-PSK can work for any positive integer M, most applications are designed with $M = 8$, as a way to fill in the gap between 4-QAM and 16-QAM. M-PSK has similar characteristics to QPSK, for example, it may be differentially encoded in a similar manner to create M-DPSK. In 60 GHz systems, $\pi/2$ 8-PSK is employed where the symbols are rotated by $j^n s[n]$ to maintain compatibility with the other $\pi/2$ rotated modulations.

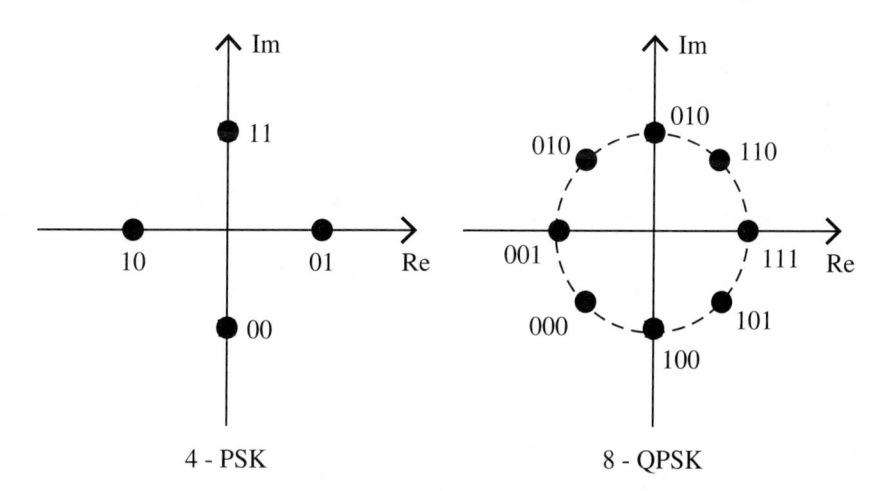

Figure 2.8 Left, 4-PSK with the same bit labeling as $\pi/2$–QPSK from IEEE 802.15.3c. Right, 8-PSK with bit labeling from ECMA 387.

2.3.7 Quadrature Amplitude Modulation

Quadrature amplitude modulation (QAM) provides a generalization of QPSK to cardinality $M = 2^b$ for $b > 2$ through phase *and* amplitude shifts [CG62] using the same complex pulse amplitude waveform representation (2.18). Several QAM constellations and their bit-to-symbol mappings are illustrated in Fig. 2.9. If M is a power of 4 (i.e., square QAM), the un-normalized M-QAM constellation is provided by combinations of $a + jb$ where $a, b \in \{-(M/2 - 1), \ldots, -1, 1, \ldots, (M/2 - 1)\}$, that is, the Cartesian product of an all-real $M/2$ pulse amplitude modulation ($M/2$-PAM) constellation and an all-complex $M/2$-PAM constellation. Common square QAM dimensions include $M = 4, 16, 64$, and 256.

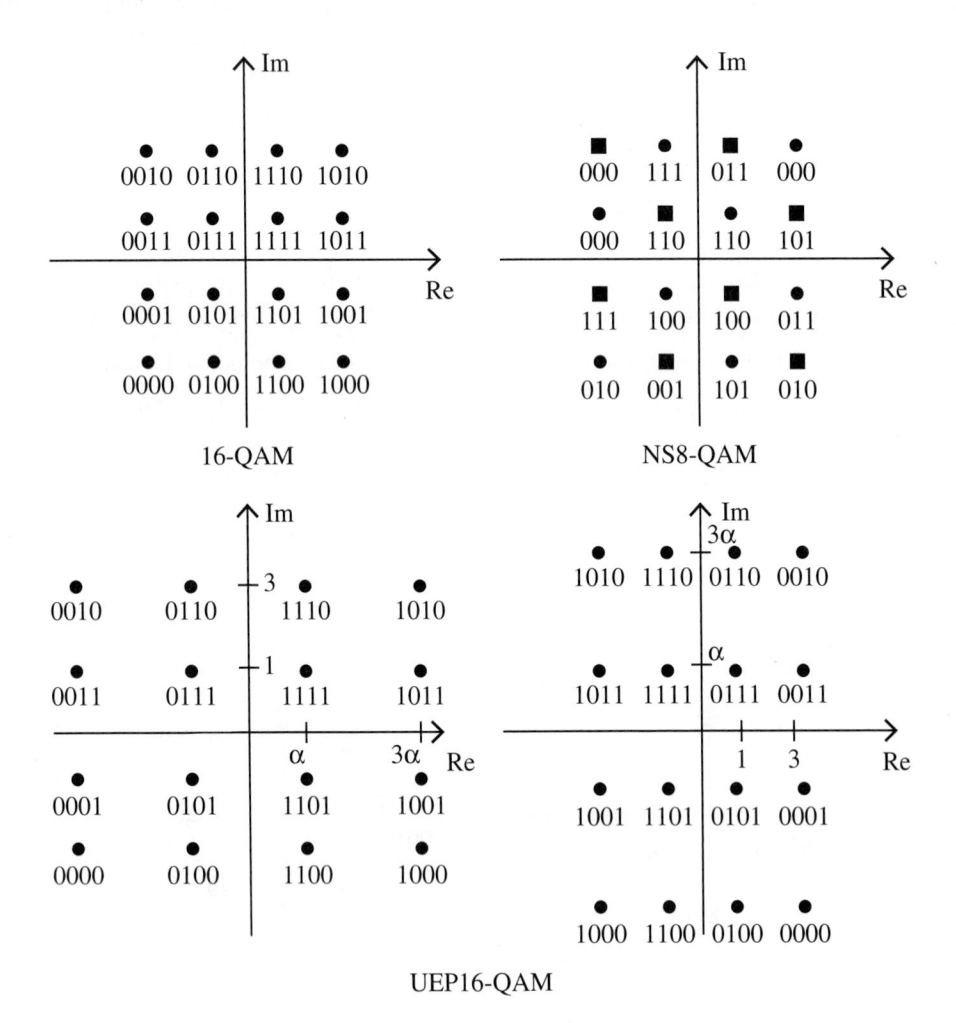

Figure 2.9 Top, left, 16-QAM with bit labeling from IEEE 802.15.3c. Top, right, NS8-QAM with bit labeling from ECMA 387. Circles correspond to mappings for even-numbered symbols, whereas squares provide the mappings for odd-numbered symbols. Bottom, UEP 16-QAM with bit labeling from ECMA 387 for even and odd transmissions. Given mappings of the form $b_1 b_2 b_3 b_4$, priority is given to bits b_3 and b_4 [ECMA08].

When M is not a power of 4, QAM constellations are not invariant to $\pi/2$ phase rotations and cannot be produced through Cartesian products of symmetric PAM constellations. Several non-square QAM constellations have been proposed, including cross constellations [CG62][Smi75] and with general non-square constellations [Tor02][WZL10]. The non-square 8-QAM (NS8-QAM) constellation, for example, is present in ECMA 387. NS8-QAM is constructed from 16-QAM by removing every other point, for example, with an unnormalized constellation given by $\{-1+3j, 3+3j, -3+j, 1+j, -1-j, 3-j, -3-3j, 1-3j\}$.

2.4 Equalization in the Time Domain

MmWave wireless transmissions will encounter frequency selective fading in the propagation channel. Frequency selective channels smear the transmitted signal, resulting in what is known as *intersymbol interference* (ISI) at the receiver. This impairment must be dealt with at the receiver along with additive noise, complicating the detection procedure. The general term for removing the effects of frequency selectivity is known as *equalizing* the received signal.

To understand the challenge of equalization, it is useful to build a mathematical model for the received signal that combines the effects of the propagation channel and the modulation. We focus the description on modulations that can be described using the complex pulse amplitude modulation in (2.18), which includes all of the modulations in Section 2.3. A basic received signal model including a frequency selective channel and additive noise is given by augmenting (2.7) with an additive noise term as

$$y(t) = \int_{-\infty}^{\infty} h(t-\tau)x(\tau)d\tau + v(t) \tag{2.28}$$

where $v(t)$ is filtered noise. Inserting (2.18) and simplifying gives

$$y(t) = \int_{-\infty}^{\infty} h(t-\tau)\sqrt{E_\text{s}} \sum_{n=-\infty}^{\infty} s[n]g_\text{tx}(\tau - nT_\text{s})d\tau + v(t) \tag{2.29}$$

$$= \sum_{n=-\infty}^{\infty} s[n] \int_{-\infty}^{\infty} h(t-\tau)\sqrt{E_\text{s}}g_\text{tx}(\tau - nT_\text{s})d\tau + v(t) \tag{2.30}$$

$$= \sum_{n=-\infty}^{\infty} s[n] \int_{-\infty}^{\infty} h_\text{g}(t - nT_\text{s})d\tau + v(t) \tag{2.31}$$

where $h_\text{g}(\tau)$ is the scaled and filtered convolution of the baseband equivalent channel $h(\tau)$ and the transmit pulse shape $g_\text{tx}(\tau)$ including the scaling factor $\sqrt{E_\text{s}}$. Sampling at the symbol rate gives the equivalent system

$$y[n] = \sum_{k=-\infty}^{\infty} s[k]h[n-k] + v[n]$$

$$= s[n]h[0] + \underbrace{\sum_{k=-\infty, k\neq 0}^{\infty} s[k]h[n-k]}_{\text{intersymbol interference}} + v[n] \tag{2.32}$$

where $h[n] = Th_{\mathrm{g}}(nT_{\mathrm{s}})$ with some abuse of notation. Frequency selective channels smear the symbols creating the form of "self" interference known as ISI. We refer to $\{h[\ell]\}$ simply as *the channel*, even though it includes multipath propagation, transmit and receive filtering, the pulse shaping function, and the symbol energy.

In practical systems, frequency selective channels are causal and have finite memory. A somewhat simpler model is

$$y[n] = \sum_{\ell=0}^{L} h[\ell] s[n-\ell] + v[n] \qquad (2.33)$$

where L is the order of the filter ($L+1$ is the number of channel taps). The size of L is a function of the propagation environment and is related to the maximum excess delay in the channel. Generally, the value of L is determined through measurements of the power delay profile, root mean square delay spread, and mean excess delay. These are statistical characterizations of multipath fading that are used to quantify the severity of the channel and are described in more detail in Chapter 3. The eventually selected value of L is chosen to be sufficient for most operating environments. As a concrete example, measurement results showed that 60 GHz systems experience root mean delay spreads of 8 ns to 80 ns depending on the environment [WRBD91][XKR02]. To a first order, L can be determined by dividing the root mean square delay spread by the symbol period (a better approximation would use the mean excess delay spread). For example, for a system with a bandwidth of 500 MHz and 25% excess bandwidth, among the smaller bandwidths under consideration, the symbol period with a pulse amplitude modulation would be 2.5 ns. In this case, the value of L would range from about 3 to 32. The value of L would be higher for larger bandwidths; for example, an RF channel bandwidth of $1,815$ MHz is used for IEEE 802.15.3c. Directional antennas can be used to reduce the effective delay spread [WAN97] but will not eliminate it completely.

Equalization is a general term for removing the effects of ISI [Rap02]. There are several different strategies for equalization that involve signal processing at the transmitter and/or at the receiver. In this section, we discuss various alternatives for equalizing the channel in the time domain. Linear equalization inverts the effects of the channel by filtering the received signal with a specially designed filter. Decision feedback equalization uses tentative decisions to feed back and subtract out previously detected symbols. Maximum likelihood sequence estimation performs a joint detection of all the symbols. In this next section, we discuss options for performing equalization in the frequency domain through what is known as single carrier frequency domain equalization (SC-FDE) and orthogonal frequency division multiplexing (OFDM).

The equalization description in this chapter assumes that the equalization is performed in discrete time. This is not necessary for implementation. As long as the Nyquist sampling theorem is satisfied, the equalization can be performed before sampling (analog equalization) or after sampling (digital). Analog equalization is not a new topic; it was the primary equalization strategy before digital circuits became feasible. It is of interest in mmWave because ADCs for the large bandwidth signals found in mmWave require high speed and precision for many kinds of modulation. For example, recent work [HRA10] makes the case that analog equalization can reduce the required resolution and thus the power consumed by the wideband ADC. Hybrids of analog and digital equalization are

2.4 Equalization in the Time Domain

possible. Recent work in [SB08] describes a decision feedback architecture in which decisions made in the digital are subtracted off in analog domain prior to sampling. Analog equalization has been used in many wireline standards, including gigabit Ethernet [HS03]. Although we focus on digital equalization, the principles can be applied to design analog equalizers as well, with additional thought required for efficient circuit implementation.

2.4.1 Linear Equalization

Linear equalization is one simple approach to remove the effects of frequency selective fading. The idea is to find a filter that removes the effects of the frequency selectivity. As illustrated in Fig. 2.10, an equalizer is followed by a symbol-by-symbol detector as in (2.22). Linear equalization is not a substitute for detection; rather, it permits application of the same detector as would be applied with just noise, at the expense of performance compared with a more optimum detector.

We consider a finite impulse response equalizer, but an infinite impulse response could also be implemented, though more complex in practice since it requires feedback. Expressed mathematically, let $\{f[\ell]\}_{\ell=0}^{L_{\mathrm{f}}}$ denote the taps of the finite impulse response (FIR) equalizing filter. Applying the equalizer to the sampled received signal in (2.33) gives

$$z[n] = \sum_{\ell=0}^{L_{\mathrm{f}}} f[\ell] y[n-\ell] \qquad (2.34)$$

$$= \sum_{k=-\infty}^{\infty} s[k] \sum_{\ell=0}^{L_{\mathrm{f}}} \sum_{m=0}^{L} f[\ell] h[n-\ell-m] + \sum_{m=0}^{L} f[\ell] v[k-\ell]. \qquad (2.35)$$

To remove intersymbol interference (ISI), the equalized channel should give a $\delta[n]$. We can relax this somewhat and allow the filter to give a delayed output $\delta[n-\Delta]$ where Δ is a design parameter. The equalizer coefficients can be optimized over Δ to give better performance.

The choice of L_{f} is a design decision that depends on the memory in the channel given by L, which represents the extent of the multipath in the channel. It is determined by the bandwidth of the signal as well as the maximum delay spread derived from propagation channel measurements. The equalizer $\{f[l]\}_{l=0}^{L_{\mathrm{f}}}$ is an FIR inverse of an FIR filter. As a consequence, the results will improve if L_{f} is large. The complexity required per symbol, however, also grows with L_{f}. Therefore, there is a tradeoff between choosing large L_{f} to have better equalizer performance and choosing a smaller L_{f} to have a more efficient receiver implementation.

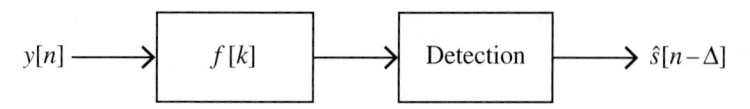

$y[n] \longrightarrow \boxed{f[k]} \longrightarrow \boxed{\text{Detection}} \longrightarrow \hat{s}[n-\Delta]$

Figure 2.10 Linear equalization filters the received signal, as shown by filter $\{f[k]\}$. It is followed by detection that operates on a symbol-by-symbol basis. Allowing for delay Δ in the output may improve performance.

There are different criteria for designing a good equalizer. Zero-forcing is a well-known approach, but this involves designing the equalizer coefficients to exactly invert the filter, requiring an infinite impulse response (IIR) [Rap02][OS09]. With a finite L_f, a related design is the least-squares equalizer. The objective is to find the filter coefficients $\{f[\ell]\}_{\ell=0}^{L_f}$ such that the squared error

$$\sum_{n=0}^{L+L_f} \left| \sum_{\ell=0}^{L_f} f[\ell] h[n-\ell] - \delta[n-\Delta] \right|^2 \qquad (2.36)$$

is minimized. A solution can be found by writing in matrix form. Let

$$\underbrace{\begin{bmatrix} h[0] & 0 & \cdots & \cdots \\ h[1] & h[0] & 0 & \cdots \\ \vdots & & \ddots & \\ h[L] & & & \\ 0 & h[L] & \cdots & \\ \vdots & & & \end{bmatrix}}_{\bar{\mathbf{H}}} \underbrace{\begin{bmatrix} f[0] \\ f[1] \\ \vdots \\ \vdots \\ f[L_f] \end{bmatrix}}_{\mathbf{f}} = \underbrace{\begin{bmatrix} 0 \\ \vdots \\ 1 \\ \vdots \\ 0 \end{bmatrix}}_{\mathbf{e}_\Delta} \leftarrow \Delta + 1 \quad . \qquad (2.37)$$

The matrix $\bar{\mathbf{H}}$ is a $(L + L_f + 1) \times (L_f + 1)$ Toepltiz matrix. Assuming that $\bar{\mathbf{H}}$ is full rank, which is guaranteed if one of the channel coefficients is nonzero, then the least-squares filter coefficients are given by $\widehat{\mathbf{f}}_\Delta = (\bar{\mathbf{H}}^* \bar{\mathbf{H}})^{-1} \bar{\mathbf{H}}^* \mathbf{e}_\Delta$. The least-squares solution can be computed efficiently using recursive least squares (see [KSH00] and the references therein).

The least-squares solution does not take into account the coloring of the white Gaussian noise that occurs because of the filtering in (2.35). Another approach to equalization that considers the impact of noise is minimizing the mean squared error, a type of Wiener filter. The minimum mean squared error (MMSE) FIR equalizer minimizes

$$\mathbb{E} \left| \sum_{\ell=0}^{L_f} f[\ell] y[n-\ell] - s[n-\Delta] \right|^2 \qquad (2.38)$$

where $\mathbb{E}$ is the expectation operator. Assuming that the transmitted symbols $s[n]$ and $v[n]$ are independent, and that $s[n]$ is modeled as an independent identically distributed random process while $v[n]$ is a zero-mean wide sense stationary random process with correlation function $R_v[k] = \mathbb{E}v[n]v^*[n+k]$, then the MMSE equalizer is given by $\widehat{\mathbf{f}}_\Delta = \mathbf{R}_y^{-1} \mathbf{H}^* \mathbf{e}_\Delta$ where $\mathbf{R}_y = \mathbf{H}\mathbf{H}^* + \mathbf{R}_v$ is known as the received signal covariance matrix,

$$\mathbf{H} = \begin{bmatrix} h[0] & h[1] & \cdots & h[L] & 0 & \cdots \\ 0 & h[0] & & \cdots & h[L] & 0 & \cdots \\ \vdots & & \ddots & & & & \\ 0 & \cdots & h[0] & h[1] & \cdots & h[L] \end{bmatrix} \qquad (2.39)$$

2.4 Equalization in the Time Domain

is the $(L_f + L + 1 \times L_f + 1)$ channel matrix, and

$$\mathbf{R}_v = \begin{bmatrix} R_v[0] & R_v[1] & \cdots & R_v[L_f] \\ R_v^*[1] & R_v[0] & \cdots & R_v[Lf-1] \\ \vdots & \ddots & \ddots & \vdots \\ R_v^*[L_f] & \cdots & \cdots & R_v[0] \end{bmatrix} \quad (2.40)$$

is the noise covariance matrix. The received signal covariance matrix can be computed directly from $y[n]$ through sample averaging or can be computed from the expression for $\mathbf{R}_y$ given $\mathbf{H}$ and $\mathbf{R}_v$. If the noise is additive white Gaussian noise (AWGN) then $\mathbf{R}_v = \sigma_v^2 \mathbf{I}$. The MMSE solution can be computed efficiently using the Levison-Durbin recursion without requiring an explicit inverse of $\mathbf{R}_y$.

There are many variations of the MMSE solution including causal and non-causal IIR filters, various kinds of adaptive filters, as well as generalization like the Kalman filter (described in many texts on statistical signal processing including [KSH00][Hay96][Say08]). Low-complexity adaptive equalization, as illustrated in Fig. 2.11, is of practical interest. The detected symbols are used to update the filter coefficients for future symbols. We give an implementation here based on the least mean squares (LMS) algorithm[Wid65][Qur85][Say08]. Let $\mathbf{f}_n$ denote the vector filter coefficients applied to the discrete-time signal at time n. We omit the Δ notation. Let $\mathbf{y}[n] = [y[n], y[n-1], \ldots, y[n-L_f]]^T$. The LMS weight vector update would compute

$$\mathbf{f}_{n+1} = \mathbf{f}_n + \mu(\mathbf{f}_n^*\mathbf{y}[n] - \widehat{s}[n-\Delta])\mathbf{y}[n]. \quad (2.41)$$

The term $\mathbf{f}_n^*\mathbf{y}[n] - \widehat{s}[n-\Delta]$ is known as an error signal. The parameter μ is the step-size and governs the convergence properties of the algorithm. The adaptive equalizer could be trained using known training data, tentatively detected symbols, or re-encoded symbols after error control decoding for more robustness.

2.4.2 Decision Feedback Equalization

A decision feedback equalizer (DFE) is a way to deal with large memory in the channel (large L). The core idea of decision feedback is to subtract past decisions from the

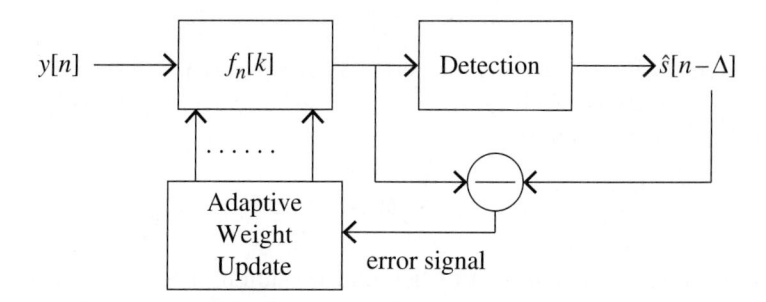

Figure 2.11 A linear equalizer filters the received signal. It is followed by detection that operates on a symbol-by-symbol basis. The input and output of the detector are used to generate an error signal, which is used to update the filter weights.

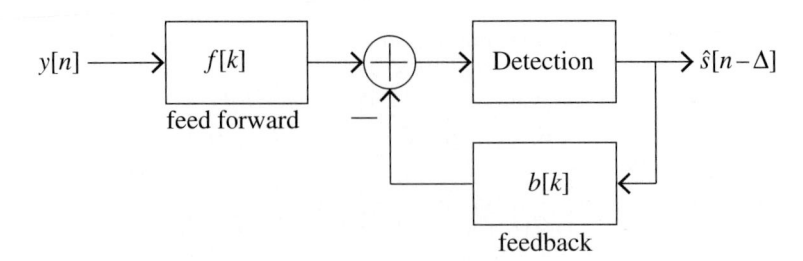

Figure 2.12 A decision feedback equalizer (DFE) subtracts previous decisions to improve equalizer performance. The feedback filtered feedback symbols are subtracted from feedforward filtered input and passed to the detector.

filtered input signal, reducing the effective amount of intersymbol interference (ISI) that must be equalized. A typical DFE structure is illustrated in Fig. 2.12. Because errors in the past decisions may cause error propagation, decision feedback works the best with high-quality decisions obtained after decoding, which may incur too much delay. Decision feedback equalization in 60 GHz has been proposed using a hybrid of analog feedforward and digital feedback [SB08]. DFEs are information lossless under certain assumptions in the sense that they do not result in a loss in channel capacity as may be found with linear equalizers [GV05]. There are many variations of decision feedback equalization including zero forcing [KT73] and MMSE [Mon71], with different degrees of causality. A wide range of algorithms, such as least mean squares (LMS) and recursive least squares (RLS), are effective at implementing time domain equalizers such as DFEs as long as the computing platform is sufficiently able to adapt at the transmitted data rates. Work in [RF91] and [RHF93] illustrates state-of-the-art simulation and implementation methods used for early second- and third-generation digital cellular standards, where data rates were on the order of a megabit per second. Computing architectures may be able to successfully implement DFEs at Gbps data rates in future wireless devices. In this section we summarize the construction from [VLC96] for an MMSE decision feedback equalizer (DFE) with finite length and delay Δ.

Let $\{f[\ell]\}_{\ell=0}^{L_f}$ denote the coefficients of the feedforward filter and let $\{b[\ell]\}_{\ell=1}^{L_b}$ denote the coefficients of the feedback filter. The following signal is input into the symbol detector

$$\sum_{\ell=0}^{L_f} f[\ell]y[n-\ell] - \sum_{\ell=1}^{L_b} b[\ell]\hat{s}[n-\Delta-\ell], \qquad (2.42)$$

which then produces the detected symbol $\hat{s}[n-\Delta]$. For derivation of the DFE coefficients and analysis, the detected symbols are assumed to be correct, thus $\hat{s}[n] = s[n]$. The MMSE DFE coefficients minimize

$$\mathbb{E}\left|\sum_{\ell=0}^{L_f} f[\ell]y[n-\ell] - \sum_{\ell=1}^{L_b} b[\ell]s[n-\Delta-\ell] - s[n]\right|^2. \qquad (2.43)$$

The detailed derivation is found in [VLC96], and we summarize the result here. Let

$$\mathbf{Q} = \begin{bmatrix} \mathbf{0}_{L_b+1,\Delta} & \mathbf{I}_{L_b+1} & \mathbf{0}_{L_b+1,L_f-L_b-\Delta} \end{bmatrix}$$
$$\times \begin{bmatrix} \mathbf{I}_{L_f+L_b+1} + \mathbf{H}^*\mathbf{R}_v^{-1}\mathbf{H} \end{bmatrix}^{-1} \begin{bmatrix} \mathbf{0}_{\Delta,L_b+1} & \mathbf{I}_{L_b+1} & \mathbf{0}_{L_f-L_b-\Delta,L_b+1} \end{bmatrix} \qquad (2.44)$$

2.4 Equalization in the Time Domain

and then partition the matrix as

$$\mathbf{Q} = \begin{bmatrix} p & \mathbf{q}^* \\ \mathbf{q} & \mathbf{P} \end{bmatrix}. \tag{2.45}$$

Then vector of feedback coefficients $\mathbf{b} = [b[1], \ldots, b[L_b]]^T$ is given by

$$\mathbf{b} = -\mathbf{P}^{-1}\mathbf{q} \tag{2.46}$$

and

$$\mathbf{f} = \left[\mathbf{I}_{L_f+L_b+1} + \mathbf{H}^* \mathbf{R}_v^{-1} \mathbf{H}\right]^{-1} \mathbf{H} \begin{bmatrix} \mathbf{0}_{\Delta,1} \\ 1 \\ \mathbf{b} \\ \mathbf{0}_{L_f-L_b-\Delta} \end{bmatrix}. \tag{2.47}$$

Adaptive versions of the decision feedback equalizer are also possible [GBS71].

2.4.3 Maximum Likelihood Sequence Estimation

The optimum receiver from the perspective of sequence detection given the received signal model in (2.33), and assuming equally likely sequences, is known as the *maximum likelihood sequence estimator* (MLSE).[11] Formulating the maximum likelihood problem and simplifying, the maximum likelihood solution is found by solving

$$\{\widehat{s}[n]\}_{n=0}^{N-1} = \arg\min_{\{s[n]\}_{n=0}^{N-1}} \sum_{n=0}^{N-1} \left| y[n] - \sum_{\ell=0}^{L} h[\ell]s[n-\ell] \right|^2. \tag{2.48}$$

The metric in (2.48) can be simplified by expanding the norm, leading to the Ungerboek form [Ung74].

The optimization in (2.48) can be performed efficiently using the Viterbi algorithm [FJ73]. There are two common approaches for implementing the Viterbi algorithm: Forney's method [FJ72] and Ungerboek's method [Ung74]. Both approaches start the problem from a continuous-time formulation for pulse amplitude modulations. The Forney receiver involves an analog matched filter, followed by sampling, noise whitening, and then sequence estimation implemented efficiently using the Viterbi algorithm. The Ungerboek receiver involves sampling after an analog matched filter and then sequence estimation with a modified metric (to account for the colored noise) implemented efficiently using the Viterbi algorithm. The Forney and Ungerboek methods are equivalent, and one can be derived from the other [BC98].

The analog matched filter (matched to the combined propagation channel and transmit pulse shape) as suggested in the Forney and Ungerboek methods is challenging to implement in practice. One approach is to perform the matched filtering in the digital domain. Then the matched filter can be implemented as a digital filter after the ADC.

---

[11]. It should really be known as the maximum likelihood sequence detector, but the term *MLSE* has gained wide acceptance.

The oversampling rate needs to be chosen to ensure that the Nyquist sampling theorem is satisfied. It does not matter whether the matched filter is performed in digital or analog as long as the front-end filtering bandwidth is matched to the oversample rate. Filtering in the oversample domain is followed by downsampling to arrive at the symbol rate sampled system (see, e.g., [BC98] for more details on the use of oversampling for receiver implementation).

One approach would be to oversample, perform the matched filter at the oversampled rate, then downsample and proceed with the Ungerboek or Forney approaches. This is still optimum provided that front-end filtering is such that the baseband noise after the first round of sampling is still white, which occurs after sampling at precisely the Nyquist rate. An alternative approach is to start directly from (2.33). If the sampling is such that the noise is sampled at exactly the Nyquist rate, then these approaches are identical. Otherwise oversampling would be employed, requiring a matched filter in the oversample domain and downsampling [BC98].

Thus far, MLSE does not seem to have been employed in mmWave implementations. The reason is complexity and memory requirements. Given a constellation size of M, there are M^L states. During each discrete-time n, M comparisons must be made at every step to determine the optimum survivor. This is a challenging requirement when the receiver must operate on gigasamples per second. There are, however, many different lower complexity, suboptimum MLSE algorithms that could be employed in future mmWave systems, for example, using a reduced number of states [Fos77][Mcl80][EQ89], sparsity in the channel [BM96], or combining with linear equalizers [Bea78], or combining with a decision feedback equalizer (DFE) [LH77], or it could be implemented as delayed decision feedback estimation [DHH89].

2.5 Equalization in the Frequency Domain

Equalization in the time domain can be challenging for channels where the delay spread is large relative to the symbol period. This means that the number of channel taps in the discrete-time channel — or the memory in the channel represented by L — is large. Using a linear equalizer, a large number of channel taps will be required to mitigate the effects of the channel. Using the MLSE solution, there will be a large number of statistics and high complexity. Note that the convolution operation becomes exceedingly complex when performing time domain equalization, but simple multiplications can be done when in the frequency domain. This section considers efficient ways to perform equalization directly in the frequency domain.

Recall that *equalization* is a term for removing the smearing of the channel on the transmitted symbols. Mathematically, this means removing the effect of $\{h[\ell]\}_{\ell=0}^{L}$ on the transmitted symbol sequence $\{s[n]\}$ in (2.33). From basic signals and systems theory, it is well known that convolution in the time domain appears as a product in the frequency domain [OS09]. Conceptually then, it should be possible to equalize the effects of the channel by taking the Fourier transform of $y[n]$ and dividing by the Fourier transform of the channel. In practice, this exact operation is challenging. The reason is that the discrete-time Fourier transform of $\{y[n]\}$ is defined as $\sum_k y[k]\exp(-j\omega k)$ where ω is the

2.5 Equalization in the Frequency Domain

discrete-time frequency with values $\omega \in [-\pi, \pi)$. The fact that the frequency domain is continuous makes this direct signals-and-systems approach not viable. An alternative is to leverage a different frequency domain transform known as the *discrete Fourier transform* (DFT).

The DFT is defined for finite-length signals as

$$\mathsf{Y}[k] = \sum_{n=0}^{N-1} y[n] e^{-j\frac{2\pi}{N}kn} \qquad (2.49)$$

where $k = 0, 1, \ldots, N-1$. The DFT can be computed efficiently using the fast Fourier transform (FFT) with complexity $\mathcal{O}(N \log N)$. Unfortunately, there is a small problem. With finite length signals, the inverse of the product $\mathsf{H}[k]\mathsf{S}[k]$ is the circular convolution between $\{h[\ell]\}$ and $\{s[n]\}$, not the linear convolution in (2.33). While it seems like an irrelevant mathematical detail, this means that the DFT cannot be applied directly to equalize the received signal. One solution to this problem is to use block processing where a specially designed guard interval between blocks allows the linear convolution to become a circular convolution. It is called a *cyclic prefix*. Essentially, extra information is prepended to the block of symbols to allow the use of the DFT operation in a straightforward way to simplify equalization.

In this section, two approaches are described for frequency domain equalization. The first is called single carrier frequency domain equalization (SC-FDE). With this approach, a block of complex pulse amplitude modulated symbols is prepended with a cyclic prefix. The received signal is transformed to the frequency domain via the DFT, equalized, and transformed back to the time domain using the inverse DFT (IDFT). The second is called orthogonal frequency division multiplexing (OFDM). With this approach, the IDFT of a block of complex pulse amplitude modulated symbols is computed at the transmitter and prepended with a cyclic prefix. At the receiver, frequency domain equalization is applied after taking a DFT. Both approaches have advantages and disadvantages relative to each other. Frequency domain equalization can also be applied to other modulation techniques, for example, continuous phase modulation [TS05][PHR09], but a detailed description is not provided since most commercial mmWave systems (those based on 60 GHz) with higher spectral efficiency employ complex pulse amplitude modulations. Zero padding can be used as an alternative to a cyclic prefix [MWG+02], as has been used in multi-band ultrawideband systems [LMR08], but this has not received much attention for mmWave thus far due to the popularity of the comparable pilot word insertion [DGE01][HH09].

2.5.1 Single Carrier Frequency Domain Equalization

Single carrier frequency domain equalization (SC-FDE) is the general term for cyclic prefix aided complex pulse amplitude modulation [SKJ95][FABSE02]. In mmWave systems, SC-FDE is one possible physical layer (PHY) for IEEE 802.15.3c, IEEE 802.11ad, and ECMA 387, as discussed in more detail in Chapter 9. In IEEE 802.15.3c, SC-FDE is used in the Single Carrier Mode. In IEEE 802.11ad, SC-FDE is used in the mmWave SC PHY and the mmWave Low Power PHY. In ECMA it is used in the single-carrier block transmission (SCBT) PHY. The term *SC-FDE* is used to distinguish a standard SC PHY from one that uses a cyclic prefix of some form to aid frequency domain equalization, but often just SC is used.

Consider the transmission of a block of symbols $\{s[n]\}_{n=0}^{N-1}$. The length of the block is given by N, which will also become the dimension of the DFT and IDFT operations. The block size should be larger than the channel memory, that is, $N > L$. Instead of transmitting the block of symbols directly, multiple blocks are separated using a guard interval of length $L_c \geq L$. Two kinds of guard intervals are common, as illustrated in Fig. 2.13. With the cyclic prefix, part of the end of each block is prepended onto the beginning of the block. With the pilot word [DGE01], the guard intervals are filled using the identical training data. Both cyclic prefixes and pilot words serve to separate out blocks of data and make the linear convolution look circular, facilitating frequency domain equalization. Using pilot words instead of a cyclic prefix facilitates adaptive frame synchronization, frequency offset estimation, and compensation for clock drift.

Now we show how the cyclic prefix leads to a convenient form of frequency domain equalization, first employed for this purpose in [PR80] for OFDM systems (however, note that OFDM systems had already been used for years prior without a cyclic prefix [Cha66]). The same approach also applies to using a guard interval with pilot words, with a slight change in notation. Let the new block of symbols be given by $\{w[n]\}_{n=0}^{N+L_c-1}$, where for the first L_c symbols, $w[n] = s[n + N - L_c]$ for $n = 0, 1, \ldots, L_c - 1$. Then for the next N symbols, $w[n] = s[n - L_c]$ $n = L_c, L_c + 1, \ldots, L_c + N - 1$. Now consider the received signal model in (2.33).

$$y[n] = \sum_{\ell=0}^{L} h[\ell] w[n - \ell] + v[n] \qquad (2.50)$$

for $n = 0, 1, \ldots, N + L_c - 1$. Discarding the first L_c samples, form the new signal

$$\bar{y}[n] = y[n + L_c] = \sum_{\ell=0}^{L} h[\ell] w[n + L_c - \ell] + v[n + L_c]. \qquad (2.51)$$

Cyclic Prefix:

| ... | $s[N-L_c] \ldots s[N-1]$ | $s[0]\ s[1] \ldots s[N-1]$ | $s[2N-L_c] \ldots s[2N-1]$ | $s[N] \ldots$ |

L_c samples copied

Guard Interval with Pilot Word:

| ... | $t[0] \cdots t[N_t-1]$ | $s[0]\ s[1] \cdots s[N-1]$ | $t[0] \cdots t[N_t-1]$ | $s[N] \cdots$ |

pilot word

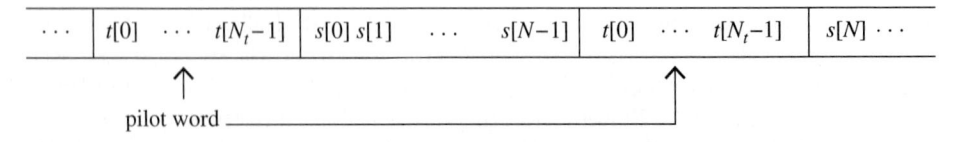

Figure 2.13 The concept of a cyclic prefix is that the last L_c samples are replicated at the beginning of the block. Alternatively, a training sequence or pilot word ($t[n]$) may be used as a guard interval.

2.5 Equalization in the Frequency Domain

Looking at the definition of $w[n]$

$$\bar{y}[n] = \sum_{\ell=0}^{L} h[\ell]s[((n-\ell))_N] + \bar{v}[n] \qquad (2.52)$$

where $((\cdot))_N$ is the integer *modulo* operation. It occurs because of the presence of the cyclic prefix and the fact that $L \leq L_c$. Because $L < N$, $h[n]$ can be treated as a length N sequence by defining $h[n] = 0$ for $n = L+1, \ldots, N-1$. This is called zero-padding. Now taking the DFT of (2.52)

$$\bar{Y}[k] = H[k]S[k] + V[k] \qquad (2.53)$$

for $k = 0, 1, \ldots, N-1$. This leads to a simple relationship between the DFT of the observed sequence $\bar{Y}[k]$ and the unknown sequence $S[k]$. The channel can be equalized by dividing $\bar{Y}[k]$ by $H[k]$, which is simply a single-tap zero-forcing equalizer. Alternatively, a single tap MMSE solution could also be implemented. Taking the IDFT of $\bar{Y}[k]/H[k]$ gives the result

$$r[n] = s[n] + \widetilde{v}[n]. \qquad (2.54)$$

The equalized signal $r[n]$ is fed to a symbol-by-symbol detector as in Fig. 2.14. With some algebra [WBF$^+$09], it can be shown that if $v[n]$ is i.i.d. $\mathcal{N}(0, \sigma_v^2)$ then the equalized noise is also i.i.d. with $\mathcal{N}(0, \sigma_v^2/N \sum_{k=0}^{N-1} |H[k]|^{-2})$. This means that the symbol-by-symbol detectors for AWGN channels as described in Section 2.3.2 can be employed, and possibly followed by more advanced coding.

With SC-FDE, linear equalization can be done perfectly. This means that the channel can be completely equalized, unlike the zero-forcing or MMSE equalizers in Section 2.4.1, which try to approximate the inverse. It is still an equalizer, though, and does not give the maximum likelihood solution. The complexity of SC-FDE is what makes it so attractive. With a linear equalizer, the length of the equalizer grows as the length of L. With SC-FDE, the complexity is fixed, completely determined by the N-DFT, N-IDFT, and the one-tap equalization. Of course, there is an overhead penalty to be paid, as L is overhead, thus N would likely increase to keep the fraction of overhead $L/(N+L)$ from becoming too large. Exactly when SC-FDE becomes more efficient than linear equalization depends

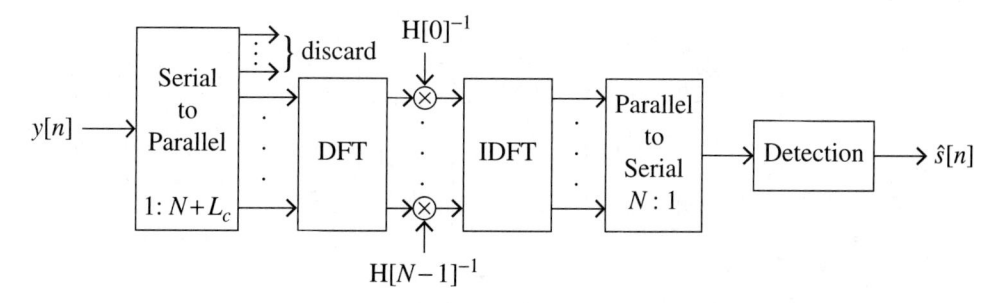

Figure 2.14 A SC-FDE operates on blocks of $N + L_c$ symbols. The first L_c symbols are discarded. The remaining symbols are input to DFT operation. The results of one-tap frequency domain equalization are applied to an IDFT operation. The output is then serialized and passed to a detector.

on many system parameters. As a general rule-of-thumb it becomes more efficient to equalize in the frequency domain for a channel with an L of 5 or higher.

2.5.2 OFDM Modulation

OFDM is a transmission technique that permits frequency domain equalization. It is used in many commercial wireless systems including IEEE 802.11 wireless local area networks [802.11-12], IEEE 802.16 broadband wireless networks, and 3GPP Long Term Evolution (LTE) cellular networks. OFDM has a long and interesting history in wireless communications [Wei09]. In 60 GHz systems, OFDM has been investigated as a modulation technique for several years [KH96][DT99][Smu02]. OFDM is the physical layer (PHY) for the High Speed Interface mode and the Audio/Visual mode in IEEE 802.15.3c, the OFDM PHY in IEEE 802.11ad, and the OFDM mode for Type A devices in ECMA [ECMA10].

OFDM can be viewed as SC-FDE with the IDFT moved from the receiver to the transmitter, as illustrated in Fig. 2.15. The size of the DFT and IDFT is given by N and is known as the (maximum) number of subcarriers. Like SC-FDE, OFDM systems append a cyclic prefix of length L_c where $L_c \geq L$ and $N \geq L_c$. To explain the concept of OFDM, consider the transmission of $\{S[n]\}_{n=0}^{N-1}$. Because the symbols are input into the IDFT, conceptually they start out in the frequency domain, thus our notation. A cyclic prefix is prepended to the output of the IDFT to give

$$w[n] = \frac{1}{N} \sum_{n=0}^{N-1} S[m] e^{j2\pi \frac{m(n-L_c)}{N}} \qquad (2.55)$$

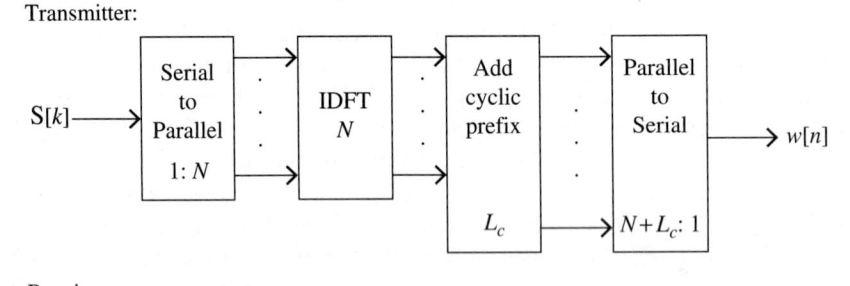

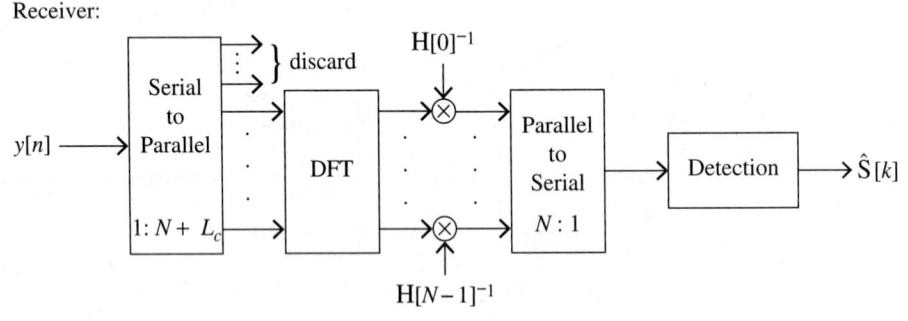

Figure 2.15 The transmitted samples in OFDM are constructed by taking the IDFT of N symbols and adding a cyclic prefix. The receiver discards the cyclic prefix, then takes the DFT of the resulting signal, followed by per-subcarrier equalization.

2.5 Equalization in the Frequency Domain

for $n = 0, 1, \ldots, N + L_c - 1$. It can be verified (due to the periodicity of discrete-time complex exponentials), that $w[n] = w[n + N]$ for $n = 0, 1, \ldots, L_c - 1$. It is common to refer to $\{S[n]\}_{n=0}^{N-1}$ as an *OFDM symbol*, and the components $S[n]$ as *subsymbols*. From (2.55), $S[n]$ rides along the complex exponential with frequency n/N, thus n is often called the subcarrier index or simply the subcarrier. The outputs $w[n]$ are referred to as samples, not symbols, since they contain a piece of every symbol being transmitted. If T_s is the sample time, then the OFDM symbol time is $(N + L_c)T_s$ and the subcarrier bandwidth is given by $1/NT_s$, assuming that a sinc pulse shaping filter is used (this is common, though other pulse shaping filters are possible).

The received signal is the same as in (2.50). Discarding the first L_c samples gives the same signal as in (2.52). Taking the DFT with $\{w[n]\}$ in (2.55) gives

$$\bar{Y}[k] = H[k]S[k] + V[k] \qquad (2.56)$$

for $k = 0, 1, \ldots, N - 1$. Thus, the channel effects can be equalized by dividing $\bar{Y}[k]$ by $H[k]$, which is simply a single-tap zero-forcing equalizer. The resulting symbols are fed directly to the symbol-by-symbol detector, with no IDFT required as in the SC-FDE case. It can be shown that in fact (assuming the samples corresponding to the cyclic prefix are discarded) applying symbol-by-symbol detection to (2.56) is also the maximum likelihood solution. This has much lower complexity than the MLSE solution, though the performance of the MLSE is better unless error control coding is used. The effective noise power per OFDM subsymbol after the one-tap equalization is $\sigma_v^2/|H[k]|^2$. With SC-FDE, the SNR performance is the same for all the subcarriers, whereas for OFDM, performance varies per subcarrier. Consequently, coding and interleaving are used in OFDM systems to provide robustness against frequency selective channel fades, but the adaptation may be more complex [DCH10].

As an aside, IEEE 802.11ad uses several variations of mapping symbols to subcarriers that are unlike other wireless systems. One example is what is called *spread QPSK*, in which both a set of QPSK symbols and their conjugates are sent on different subcarriers. Another example is a non-standard QPSK mapping, in which two QPSK constellation points are combined together to create two 4-bit constellations that are different from 16-QAM. The resulting constellations are then transmitted on two different subcarriers as another way to provide diversity with the overhead that results from spreading. More information is provided in Chapter 9.

There have been many comparisons between OFDM and SC-FDE [SKJ95][FABSE02]. Each has advantages and disadvantages in certain cases. The main drawback of OFDM modulation is its sensitivity to RF impairments. Because the transmitter takes the IDFT of a block of signals, the samples used for modulation in (2.27) have a much larger fluctuation. The resulting analog signal has a high peak-to-average power ratio. Essentially, the peak value of the signal can be much larger than the average value of the signal. This means that a larger power amplifier (PA) backoff is required (see Chapter 5 for more information), implying that either a more expensive (power-hungry) power amplifier would be required to maintain the same range or additional nonlinear distortion would have to be tolerated. There are peak-to-average power ratio reduction techniques, which can be used to partially solve this problem (see [JW08] and the papers therein). The OFDM waveform is sensitive to other nonlinearities [Shi96]. OFDM is more sensitive to carrier frequency offset [CM94] since offsets create what is called *intercarrier interference* between symbols. It is also sensitive to gain and phase imbalance [PH02] and phase noise [PVBM95]. In terms of complexity, it depends on the application. Complexity is more symmetric in OFDM, whereas in SC-FDE the complexity is primarily at the receiver. In 3GPP LTE, for example, OFDM is used on the downlink and a SC-FDE-like modulation is used on the uplink. Although there appear to be several disadvantages, there are a number of important advantages as well. OFDM gives better performance with adaptive modulation and coding since the information rate can be adjusted to current channel conditions based on the frequency selectivity of the channel. OFDM allows for easy spectral shaping by zeroing or reducing the power on some subcarriers. It is also possible to assign different OFDM subcarriers to different users in what is called *orthogonal frequency division multiple access* (OFDMA).

In the application of OFDM to mmWave systems, there have been several specific comparisons between SC-FDE and OFDM for 60 GHz [SB07][LCF$^+$07][RJW08]. The general conclusions of these studies are summarized as follows. SC-FDE allows lower-resolution DACs and ADCs to achieve the same performance as OFDM. SC-FDE is less sensitive to nonlinearity from the power amplifier (PA) and thus requires less power amplifier backoff, resulting in higher range. SC-FDE can provide higher throughputs because it is more robust to various RF impairments. The performance differences in many cases are on the order of 10% throughput or 1 dB in SNR. Given the substantial industry already invested in producing OFDM radios, these differences are not necessarily enough to make SC-FDE a clear winner. Indeed, most WPAN and WLAN standards at 60 GHz have adopted both SC and OFDM transmission models. More discussion on the comparison between OFDM and SC transmission and a more detailed treatment of the impact of impairments are found in Chapter 7.

2.6 Error Control Coding

Communication is not perfect, and wireless is no exception. Many impairments are introduced into the communication signal at the transmitter, by the channel, and at the receiver. Common impairments include nonlinearities from active circuit components (amplifiers, mixers), frequency selective fading and time selective fading introduced by the propagation channel, noise from the circuits components (thermal noise, phase noise, quantization noise), as well as interference (caused by other transmissions in the unlicensed bands). Wireless systems are designed so that the impairments are managed through the careful choice of components, selection of transmission strategies, receiver processing algorithms, and protocols.

Error control coding is an important component of any communication system. The idea of an error control code is to introduce redundancy in the transmitted signal that can be used at the receiver to deal with bit errors. In many systems, multiple kinds of error coding techniques are concatenated and used together. Error detection codes add redundancy, for example, a parity check, to detect whether a block of bits was transmitted correctly. Error detection is used so that erroneous packets are not further processed by higher layers and also may be used as part of a radio link protocol to request a retransmission. Error correction codes add redundancy that permits bit errors to be

2.6 Error Control Coding

corrected. Some codes can correct groups of bit errors, others work better with more random error patterns. In general, the higher the amount of redundancy, the better the error detecting or correcting capability of a code will be, at the expense of more overhead.

There are many different kinds of error control codes. A broad classification can be made into block codes and trellis codes. Block codes generally map a block of information bits onto a longer block of information bits. Block codes are often systematic, where the output block of bits consists of the input bits concatenated with some parity bits. Trellis codes are a type of memory based encoding in which the encoded bits or symbols depend on the values of a finite number of bits; they are often represented using a trellis diagram that captures the state (determined by the previous bits) and the transitions from one state to another (determined by the current bits). The mathematics behind traditional error control coding was finite field theory. In the past 15 years, classical code constructions have been combined with sophisticated iterative decoding algorithms to achieve performance close to fundamental limits.

MmWave systems make use of many different kinds of error control coding techniques. A main consideration, besides good performance, is the ability to implement error control decoding techniques efficiently in hardware. This section summarizes several important error control codes proposed for early mmWave (e.g., 60 GHz) wireless standards. These strategies are likely to find use in other mmWave systems, as well. More detail on error control coding can be found in various texts devoted to the subject [Bla03][LC04][RU08].

2.6.1 Block Codes for Error Detection

Error detection is an important part of establishing a reliable communication link. Error detecting codes are used to check that a packet (block of bits) was decoded correctly; even with strong error control codes, it is not possible to correct for all possible bit errors. There are different ways a system may deal with packet errors. If the application is real-time voice, for example, the packet may be flagged as an error and then the source decoder would apply a concealment algorithm to mitigate the effects of the loss. Alternatively, if the wireless link is used to transmit a binary file, the receiver may ask the transmitter to resend the packet until it is received correctly. Error detection may be performed at the physical layer (PHY), at the medium access control (MAC) layer, or at higher layers (see Section 2.10 for a discussion on abstraction layers).

Block codes operate on fixed length blocks of the data to be transmitted. Block codes may operate on binary data or non-binary data. In the binary case, the data symbols are bits with values either 0 or 1 and the operations are performed in a finite field known as a *Galois field* (GF), specifically $GF(2)$. Non-binary block codes operate on symbols that consist of a collection of $m > 1$ bits. The term *symbol* is used in a more general fashion than in Section 2.3.1; the m bits may be mapped to multiple constellation symbols. Non-binary block codes operate in a higher dimension field, for example, Reed-Solomon codes work in $GF(2^m)$. A binary block code with block length K and encoded length N takes K data symbols and produces N encoded symbols. A block code is linear if it satisfies the property that any sum of the output codewords in the appropriate finite field is also a codeword. Systematic block codes satisfy the property where the first K symbols of the encoded data correspond to the uncoded K data symbols to be transmitted; the remaining $P = N - K$ symbols are parity symbols.

The most common type of error detecting code is a block code known as the *cyclic redundancy check* (CRC). Essentially, a CRC takes a block of K data bits and computes $P = N - K$ parity check bits that are appended to the K data bits to create an encoded block length of N. The parity bits are all different functions of the data bits. At the receiver, if the decoded data bits do not give the same K parity bits as transmitted, then an error is declared. In general, CRC codes can detect $1 - 2^{-P}$ percent of possible errors. For example, with the common value of $P = 32$, the receiver detects 99.9999999767% of the errors.

Mathematically, CRC codes work as follows through polynomial division in the binary field. A particular length CRC code is associated with a generator polynomial $g(D) = g[0]D^P + g[1]D^{P-1} + g[P]D + g[P+1]$ where $g[0] = g[P+1] = 1$. The generator polynomial will be used to generate P parity bits. Let $b(D) = b[0]D^{K-1} + \ldots + b[K-2]D + b[K-1]$ denote the data polynomial corresponding to the bit sequence $\{b[n]\}_{n=0}^{K-1}$. The parity bits are computed from the division of the generator polynomial as

$$r(D) = \text{remainder}\left(\frac{D^P b(D)}{g(D)}\right). \tag{2.57}$$

Then the coefficients of the resulting polynomial $d(D) = D^P b(D) + r(D)$ are sent. Thus, $d[k] = b[k]$ for $k = 0, 1, \ldots, K-1$, and $d[k+K] = r[k]$ for $k = 0, \ldots, P-1$. Practically, the remainder is computed using a linear feedback shift register, which can be implemented very efficiently both in hardware and software. A similar operation is applied at the receiver to check the parity.

2.6.2 Reed-Solomon Code

Reed-Solomon (RS) codes are a type of non-binary block code [RS60] used primarily for burst error correction. Burst errors occur, for example, when a symbol is decoded incorrectly, or when multiple symbols are fading in OFDM, or in the output of some other kinds of error control codes like convolutional codes when there is a decoding error. RS codes are non-binary in the sense that they operate on symbols that consist of a collection of $m > 1$ bits. And although they have many interesting mathematical properties, perhaps the most well known is that they have a special property known as being maximum distance separable. This means that RS codes achieve the largest possible minimum distance compared with other linear codes with the same input and output block lengths. In other words, they are good codes.

A typical notation for an RS code is $RS(N, K)$, where $K < N$ is the number of input symbols and N is the number of output symbols. The number of output symbols may be either $N = 2^m - 1$ for traditional codes or $N = 2^m$ or $N = 2^m + 1$ for extended RS codes. The RS code can correct up to $\lfloor (N-K)/2 \rfloor$ symbol errors using hard decoding techniques. If some symbols are known to be likely in error, as can be obtained from soft information, these symbols can be considered to be an erasure and the code can correct E errors and S erasures as long as $2E + S \leq N - K$. Thus, side information can be used to improve the decoding capability of the RS code. Encoding and decoding of RS codes is usually viewed through the lens of Bose, Ray-Chaudhuri, and Hocquenghem (BCH) codes [BRC60][Hoc59][LC04], as a non-binary cyclic code with an encoding similar to that described in Section 2.6.1.

In 60 GHz systems, RS coding is used in IEEE 802.15.3c [802.15.3-03], ECMA 387 [ECMA10], and IEEE 802.11ad [802.11-12]. The code $RS(255, 239)$ is used in the SC PHY of IEEE 802.15.3c and in all the modes of ECMA 387. It encodes groups of $m = 8$ bits and can correct $(N - K)/2 = 8$ symbol errors. The code $RS(224, 208)$ is used in IEEE 802.11ad. It encodes groups of $m = 8$ bits and can also correct 8 symbol errors. IEEE 802.15.3c and ECMA 387 allow for shortening through zero padding so that the RS code can be applied to shorter block sizes, useful for protecting header information.

As a concrete example, consider the $RS(255, 239)$ code used in IEEE 802.15.3c, ECMA 387, and IEEE 802.11ad. The description comes from IEEE 802.15.3c [802.15.3-03]. Note that the $RS(224, 208)$ code used in IEEE 802.11ad [802.11-12] is a shortened version of this code, thus it also has the same parameters. The code uses a generator polynomial $g(D) = \prod_{k=1}^{1} 6(D + \alpha^k)$ where α is a root of the binary primitive polynomial $p(D) = 1 + D^2 + D^3 + D^4 + D^8$ given by $\alpha = 0x02 = 0b00000010$. Groups of 8 bits are mapped to $m = 8$ bit symbols elements as $b_7 D^7 + b_6 D^6 + \cdots + b_1 D + b_0$ where b_7 is the most significant bit and b_0 is the least significant bit. The symbols are mapped to the data polynomial $m(D) = m[238]D^{238} + \cdots + m[1]D + m[0]$. Then the parity bits are computed from the division of the generator polynomial as

$$r(D) = \text{remainder}\left(\frac{D^{16} m(D)}{g(D)}\right). \tag{2.58}$$

Then the coefficients of the resulting polynomial $d(D) = D^P m(D) + r(D)$ make up the encoded sequence of symbols. The symbols are then converted back to bits for transmission. The encoding operation can be performed efficiently using a linear feedback shift register [LC04].

There are many possible types of decoders for RS codes; it is an ongoing topic of research. Perhaps the most well known hard decoding algorithm takes an approach leveraging the connection to BCH codes. First, the syndrome is calculated by evaluating the received polynomial $\widehat{d}(D)$ at the 16 roots of the generator polynomial $g(D)$. The symbol error locations are found by finding an error locator polynomial, using, for example, the Berlekamp-Massey algorithm [Ber65][Mas65], and then finding the roots of that polynomial, using, for example, the Chien algorithm [Chi64]. Then the symbol error values are found by solving another set of simultaneous equations with $N - K$ unknowns, using, for example, the Forney algorithm [For65]. Better performance can be achieved using what is called *soft decision decoding*. In that type of decoding, bit reliabilities (instead of hard decisions) are incorporated into the decoding process [GS99][KV03]. For example, a bit could be assigned a -5 if it is very likely to be a 0, a 5 if it is very likely to be a 1, and something in the middle, say 2, if the value of the bit is unclear but it is a little more 1 than 0. There are many hardware implementations of RS decoders including both hard decoding [Liu84][SLNB90] and, more recently, soft decoding [AKS11].

2.6.3 Low Density Parity Check Codes

Low density parity check (LDPC) codes are linear block codes with some special properties that permit very efficient decoding near Shannon limits [MN96] within 0.0045 dB [CFRU01]. They were first proposed in [Gal62], but interest did not become intense until the development of turbo codes [RU03]. LDPC codes may be binary or non-binary; only the binary ones have been employed in 60 GHz, thus we focus on the binary case here. LDPC codes are high-performance codes correcting bit errors due to noise.

Linear block codes with parameters K (the amount of parity) and N are associated with a $K \times N$ generator matrix $\mathbf{G}$, which creates the encoded sequence from the $K \times 1$ vector of uncoded bits $\mathbf{u}$ as $\mathbf{u}^T\mathbf{G}$. The general convention in the coding theory community is to use row vectors for encoding and decoding operations. Every generator matrix is associated with an $N \times K$ parity check matrix $\mathbf{H}$ that satisfies the property $\mathbf{GH}^T = \mathbf{0}$. If $\mathbf{x}^T = \mathbf{u}^T\mathbf{G}$ then $\mathbf{x}^T\mathbf{H}^T = \mathbf{0}$, meaning that the codeword was received correctly. LDPC codes are called low density because the parity check matrix $\mathbf{H}$ is sparse, meaning that it contains many more zeros than ones. For efficient LDPC decoding, the parity check matrix must have a special structure. The structure can be employed to make the encoding process more efficient, as well [RU01a]. Like most random-like coding strategies, better code performance is generally achieved with larger block length N.

LDPC codes are used in certain operating modes of IEEE 802.15.3c and IEEE 802.11ad. For example, for IEEE 802.11ad, RS is used in some of the SC modulation and coding schemes. LDPC is used for all of the OFDM modulation coding schemes and most of the SC modulation and coding schemes. There is some perception that RS logic may be lower power than LDPC logic, an important point when considering the total radio power budget in personal-portable devices such as cellphones. However, a detailed comparison of power requirements depends on many implementation decisions such as the semiconductor process and the parameters of the codes.

The notation LDPC(N,K) means a LDPC code with an input size of K and an output size of N. The following codes are used in those systems. The rate 1/2, 3/4, and 7/8 codes have an encoded block length of $N = 672$ bits and are denoted LDPC(672,336), LDPC(672,504), and LDPC(672,588). They have a parity check matrix that can be partitioned into square submatrices of size 21×21 and are either a cyclic permutation of the identity matrix or the zeros matrix. The rate 14/15 code has an encoded block length of $N = 1440$ and is denoted LDPC(1440,1344). It is a quasi-cyclic code with period of 15. It uses a particular parity check matrix with a block circular shift structure, meaning that every 15th block of rows is obtained by a circular shift of the previous 15 rows. The LDPC(1440,1344) code may be shortened by appending zeros to the incoming message and not transmitting the systematic part of the data corresponding to those zero bits. IEEE 802.11ad has mandatory support for LDPC codes with rates 1/2, 5/8, 3/4, and 13/16, also using a block length of 672 and a parity check matrix with a cyclic permutation structure.

The power of an LDPC code is realized through iterative decoding algorithms. Most algorithms employ an interpretation of an LDPC parity check matrix as a Tanner graph [Tan81], which falls in the general framework known as *factor graphs* [KFL01]. The connections between graph theory and LDPC codes have been used to develop many different kinds of iterative decoding algorithms. One class of algorithms is bit flipping [Gal62] and recent variations, for example [MF05]. Another class of algorithms is based on a concept called *message passing* [Gal62][RU01b]. Bit flipping algorithms generally facilitate higher speed implementation at the expense of worse error correction performance. In the past several years there has been substantial interest in hardware implementations of LDPC encoders and decoders, for example, [BH02][DCCK08][MS03][ZZH07]. A specific hardware implementation for some of the LDPC codes from IEEE 802.15.3c is found in [SLW+09].

2.6.4 Convolutional Codes

Convolutional codes are a type of error control code, usually considered to be a special case of trellis codes. They are widely used in many communication systems for their ability to correct non-burst errors. Convolutional codes are usually employed in conjunction with interleaving to spread out burst errors. They are also commonly used in a concatenated coding technique where it is used as an inner code and a RS code is used as an outer code. Convolutional codes are most powerful when used with a good soft input decoding algorithm.

A convolutional code is characterized by the generator polynomials used for the code. The generator polynomials give the code memory. With order L polynomials, the code is said to have constraint length of $L + 1$. A larger amount of memory in the generator polynomials generally gives better performance at the expense of higher decoding complexity. The rate of a convolutional code is given by K/N where K is the number of input bits and N is the number of output bits per K input bits. In general, a rate K/N convolutional code will have KN corresponding generator polynomials. In many implementations, especially those that employ rate adaptation, a rate $1/N$ convolutional code is used as a mother code. Different rates are created through the process of periodic puncturing of the output.

Convolutional codes are widely used in wireless systems. In mmWave, they are used in the Audio/Visual mode of IEEE 802.15.3c [802.15.3-03] and in ECMA 387 [ECMA10]. IEEE 802.15.3c uses a rate 1/3 mother code with a constraint length of 7 with generator polynomials $g_0(D) = 1 + D^2 + D^3 + D^5 + D^6$, $g_1(D) = 1 + D + D^2 + D^3 + D^4$ and $g_2(D) = 1 + D + D^2 + D^4 + D^6$. Puncturing is used to create effective rates of 1/2, 4/7, 2/3, and 4/5. ECMA uses a rate 1/2 mother code with a constraint length of 5 with generator polynomials $g_0(D) = 1 + D^3 + D^4$ and $g_1(D) = 1 + D + D^2 + D^4$. Puncturing is used to create effective rates of 4/7, 2/3, 4/5, 5/6, or 6/7. The ECMA encoder is illustrated in Fig. 2.16.

Like it sounds, the convolutional encoder involves the binary convolution of a data sequence with multiple generator polynomials. The outputs are combined together in a

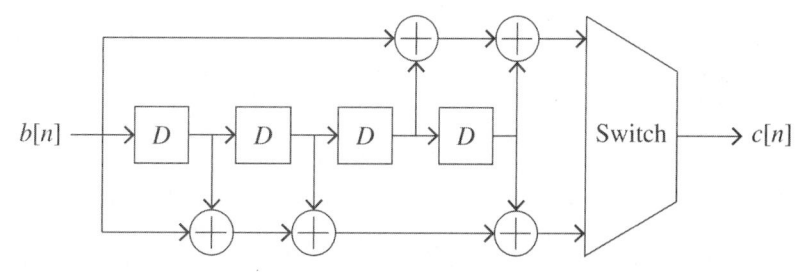

Figure 2.16 The convolutional encoder from ECMA [ECMA10]. The outputs of the encoder are alternately switched to create 2 output bits for every 1 input bits.

round-robin fashion. To make this more clear, we consider an example with the ECMA encoder. Let $\{m[n]\}$ denote a sequence of input bits. Let $g_1[k]$ and $g_2[k]$ denote the kth coefficients of the generator polynomials. Then the output sequence is given by (assuming binary additions)

$$c[2n] = \sum_{k=0}^{4} g_0[k]m[n-k] = m[n] + m[n-3] + m[n-4]$$

$$c[2n+1] = \sum_{k=0}^{4} g_1[k]m[n-k] = m[n] + m[n-1] + m[n-2] + m[n-4].$$

The even outputs are the input bit sequence convolved with the first generator polynomial $g_0(D)$, and the odd outputs are the input bit sequence convolved with the second generator polynomial $g_1(D)$. Normally, a convolutional code is used to encode a block of data. The state of the encoder is typically initialized to zero; this means that it is assumed that $m[n] = 0$ for $n < 0$. To facilitate decoding, the incoming bit sequence is padded with L zeros where the constraint length is $L + 1$. It forces the state of the encoder back to zero, which aids in the decoding process.

The idea of puncturing is to remove periodically certain bits at the transmitter to reduce the redundancy in the transmitted signal and thus increase the rate. At the receiver, the punctured bits are related with dummy bits or erasures. The puncturing patterns are optimized offline. For example, ECMA creates a rate 2/3 code by taking every four outputs of the encoder and puncturing (or stealing) the last bit. Essentially with the output $\widetilde{c}[n]$, then $\widetilde{c}[3n] = c[4n]$, $\widetilde{c}[3n+1] = c[4n+1]$, and $\widetilde{c}[3n+2] = c[4n+2]$. More complex puncturing patterns are possible [Hag88][LC04].

Interleaving is used in convolutional codes to spread out burst errors, which the convolutional decoder is not able to correct very well. Interleaving with convolutional codes is usually applied at the bit level. For other types of codes like trellis coded modulation, it may be performed at the symbol level. Convolutional coding followed by interleaving and symbol mapping is also called bit interleaved coded modulation (BICM), see for example [CTB98]. There are different kinds of interleavers. For example, ECMA uses a length 48 bit interleaver in which a block of 48 bits $c[n]$ is mapped to an interleaved sequence $i[\ell]$ where

$$\ell = [6\lfloor n/6 \rfloor + 7(n \bmod 6)] \bmod 48. \qquad (2.59)$$

The performance of a convolutional code depends to a large extent on the type of decoder implemented at the receiver. Like most receiver algorithms, the decoder is not normally specified in a standard. The most common decoding algorithm is the maximum likelihood sequence estimator (MLSE), commonly implemented using the well-known Viterbi algorithm [FJ73]. There are different algorithms depending on whether the input is hard (the bit decisions have already been determined) or soft (likelihoods or reliability metrics are given). The best performance is achieved when there is a soft input. The Viterbi algorithm operates by exploiting the Markov property of the output of the convolutionally encoded source. Specifically, the fact that $c[n]$ depends only on $m[n]$, $m[n-1], \ldots, m[n-L]$ where L is the memory of the code assuming a rate $1/N$ code.

2.6 Error Control Coding

The core idea of the decoder is to define a state-space that consists of all possible values of $m[n-1], \ldots, m[n-L]$. The state at time $n-1$ is connected to the state at time n through a branch, which depends on the value of $m[n]$. The collection of states and transitions creates what is known as a *trellis*. The Viterbi algorithm in the forward step at each time n, for every state k, determines the most likely previous state considering $m[n] = 0$ or $m[n] = 1$. Then in the traceback step, the decoder starts from the end state and traces back the most likely path, generating the corresponding bit values that lead to those transitions.

If a convolutional code is used as an inner code, it is followed by an outer code that allows for soft inputs, then the decoder needs to generate soft information. With soft information, the output $\hat{m}[n]$ is a real number instead of a bit. For example, 100 might mean that the decoder is sure that the output is a 1, -100 might mean that the decoder is sure that the output is a -1, and a small value like 10 might be used to mean that the decoder thinks the output is a 1 but does not have much confidence. A classic modification of the Viterbi algorithm is the list output Viterbi decoder algorithm (LOVA) [SS94], which computes the K best paths instead of just a single best path. One of the best known algorithms for generating soft outputs is the soft output Viterbi algorithm (SOVA), which computes the maximum-likelihood sequence and delivers an approximation of the symbol (or bit) likelihood values [HH89]. The optimum bit error probability can be achieved by the trellis-based maximum a posteriori (MAP) decoding algorithm proposed in [BCJR74], at the expense of added complexity over the SOVA. There are hardware implementations of all of these algorithms, see for example [CR95][WP03][MPRZ99]. Convolutional codes in the form of BICM can be combined with iterative turbo decoders for further improved performance [LR99a], taking into account the interleaving and modulation in the decoding process.

2.6.5 Trellis Coded Modulation

Trellis coded modulation (TCM) is a type of trellis code generally considered to be a generalization of convolutional codes. With convolutional codes, the bits are mapped to symbols after the encoding operation. The code polynomials for a convolutional code are designed so that the code has good Hamming distance properties and are not necessarily optimized for a specific modulation technique. When employed with soft decoding Viterbi receivers, Euclidean distance in the symbol space is a more important measure of code performance in additive white Gaussian noise (AWGN). TCM generalizes convolutional coding by concatenating a convolutional encoding and symbol mapping operations together, resulting in better Euclidean distance properties when the encoder is optimized. The receiver structure is similar to a soft-input Viterbi decoder for a convolutional code. TCM performs better than BICM in additive white Gaussian noise channels but may have worse performance in fading channels without further modification [CR03] especially when BICM is used with iterative decoding.

TCM is not widely used in wireless systems. It is found, though, in one 60 GHz standard. ECMA 387 uses a type of TCM in addition to several modulations and code rates that use BICM [ECMA10]. ECMA uses a pragmatic form of TCM, where the concept of pragmatic means that a conventional rate 1/2 convolutional code is used to

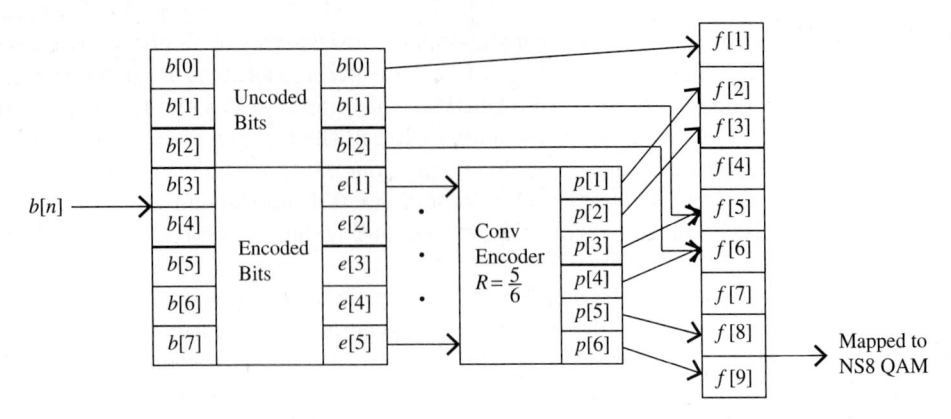

Figure 2.17 An example pragmatic trellis encoder from ECMA [ECMA10]. For every 8 input bits, 3 are uncoded and 5 are coded. The output bits are mapped to NS8 QAM symbols. The uncoded bit determines whether the inner 4 points or outer 4 points are used.

create the trellis coded modulation [VWZP89]. The input bits are partitioned so that some bits are convolutionally encoded and other bits are uncoded. The uncoded bits may, for example, select a quadrant for the constellation point, whereas the coded bits may select the specific point in that quadrant. Pragmatic codes use the existing convolutional code infrastructure and achieve performance that is close to TCM. One of the mappings used in ECMA is illustrated in Fig. 2.17, which uses the NS8 constellation; the other encoder for 16-QAM is not shown.

The encoder and decoder for pragmatic TCM are similar to convolutional encoding, thus the implementation challenges are similar. TCM is also likely to be used as an inner code with an outer RS code. TCM for mmWave systems in the presence of hardware impairments is analyzed in [SBL+08]. It was found that a certain TCM code with an RS outer code offers around a 1.5 dB improvement over QPSK with an RS code. TCM can use a soft-input hard-output or soft-input soft-output decoder, and can be used with iterative decoding techniques.

2.6.6 Time Domain Spreading

Time domain spreading is a simple way of adding redundancy through repetition. The idea is to repeat every coded symbol L_{TDS} times to improve the reliability of reception, at the expense of losing efficiency by $1/L_{\text{TDS}}$. The number of repetitions L_{TDS} is also known as the *spreading factor*. For example, with a spreading factor of 2, an input sequence of symbols s_1, s_2 would become s_1, s_1, s_2, s_2. Time domain spreading is used in mmWave systems in two common ways: to provide additional error protection and redundancy to header information and to permit reliable operation before a beamforming link is established. For example, ECMA uses a spreading factor of two only on one coding and modulation mode, considered the lowest (the one with the smallest rate). IEEE 802.15.3c allows spreading factors of 1, 2, 4, 6, and 64, providing several lower modes of operation especially useful for communication prior to and during beam training. High spreading factors do not always provide better resilience in the presence of interference, depending on how they are used [BAS+10].

2.6.7 Unequal Error Protection

The concept of unequal error protection (UEP) is to provide different levels of error correction and/or detection capability to different classes of bits. UEP is of particular interest for real-time multimedia transmission including audio and video. For example, with UEP an audio signal might be divided into a more perceptually important block and a less perceptually important block. The audio might sound great if both blocks are received, average if just the most important block is received, but poor if only the less important block is received. By partitioning bits into different classes, a multimedia source can be better matched to a wireless channel, whose quality is varying over time. UEP is employed for audio coding in cellular systems, digital audio broadcasting, digital video broadcasting (see, e.g., [HS99] and the references therein), and in 60 GHz systems (see, e.g., [SQrS+08][LWS+08] video and also the discussion in Chapter 8, Section 8.5).

UEP can be achieved through the use of different constellations; see the examples in Section 2.3.1. It can also be achieved by coding the different priority classes differently, for example, with a different rate convolutional code [HS99]. The precise partitioning of bits (what is significant) depends on the type of source, the type of source compression, and the perceptual distortion metric. The extent of the differences in error protection between the two classes is also empirically determined.

Unequal error protection (UEP) is used in the Audio/Visual mode of IEEE 802.15.3c and in ECMA 387. IEEE 802.15.3c supports three types of UEP to protect two classes of bits. Type 1 uses different error control codes, Type 2 uses different coding and modulation schemes, and Type 3 uses different error control code rates and/or a skewed UEP constellation. ECMA supports UEP for uncompressed video transmission (HDMI) for two classes of bits. The partitioning is applied to each pixel by color in two partitions: most significant bits and least significant bits. The bits are coded by an outer RS code and protected unequally using a convolutional code with two different rates.

Various approaches to incorporating UEP with mmWave have been proposed in the literature for both uncompressed [SQrS+08] and compressed [LWS+08] video. In [SQrS+08] a 60 GHz system is proposed for supporting uncompressed video that protects partitions of pixels unequally by color into two partitions. It uses a combination of CRC and RS codes for each partition to allow for error detection and error concealment. Two different modulation and coding modes support UEP. One uses UEP-16-QAM whereas the other uses 16-QAM with two different rate RS codes. A main conclusion in the paper is that UEP provides less fluctuation in quality, though it is sometimes outperformed by equal error protection. In [LWS+08], a 60 GHz system is described for transmission of MPEG video. All the bits are encoded with the same RS code and a UEP-QPSK constellation is used to give different priority (I-frames of the video receive high priority whereas P- and B-frames receive low priority). The conclusion is that quality improves at lower signal-to-noise ratios (SNRs) with UEP but is similar at higher signal-to-noise ratios. The configuration proposed in [LWS+08] is similar to the Type 3 mode of operation in IEEE 802.15.3c.

2.7 Estimation and Synchronization

There are many different kinds of uncertainty in communication systems, and even more algorithms for dealing with them. For example, the presence of additive noise is incorporated into the detection algorithm; errors due to noise are corrected through the use of error control coding. This section describes other types of uncertainty in communication, reviews the structure included in communication signals to combat uncertainty, and summarizes some algorithmic approaches for estimating unknown parameters. The main topics are frame synchronization (finding the start of the waveform), frequency offset synchronization (determining and correcting the frequency offset), and channel estimation. More specific algorithms are described in Chapter 7.

Frame synchronization deals with a form of uncertainty in time. In wireless systems, communication involves the transmission of a group of bits in what may be known as a frame, burst, or packet, depending on the system. To decode the frame, it is necessary to know when it begins. In random access transmission, the receiver may not know when the transmitter is sending a packet of information. Therefore, the receiver will monitor the spectrum and look for an appropriate signal. Even if a transmission is scheduled, there will be propagation delays, which may be unknown and also contribute to the delay. Recall that it takes radio waves 3 ns to travel 1 meter. Given that a symbol period in a mmWave system may be less than 1 ns, even a distance of a few meters will incur tens of symbols worth of propagation delay. *Frame synchronization* is the general term for finding the starting point of a frame, the frame offset, and correcting for it by advancing the received signal. The concept of frame synchronization is illustrated in Fig. 2.18. In a typical system, frame synchronization is performed by looking for a specially designed header sequence that has good correlation properties.

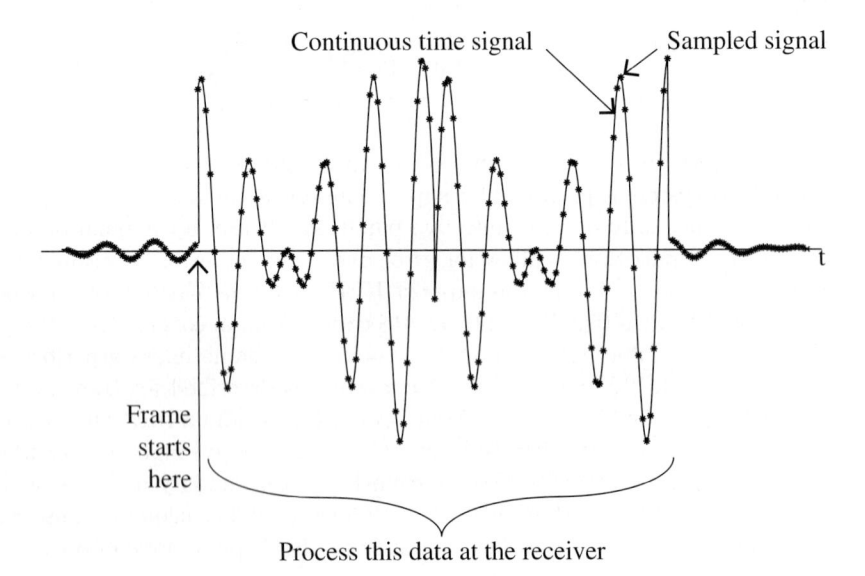

Figure 2.18 A communication waveform in noise. Frame synchronization involves finding the starting point of the frame.

2.7 Estimation and Synchronization

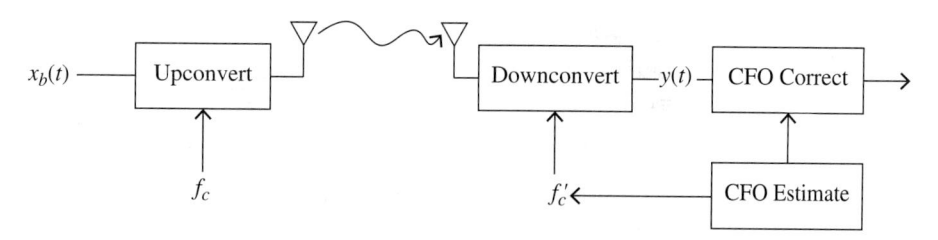

Figure 2.19 Carrier frequency offset (CFO) is created by using a different frequency to demodulate the signal than to modulate it. Frequency offset is often estimated based on structure in the transmitted signal. Frequency offset can be corrected in digital, analog, or both.

Frequency synchronization is a form of uncertainty. In Section 2.2, the concept of the complex envelope of a signal was explained. Creating the transmitted passband signal involves upconverting the baseband signal via modulation onto the carrier frequency f_c, as illustrated in Fig. 2.2. Creating the received baseband signal involves downconverting the received passband signal, effectively removing the carrier f_c. Unfortunately, it is impossible to exactly generate the same clock frequency at the receiver and the transmitter. The clocks can be generated from the same reference signal, for example through GPS, but this still results in some uncertainty over time. Because of the uncertainty, the receiver downconverts the signal with a frequency of f_c', creating what is known as a carrier frequency offset $f_c - f_c'$. Estimating and removing the carrier frequency offset is known generally as frequency synchronization, or more specifically as carrier frequency offset estimation and correction, as illustrated in Fig. 2.19. In a typical system, the frequency offset is estimated in digital using structure in the transmitted signal and is also corrected in digital. Hybrid approaches are possible. There may be additional analog carrier frequency tracking loops. Frequency offset estimation is often performed in two steps: coarse estimation in conjunction with frame synchronization and fine estimation in conjunction with channel estimation.

Channel estimation is needed to reduce uncertainty in the channel impulse response. The equalization algorithms described in Sections 2.4 and 2.5 remove the effects of the channel on the received signal. The baseband equivalent channel response, however, is unknown to the receiver. Therefore, information known to both the receiver and transmitter is inserted into the transmitted waveform. This known information — usually called *training signals* or *pilot signals* — is used by receiver algorithms to estimate the channel response. Training signals are known sequences of bits (or known waveforms) that are inserted into the transmit signal for many purposes including channel estimation. Pilots are also a type of training signal, typically a shorter but more frequent signal that may happen in time and/or frequency if used in an OFDM system. Training signals and pilots are used to estimate and track the propagation channel to facilitate equalization. The same training signals may also be used to facilitate frame synchronization and carrier frequency offset estimation. Note that channel estimation, as illustrated in Fig. 2.20, is performed after the frame synchronization and frequency offset correction blocks.

2.7.1 Structure to Facilitate Communication

Structure in the transmit signal that is known by the receiver can be exploited to deal with different forms of uncertainty. This section summarizes some common forms of

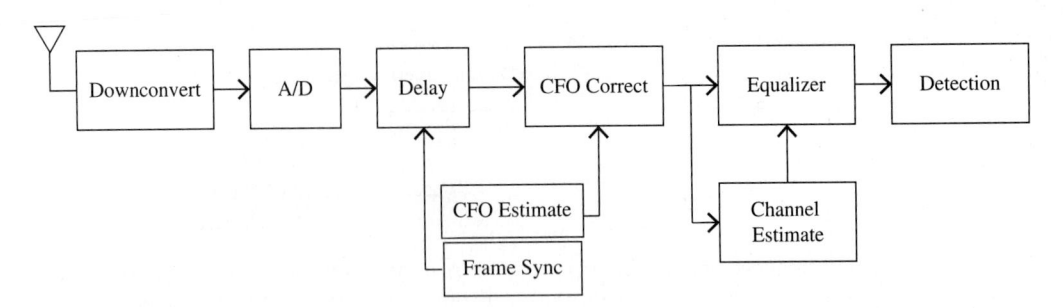

Figure 2.20 A communication receiver with frame synchronization, carrier frequency offset (CFO) synchronization, and channel estimation. The frame offset and carrier frequency offset may be estimated jointly.

structure that facilitate the implementation of signal processing algorithms to deal with synchronization and estimation.

The physical layer (PHY) frame in a wireless system often includes a preamble, inserted at the transmitter but entirely known to the receiver, to facilitate packet detection, frame synchronization, frequency synchronization, automatic gain control, timing synchronization, channel estimation, and beamforming. The receiver may also exploit known transmitted signals, such as pilot signals and other periodic or specially constructed transmissions that enable the receiver to estimate the channel, and to synchronize frame timing, carrier offsets, and other transient characteristics over time. Each standard uses a very specific structure for the known information, as described in more detail in Chapter 9. Here we summarize general structural features that may be present in a mmWave wireless system.

Training sequences are employed for various purposes. A training sequence consists of a sequence of known symbols. In non-OFDM systems, a training sequence consists of N_{tr} known symbols $\{t[n]\}_{n=0}^{N_{tr}-1}$. These known symbols would be modulated in the usual way. In an OFDM system, the training symbols would be sent in the frequency domain, typically on a single OFDM symbol. The training for several subcarriers may be zero. The values of the training symbols are fixed by the standard. Often the training symbols are chosen from classes of sequences with known good properties. For example, ECMA uses Frank-Zadoff-Chu sequences, which are known to have ideal periodic autocorrelation properties (zero sidelobes) [FZ62][Chu72] with PSK constellation symbols. Golay sequences are used in IEEE 802.15.3c and IEEE 802.11ad. Golay sequences are complementary, meaning that the sums of the autocorrelations of two complementary sequences with BPSK modulation have an ideal correlation with a peak and zero sidelobes [Gol61].

The preamble may use a repeated training structure in which the training sequence is repeated several times. Multiple sets of repeated sequences, as illustrated in Fig. 2.21, are also common. Repeated training sequences introduce periodic correlation structure into the transmitted signal. The periodic structure is useful for frame detection, frame synchronization, and frequency offset estimation. Sign inversions, for example, $T, T, T, -T$, or the use of complementary sequences, help sharpen frame synchronization algorithms. Cyclic prefixes may also be applied to the training data to facilitate frequency domain

2.7 Estimation and Synchronization

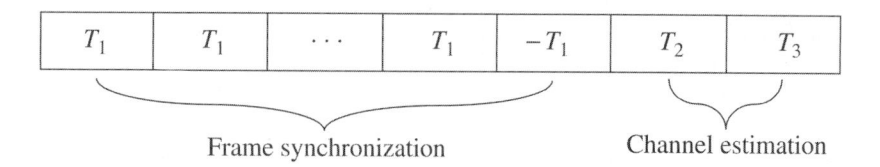

| T_1 | T_1 | $\cdots$ | T_1 | $-T_1$ | T_2 | T_3 |

$\underbrace{\hspace{4cm}}_{\text{Frame synchronization}}$ $\underbrace{\hspace{2cm}}_{\text{Channel estimation}}$

Figure 2.21 A typical preamble structure with multiple repeated sequences. The initial sequences may be used for automatic gain control, frame synchronization, and coarse frequency offset estimation. The later sequences may be used for channel estimation, fine frequency offset estimation, and fine frame synchronization.

channel estimation, useful with SC-FDE or OFDM modulations. The common SC and OFDM preamble in IEEE 802.11ad uses a short training field with 14 repetitions of a Golay sequence, followed by a negated Golay sequence, then a channel estimation field that consists of two sets of four Golay sequences (with different signs). The common mode signaling preamble in IEEE 802.15.3c consists of a channel estimation sequence composed of multiple Golay sequences, a synchronization field that consists of a Golay spread sequence, and a synchronization field that consists of 48 repetitions of a Golay sequence. The ECMA-387 preamble uses a Frank-Zadoff-Chu sequence where the Kronecker product is taken with itself. The frame synchronization sequence uses 8 such repetitions where the last one is negated and 3 repetitions of a different Frank-Zadoff-Chu sequence for channel estimation.

Pilots are also used to facilitate receiver processing. With OFDM modulation, pilots refer to subcarriers with known symbol values, also called *pilot carriers*. They are used to correct common gain and phase errors and are typically spread out in frequency, that is, they are used on subcarriers that are far apart. In the high-speed interface (HSI) mode of IEEE 802.15.3c and the OFDM mode of IEEE 802.11ad, there are 16 pilot subcarriers and 336 data subcarriers in every OFDM symbol with $N = 512$ subcarriers. ECMA 387 also has 16 pilot subcarriers but 360 data subcarriers. Pilots can also be used for channel estimation when there are enough pilots per OFDM system (at least one per coherence bandwidth). This approach is used in cellular communication systems like 3GPP Long Term Evolution (LTE).

Pilot words are a name for pilots used in the SC mode of operation. In IEEE 802.15.3c, for example, the pilot word serves the purpose of a cyclic prefix and a guard interval. Normally the cyclic prefix consists of unknown data, but in this case it consists of known data. Pilot words used in this fashion allow frequency domain equalization. Because their values are known, they can also be used for timing and frequency offset synchronization. The SC mode of IEEE 802.15.3c allows a pilot word length of 0, 8, or 64. The guard interval in the SC mode of IEEE 802.11ad has length 64.

2.7.2 Frequency Offset Synchronization

In this section we consider the problem of frequency offset estimation and correction. The resulting algorithms will give some insight into frame detection and synchronization. Frequency offset is due to differences in the oscillators at the transmitter and receiver.

Though very small, the residual frequency has a great impact on the observed error rate at the receiver. High-quality RF components and analog control units are used to make the frequency offset small. Subsequently, digital receiver algorithms are used to estimate and correct for the remaining offset.

A good mathematical model for the received signal with carrier frequency offset is

$$y(t) = e^{j2\pi f_o t} \sum_{n=-\infty}^{\infty} s[n] h_\text{g}(t - nT_\text{s}) + v(t) \tag{2.60}$$

where the first term contains the frequency offset $f_o = f_\text{c} - f'_\text{c}$. This model is valid for small offsets, meaning that the received signal is still present after the various stages of filtering and downconversion. We will describe algorithms for carrier frequency offset estimation that operate on the sampled signal. Here we consider the symbol-rate sampled signal, but the algorithms can be generalized to operate on the over-sampled signal. Sampling at the symbol rate, assuming the system is causal, and using the notation from Section 2.4 gives the equivalent system:

$$y[n] = e^{j2\pi \epsilon n} \sum_{\ell=0}^{L} h[\ell] s[n - \ell] + v[n] \tag{2.61}$$

where $\epsilon = f_o T_\text{s}$ is the normalized frequency offset. The effect of carrier frequency offset is to rotate the sampled received signal by a complex exponential with *unknown* frequency ϵ. If the offset was known, then it can be removed simply by the operation $e^{-j2\pi \epsilon n} y[n]$. The challenge with carrier frequency offset synchronization is to estimate ϵ, possibly using known training information in $\{s[k]\}$, but when the channel impulse response is as of yet unknown.

One approach for estimating the carrier frequency offset is to employ multiple training repetitions, as described in Section 2.7.1. Consider a training sequence of length N_t given by $\{t[n]\}_{n=0}^{N_\text{t}-1}$, and suppose that there are P repetitions of this training sequence. This means that

$$s[n] = t[n], \quad n = 0, 1, \ldots, N_\text{t} - 1$$
$$s[n] = t[n - N_\text{t}], \quad n = N_\text{t}, N_\text{t} + 1, \ldots, 2N_\text{t} - 1$$
$$\vdots$$
$$s[n] = t[n - (P-1)N_\text{t}], \quad n = (P-1)N_\text{t}, N_\text{t} + 1, \ldots, PN_\text{t} - 1.$$

Now assuming that frame synchronization has already been performed, notice from (2.61) that for

$$y[n] = e^{j2\pi \epsilon n} \sum_{\ell=0}^{L} h[\ell] t[n - \ell] + v[n], \quad n = L, L+1, \ldots, N_\text{t} - 1$$

$$y[n] = e^{j2\pi \epsilon (n + N_\text{t})} \sum_{\ell=0}^{L} h[\ell] t[n - \ell] + v[n], \quad n = N_\text{t} + L, N_\text{t} + L + 1, \ldots, L + 2N_\text{t} - 1$$

$$\vdots$$

2.7 Estimation and Synchronization

$$y[n] = e^{j2\pi\epsilon(n+(P-1)N_t)} \sum_{\ell=0}^{L} h[\ell]t[n-\ell] + v[n], \quad n = (P-1)N_t + L, (P-1)N_t + L + 1,$$

where the values of n start at L to avoid the edge effect.[12] Thus, $y^*[n + pN_t]y[n + (p-1)N_t] \approx \exp(j2\pi\epsilon N_t)|\alpha[n]|^2$ for $n = L, L+1, \ldots, N_t - 1$, $p = 1, \ldots, P$, and $\alpha[n]$ is unknown and the approximation is used because of the presence of noise. One simple approach, inspired from a least-squares problem, is

$$\widehat{\epsilon} = \frac{\text{phase}\left(\sum_{p=1}^{P-1} \sum_{n=L}^{N_t - 1} y^*[n + pN_t]y[n + (p-1)N_t]\right)}{2\pi N_t} \quad (2.62)$$

where phase(x) gives the phase value of the complex number x. The range of offsets that can be estimated without further correction using this approach is $|\epsilon| < 1/2N_t$ or in terms of frequency $|f_o| < 1/2T_s N_t$. Large values of P give better averaging. Longer values of N_t give better averaging but reduce the range. It should be clear from the development that the oversampled received signal can also be employed in (2.62) using the same approach.

Larger numbers of packets P also have several other advantages. For example, multiple packets are also useful in real radios where the signal may not be stable during the first few repetitions. The start of any pulsed waveform is usually the most banged up as the PAs are warming up and the local oscillators are stabilizing after having been shut down to conserve power.

The approach as described in this section seems to have been presented first in [Moo94] for OFDM systems, though the core concept of deriving the frequency from the difference in phase is credited to other papers. With $P = 2$, [Moo94] shows that the estimator in (2.62) is a maximum likelihood estimator; essentially it is a good estimator. The use of multiple training packets in a more efficient manner than employed in (2.62) is described in [MM99]. There, multiple correlations — not just the average of the pairwise correlations in (2.62) — are combined to find the best linear unbiased estimator under certain assumptions, with somewhat higher complexity. Other multiple training packet algorithms with advantages in terms of frame synchronization are described in [MZB00][SS04] and also in Section 2.7.3.

The extension to OFDM is further developed in [SC97a], where it is suggested to use two specially designed OFDM symbols. The first OFDM symbol has all the odd subcarriers with the zero value, making it periodic with a period of $N/2$ where N is the size of the FFT algorithm. The second OFDM symbol contains training information on all the subcarriers; the training on the even subcarriers is differentially encoded with the previous OFDM symbol. This algorithm first estimates and corrects for the fine frequency offset with a range of $|\epsilon| < 1/N$. After correction, there is an ambiguity of the form $\exp(j2\pi k/(N/2))$ where k is the integer frequency offset. The offset is estimated using the second differentially encoded training packet.

---

12. At the edge, unknown symbols from the previous packet are smeared with the current packet. By looking only at the later samples, this effect can be avoided.

2.7.3 Frame Synchronization

Frame synchronization involves estimating the beginning of a frame of data. Equivalently, frame synchronization involves estimating and correcting for an unknown timing offset. Frame synchronization is performed prior to frequency offset synchronization but may also be performed in concert with frequency offset synchronization. A reasonable model for timing offset in discrete-time is

$$y[n] = e^{j2\pi\epsilon n} \sum_{\ell=0}^{L} h[\ell]s[n - \ell - \Delta] + v[n] \qquad (2.63)$$

where Δ is the unknown integer offset. Assuming, for example, that the observed data starts at $n = 0$, the offset Δ should be greater than 0, otherwise the beginning of the frame is missed. Given an estimate of the offset $\hat{\Delta}$, correction simply involves shifting the data to form $\widetilde{y}[n] = y[n + \hat{\Delta}]$. Essentially the first $\hat{\Delta} - 1$ samples are removed.

Timing offset algorithms for many wireless systems exploit the same preamble structure used for frequency synchronization. In fact, the correlation used to compute (2.62) can be used as a basis for frame synchronization. For example, the proposed algorithm in [SC97a] with $P = 2$ repetitions would look for a peak in a function

$$J[d] = \frac{\left|\sum_{n=L}^{N_t-1} y^*[n + d + N_t]y[n + d]\right|^2}{\sum_{n=L}^{N_t-1} |y[n + d]|^2 + |y[n + N_t + d]|^2}. \qquad (2.64)$$

The numerator of (2.64) will be large when the two replicas of the signal are found. The normalization by the denominator is used to reduce the effects of noise when there is no signal present. Different normalizations are possible; the one in (2.64) comes from [MZB00].

The function in (2.64) may not have a sharp peak due to the presence of the cyclic prefix. In general, the system is resilient to small errors in $\hat{\Delta}$, which can be incorporated into the the channel impulse response. Better timing performance can be improved through the use of multiple repetitions including a sign change in at least one repetition, as proposed in [MZB00] and analyzed in more detail in [SS04]. For example, as described in Section 2.7.1, the last repetition may be negated. Following the development in [SS04], define a function

$$\begin{aligned}P[d] = &\left|\sum_{n=L}^{N_t-1} y^*[n + d + N_t]y[n + d] + y^*[n + d + 2N_t]y[n + d + N_t]\right.\\ &\left. - y^*[n + d + 3N_t]y[n + d + 2N_t]\right| \\ &+ \left|\sum_{n=L}^{N_t-1} y^*[n + d + 3N_t]y[n + d] - y^*[n + d + 4N_t]y[n + d + 2N_t]\right| \\ &+ \left|\sum_{n=L}^{N_t-1} y^*[n + d + 4N_t]y[n + d]\right|.\end{aligned} \qquad (2.65)$$

Then timing could be performed based on the modified metric

$$\frac{P[d]}{\sum_{n=L}^{N_t-1} |y[n+d]|^2 + |y[n+d+N_t]|^2 + |y[n+d+2N_t]|^2 + |y[n+d+3N_t]|^2}. \quad (2.66)$$

The sign change creates differences in the metric that improves performance. The concept can be generalized to more than $P = 4$. A better frequency offset can be estimated from this timing estimator using the term inside the absolute value in (2.65).

Other structures in the preamble may also be exploited for timing estimation, for example, the correlation properties of the Frank-Zadoff-Chu sequence [YLLK10] to reduce the number of autocorrelations that must be performed. Autocorrelations convolve the received data with itself, while correlation with a known sequence can be implemented using convolution. Not all the available repetitions have to be exploited. For example, the approach in [PG07b] uses only one of the negated sequences among several.

A related concept of frame synchronization is packet detection. The objective of packet detection is to determine whether a frame (or packet) is present or not in a given buffer of data. This is a necessary part of the frame synchronization operation and is also used as part of a carrier sense operation in certain medium access control (MAC) protocols. The classic approach to packet detection is to perform a windowed estimate of the received power and compare it to a threshold. The idea is that if the power increases then there must be a packet present. It is more robust to exploit structure found in the preamble; this also helps avoid false detections in shared unlicensed bands. A simple approach would be to compute the metric in (2.64) or (2.66) and compare to a threshold that is implementation specific.

2.7.4 Channel Estimation

After frame synchronization and carrier frequency offset correction, the next major receiver task is equalization. The strategies outlined in Sections 2.4 and 2.5 require knowledge of the coefficients of the channel impulse response to implement equalization. After performing frame synchronization and frequency offset correction, the received signal can be modeled as

$$y[n] = \sum_{\ell=0}^{L} h[\ell]s[n-\ell] + v[n]. \quad (2.67)$$

The objective of channel estimation is to generate an estimate $\{\widehat{h}[\ell]\}$ given known information in the transmitted signal. This section reviews channel estimation from a discrete-time perspective; an analog perspective may be more suitable for analog equalization [SGL+09].

There are different ways to estimate the channel coefficients. Perhaps the most popular approach is the least squares estimate, which is also the maximum likelihood estimate if the additive noise is white and Gaussian. To formulate the estimator, suppose that $\{t[n]\}_{n=0}^{N_t-1}$ is a sequence of known training data and suppose that $s[n] = t[n]$ for

$n = 0, 1, \ldots, N_t - 1$. A general approach is to write the input-output relationship in a matrix form neglecting noise

$$\underbrace{\begin{bmatrix} y[L] \\ y[L+1] \\ \vdots \\ y[N_t - 1] \end{bmatrix}}_{\mathbf{y}} = \underbrace{\begin{bmatrix} t[L] & \cdots & t[0] \\ t[L+1] & \ddots & \vdots \\ \vdots & & \vdots \\ t[N_t - 1] & \cdots & t[N_t - 1 - L] \end{bmatrix}}_{\mathbf{T}} \underbrace{\begin{bmatrix} h[0] \\ h[1] \\ \vdots \\ h[L] \end{bmatrix}}_{\mathbf{h}}. \quad (2.68)$$

The observations start from $y[L]$ since the values of $s[n]$ for $n < 0$ may be unknown. The matrix $\mathbf{T}$ is the training matrix. Normally N_t is chosen such that $\mathbf{T}$ is tall to give better performance. The least-squares channel estimate solves the following optimization problem

$$\widehat{\mathbf{h}} = \arg\min_{\mathbf{a}} \|\mathbf{y} - \mathbf{T}\mathbf{a}\|^2. \quad (2.69)$$

Assuming that the training matrix is full-rank, which is achieved, for example, by a training sequence with good correlation properties, the least squares estimate has a simple form: $\widehat{\mathbf{h}} = (\mathbf{T}^*\mathbf{T})^{-1}\mathbf{T}^*\mathbf{y}$. The squared error of the least-squares estimate is given by $\mathbf{y}^*\mathbf{y} - \mathbf{y}^*\mathbf{T}(\mathbf{T}^*\mathbf{T})^{-1}\mathbf{T}^*\mathbf{y}$. Note that the operation $\mathbf{T}^*\mathbf{y}$ is a correlation of part of the training data with the observed data. If the training sequence has good correlation properties, $\mathbf{T}^*\mathbf{T} \approx \mathbf{I}$ and thus the least squares operation can be interpreted as taking the outputs of the correlation with a known training signal. When there are multiple repeated training sequences, and the training has good periodic correlation properties as with the Frank-Zadoff-Chu sequences, this interpretation can be made exact.

Channel estimation using least squares can be performed with training in the time domain or the frequency domain. Consider now an OFDM system with training in the frequency domain for one OFDM symbol. It helps to write an equivalent system like in (2.68) where the unknown channel coefficients are written in a vector form. Assuming training on all the subcarriers

$$\underbrace{\begin{bmatrix} \bar{Y}[0] \\ \bar{Y}[1] \\ \vdots \\ \bar{Y}[N-1] \end{bmatrix}}_{\bar{\mathbf{y}}} = \underbrace{\begin{bmatrix} S[0] & 0 & \cdots & 0 \\ 0 & S[1] & 0 & \cdots & 0 \\ \vdots & & \ddots & \ddots \\ 0 & \cdots & 0 & S[N-1] \end{bmatrix}}_{\mathbf{S}} \underbrace{\begin{bmatrix} H[0] \\ H[1] \\ \vdots \\ H[N-1] \end{bmatrix}}_{\bar{\mathbf{h}}}. \quad (2.70)$$

A least squares version of (2.70) can be solved directly, but the result is quite noisy. Better performance is achieved by recognizing that $\bar{\mathbf{h}} = \mathbf{F}\mathbf{h}$ where $\mathbf{F}$ consists of the first $L+1$ columns of the $N \times N$ DFT matrix. Then the estimate is performed in the time domain to obtain $\widehat{\mathbf{h}} = (\mathbf{F}^*\mathbf{S}^*\mathbf{S}\mathbf{F})^{-1}\mathbf{F}^*\mathbf{S}^*\bar{\mathbf{y}}$. The frequency domain channel coefficients can be found as $\widehat{\bar{\mathbf{h}}} = \mathbf{F}\widehat{\mathbf{h}}$. Though the explanation here assumes that all N subcarriers are used for data, it is easy to accommodate only a partial number of active subcarriers by removing rows from (2.70) and $\mathbf{F}$.

Channel estimation is performed in most wireless communication systems — the concept is not particular to mmWave. The performance of the channel estimator, however, is tied to the design of the training structure in the signal (training signals or pilots) and more generally in the selection of a preamble sequence. Related work on channel estimation for 60 GHz systems is reported in [YLLK10][LH09]. A correlation based approach using Frank-Zadoff-Chu sequences is proposed in [YLLK10] for a 60 GHz SC system where the channels are estimated in the time domain but then smoothed in the frequency domain. Another correlation based approach using Golay sequences is suggested in [LH09] for a 60 GHz SC system where multiple estimates averaged over time are used to improve accuracy.

The channel estimate can be updated using adaptive estimation techniques. For example, the approaches mentioned in Section 2.4.1 can be used to derive recursive least squares channel estimators or least mean squares channel estimators, using pilots, pilot words, or decision directed estimates. Adaptive estimators can track channel variations over time, improving performance.

One specific application of pilots, versus training sequences, is tracking of common phase and gain errors. Phase noise and clock drive create common phase error, which manifests in OFDM systems as a constant phase shift across all subcarriers [RK95]. Common gain errors manifest if the automatic gain control is not settled. Common phase errors in OFDM can be modeled as $\mathsf{H}[n,k] = e^{j\theta[n]}\mathsf{H}[n,0]$ where $\mathsf{H}[n,k]$ is the value of the channel in the frequency domain at subcarrier n and for OFDM symbol k, and $\mathsf{H}[n,0]$ is the channel estimated from the preamble. By sending known pilots on certain subcarriers $n_1, n_2, \ldots, n_p$, the phase error can be estimated as phase($\mathsf{H}^*[n_1,k]\mathsf{H}[n_1,0] + \mathsf{H}^*[n_2,k]\mathsf{H}[n_2,0] + \cdots + \mathsf{H}^*[n_p,k]\mathsf{H}[n_p,0]$). In IEEE 802.11ad and IEEE 802.15.3c, 16 pilots are used for this purpose.

When estimating the channel, the parameter L known as the *channel order* defines the maximum propagation delay in the channel, in terms of the numbers of symbols spanning such delay, and must be assumed or discovered. In traditional estimation this parameter would be estimated using, for example, a maximum description length criterion [Aka74][Ris78]; see [GT06] for a recent example applied to wireless channels. Wireless systems are typically designed to deal with a maximum size of L, as determined from multi-path delay spread propagation measurement studies; see Chapter 3 for more information. The length of the cyclic prefix or guard interval, for example, provides an upper bound on L. Consequently in existing systems, the value of L can be taken a priori from the features of the system as described in the standards specification. Better order estimation, when the channel is not very frequency selective, will give better performance at the expense of higher complexity in the receiver. Sparsity can also be employed to improve performance [BHSN10].

2.8 Multiple Input Multiple Output (MIMO) Communication

Multiple antennas are widely used in wireless communication systems like IEEE 802.11n [802.11-12a], IEEE 802.11ac, 3GPP LTE, and 3GPP LTE Advanced. Multiple antennas may be located at the transmitter, receiver, or both. Communication systems that use multiple transmit and receive antennas, as illustrated in Fig. 2.22, are broadly known

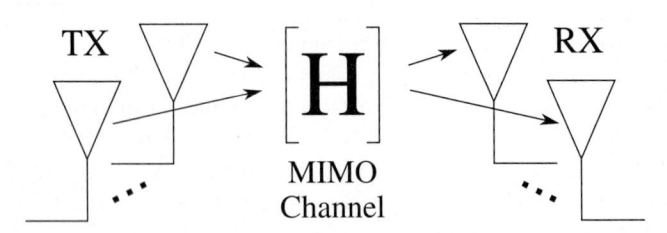

Figure 2.22 A generic MIMO communication system with a MIMO channel. The inputs of the channel correspond to the outputs of the transmit antennas. The outputs of the channel are inputs to the receive antennas.

as MIMO communication systems because the propagation channel has multiple inputs (from different transmit antennas) and multiple outputs (from different receive antennas) [PNG03].[13] MIMO communication offers many advantages in communication including diversity against small-scale fading, higher data rates, and the ability to cancel interference.

MIMO communication is important and relevant for mmWave systems. The small wavelengths at mmWave frequencies make it possible to pack a large number of antennas in a small area. This permits the use of large antenna arrays at the transmitter and receiver and allows mmWave systems to exploit the potential benefits of MIMO communication. MIMO is found in IEEE 802.15.3c [802.15.3-03], ECMA 387 [ECMA10], and IEEE 802.11ad [802.11-12]. More details on the specific supported transmission modes are discussed in Chapter 9.

This section provides an overview of the fundamentals of MIMO communication. It explains the different transmission modes that are common in MIMO communication. Spatial multiplexing is used to provide higher data rates by sending, on the same carrier frequency (channel), multiple streams of symbols. Diversity is used to provide higher reliability by sending redundant information across the antennas. Beamforming is used to adapt the shape of the transmit and receive antenna arrays to improve communication through, for example, array gain and/or diversity gain. Hybrid beamforming approaches are used to obtain both the advantages of spatial multiplexing and beamforming, using hardware constraints often found in mmWave systems. The survey of MIMO communication in this section is provided from a signal processing perspective. More details about antennas and antenna patterns are provided in Chapter 4.

2.8.1 Spatial Multiplexing

Spatial multiplexing, illustrated in Fig. 2.23, is a MIMO transmission technique in which information is demultiplexed across the transmit antennas and jointly decoded from the signals observed by multiple receive antennas [PK94][PNG03]. Under some assumptions about the richness of the propagation environment (which is yet to be fully understood at mmWave), it is possible to increase the spectral efficiency (the rate of data per unit

---

13. The term *MIMO* comes from signal processing and control theory. It is used to refer to a system that has multiple inputs and multiple outputs.

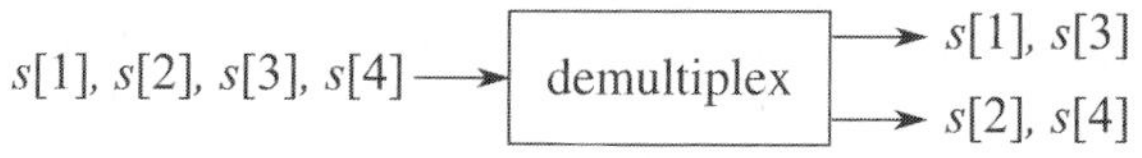

2.8 Multiple Input Multiple Output (MIMO) Communication

$s[1], s[2], s[3], s[4] \longrightarrow$ [demultiplex] $\longrightarrow s[1], s[3]$
$\longrightarrow s[2], s[4]$

Figure 2.23 Example of a MIMO spatial multiplexing transmitter sending two data streams. The spatial multiplexing operation spreads the symbols across the transmit antennas, sending the symbols in this case in half the time.

bandwidth) linearly with the smaller of the number of either transmit or receive antennas [Tel99][Fos96]. Essentially, spatial multiplexing allows multiple symbols to be transmitted at the same time on the same carrier, with the same total power as if one symbol were sent.

A complex baseband MIMO input-output relationship can be obtained by following the techniques from Section 2.2. Similarly, an input-output relationship can be derived by generalizing the results in Section 2.3 to the case of spatial multiplexing.

Consider a MIMO communication system with N_t transmit antennas and N_r receive antennas. Let $s_p[n]$ denote the symbol sent on the pth transmit antenna. The symbol $s_p[n]$ could be obtained by demultiplexing a symbol stream $s[n]$ to create $s_p[n] = s[N_t n + p - 1]$. Let $\mathbf{s}^T[n] = [s_1[n], s_2[n], \ldots, s_{N_t}[n]]^T$ denote the vector of transmit symbols. Let $\{\mathbf{H}[\ell]\}_{\ell=0}^{L}$ denote the matrix impulse response of the channel. Each entry of the channel response, $\{h_{m,p}[\ell]\}_{\ell=0}^{L}$, is the impulse response between the pth transmitter antenna and the mth receiver antenna. Let $\mathbf{y}[n]$ be the vector of sampled observations from each antenna. Then, the input-output relationship for spatial multiplexing is

$$\mathbf{y}[n] = \sum_{\ell=0}^{L} \mathbf{H}[\ell]\mathbf{s}[n-\ell] + \mathbf{v}[n]. \tag{2.71}$$

Note that the normalization of the transmitted signal vector is chosen so that the sum symbol energy is given by $\mathbb{E}\|\mathbf{s}[n]\|^2 = E_s$ to make comparisons fair with single antenna systems.

All of the techniques of linear equalization, frequency domain equalization, decision feedback equalization, and maximum likelihood sequence detection can be extended to MIMO systems. Because the impulse response of the system is a matrix, equalization is generally more complex in a MIMO system. The result is that frequency domain techniques have received the most emphasis, as DFTs are easy to implement in signal processing hardware and convolution in the time domain is simple multiplication in the frequency domain.

Using the OFDM notation from Section 2.5.2, the received MIMO communication signal after the DFT operation can be written as a multiplication of the signal matrix and the channel matrix

$$\mathbf{y}[k] = \mathbf{H}[k]\mathbf{s}[k] + \mathbf{v}[k]. \tag{2.72}$$

The equivalent channel $\mathbf{H}[k] = \sum_{\ell=0}^{L} \mathbf{H}[\ell]\exp(-j2\pi k\ell/N)$ is now a matrix instead of a scalar as in (2.56). The decoding complexity is reduced in (2.72) versus (2.71) since there is only a single matrix representing the received frequency samples.

Detecting the transmitted symbols from $\mathbf{y}[k]$ in (2.72) requires removing the effects of $\mathbf{H}[k]$. Because the MIMO channel is now a matrix, this requires equalizers that operate in the spatial domain to remove the co-antenna interference created from multiple simultaneous symbol streams. Many of the other techniques described in this chapter can be applied to the simplified model in (2.72). For example, a zero-forcing equalizer would compute the pseudo inverse of the channel given by $\mathbf{H}^{\dagger}[k]$, apply it to the observed vector to create $\mathbf{H}^{\dagger}[k]\mathbf{y}[k]$, and then perform symbol detection on each resulting entry separately.

Much of the theoretical work on MIMO communication focuses on narrowband (frequency-flat) channels, essentially a special case of (2.71) in which there is no time delay and the convolution becomes a multiplication where $L = 0$ [Rap02]. OFDM and similar modulations exploit MIMO by essentially treating a wideband RF channel as an orthogonal linear sum of narrowband flat fading channels, where each narrowband channel is a subband of the wideband channel. Let $\mathbf{H}$ denote the matrix channel assuming that $L = 0$ (there is flat fading and only one discrete-time tap). Then, the input-output relationship becomes

$$\mathbf{y}[n] = \mathbf{H}\mathbf{s}[n] + \mathbf{v}[n], \tag{2.73}$$

which shares similarities with (2.72). The main difference in the flat fading case is that the channel is the same for all subcarriers (effectively $\mathbf{H}[k] = \mathbf{H}$). It is useful to explain other MIMO techniques using the simplified model in (2.73), though it should be understood that a more practical transceiver will be built around the more complete signal models in (2.71) or (2.72).

There is limited work on the direct application of spatial multiplexing to mmWave systems. MmWave channels are not yet fully understood as to the extent of rich scattering, making the resulting MIMO channels possibly ill conditioned (not very invertible, meaning that equalization may not work well). However, as shown in Chapter 3, rich scattering does appear to some extent in outdoor urban mmWave channels. Spatial multiplexing has some potential for mmWave communication. One case that has been studied is line-of-sight MIMO, where the transmitter-to-receiver range is small such that the phases from different antenna pairs are different enough to create well conditioned channels [MSN10]. More details about line-of-sight MIMO are provided in Chapter 7. Other applications of spatial multiplexing in mmWave systems combine with some form of beamforming, discussed in more detail in Section 2.8.4.

2.8.2 Spatial Diversity

Another application of multiple antennas in a wireless system is to obtain diversity advantage in the presence of small-scale fading. The idea is to exploit the presence of multiple propagation paths between the transmitter and receiver. Such paths can be created through the presence of multiple transmit and multiple receive antennas. For example, with a single transmit antenna, multiple receive antennas can be used to obtain different "looks" at the same signal. These observations can be combined together to better detect the transmitted symbols. With multiple receive antennas, diversity can be obtained through different combining techniques [Kah54][Jak94, Chapter 5][Rap02]. With multiple transmit antennas, obtaining diversity is more complicated. One approach is to spread the symbols in a smart way across the transmit antennas using delays [Wit91][SW94a]

2.8 Multiple Input Multiple Output (MIMO) Communication

$s[1], s[2], s[3], s[4] \rightarrow$ delay diversity $\rightarrow s[1], s[2], s[3], s[4], 0$
$\rightarrow 0, s[1], s[2], s[3], s[4]$

Figure 2.24 A transmitter that employs delay diversity. The symbols sent on the first antenna are delayed then sent on to the second antenna.

(see Fig. 2.24), space-time trellis codes [TSC98], or space-time block codes [JT01], of which the Alamouti code is most famous [Ala98]. Another approach is to use feedback (control data from the receiver to the transmitter) to steer the transmitted signal in directions that take advantage of multiple signal paths (known as *limited feedback communications*) [LH03][LHS03][MSEA03][LHSH04].

Spatial diversity techniques for fading mitigation have not been widely studied in mmWave systems, except in a few cases [JLGH09][RSM+13]. There are several possible reasons. For example, because there is a misconception that there is a shortage of nearby scatterers, the coherence distance in mmWave systems is thought to be large and the impulse responses between distinct antenna pairs are thought to be correlated (making beamforming useful, but mitigating the value of diversity techniques). Additionally, the bandwidth used in mmWave communication is extremely large, often much larger than the coherence bandwidth, meaning that delay spread is significant, producing many discrete-time channel taps. Channels with delay spread are a good source of frequency and time diversity. Further discussion on mmWave channels may be found in Chapter 3.

2.8.3 Beamforming in MIMO Systems

Classical beamforming uses multiple antennas at the transmitter and/or the receiver. To explain this concept, we consider transmit beamforming. The essential idea of beamforming is to send the same information on each antenna, but with varying amplitude and/or phase for the signal on each antenna. Beamforming is effectively a form of spatial filtering. By varying the amplitude or phase, the effective radiation pattern of the overall antenna array can be shaped and directed. For example, the antenna pattern can be steered in the direction of the most favorable propagation path [LR99b][Dur03][CR01a][SNS+14]. In more sophisticated beamforming strategies, as implemented by smart antennas, antenna array patterns may also be adjusted to steer nulls to nearby interferers. Canceling interference further improves system performance [LR99b][AAS+14].

To explain the concept of beamforming, consider a MIMO communication system. Let $\mathbf{f}$ denote the transmit beamforming vector and $s[n]$ the symbol to be transmitted. Further, let $\mathbf{w}$ denote the receive beamforming vector, often known as the *combining vector*. Using these definitions, a narrowband MIMO system with beamforming (see Fig. 2.25) has an input-output relationship

$$y[n] = \mathbf{w}^*\mathbf{H}\mathbf{f}s[n] + \mathbf{w}^*\mathbf{v}[n]. \tag{2.74}$$

Compared to (2.73), only a single stream is sent; there is no spatial multiplexing rate improvement. The effective channel includes the effects of the channel, beamformer, and combining vectors in $\mathbf{w}^*\mathbf{H}\mathbf{f}$. Intelligent selection of $\mathbf{f}$ and $\mathbf{w}$ yields a large channel gain

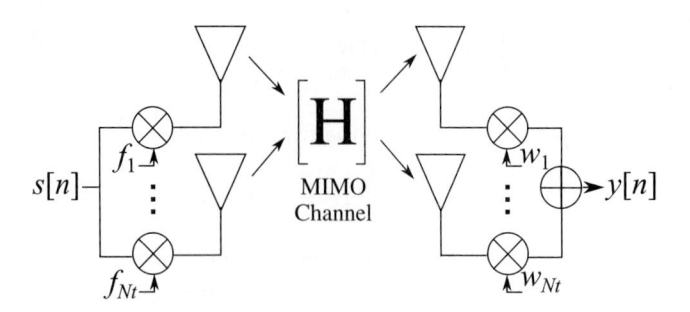

Figure 2.25 A MIMO system with transmit and receive beamforming.

$|\mathbf{w}^*\mathbf{Hf}|$. The maximum gain solution is given by setting $\mathbf{f}$ to the dominant right singular vector of $\mathbf{H}$ and by setting $\mathbf{w}$ to the dominant left singular vector of $\mathbf{H}$. Note that the transmitter's beamforming vector $\mathbf{f}$ depends on the channel $\mathbf{H}$, which is typically only measured at the receiver. Consequently, some feedback information may be needed from the receiver to help the transmitter select the best beamforming vector. This is known as *limited feedback* and is widely deployed in commercial wireless systems (see, for example, [LHL+08] for an overview and the references therein).

Beamforming in mmWave systems is different from beamforming in microwave systems due to practical constraints. For example, in UHF and microwave MIMO systems, the beamforming is often performed digitally at baseband, which enables controlling both the signal's phase and amplitude. Digital processing, however, requires dedicated baseband and RF hardware for each antenna element. Unfortunately, the high cost and power consumption of mmWave mixed-signal hardware precludes such a transceiver architecture at present, and forces mmWave systems to rely heavily on analog or RF processing [DES+04][Gho14][PK11] or, alternatively, very low resolution ADCs [MH14]. Analog beamforming is often implemented using phase shifters [PK11][DES+04][VGNL+10], which places constant modulus constraints on the elements of the beamformers. Designing MIMO beamforming solutions with constraints has been an active area of research [LH03][SMMZ06][ZXLS07][NBH10][GVS11][Pi12] [ZMK05][AAS+14]. Designing good beamforming solutions is tightly coupled to higher layer protocols, for example, beam training as discussed in Chapter 8. As shown in Chapter 3, mmWave systems benefit from significant range extension through the use of beamforming [SR13][SR14][SMS+14].

In mmWave systems, the antennas are often tightly packed together (antenna element spacings are much less than the coherence distance of the channel). This means that there is a deterministic relationship between the signal arriving at one antenna element and the delayed version of the signal arriving on another antenna element, which is a function of the angle-of-arrival. This leads to sparsity and mathematical structure in the channel. These concepts are explained in more detail in Chapter 4. Here we consider a specific example. Consider an example of a single path and a uniform linear array (where the antenna elements are uniformly spaced with spacing distance d/λ, where λ is the wavelength). The spacing could be different at the transmitter and receiver, but for this example we assume it is the same. Let θ_t denote the angle of departure and θ_r denote the angle of arrival. The array response for a uniform linear array is given by

$$\mathbf{a}(\theta)^T = [1, e^{j\pi \frac{d\cos(\theta)}{\lambda}}, \cdots, e^{jpi \frac{(N_r-1)d\cos(\theta)}{\lambda}}]^T. \qquad (2.75)$$

Now, assuming a single propagation path between the transmitter and receiver with complex gain α, the channel can be modeled as $\mathbf{H} = \alpha \mathbf{a}(\theta_r)\mathbf{a}(\theta_t)^*$. This channel is sparse and has low rank because there is only one propagation path (in fact, the channel has a rank of 1 since there is only one propagation path). The optimum beamformer in this example is just $\mathbf{f} = \frac{1}{N_t}\mathbf{a}(\theta_t)$ (normalized to maintain constant transmit energy). The receive combining vector is similarly $\mathbf{w} = \mathbf{a}(\theta_r)$ (normalization is not required at the receiver). The entries are just phase shifts, thus the optimum beamformer can be implemented using phase shifters. If the phase shifts are quantized, as is usually the case in analog beamforming, then this will require optimization to select the best phase shift for each antenna; see, for example, the algorithms in [LH03][SMMZ06][ZXLS07][NBH10][GVS11][Pi12]. Design of transmit and receive beamformers remains an important research area for mmWave communication [LR99b][BAN14][SR14].

2.8.4 Hybrid Precoding

Hybrid precoding has been developed to support spatial multiplexing (with only a few data streams) while still accounting for the hardware limitations in mmWave beamforming [PK11][TSMR09][TAMR06][XYON08][ERAS+14][AELH14]. The term *precoding* essentially means that multiple beamforming vectors are used, one for each stream to be transmitted.

A mathematical model of a narrowband MIMO system with precoding at the transmitter and combing at the receiver is given by

$$\mathbf{y}[n] = \mathbf{W}^*\mathbf{H}\mathbf{F}\mathbf{s}[n] + \mathbf{v}[n] \qquad (2.76)$$

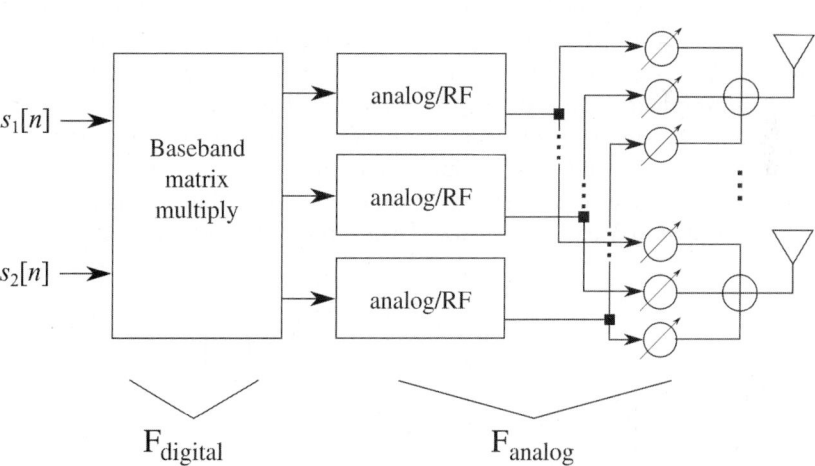

Figure 2.26 An example of hybrid precoding at the transmitter with $N_s = 2$ and $N_m = 3$. The digital outputs are combined with different weights and sent on each antenna.

where $\mathbf{F}$ is the $N_t \times N_s$ precoding matrix used to send N_s symbols and $\mathbf{W}$ is the $N_r \times N_s$ combining matrix.

In hybrid precoding, as shown in Fig. 2.26 [PK11][TSMR09][TAMR06][XYON08] [ERAS+14], some precoding is performed in digital and some precoding is performed in analog. For example, the precoding matrix $\mathbf{F}$ may be decomposed as $\mathbf{F} = \mathbf{F}_{\text{analog}} \mathbf{F}_{\text{digital}}$ where $\mathbf{F}_{\text{analog}}$ is a $N_t \times N_m$ matrix corresponding to the analog beamforming coefficients and $\mathbf{F}_{\text{digital}}$ is a $N_m \times N_s$ matrix of digital precoding coefficients. The parameter N_m corresponds to the number of outputs of the digital transmit beamformer and should satisfy $N_m \geq N_s$. The design of $\mathbf{F}_{\text{analog}}$ and $\mathbf{F}_{\text{digital}}$ is complicated, but if done well results in little loss over the unconstrained solution. Some interesting approaches have been developed that exploit sparsity in the channel as described in [KHR+13][ERAS+14][AELH14][AAS+14]. Research on hybrid precoding is still ongoing. There are several advantages of the hybrid architecture including easy support of multi-user transmission and reception (as in cellular systems) [GKH+07].

2.9 Hardware Architectures

The illustrations of upconversion and downconversion in Fig. 2.2 are idealized representations of the actual hardware architectures. Practical implementations may differ due to factors relating to the non-ideality of circuit components, for example, the difficulty in realizing very sharp filters. When filters are non-ideal, signals occupying other carriers (interferers) are not completely eliminated and may fold into the desired signal during the mixing operation. This section describes several common hardware architectures for realizing upconversion and downconversion from a circuit perspective. A more comprehensive review is given in [Raz98][LR99b].

Consider downconversion as illustrated in Fig. 2.2. Essentially, the downconversion operation involves mixing the received signal with a cosine or sine and filtering to reject the higher frequency products. The closest hardware architecture to that in Fig. 2.2 is the homodyne receiver, also called the direct conversion receiver as illustrated in Fig. 2.27. In this receiver, the received signal is (possibly) filtered with a band-select filter followed by a low-noise amplifier. The resulting signal is mixed in quadrature with a cosine and sine wave, where the sine is created by phase shifting the cosine. The received signal is low-pass filtered to reject interference, amplified, and sampled with an ADC.

The homodyne receiver has a number of limitations that make it challenging to implement, especially at mmWave frequencies. Two important limitations are DC offset and

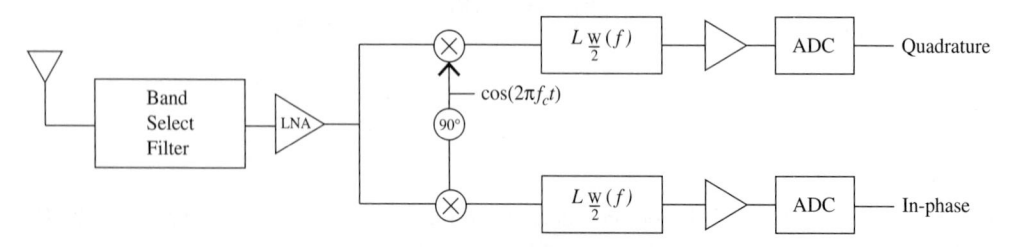

Figure 2.27 Illustration of a homodyne or direct conversion receiver. There is only one stage of downconversion from the carrier frequency to baseband. A sharp band-select filter is not required as most of the filtering is taken care of by the baseband low pass filter.

2.9 Hardware Architectures

in-phase/quadrature (IQ) mismatch. DC offset needs either a special zero-DC transmit signal or some kind of receiver correction circuitry. DC offset is created in a number of ways. For example, because of imperfect port isolation, some of the carrier makes its way to the input to the low-noise amplifier (called *carrier feedthrough*). This signal gets converted down to 0 Hz (DC). DC offset needs either a special zero-DC transmit signal or some kind of correction circuitry. IQ mismatch distorts the received constellations with SC modulations and, if severe, requires more complex signal processing at the receiver to compensate for the added distortion. IQ mismatch creates inter-carrier interference in OFDM systems, which creates a noise-like distortion that reduces performance (e.g., increases bit errors). IQ mismatch is created by phase and gain differences in the different paths between the in-phase and quadrature signal branches. Other impairments found in homodyne receivers include flicker noise and local oscillator leakage.

Note that DC offset and IQ mismatch have other sources as well at mmWave bandwidths, which are different than narrowband designs. For example, in narrow-band devices, the IQ mismatch is usually the result of the local oscillator outputs not being in perfect quadrature. In wideband circuits, not only do the two filters in Fig. 2.27 have to match, but also the electrical path lengths of the interconnects and the clock lines of the ADCs. ADC sample skew is rarely a problem in narrowband designs. ADCs are discussed in Chapter 6.

The heterodyne receiver is one of the most common downconversion architectures. The idea of the heterodyne receiver is to filter and mix successively through multiple intermediate frequencies (IFs) to shift the signal to baseband. An example known as a *single IF* receiver is illustrated in Fig. 2.28. In this example, the signal is first downconverted from carrier f_c to $f_c - f_{LO}$. The carrier f_{LO} is the first IF frequency. The filtered signal is then converted to baseband by a quadrature demodulation that multiplies by a carrier of frequency $f_{LO2} = f_c - f_{LO}$. Compared with the homodyne receiver, the problem of DC offset is avoided because the first stage is not at DC and the signal level at the second stage is much higher, making the carrier feedthrough less significant. IQ mismatch is less severe because the gain and phase differences are much smaller for a stronger and lower frequency signal. A main challenge in designing a heterodyne receiver is the choice of image reject filter and the IF frequencies. After each mixing with an IF, signals found at a carrier frequency of $2f_{LO}$ (called *images*) mix into the received signal. The image reject filter tries to attenuate these signals as much as possible, but very sharp filters are hard to implement, pushing f_{LO} to be lower in frequency.

An alternative heterodyne receiver structure that reduces the required baseband hardware is the sampling IF architecture. In the sampling IF architecture, the final stage of

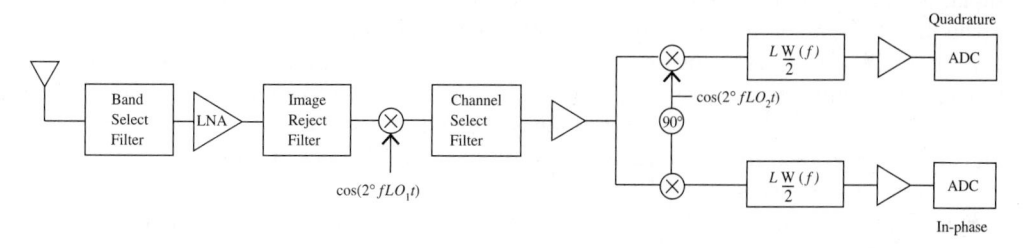

Figure 2.28 Illustration of a single IF heterodyne receiver. The signal is converted to baseband through two stages of mixing. In the second stage, the in-phase and quadrature components are extracted.

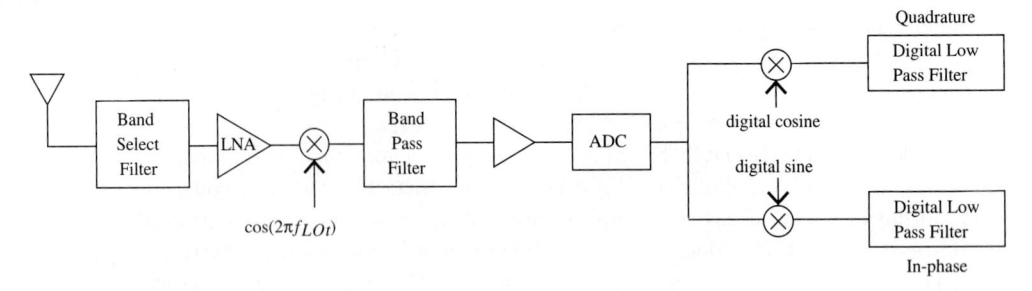

Figure 2.29 Illustration of a sampling IF receiver. Instead of a second stage of downconversion in analog, the received signal is sampled and then the subsequent downconversion is performed in digital in baseband.

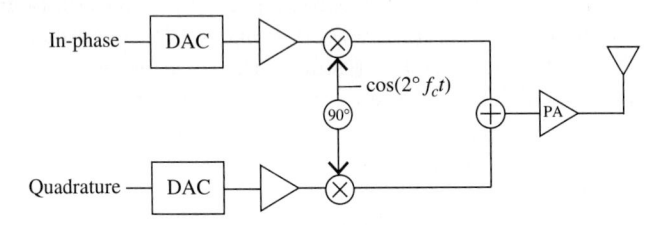

Figure 2.30 Illustration of a direct conversion transmitter. The in-phase and quadrature signals and mixed, added together, and amplified prior to transmission.

conversion to baseband is replaced with a sampling operation. Then the conversion to baseband is performed in digital. An example is illustrated in Fig. 2.29. This architecture reduces the number of ADCs required and eliminates IQ mismatch, but requires sampling at a higher rate [LR99b].

There are some important variations of the sampling IF architecture. For example, the wide bandwidths of mmWave signals favor high IF frequencies as it is always hard to work near DC. It not uncommon to subsample, that is, to place the IF signal in a higher Nyquist interval (greater than the Nyquist frequency but less than the Nyquist rate). Also, for efficiency reasons, the digital sine and cosine generators shown in Fig. 2.29 may be run at one fourth the Nyquist rate for very efficient, low-complexity implementations. See [GFK12] for an example of a subsampling heterodyne receiver.

Now consider upconversion as illustrated in Fig. 2.2. Essentially, the upconversion operation involves mixing the baseband signal with a cosine or sine and amplifying the transmit signal with a power amplifier (PA). The closest hardware architecture to that in Fig. 2.30 is the direct conversion transmitter, as illustrated in Fig. 2.27. The transmitter operation is virtually identical to the theoretical description. The main drawback in this design is the leakage of the amplified signal back into the mixing carrier, a phenomenon known as *local oscillator pull*. There are various techniques to alleviate this problem, including the use of multiple oscillators to maintain the desired frequency in the face of pull.

A multi-step conversion is also possible at the transmitter. The use of multiple steps avoids the disadvantages of the direct conversion transmitter. A two-step conversion is illustrated in Fig. 2.31. The filter before the second mixing operation rejects harmonics from the first mixing and reduces noise. The filter after the second mixing operation

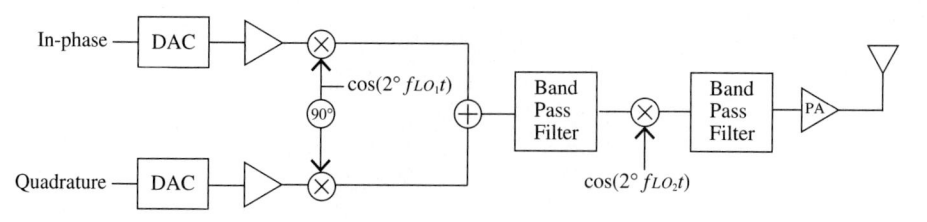

Figure 2.31 Illustration of a two-step conversion transmitter. The first passband filter rejects harmonics created by the mixing operation and reduces noise, while the second filter eliminates images from the second mixing.

eliminates images from the second mixing. The second filter must be good enough to avoid amplifying the images and creating unwanted interference in other frequency bands for other communication devices.

This review of hardware architectures for upconversion and downconversion provides a high-level review of some of the issues involved in the translation of signals between baseband and passband. Of course, the actual design of hardware for these purposes involves making many design tradeoffs and requires sophisticated circuit design as discussed in Chapter 5. Because of commercial interests, a number of designs have been proposed in the literature for 60 GHz. A homodyne upconverter was presented in [FOV11], whereas a corresponding downconverter was presented in [OFVO11] for a WLAN application. A reconfigurable subsampling heterodyne receiver was presented in [GFK12]. A heterodyne receiver architecture was presented in [PR08]. A three-stage super heterodyne transceiver architecture was presented in [ZYY+10]. A single-chip integrated transmitter and receiver was presented in [PSS+08] based on a wideband super-heterodyne architecture. This work shows that realizing practical circuits for upconversion and downconversion is possible for mmWave wireless systems.

2.10 System Architecture

In this section, we describe the reference system architecture used in communication networks. It is based on the Open Systems Interconnection (OSI) model and was developed by the International Organization for Standardization [ISO]. The OSI model is an abstraction of network functions that is used to help design and understand communication networks. The model is presented in this section to provide some context about how and where mmWave communication fits in with the larger view of networking. More details about the OSI model are provided in basic textbooks on networking [BG92][Tan02].

A standard illustration of the OSI model is provided in Fig. 2.32. The OSI model categorizes communication into seven layers. The highest layer is called the application layer; the lowest layer is the physical layer (PHY). In the conventional OSI model, layers are specified in terms of a communication protocol and an interface. The protocol governs the interactions between different nodes of a network. For example, the networking protocol is concerned with the routing of packets among multiple nodes in the network. The layer of one node is said to communicate with the same layer of another node. The interface defines what control and data information is communicated between layers. A given layer n requests service from layer $n-1$. Layer n in return provides service for

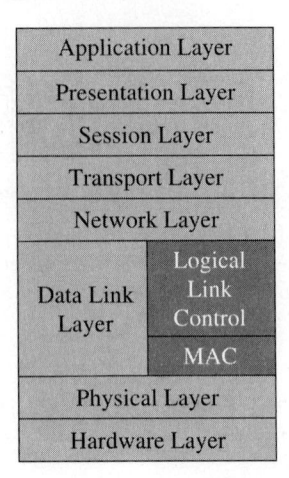

Figure 2.32 Reference system architecture for a communication network. The Physical Layer (PHY) is considered the lowest layer, and the Application Layer is the highest layer. We propose a new layer, called the *Hardware Layer,* that is below PHY, in order to account for complexities involved with the creation of new hardware and devices for mmWave communications.

layer $n+1$ above it. In this way of abstraction, the design of layers can be performed independently, without the engineering of one layer knowing the details of another layer. Many of the layers in Fig. 2.32 can be divided into sublayers. In wireless systems, the division of the Data Link Layer into Logical Link Control and a Medium Access Control (MAC) layer is common.

The layered system architecture is facing increasing scrutiny from communication engineers. A recent topic of research has been cross-layer design [SRK03], in which effort is focused on the joint design of functions that are abstracted in different layers. For example, multiple antennas can be used at the physical layer to cancel interference, which the MAC in a wireless system is tasked with avoiding. A cross-layer MAC design might design the physical layer algorithms and the MAC protocol jointly to manage interference. Though research proceeds on cross-layer design and redefining network boundaries, the layers are still useful for reviewing different network functions.

The focus of mmWave in this book is primarily at the data link layer and below. In fact, most chapters deal with the physical layer and below, except for Chapters 8 and 9. It is useful, though, to have some understanding of the higher layer operations. In early 60 GHz systems, applications have played an important role in the design of the lower layers. Perhaps the most striking example discussed in this chapter is multimedia. The video application has led to various physical layer modulation and coding decisions, including the use of unequal error protection (UEP) coding and unequal error protection modulations. The rest of this section provides some background on each layer and postulates the existence of an additional lower layer.

Application Layer The Application Layer is the highest layer, layer 7, in the OSI model. It is not the application itself (e.g., a software program); rather, it is the communication protocol that serves the application. The application layer facilitates communication between applications. Common example of protocols at the application layer include email, Hypertext Transfer Protocol (HTTP), Session Initiated Protocol (SIP), and File Transfer Protocol (FTP).

Presentation Layer   The Presentation Layer is layer 6 in the OSI model. One of the main functions of the presentation layer is data formatting or translation, so that data can be interchanged between different types of systems — for example, converting between the Western (ISO Latin 1) character set and the Unicode (UTF-8) character set. Other functions of the presentation layer are encryption and decryption, ensuring secure transmission of information, and compression. An example protocol that incorporates these different facets is the Multipurpose Internet Mail Extensions (MIME) protocol, which adds headers to identify content in email programs such as the content type, character set, and type of encryption.

Session Layer   The Session Layer is layer 5 in the OSI model. It deals primarily with connections, which are the association between two communicating end points, and sessions, which are conversations between two end points that result in the exchange of multiple messages. The Session Layer is concerned with setting up, coordinating, closing, and, in general, managing sessions and their corresponding connections. Examples of protocols at the session layer include Remote Procedure Call (RPC) Protocol, Network Basic Input Output System (NetBIOS), and Session Control Protocol (SCP).

Transport Layer   The Transport Layer is layer 4 in the OSI model. Transport protocols provide transparent end-to-end communication. It is responsible for providing reliability, flow control, and ordering. An example of reliability is checking that data is not corrupt, for example, through the use of error detection codes, and asking for data to be retransmitted. Flow control refers to management of the aggregate rate of the connection, for example, to avoid buffer overflow at the receiver. Ordering is a mechanism for reassembling data from the network layer, which may provide packets that were received in a different order than they were sent. Examples of Transport Layer protocols include Transmission Control Protocol (TCP), User Datagram Protocol (UDP), and Stream Control Transmission Protocol (SCTP).

Network Layer   The Network Layer is layer 3 in the OSI model. The Network Layer has knowledge about multiple nodes in the network. Network protocols include functions such as routing (or switching), forwarding, addressing, sequencing, and congestion control. Routing involves selecting the best path between the source and the destination. Forwarding is the operation of receiving a packet and sending it to the next node in the route. Addressing involves giving unique addresses to nodes in the network, useful as part of routing. Sequencing is the labeling of packets so they can be sent along different paths in the network. Congestion control refers to managing traffic on links between nodes to avoid excessive delay or dropped packets. A well known example of a Network Layer protocol is the Internet Protocol (IP).

Data Link Layer   The Data Link Layer is layer 2 of the OSI model. Data Link protocols facilitate communication between directly connected nodes in a network. It is common for the Data Link Layer to be divided into two sublayers, as illustrated in Fig. 2.32. The higher of the two sublayers is known as the *Logical Link Control* (LLC). The LLC is concerned with flow control and error checking. Flow control is the process of controlling the transmission rate to prevent the sending node from overwhelming the receiving node. Error checking involves using an error control code to detect errors in the packets, and possibly acknowledging packets. For example, the standard IEEE 802.2

specifics a LLC that is employed in conjunction with the IEEE 802.15.3c and IEEE 802.11ad 60 GHz standards.

The MAC sublayer is concerned with physical addressing and channel access. A MAC address is associated with a physical piece of hardware, like a serial number. Channel access refers to allowing multiple physical devices to access the same physical communication medium. The MAC protocol coordinates channel access in either a centralized or distributed fashion, depending on the kind of network. In a centralized protocol, a controller node might query neighboring nodes if they would like to use the medium, whereas in a distributed protocol, rules determine how to resolve conflicts and collisions when multiple nodes want to access the same channel. In IEEE 802.11, the point coordination function (PCF) is an example of a centralized protocol whereas the distributed coordination function (DCF) is an example of a distributed protocol. IEEE 802.15.3c and IEEE 802.11ad specify 60 GHz MAC protocols.

The MAC protocol is perhaps the most relevant for mmWave communication engineers. An entire chapter in this book is devoted to it: Chapter 8. The MAC protocol is interesting because the communication medium at mmWave frequencies has different properties than at lower frequencies. An acute difference is the importance of beamforming, using multiple antennas to direct the transmission and reception of a communication signal. Beamforming is used to provide array gain to overcome loss in the channel and attenuation, and to reduce interference. Multiple beams can be a source of path diversity, providing benefits against line-of-sight blockage. MmWave systems need a MAC protocol that works with beamforming.

Physical Layer The Physical Layer (PHY) is layer 1 in the OSI model. The main task of the PHY protocol is to convey digital information (bits) over a physical communication link. A specification for a PHY describes the interface to the communication medium and how the physical waveform is created. For example, a PHY protocol may specify features like the carrier frequency, bandwidth, symbol constellation, type of modulation, location and type of training data, and error control coding. Algorithms used to decode information sent by the PHY are not usually provided directly in a PHY specification and are left up to the designer. Much of this book deals with the physical layer and PHY protocols. This chapter reviewed the key features of the PHY of a digital communication system. Chapter 3 delves into the details of the mmWave communication medium. Chapter 7 describes mmWave-specific receiver signal processing algorithms in more detail, and Chapter 9 describes PHY and MAC specifications.

The Hardware Layer 0 In a traditional description of the OSI model, the PHY is the lowest layer. In practice, however, there is another perceived layer that we call layer 0 for lack of a better term. Layer 0 describes the physical hardware implementations and capabilities. A physical layer protocol specification does not typically specify the exact hardware components, at least for wireless systems (in wired systems there may be detailed specifications on the cabling, for example). For example, the type of antenna, the semiconductor technology, circuit implementation, or the configuration of the analog front end is not normally a part of the specification. However, we observe that the expertise associated with engineering the physical layer (signal processing, digital communication, and control theory) is quite different from that required to implement the hardware layer (electromagnetics, antenna design, RF circuits, mixed-signal). Consequently, several chapters are devoted to these topics to provide specific background in the areas of antennas in Chapter 4, circuits and devices in Chapter 5, and baseband processing in Chapter 6. The importance of the hardware layer 0 underlies the emphasis on both communications and circuits in this book.

2.11 Chapter Summary

Digital communication is the foundation of mmWave communication systems. This chapter reviewed key components of wireless digital communication systems, creating a foundation for the remainder of the book. The complex baseband representation is used to mathematically represent the analog signals that are transmitted and received in a digital communication system. Sampling those signals leads to discrete-time relationships, which exposes the connection between communication and digital signal processing. Digital modulation involves mapping sequences of bits onto waveforms, and is usually a two-step process that involves symbol mapping and pulse-shaping. Digital demodulation involves extracting a guess as to which bits were sent from the transmitter. The type of demodulation applied is a function of the noise and the channel impairments. Equalizing the effects of the channel is one of the most important functions of the receiver. It can be performed in the time domain or the frequency domain, each with different tradeoffs in terms of performance and complexity. Error control coding is used to either detect or correct errors, through the careful insertion of redundancy in the transmitted waveform. There are many different types of error control codes, some suitable for correcting single errors and some for block errors. Effective communication involves many other functions to help the receiver cope with uncertainty in the transmitted waveform. Exemplary forms of uncertainty include frame synchronization, carrier frequency offset synchronization, and channel estimation. The process of up conversion and down conversion is abstracted in the signal processing description of a communication system. In practice, the operations are often performed with multiple stages of oscillators. Digital communication is only one piece of an entire communication system architecture. There are many layers of a communication protocol, each serving a different purpose: application, presentation, session, transport, network, data link, and physical layer. Circuits have a critical influence on communication and could even become a layer in their own right.

Other applications of mmWave are coming. Some of the examples in this chapter were drawn from current practice in 60 GHz systems, but 60 GHz WLAN/WPAN is only one potential application of mmWave. There is a great future in applications of mmWave to other wireless systems, for example, to cellular systems [RMGJ11][RSM+13][PK11] [MBDQ+12][RRE14][AEAH12][ALRE13][BAH14][BH13b][BH13c][BH14]. MmWave cellular offers the new dimension of space along with massive bandwidth allocations far beyond today's systems to revolutionize methods for distributing content [Rap09][RMGJ11][Rap12a][Rap12b][Rap11][RMGJ11][BH14][AELH14][EAHAS+12] [EAHAS+12a][ERAS+14][KHR+13].

A challenge in cellular systems is how to support multiple users over a large geographic area. Because of the need to employ frequency reuse for increased capacity within a finite spectrum allocation [Rap02], interference becomes an important impairment in cellular systems. Early work shows that mmWave cellular systems that employ directional steerable antennas are different from all previous cellular systems because they are limited by noise rather than by co-channel interference, as in today's cellular systems [SBM92][CR01a][CR01b][ALS+14][RRE14][Gho14][BAH14][SNS+14]. Interference can be treated as additional noise that is statistically combined due to the propagation environment [CR01b], or its structure can be exploited by receiver algorithms to further improve performance. Cellular systems also need to support many other functions including handoff, in which a user is passed from one base station to another. Resources will need to be efficiently shared among users in cellular systems, thus multiple access strategies are also important. MmWave systems may be the first wireless networks to incorporate site-specific knowledge for real-time frequency allocations and load-balancing, due to the predictable nature of propagation when directional antennas are used [CRdV06][CRdV07]. These concepts are described in Chapter 3. Users may have different time slots or may share frequency resources by using different subcarriers in an OFDM, or single carrier symbol, for example. Orthogonal frequency division multiple access is used in today's fourth generation (4G) cellular systems is based on the 3GPP Long Term Evolution (LTE) standard [3GPP09].

The system models illustrated in this chapter are most suitable for transmission from a single transmit antenna and reception from a single receive antenna. To provide the ability for mmWave cellular systems to flexibly close the link margin for any particular operating environment or location, mmWave systems will make use of antenna arrays and beamforming. This beamforming will likely be handled at least partially in the analog domain at RF but will be controlled by the digital portions of the radio. This requires steering the transmit and receive beams in dominant directions of propagation in the channel, or possibly a joint design of beams to support multiple users or to cancel interference. The use of beamforming also has an impact on the receiver algorithms. A sharp beam pointed along a dominant propagation path can reduce the amount of multipath and thus the subsequent extent of equalization required. Beamforming may be performed in a combination of analog and digital domains, in what is known as a hybrid beamformer (see, for example, [Gho14], [ERAS+14], and the references therein). This gives more control over the transmit and receiver beamforming design at the expense of higher complexity. Beamforming for mmWave is an active area of investigation. Channel measurements that illustrate the dimensionality and potential link margin improvements for beam forming are given in Chapter 3, and beam patterns and antennas are discussed in more detail in Chapter 4.

Part II

Fundamentals

Part II

Fundamentals

Chapter 3

Radio Wave Propagation for MmWave

3.1 Introduction

Radio wave propagation holds the key to understanding receiver design, the transmitter power requirements, antenna requirements, interference levels, and expected distances for wireless communication links. At mmWave frequencies, where the wavelengths are smaller than a centimeter — even smaller than the size of a human fingernail — most objects in the physical environment are very large relative to the wavelength. Lampposts, walls, and people are large relative to the wavelength, and this causes very pronounced propagation phenomena, such as signal blockage (e.g., shadowing) when an obstacle is in the way of the path between the transmitter and receiver. However, reflection and scattering allow wireless links to be made between the transmitter and receiver, even when there are physical obstructions that block the line-of-sight (LOS) paths, as long as steerable antennas are used to "find" objects that bounce or scatter energy. Fortunately, highly directional multiple-element antennas, capable of being electrically steered, can be made in very small form factors and integrated inexpensively, as we describe in Chapter 4.

The wavelengths at mmWave frequencies are so small, in fact, that the molecular constituency of air and water play a major role in defining the free space distances achievable across the sub-terahertz spectrum. Fig. 3.1 illustrates the excess attenuation (that is, in addition to the well-known Friis distant-dependent free space loss discussed next) in air across the electromagnetic (EM) spectrum up to 400 GHz, and shows how EM waves are dramatically attenuated by atmospheric absorption caused by the oxygen molecule at 60 GHz and the water molecule at 180 and 320 GHz [RMGJ11]. Temperature and humidity greatly impact the actual excess attenuation caused by absorption [FCC88].

It can be seen from Fig. 3.1 that certain spectrum bands, such as frequencies in the 60 and 180 GHz band, and particularly in the 380 GHz band, have very high excess attenuation over distance. These frequencies are well suited for unlicensed networks in and around homes and buildings (say, within a few tens of meters coverage distances, and even much smaller distances at 380 GHz) where the radiated signals will rapidly

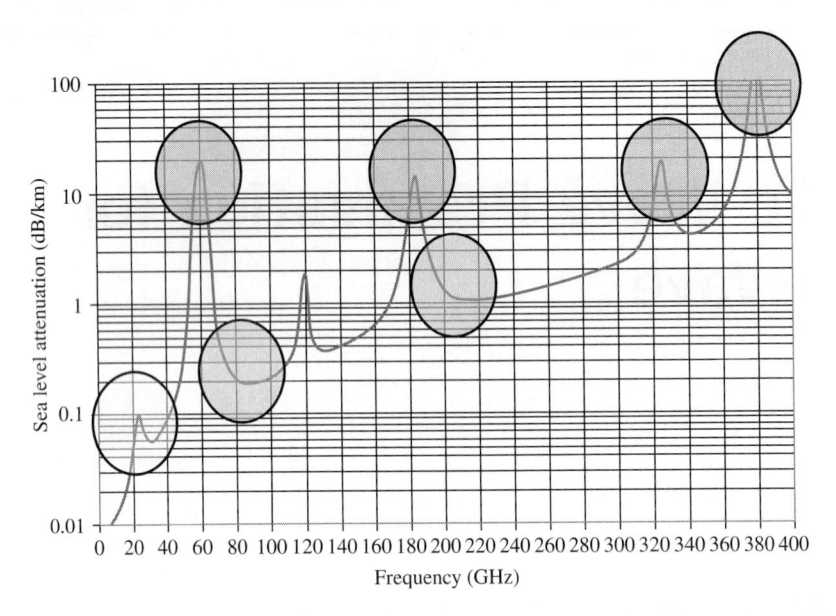

Figure 3.1 The attenuation (dB/km) in excess of free space propagation due to absorption in air at sea level across the sub-terahertz frequency bands. The far left (unshaded) bubble shows extremely small excess attenuation in air for today's UHF and microwave consumer wireless networks, and other bubbles show interesting excess attenuation characteristics that are dependent on carrier frequency [from [RMGJ11], © IEEE].

decay and not interfere with other nearby networks. Other frequencies, such as in the 0-50 GHz or 200-280 GHz bands, have very little excess atmospheric attenuation beyond free space propagation loss, making them strong candidates for future cellular and mobile communication in both indoor and outdoor wireless networks, where long distances, on the order of hundreds of meters, or even kilometers, must be covered (e.g., for femtocells, microcells, and backhaul links in future cellular networks). The height above ground also impacts the atmospheric absorption, and thus the attenuation of radio signals, as elevation substantially above sea level has lower temperatures and often less humidity, and this alters the attenuation due to changes in the molecular constituency of air. For example, at 1 km above sea level, the attenuation may be 10-20 dB per km less than at ground level for certain frequencies where the oxygen and water molecules create huge absorption closer to ground level. Thus, in arid or high-altitude locations, there may be substantially less excess attenuation at certain lossy mmWave bands.

Note that atmospheric attenuation is not the only issue that impacts coverage distance or interference seen by co-channel transceivers. The impact of weather, such as rain, hail, sleet, or snow, also has a great impact on the attenuation of particular frequency bands over distance. It is worth noting that very high gain directional antennas will play a key role in overcoming the losses due to both atmospheric and weather-related attenuation, while simultaneously providing inherent interference protection based on the very narrow beamwidths used to transmit (or receive) mmWave signals.

In future mmWave systems, antenna gain, beamwidth, and beam pointing directions will be adaptable to offset and adjust to specific levels of interference in low-attenuation mmWave bands, and they may be used to adaptively compensate for particular loss characteristics of weather or atmosphere. Fig. 3.1 compels us to contemplate the use of "whisper radios," for which massive bandwidth communication links using tiny, high gain miniature antennas could be confined to only a meter or so of coverage distance, thereby replacing wired connectivity and ensuring privacy for critical military or consumer applications, at frequencies near 380 GHz. At such a carrier frequency, many tens of gigahertz of spectrum could be used to provide circuit board and wire replacements in automobiles, consumer appliances, and wiring within buildings [Rap12c].

Wireless engineers familiar with cellular radio or WLAN communications in the 1-5 GHz bands will find frequencies at the mmWave band, 20 GHz and above, to be vastly different in many ways, but surprisingly similar in other ways. MmWave frequencies are more susceptible to increased path loss from weather and atmospheric effects, experience increased reflectivity from common objects, and have increased sensitivity to the channel's environment-dependent characteristics. Fig. 3.1 shows, however, that for paths of less than 1 km in free space, the excess attenuation at many mmWave frequencies is quite small and, in fact, comparable to today's UHF and low microwave wireless systems. Because the mmWave radio channel, as experienced through highly directional transmit and receive antennas, is different than traditional wireless channels, this chapter has special importance for the design and/or evaluation of mmWave wireless communication systems. In this chapter we first characterize the macroscopic channel properties, known as *large-scale channel effects*, by describing the expected signal propagation power loss as a function of distance between the transmitter and receiver. This chapter describes both traditional path loss due to electromagnetic wave transmission and the impact of antenna gain, as well as atmospheric loss due to molecular resonances. Rain and fog also impact large-scale path loss because mmWave frequencies are attenuated over distance much more severely than today's 1-5 GHz cellular and WLAN bands, due to the physical size of raindrops being on the order of the propagating wavelength. After we analyze these path loss effects on mmWave links, we also look at the microscopic view of the received signal fluctuation, also known as *small-scale channel effects*. At frequencies at or above 20 GHz, small-scale channel effects are determined by myriad environmental propagation properties. For example, with indoor communication this could include the speed of moving objects, room dimensions, building materials and their surface roughness, furniture, path obstructions, antenna radiation patterns, the polarization of the radiated waves, and the presence of people. We conclude the chapter with a summary of indoor and outdoor mmWave channel models and measurements of material properties, and we refer often to state-of-the-art measurements that describe radio propagation at 28, 38, 60, and 73 GHz.

3.2 Large-Scale Propagation Channel Effects

The free space propagation of electromagnetic waves is a useful starting point for the evaluation of large-scale wireless channel characteristics, where the propagation loss of radiated signal power is characterized over several orders of magnitude change in distance between the transmitter and receiver, from meters to hundreds or thousands of meters. The propagation of electromagnetic waves in free space, with no obstructions, reflections, or scatterers, is modeled mathematically by the free space path loss equation attributed to Harold T. Friis [Fri46]. As shown in [Rap02], the Friis free space propagation theory describes how the effective isotropic radiated power (EIRP) of a transmitter is given by the product of its transmitted power P_t and its transmitter antenna gain G_t, where antenna gains are specified relative to an isotropic radiator. Antenna gains are described by their maximum gain (boresight) direction, relative to a unity gain, 0 dB omnidirectional isotropic antenna.

The power flux density, measured in units of Watts/square meter, is the amount of power per area (e.g., the radiated power spread out over the surface area of a sphere) that is radiated into free space. The receiver is located at a particular transmitter-receiver (transmit-receive or TX-RX) separation distance d from the transmitter, and the receiving antenna captures a small portion of the radiated surface area. If two receiver antennas have different physical sizes that correspond to different electrical sizes with respect to wavelength, the larger receiver antenna will have greater capture area, and thus will have greater gain and will have more directionality (a narrower beamwidth) than a smaller antenna. The power flux density of a transmitted signal, propagating over a distance d in free space (with units of Watts per square meter) is given by the EIRP divided by the surface area of a sphere with radius d (see [Rap02] and (3.2) for further details).

The gain of any antenna may be expressed in terms of its effective area and the operating frequency, where the effective area of any antenna is estimated based on its physical size and operating frequency according to its maximum effective area (or aperture area) [Fri46]

$$G_{\max} = e_{\max} A_{\max} \left(\frac{4\pi}{\lambda^2} \right) \tag{3.1}$$

where $e_{\max}$ is the maximum efficiency of the antenna (which is always less than 1, and may be substantially smaller than 1 for inefficient antennas), λ is the operating wavelength, and $A_{\max}$ is the maximum effective aperture (a measure of the electrical area of an antenna). The maximum effective aperture has units of m^2. That is, $A_{\max}$ represents the physical area of the antenna that captures and delivers useful energy. Because antenna gain $G_{\max}$ denotes the directionality of an antenna in its maximum (e.g., boresight) direction compared with an omnidirectional isotropic reference, it is clear from (3.1) that as either the carrier wavelength shrinks, or the aperture of the antenna increases, antennas have much greater directionality and gain. Antenna gain also increases with frequency for a fixed antenna aperture (i.e., physical size), where the gain increases by the square of the frequency.

The received power in free space, in Watts, is proportional to the product of the *effective isotropic radiated power* (EIRP) given by $P_t G_t$, the effective area $A_{eff} = e_{\max} A_{\max}$ of the receiver antenna from (3.1), and the inverse square of the propagation distance d that follows from Friis' free space path loss law

$$P_r = \frac{EIRP}{4\pi d^2}(A_{eff}) = \frac{P_t G_t G_r}{L}\left(\frac{\lambda}{4\pi d}\right)^2 \tag{3.2}$$

where P_r and P_t are the received and transmitted power, respectively, given in absolute linear units (usually Watts or milliwatts); G_r and G_t are the linear (not dB) gains of the receiver and transmitter antennas, respectively, compared to an isotropic antenna;

and λ is the operating wavelength of transmission (in meters). The unitless loss factor, L, in the denominator of (3.2) is greater than unity and accounts for all losses associated with the antennas and components. Because path loss is the reciprocal of path gain [the term $\left(\frac{\lambda}{4\pi d}\right)^2$ in (3.2)], it can be seen that path loss increases as the transmission distance is increased or as the wavelength is shortened. Because the frequency (f, in Hz) and wavelength (λ, in meters) of a propagating wave are related by the speed of light through $c = \lambda f$, (3.2) shows that for a fixed transmit-receive separation distance and a fixed antenna gain at the transmitter and receiver, free space path loss is proportional to the square of the operating frequency.

To illustrate the impact of frequency on path loss, we compare free space path loss at traditional wireless frequencies of 460 MHz (early cellular), 2.4 GHz (early WLAN and modern connectivity standards such as Bluetooth and BLE), 5.0 GHz (modern WLAN), and a mmWave operating frequency of 60 GHz. Assuming equal transmitter power levels, omnidirectional antennas, and no system losses ($L = 1$), the path loss (in decibels) for distances of $d = 1, 10, 100$, and $1,000$ m, given that $\lambda = \frac{c}{f_c}$ with $c = 3 \times 10^8$ m/s (speed of light), are shown in Table 3.1. These simple calculations show that moving up to the mmWave frequencies will be no trivial affair if omnidirectional antennas are used. Clearly, we must compensate for 20-40 dB received power loss when compared with current unlicensed UHF/microwave communication environments. Increased path loss between 60 GHz and 2.45 GHz has been confirmed with many measurement campaigns [BMB91][SR92][ARY95][AR04][AR+02][BDRQL11].

However, Table 3.1 only tells part of the story because there is a hidden benefit to propagation at mmWave frequencies that may not be obvious on first consideration of the Friis free space equation. This benefit is evident when we begin to consider that mmWave frequencies will permit small directional antennas that have substantial gain that can offset, and even reduce, the path loss compared to UHF and microwave frequencies. This is easily seen in (3.2). Consider antenna arrays, which are able to offer substantial gain at mmWave frequencies in a very small physical form factor. Adaptive arrays may be used to form narrow beams (high gain antennas) that are physically small at mmWave frequencies. These high gain, steerable adaptive antennas allow mmWave communication systems to steer beams within the environment and to bounce energy off the surrounding scatterers and reflectors in a real-world propagation environment, while concentrating radiated energy in only those directions that prove fruitful in making a viable link for the communication path. Furthermore, using MIMO and beam-combining techniques, the path loss in the channel can be decreased dramatically by forming simultaneous beams in many different directions, and different spatial paths can be used to support MIMO and spatial multiplexing where multiple data streams are sent in parallel to increase capacity.

Table 3.1 Path loss calculations for various mobile communication frequencies in free space.

	$f_c = 460$ MHz	$f_c = 2.4$ GHz	$f_c = 5$ GHz	$f_c = 60$ GHz
$d = 1$ m	-25.7 dB	-40 dB	-46.4 dB	-68 dB
$d = 10$ m	-45.7 dB	-60 dB	-66.4 dB	-88 dB
$d = 100$ m	-65.7 dB	-80 dB	-86.4 dB	-108 dB
$d = 1,000$ m	-85.7 dB	-100 dB	-106.4 dB	-128 dB

At frequencies above 28 GHz, the wavelengths are about 10 mm or smaller, thus allowing for half-wave dipole antennas that are only a few millimeters long, or even smaller, when fabricated on high dielectric constant materials [GAPR09]. Thus, a very large array of dipoles could be put in a form factor much smaller than the size of today's cellphone or tablet computer [GAPR09][RMGJ11].

Consider an adaptive antenna array made up of identical antenna elements each having maximum length dimension D, noting the antenna gain of each antenna element is given by (3.1). For a properly impedance-matched antenna, [Bal05, Chapter 2] shows the maximum gain (see (3.1)) of any antenna element is proportional to

$$G_{\max} \propto \frac{4\pi e_{\max} D^2}{\lambda^2}. \tag{3.3}$$

If there is a linear array of N antenna elements, then (3.3) will still hold approximately for each antenna, so that the effective aperture of an array is the physical area of each antenna multiplied by the number of antennas in the array given by

$$D_{\text{array}} = N D_{\text{ant}}. \tag{3.4}$$

If the antenna array is a square, two-dimensional (quadratic) array, with N^2 elements and N elements per side, then the maximum linear dimension of the array given by (3.4) still holds, but the effective area, and thus the array gain, increases even more due to the greater number of antennas. For a linear array, the maximum gain is

$$G_{\max} = \frac{4\pi e_{\max} D_{\text{array}}^2}{\lambda^2}. \tag{3.5}$$

Inserting (3.5) into the Friis path loss equation of (3.2), and assuming antenna arrays are used for both the transmitting and receiving antenna, then

$$P_r = \frac{\left(P_t e_t e_r (D_r D_t)^2\right)}{L(\lambda d)^2} \tag{3.6}$$

where D_r and D_t are for the receiver and transmitter antenna arrays, respectively. There are two key points to glean from equation (3.6).

1. As the carrier frequencies increase (e.g., at 28 GHz and higher), engineering antenna arrays with an area that is a substantial portion of the operating wavelength becomes easier to implement on handheld structures such as cellphones, since the smaller wavelength allows high gain antennas to be physically smaller and to use dimensions that are on the order of or larger than the wavelength compared with conventional microwave frequencies (where the wavelength is often larger than the form factor of a handheld device.) At 60 GHz, the free space wavelength is 5 mm, and at 1,000 GHz (1 THz) the wavelength is 0.3 mm. These small wavelengths mean that as carrier frequencies increase, it will be possible to fit more and more antennas into a small printed circuit board, small package, or on a chip, resulting in amazing antenna gains compared with today's nearly omnidirectional cellphone antennas. By combining steerable high gain antennas at both a

3.2 Large-Scale Propagation Channel Effects

base station and a mobile device, the enormous antenna gains can overcome much greater propagation path loss than in today's systems where low gain antennas are used. In short, (3.6) shows that the antenna array dimensions at the transmitter and receiver (in the numerator) can overcome the distance-dependent propagation path loss (given in the denominator).

2. The ratio of array area to operating frequency becomes a completely new design metric that can be used to engineer the size of a communication device for small handheld applications. Gains (and thus antenna beamwidths) may be rapidly adjusted by turning on and off different antenna elements in an array, and a small physical form factor can allow a widely varying steerable beam antenna system through electronic phasing and selective excitement of array elements.

Exploiting physically small, high gain, steerable antennas is a key aspect of mmWave communications. In fact, (3.6) shows that the physical length of the antennas provides a fourth-order increase in received power that can overcome the second-order loss of power with distance in free space. As we show in Chapter 4, such gains and abilities will allow massive multiple input multiple output (MIMO) and steerable antenna arrays in a very small form factor, through the use of printed circuit board, on-chip, and in-package antennas [GAPR09], and completely new wireless architectures and future wireless systems will be enabled by these capabilities [RMGJ11][RSM+13][R+11][EAHAS+12a][RRE14][ALS+14][Gho14].

3.2.1 Log-Distance Path Loss Models

Free space path loss does not always hold true in practice. Early work shows that it may only work well for mmWave systems when there is a line-of-sight path and when antennas are perfectly aligned on boresight [BDRQL11][RBDMQ12][RGBD+13][RSM+13][AWW+13][SWA+13]. Observe in (3.2) that the received power in free space decays 20 dB per decade in distance due to the distance squared term in the denominator. By generalizing the log-distance slope in the far field, a better fit to path loss measurements can be made using the log-distance path loss model

$$P_r(d) = P_t K_{\text{fs}} \left(\frac{d_o}{d}\right)^\alpha \text{ for } d \geq d_o \quad (3.7)$$

where d is the propagation distance and $d_0 \gg \lambda$ is a close-in free space path loss reference distance in the far field, such that the dimensionless constant K_{fs} and path loss exponent (PLE) α are adjusted to fit field measurements [Rap02]. In communications and propagation analysis, it is customary to represent propagation path loss using decibel values because decibel values are much easier to use for "back of the envelope" calculations, since the multiplication of linear (absolute) values results in the simple addition of decibel values. Decibels are also preferred because the dynamic range of propagating signal powers changes by many orders of magnitude over relatively small distances. By converting (3.7) into decibel values, we see that the path loss is a function of the logarithm of the propagation distance and the PLE, as given by

$$P_r[\text{dBm}](d) = P_t[\text{dBm}] + 10 \log_{10} K_{\text{fs}} - \underbrace{10\alpha \log_{10}(d/d_o)}_{\text{path loss}} \text{ for } d \geq d_o \quad (3.8)$$

By considering (3.2) and (3.7), the received power P_r of a wireless system using directional antennas may be related to the transmitted power P_t (in units of milliwatts or dBm), the antenna gains G_t and G_r, respectively, and the propagation path loss (PL) in the channel. Received power may then be simply expressed using the path loss exponent, alpha, and antenna gains, in a distance-dependent form using decibel units:

$$P_r(d)[\text{dBm}] = P_t(d)[\text{dBm}] + G_t + G_r + PL(d) \qquad (3.9)$$

where PL is expressed as equation (3.8). Although values of $\alpha \leq 2$ (e.g., better than or equal to free space path loss, where (3.2) shows that $\alpha = 2$ is the PLE in free space) are possible in certain situations where waveguiding or particular constructive interference exists, typically $\alpha > 2$, because free space is an optimistic environment.

Measurements of 60 GHz indoor wireless channels with a clear line-of-sight path have shown varying α values in different line-of-sight (LOS) environments, including open space, hallways, and laboratories [TM88][BMB91][SR92][ARY95][FLC$^+$02]. For open space scenarios of limited distance such that atmospheric absorption has minimal impact and where few reflectors are present, as expected, $\alpha \approx 2.0$. In hallways, where waveguiding occurs, or where reflections from walls, floors, and ceilings can coherently combine, it was found that $1.2 \leq \alpha \leq 1.32$. Note that α must always be greater than unity, else (3.7) would provide gain in the channel with increasing distance. For the laboratory environment, both constructive and destructive reflections contribute to the estimated values of $1.71 \leq \alpha \leq 2.71$. Measurements with an obstructed line-of-sight path have shown that $2 \leq \alpha \leq 10$ [Yon07]. In outdoor LOS measurements for mmWave cellular at 28, 38, 60, and 73 GHz, α has been found to be between 1.8 and 2.2, but changes to about 4 or 5 in non-line-of-sight (NLOS) conditions or LOS conditions when the highly directional antenna beams from the transmitter and receiver are not pointing at each other (see Section 3.3 for actual measured data). Surprisingly, however, as we show in Section 3.7.1, there is little difference between the omnidirectional channel path loss at mmWave frequencies when compared with today's UHF/microwave channels. Hence, we can assume that for line-of-sight conditions, the Friis free space model of (3.2) provides an adequate if not conservative model of large-scale path loss. Without a clear line-of-sight path, however, the Friis model is too generous to represent reality, just as is the case with today's UHF/microwave wireless systems.

3.2.2 Atmospheric Effects

Unfortunately, increased path loss at mmWave frequencies is not only the result of the increased transmission frequency. As reported in [Rog85], attributes such as atmospheric attenuation, rain attenuation, and path depolarization also contribute to degrade mmWave received signal power, more so at 60 GHz than at the 28-38 GHz bands, where there is relatively little oxygen absorption (only a fraction of a decibel per kilometer, as shown in Fig. 3.1). Atmospheric attenuation creates additional path loss beyond the propagation loss of (3.2) or (3.7) and is multiplicative in absolute terms (additive in decibel terms). Atmospheric loss is not unique to 60 GHz systems, as all electromagnetic waves are absorbed to some degree by gaseous molecules such as oxygen and water

3.2 Large-Scale Propagation Channel Effects

vapor. This effect, however, is greatly magnified at certain mmWave frequencies such as 60 GHz. Under typical atmospheric conditions (temperature = 20° C, atmospheric pressure = 1 atm, and water vapor density = 7.5 g/cm^3) atmospheric attenuation does not become significant until carrier frequencies exceed 50 GHz. This effect, which is characterized by a logarithmic power decrease per kilometer of transmission distance (dB/km), can be seen in Fig. 3.1 [RMGJ11][GLR99][OET97]; the dominant effects of atmospheric attenuation are due to water vapor and oxygen. Fig. 3.1 shows O_2 absorption dominates with a peak at 60 GHz. This quality, under normal atmospheric conditions, accounts for 7-15.5 dB/km in the received signal at 57-64 GHz.

Although water vapor does not contribute greatly to signal attenuation at normal concentrations, rain droplets that form when the atmosphere becomes saturated can further attenuate the signal, as shown in Fig. 3.2. For example, for a rainfall rate of 50 mm/hr (heavy rain), different models predict between 8 and 18 dB/km additional atmospheric attenuation [GLR99]. Therefore, outdoor cellular or backhaul systems will need to overcome rain conditions through adaptive beamforming to achieve more gain. Precipitation effects are not limited to the attenuation of the signal, as depolarization can also occur [Rog85]. This is particularly troublesome for systems utilizing cross-polarized signals for RF isolation.

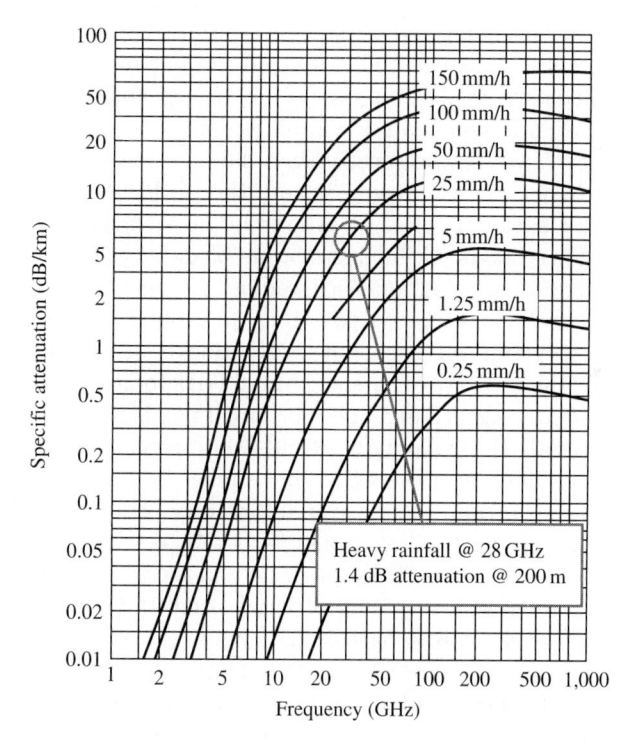

Figure 3.2 Rain attenuation as a function of frequency and rain rate in the mmWave spectrum [from [AWW$^+$13][RSM$^+$13][ZL06] © IEEE].

3.2.3 Weather Effects on MmWave Propagation

Weather has a dramatic impact on the attenuation of mmWaves because the physical dimensions of raindrops, hail stones, and snowflakes are on the order of a wavelength of the propagating radio frequency. In the 1970s and 1980s, a great deal of research focused on the characterization of weather on the attenuation of slant-path satellite communication links [PBA03]. This body of knowledge has helped foster understanding of mmWave propagation in various weather conditions. Indeed, a commonly used rain attenuation model based on work by R. K. Crane [Cra80] has proved useful for estimating rain attenuation that results in excess path loss, beyond that caused by free space propagation and atmospheric losses. Basically, the impact of weather is properly modeled as an additional path loss factor that is simply added to the propagation loss, where the addition is done in decibel units. The weather attenuation is a function of distance, rainfall rate, and the mean size or shape of the raindrops. Due to the statistical nature of rainfall rate in a particular region, different parts of the world will have different likelihoods of coverage outages [PBA03].

From the viewpoint of the satellite industry, the belief that rain attenuation makes mmWaves unusable for reliable mobile communications is understandable because the slant path distances for satellite links can be several and even tens of kilometers. However, this conclusion must not be made for future mmWave mobile communication systems because of the trend for shrinking cell sizes below 1 km [RMGJ11][ACD$^+$12].

In recent years, several researchers have studied the impact of weather on terrestrial mmWave communications from the standpoint of backhaul and mobile communications. For backhaul, mmWave wireless enables a rapidly deployable *wireless fiber* radio connection, which is often much less expensive than having to pay for leased lines or connections from a third party. Governments around the world have allowed operations in the mmWave bands for backhaul, often at little or no licensing cost. This is an emerging industry in the mmWave bands. Cellular operators often find that their backhaul costs are much less expensive when using mmWave radios. The move to mmWave wireless for backhaul usage will surely continue, as governments around the world have promoted the use of mmWave spectrum for backhaul [Fr11]. Backhaul communications are typically used throughout the microwave bands (2 and 6 GHz today), but they are becoming increasingly popular at various licensed and unlicensed bands in the 18, 22, 28, 33, 38-40, 42, 50, and 60 GHz bands throughout the world. More recently, the mmWave E-bands at 71-76, 81-86, and 92-95 GHz in the USA (and similar bands in other countries) have become popular, as the spectrum is made available at very little cost to carriers (a "light" licensing model is used, in which a license can be obtained in minutes over the Internet for a particular location and path). These spectrum bands offer massive amounts of bandwidth, and when combined with the small physical size of high gain steerable antennas (as we discuss in Chapter 4), it is apparent that small cells will be more likely to move to mmWave backhaul for interconnection of base stations and access points, as well as for mobile connectivity to future cellphones and other subscriber devices [RSM$^+$13].

Outdoor mobile applications for mmWave communications have only been considered by a handful of researchers. Given the vast amount of spectrum available in the mmWave bands, and with semiconductor capabilities now supporting mmWave RF and antenna systems (as we describe in Chapters 4 and 5), the potential for mmWave systems to carry

3.2 Large-Scale Propagation Channel Effects

massive data rates to mobile users is clear. This section now outlines some of the key factors to date that impact the large-scale path loss due to weather in both backhaul and mobile scenarios.

Rain and hail present path loss attenuation, based on the fall rate. Different parts of the world have different average rainfall rates, as well as small-scale peak rainfall rates, all of which impact the statistical variation of path loss. Heavy rainfall is generally defined as being greater than 25 mm/hr (approximately 1 in/hr). As seen in Fig. 3.2, the rain attenuation at 28 GHz is about 7 dB/km, which is only 1.4 dB over 200 m. Fig. 3.2 also shows that rain attenuation flattens out and is approximately constant at frequencies above 90 GHz [ZL06]. For small distances, it is evident that rain attenuation is not as severe as might be intuitively expected, especially when considering that high gain antennas could be used to overcome the rain attenuation using gains that vary with instantaneous rainfall rates.

Work in [XKR00], [XRBS00], and [HRBS00] studied two 38 GHz backhaul links: 265 m in a partially obstructed path, and a 605 m line-of-sight (LOS) path over a summer in Virginia. Rain and hail events were captured. As shown in Fig. 3.3, the attenuation over the two paths at 38 GHz is directly a function of rain rate, and is a few decibels more pessimistic than the Crane model. Note, however, that instantaneous rain rates

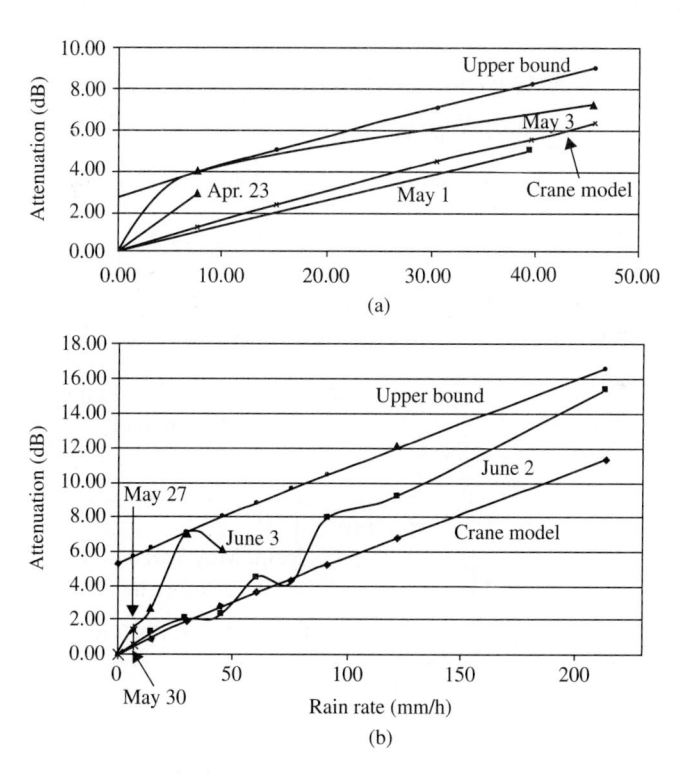

Figure 3.3 Rain attenuation summary and upper bounds for two 38 GHz backhaul links measured over different rain rates. (a) 605 m unobstructed link and (b) 265 m partially obstructed and obstructed links [from [XKR00] © IEEE].

can approach 200 mm/hr during extreme cloudbursts (e.g., flash flooding rain rates), and such events caused approximately 16 dB of loss over a 265 m path. Data in [XRBS00] and [HRBS00] showed that rare summer hail events can cause even greater loss, up to 25 dB or more in extreme cases. Such weather events pose difficulties for the use of mmWave links over significant distances, yet they can be overcome with adjustable gain antennas and transmit power compensation techniques over ranges of a few hundred meters. A mesh-like architecture could be used to link infrastructure backhaul networks together, cooperatively, thus enabling diversity paths as spot rainfall rates spike at particular base stations [Gho14].

Interestingly, the longer 605 m LOS link in Virginia suffered from multipath at higher rain rates, due to wet surfaces and moisture in the air, making the environment more reflective. As shown in Fig. 3.4, higher rain rates caused a ground reflected path to increase while the direct LOS path was attenuated from rain [XKR00].

The impact of rain is an important consideration for outdoor mmWave systems. Rain attenuation is worse at mmWave frequencies than at lower frequencies due to the larger electrical size of raindrops. As a rule of thumb, worst-case rain provides approximately an extra 15 dB/km of attenuation throughout the mmWave bands over 1 km distances. Maximum rain rates in Europe are approximately 50 mm/hr [SL90]. The ITU approximates rain attenuation according to the rain-rate R (mm/h) as given in [SC97b]:

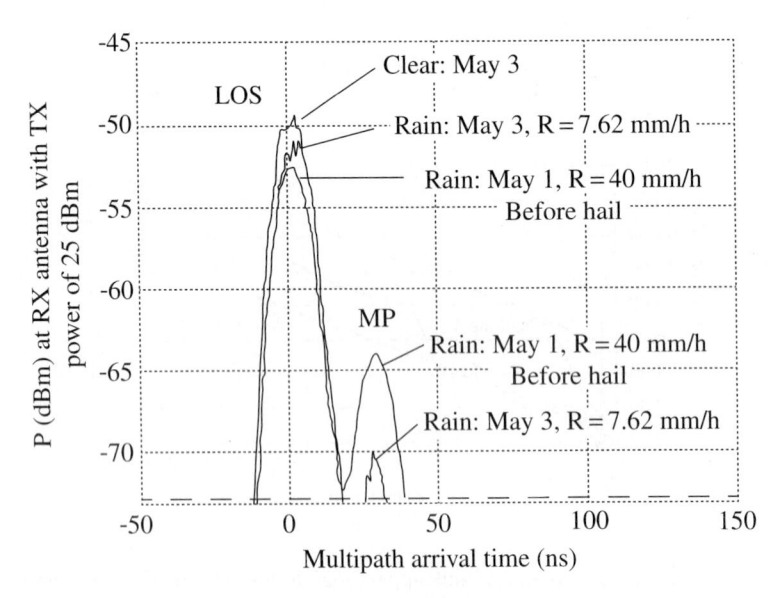

$$\lambda_r \left(f \left[\text{GHz} \right], R \right) = k \left(f \right) R^{a(f)} \left[\frac{\text{dB}}{\text{km}} \right]$$

$$k \left(f \right) = 10^{1.203 \log(f) - 2.290}$$

$$a \left(f \right) = 1.703 - 0.493 \log(f).$$

Figure 3.4 The power delay profile (PDP) of a 38 GHz LOS 605 m backhaul link during different weather conditions [from [XKR00][XRBS00][HRBS00] © IEEE].

3.2.4 Diffraction

The mmWave propagation environment also provides new approaches to combat the greater losses induced by diffraction. Diffraction is the propagation of radio signals around an object and is the mechanism that supports radio communications when a mobile device is blocked or shadowed by an obstruction, or when a wireless terminal turns the corner and moves from a line-of-sight (LOS) propagation condition to a non-LOS (NLOS) propagation condition. While diffraction is a powerful propagation mechanism in today's 2 GHz cellular systems, diffraction becomes very lossy with movements of only a few centimeters at frequencies in the mmWave bands and cannot be relied upon for mmWave propagation. For example, one of the authors has observed that the difference in received signal level before and after a mobile receiver goes around a building corner is more than 40 dB at 28 and 73 GHz. Future steerable beams will need to work with the PHY and MAC protocols to quickly adapt to the change in propagation conditions. Highly directional steerable antennas at both the base station and receiver will overcome diffraction by steering away from the original LOS signal, finding reflections and scattered paths from nearby buildings or surfaces. Similarly, inside buildings, the attenuation of diffracted signals is quite severe at mmWave frequencies, resulting in 10 dB attenuation when a receiver moves around a hallway corner, and more than 40 dB when a receiver moves behind an elevator shaft [ZMS+13]. It is clear that because of the very small wavelengths, diffraction will be the weakest and least reliable propagation mechanism for mmWave mobile systems, whereas scattering and reflection will become the most dominant. This is in contrast to today's UHF and microwave systems in which scattering is the weakest mechanism and diffraction offers appreciable signal propagation [Rap02].

3.2.5 Reflection and Penetration

Although diffraction creates large dynamic range changes in received signal levels at mmWave frequencies over relatively small distance movements, the reflective properties of many materials at mmWave frequencies are surprisingly good in both indoor and outdoor environments. Recent work at 28, 38, 60, and 73 GHz show that human beings, building walls, lampposts, and trees can be very reflective, allowing multipath signals to propagate by bouncing off natural and man-made objects. Despite many propagation paths, obstructions such as foliage, metal walls, elevator shafts, external buildings, and modern tinted external windows can attenuate individual signal multipath components more than 40 dB at mmWave frequencies, which is much larger than today's microwave cellular systems at 2 GHz [ZMS+13][RSM+13][LFR93][SSTR93]. The key to future mmWave systems will be to find and process the strongest direct, reflected, and scattered multipath components to create viable links in different operating environments.

Measurements conducted for indoor and outdoor environments from 28 to 60 GHz illustrate how reflective objects can be when compared with today's microwave (1-5 GHz) bands. In [BDRQL11] and [AWW+13], lampposts, metal garbage cans, the heads of human beings, as well as the exterior walls of buildings were found to be highly reflective, and careful studies at different incident angles show that many outdoor objects have reflective coefficients exceeding 0.7 [ZMS+13][BDRQL11][LFR93]. Alternative propagation paths resulting from the highly reflective (and highly scattering) nature of mmWave channels offer hope for overcoming path loss and attenuation caused by diffraction, provided that directional antennas are used to offer gains that overcome the free space path loss.

Given the remarkable reflectivity of indoor and outdoor channels, site-specific ray tracing methods will likely be the most important propagation modeling approach to provide first-order designs for deployment of infrastructure for future mmWave wireless networks in both indoor and outdoor environments [RGBD+13][HR93][SSTR93][SR94][RBR93][DPR97b][DPR97a][BDRQL11][RSM+13]. Adaptive, steerable antennas will then be used to provide proper performance during network operation.

Measurements in and around buildings at 28 and 72 GHz provide insights into the penetration and reflection characteristics of common building materials. Table 3.2 shows that the angle-dependent reflection coefficients are surprisingly high for common building materials as measured by a broadband channel sounder [ZMS+13][NMSR13]. Table 3.3 shows measured attenuation (e.g., penetration) values for common indoor and outdoor

Table 3.2 Comparison of reflectivity for different materials at 28 GHz. Concrete and drywall measurements were conducted with 10 degree and 45 degree incident angles, and tinted and clear glass reflectivity were measured at 10 degrees. Both of the horn antennas have 24.5 dBi gains with 10 degree half power beamwidth [reproduced from [ZMS+13] © IEEE].

Environment	Location	Material	Angle (°)	Reflection Coefficient ($\|\Gamma_{\|\|}\|$)
Outdoor	ORH	Tinted Glass	10	0.896
		Concrete	10	0.815
			45	0.623
Indoor	MTC	Clear Glass	10	0.740
		Drywall	10	0.704
			45	0.628

Table 3.3 Comparison of penetration losses for different environments at 28 GHz. Thicknesses of different common building materials are listed. Both of the horn antennas have 24.5 dBi gains with 10 degree half power beamwidth [reproduced from [ZMS+13] © IEEE].

Environment	Location	Material	Thickness (cm)	Received Power − Free Space (dBm)	Received Power − Material (dBm)	Penetration Loss (dB)
Outdoor	ORH	Tinted Glass	3.8	−34.9	−75.0	40.1
	WWH	Brick	185.4	−34.7	−63.1	28.3
Indoor	MTC	Clear Glass	<1.3	−35.0	−38.9	3.9
		Tinted Glass	<1.3	−34.7	−59.2	24.5
	WWH	Clear Glass	<1.3	−34.7	−38.3	3.6
		Wall	38.1	−34.0	−40.9	6.8

3.2 Large-Scale Propagation Channel Effects

wall materials. Notice from Table 3.3 that outdoor materials such as tinted glass and brick offer 28 to 40 dB of attenuation over free space, indicating that penetration into buildings will be difficult for outdoor mmWave systems. This may prove to be a valuable insulator of interference, allowing indoor networks to operate simultaneously and without interference from co-channel outdoor networks. Indoor partition measurements at 72 GHz are given in Table 3.4, measured at locations shown in Fig. 3.5. The results show that most interior walls and furniture are not terribly attenuative, allowing relatively low loss and good penetration, for example, 2-6 dB loss per partition. Metal objects such as elevator shafts deeply attenuate signals by 40 dB when the receiver is moved from a hallway to behind the elevator [NMSR13].

For completeness, we now provide a summary of material penetration measurements at 60 GHz as well as fundamental propagation theory that may estimate loss when measurements are scarce or unavailable. Propagating electromagnetic waves across different mediums are altered at the junction of materials that are different in composition from the source medium, due to the difference in electromagnetic impedance of the two media. For the purposes of this chapter, the medium is air (free space) and the junctions with which we will be concerned are materials in the physical environment. At such junctions we say that the electromagnetic wave is partially reflected and partially transmitted. The reflected portion of the electromagnetic wave bounces off the junction material and back into the source medium, whereas the transmitted portion propagates through the junction material. For a wireless communications system, we are primarily concerned with the magnitude of energy transmitted and reflected. To calculate such values without empirical measurements, we must have knowledge of the material complex permittivity $\epsilon = \epsilon' + j\epsilon''$ (where ϵ', $\epsilon'' \in \mathbb{R}$), the material thickness and surface roughness, the electromagnetic polarization, and the incident angle of the electromagnetic wave on the junction material [LLH94].

This is easily understood from a brief review of the loss mechanisms that affect a wave traveling through a material. Penetration loss can be understood through consideration of propagation characteristics of waves in dielectrics, conductors, and semiconductors. Dielectrics may become polarized (i.e., bound charges may be moved slightly apart from one another according to the polarization of the exciting wave and properties of the material) to create dipole moments, and free charges in conductors will move under the influence of electromagnetic waves. These topics are covered at great length in many basic and advanced texts on electromagnetic field theory [Bal89, Chapter 2]. As a basic reminder, the time harmonic form of Ampere's equation may be written as

$$\nabla \times H = j\omega\epsilon E + J + P \qquad (3.10)$$

where H is the magnetic field [A/m], E is the electric field [V/m], J is conduction current [A/m^2], ϵ is the electric permittivity, ω is the angular frequency ($2\pi f$), and P is the polarization vector [A/m^2] for a time harmonic charge movement. Both J and P are proportional to applied electric field: $J = \sigma E$, and $P = \chi\epsilon_o E$, where σ is the conductivity, ϵ_o is the permittivity of free space, and χ is the electric susceptibility. These result in a new form for (3.10):

$$\nabla \times H = j\omega\epsilon \left(1 - \frac{j}{\omega}\left(\frac{\sigma}{\epsilon} + \frac{\chi}{\epsilon_r}\right)\right) E \qquad (3.11)$$

Table 3.4 Penetration losses at different indoor receiver locations at 73.5 GHz. Numbers and types of obstructions are listed. Both the transmit and receive antennas are 20 dBi gain with 15 degree half power beamwidth. Multiple obstructions exist between TX and RX [reproduced from [NMSR13] © IEEE].

RX ID	TX-RX Separation (m)	# of Partitions					Received Power for Free Space (dBm)	Received Power for Test Material (dBm)	Penetration Loss (dB)
		Cubicle Wall	Metal Cabinet	Dry Wall	Wood Door				
1	6.8	1	0	0	0	−34.1	−39.4	5.3
2	8.0	1	1	0	0	−35.6	−52.8	17.2
3	10.1	2	2	0	0	−37.6	−61.4	23.8
4	11.5	1	2	1	1	−38.7	−75.5	36.8
5	8.6	0	2	0	0	−36.2	−50.3	14.1
6	8.1	0	2	0	0	−35.7	−45.4	9.7
7	8.8	1	2	0	0	−36.4	−63.0	26.6
8	14.0	0	2	1	1	−40.4	−55.6	15.2
9	13.0	1	3	0	0	−39.7	−53.0	13.3
10	15.2	1	2	1	0	−41.1	−60.4	19.3
11	15.2	1	2	1	0	−41.1	−59.0	17.9

3.2 Large-Scale Propagation Channel Effects

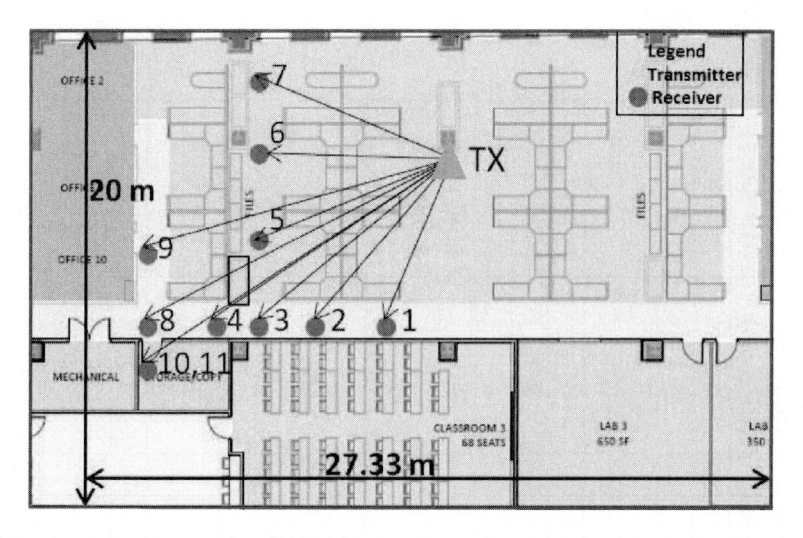

Figure 3.5 Indoor penetration measurements at 72 GHz in a building in Brooklyn, New York. The TX location is marked by a triangle, the RX locations are shown as numbered dots. The primary ray paths for signal penetration are shown with arrows [reproduced from [NMSR13] © IEEE].

where we have used $\epsilon = \epsilon_r \epsilon_o$. Equation (3.11) indicates that the permittivity of the material when loss is considered is given by

$$\epsilon_{\text{loss}} = \epsilon_{\text{no\_loss}} \left(1 - \frac{j}{\omega}\left(\frac{\sigma}{\epsilon} + \frac{\chi}{\epsilon_r}\right)\right). \qquad (3.12)$$

This results in a new wave number for the electromagnetic wave that contains both real and imaginary parts:

$$E = E_o e^{jkx}, \qquad k = \omega\sqrt{\epsilon_{\text{loss}}\mu} \qquad (3.13)$$

where k is the wave number and μ is the permeability, equal to the permeability of free space for most materials. Because (3.12) will result in a negative imaginary component for the wave number in (3.13), it is clear that the wave will experience exponential damping as it propagates into the object.

A comprehensive discussion of indoor material characteristics as well as transmission and reflection properties are summarized in [HR93], [LLH94], [CF94], [ATB+98], [HA95], [AR04], [MC04], and [Rap02]. First, we consider the measured reflected and transmitted energy of different materials of interest at 60 GHz. In [LLH94], a comprehensive table of such measurements for different angles of incidence is presented. Here we include representative data from [LLH94] in Table 3.5. The reflected and transmitted power losses summarized in Table 3.5 represent the relative measured power reflected or transmitted for different materials at different angles of incidence. From these measurements, it is possible to numerically estimate the permittivity, the loss tangent, and attenuation coefficient property of materials at 60 GHz. This was conducted in [CF94], and the results are summarized in Table 3.6. Although we focus here on 60 GHz waves, the same basic principles that apply to 60 GHz will also apply to other mmWave frequencies.

Table 3.5 Reflection and transmission power loss for different incidence angles on indoor materials at 60.2 GHz [from [LLH94] © IEEE].

Material	Thickness (cm)	Roughness (mm)	Refl. 10° (dB loss)	Refl. 40° (dB loss)	Refl. 70° (dB loss)	Trans. 0° (dB Loss)
Granite	3.0	0.6	17.5	11.7	3.4	$\geq$ 30.0
Quartzite	2.0	0	5.8	24.1	4.4	3.4
Marble	1.7	0	3.8	5.5	0.8	5.2
Limestone	3.0	0	6.5	5.1	0.8	5.2
Aerated concrete	5.0	0.2	14.1	11.0	5.1	18.9
Concrete	5.0	0.1	7.5	6.2	2.0	$\geq$ 30.0
Brick	11.0	0.3	14.8	17.5	4.8	16.9
Breeze block	5.0	0.5	17.5	12.7	5.1	$\geq$ 30.0
Tiles	0.5	0.1	4.1	3.8	2.1	$\geq$ 30.0
Plasterboard	1.0	0	23.8	4.5	6.9	2.1
Plasterwork	1.0	1.0	27.9	30.0	6.9	$\geq$ 30.0
Polyfoam	$\geq$ 30.0	$\geq$ 30.0	$\geq$ 30.0	$\geq$ 30.0	22.7	0
Rockwool	3.5	0.9	28.9	$\geq$ 30.0	$\geq$ 30.0	0.5
Wood fiberboard	1.2	0.2	21.0	15.5	5.5	3.4
Pertinax	0.8	0	9.1	9.1	2.6	6.9
Wooden panels	1.2	0	6.4	14.5	4.8	7.6
Wooden chipboard	1.3	0.2	13.4	11.7	5.3	6.2
Acrylic glass	0.4	0	0.2	5.5	13.1	1.7
Glass	0.4	0.3	6.7	3.8	0.8	4.5
Glass	0.4	0	17.6	7.6	2.9	2.4
Glass	0.8	0	8.8	9.1	2.6	3.1

Table 3.6 Average calculated electrical characteristics from reflected/transmitted power measurements [from [CF94] © IEEE].

Material	Relative Permittivity ϵ_r	Loss Tangent $\tan(\delta)$	Atten. Coeff. $\alpha\left(\frac{dB}{cm}\right)$
Stone	6.81	0.0401	5.73
Marble	11.56	0.0067	1.25
Concrete	6.14	0.0491	6.67
Aerated concrete	2.26	0.0449	3.7
Tiles	6.30	0.0568	7.81
Glass	5.29	0.0480	6.05
Acrylic glass	2.53	0.0119	1.03
Plasterboard	2.81	0.0164	1.51
Wood	1.57	0.0614	4.22
Chipboard	2.86	0.0556	5.15

3.2.6 Scattering and Radar Cross Section Modeling

Scattering at mmWave frequencies is an important propagation mechanism, since physical objects such as walls, people, and lampposts are larger than a wavelength. Scattering causes propagated power (in free space) to be inversely proportional to d^4 while free space is proportional to d^2. Scattering, being a weaker propagation phenomenon, is negligible at today's 1-5 GHz cellular and WLAN systems, but at mmWave frequencies, the large relative size of all objects in the channel implies that illuminated scatterers may actually create signal paths that are as substantial as (or even occasionally stronger than) reflected paths [RSM+13][AWW+13][RGBD+13][S+91][S+92][SSTR93]. One method to estimate the impact of scattering from the sides of buildings and other large objects relative to a wavelength is to apply a radar cross section (RCS) model to estimate the impact of scattering in a propagation environment. By multiplying the RCS (units of m^2) by the scattered field, it is possible to estimate received power.

Scattering models, when combined with ray tracing, may accurately predict the large-scale variations of coverage and interference. The most important scattering effects that impact propagation in an outdoor environment are the radar cross section and surface roughness of the objects in the channel. The radar cross section of an object describes how the electromagnetic fields are scattered from that object and represents that object in terms of an aperture with a particular area. The RCS area does not necessarily relate to the physical area of the object.

The monostatic cross section describes how the field is scattered in the direction of the transmitter when the transmitter and receiver occupy the same point in space, such as when a police officer uses a radar gun that serves as a transceiver. The bistatic cross section is applicable to mmWave propagation for communication systems. It describes how the electric field is scattered in the direction of a receiver that is not collocated with the transmitter [Bal89, Chapter 2]. The cross section of an obstacle is defined as [Bal89, Chapter 2]

$$\sigma_{3D} = \lim_{r \to \infty} \left[\frac{4\pi r^2 S_s}{S_i} \right]$$
$$= \lim_{r \to \infty} \left[\frac{4\pi r^2 |E_s|^2}{|E_i|^2} \right] \quad (3.14)$$

where σ_{3D} is the 3-dimensional radar cross section, S_s is the scattered power density (in the direction of observation), S_i is the incident power density, E_s is the scattered electric field, E_i is the incident electric field, and r is the distance between the target and the observation point. Note that the electric field is proportional to the square root of power, and thus falls off as $1/d$ in free space [Rap02]. The exact mathematical formulation of the scattering cross section depends on the nature of the object and the polarization of the incident field, and this is an open area of research for a mmWave wireless communication design system. Intuition can be gained, however, by considering the monostatic (i.e., backscattering) cross section for a smooth plate of length a and width b for the polarized magnetic field parallel to the surface (i.e., transverse electric polarization relative to the surface). As shown by [S+91], [S+92], and [Bal89, Chapter 2], the monostatic cross section for the slab with this incident field is equal to

$$\sigma_{3D}^{\text{monostatic}} = 4\pi \left(\frac{ab}{\lambda} \right)^2 \cos^2(\theta_i) \left[\left(\frac{\sin(k_o b \sin(\theta_i))}{k_o b \sin(\theta_i)} \right) \right]^2 \quad (3.15)$$

where θ_i is the angle of incidence, λ is the wavelength of operation, and $k_o = 2\pi/\lambda$ is the free space wave number. Equation (3.15) indicates that as the frequency increases, the radar cross section of an object will become sharper and stronger.

The popular radar cross section (RCS) model for received power from a scattered ray indicates that as the cross section increases, we can expect a higher received power from the scattered ray [Rap02][S+91][S+92]

$$P_r[\text{dBm}] = P_t[\text{dBm}] + G_t[\text{dBi}] + 20\log_{10}\lambda + RCS[\text{dBm}^2]$$
$$- 30\log_{10}(4\pi) - 20\log_{10}d_t - 20\log_{10}d_r. \quad (3.16)$$

Equation (3.16) indicates that a larger RCS in the direction of the receiver is associated with higher received power. In (3.16), P_t is the transmit power, G_t is the transmitter antenna gain, RCS is the radar cross section given by (3.14) and (3.15), d_r is the distance from the scattering object to the receiving antenna, d_t is the distance from the scattering object to the transmitting antenna, and P_r is the power into the receiving antenna. We see from (3.15) and (3.16) that a higher RCS can partially compensate for the decreased received power due simply to the shorter wavelength of mmWave frequencies. Equation (3.15) assumes that the scattering slab is perfectly smooth. In an outdoor environment, this is sometimes not likely to be the case, as propagating signals are likely to encounter such materials as brick or tree bark. Therefore, the impact of surface roughness on the scattering characteristics of objects should also be considered. As discussed

3.2 Large-Scale Propagation Channel Effects

by [DRU96] and [BC99], when a surface is rough, its total scattering cross section may be defined as a weighted-sum of the scattering that would result from a smooth surface, and the cross sectional scattering area attributable to the surface roughness:

$$\sigma_{\text{tot}} = \sigma_{\text{rough}} + |\chi_s|^2 \sigma_{\text{smooth}} \qquad (3.17)$$

where σ_{rough} is the scattering area attributable to the surface roughness and σ_{smooth} is the scattering due to the large-scale features of the object (e.g., for a slab with TE incident field we may apply (3.15) in the backscatter direction). In (3.17), χ_s accounts for the surface roughness [DRU96][BC99], given as

$$\chi_s = e^{-k_o^2 \langle h_s^2 \rangle \cos^2(\theta_i)} = e^{-\left(\frac{2\pi}{\lambda}\right)^2 \langle h_s^2 \rangle \cos^2(\theta_i)} \qquad (3.18)$$

where $\langle h_s \rangle$ is the average surface roughness (i.e., the mean square height of the small-scale features of the surface). Equations (3.17) and (3.18) indicate that as the surface becomes rougher relative to the wavelength, the "smooth" cross section of (3.15) will play less of a role in determining the total radar cross section of the scatterer. The scattering cross section attributable to the surface roughness has a complicated relationship to the nature of the surface, including the periodicity of its variations, the slopes of its surface variations, and the height of its surface variations [PLT86]. The value of σ_{rough} is related to the reflection coefficient for a rough surface, and this reflection coefficient gives more physical insight into the pertinent effects. For a smooth surface with a specular reflection coefficient (i.e, the proper direction given by Snell's law, as shown by the reflected rays r_1 and r_2 in Fig. 3.7) R, the rough surface reflection coefficient R_{rough}, is approximated by

$$R_{\text{rough}} = R e^{-2k_o^2 \langle h \rangle^2 \cos(\theta_i)}, \qquad (3.19)$$

which is obtained from a physical optics model [DRU96]. Therefore, surfaces and scatterers will appear rougher at mmWave frequencies unless they are perfectly smooth, but they will also be electrically much larger than at lower frequencies. Smooth surfaces, such as outdoor lampposts or other metallic objects, serve as strong sources of multipath in outdoor environments, much stronger than at lower RF frequencies [RGBD+13][RSM+13][AWW+13][RQT+12]. In contrast, rough surfaces such as brick or tree bark are less effective as sources of multipath. Yet, because of the very large surface area relative to a wavelength, even rough surface scatterers such as large building facades or rows of trees seem to produce scattering energy that is useful at mmWave bands. Indeed, at 60 and 38 GHz, one of the authors and his students [RBDMQ12][RQT+12][BDRQL11] found that lampposts and other metallic objects commonly found outdoors may be used to form NLOS links between a transmitter and a receiver, whereas rough surfaces were much less effective in this regard. Buildings in New York City, however, even when rough on the surface, provided strong scattering in many instances in NLOS conditions [AWW+13][SWA+13][RSM+13].

3.2.7 Influence of Surrounding Objects, Humans, and Foliage

The presence of furniture in a room (compared with the equivalent empty room) has inconsistent effects on the multipath [CZZ03b]. For the LOS scenario, the multipath root-mean-square (RMS) delay spread (a measure of multipath propagation time dispersion described in Section 3.3.1) is increased due to *spreading* of the energy by the furniture. In the NLOS scenario, the RMS delay spread is decreased because weaker reflections are more highly attenuated. The presence of humans has a more pronounced effect. In [MC04], it was discovered that fading attributed to humans who shadow a signal is continuously changing with a dynamic range of 35 dB. [CZZ04] carefully studied the performance of a system using outage calculations through measurements of an indoor environment with varying human numbers and speed. Not only are humans significant obstacles, they are also significant reflectors and scatterers. In general, when more humans populate a room, the RMS delay spread becomes larger, and outages occur more often. In addition, the coherence time of a channel decreases if humans move quickly in a room while the transmitter and receiver are stationary [CZZ04].

There have been relatively few studies of the impact of foliage and vegetation on mmWave propagation. Outdoor mmWave wireless systems will someday be used for both mobility and fixed backhaul applications throughout a wide range of sub-terahertz bands. Early work, however, has focused primarily on the 28 and 38 GHz Local Multipoint Distribution Service (LMDS). It is instructive to review studies of foliage in LMDS channels because this provides insight into gross propagation effects that are likely to impact many mmWave systems of the future.

[Kaj00] studied the impact of foliage on 29 GHz propagation and found that foliage results in attenuation well described by a Rician distribution, and that moving foliage can result in fading up to 10 dB, whereas at 5 GHz they found moving foliage only resulted in 2 dB fades. [XKR00] studied the 38 GHz outdoor channel and found that wet foliage, in particular, can serve as a source of multipath reflection. Early mmWave researchers [JEV89] conducted extensive measurements at 9.6, 28.8, 57.6, and 96.1 GHz in a conifer tree orchard in order to study propagation loss caused by foliage using horn antennas. They studied links ranging from 55 to 177 m. One of their more surprising findings is that highly vegetated areas can support NLOS paths (e.g., in which the transmitter and receiver horn antennas do not point directly at one another, and scattering occurs from the foliage) consistently stronger than LOS paths, even when there is not total obstruction between the transmitter and receiver (e.g., a quasi-LOS path). [JEV89] found that paths that are $2°$ to $5°$ off boresight were consistently strongest, suggesting scattering from the foliage. At 57.6 GHz, the researchers found foliage to contribute approximately 40 dB of extra attenuation compared with 9.6 GHz for co-polarized measurements (i.e., when the receiving and transmitting antenna have the same polarization) over a 5 m distance. This important study by the National Telecommunications and Information Administration (NTIA) [JEV89] also found that, at 57.6 GHz, signals can experience from 50 to 80 dB of additional attenuation for propagation through 20-80 m of foliage depth (or for an equivalent number of trees ranging from 4 to 14) for areas with heavy foliage. [JEV89] also found that conifer foliage can result in substantial de-polarization of 57.5 GHz signals, resulting in less effective orthogonal frequency reuse [PHAH97].

For outdoor communication scenarios, the foliage effects of vegetation help to better illustrate properties of propagation of 60 GHz and other mmMave frequency electromagnetic waves. [PB02] analyzed the foliage attenuation effects of trees at frequencies from 2.45 to 60 GHz. Contrary to what might be expected, lower frequencies can actually display larger mean attenuation of signals propagating through leaves. It was determined, however, that the variance of attenuation increased as the carrier frequency and ambient wind speed increased.

The impact of foliage on mmWave propagation versus lower frequency propagation can be understood in part from a consideration of Fresnel zones, as discussed in [Rap02]. Fresnel zones are circles drawn around an obstacle between a receiver and transmitter that represent path length increases by increments of one half wavelength and are illustrated in Fig. 3.6. The distances between the transmitter and the obstacle, and the receiver and the obstacle may be denoted d_1 and d_2, respectively. The radius of the nth Fresnel zone is then found according to

$$r_n = \sqrt{\frac{n\lambda d_1 d_2}{d_1 + d_2}}. \qquad (3.20)$$

As discussed in [Rap02], when the volume contained by the first Fresnel zone is not blocked, there will be minimal diffraction loss between the transmitter and receiver. In general, if the volume contained within the first Fresnel zone is blocked, then the signal will be heavily attenuated, and further Fresnel zones being blocked induces even greater diffraction loss. For mmWave communications, since the wavelength is so small (on the order of millimeters), a useful rule of thumb is to simply consider only if the first Fresnel zone has clearance, and to assume complete outage whenever the first zone is completely blocked. Note that the small wavelength of mmWave signals results in the Fresnel zones being much smaller than at traditional RF frequencies. In outdoor situations in which vegetation is likely to be an issue, it is possible for bushes and trees to block the majority of the Fresnel zone. For example, if a tree is 100 m from the transmitter and 100 m from the receiver, the first Fresnel zone has a volume of 0.52 m^3 at 60 GHz, while at 2 GHz the first Fresnel zone has a volume of 86 m^3. With only 10 m between the transmitter/receiver and the obstacle, the volume of the first Fresnel zone

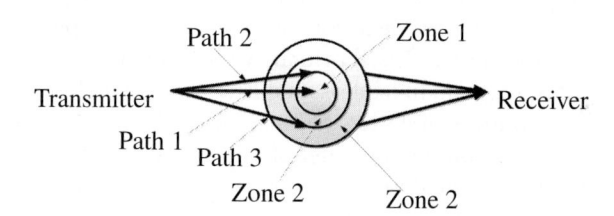

Figure 3.6 Example of a diffraction object blocking the line-of-sight (LOS) path between transmitter and receiver. At millimeter wave frequencies, objects such as trees and people may induce fading and scattering as they move.

is even smaller at 0.005 m³. As a result, when a mmWave signal encounters a tree or bush that lies between the transmitter and receiver, the signal is likely to be much more highly attenuated than at lower RF frequencies. This is why we stated in Section 3.2.4 that diffraction cannot be relied upon for large-scale propagation at mmWave frequencies. [PHAH97] used Fresnel zone analysis to understand the impact of foliage and vegetation on the LMDS channel from 27.5 to 29.5 GHz. They suggested that vegetation can result in fading over short time intervals as the amount of the first Fresnel zone that is blocked changes with time as tree foliage sways and moves.

Although foliage can severely attenuate a signal between the transmitter and receiver at mmWave frequencies, it also can serve as a source of multipath reflections that can be used to form NLOS links. For example, [RBDMQ12], [RQT+12], [BDRQL11], and [MBDQ+12] all studied the mmWave channel at either 60 or 38 GHz for outdoor urban peer-to-peer or cellular type applications. In their study, they used highly directional and rotatable transmit and receive antennas (7° beamwidths) that enabled identification of the object in the environment that was the source of multipath energy. They found that tree trunks were moderately reflective and that people's heads are highly reflective.

3.2.8 Ray Tracing and Site-Specific Propagation Prediction

Using the theory of ray tracing, it is possible to use a computer to predict with remarkable accuracy the temporal and spatial characteristics of the multipath channel and to model the impact of reflectors and scatterers, as well as diffraction objects [BMB91] [SDR92][SR92][SSTR93][SR94][RBR93][YMM+94][ARY95][DPR97a][DPR97b][ANM00] [WR05]. Ray tracers also allow simple implementation of atmospheric models, distant-dependent path loss, diffraction, and reflection models, such as those given in Sections 3.2.1–3.2.5 and 3.2.7, and scattering models, such as those described in Section 3.2.6. As mmWave wireless systems proliferate, the use of ray tracing will become critical for proper site deployment, infrastructure placement, and research into spatial processing and antenna characteristics, as the large-scale propagation mechanisms that impact coverage and interference at mmWave frequencies are based primarily on specular reflections and radar cross section (e.g., large scattering) and much less on diffraction or small scattering.

Ray tracing uses computer simulation to model and discretize the energy radiated in space as it interacts with a computer model of the physical environment. As shown in Fig. 3.7, a radiating source is modeled as a discrete source having radiation of varying amplitudes over a sphere. A *ray tube* is used to represent a fractional spatial portion of radiated energy from (or to) an antenna, and hundreds of ray tubes are launched by a computer program, as shown in Fig. 3.8, to trace all of the possible discretized physical paths of propagation. This is sometimes called brute force ray tracing. The ray tracer also simulates the receiver and seeks to find the intersection of the radiated rays with *reception spheres*, as shown in Fig. 3.9. Ray tracers, also known as *ray tracing engines*, may be improved for computational speed as well as accuracy by considering only those physical paths that contribute the most energy. Databases that represent the physical environment are a critical component of ray tracing engines, and the accuracy of the physical dimensions and electromagnetic properties of the environment play a major role in their utility. For relatively simple physical environments, the method of images may be used instead of "brute force" ray tracing, where specific reflection points are identiied in the physical model [HR93][H+94].

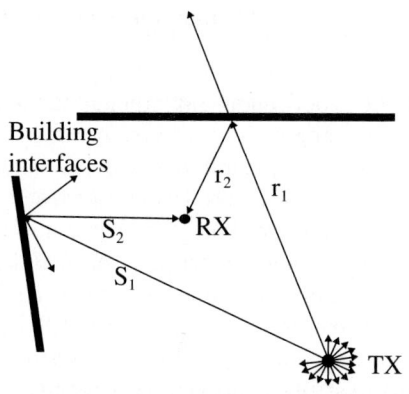

Transmitted, reflected, and scattered rays at a planar interface.

Figure 3.7 Ray tracing uses a computer to discretize the radiation from a transmitter and models the interaction of propagating waves with a computerized model of the physical environment [from [SDR92] © IEEE].

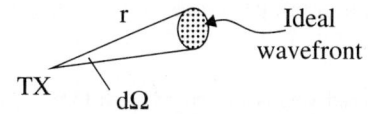

Ideal wavefront represented by each source ray.

Figure 3.8 Ray tubes are used to represent each of the rays shown in Fig. 3.7, where each ray represents a portion of the radiated wavefront [from [SDR92] © IEEE].

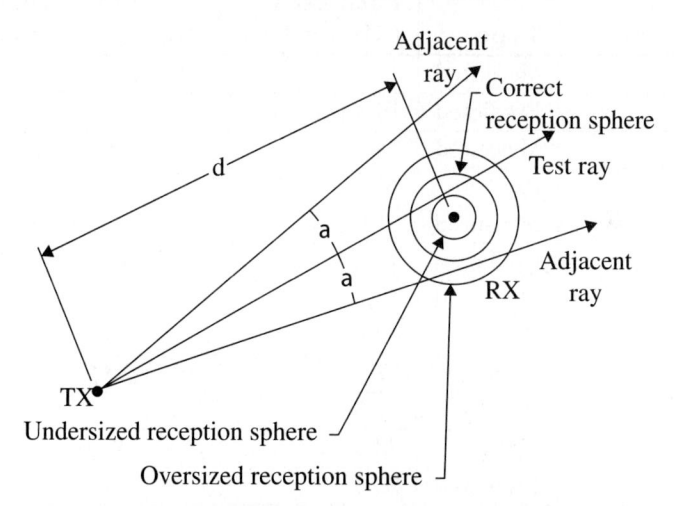

Two dimensional view of the reception sphere. The total ray path length is d, producing a reception sphere radius that varies with α and d.

Figure 3.9 A 3-D reception sphere is used to find the intersections of the launched rays at a particular receiver location. The radius of the reception sphere is determined from the propagated distance of the ray and the angular resolution of the rays launched in Fig. 3.7 [from [SDR92] © IEEE].

Accuracy can be remarkably good using ray tracing, as shown for measurements in Table 3.7. The power of a ray tracing engine is that once it is constructed, it is a simple matter to change carrier frequency, transmit and receive antennas, and physical environmental models. Work in [JPM+11] showed that ray tracing can accurately predict random human shadowing effects, and work in [YMM+94] and [MEP+08] have predicted 60 GHz indoor channel responses using ray tracing. As new knowledge is developed at higher frequencies for a wide range of environments, ray tracing is sure to be a fundamental planning, deployment, and research tool [SDR92][SSTR93][SR94][He+04][RGBD+13] [DPR97a][DPR97b]. As computing power and connectivity to personal devices continue to expand, it will also become possible to implement site-specific prediction for real-time wireless network control of coverage and capacity. [WR05] suggested real-time methods for computing diffraction, and rapid methods for computing the impact of beamforming on mmWave signal strength with the use of environmental maps could become a valuable real-time control approach for mobile users [CRdV06][CRdV07][SSTR93].

Recent research shows that exhaustive ray tracing may not be needed to obtain reasonably accurate models of mmWave path loss (although this approach might not work for small scale spatial or temporal modeling). Work at 5.8 GHz reported in [DRX98b][SRA96]

Table 3.7 Comparison of predicted versus measured path loss (with respect to a 1 m free space path loss distance d) and multipath RMS delay spread for the locations indicated in a microcell environment at 1 GHz. As frequencies increase, ray tracing is likely to be a more important deployment and research technique [from [SDR92] © IEEE].

Location	Type	Path Loss (dB wrt 1 m FSPL)	RMS Delay Spread (ns)
A	Measured	17.3	15.0
	Predicted	16.1	5.8
B	Measured	19.4	8.2
	Predicted	14.9	3.8
C	Measured	41.9	24.0
	Predicted	41.6	19.4
D	Measured	43.5	19.8
	Predicted	42.5	13.3
E	Measured	17.7	38.1
	Predicted	17.3	21.0
F	Measured	23.9	43.4
	Predicted	20.9	23.8
G	Measured	22.3	16.4
	Predicted	25.2	30.8
H	Measured	42.9	35.4
	Predicted	40.9	22.9
I	Measured	47.1	34.0
	Predicted	42.0	13.9

3.2 Large-Scale Propagation Channel Effects

showed that using a very simple "primary ray-tracing" technique, in which a single line (the primary ray) is drawn between the transmitter and receiver, it is possible to use a simple large-scale distance-dependent path loss model such as given in (3.7), combined with a systematic partition loss factor for each "object" that is encountered by the ray, to come up with remarkably accurate path loss predictions. While this approach may not always be accurate for specific situations, work in [DRX98b] and [SRA96] as well as work in [BMB91], [SR92], and [ARY95] shows that over a large number of environments, the "primary ray" partition loss model works surprisingly well for real-world system deployments. Fig. 3.10 shows a potential mmWave cellular or last-mile wireless scenario, in which an outdoor transmitter at modest height (5.5 m) is used to transmit into and around a nearby home. Measurements in [DRX98b] showed that path loss partition values may be obtained for various common obstructions, and when used for prediction in new environments, they can yield excellent accuracy for determining path loss, coverage, interference, and other large-scale propagation issues. Table 3.8 shows

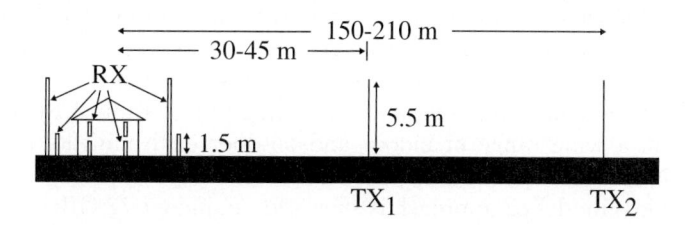

Figure 3.10 A possible scenario for mmWave wireless communications as a last-mile solution. Measurements in [DRX98b] and [SRA96] demonstrated that a simplified "primary" ray tracing method may be used in conjunction with partition losses to determine large-scale path loss at many microwave frequencies [from [DRX98b] © IEEE]. Research is needed to determine whether such an approach will work for small cells at mmWave frequencies. Early work shows that such an approach may be used for in-building mmWave systems [NMSR13][RSM+13].

Table 3.8 Attenuation (in decibels) into three homes due to shadowing (receiver located behind the homes) and penetration into homes at 5.85 GHz using 5.5 m transmitter height [DRX98a]. Attenuation is given for close-in shadowing of a single house as well as aggregate penetration loss (APL) values for all homes. Work in [ZMS+13] shows that penetration into urban buildings is much more difficult, with penetration losses on the order of 40 dB at 28 GHz. "N/A" denotes locations that were not measured for shadowing loss with external receivers [from [DRX98a] © IEEE].

Home	TR sep	Shadowing Loss		APL (dB)
		5.5 m RX (dB)	1.5 m RX (dB)	
Rappaport	30 m	19.1	23.2	13.3
	150 m	10.8	11.9	16.4
Woerner	30 m	14.1	27.8	13.1
	210 m	N/A	N/A	7.2
Tranter	48 m	17.2	19.0	21.1
	160 m	N/A	N/A	15.3
Linear Average		16.3	23.6	16.3
dB Average		15.3	20.5	14.4

the shadowing loss at 5.8 GHz when a receiver is moved to behind the home, so that the home acts as a shadowing object, and also shows the average penetration loss into three homes. Recent work in [ZMS+13] shows that for office buildings in Brooklyn, New York, 28 GHz penetration losses into buildings are much greater than the homes measured in rural Virginia at 5.8 GHz [DRX98b], with penetration losses ranging between 28 and 40 dB [ZMS+13]. More data are required to understand the penetration losses for mmWave cellular applications.

To use primary ray tracing, an optimization algorithm is generally used to form a best-fit (minimum mean square error, with least standard deviation) among many measurement campaigns and coverage distances. Table 3.8 shows the path loss (shadowing loss) due to the building obstructions for receiver antennas located behind houses as shown in Fig. 3.10, and Table 3.9 shows a tabulation of the various partitions encountered when a single, primary ray is drawn between a transmitter and receiver in a computer model of the physical environment (also, see Figure 3.5). By counting the types and number of partitions, as shown in Table 3.9, and noting the physical distance traveled by the primary ray, it is easy to develop an optimized path loss model that accounts both for free space distance and the cumulative effect of partitions of differing attenuations [SRA96][DRX98b]. The resulting optimized partition loss values and standard deviations for a wide range of residential building materials are given in Table 3.10. This technique may be applied in a wide range of indoor and outdoor environments for determining rapid accurate signal or interference levels as long a primary ray path is known.

The partition loss modeling approach is shown for an indoor 72 GHz wireless network (see Fig. 3.5 and Table 3.4). Early work at mmWave frequencies indicates that, while effective at some locations, primary ray tracing may not be as effective as more comprehensive ray tracing for accurately predicting power over wide indoor coverage areas, possibly due to the lack of diffraction at the mmWave bands. More work is required to develop reliable and easy-to-use modeling tools. This is clearly seen by the wide range of partition losses over the measurements. At location 4 in Fig. 3.5, it is possible that edge diffraction from the nearby cube wall is causing a much greater attenuation than other locations measured at comparable distances and with a comparable number of partitions.

Much work is needed to determine the most suitable ways to model the large-scale propagation effects of mmWave communication systems in efficient and intuitive ways, yet the use of physical maps and ray tracing promise to be important tools in the design and management of mmWave networks and may also be used to assist in the real time control of such networks.

3.3 Small-Scale Channel Effects

The complex baseband linear system impulse response of a multipath channel was developed in Chapter 2. For the purpose of propagation channel characterization, it is useful to consider the pseudo-complex baseband equivalent channel $h_e(t)$ described in (2.14) written as

$$h_e(t) = \sum_{\ell=1}^{L} h_e[\ell]\delta(t - \tau[\ell]) \qquad (3.21)$$

3.3 Small-Scale Channel Effects

Table 3.9 A systematic way to compute the overall path loss, and additional path loss due to partitions, using a primary ray tracing method. Partition frequency, distance, and 5.85 GHz path loss W.R.T. 1 m free space are given for the 30 m transmitter at the Rappaport home using an outdoor 5.5 m transmitter height. Early work shows that this method is potentially useful for in-building at 28 and 73 GHz [SRA96][NMSR13][RSM+13] [from [DRX98b] © IEEE].

Location	Small Tree	Brick Ext.	Int. Wall	TR Sep. (m)	PL (dB)
1	1	0	0	22	31.3
2 *Outdoors*	1	0	0	22	33.4
3 *Front Side*	0	0	0	23	32.4
4 *5.5m height*	0	0	0	25	33.7
5	0	0	0	27	31.8
6	0	0	0	29	32.0
1	1	0	0	22	31.3
2 *Outdoors*	0	0	0	23	25.9
3 *Front Side*	0	0	0	25	27.3
4 *1.5m height*	0	0	0	27	32.1
5	0	0	0	29	32.0
1st Floor					
Living Room	1	1	0	32	40.1
Front Hall	0	1	0	30	39.6
Office	0	1	0	32	41.6
Stairs	0	1	0	31	45.8
Bathroom	0	1	1	35	46.7
Laundry	0	1	1	35	43.7
Kitchen	0	1	2	38	51.2
Dining Room	1	1	0	38	42.5
Family Room	0	1	2	41	51.9
2nd Floor					
Front Bed	1	1	0	32	44.4
Rear Bed 1	1	1	1	38	51.2
Bathroom	0	1	2	38	51.7
Rear Bed 2	0	1	1	42	46.6
Master Bed	0	1	1	34	40.6
		$\underbrace{\qquad}_{A}$		$\underbrace{\qquad}_{\vec{d}}$	$\underbrace{\qquad}_{\vec{p}}$

where we define each multipath component of the impulse response arriving at the τ_ℓth time as having a complex voltage $\bar{h}_e[\ell] = \alpha_\ell e^{-j2\pi f_c \tau[\ell]}$ and $L > 0$ is the number of excessive multipath components. The complex coefficients of each multipath component $h_e[\ell]$ incorporate the large-scale propagation path loss effects from the previous section. Conceptually, the impulse response model shows the different arrivals of the transmitted signal due to reflection, scattering, or diffraction in the environment, and while (3.21)

Table 3.10 Partition losses, and their standard deviation, for a minimum-mean square fit for common building materials in modern residential homes. Summary of all attenuation values (loss in excess of free space) is given at 5.85 GHz with outdoor transmitters at 5.5 m height above ground. Such data does not yet exist publicly for most mmWave frequencies [NMSR13] [from [DRX98b] © IEEE].

Partition	Loss (dB)	σ (dB)	$\Delta\sigma$ (dB)
Home exteriors			
Brick†	12.5		
Rappaport Home, 30m TX	10.2	2.6	3.1
Rappaport Home, 150m TX	14.8	2.1	4.5
Brick*	16.4		
Tranter Home, 48m TX	16.1	3.4	3.9
Tranter Home, 160m TX	16.6	3.2	4.5
Wood Siding†	8.8	3.5	0.9
Cinderblock wall	22.0	3.5	6.4
Subterranean basement	31.0		
Tranter Home, 48m TX	34.0	3.4	3.7
Tranter Home, 160m TX	29.0	3.2	2.7
Home Interior			
Plaster walls	4.7		
Rappaport Home 30m TX	4.7	2.6	1.1
Rappaport Home, 150m TX	4.6	2.1	0.8
Plasterboard walls	4.6		
Tranter home, 48m TX	3.6	3.4	1.9
Woerner Home, 30m TX	5.6	3.5	1.2
Foliage Shadow			
Small deciduous tree	3.5	2.6	0.5
Large deciduous tree	10.7		
Woerner Home, 30m TX	9.0	3.5	1.7
Woerner Home, 210m TX	12.3	3.3	2.4
tree line, 5.5m RX	12.4	–	–
Large coniferous tree	13.7		
tree line, 5.5m RX	16.4	–	–
tree line, 1.5m RX	11.0	–	–

†paper-backed insulation
*foil-backed insulation

does not show spatial angles of arrival or departure (AoA or AoD), it is understood that individual multipath components may arrive from different spatial directions as well. The delays correspond to path length differences of the scattered or reflected path or time length differences when propagating through different mediums. MmWave frequency signals also observe a large percentage of multipath contributions due to large surface

3.3 Small-Scale Channel Effects

scattering (i.e., scattering from large objects). Scattering occurs when objects similar in dimension to λ, the operating wavelength, act as point sources when they obstruct the propagation of electromagnetic waves. Due to the reduced diffraction experienced by mmWave and even higher frequency waves, the channel is also characterized by more severe shadowing [AR04]. Reflection occurs on objects much larger than the dimensions of λ. Therefore, objects traditionally acting as scattering objects now become reflectors at mmWave frequencies, and may induce significant multipath effects in mmWave systems.

3.3.1 Delay Spread Characteristics

The time delay spread of a wireless channel represents the propagation times and temporal spread of multipath energy of a received signal propagating through a channel. Given the multipath channel model in (3.21), the delay of the channel $\tau[0]$ is the absolute propagation time from the transmitter to the receiver of the first arriving signal component (or, in some cases, $\tau[0]$ is called the *minimum excess delay time* and is used to denote the first arriving multipath component with a reference delay of zero), while $\tau[L]$ is the absolute maximum propagation time of a multipath signal component. We define the *maximum excess delay spread* as the difference $T_{\max} = \tau[l] - \tau[0]$. The root-mean-square (RMS) delay spread is defined as [Rap02]

$$\tau_{\text{RMS}} \equiv \sqrt{\left(\frac{\sum_{\ell=0}^{L}(\tau[\ell] - \tau[0])^2 \mathbb{E}\,|h_{\text{e}}[\ell]|^2}{\sum_{\ell=0}^{L} \mathbb{E}\,|h_{\text{e}}[\ell]|^2}\right) - \left(\frac{\sum_{\ell=0}^{L}(\tau[\ell] - \tau[0]) \mathbb{E}\,|h_{\text{e}}[\ell]|}{\sum_{\ell=0}^{L} \mathbb{E}\,|h_{\text{e}}[\ell]|^2}\right)^2} \quad (3.22)$$

where $\mathbb{E}\,|h_{\text{e}}[\ell]|^2$ denotes the average of the channel impulse response over a local area. This is often called a power delay profile (PDP), where the area under the curve of the PDP represents the average received power [Rap02, Chapter 5].

Intuitively, the RMS delay spread quantifies the temporal spreading (multipath time dispersion) effect of the wireless channel and is a simple channel parameter that loosely defines the spread of propagation delays of a multipath channel. As a rule of thumb, for a given τ_{RMS} the number of symbols that need to be equalized to remove intersymbol interference effects is $\left\lceil \frac{\tau_{\text{RMS}}}{T_{\text{s}}} \right\rceil - 1$ for symbol rate $1/T_{\text{s}}$. Numerous measurements studies have been completed to better understand the RMS delay spread of mmWave channels in both indoor and outdoor environments. Here, we reproduce the approximated measurements from [SW92] in Table 3.11, as this reference is in agreement with other measurements and nicely summarizes the RMS delay spreads that can be expected for indoor scenarios across a wide range of mmWave frequencies.

As reported in [SW92], the delay spread has been observed to be relatively independent of the location of the transmitter and receiver in a room. The measurements [SW92] were taken with bi-conical horn antennas [SW94b] that exhibit omnidirectional characteristics on the azimuthal (horizontal plane). The transmitter was elevated at a level of 3 m, and the receiver varied its elevation from 1 to 4 m. As shown in Table 3.11, delay spread increases when the dimensions of the room increase and when the wall reflection coefficient increases. Measurements by other researchers, such as [AR04], [Yon07], [ZBN05], [LBVM06], [M+09], [RH92], [RH91], [H+94], [XKR02], [HR92] and [AR+02], provide additional observations for indoor mmWave channels.

Table 3.11 Approximate T_{RMS} for different indoor scenarios at 60 GHz [reproduced from [SW92] © IEEE].

Dimensions (m³)	Wall Material	$T_{RMS,\ 90\%}$	$T_{RMS,\ 10\%}$ (ns)
24.5 × 11.2 × 4.5	Wood	40	45
30 × 21 × 6	Rock wool	30	35
43 × 41 × 7	Concrete	40	60
33.5 × 32.2 × 3.1	Concrete	40	70
44.7 × 2.4 × 3.1	Metal	60	80
9.9 × 8.7 × 3.1	Metal	40	45
12.9 × 8.9 × 4.0	Wood	15	25
11.3 × 7.3 × 3.1	Concrete	25	30

In outdoor channels at 28 and 38 GHz, measurements in Austin, Texas, and New York City have shown the impact of different antenna beamwidths on multipath delay spread. For 38 GHz outdoor channels in Austin, works in [BDRQL11], [RQT+12], [RBDMQ12], [MBDQ+12] and [RGBD+13] show how wider beamwidth steerable antennas (or equivalently, smaller gain antennas) provide larger multipath delay spreads and better signal coverage (due to smaller path loss exponent values) over short distances when compared with narrower beam antennas, yet the smaller antenna gains (i.e., wider beamwidths) may not be effective over longer distances, for which LOS links achieved through higher gain antennas will be superior. Similar effects were seen in New York City at 28 GHz [SWA+13][AWW+13][RSM+13], where wider beam (i.e., smaller gain) antennas offered smaller path loss exponents and greater multipath delay spreads, but over smaller distances. This indicates the importance of adaptive beam antennas for contouring the particular multipath and received power responses of the channel to the most desirable operating setpoint for future wireless networks, depending on the distance between the TX and RX.

As described in Section 3.7, where small-scale propagation measurements for outdoor channels are analyzed and modeled, the extent of the multipath time delay of a channel determines the amount of bandwidth over which the spectrum appears to be flat in gain (this is called a *frequency flat* channel). RMS delay spread increases when reflections and scattering induce larger propagation delay times in a channel, in which case the channel becomes more *frequency selective* in nature. In general, as shown in Section 3.7, directional antennas and the higher frequencies that will be used in mmWave communication systems will offer smaller multipath delay spreads and less severe frequency selective fading than conventional UHF/microwave systems that use omnidirectional antennas.

3.3.2 Doppler Effects

The Doppler effect of a traveling wave can be explained as follows. For a receiver with velocity v and a transmitter with velocity v_0 traveling toward each other, the perceived frequency f' of the traveling wave at the receiver is given by

$$f' = \frac{c+v}{c-v_0} f \qquad (3.23)$$

3.3 Small-Scale Channel Effects

where f is the frequency of the signal transmitted. Hence, from (3.23), the Doppler frequency (change in frequency at the receiver), $f_d = f' - f = f\left(\frac{c+v}{c-v_0} - 1\right)$, is proportional to the transmitted wave frequency. Therefore, as the carrier frequency is increased in wireless systems, motion causes magnified Doppler effects. As a result, we anticipate that the Doppler effect might be 15-30 times greater at 28-60 GHz compared with microwave wireless systems. This effect was verified in [TM88], in which a receiver moving at speeds of 1 m/s in a stable, indoor environment observed Doppler frequencies around 250 Hz for 60 GHz carrier frequencies. The Doppler frequency calculation in (3.23) results in $f_d = 60 \times 10^9 \left(\frac{3 \times 10^8 + 1}{3 \times 10^8}\right) = 200$ Hz. For bullet train and airplane speeds, Doppler shifts will exceed $\pm$ 10 KHz about a 60 GHz carrier frequency. Note that historic models for Doppler spread are based on omnidirectional antenna assumptions derived by Clarke and Gans [Rap02], but the use of highly directional antennas at mmWave frequencies provide fading characteristics that are highly dependent on specific angles of arrival [DR98][DRD99][DR99a][DR99b][DR99c][DR00][DRD02][Dur03]. Highly directional antennas provide a small number of specular signal paths with diffuse multipath as described by the *Two Wave with Diffuse Power* (TWDP) distribution [DRD99][DRD02][SB13]. See also Section 3.8.2.

The Doppler effect results in new challenges for the design of the physical layer. There is an approximately proportional relationship between the Doppler effect and the time-varying nature of the wireless channel [Rap02], but the precise signal envelope fading distribution is no longer Rayleigh, but depends on the particular beamwidths and finite multipath signals, and thus the fading and time coherence of mmWave channels will depend on beamwidth, velocity, frequency, and bandwidth [Dur98][Dur99a][Dur99b][Dur09c][Dur00][DRD99][DRD02][Dur03]. Early work suggests a fading probability distribution of signal envelope that is bi-modal, as reported in [DRD02] and [SB13] using Durgin's TWDP distribution. While more research is needed for the directional antenna case, a good first order estimate for the impact of Doppler spread on mmWave channels is that if the carrier frequency, and thus the Doppler frequency, increases by a factor of 10, then the wireless channel changes 10 times faster and requires channel retraining time and frame sizes to be reduced proportionally. As a result, the time-varying nature is more rapid for mmWave channels, requiring shorter frame times or packet durations in future wireless systems [RGBD+13], although research is needed to determine fundamental relationships that vary with antenna beamwidth and/or directional channel impulse response. Shorter synchronization times are not expected to be an insurmountable problem, symbol times for future mmWave systems will be smaller than today's systems, and Doppler spreads at mmWave frequencies are still slow for modern signal processing computing devices. Furthermore, the commensurate low-latency offered by shorter time slots and frames will be a needed feature for future mmWave systems [RSM+13]. In short, the time slots and frames will simply shrink linearly in time as the carrier frequency increases from today's UHF and microwave networks to the mmWave region, with the impact of directional antennas offering improved time coherence to a presently unknown degree.

As described in Sections 3.7 and 3.8, where small-scale propagation measurements for outdoor and indoor channels are analyzed and modeled, the extent of the Doppler spread of a channel is shown to govern the interval of time over which a channel appears static. The coherence time is inversely proportional to the Doppler shift, and as discussed in Section 3.7, there appears to be a distant-dependent relationship on the Doppler spectrum, and thus on the coherence time, in outdoor mmWave scenarios. The impact of antenna beamwidth promises to be an important factor for reducing signal fading and increasing time coherence. Determining the time intervals over which the channel is static is critical for properly sizing the frames and update rates for equalizers, coders, and beam steerers. Thus, the Doppler shift of wireless channels is critical information for exploiting small-scale fading effects, and the much smaller wavelength of mmWave signals results in a much smaller coherence time than conventional UHF/microwave systems.

3.4 Spatial Characterization of Multipath and Beam Combining

Works in [TM88] and [XKR00] were among the first to demonstrate the viability of directional, steerable antennas to find viable communication links at 60 GHz, and work in [SBM92], [XKR00], [LR99b], [LR96], [CR01a], and [CR01b] showed the value of using spatial division multiplexing to increase capacity in wireless systems. Recent work further confirms that adaptive antennas do indeed create viable links at mmWave frequencies [RBDMQ12][RQT+12][BDRQL11][RMGJ11][M+09] [MBDQ+12] [RGBD+13] [AWW+13] [SWA+13][BAN13][BAN14]. The spatial discrimination aspect of mmWave communications creates an entirely new dimension for wireless network design. Throughout this section we refer to non-line-of-sight (NLOS) and line-of-sight (LOS) environments or paths. An LOS environment is characterized by a direct non-obstructed path between the transmitter and receiver. An NLOS environment has the characteristic of having no direct unobstructed path from the receiver to the transmitter.

Consider the following experimental data, collected in New York City at 28 GHz. Using directional antennas and beamforming approaches at both the handset (RX) and transmitter (TX), carriers may hope to ensure that sufficient link margin is provided in even the most hostile NLOS environments in outdoor urban environments. Measurements made by one of the authors in New York City during summer 2012 with 175 dB of maximum measurable path loss [AWW+13] reveal the impact of beam combining, wherein the powers contained in various lobes or angular segments at the receiver are combined to obtain a higher received power level. By using "angular segment combining," the receiver may receive considerably more signal power than just utilizing one single angular segment seen by an antenna, thus fully exploiting the rich multipath diversity that exists in mmWave channels in the outdoor urban channel. These measurements provide a glimpse into achievable improvements of path loss (e.g., link budget) through the use of beam combining. Fundamentals of array steering are given in Chapter 4.

3.4.1 Beam-Combining Procedure

The success of beam combining at improving link budgets suggests that future mmWave cellular handsets are likely to have at least two antenna arrays, or single arrays that can be broken down to functional and independent sub-arrays. Beam combining was analyzed for steerable high gain antennas (24.5 dBi antennas at both the TX and RX, and 30 dBm applied to the TX antenna) at 28 GHz in NLOS channels in New York City, using spatial scanning over all 3-D antenna pointing directions at both the base

3.4 Spatial Characterization of Multipath and Beam Combining

station and mobile receiver. The RX locations were set on New York City streets, with a rotatable horn antenna situated at ear level for a typical person. The powers that were received when pointing at all of the different 10° angular segments (the 3 dB beamwidth of the steerable antennas), measured at 75 different receiver (RX) locations from 3 typical microcell base station transmitter (TX) locations in downtown New York City as part of a 28 GHz outdoor cellular measurement campaign in deep urban environments [AWW+13][RSM+13][SWA+13], were compared. At each RX location, we considered the improvement in received power that could be obtained if the strongest 10° segments could be combined, assuming that a RX beamforming antenna system could combine the strongest, two strongest, and three strongest signals that were received over 3-D space at all RX locations. As shown in Section 4.7, well-known methods exist to find the strongest directions of arrival at an antenna array. This study considered both coherent reception, in which the received powers are combined using known phase information from each formed beam (the optimal/maximum amount of power that could be combined), and also non-coherent beamforming, in which the powers in each of the received angular segments are combined non-coherently (i.e., the powers from the strongest angular segments were simply added, with no knowledge of phase, and with the assumption that the phases of each received signal are uniform and i.i.d. such that the powers may be simply added). This allowed us to compare the path loss scatter plots and resulting link improvement for different types of receivers [SMS+14][SR14].

3.4.2 Beam-Combining Results

Measurements in New York City, made using an 800 MHz first-null RF bandwidth pseudo noise (PN) sequence channel sounder with 10° rotatable horn antennas and 54.5 dBm EIRP, showed that 86% of all links (using just a single directional beam at both the transmitter and receiver) with a transmit-receive separation distance less than 200 m could be completed in heavy urban NLOS environments. Similar results were also found in Austin, Texas. Scatter plots of measured 28 GHz path loss are provided in Fig. 3.11. The baseline results are presented in Fig. 3.11a where the strongest signal measured from a single 10° beam antenna at the best pointing angle was determined over all RX locations. The circles and crosses denote the measured path loss values in LOS and NLOS environments, respectively. The values in the legend represent the path loss exponent and standard deviation (shadowing factor) in each environment. Note that LOS locations have higher path loss than free space (i.e., path loss exponent greater than 2.0) because the directional antennas were often not pointed directly at each other. Figs. 3.11c and 3.11d show the improvement in received power when the strongest two and strongest three pointing angles, respectively, are used to combine the received powers using coherent detection where the square root of the powers from each of the best beams were added, and the total power was then determined as the square of the resulting sum [SR13][SR14]. Fig. 3.11b shows the result for non-coherent combining of the two strongest beams, where the powers from each beam are summed.

Table 3.12 shows that the path loss exponent (PLE) drops dramatically as the number of combined signals increases from one to three, for both non-coherent and coherent combinations of beams at the mobile RX. It is worth noting that, as would be expected for the same number of combined signals, the PLE of coherent combining is substantially lower than the non-coherent case, about 3.4 dB/decade in distance for the case of

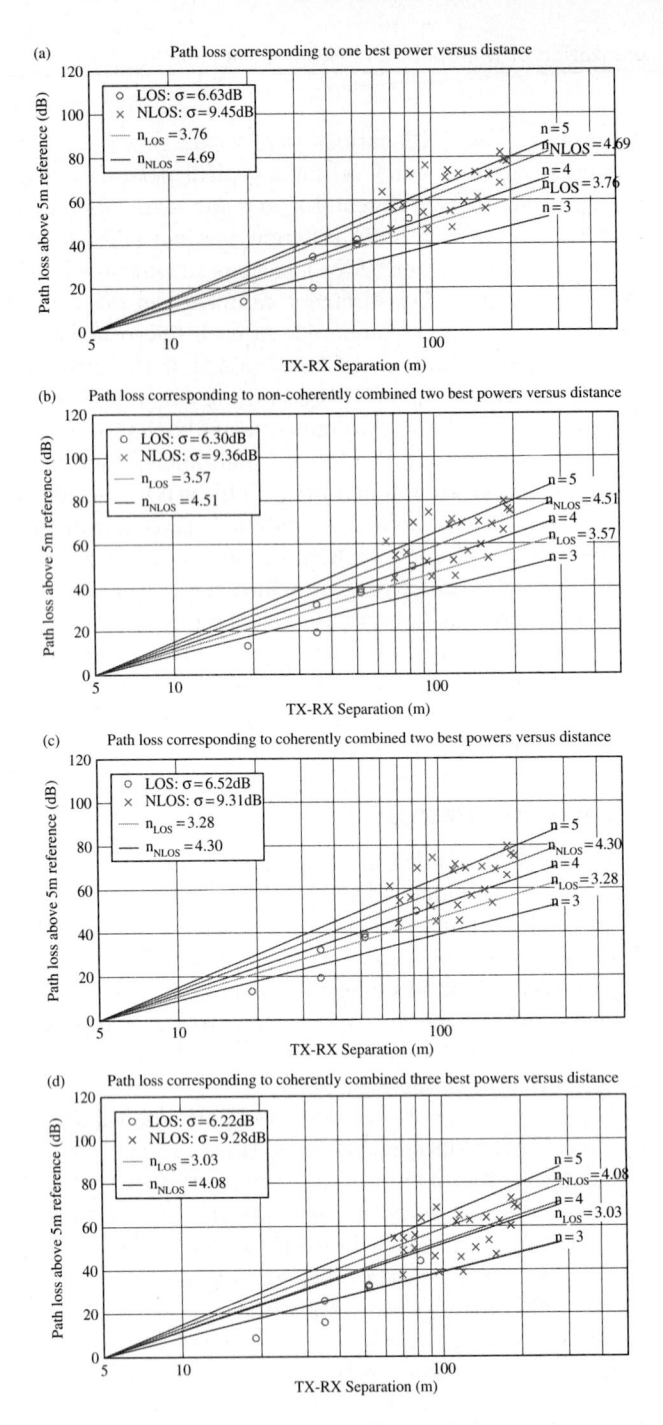

Figure 3.11 Scatter plots of measured 28 GHz cellular path loss in New York City [SR13][RSM+13][AWW+13][SWA+13]. The plots illustrate the reduction in path loss that can be achieved when a mobile handset using 10° steerable beams combines individual multipath signals arriving at different angles from the same transmitter. In (a), the single best beam pointing direction is used to make a link at each RX location. In (b), the two best beam pointing directions are non-coherently combined (where the powers in each unique beam are simultaneously added). In (c) and (d), the two and three best beams, respectively, are coherently added (where the total voltage in each unique beam is simultaneously added and then squared to produce power).

Table 3.12 Path loss exponents (PLEs) in 28 GHz urban channels decrease along with shadowing as the signals of different beams are combined at a receiver in both LOS and NLOS environments [from [SR13] © IEEE].

	Number of Combined Signals	Path Loss Exponent n	Decrease	Standard Deviation σ (dB)
	1	3.76	N/A	6.63
	2 (Noncoherent)	3.57	5.1%	6.30
	3 (Noncoherent)	3.48	7.4%	6.02
	2 (Coherent)	3.28	12.8%	6.52
LOS	3 (Coherent)	3.03	19.4%	6.22
	1	4.69	N/A	9.45
	2 (Noncoherent)	4.51	3.8%	9.36
NLOS	3 (Noncoherent)	4.42	5.8%	9.34
	2 (Coherent)	4.30	8.3%	9.31
	3 (Coherent)	4.08	13.0%	9.28

three-beam combining (where the PLE drops from 4.42 to 4.08 in NLOS channels), and 6.1 dB/decade in distance better than the case in which a single beam is used (where PLE is 4.69 for just a single beam). From (3.2), 6.1 dB/decade decrease in PL is shown to offer $\sim 40\%$ more coverage distance, and simulations have confirmed this [SNS+14]. It is interesting to note that 28 GHz path loss, while more lossy than pure $\alpha = 2$ free space conditions, is not drastically different from today's outdoor urban microcellular systems deployed at 1-5 GHz [BFR+92][HR92][RH92][H+94][BFR+94][RH91]. Not surprisingly, coherent combination of in-phase signals from different spatial angles gives rise to stronger received powers and offers considerable *range extension* compared to the use of a single beam antenna [SR14].

These measurement results show how beam combining can significantly improve the coverage distance of cells by improving the signal level of received signals. With lower PLEs and lower standard deviations about the mean path loss, better signal coverage and improved link margins may be provided for carriers, thus demonstrating the promise of beamforming antennas for future mmWave communications for both indoor and outdoor PAN, WLAN, and cellular applications.

3.5 Angle Spread and Multipath Angle of Arrival

Angular characterization of multipath propagation will be critical for mmWave systems that exploit adaptive arrays and high gain steerable antennas. To properly model the spatial characteristics of wireless channels, measurements must be made using antennas comparable to eventual system implementations. Work in [SWA+13], [RSM+13], [SR13], and [SR14] used rotational horn antennas to emulate phased arrays in future mmWave systems. There are different ways to characterize the spatial characteristics of multipath as a function of angle. Measurements in urban outdoor channels [SWA+13] [SR14a] show that there are generally a few distinct directions of arrival with significant energy, even in NLOS environments.[1] These directions of arrival (DOA) have distinct angular spread about the main directions, as illustrated in Fig. 3.12. Each unique direction has a lobe of energy that has a particular spread (in angle). Different locations provide a different number of lobes, different angle spreads, and varying power levels. It is useful to create statistics to describe the angular characteristics of the mmWave channel. Table 3.13 provides several examples of statistics that may be used to describe angular propagation

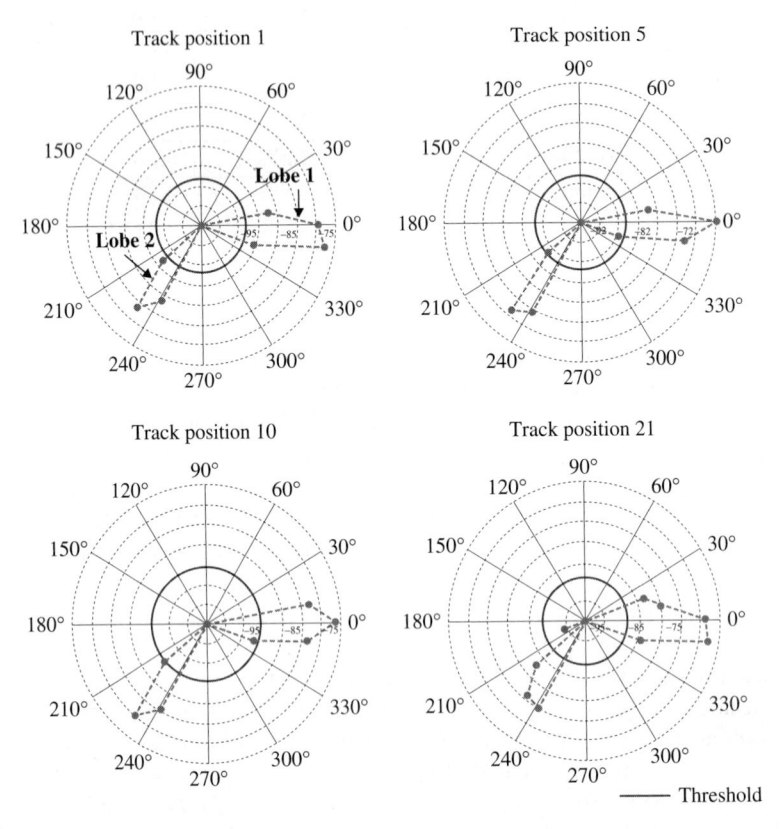

Figure 3.12 Four polar plots of 28 GHz propagation at track positions 1, 5, 10, and 21 along a 21-step linear track with $\lambda/2$ step sizes show two lobes of received power across azimuth. Measurements are for a partially obstructed NLOS RX environment in downtown Brooklyn using 24.5 dBi horn antennas at both the TX and RX. The TX was placed on the rooftop of NYU's Rogers Hall 135 m away from the RX. Each dot represents the received power level at a particular RX azimuth angle. For NLOS RX locations, a threshold of 20 dB below maximum power level was defined for a threshold (shown as a solid-line circle) to determine lobe statistics, whereas 10 dB was used for the LOS threshold [reproduced from [SWA+13] © IEEE].

---

1. See Table 3-13, where it is shown that there is an average of 2.5 distinct lobes in urban mmWave channels.

3.5 Angle Spread and Multipath Angle of Arrival

as well as measured results for microcellular mmWave 28 GHz systems in New York City [SWA+13]. Section 3.7 presents a complete spatial-temporal statistical channel model that incorporates angle spread information for the various lobes [SR14a].

Table 3.13 Summary of 28 GHz angular propagation statistics, the procedure used to compute angular statistics, their physical meaning, and the initial empirical distributions from New York City RX locations [reproduced from [SWA+13] © IEEE].

Statistics	Computation Procedure	Physical Meaning	Empirical Distribution in Manhattan
AOA	$\bar{\theta} = \frac{\sum_k P(\theta_k)\theta_k}{\sum_k P(\theta_k)}$	The mean direction of arrival of a lobe	Uniform $[0°, 360°]$
Lobe angle spread (LAS)	Apply threshold to polar plot, e.g., 10 dB of peak at LOS RX location, 20 dB for NLOS	Angle span of a lobe above a predefined threshold	Exponential $\mu = 40.3°$ $\sigma = 42.5°$
RMS LAS (standard deviation of lobe angle spread)	RMS LAS = $\sqrt{\bar{\theta^2} - (\bar{\theta})^2}$ where $\bar{\theta^2} = \frac{\sum_k P(\theta_k)\theta_k^2}{\sum_k P(\theta_k)}$	Angle span of lobe in which most power is received	Exponential $\mu = 7.8°$ $\sigma = 10.7°$
# of lobes for a particular RX location/antenna configuration	Number of lobes above predefined threshold	# of spatial directions from TX to RX	Exponential $\mu = 2.5$ $\sigma = 1.7$
Total power in a lobe for a particular RX location/antenna configuration	$\sum_k P(\theta_k)$ over consecutive k values where $P(\theta_k)$ above threshold	Total power in a lobe	Applies to each lobe at each RX for a particular antenna configuration at TX and RX
Average power in a lobe for a particular RX location/antenna configuration	$\frac{\sum_k P(\theta_k)}{k_{max}}$ over consecutive k values where $P(\theta_k)$ above threshold	Average power in a lobe	Applies to each lobe at each RX for a particular antenna configuration at TX and RX
Max power in a lobe for a particular RX location/antenna configuration	$\max_k P(\theta_k)$ over consecutive k values where $P(\theta_k)$ above threshold	Max power in a lobe	Applies to each lobe at each RX for a particular antenna configuration at TX and RX

Angle spread is a propagation metric that promises to have important relevance as future mmWave wireless systems exploit directional, adaptive antennas. Angle spread represents the spatial spread of the received signals that arrive through multipath propagation, as measured with respect to a mean angle of arrival or departure. As developed in [DR00], an angular spread of value 1 indicates a channel that does not favor any angle in space (i.e., propagation is received over the entire, omnidirectional, azimuthal directions). An angular spread of 0 indicates a channel is received in a single narrow beam. In [XKR02] and [CZZ03a], the authors characterized spatial properties of angular spread. It was discovered that the angular spread varied from 0.3 to 0.8 in indoor environments, meaning that reflections contribute a significant portion of the signal in a variety of directions. Outdoor scenarios had a reduced angular spread of 0.1-0.5. Power angle profiles show that reflections and large scatterers indeed contribute to the majority of received multipath.

Durgin made the important discovery that the direction of arrival, as well as the angle spread, of multipath energy at a receiver can be determined by using a relatively simple method of measuring and cross correlating the received signals at two closely spaced omnidirectional antennas [DR98][DR99a][DR99b][DR99c][DR00]. These results have implications for future blind beamforming algorithms, where narrowband pilot tones in a transmitted signal could serve as the transmission beacon by which adaptive beams at the receivers could constantly point toward the strongest multipath signals, while omnidirectional antennas at the receivers could be used to cross correlate the pilot signals in parallel, to provide pointing updates to the adaptive beam receiving antennas. Simple non-coherent envelope detection could be used on very small/inexpensive omnidirectional antenna arrays to provide direction steering for array antennas in real time using this theory. New models such as Durgin's TWDP distribution provide physical reasoning for the new distributions that are beginning to be found in practice [DRD09][DRD02][SB13].

Researchers have found the materials of the reflective elements (e.g., walls, building surfaces) are critical in determining the angular spread, and as mentioned above, at mmWave frequencies, the propagation environment is reflective and likely to produce stronger scattering, while providing more severe diffraction if large objects block signal propagation. Thus, the site-specific nature of the environment will give rise to a particular angle spread at a particular location [SWA+13][XKR00]. Many typical angle spread results are given in [XKR00] for typical office buildings. Recent research in outdoor 28 and 38 GHz channels show that base station microcells will only need a +/−30 degree span, but the mobile devices should have nearly omnidirectional search capabilities [RGBD+13]. The average angular spread in New York City is about 40° at the receiver, with approximately two to three main directions of arrival ("lobes") received occurring at random locations using receivers at typical human head height above ground [RSM+13][AWW+13][SWA+13].

3.6 Antenna Polarization

Polarization refers to the orientation of the electric and magnetic fields of the emitted or received electromagnetic waves. Circular polarization at mmWave frequencies can be used as an effective method to reduce multipath contributions, thus making circular antennas advantageous from the perspective of reducing, for example, equalization requirements while possibly also reducing the effectiveness of multi-beam combining. In [RH91], [HR92], [HR93], [H+94], [MMI96], and [SMI+97] and in later work [FLC+02], it was observed that circularly polarized signals may reduce RMS delay spread values by a factor of two over linearly polarized signals, although there is virtually no distinction between the link performance of various linear polarizations [ZGV+03]. This can be attributed to the change from left-handed to right-handed polarization that a circularly polarized wave experiences upon reflection from a material, reducing susceptibility to even order reflections, as well as lower overall reflection coefficients for circularly polarized waves.

Although polarization mismatches between antennas and the waves propagating in the channel may serve to reduce RMS delay spreads, they can also substantially reduce the strength of LOS reception. For example, in [MMS+10], [MMSKL09a], [MPMSKL10], and [Mal10], Intel Russia found that polarization mismatches can result in lower received power by 10-20 dB.

Work suggests that in dry weather, or inside buildings, it may be possible to use both horizontal and vertical polarization on the same propagation path and obtain sufficient discrimination so that two orthogonal signals could exist [RH91][RH92][H+94][AWW+13][RGBD+13]. More than 20-30 dB of isolation has been observed by the authors on the same propagation links in outdoor urban channels when using high gain horn antennas with simultaneous vertical and horizontal polarization in outdoor LOS mmWave channels at 28 and 73 GHz. It may be possible to exploit MIMO concepts in concert with polarization agile antennas to achieve spatial multiplexing gains. The impact of moisture (see Figure 3.4), moving crowds, and surface roughness of common objects and materials must be fully understood to determine worst-case situations and the potential for antenna polarization diversity in future mmWave networks, and this is an open research area.

3.7 Outdoor Channel Models

There has been comparatively little study of mmWave outdoor propagation compared to study of indoor propagation scenarios, but researchers in [BDRQL11], [RQT+12], [MBDQ+12], [RBDMQ12], [AWW+13], [SWA+13], [RGBD+13], and [RSM+13] have used a sliding correlator channel sounder to capture small-scale (i.e., closely spaced impulse responses) and large-scale channel data (i.e., path loss and beamforming link outage data) in outdoor environments.

The sliding correlator system essentially uses spread spectrum techniques and does not require a cable to synchronize the phase of the transmitter and receiver, as is required for Vector Network Analyzer (VNA) measurement systems that are generally confined to indoor locations [BDRQL11]. This makes the sliding correlator system easy to use in outdoor urban environments, where providing a phasing cable would be impractical. The sliding correlator requires a sufficiently small chip duration (i.e., sufficiently high chipping code rate, where the chip code is the pseudo-random code that is mixed with the RF carrier to spread the transmitted signal in the frequency domain) to obtain sufficient temporal multipath resolution, and post detection averaging may be used with the processing gain to provide substantial link margin improvement at the expense of real-time measurements (e.g., typically, a power delay profile is measured over a couple of seconds, to provide for sufficient averaging). Works cited earlier used a chip code of 400-800 Megachips/s, providing temporal multipath resolution of 1.25-2.5 ns [BDRQL11] [RBDMQ12] [AWW+13] [SWA+13] [MBDQ+12] [RQT+12] [RGBD+13] [RSM+13]. The block diagram of the popular sliding correlator, and its method of operation, are described in [Rap02] and [NRS96].

Knowledge of the outdoor mmWave channel will be increasingly important as industry seeks to take advantage of the massive amounts of untapped mmWave spectrum that can support very high data rates to mobile users in outdoor environments, for example, for cellular and peer-to-peer applications. Some of our discussion is based on the results of extensive channel sounding campaigns at 28, 38, and 60 GHz conducted on the campus of the University of Texas at Austin over the summer of 2011 and extensive measurements on the campus of New York University in New York City during the summers of 2012 and 2013 (and still ongoing). In practice, the 28 and 38 GHz results should apply fairly well to 60 GHz channels because there is less than an octave difference between the two frequencies [AH91], except that the oxygen absorption attenuation factor must be applied, such that 60 GHz waves will be more heavily attenuated than 28 or 38 GHz waves due to oxygen absorption. We refer the reader to [RBDMQ12], [RQT+12], [BDRQL11], and [MBDQ+12] for an overview of the channel measurements and hardware used in the Texas measurements. At a high level, the 38 GHz measurement campaigns used 13.3 dBi 49.4° half power beamwidth (HPBW) or 25 dBi 7.8° HPBW receive antennas and 25 dBi RX antennas, with an EIRP of approximately 47 dBm. At 38 GHz, a BPSK spread-spectrum 800 MHz RF-bandwidth sequence was used to sound the channel, while at 60 GHz a 1.5 GHz RF bandwidth signal was used. For the 38 GHz campaign, both the cellular and peer-to-peer channels were studied. The channel sounder used in the campaign was fully quadrature, allowing for determination of both phase and amplitude information, whereas for the 60 GHz campaign primarily peer-to-peer outdoor channels and measurements into vehicles were studied. In both cases, the impacts of antenna pointing angle and transmitter-receiver separation were studied.

Indoor and outdoor environments at 60 GHz were originally found to be moderately frequency selective, with fades of 15-20 dB occurring over a wide bandwidth [PPH93]. Frequency selectivity indicates that the channel's frequency response is not flat when swept across the band of interest. Indeed, measurements performed at the University of Texas at Austin for 38 GHz cellular channels, in which the transmitter is substantially elevated relative to the receiver, indicate that deep fades occasionally occur in the passband. For example, Fig. 3.13 shows the spectral representation of a non-line-of-sight (NLOS) link measured for the 38 GHz cellular channel over a distance of 70 m and with the transmitter (25 dBi) and receiver (13.3 dBi) antennas pointing off bore sight (the azimuth/elevation angles for each antenna are indicated in the figure). Although deep fades can occur, in general the channel is not severely frequency selective. This is indicated by the cumulative distribution function (CDF) of the individual gains of 1 MHz sub-carriers ± 222 MHz from the carrier frequency at 38 GHz relative to the average channel gain for each measurement. Such a distribution is given for both 13.3 dBi and 25 dBi RX and 25 dBi TX antennas. A more frequency selective channel would have a more gradual slope (i.e., a longer tail in the fading distribution) in Fig. 3.14 on the transition from 0% probability to 100% probability.

3.7 Outdoor Channel Models

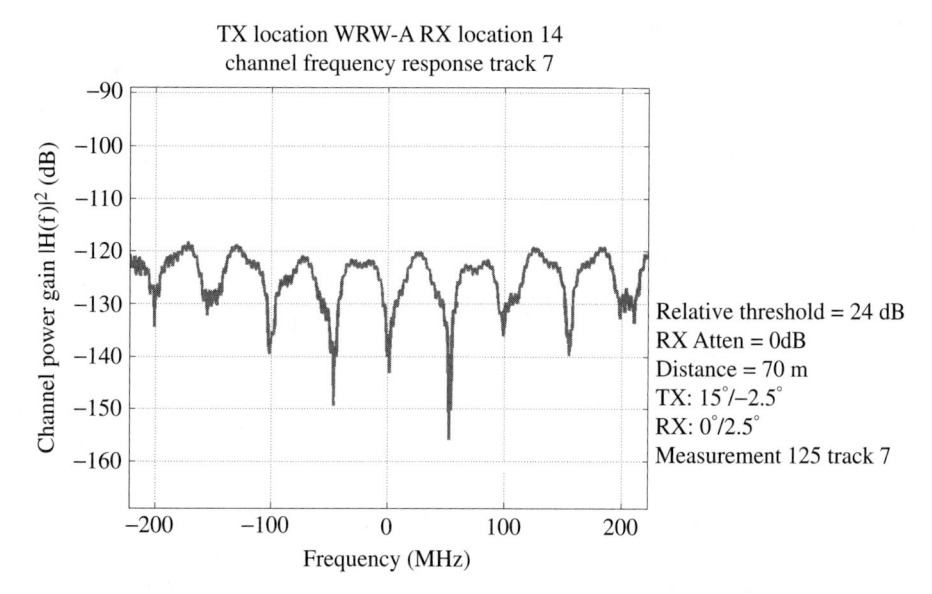

Figure 3.13 Frequency-selective fading occurs about the 38 GHz carrier frequency in outdoor urban NLOS channels. Note the periodic 50 MHz fades in frequency about the carrier correspond to a RMS delay spread that is approximately 20 ns. Here we see a channel that has deep fades as low as 30 dB from the peak channel gain.

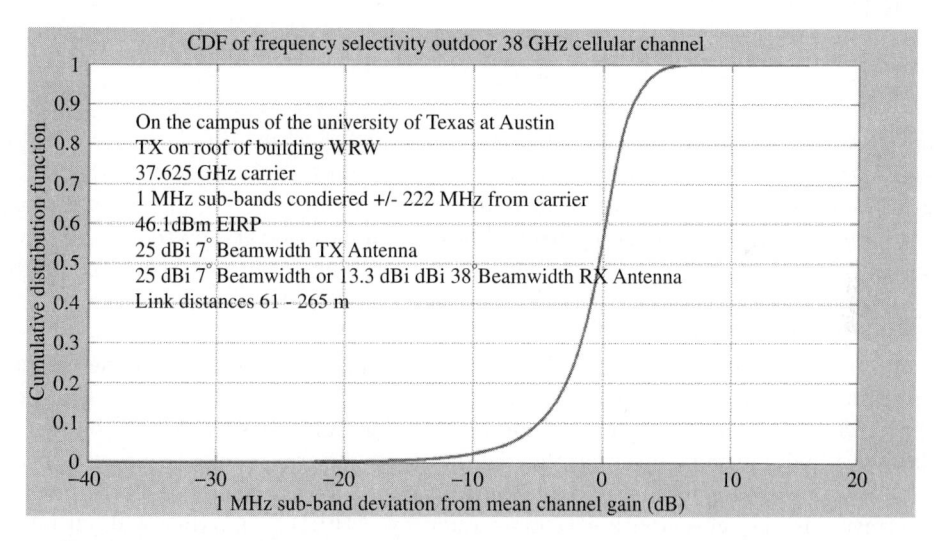

Figure 3.14 When a channel frequency representation such as that shown in Fig. 3.13 is considered over 1 MHz subbands (i.e., we evaluate the average channel gain at 1 MHz intervals and compare these small intervals to the overall average channel gain across the band), we see that fading is not severe for outdoor urban cellular mmWave channels. The time delay spread and the number of resolvable multipath components directly contribute to the fading characteristics across the occupied spectrum. Directional antennas change small-scale fading from today's common Rayleigh fading characteristics (for omnidirectional antennas) into much narrower fade depths over much wider frequency bands.

Fig. 3.14 indicates that most 1 MHz sub-carriers are within 5 dB of the mean channel gain. The transmitter location for the measurements corresponding to Fig. 3.14 was on the building WRW (W. R. Woolrich laboratory of The University of Texas at Austin), approximately 18 m above the ground with the receiver at ground level. Note that the height of the transmitter used in this measurement campaign may reduce the frequency selectivity of the channel. Indeed, [RBDMQ12], [RQT+12], [BDRQL11], and [MBDQ+12] found that RMS delay spreads increase with lower transmitter antenna heights, which in turn indicates that the channel will be more frequency selective, as indicated by the reduced coherence bandwidth [Rap02]. The coherence bandwidth is defined as the frequency bandwidth over which two continuous waves (CW waves) are highly correlated, and is approximated by

$$B_c \approx \frac{1}{5\tau_{\text{RMS}}}. \quad (3.24)$$

Note that (3.24) is only approximate. All else being equal, moving to mmWave carrier frequencies will result in less frequency selectivity due to the smaller RMS delay spreads resulting from highly directional antennas and much less diffraction. Note modern modulation techniques such as OFDM can greatly reduce sensitivity to frequency selectivity by judiciously selecting portions of the channel that are not severely faded.

Temporal fading is different from frequency selectivity and is the result of movement by the transmitter, receiver, or objects in the channel. The Doppler shift discussed in Section 3.3.2 is related to temporal variations in the channel according to the coherence time [Rap02] defined as

$$T_c \approx \frac{1}{f_d}, \quad (3.25)$$

where f_d is the maximum Doppler shift and T_c is the coherence time, and indicates the longest interval over which the channel is relatively constant in time. The short wavelength of mmWave signals results in increased Doppler shift, that is, increased temporal fading. Temporal fading is generally studied by examining spectral representations of variations in received power versus time. Such a plot will have frequency components that extend from $-v/\lambda$ to v/λ, where v is the speed of movement of the receiver/transmitter or objects in the channel and λ is the carrier wavelength. [AH91] studied narrowband fading characteristics and found that fading is strongly impacted by the ratio of the strongest received ray (generally from the LOS path) to the next strongest ray — the ratio should be within 5 dB in order for significant fading to occur. As a consequence of this, temporal fading is highly dependent upon the distance from the transmitter/receiver to nearby reflectors and will also be highly dependent on antenna beamwidth. If the transmitter and receiver are very close (within 17 m as found by [AH91]) then fading will not be very severe. If the transmitter and receiver are more distantly spaced and near large walls or obstacles such as buildings, then there is much more severe fading. This conventional fading is generally well described by a Rayleigh distribution, discussed in Section 3.8.2, but with the advent of directional antennas that receive only a few specular components, the TWDP distribution is more realistic [DRD02]. Note that if the signal bandwidth is significantly greater than the Doppler spread, then channel temporal fading will be

3.7 Outdoor Channel Models

negligible and the channel may be assumed to be static over many transmitted symbols (e.g., slow fading) [Rap02]. This is one advantage of wideband systems. The increased Doppler shift at mmWave frequencies offers an incentive to use very wide bandwidths and directional antennas at these frequencies in order to avoid excessive fading and temporal selectivity over thousands of consecutive symbols.

Spatial fading results from movement by the receiver or transmitter and is closely linked to temporal fading. Spatial fading is experienced as variations in the received power at the receiver as it moves over short distances. As discussed by [Rap02], very wideband systems will generally experience less spatial fading than narrowband systems. It is worth repeating a derivation found in [Rap02] that describes the impact of signal bandwidth on the fading characteristics experienced by a signal in the channel. First, as in [Rap02], we assume that the channel is excited by a repetitive pulse train $p(t)$ with pulse duration T_{bb} whose repetition period is much greater than the maximum excess delay of the channel $\tau_{\max} = \tau_L$

$$p(t) = \sqrt{\frac{\tau_{\max}}{T_{bb}}} \sum_{i=-\infty}^{\infty} \Pi\left(\frac{t - k_i T_{\text{rep}}}{2T_{bb}}\right) \qquad (3.26)$$

where T_{rep} is the repetition period and is much greater than the length of the channel impulse response, and the amplitude of $p(t)$ is chosen to normalize the energy relative to a CW signal [Rap02] and

$$\Pi(x) = \begin{cases} 1 & -\frac{1}{2} \leq x \leq \frac{1}{2} \\ 0 & \text{otherwise} \end{cases} \qquad (3.27)$$

is the rectangular function. As in [Rap02], we convolve (3.21) and (3.26) to find the received signal $r(t)$ as passed through the channel

$$r(t) = \sum_{\ell=0}^{L} \alpha_\ell e^{-j\theta_\ell} p(t - \tau_\ell). \qquad (3.28)$$

Over a single repetition period of the pulse train $p(t)$, (3.28) may be written as

$$r(t) = \sqrt{\frac{\tau_{\max}}{T_{bb}}} \sum_{\ell=0}^{L} \alpha_\ell e^{-j\theta_\ell} \Pi\left(\frac{t - \tau_\ell}{2T_{bb}}\right). \qquad (3.29)$$

The instantaneous power delay profile is then found by correlating $r(t)$ to find the squared amplitude of $r(t)$ evaluated at a particular time t_0 from time t_0 to time $t_0 + \tau_{\max}$ and dividing by $\tau_{\max}$ to give [Rap02]

$$|r(t_0)|^2 = \frac{1}{\tau_{\max}} \int_{t_0}^{t_0+\tau_{\max}} r(t) r(t)^* dt \qquad (3.30)$$

$$= \frac{1}{\tau_{\max}} \frac{\tau_{\max}}{T_{bb}} \int_{t_0}^{t_0+\tau_{\max}} \sum_{k=1}^{L} \sum_{i=1}^{L} \alpha_i \alpha_k e^{-j\theta_i + j\theta_k} \qquad (3.31)$$

$$\cdot \Pi\left(\frac{t - \tau_i}{2T_{bb}}\right) \Pi\left(\frac{t - \tau_k}{2T_{bb}}\right) dt.$$

If the rectangular pulses are sufficiently narrow in time, that is, sufficiently broad in bandwidth relative to the bandwidth of the channel, then two rectangular pulses never overlap unless $i = k$, revealing that the overall received power at a given instant in time is

$$|r(t_0)|^2 = \frac{1}{T_{bb}} \int_{t_0}^{t_0+\tau_{\max}} \sum_{i=0}^{L} \alpha_i^2 \Pi\left(\frac{t-\tau_i}{2T_{bb}}\right) dt$$

$$= \frac{1}{T_{bb}} \sum_{i=0}^{L} \int_{\tau_i - \frac{T_{bb}}{2}}^{\tau_i + \frac{T_{bb}}{2}} \alpha_i^2 \, dt \quad (3.32)$$

$$= \sum_{i=1}^{L} \alpha_i^2.$$

Now, the average power of a narrowband CW signal in a local area is equal to the ensemble average of the sum of the powers of the individual resolvable multipath components, each having width (time bins) T_{bb} as found in a PDP. In other words, when $p(t)$ is DC in (3.28),

$$\mathbb{E}_{a,\theta}\left[|r(t)|^2\right] = \mathbb{E}\left[\sum_{i=0}^{L} \left|\alpha_i e^{-j\theta_i}\right|^2\right] \quad (3.33)$$

$$\approx \sum_{i=0}^{L} \overline{\alpha_i^2}. \quad (3.34)$$

Equation (3.34) shows that the average received power for a narrowband CW signal and the ensemble average of the received power as computed by the sum of the powers of the individual multipath components in a wideband channel are approximately equal over a local area (provided the individual multipath components have sufficient time resolution such that the channel multipath is resolved and the multipath components do not fluctuate greatly). This suggests that wideband systems will not experience rapid spatial fading over small distances. In fact, this is confirmed by small-scale mmWave measurements in [RSM+13]. When (3.28) is used for a narrowband signal that cannot resolve the channel, the result indicates that narrowband systems will experience much more severe local area fading [Rap02]. However, if the channel is rank one (i.e., $L = 1$), (3.34) shows that no fading occurs for the narrowband signal.

To summarize the discussion on fading: Movement toward higher frequencies will in general result in increased temporal and spatial fading due to Doppler, but these effects are mitigated by using wider-band systems and directional antennas. Wideband systems will in turn experience more frequency selectivity over the wider bandwidths, but modern modulation techniques such as OFDM are designed to counteract or exploit these effects. Furthermore, outdoor mmWave systems are likely to use highly directional antennas that provide spatial filtering that can be used to mitigate frequency selectivity that results from interference between multipath components.

RMS delay spreads for outdoor mmWave channels have been found to be moderate in most cases. [DCF94] found spreads under 100 ns for 88% of measurements. [RBDMQ12], [RQT+12], [BDRQL11], and [MBDQ+12] studied RMS delay spreads for outdoor cellular

3.7 Outdoor Channel Models

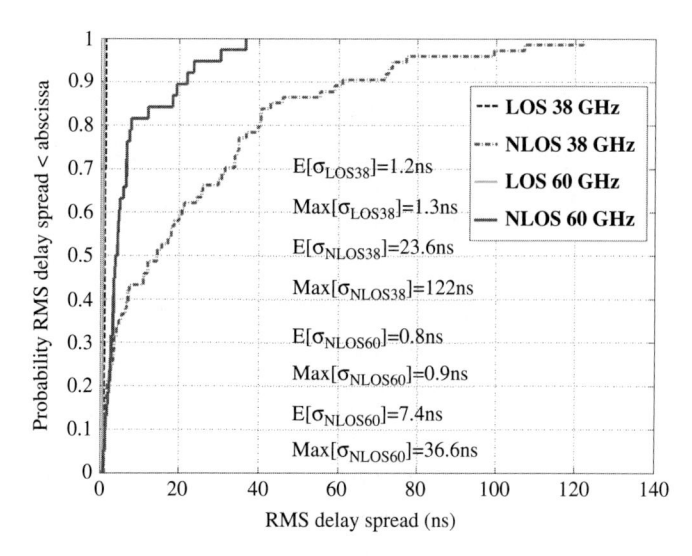

Figure 3.15 Differences in RMS delay spread and their distribution at 38 and 60 GHz in various outdoor environments [from [RBDMQ12] © IEEE].

and peer-to-peer applications at 38 and 60 GHz. Fig. 3.15 shows results for peer-to-peer applications with link distances ranging from 19 to 129 m measured with highly directional rotatable 7° 25 dBi horn antennas at the transmitter and receiver. The plot includes both line-of-sight (LOS) links and NLOS links. Fig. 3.15 confirms that mmWave systems with highly directional antennas will experience only moderate RMS delay spread (typically not greater than 130 ns for NLOS paths, and only a few ns for LOS paths). The difference in the RMS delay spread seen at 38 and 60 GHz reflects the difference in scattering capabilities of environmental objects at the two frequencies, as well as the oxygen absorption at 60 GHz.

To highlight the importance of transmitter antenna height, Fig. 3.16 shows results for cellular type measurements at 38 GHz with the transmitter ranging in height from 8 to 36 m using a 25 dBi gain TX antenna and a 13.3 dBi or 25 dBi RX antenna. Comparing Fig. 3.15 to Fig. 3.16 at the 90% likelihood level indicates that greater transmitter antenna heights will result in decreased RMS delay spread, although tall TX antenna heights produced the overall worst-case delay spreads. Interestingly, Fig. 3.16 also demonstrates that for highly elevated transmitter antennas, there is little impact of the beamwidth of the receiver antenna on the RMS delay spread for RX beamwidths less than approximately 40° [RGBD+13].

When highly directional antennas are used in outdoor scenarios, we must study how antenna pointing angle impacts link performance [RBDMQ12]. Fig. 3.17 shows a scatter plot of azimuth pointing angles for both 60 and 38 GHz outdoor *peer-to-peer* channels found using highly directional steerable antennas. The wide range of transmitter and receiver azimuth angles indicates that applications in which the transmitter and receiver are close to the ground will provide a wider range of incoming ray directions at the receiver than applications in which the transmitter is substantially elevated. This is easily seen by comparing Fig. 3.17 to Fig. 3.18, which shows the antenna pointing angles that were used

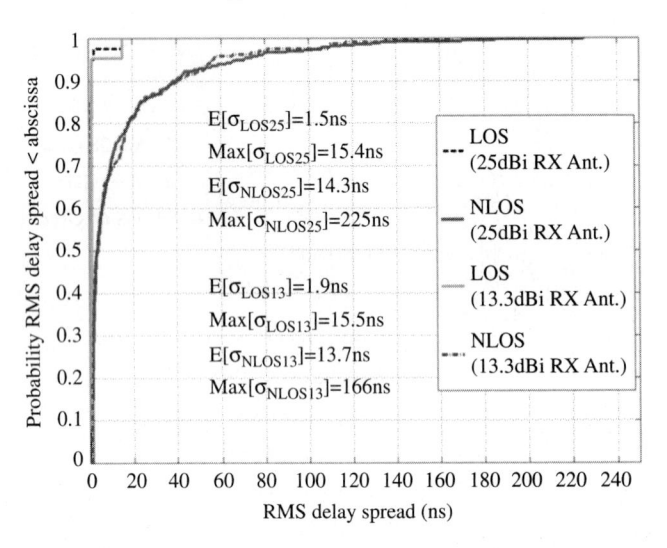

Figure 3.16 Greater transmitter antenna heights resulted in decreased 90% RMS delay spread compared with situations in which the 38 GHz transmitter is near the ground, and the worst-case RMS delay spread was found to be 225 ns on a Texas college campus using the tallest transmitter antenna height [from [RSM$^+$13] [RGBD$^+$13] © IEEE].

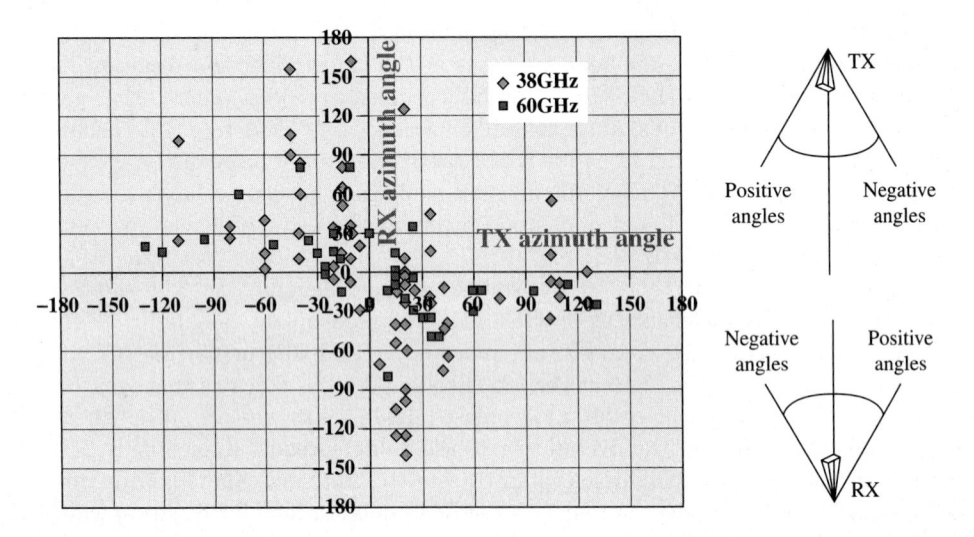

Figure 3.17 MmWave applications in which the transmitter and receiver are close to the ground (such as peer-to-peer or vehicle-to-vehicle) will provide a wide distribution of angles at which links may be established [from [RBDMQ12][RGBD$^+$13] © IEEE].

for a single transmitter location for the 38 GHz cellular channel in which the transmitter was 18 m above the ground for link distances from 61 to 265 m. The plots indicate that the range of transmitter angles at which a link may be formed decreases as the height of the transmitter increases [RBDMQ12][RGBD$^+$13].

3.7 Outdoor Channel Models   **147**

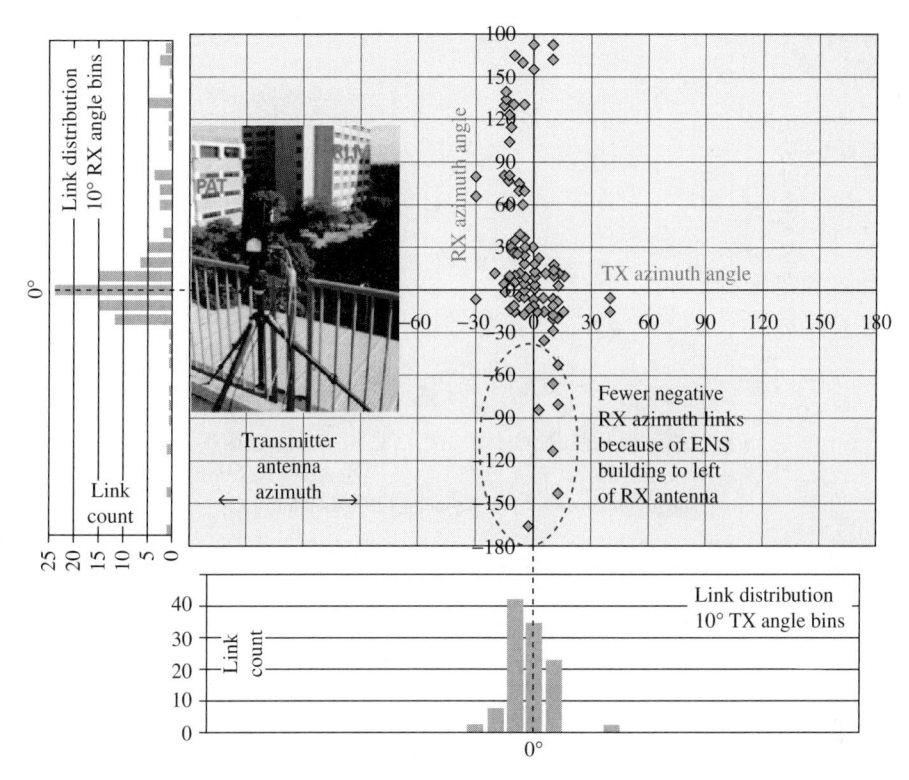

Figure 3.18 The antenna pointing angles found with a 37.625 GHz carrier and highly directional antennas at the receiver and transmitter. The transmitter was elevated to 18 m [from [RBDMQ12] [RGBD+13] © IEEE].

It can be seen in Figs. 3.17 and Fig. 3.18 how the height of the antenna impacts the angular spread of multipath energy for mmWave cellular. Fig. 3.18 shows that the base station antenna, located 18 m above street level, provides most of its energy within a ±30° beamwidth relative to boresight to the RX location. Similar angle spreads were found for 28 GHz transmitters ranging in height from 7 to 17 m in New York City [RSM+13]. This means that base station antennas in future outdoor mmWave cellular systems will not need to consider beamforming more than a typical 60° sector width. However, Fig. 3.17 shows that as the base station is lowered closer to street level, the sector width will need to increase.

The antenna pointing angle has an important impact on the RMS delay spread measured with mmWave systems. In general, steeper (i.e., away from boresight) pointing angles will result in higher RMS delay spreads. This was found at both 38 and 60 GHz by [RBDMQ12], [RQT+12], [RGBD+13], and [BDRQL11]. Fig. 3.19 illustrates the impact of the total pointing angle (sum of the absolute azimuth angles off boresight for both the transmitter and receiver, where the boresight of an antenna corresponds to the antenna pointing directly to the other antenna). Fig. 3.20 shows their results for the 38 GHz cellular channel, including the impact of the elevation and azimuth angles. Figs. 3.17, 3.18, 3.19, and 3.20 demonstrate two very important effects for mmWave outdoor channels: A

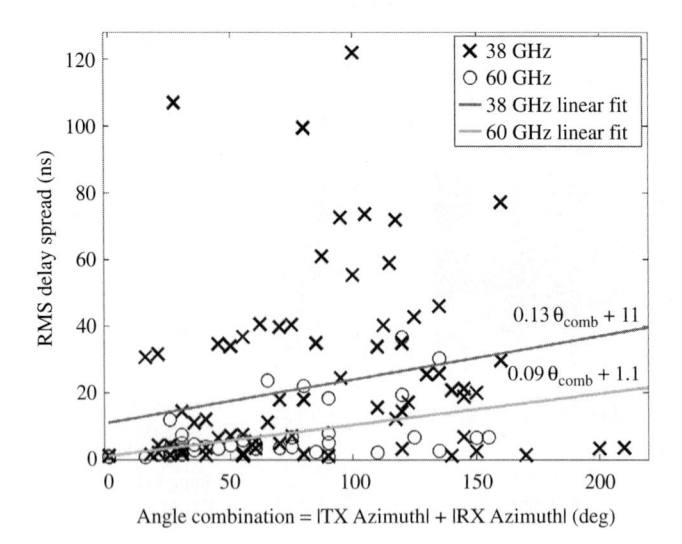

Figure 3.19 Steeper azimuth pointing angles are associated with higher RMS delay spreads for outdoor peer-to-peer channels. The measurements from this plot were taken with 25 dBi 7° beamwidth horns at the transmitter and receiver, and with link distances from 19 to 129 m [from [RBDMQ12] © IEEE].

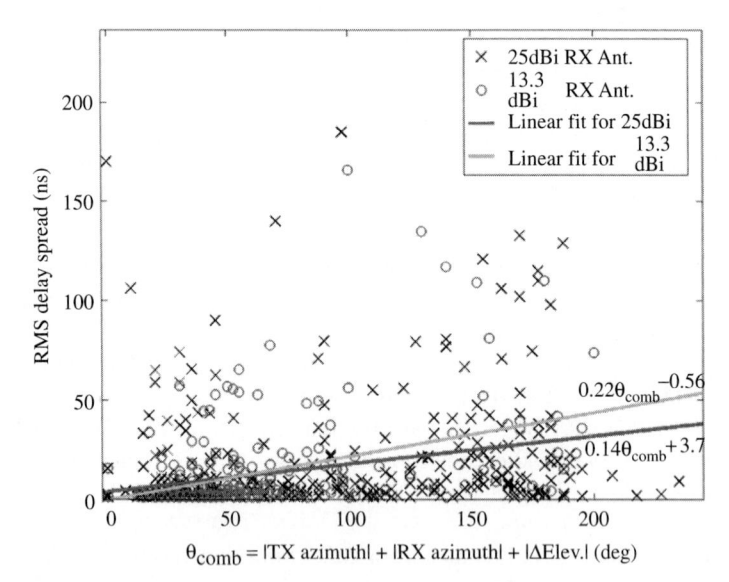

Figure 3.20 Steeper antenna pointing angles are associated with higher RMS delay spreads. These measurements were taken at 38 GHz with at 25 dBi TX antennas, and a 25 dBi or 13.3 dBi RX antennas. Link distances ranged just beyond 900 m [from [RQT+12] © IEEE].

diversity of paths can be found in most cases, indicating that beam steering and beam combining can enable reliable communication, and RMS delay spreads will be higher for steeper antenna pointing angles. This motivates the development of beam-steering algorithms.

3.7 Outdoor Channel Models

There are several popular methods to model outdoor mmWave propagation. One of the most popular methods is ray tracing, which uses geometric optics to simulate the propagation and reflection of waves in an environment. Generally, only two to four reflections per wave are considered, and if the transmitter and receiver are at the same height and surrounded by much taller buildings, then a two-dimensional approach may be taken [CR96] [SC97b]. When simulating wave propagation, it is important to consider the path loss exponent, which describes how received power diminishes as a function of the distance between the transmitter and receiver. In many current cellular systems, a break-point model is used for LOS path loss in which the path loss exponent is 2 (free space) before what is called the *break-point distance*, and 4 afterward [Rap02]. For a transmitter at height h_t and a receiver at height h_r, the break-point distance d_bp is defined as [Rap02][FBRSX94]

$$d_\mathrm{bp} = \frac{20 h_\mathrm{t} h_r}{\lambda}. \tag{3.35}$$

Equation (3.35) indicates that the break-point distance decreases with increasing frequency. For example, if the transmitter and receiver are 1.5 m above the ground, then at 60 GHz the break-point distance is 9 km, far longer than the link distance likely to be in practice. Therefore, for LOS links at mmWave frequencies (here antennas are pointing at each other, or for omnidirectional models) we anticipate that the path loss will almost always be near that of free space. This was found by [RBDMQ12], [RQT+12], and [SC97b], and earlier by Feurestein and Blackard [BFR+92] [FBRSX94]. Fig. 3.21 from [RBDMQ12] shows the measured path loss for the outdoor 60 GHz peer-to-peer channel, and shows that for LOS paths the path loss was very close to that of free space. Fig. 3.22 from [RBDMQ12] for outdoor 38 GHz peer-to-peer applications shows that this is also true at 38 GHz. Figs. 3.23 and 3.24 from [RQT+12] show measured path loss values for 38 GHz outdoor cellular channels, confirming the near-free space-quality of LOS links at these frequencies as well. Comparing Figs. 3.23 and 3.24, we see that NLOS links are often stronger (i.e., have lower path loss exponents) with lower-gain antennas. This is due to the fact that more energy is captured from many angles with a wider beam antenna than with a very narrow beam antenna. The plots also indicate the path loss exponents that would be found if only the strongest NLOS paths (e.g., single best pointing angle) at particular transmitter-receiver distances are considered. The lower path loss values for these paths indicate the great benefit that can be derived from intelligently steering the transmitter and/or receiver antenna or adapting the gain/beamwidth of an antenna to accommodate a wider field of view (less gain) when the link has sufficient signal, and using a narrower field of view (more gain) when rain or obstructions require greater power from the link. [SC97b] found that antenna steering can significantly improve link quality, by as much as 8.4 dB improvement in signal strength.

3.7.1 3GPP-Style Outdoor Propagation Models

To properly develop global standards for future mmWave wireless systems, the engineering community requires propagation models that allow engineers to compare different system aspects of wireless communications. While indoor mmWave wireless standards have

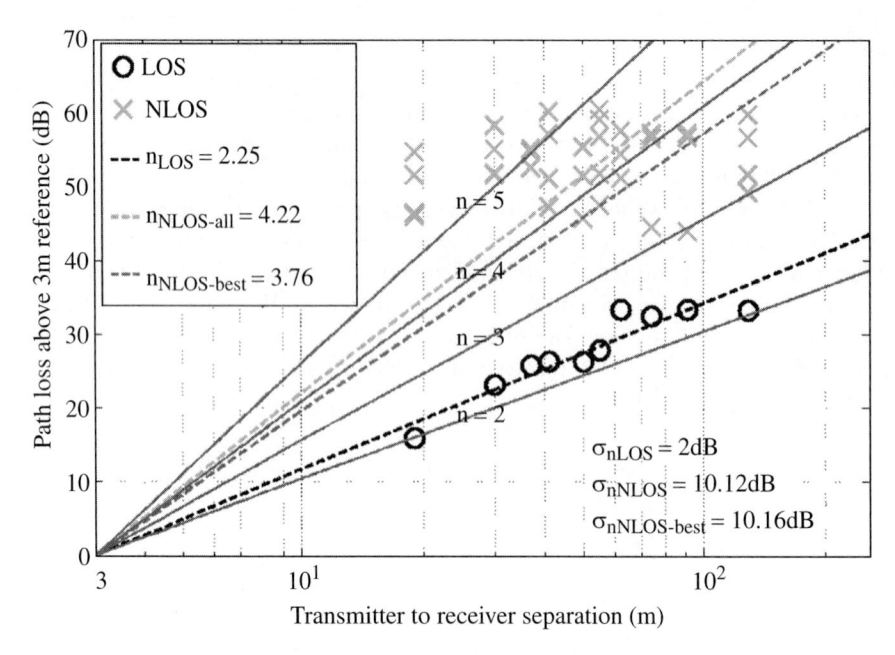

Figure 3.21 Due to the very high value for the break-point distance, LOS links at mmWave frequencies are very close to free space in terms of path loss. This plot was generated with highly directional antennas at the receiver and transmitter with 25 dBi gain and 7° beamwidths at 60 GHz [from [RBDMQ12] © IEEE]. Note that the oxygen absorption causes the path loss exponent to be slightly greater than 2.0.

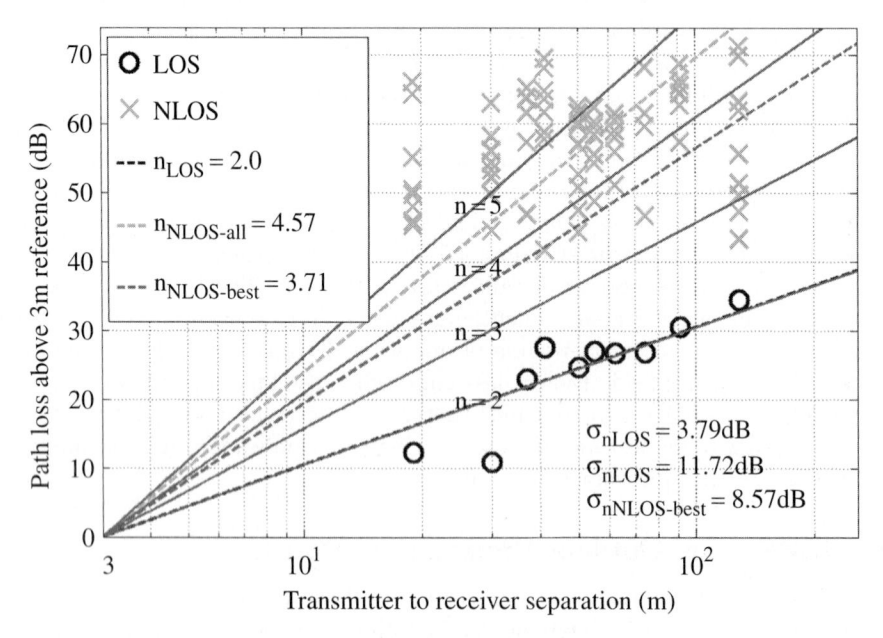

Figure 3.22 This plot shows measured path loss values for 38 GHz peer-to-peer applications with highly directional 25 dBi 7° beamwidth horn antennas [from [RBDMQ12] © IEEE].

3.7 Outdoor Channel Models 151

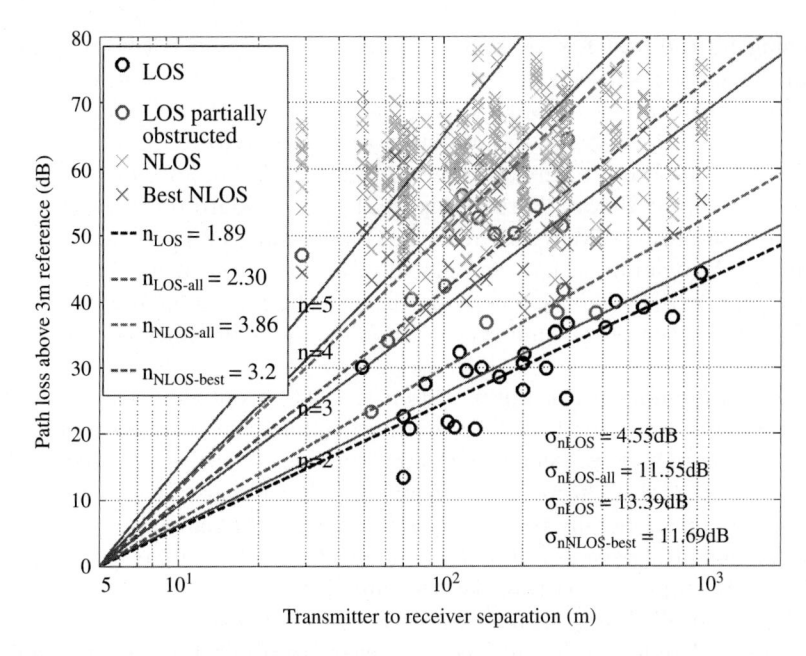

Figure 3.23 When a highly directional antenna is used at the receiver, LOS links will be very close to free space but NLOS links may be more heavily attenuated. This plot is for 38 GHz and the measurements used the same highly directional antennas at both the transmitter and receiver [from [RQT+12] © IEEE].

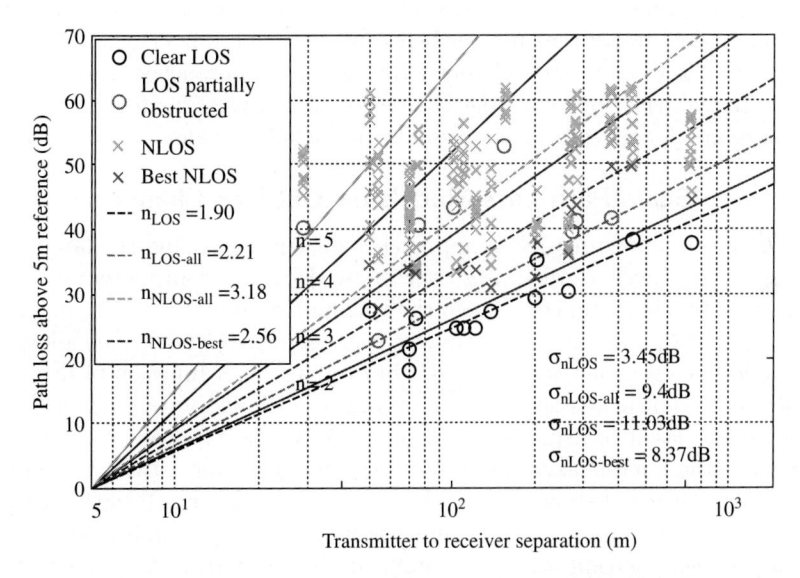

Figure 3.24 This plot was generated from measurements using a 25 dBi 7° beamwidth horn TX antenna and a less directional 13.3 dBi 40° beamwidth horn at the receiver. NLOS paths are significantly stronger as the receiver cannot filter out multipath as effectively as when a more directional antenna is used [RQT+12] © IEEE].

been ongoing for the past few years (e.g., IEEE 802.11ad and IEEE 802.15.3.c channel models are discussed subsequently in Section 3.8.3), there are not yet any accepted outdoor mmWave channel models. This is because the technical community is in the very early stages of formation, and standard-setting activities are in their infancy but will likely evolve in Europe through the existing COST bodies and the newly formed 5th Generation Public Private Partnership (5GPPP) standard organization, in the USA through the International Wireless Industry Consortium (IWPC) and Telecommunications Industry Association (TIA), and globally through continued 3rd Generation Partnership Project (3GPP) efforts. There are likely to be new standards bodies that arise in the IEEE and elsewhere as mmWave communications becomes popular in the years to come.

This section provides various propagation models for outdoor channels that may be used by mmWave standards bodies for outdoor cellular and wireless access systems in urban areas. The models presented here are based on extensive field data collected by one of the authors and his students at 28 and 73 GHz in downtown New York City.

Standards bodies generally require omnidirectional antenna pattern models where the total path loss and the path loss of individual multipath components, as well as mutipath time delays, directions of departure (DODs), and directions of arrival (DOAs) of all multipath energy over all directions are captured in a statistical model [SR14a]. The use of omnidirectional channel models allows researchers to apply the channel model to any particular system or design concept, such as arbitrary MIMO or beamforming antenna systems. The channel models are also useful for determining capacity and coverage, usually by simulation or analysis [RRE14] [Gho14] [SNS$^+$14] [BDH14]. Such omnidirectional channel models allow the research and product development communities to conduct research and experimentation with new ideas without having to make expensive and time-consuming field measurements on their own. By using universally accepted models, it becomes possible to compare different PHY or MAC approaches with the same baseline channel model, thereby providing fair technical comparisons for competing modem designs and research concepts among vendors and academicians.

3.7.1.1 Large-Scale Path Loss Models for Outdoor Channels

There are two well-known modeling approaches for characterizing the large-scale coverage distances (i.e., coverage distances over many decades, from meters to hundreds or thousands of meters of distance between the transmitter and receiver). One approach, as discussed in [Rap02], is to use a *close-in free space reference distance* as a leverage point to fit a simple line on a plot for path loss (in decibels) as a function of distance (in log scale), such as is shown in Figs. 3.25, 3.26, and 3.27. This model provides a simple fit and is physically based on equation (3.2) or (3.8), where $n = 2$ denotes free space and $n = 4$ provides the famous asymptotic path loss exponent for a two-ray ground-bounce channel model at great distances beyond the first Fresnel zone [Rap02]. Another modeling approach, used in prior 3GPP standards bodies, has a similar mathematical form but uses a *floating intercept* where measured path loss data is modeled to a line with a least-squares fit, using an arbitrary leverage point (and not a physically based close-in free space reference distance). This latter approach provides a bit smaller standard deviation to measured data (e.g., a better fit, by about 0.5-1.0 dB typically, as shown in

3.7 Outdoor Channel Models

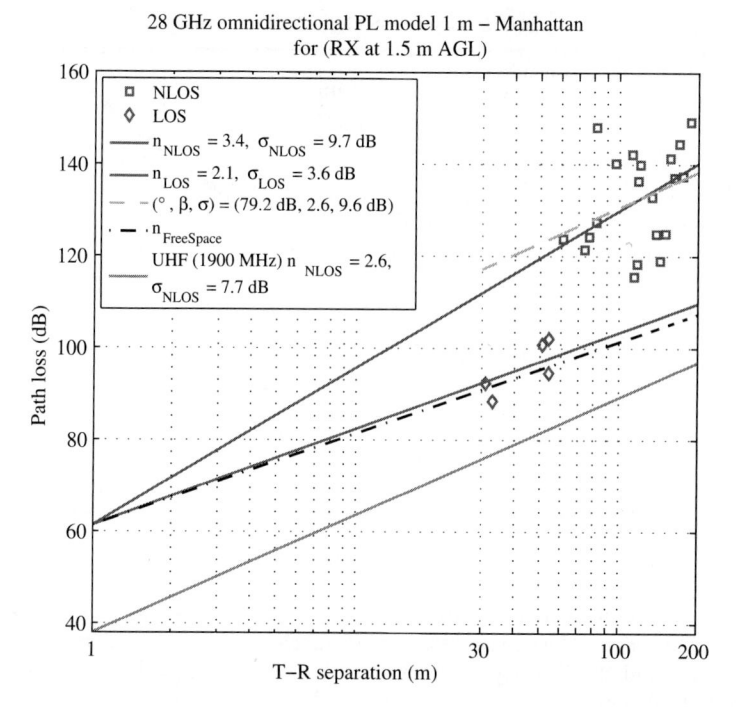

Figure 3.25 28 GHz omnidirectional close-in free space reference distance ($d_0 = 1$ m) and floating intercept path loss models for a non-line of sight (NLOS) urban environment with a receiver antenna 1.5 m above ground. A comparison is made to path loss in a 1.9 GHz urban NLOS environment as reported in [BFR+92] [FBRSX94].

[MZNR13]), but yields a model that has no physical basis (e.g., the slope of the line cannot be related to anything physical because of the arbitrary y-intercept, and thus cannot be used accurately beyond the observed measurement points [MR14a][RRE14]). In other words, the close-in free space reference distance path loss model uses a known close-in free space anchor point based on free space propagation physics at a close-in distance to explain a channel's path loss over distance, whereas the floating intercept model simply fits measured data to the best possible model without regard for any physical rationale, and only over the distances that were originally measured [MZNR13][MR14a][MSR14].

The close-in free space reference distance model is given by

$$PL[dB](d) = PL(d_0) + 10\bar{n} \log_{10}\left(\frac{d}{d_0}\right) + \chi_\sigma \quad (3.36)$$

where,

$$PL[dB](d_0) = 20 \log_{10}\left(\frac{4\pi d_0}{\lambda}\right) \quad (3.37)$$

$$\lambda = \frac{3 \times 10^8 (\text{m/s})}{f_c(\text{Hz})} \quad (3.38)$$

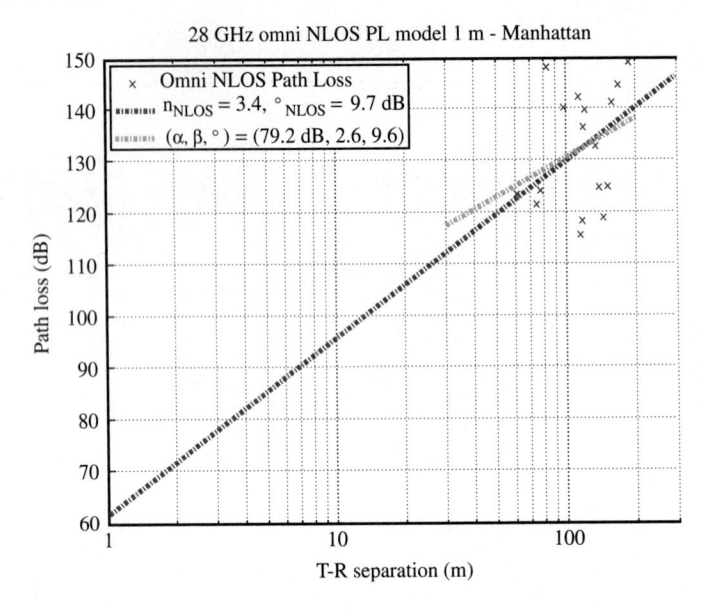

Figure 3.26 28 GHz omnidirectional path loss model from which the TX and RX antenna gains have been removed. The close-in free space reference distance model with respect to a 1 m free space reference distance, and the floating intercept (α, β) model from [RRE14] is shown for distances ranging from 30 to 200 m.

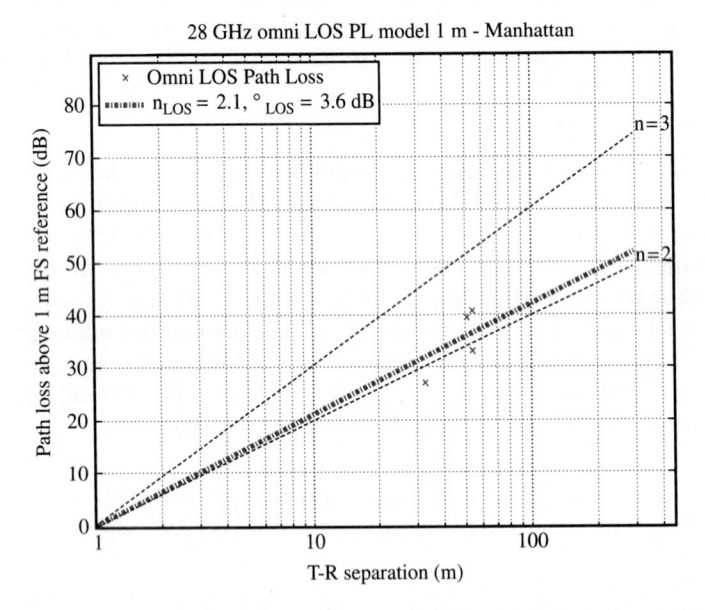

Figure 3.27 28 GHz omnidirectional path loss model from which the TX and RX antenna gains have been removed. The close-in free space reference distance model with respect to a 1 m free space reference distance is shown. Note that one point at 100 m had excessive path loss due to the fact that the antennas were not aligned on boresight at this location. By removing this single point, it is evident that the LOS path loss exponent is very close to 2.

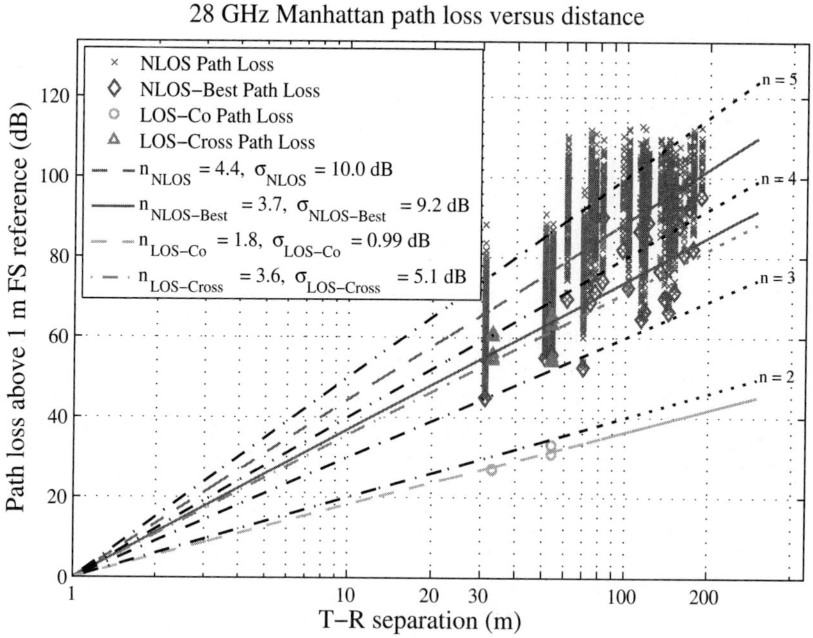

3.7 Outdoor Channel Models

Figure 3.28 28 GHz Manhattan single beam path loss measurements as a function of T-R separation distance using 24.5 dBi horn antennas with 10.9° half-power beam width at both the TX and RX and 15 dBi (28.8 degree HPBW) horn antennas at both the TX and RX. NLOS path losses include LOS non-boresight and truly NLOS measurements. Co-polarized and cross-polarized LOS measured path losses are also shown. The close-in free space reference distance model with respect to a 1 m free space reference distance is shown. All data points represent path loss values calculated from recorded PDP measurements.

and d_0 (m) is the free space reference distance in meters, λ (m) is the carrier wavelength, d (m) is the T-R separation distance between the transmitter and receiver, $\bar{n}$ is the path loss exponent shown subsequently on the scatter plot graphs (see Figs. 3.26 and 3.27), and χ_σ is the typical log-normal random variable with 0 dB mean and standard deviation σ in dB for modeling the large-scale shadow fading [CR01b]. For creating a simple modeling approach for use by standard bodies, we note that as long as the original measurements were made with the antennas in the far field, we can specify $d_0 = 1$ m for large-scale path loss models. For measurements made in [RSM+13][MR14a][SWA+13] [AWW+13][NMSR13][RRE14][SMS+14][MZNR13] [Gho14][SNS+14][MSR14] at 28 and 73 GHz, we observe from (3.37) and (3.38) that $PL(d_0 = 1$ m$)$ is 61.38 dB for 28 GHz and is 69.77 dB for 73 GHz, compared with a value of only ∼40 dB at 2 GHz.

Remarkably, early work in outdoor mmWave propagation shows that the difference in path loss, within the very first meter of propagation, is the *primary* difference between UHF/microwave path loss and the omnidirectional path loss at mmWave frequencies [MSR14][RRE14][SNS+14]. When comparing omnidirectional NLOS channels at 1.9 GHz and 28 GHz in urban environments, early work in Fig. 3.25 shows the path loss exponent changes from a value of 2.6 at 1,900 MHz [BFR+92] [FBRSX94] to a value of 3.4 at 28 GHz (only 8 dB more loss per decade in distance at mmWave, with only a dB or two more shadow fading) [MSR14].

The floating intercept model, simply called an (α, β) *model*, is merely a best curve fit to measured data on a scatter plot of received power versus distance [MZNR13]. The floating intercept PL model is given by

$$PL[\text{dB}] = \alpha + 10\beta \log_{10}(d) + \chi_\sigma \quad (3.39)$$

where α is a floating intercept in dB, β is the slope, and χ_σ is the typical log-normal random variable with 0 dB mean and standard deviation σ in decibels to model shadow fading.

Recent work by one of the authors in mmWave propagation modeling has considered both types of large scale PL models using steerable directional horn antennas, but previously published works did not process the mmWave measurements as omnidirectional channel models, and they did not consider a simplified reference distance of $d_0 = 1$ m for the close-in free-space reference path loss model [RSM+13][MR14a][SWA+13][AWW+13][NMSR13][SMS+14][MZNR13]. Those earlier works documented mmWave measurements in urban outdoor environments in downtown New York City in 2012 (28 GHz) and 2013 (73 GHz) using highly directional rotatable horn antennas with a sliding correlator spread spectrum channel sounder, and using several different TX heights representative of cellular mmWave deployments. Multipath time resolution was 2.5 ns and spatial resolution was between 7° and 10°, depending on the particular horn antennas and frequencies measured [RSM+13][MR14a][SWA+13][AWW+13][NMSR13][RRE14][SMS+14][MZNR13]. The 73 GHz measurements conducted in 2013 considered both base station to mobile measurements as well as base station to base station measurements, in order to model how lamppost-tall base stations could communicate with each other in urban channels (for fronthaul or backhaul use).

Here, we present the omnidirectional path loss data for the measurement campaigns in 2012 and 2013 [MSR14]. The omnidirectional models were formed by computing the individual received power at every unique pointing angle at the TX and RX (e.g., computing the area under the power delay profile for each pointing angle) and by subtracting the directional TX and RX antenna gains for each of the measured power levels at each unique pointing angle, and then summing up all received powers over all unique angles at the TX and RX to obtain an omnidirectional path loss value at each RX location. Simplified path loss models with respect to a 1 m free-space reference are provided here for the omnidirectional channels, suitable for use in 3GPP-like standards [MSR14].

Table 3.14 and equations (3.40) and (3.41), respectively, provide omnidirectional large-scale path loss models for 28 GHz NLOS and LOS channels [MSR14]. Table 3.15 provides the parameters for the 73 GHz models, and equations (3.42) and (3.43) provide models for hybrid 73 GHz measurement that combine all mobile-height (2.0 m height) RX locations and all lamppost (4.06 m height) RX locations for NLOS and LOS locations, respectively [MSR14]. Equations (3.44) and (3.45) show 73 GHz path loss models for the NLOS and LOS situations for mobile height receivers, and equations (3.46) and

3.7 Outdoor Channel Models

Table 3.14 Omnidirectional large-scale path loss models for the close-in free-space reference distance model with $d_0 = 1$ m, and floating point (α, β) models for 28 GHz. The (α, β) model from [RRE14] is only valid for distances ranging from 30 to 200 m.

	28 GHz Path Loss Models		
NLOS	(α, β) model $30 < d < 200$ m	α[dB]	79.2
		β	2.6
		σ[dB]	9.6
	close in ref. 1 m	PLE	3.4
		σ[dB]	9.7
LOS	close in ref. 1 m	PLE	2.1
		σ[dB]	3.6

Table 3.15 Omnidirectional path loss models for the close-in free-space reference distance model with $d_0 = 1$ m and (α, β) models for 73 GHz [MSR14]. The (α, β) model from [RRE14] is only valid for distances between 30 and 200 m.

			73 GHz Path Loss Models		
			RX: Height of 2 and 4.06 m	RX: Height of 4.06 m	RX: Height of 2 m
NLOS	(α, β) model $30 < d < 200$ m	α[dB]	80.6	84.0	81.9
		β	2.9	2.8	2.7
		σ[dB]	7.8	7.8	7.5
	close in ref. 1 m	PLE	3.4	3.5	3.3
		σ[dB]	7.9	7.9	7.6
LOS	close in ref. 1 m	PLE	2.0	2.0	2.0
		σ[dB]	4.8	4.2	5.2

(3.47) provide 73 GHz path loss models for base station to base station (e.g., backhaul) NLOS and LOS situations, respectively [MSR14]. These path loss models are

$$PL_{28GHz}(LOS)[\text{dB}](d) = 61.4 \text{ dB} + 21\log(d) + \chi_\sigma \quad [\sigma = 3.6 \text{ dB}] \quad (3.40)$$

$$PL_{28GHz}(NLOS)[\text{dB}](d) = 61.4 \text{ dB} + 34\log(d) + \chi_\sigma \quad [\sigma = 9.7 \text{ dB}] \quad (3.41)$$

$$PL_{73GHz-Hybrid}(LOS)[\text{dB}](d) = 69.8 \text{ dB} + 20\log(d) + \chi_\sigma \quad [\sigma = 4.8 \text{ dB}] \quad (3.42)$$

$$PL_{73GHz-Hybrid}(NLOS)[\text{dB}](d) = 69.8 \text{ dB} + 34\log(d) + \chi_\sigma \quad [\sigma = 7.9 \text{ dB}] \quad (3.43)$$

$$PL_{73GHz-Access}(LOS)[\text{dB}](d) = 69.8 \text{ dB} + 20\log(d) + \chi_\sigma \quad [\sigma = 5.2 \text{ dB}] \quad (3.44)$$

$$PL_{73GHz-Access}(NLOS)[\text{dB}](d) = 69.8 \text{ dB} + 33\log(d) + \chi_\sigma \quad [\sigma = 7.6 \text{ dB}] \quad (3.45)$$

$$PL_{73GHz-Backhaul}(LOS)[\text{dB}](d) = 69.8 \text{ dB} + 20\log(d) + \chi_\sigma \quad [\sigma = 4.2 \text{ dB}] \quad (3.46)$$

$$PL_{73GHz-Backhaul}(NLOS)[\text{dB}](d) = 69.8 \text{ dB} + 35\log(d) + \chi_\sigma \quad [\sigma = 7.9 \text{ dB}] \quad (3.47)$$

where $\chi\sigma$ is a zero mean Gaussian random variable with a standard deviation of σ in decibels. The 73 GHz hybrid models include measurements for RX antenna at mobile (2 m Above Ground Level [AGL]) and backhaul (4.06 m AGL) heights.

The scatter plots shown in Figs. 3.26 and 3.27 show the best fit to measured field data for omnidirectional path loss at 28 GHz in NLOS (Fig. 3.26) and LOS (Fig. 3.27) channels. Fig. 3.26 also shows the floating-intercept path loss model, computed in [SMS+14] and in equation (3.39), to illustrate how the floating intercept model agrees with the close-in free-space reference distance model over a limited range of distances. Fig. 3.28 shows all 28 GHz Manhattan path loss values as measured with 24.5 dBi TX and RX antennas [RSM+13][MR14a][SWA+13]. It is worth noting that the omnidirectional path loss model has a smaller path loss exponent (e.g., less loss) than directional single beam measurements as can be seen in Fig. 3.28, but the omnidirectional nature of the model implies 0 dB antennas, meaning that the radio link margin (e.g., overall coverage distance) will be smaller for omnidirectional antennas as opposed to directional antennas, due to smaller antenna gain. This is a phenomenon reported consistently over the past few years, indicating that wide beam (smaller gain) communication systems capture more energy but have less overall link margin compared with narrow beam (high gain) systems. The directional path losses shown on the scatter plot in Fig. 3.28 should not be confused with the omnidirectional path losses shown in Figs. 3.26 and 3.27, where all received power was summed over all TX and RX azimuth and elevation angles.

The scatter plot shown in Fig. 3.29 shows the best fit to measured field data for omnidirectional path loss at 73 GHz in both NLOS and LOS channels for a hybrid combination of mobile height and base station height measurements. Fig. 3.29 also shows the floating-intercept model, computed in [SMS+14] and in equation (3.39) above, to illustrate how the floating intercept model somewhat agrees with the close-in free space reference distance model over a limited range of distances. Fig. 3.30 shows all 73 GHz Manhattan omnidirectional path loss values measured for mobile height (2.0 m RX height), as well as the floating point model from 30 to 200 m.

Fig. 3.31 shows the scatter plot of omnidirectional path loss values and the two different large-scale path loss models for all backhaul (base station to base station) measurements in downtown New York City. Figs. 3.32 and 3.33 show all 73 GHz Manhattan path loss values measured with 27 dBi TX and RX antennas for both backhaul and mobile RX antenna heights, obtained at each measured RX azimuth and elevation angle combination using the ten RX antenna-pointing combinations, and the two TX antenna-pointing combinations at each T-R separation, for NLOS and LOS environments. Since each antenna-pointing combination consisted of rotating the RX antenna in 10° increments for a LOS environment and 8° increments for a NLOS environment, a maximum of 45 possible measurements could be performed for each antenna pointing combination, with up to 540 measured channel impulses when considering all 12 antenna-pointing combinations for a given RX location. These results were published in [MR14a] for a 4 m free-space reference and are now shown here for a 1 m free-space reference. It is worth noting that beamforming and beam-combining technologies using electrically phased on-chip antennas will combine received energy from multiple incoming directions to improve signal quality, as demonstrated in [SMS+14]. The directional path loss models shown in Figs. 3.28, 3.32, and 3.33, although not in use today in standards work, are applicable

3.7 Outdoor Channel Models

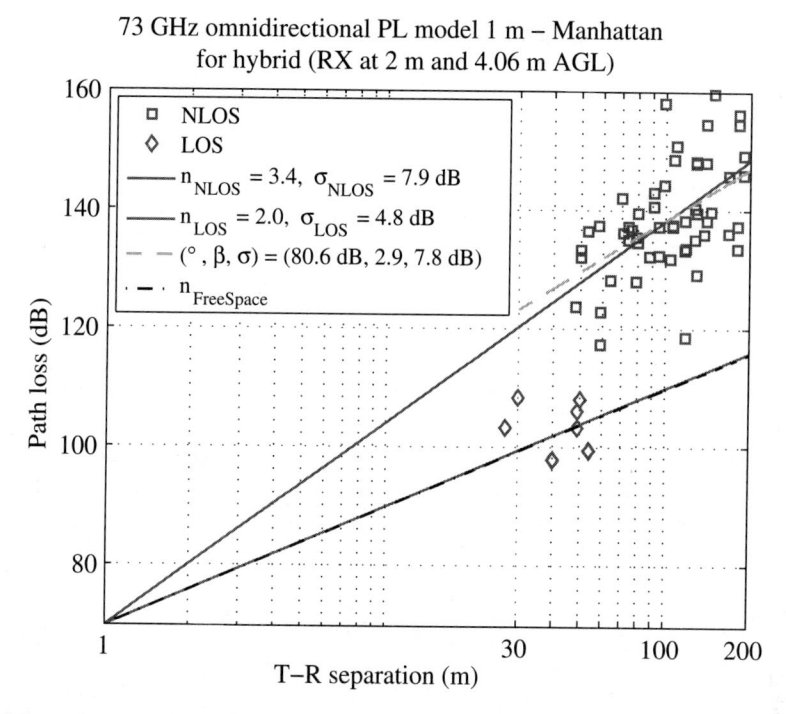

Figure 3.29 73 GHz omnidirectional path loss model from which the TX and RX antenna gains have been removed for a combination of cellular and backhaul (hybrid) RX antenna heights. The close-in free-space reference distance model for $d_0 = 1$ m and the floating intercept (α, β) model from [RRE14] over 30-200 m are shown.

for estimating the received power at various RX antenna azimuth and elevation angle combinations, for a fixed TX azimuth and downtilt angle. As in Fig. 3.28, it can be seen in Figs. 3.32 and 3.33 that the omnidirectional path loss model has a smaller path loss exponent (e.g., less loss) than directional single beams, but the omnidirectional channel model implies 0 dB antennas, meaning that the radio link margin (e.g., overall coverage distance) will be smaller for omnidirectional antennas as opposed to directional antennas, due to smaller antenna gain. The directional path losses shown on the scatter plot in Figs. 3.32 and 3.33 should not be confused with the omnidirectional path losses shown in Figs. 3.29, 3.30, and 3.31, where all received power was summed over all TX and RX azimuth and elevation angles at 73 GHz.

It is a relatively straightforward procedure to process the measured data, collected from directional antennas as shown in Figs. 3.28, 3.32, and 3.33, to create models for angles of arrival and departure as well as omnidirectional path loss. Such approaches, described in [RRE14][ALS+14] and [SR14a], use clustering algorithms and detailed statistical analysis in order to fit the measured data to appropriate statistical models. In [RRE14], such angular models are combined with the path loss models described above to provide capacity analyses that show future mmWave systems could have 20 times the

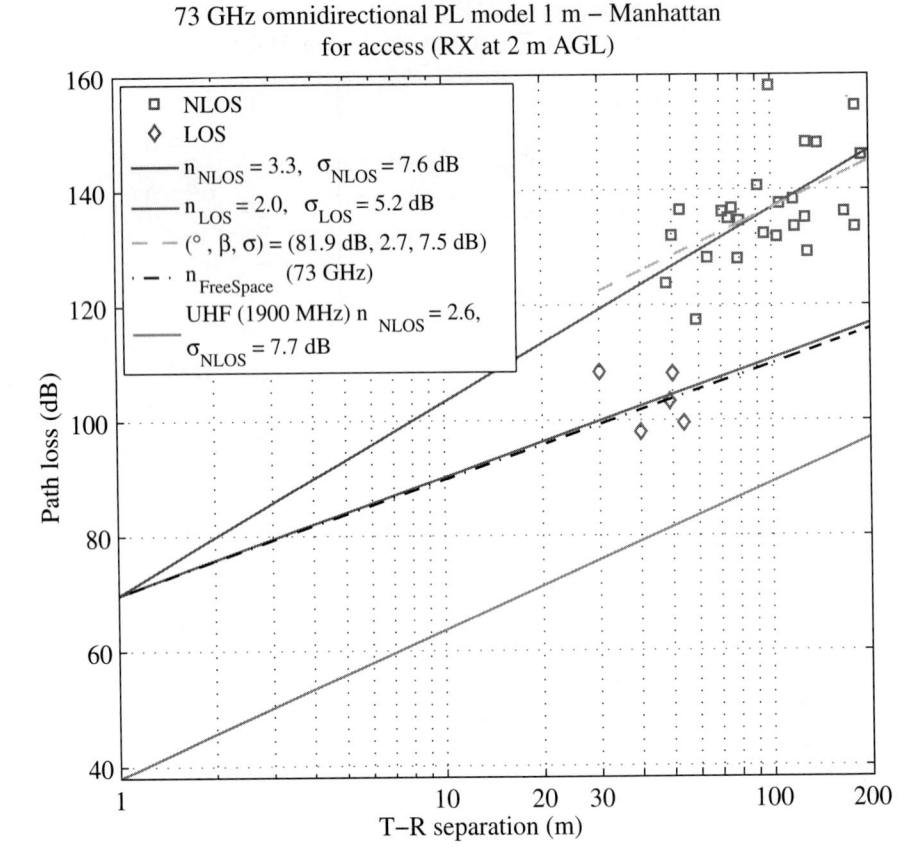

Figure 3.30 73 GHz omnidirectional path loss model from which the TX and RX antenna gains have been removed for mobile RX antenna heights of 2 m. The close-in free-space reference distance model for $d_0 = 1$ m and the floating intercept model (α, β) model from [RRE14] over 30-200 m are shown. A comparison is made to path loss in a 1.9 GHz urban NLOS environment as reported in [BFR+92].

amount of average data rate over modern 4G LTE systems. Work in [SR14a] provides a detailed statistical channel model for wideband channels, very similar to past 3GPP models, and exploits the greater temporal and spatial resolution of the field measurements to provide new wideband models for NLOS urban channels. These types of standardized models for outdoor mmWave channels will aid researchers in equipment design and simulation, as well as contributing to network installations in the future.

3.7.1.2 Small-Scale Temporal and Spatial Multipath Models for Outdoor mmWave Channels

Temporal and spatial multipath channel models such as those created by 3GPP and WINNER II are used by standard bodies for current UHF and microwave bands [3GPP][Winner2]. These statistical spatial channel models (SSCMs) are based on empirical studies between 1 and 6 GHz, and RF bandwidths between 5 and 100 MHz, and they

3.7 Outdoor Channel Models

73 GHz omnidirectional PL model 1 m – Manhattan for backhaul (RX at 4.06 m AGL)

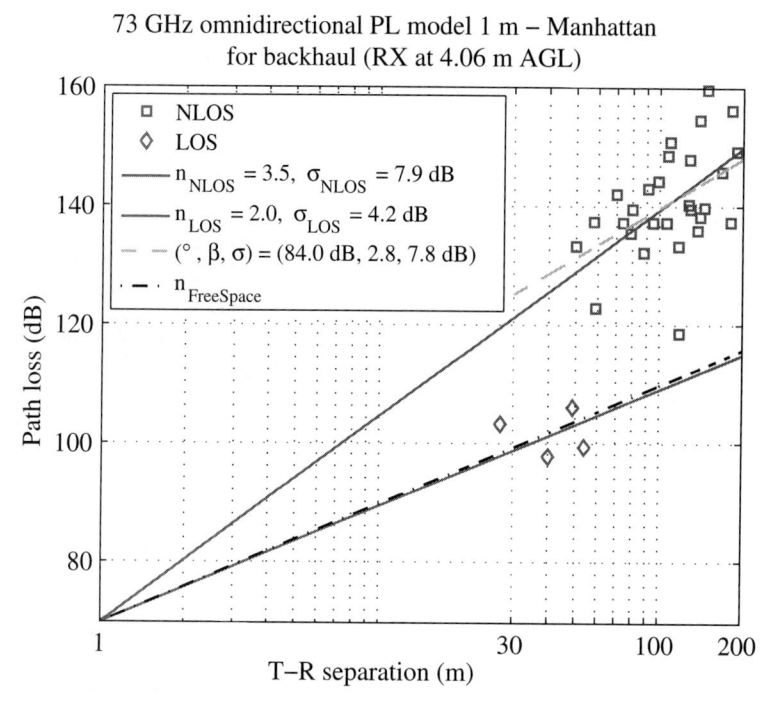

Legend:
- □ NLOS
- ◇ LOS
- $n_{NLOS} = 3.5$, $\sigma_{NLOS} = 7.9$ dB
- $n_{LOS} = 2.0$, $\sigma_{LOS} = 4.2$ dB
- $(\alpha, \beta, \sigma) = (84.0$ dB, $2.8, 7.8$ dB$)$
- $n_{FreeSpace}$

Figure 3.31 73 GHz omnidirectional path loss model from which the TX and RX antenna gains have been removed for backhaul RX antenna heights of 4.06 m. The close-in free-space reference distance model for $d_0 = 1$ m and the floating intercept (α, β) model from [RRE14] over 30-200 m are shown.

provide important statistical channel parameters such as multipath delays, cluster powers, angle of arrival (AOA) and angle of departure (AOD) information, as well as large-scale path loss models based on real-world measurements. These models also produce complex channel coefficients for simulating channel impulse responses [3GPP][Winner2]. However, while these two channel models have been successful in describing the stochastic nature of the low frequency (UHF/microwave) wideband channels, they are limited in temporal resolution (e.g., the 100 MHz bandwidth only allows baseband temporal resolution to 20 ns), and they make a simplifying assumption that all clusters of multipath energy travel closely together in both time and space. This past modeling approach does not accurately portray the urban mmWave channels, where the bandwidths will be much wider, and where distribution of power over the spatial domain has been observed to behave such that multiple multipath clusters can arrive within a particular spatial direction [SR14].

Recent wideband mmWave measurements, using much greater multipath time resolutions of 2.5 ns and highly directional rotatable horn antennas [RSM+13][MR14a][SWA+13][AWW+13][NMSR13][RRE14][SMS+14][MZNR13][SR14] show that, in urban outdoor mmWave channels, there are some angles of arrival (AOAs) that have energy arriving from multiple propagation time clusters. This occurs from reflections off the ground, and from reflections off buildings behind and in front of a receiver in an urban

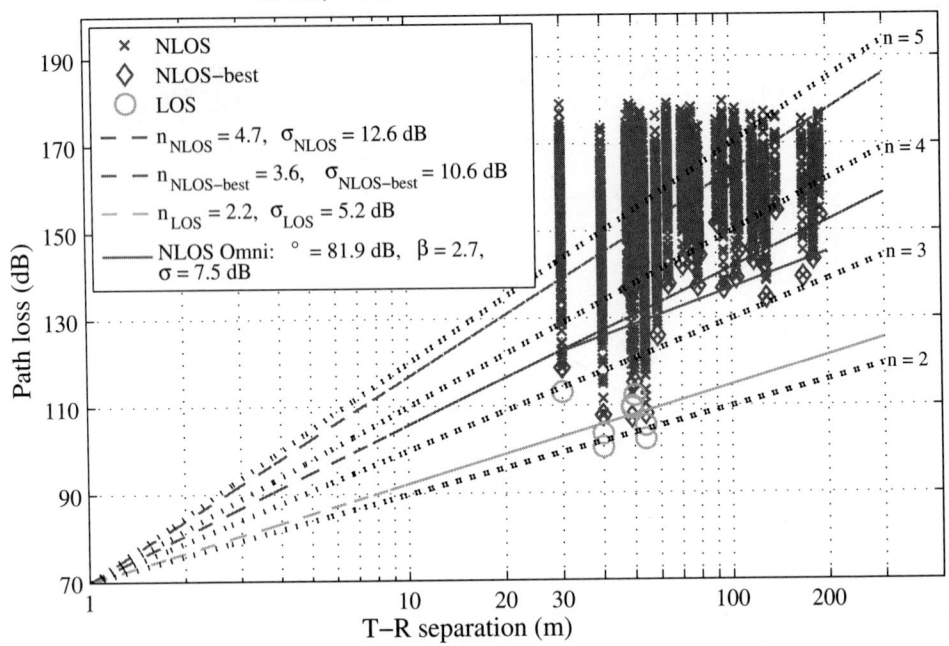

Figure 3.32 New York City cellular RX height (2 m) path loss measurements at 73 GHz as a function of T-R separation distance using vertically polarized 27 dBi, 7° half-power beam width TX and RX antennas. All data points represent path loss values calculated from recorded PDP measurements. Crosses indicate all NLOS pointing angle data points, diamonds indicate best NLOS pointing angle data points for each RX location and each T-R combination, and circles indicate LOS data points. The measured path loss values are relative to a 1 m free-space close-in reference distance. NLOS PLEs are calculated for the entire data set and also for the best recorded link. LOS PLEs are calculated for strictly boresight-to-boresight scenarios. n values are PLEs and σ values are shadow factors. The solid line spanning 30 to 200 m is the omnidirectional (α, β) model from [RRE14] [ALS+14] depicted in Fig. 3.30.

corridor, as well as very strong reflections that can arrive on the back lobe of an antenna. Thus, mmWave channel models will require the concept of *spatial lobes* to allow energy to be modeled in space as well as time, where it is understood from recent measurements that multipath energy from different time clusters may arrive within a single lobe. Such an approach allows a channel model to consider multipath energy having many different time delays to be due to a particular direction of arrival or departure for a receiver or transmitters.

To properly model the arrival and departure angles of multipath energy in an omnidirectional fashion for mmWave channels, the absolute time delays, power levels, and angles of arrival and departure for each multipath component must be measured [SR14a].

3.7 Outdoor Channel Models

73 GHz unique pointing angle path loss vs. distance with
RX height: 4.06 m using 27 dBi, 7° 3dB BW TX antennas and
27 dBi, 7° 3dB BW RX antennas in Manhattan

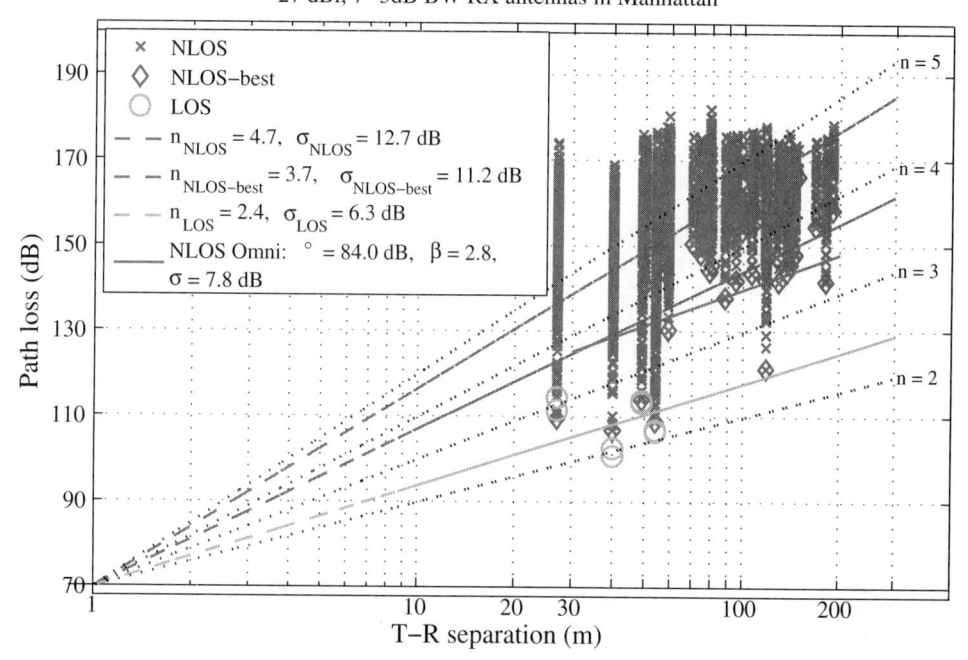

Legend:
- × NLOS
- ◇ NLOS–best
- ○ LOS
- – – $n_{NLOS} = 4.7$, $\sigma_{NLOS} = 12.7$ dB
- – – $n_{NLOS-best} = 3.7$, $\sigma_{NLOS-best} = 11.2$ dB
- – – $n_{LOS} = 2.4$, $\sigma_{LOS} = 6.3$ dB
- —— NLOS Omni: $\alpha = 84.0$ dB, $\beta = 2.8$, $\sigma = 7.8$ dB

Figure 3.33 New York City backhaul measurements with RX heights of 4.06 m path losses at 73 GHz as a function of T-R separation distance using vertically polarized 27 dBi, 7° half-power beam width TX and RX antennas. All data points represent path loss values calculated from recorded PDP measurements. Crosses indicate all NLOS pointing angle data points, diamonds indicate best NLOS pointing angle data points for each RX location and each T-R combination, and circles indicate LOS data points. The measured path loss values are relative to a 1 m free-space close-in reference distance. NLOS PLEs are calculated for the entire data set and also for the best recorded link. LOS PLEs are calculated for strictly boresight-to-boresight scenarios. n values are PLEs and σ values are shadow factors. The solid line spanning 30 to 200 m is the omnidirectional (α, β) model from [RRE14] depicted in Fig. 3.31.

By using the concept of temporal clusters and lobes, Figs. 3.34 and 3.35 show how an omnidirectional mmWave impulse response may be represented simultaneously in the time domain and spatial domain. Fig. 3.34 illustrates the modeling of a power delay profile based on multipath clusters in the time domain, and Fig. 3.35 shows the equivalent channel as represented by spatial clusters, or lobes, in the spatial domain, where it is understood from recent high-resolution mmWave measurements that one or more temporal clusters may exist within a spatial lobe [SR14a].

For omnidirectional multipath models in the time domains, the concept of temporal clusters (for recreation of statistical wideband time domain impulse responses, or PDPs) is used, where the key statistical parameters needed in a channel model include the number of multipath clusters, the received power in each multipath cluster, the time

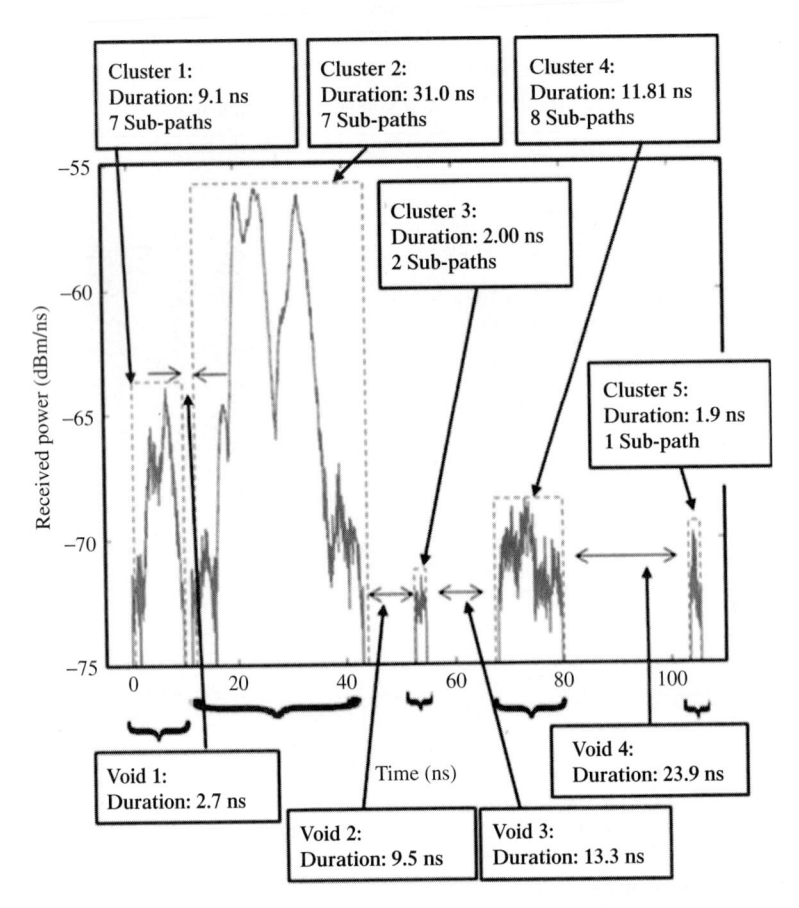

Figure 3.34 Illustration of some of the key temporal modeling parameters used for modeling the temporal clusters in an omnidirectional SSCM wideband mmWave channel. This example shows five time clusters, with time durations ranging from 2 to 31 ns, and voids between clusters ranging from 2.7 to 23.9 ns [SR14a].

durations of each cluster, the number of sub-paths within a cluster (e.g., the number of multipath components within a particular cluster of multipath energy), the sub-path energy distribution over time, and the distribution of temporal voids between clusters.

For modeling omnidirectional outdoor mmWave channels in the spatial domain, the concept of lobes incorporates key statistical parameters such as the number of lobes, the number of segments within a lobe (and corresponding sampling beam width of each lobe segment), the directional pointing angle of a lobe, a lobe angle spread (LAS) for each lobe, and the threshold of each lobe with respect to the strongest lobe segment [SR14a]. A statistical simulator based on Figs. 3.34 and 3.35 has been developed [SR14a] and shows strong agreement for wideband mmWave channels in outdoor urban environments at 28 GHz. The key (e.g., primary) parameters used in the SSCM for both NLOS and LOS urban outdoor channels in [SR14a] are shown in Tables 3.16 and 3.17, and secondary parameters that give gross validations of an ensemble as compared to field measurements are also shown.

3.7 Outdoor Channel Models

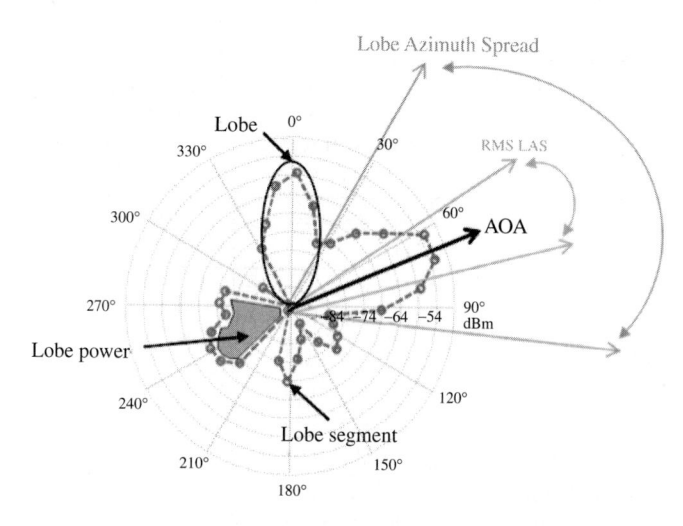

Figure 3.35 Illustration of some of the key spatial modeling parameters used to model the spatial lobes in an omnidirectional SSCM wideband mmWave channel. The polar plot (in the azimuthal/horizontal plane only) shows five distinct lobes with various lobe azimuth spreads and AOAs. Each dot is a lobe angular segment simulated for a particular discrete pointing angle and represents the total integrated received power over a particular beam width (and corresponds to the area under a PDP for the particular RX pointing angle). The lobe power is the sum of powers from each segment within the lobe (e.g., the sum of powers from each lobe segment in a lobe).

3.7.2 Vehicle-to-Vehicle Models

One of the most commonly discussed applications for outdoor mmWave signals is for vehicle-to-vehicle applications. [SL90] investigated intervehicle communications at 60 GHz, and found that link distances of up to 500 m with transmission powers of approximately 2 W allowed for reliable communications at these frequencies. [BDRQL11] studied the impact of vehicles on received signal strength at 60 GHz. Fig. 3.36 from [BDRQL11] shows the path loss measurements when the receiver was in a vehicle and also the path loss measurements for peer-to-peer applications in which the transmitter and receiver were in the open. The plot indicates that LOS communications to vehicles from an outdoor transmitter at the level of the vehicle is possible at 60 GHz using highly directional antennas at the receiver and transmitter, but that the received signal will be slightly attenuated. NLOS paths are greatly attenuated. Fig. 3.37 shows that RMS delay spreads for communication to a receiver inside a vehicle will be much lower than for situations in which the transmitter and receiver are both out in the open.

In modeling outdoor vehicle communication channels, the impact of neighboring cars must be accounted for, as well as traffic density, number of lanes of traffic, and potential interference sources. Work in [RDA11] explored co-channel interference and out of band emission (OOBE) scenarios between satellite radio and cellular users in the 2 GHz band, and similar techniques could be used for determining coverage, capacity, and interference for vehicle-to-vehicle communication systems of the future. For additional information on vehicle-to-vehicle channel models, we refer the reader to a recent book [Zaj13] that focuses on vehicle-to-vehicle and underwater vehicle to underwater vehicle communications.

Table 3.16 Summary of 28 GHz wideband channel statistics for NLOS urban outdoor channels from omnidirectional wideband PDPs, and the simulated statistics using 10,000 PDPs and Power Angle Spectrum (PAS) plots using the simulation procedure in [SR14a]. "P" denotes a primary modeling statistic, and "S" denotes a secondary statistic as described in [SR14a]. A 20 dB lobe threshold is used.

Type of Statistic	Quantity	Measured (μ, σ)	Simulated (μ, σ)	Error (%)
Temporal	Number of clusters (P)	Poisson (3.4, 2.1)	(3.2, 2.1)	(5.9, 0)
	Number of cluster sub-paths (P)	Exponential (2.1, 1.6)	(2.2, 1.7)	(4.7, 6.3)
	Cluster excess time delay (ns) (P)	Exponential (66.3, 68.0)	(71.8, 62.1)	(8.3, 8.7)
	Cluster sub-path excess time delay (ns) (P)	Exponential (8.1, 8.8)	(8.6, 8.0)	(6.2, 9.1)
	RMS delay spread (ns) (S)	Exponential (13.4, 11.5)	(12.9, 11.3)	(3.7, 1.7)
	Cluster RMS delay spread (ns) (S)	Exponential (2.0, 2.0)	(2.4, 1.7)	(20.0, 15.0)
	Cluster duration (ns) (S)	Exponential (8.9, 8.7)	(10.7, 8.4)	(20.2, 3.5)
	Inter-cluster void duration (ns) (S)	Exponential (16.8, 17.2)	(21.5, 15.9)	(28.0, 7.5)
Spatial	Number of lobes (P)	Poisson (2.4, 1.3)	(2.3, 1.1)	(4.2, 15.4)
	AOA (°) (P)	Uniform (0, 360)	Uniform (0, 360)	(0)
	Lobe azimuth spread (°) (P)	Normal (34.8, 25.7)	(34.6, 27.8)	(0.5, 8.1)
	RMS lobe azimuth spread (°) (S)	Exponential (6.1, 5.8)	(8.3, 6.8)	(36.0, 17.0)

3.8 Indoor Channel Models

For proper link budget and coverage/capacity designs, indoor channel models will be needed to deploy mmWave communication systems. For deployment and system analysis, frequencies above 5 GHz lend themselves well to site-specific modeling, where the accuracy of such methods is remarkably good for estimating both the received signal levels, as well as multipath components and their angles of arrival. For the development of wireless standards and for having a unified approach to simulating bit error rates, PHY and MAC layer improvements, as well as beam steering and cooperative communication methods, it is more desirable to have a statistical channel model, which accounts for both time delays and spatial multipath, for simulations and bench testing of hardware.

3.8 Indoor Channel Models

Table 3.17 Summary of 28 GHz wideband channel statistics for LOS urban outdoor channels from omnidirectional wideband PDPs, and the simulated statistics using 10,000 PDPs and Power Angle Spectrum (PAS) plots using the simulation procedure in [SR14a] for the NLOS environment. "P" denotes a primary modeling statistic, and "S" denotes a secondary statistic as described in [SR14a]. A 10 dB lobe threshold is used.

Type of Statistic	Quantity	Measured (μ, σ)	Simulated (μ, σ)	Error (%)
Temporal	Number of clusters (P)	Poisson (4.1, 2.3)	(4.0, 2.4)	(2.1, 1.7)
	Number of cluster sub-paths (P)	Exponential (2.0, 1.7)	(2.1, 1.6)	(5.0, 5.9)
	Cluster excess time delay (ns) (P)	Exponential (161.8, 189.1)	(172.9, 170.7)	(6.7, 9.6)
	Cluster sub-path excess time delay (ns) (P)	Exponential (8.0, 8.3)	(8.3, 7.8)	(3.8, 6.0)
	RMS delay spread (ns) (S)	Exponential (60.5, 80.7)	(25.0, 20.3)	(58.7, 74.6)
	Cluster RMS delay spread (ns) (S)	Exponential (1.8, 1.9)	(2.3, 1.6)	(27.8, 11.0)
	Cluster duration (ns) (S)	Exponential (8.6, 8.4)	(10.2, 8.1)	(18.6, 3.6)
	Inter-cluster void duration (ns) (S)	Exponential (14.8, 17.0)	(41.6, 26.0)	(180, 53.0)
Spatial	Number of lobes (P)	Exponential (2.9, 1.5)	(3.1, 1.3)	(7.0, 13.3)
	AOA (°) (P)	Uniform (0, 360)	Uniform (0, 360)	(0)
	Lobe azimuth spread (°) (P)	Normal (39.9, 31.4)	(39.5, 30.5)	(1.0, 2.9)
	RMS lobe azimuth spread (°) (S)	Exponential (8.9, 8.7)	(9.0, 8.3)	(1.3, 4.6)

Two dominant characteristics of antennas in the indoor environment influence the observed multipath at the receiver: *antenna directivity* and electromagnetic field polarization. The first characteristic, antenna directivity, refers to ratio of the power radiated in a given direction to the total radiated power average over all directions. Because directive transmit antennas favor certain directions, less space is excited. Similarly, directive receive antennas capture energy from less space. As a result, directive antennas can be used to reduce multipath contributions by limiting the space from which radiation is captured [YHS05]. This effect can be drastic. In [MMI96], a scenario was shown where the RMS delay spread was reduced from 18 ns to 1 ns by using a highly directive antenna. Additionally, [WAN97] shows a reduction from 23 to 10 ns. Similar results are shown in [RH91], [HR92], and [H+94] for different environments. However, as

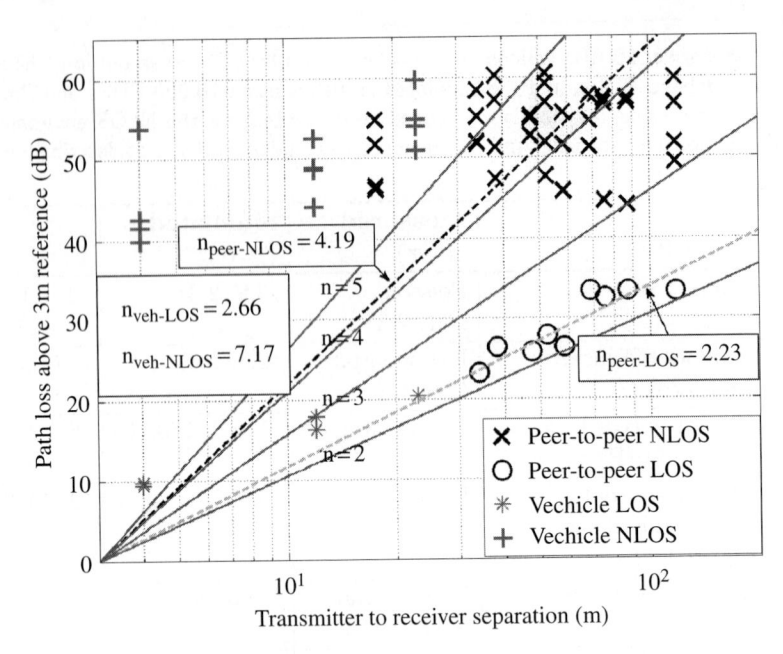

Figure 3.36 Path loss for 60 GHz for peer-to-peer applications and communication from a ground-based transmitter to a receiver in a vehicle. These measurements used highly directional 25 dBi 7° beamwidth antennas as the transmitter and receiver [from [BDRQL11] © IEEE].

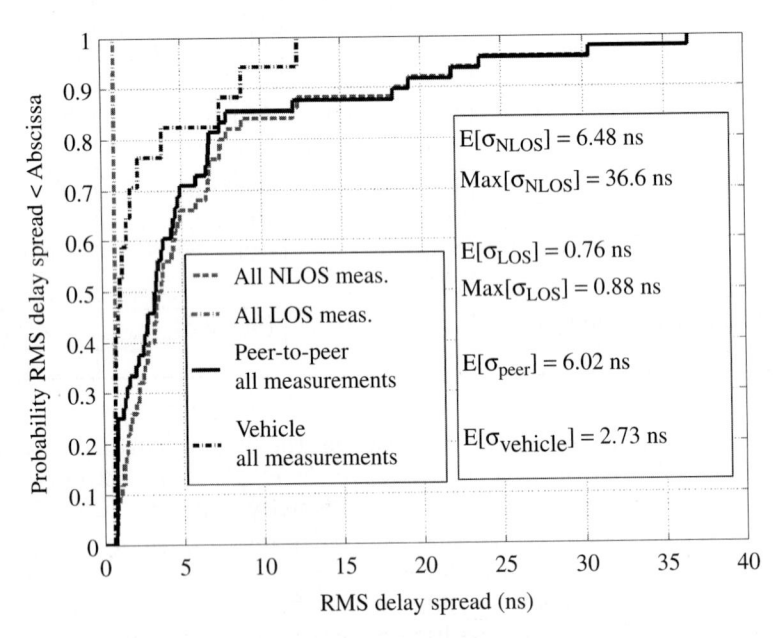

Figure 3.37 These measurements used highly directional 25 dBi 7° beamwidth antennas at the transmitter and receiver. When the transmitter communicates to a receiver inside a vehicle, much lower RMS delay spreads result than when the transmitter and receiver are in the open [from [BDRQL11] © IEEE].

3.8 Indoor Channel Models

will be discussed in Chapters 7 and 8, directivity presents problems since transmitters and receivers must "point" to each other in order to communicate.

As shown earlier in Sections 3.2.5 and 3.2.8, and in Fig. 2.5 and Table 2.4, the penetration loss of partitions, as well as reflection coefficients, play a large role in properly modeling mmWave indoor channels [SRA96][AR04][AR+02][NMSR13][XKR02]. We now present some promising techniques and early work for mmWave indoor channel modeling for both large-scale effects and small-scale statistical effects.

3.8.1 Ray-Tracing Models for Indoor Channels

Due to the dominance of reflection and scattering in the mmWave wireless environment and the lack of diffraction contributions, ray tracing is a popular and accurate method to reproduce wireless channels. The main limitation of ray-tracing models is the lack of flexibility. Ray-tracing models must be applied to a specific environment and are, in general, deterministic in nature [SDR92][SR94][HR93][H+94][SSTR93][DPR97b]. That means that it is harder to characterize channel models for different environments (e.g., indoor, outdoor) without an accurate physical model of the environment, although such a site-specific approach is highly desirable for individual installations of wireless networks (say, within a particular city or enterprise). Statistical-based models, such as Rician fading (discussed next) and the Saleh-Valenzuela cluster model (discussed in Section 3.8.3), are more analytically tractable for research and can reference general environment scenarios through stochastic parameterization. As shown in Section 3.2.8, [SDR92], [SSTR93], [SR94], [HR93], [H+94], [YMM+94], [HA95], and [DEFF+97] describe N-ray tracing models, where N is the number of rays considered for propagation from the transmitter to receiver through the environment. In general, good agreement is demonstrated between N-ray models and measurements. For most scenarios, only a few reflections are necessary to include in the ray-tracing models since the large path loss diminishes the contributions of signals from more than a few reflections [SSTR93][XKR02].

3.8.2 Rayleigh, Rician, and Multiwave Fading Models

The Rayleigh fading model, derived from a rich scattering channel response in which many multipath signal components are not resolvable in time or space [Rap02], statistically describes the received signal envelope about some average value specified by the large-scale channel model. Such models pertain generally only to narrowband channels, or to omnidirectional low gain antenna systems, since those do not resolve individual multipath components, and hence vectorially sum the randomly arriving signal energy. As tiny phase shifts are induced on each of the individual unresolved multipath components, the received signal envelope undergoes a Rayleigh characteristic. In the case of a dominant, for example, non-fading LOS multipath signal component, the signal envelope undergoes a Ricean distribution in which the dominant non-fading signal provides a baseline for the received signal characteristic.

The probability density function (PDF) of the small-scale received signal envelope voltage (not power) random variable r of these narrowband channels under Rayleigh fading with large-scale received signal power P_r is given by

$$p_R(r) = \left(\frac{2r}{P_r}\right) \exp\left(-\frac{r^2}{P_r}\right) \text{ for } 0 \leq r \leq \infty. \quad (3.48)$$

The Rayleigh fading model does not include a specular (non-fading) LOS signal component. The Rician fading model for the received envelope voltage generalizes the Rayleigh fading model to include a LOS component such that

$$p_R(r) = \left(\frac{2r}{P_{r,\text{NLOS}}}\right) \exp\left(-\frac{(r^2 + P_{r,\text{LOS}})}{P_{r,\text{NLOS}}}\right) I_0\left(\frac{2r\sqrt{P_{r,\text{NLOS}}}}{P_{r,\text{NLOS}}}\right) \text{ for } 0 \leq r \leq \infty \quad (3.49)$$

where $P_{r,\text{LOS}}$ is the average received signal power in the LOS component, $P_{r,\text{NLOS}}$ is the average received signal power in the non-line-of-sight (NLOS) component, and $I_0(\cdot)$ is the zeroth-order Bessel function. To quantify the degree of stationary LOS components in the received signal, we define the Rician K-factor

$$K_R = \frac{P_{r,\text{LOS}}}{P_{r,\text{NLOS}}} \quad (3.50)$$

such that if $K_R = 0$ there is no LOS component, then $P_{r,\text{NLOS}} = P_r$ and the received signal power is distributed as Rayleigh. Studies have measured adherence of 60 GHz systems to Rayleigh/Rician fading [TM88][AH91][DMTA96][PBC99]. In [AH91], a 60 GHz constant power signal was analyzed at 5,000 samples/sec with omnidirectional antennas in the azimuthal direction. The authors determined that, for an outdoor urban environment, the 60 GHz wireless channel is *not* well-modeled by the Rayleigh distribution. Specifically, the 60 GHz outdoor channel does not include the richness that is necessary to have a highly varied fading distribution. At larger distances between the transmitter and receiver, however, the fading distribution does become increasingly close to Rayleigh fading. In [DMTA96], a ray-tracing simulation calculated K_R averaged over different transmitter/receiver separation distances within an empty room. Results showed that K_R decreases logarithmically proportional to distance (K_R varies from 20 to 5 dB when the separation distance varies from 1 to 12 m). We can conclude that, as expected, when present, the LOS component dominates and the strongest NLOS component is often not very strong in comparison, leading to Rician fading, and *not* Rayleigh fading.

With the use of directional antennas at mmWave frequencies, a new class of fading distributions developed by Durgin, Rappaport, and de Wolf are becoming important. This distribution, awkwardly called the *Two Wave with Diffuse Power* (TWDP) distribution, properly models the fading caused by the combination of just a few strong specular multipath components above diffuse power or random noise, and are steeped in physical realities. Durgin's TWDP distribution functions have the Rayleigh and Ricean fading distributions as special cases of a more general family of distributions [DRD99][DRD02][Dur03], and properly predicts bi-modal density functions now being reported [SB13]. The Durgin distribution provides accurate closed-form expressions that model the impact of two specular waves and three specular waves that combine with diffuse multipath, and improved closed form expressions have recently been presented for Durgin's distribution [SB13]. These distributions will have particular applicability to mmWave channels, where often a few strong multipath components exist due to the use of highly directional antennas [SR14][SR14a].

3.8.3 IEEE 802.15.3c and IEEE 802.11ad Channel Models

The small-scale channel models used in the IEEE 802.15.3c [BSL+11] and IEEE 802.11ad [IEE10] standards for different PHY evaluations are clustered in both time and space. They are based on a modification of the standard Saleh-Valenzuela (S-V) propagation model [SV87][YXVG11][Yon07][MEP+10]. Their difference with the standard S-V propagation model is that they represent a LOS path between the transmitter and receiver separately — that is, the LOS path between the transmitter and receiver is not assumed to be part of a cluster. The separateness of the LOS path and its strength relative to the other components suggests that the channel may be considered Rician when the LOS path is not blocked. Indeed, a K-factor is often used to describe the channel-impulse response. In most Rician channels the K-factor describes the ratio of the strongest ray (usually the LOS ray unless it is blocked) to the next strongest ray. When the LOS path is blocked or is not present, the channel is well described by a Rayleigh distribution. The clustering nature of the model is illustrated in Fig. 3.38 and the physical environment is modeled as shown in Fig. 3.39. The model for the Channel Impulse Response (CIR) for the outdoor models is written as [YXVG11][Yon07][MEP+10]

$$h(t,\phi) = \beta\delta(t,\Phi_{\text{LOS}}) + \sum_{l=0}^{L}\sum_{k=0}^{K} a_{k,l}\delta(t - T_l - \tau_{k,l})\delta(\phi - \Phi_l - \psi_{l,k}) \quad (3.51)$$

where β is the gain of the LOS component, which is assumed to occur at zero excess delay and to arrive from azimuth angle Φ_{LOS}. The NLOS clusters are represented by the double summation that follows the LOS impulse. The first summation represents the individual clusters, of which there are L. The second summation represents the rays or multipath components (MPCs) within each cluster, and each cluster is assumed to have K such components. The clusters occur at a rate $1/T_l = \alpha$. The primary (i.e., strongest) ray of each cluster is called the *curser*. The number of minor rays that precede the primary ray is N_f and the number that follow is N_b. The rays are assumed to occur at rate λ_f or λ_b depending on whether the ray precedes or follows the main ray. In (3.51), each of the rays is assigned a unique delay of $T_l + \tau_{k,l}$. The pre-curser arrival rate is the average of the inter-arrival times of the rays that precede the primary ray. The post-curser arrival rate is the average rate at which rays within a given cluster that follow the curser ray arrive. The pre-curser rays grow exponentially at a rate determined by γ_f, while the post-curser rays decay at a rate determined by γ_b. The clusters decay in amplitude at an exponential rate determined by Γ. Equation (3.51) indicates that rays are also clustered in angle, that is, they arrive in small deviations from a few key angles given by the set of Φ_l. The impinging angle of each ray is given by $\Phi_l + \psi_{k,l}$. Although not indicated in (3.51), the rays will also be clustered in elevation angle. The average number of clusters, $\bar{L}$, is a statistical parameter. The mmWave channel is found to be Rician in many cases, although some have also found that the LOS path characteristic of a Rician channel may actually be the combination of two rays [GGSK11]. For Rician channel models, the K-factor is a primary descriptor of the channel — it tells the ratio of the LOS path to the mean power of the cluster's MPCs. Quoting [GGSK11]: "The larger the value of the Rician K-factor, the stronger the LOS component in the channel. Experimental results show that the Rician K-factor increases with the decrease of channel RMS delay spread in general." The IEEE 802.15.3c model includes a pre-curser and post-curser K-factor for

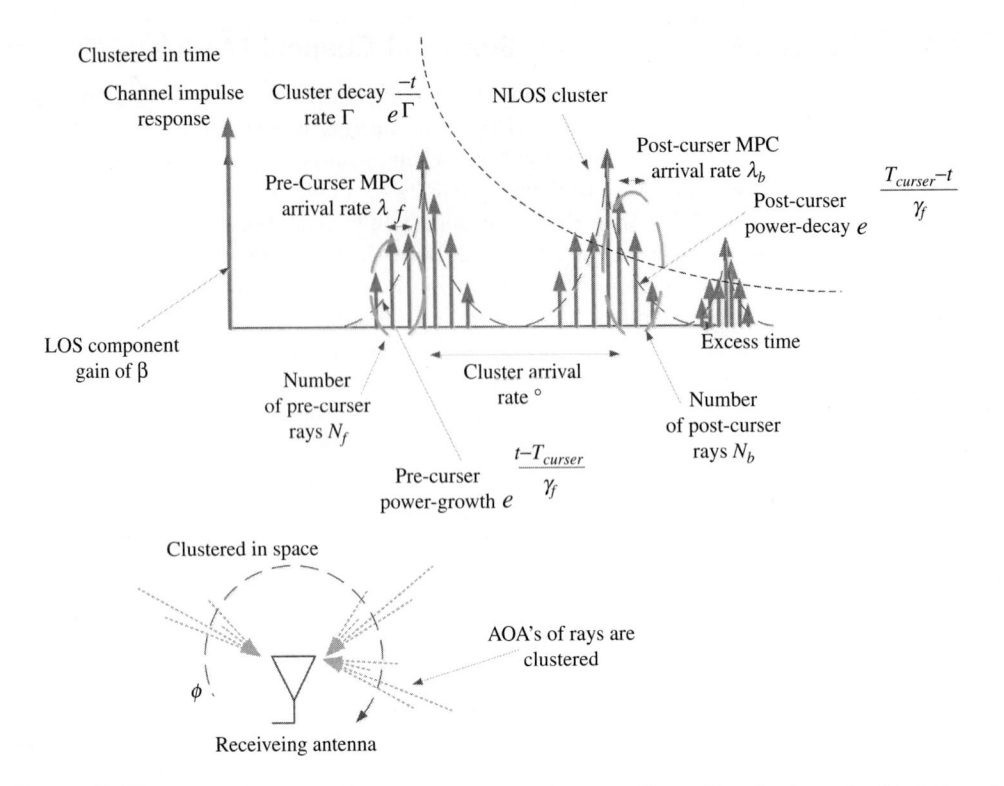

Figure 3.38 Representation of key parameters used to specify multipath channels. Statistics of the key channel parameters are generated from measured data, as collected by wideband channel sounders, to determine the temporal and spatial channel models that can be used by researchers and standard bodies for modem design and signaling protocols.

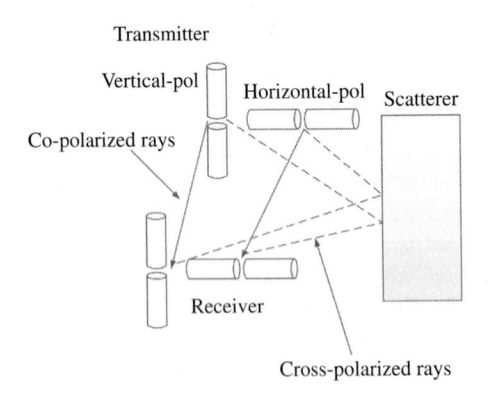

Figure 3.39 Physical assumptions for a multipath channel model.

each cluster. The pre-curser K-factor, k_b, is the ratio of the curser ray (the strongest ray in the cluster) to the strongest ray that precedes the curser ray. The post-curser K-factor, k_p, is the ratio of the curser ray to the strongest ray that follows the curser ray.

3.8 Indoor Channel Models

When no directional antennas (or directional beamforming with antenna arrays) are used, the multipath components within each cluster are exponentially decaying [GGSK11]

$$E\left\{|a_{k,l}|^2\right\} = \Omega_o e^{\frac{-T_l}{\Gamma}} e^{-\frac{\tau_{k,l}}{\gamma}} \tag{3.52}$$

where Ω_o is "the mean energy of the first path of the first cluster, Γ is the cluster decay factor, and γ is the ray decay factor." T_l is the lth cluster and k, l refers to the kth MPC of the lth cluster. This is an omnidirectional channel model.

To account for the impacts of antenna polarization, the channel model must be expanded from a simple CIR to a 2×2 time-dependent matrix that captures co- and cross-polarization characteristics of the channel. Although LOS paths will not change polarization (significantly) as they propagate from the transmitter to the receiver, scattered and reflected rays may change polarization, allowing for a vertically polarized transmitting antenna to communicate with a horizontally polarized antenna. The complete single input single output (SISO) CIR model that allows for consideration of antenna polarization effects is written

$$\mathbf{h}(t, \phi) = \beta \delta(t, \Phi_{\text{LOS}}) \mathbf{I} + \sum_{l=0}^{L} \mathbf{H}^{(l)} \sum_{k=0}^{K} a_{k,l} \delta(t - T_l - \tau_{k,l}) \delta(\phi - \Phi_l - \psi_{l,k}) \tag{3.53}$$

where $\mathbf{H}^{(l)}$ is a 2×2 matrix that describes the co- and cross-polarization characteristics of the lth cluster. The LOS component is scaled by the identity matrix $\mathbf{I}$, as the polarization of the LOS component is not expected to change between the transmitter and the receiver.

The value of β depends on the transmitter and receiver antenna gains and orientations. The IEEE 802.15.3c channel model contains an explicit formula for β based on the transmitter and receiver antenna heights and gains in the direction of ray departure/arrival.

Equations (3.51) and (3.53) only indicate angle of arrival (AOA) information. To allow for a complete simulation based on the transmitter's location, angle of departure (AOD) information must also be included in the model [MEP+10]

$$\mathbf{h}(t, \phi_{\text{TX}}, \phi_{\text{RX}}) = \beta \delta(t, \Phi_{\text{LOS}}) \mathbf{I}$$
$$+ \sum_{l=0}^{L} \mathbf{H}^{(l)} \sum_{k=0}^{K} a_{k,l} \delta(t - T_l - \tau_{k,l}) \delta(\phi_{\text{RX}} - \Phi_{l,\text{RX}} - \psi_{l,k,RX}) \delta(\phi_{\text{TX}} - \Phi_{l,\text{TX}} - \psi_{l,k,\text{TX}})$$
$$\tag{3.54}$$

where we have included a spatial delta-function to indicate the angle of arrival and the angle of departure of the k, l^{th} ray and consider only azimuth angles but it could be generalized to elevation. AOD information is needed to accurately simulate a transmitter-receiver system.

One important assumption of the channel models used for the 802.11ad and 802.15.3c standards is that the time of arrival, azimuth-angle, and elevation-angle distributions are all assumed to be independent [YXVG11]. The angle and time distributions are of course not truly independent, but this assumption is often used, so that

$$p(\phi_{\text{RX}}, \phi_{\text{TX}}, \theta_{\text{RX}}, \theta_{\text{TX}}, \tau) = p(\phi_{\text{RX}}) p(\phi_{\text{TX}}) p(\theta_{\text{RX}}) p(\theta_{\text{TX}}) p(\tau), \tag{3.55}$$

where $p(\phi_{\text{RX}})$, $p(\phi_{\text{TX}})$, $p(\theta_{\text{RX}})$, $p(\theta_{\text{TX}})$, and $p(\tau)$ are the distributions for the receiver azimuth angle, the transmitter azimuth angle, the receiver elevation angle, the transmitter elevation angle, and the time-of-arrival of an MPC component, respectively. Although each of the distributions for these separate parameters is approximated as independent in the models, there is generally a dependence of one ray or cluster on the cluster or clusters that precedes it.

The values for different distributions are summarized from [Yon07]. The azimuth angle of arrival/departure of a cluster is described by a uniform distribution conditioned on the azimuth angle of the preceding ray

$$p\left(\Phi_{\text{RX},i} | \Phi_{\text{RX},i-1}\right) = \frac{1}{2\pi} \tag{3.56}$$

where $\Phi_{\text{RX},i}$ is the average azimuth angle of the MPC rays in the ith cluster, and $\Phi_{\text{RX},i-1}$ is the average angle of the MPC rays in the $(i-1)^{\text{th}}$ cluster. The deviation of each ray's azimuth angle from the mean angle of its cluster is given by either a zero-mean Laplacian or Gaussian distribution

$$p\left(\psi_{l,k,\text{RX}}\right) = \frac{1}{\sqrt{2\pi}\sigma_\phi} \exp\left(-\frac{\psi_{l,k,\text{RX}}^2}{2\sigma_\phi^2}\right) \tag{3.57}$$

$$p\left(\psi_{k,l}\right) = \frac{1}{\sqrt{2}\sigma_\phi} \exp\left(-\left|\frac{\sqrt{2}\psi_{k,l}}{\sigma_\phi}\right|\right) \tag{3.58}$$

where (3.57) is a Gaussian distribution with variance σ_ϕ^2 and (3.58) is a Laplacian distribution with standard deviation σ_ϕ, also called the *angular spreads* of the distributions.

The cluster arrival times and MPC arrival times within each cluster are described by exponential distributions

$$p\left(T_l | T_{l-1}\right) = A\left[-A\exp\left(T_l - T_{l-1}\right)\right], l > 0 \tag{3.59}$$
$$p\left(\tau_{k,l} | \tau_{k,l-1}\right) = \lambda\left[-\lambda\exp\left(\tau_{k,l} - \tau_{k,l-1}\right)\right], k > 0 \tag{3.60}$$

and the ray and cluster amplitudes are log-normally distributed

$$p\left(a_{k,l}\right) = \frac{1}{\sqrt{(2\pi)}\sigma_r a_{k,l}} \exp\left(-\frac{(\ln(a_{k,l}) - \mu_r)^2}{2\sigma_r^2}\right) \tag{3.61}$$

where μ_r and σ_r^2 are the mean and variance, respectively, of the ray amplitudes $a_{k,l}$. Note that the ray standard deviation and the path loss shadowing variable are similar in that they both are for log-normal distributions, but they are not equal. This is because the path loss is generally computed by accounting for the energy in all of the rays of the channel impulse response. Therefore, both individual ray amplitudes and sums of ray amplitudes are well described by log-normal distributions.

The average MPC amplitude of a cluster may also be described by a log-normal distribution, with standard deviation σ_c.

The large-scale characteristics of the 60 GHz models are based on a log-distance dependence of path loss with log-normal shadowing [YXVG11]. Log-normal shadowing

3.8 Indoor Channel Models

indicates that measured values of path loss will fit a normal distribution with the values represented in a decibel scale. Due to the very wide absolute bandwidth of channels dedicated to 60 GHz communication, the frequency-dependence of the path loss is also sometimes taken into account:

$$PL(d,f) = PL_0 + 10\kappa \log_{10}\left(\frac{f}{f_0}\right) + 10n \log_{10}\left(\frac{d}{d_0}\right) + X_\sigma \qquad (3.62)$$

where PL_0 is the reference path loss at carrier frequency f_0 and transmitter-receiver separation distance d_0. κ is the frequency-loss factor (and is usually equal to 2); though further testing is needed to accurately determine the frequency-dependence of path loss [YXVG11]. Note that n is the path loss exponent, which describes how the path loss increases as the transmitter-receiver separation distance d increases above d_0. X_σ is the log-normally distributed random variable that describes the shadowing characteristics of the channel. [YXVG11] states that most indoor LOS channels have a path loss exponent from 0.4 to 2.1 (a value less than 2 indicates that the environment prevents the energy from spreading out spherically, but rather guides some of the radiated energy from the transmitter to the receiver). NLOS indoor 60 GHz channels have a path loss of 1.97-5.40 [YXVG11]. Higher antenna positions relative to the floor result in lower path loss exponents for most indoor environments [YXVG11]. If path loss is parameterized according to polarization (co-polarized path loss or cross-polarized path loss), then the cross-polarized path loss exponent is in most cases higher than the co-polarized path loss exponent.

To convert data from a given measurement with a particular set of transmitter and receiver antennas, simply add the gains (in decibel units) of the transmitter and receiver antennas to the reference path loss.

The rate at which the channel changes in time is described by the Doppler characteristics of the channel. The Doppler spread is the maximum amount by which the frequency of a signal sent into channel can be changed by movement of the transmitter, receiver, or objects in the channel. The inverse of the Doppler spread is the coherence time, and it can be thought of as the time over which the channel's properties are relatively constant. [YXVG11] gives a maximum Doppler spread of 300 Hz, corresponding to a coherence time of 0.6 ms for indoor channels. The coherence time is generally much longer than the symbol duration from IEEE 802.11ad or IEEE 802.15.3c, so that many consecutive symbols will experience approximately the same channel—that is, the channel has a block-fading characteristic (also known as slow-fading) [Rap02].

The wide bandwidth used by 60 GHz modulation and coding scheme (MCS) standards, however, results in frequency selective fading, that is, the frequency response of the channel is often not flat across the band of interest [YXVG11][Rap02]. Because of the wide bandwidth and frequency selective fading nature of these systems, they will not experience substantial received-power fading (i.e., variations in the received power) for movement over short distances. Note that the use of high gain directional antennas may result in a flattening of the channel frequency response, especially for LOS links. This is due to the spatial filtering characteristics of directional antennas, which acts to reduce RMS delay spread (recall the RMS delay spread is inversely proportional to coherence bandwidth) by blocking certain paths. Use of circularly polarized antennas also reduces RMS delay spread and hence reduces frequency selectivity [YXVG11]. Contrastingly, a flat-fading system has a channel frequency response that is approximately flat across the bandwidth of interest, but experiences substantial received-power fading for movement over short distances.

Inclusion of the effect of human-induced fades or shadows is a distinguishing feature of the 60 GHz channel models [YXVG11]. Human-induced fades can range from 18 to 36 dB in strength and can last on the order of tens to hundreds of milliseconds [YXVG11] [JPM+11]. [JPM+11] showed that humans generally do not entirely block paths, but they rather severely attenuate paths. Some researchers have shown that a knife-edge diffraction model works well to simulate blockage due to human bodies. [JPM+11] uses a multi-knife-edge model that includes the possibility of a ray diffracting over a person's head. People are modeled as cubes with six knife edges. For indoor scenarios, a reflection from the ceiling can greatly alter the impact of a human. A human blockage event was found to significantly increase RMS delay spread: Human blockage results in lower SNR and increased RMS delay spread, both of which increase BER or decrease link capacity.

To summarize, the key parameters of the IEEE 802.11ad and IEEE 802.15.3c channel models are [Yon07]:

1. PL_0, the path loss at a reference distance, usually 1 m

2. n, the path loss exponent, which describes how average path loss depends on transmit-receive distance

3. X_σ, shadowing standard deviation

4. β, gain of the first MPC component, usually assumed to be LOS

5. α, inter-cluster (cluster) arrival rate

6. λ, intra-cluster (ray) arrival rate

    (a) λ_f the intra-cluster pre-curser ray arrival rate

    (b) λ_b the intra-cluster post-curser ray arrival rate

7. Γ, inter-cluster (cluster) decay rate

8. γ, intra-cluster (ray) decay rate

    (a) γ_f the intra-cluster pre-curser growth rate

    (b) γ_b the intra-cluster post-curser decay rate

9. σ_c, cluster log-normal standard deviation

10. σ_r, ray log-normal standard deviation

11. σ_ϕ, angle spread, aka the standard deviation of the azimuth angle distribution

12. $\bar{L}$, average number of clusters

13. $P(L)$, the distribution of the number of clusters

14. d, TX-RX separation, h_1, TX height, h_2, RX height, G_T, TX gain, G_r, RX gain, Δ_k, ray Rician factor. These parameters are used in the estimation of β.

We now review the particular forms of the channel models used in IEEE 802.15.3c and IEEE 802.11ad. Due to the limited measurement available, these models were derived largely through ray tracing simulations (that were subsequently verified against measurements). As shown in Section 3.2.8, ray tracing works well provided accurate information regarding topography and type/level of clutter and material properties is available.

3.8.4 IEEE 802.15.3c

The IEEE 802.15.3c channel model includes sub-models for nine different situations/environments. These are described in Table 3.18, reproduced from [Yon07] [BSL+11]. The IEEE 802.15.3c model is applicable to LOS or NLOS channels [BSL+11]. The generalized S-V model (clustered in time and space [SV87]) is used to represent channel impulse responses, with the exception of the use of separate LOS component representation [BSL+11]. Cluster arrivals and ray arrivals within clusters are modeled with Poisson random processes, whereas amplitudes are modeled with log-normal distributions [BSL+11]. Angles of arrival are considered to take a uniform distribution and are assumed to be independent across paths [BSL+11], while individual rays within a cluster have angles of arrival that are either Gaussian or Laplacian distributed.

The path loss exponent for various IEEE 802.15.3c channels is given by Table 3.19. For those entries in Table 3.19 with frequency dependence, the path loss is given as:

$$PL = PL_0 + 20\log_{10}(f) + 10n\log_{10}\left(\frac{d}{d_0}\right) + X_\sigma \qquad (3.63)$$

while for those entries without frequency dependence, the frequency-dependent term is not included.

The value of β, which describes the gain of the LOS path in the channel impulse response, is given by the following equation for the IEEE 802.15.3c channel model in a desktop environment [Yon07]:

$$\beta\,[\text{dB}] = 20\log_{10}\left[\frac{\mu_d}{d}\left|\sqrt{G_{t1}G_{r1}} + \sqrt{G_{t2}G_{r2}}\Gamma_0 \exp\left(\frac{j\frac{2\pi}{\lambda_f}2h_1h_2}{d}\right)\right|\right] - PL_d(\mu_d) \qquad (3.64)$$

where

$$PL_d(\mu_d)\,[\text{dB}] = PL_d(d_o) + 10 \cdot n \cdot \log_{10}\left(\frac{d}{d_o}\right) \qquad (3.65)$$

$$PL_d(d_o) = 20\log_{10}\left(\frac{4\pi d_o}{\lambda_f}\right) + A_{\text{NLOS}} \qquad (3.66)$$

where λ_f is the operating wavelength in (3.64) and (3.65).

Table 3.18 IEEE 802.15.3c channel models for various indoor environments where a human is holding a portable device [reproduced from [Yon07][BSL+11] © IEEE].

Channel Model	Scenario	Environment	Description
CM1	LOS	Residential	Home with furnished rooms. Comparable in size to a small office. Wood or concrete walls and floor with carpet or paper covering. Windows and doors are present.
CM2	NLOS		
CM3	LOS	Office	Multiple office chairs and desks, computers. Bookshelves, cabinets, and cupboards. Metal/concrete walls. Cubicles or lab workstations may be present. Long corridors connect offices together.
CM4	NLOS		
CM5	LOS	Library	Multiple desks and book-filled metal/wooden shelves. Large windows may be on one or more sides. Large public doorways.
CM6	NLOS		
CM7	LOS	Desktop	Office and computer clutter. Often enclosed by a cubicle.
CM8	NLOS		
CM9	LOS	Kiosk	Station in a mall or other public venue. Users are intended to stand directly in front and close (1-2 m) to the kiosk.

The mean number of clusters in the IEEE 802.15.3c channel model is 3-14, but ranged most typically from 3 to 4 [YXVG11]. Notably, [Yon07] found that the number of clusters did not follow any one distribution but rather had a statistical nature that was highly dependent on the environment.

The other channel parameters used in the IEEE 802.15.3c model may be found in Table 3.20, Table 3.21, Table 3.22, and Table 3.23, reproduced from [Yon07]. Table 3.21 lists particular values for the IEEE 802.15.3c channel model for a kiosk environment.

3.8.5 IEEE 802.11ad

The IEEE 802.11ad channel model is very similar to the IEEE 802.15.3c channel model and is more pertinent because there is large commercial activity surrounding 60 GHz wireless communications. The IEEE 802.11ad channel modeling committee has expended considerable effort in studying the statistical nature of the 60 GHz indoor channel,

Table 3.19 Path loss values used in the development of 60 GHz channel models [top 4 entries from [YXVG11] © IEEE].

Environment	LOS/NLOS	n	PL_0 (dB)	X_{sigma} (dB)	Frequency Dependence	Antennas	Reference
Residential	LOS	1.53	75.1	1.5	None	TX 72° HPBW, RX 60° HPBW	[YXVG11]
Residential	NLOS	2.44	86.0	6.2	None	TX 72° HPBW, RX 60°	[YXVG11]
Office	LOS	1.16	84.6	5.4	None	TX Omni, RX 30° HPBW	[YXVG11]
Office	NLOS	3.74	56.1	8.6	None	TX Omni, RX 30° HPBW	[YXVG11]
Conference Room[a]	LOS	2.0	32.5	N/A	$20 \times log_{10}(f)$	N/A	[MEP+08]
Conference Room[a]	NLOS	0.6	51.5	N/A	$20 \times log_{10}(f)$	N/A	[MEP+08]

[a] Applicable also to IEEE 802.11ad models.

Table 3.20 The model parameters used in CM1 and CM2 for residential environments for use in evaluation of IEEE 802.15.3c PHYs [reproduced from [Yon07] © IEEE].

Residential	LOS (CM1)					NLOS (CM2)	Comment
	TX-360°, RX-15° NICT	TX-60°, RX-15° NICT	TX-30°, RX-15° NICT	TX-15°, RX-15° NICT	TX-360°, RX-15° NICTA		
Λ [1/ns]	0.191	0.194	0.144	0.045	0.21	N/A	
λ [1/ns]	1.22	0.90	1.17	0.93	0.77	N/A	
Γ [ns]	4.46	8.98	21.50	12.60	4.19	N/A	
γ [ns]	6.25	9.17	4.35	4.98	1.07	N/A	
σ_c [dB]	6.28	6.63	3.71	7.34	1.54	N/A	
σ_r [dB]	13.00	9.83	7.31	6.11	1.26	N/A	
σ_ϕ [degree]	49.80	119.00	46.20	107.00	8.32	N/A	
$\bar{L}$	9	11	8	4	4	N/A	
Δk [dB]	18.80	17.40	11.90	4.60	N/A	N/A	
$\Omega(d)$ [dB]	−88.70	−108.00	−111.00	−110.70	N/A	N/A	Ω_0 was derived at 3 m
n_d	2	2	2	2	N/A	N/A	
A_{NLOS}	0	0	0	0	N/A	N/A	

Table 3.21 The channel parameters used for evaluation of IEEE 802.15.3c PHYs for use in office environments [reproduced from [Yon07] © IEEE].

Office	LOS (CM3)			NLOS (CM4)		Comment
	TX-30°, RX-30° NICT	TX-60°, RX-60° NICT	TX-360°, RX-15° NICT	TX-30°, RX-15° NICT	Omni-TX, RX-15° NICTA	
Λ [1/ns]	0.041	0.027	0.032	0.028	0.07	
λ [1/ns]	0.971	0.293	3.45	0.76	1.88	
Γ [ns]	49.80	38.80	109.20	134.00	19.44	
γ [ns]	45.20	64.90	67.90	59.00	0.42	
σ_c [dB]	6.60	8.04	3.24	4.37	1.82	
σ_r [dB]	11.30	7.95	5.54	6.66	1.88	
σ_ϕ [degree]	102.00	66.40	60.20	22.20	9.10	
$\bar{L}$	6	5	5	5	6	
Δk [dB]	21.90	11.40	19.00	19.20	N/A	
$\Omega(d)$ [dB]	−3.27d −85.80	−0.303d −90.30	−109.00	−107.20	N/A	
n_d	2.00	2.00	3.35	3.35	N/A	
A_{NLOS}	0	0	5.56@3 m	5.56@3 m	N/A	

Table 3.22 The IEEE 802.15.3c channel model for library environments [reproduced from [Yon07] © IEEE].

Library	LOS (CM5)	NLOS (CM6)	Comment
Λ [1/ns]	0.25	N/A	
λ [1/ns]	4.00	N/A	
Γ [ns]	12.00	N/A	
γ [ns]	7.00	N/A	
σ_c [dB]	5.00	N/A	
σ_r [dB]	6.00	N/A	
σ_ϕ [degree]	10.00	N/A	
$\bar{L}$	9	N/A	
K_{LOS} [dB]	8	N/A	

Table 3.23 The IEEE 802.15.3c channel model for desktop environments [reproduced from [Yon07] © IEEE].

Desktop	LOS (CM7)		LOS (CM7) Omni-TX, RX-21 dBi	NLOS (CM8)
	TX-30°, RX-30°	TX-60°, RX-60°		
Λ [1/ns]	0.037	0.047	1.72	N/A
λ [1/ns]	0.641	0.373	3.14	N/A
Γ [ns]	21.10	22.30	4.01	N/A
γ [ns]	8.85	17.20	0.58	N/A
σ_c [dB]	3.01	7.27	2.70	N/A
σ_r [dB]	7.69	4.42	1.90	N/A
σ_ϕ [degree]	34.60	38.10	14.00	N/A
$\bar{L}$	3	3	14	N/A
Δk [dB]	11.00	17.20	N/A	N/A
$\Omega(d)$ [dB]	4.44d −105.4	3.46d −98.4	N/A	N/A
h_1	Uniform dist. Range: 0-0.3	Uniform dist. Range: 0-0.3	N/A	N/A
h_2	Uniform dist. Range: 0-0.3	Uniform dist. Range: 0-0.3	N/A	N/A
d	Uniform dist. Range: d±0.3	Uniform dist. Range: d±0.3	N/A	N/A
G_{T1}	GSS[a]	GSS	N/A	N/A
G_{R1}	GSS	GSS	N/A	N/A
G_{T2}	GSS	GSS	N/A	N/A
G_{R2}	GSS	GSS	N/A	N/A
n_d	2	2	N/A	N/A
A_{NLOS}	0	0	N/A	N/A

[a] GSS is a Gaussian antenna model with side lobe level discussed in [Toy06].

3.8 Indoor Channel Models

Table 3.24 The IEEE 802.15.3c channel model for a kiosk environment, where a human is holding a portable device pointed to the kiosk [reproduced from [Yon07] © IEEE].

Kiosk	LOS (CM9)		Comment
	TX-30°, RX-30° Environment 1	TX-30°, RX-30° Environment 2	
Λ [1/ns]	0.0546	0.0442	
λ [1/ns]	0.917	1.01	
Γ [ns]	30.20	64.20	
γ [ns]	36.50	61.10	
σ_c [dB]	2.23	2.66	
σ_r [dB]	6.88	4.39	
σ_ϕ [degree]	34.20	45.80	
$\bar{L}$	5	7	
Δk [dB]	11.00	9.10	
$\Omega(d)$ [dB]	−98.00	−107.80	Ω_0 was derived at 1 m
n_d	2	2	
A_{NLOS}	0	0	

Table 3.25 The average number of clusters for the IEEE 802.11ad channel model for a conference room environment [reproduced from [MEP+10] © IEEE].

Type of clusters	Number of clusters for STA-STA sub-scenario	Number of clusters for STA-AP sub-scenario
LOS path	1	1
First order reflections from walls	4	4
Second order reflections from two walls	8	8
First order reflection from ceiling	1	
Second order reflections from the walls and ceiling	4	

primarily through the use of ray-tracing simulation tools. As described by [MEP+10], the channel model describes spatial and temporal effects, in addition to capturing amplitude, phase, and polarization channel characteristics or impacts. Spatial characteristics are captured through transmit-receive separation, azimuth angle, and elevation angle.

As for the IEEE 802.15.3c channel model, the IEEE 802.11ad channel model is based on a generalization of the popular Saleh-Valenzuela channel model. Most of the results for the IEEE 802.11ad channel model were derived from study of a conference room environment as of 2007.

The IEEE 802.11ad model's representation of LOS paths is based on Friis' free space equation [YXVG11]. The gain of non-LOS paths is modeled using the following equation (for the ith path), which accounts for the wavelength λ, the reflection loss g_i of the ith path, the LOS distance between the transmitter and receiver d, and the NLOS distance excluding the LOS path distance R:

$$\beta_i(\text{dB}) = 20 \log_{10}\left(\frac{g_i \lambda}{4\pi(d+R)}\right). \tag{3.67}$$

The IEEE 802.11ad channel model classifies clusters according to two different use cases — Station-to-Station (STA-STA) or Station-to-Access Point (STA-AP) [MEP+10]. For each scenario, both LOS and NLOS paths are considered, and the impact of the number of reflections of a ray from the transmitter to the receiver is taken into account. The average number of clusters for each scenario in a small conference room setting was found using ray tracing by [MEP+10], and the results are shown in Table 3.25 (reproduced from [MEP+10]).

Table 3.26, reproduced from [MEP+10], summarizes the key factors used in the IEEE 802.11ad channel model.

3.9 Chapter Summary

The emerging world of mmWave wireless communications offers a new frontier for the engineering world. Radio propagation characteristics are quite different in many respects at this new spectral frontier, given the small wavelengths and remarkable capabilities that now exist to build highly directional antennas. Yet, in other ways, the mmWave channel is remarkably similar to UHF and microwave channels in use today. For example, the

Table 3.26 The key parameters for the conference room environment for the IEEE 802.11ad channel model [reproduced from [MEP+10] © IEEE].

Parameter	Notation	Value
Pre-cursor rays K-factor	K_f	5 dB
Pre-cursor rays power decay time	g_f	1.3 ns
Pre-cursor rate arrival rate	I_f	0.20 ns^{-1}
Pre-cursor rays amplitude distribution		Rayleigh
Number of pre-cursor rays	N_f	2
Post-cursor rays K-factor	K_b	10 dB
Post-cursor rays power decay time	g_b	2.8 ns
Post-cursor rate arrival rate	I_b	0.12 ns^{-1}
Post-cursor rays amplitude distribution		Rayleigh
Number of post-cursor rays	N_b	4

3.9 Chapter Summary

omnidirectional channel models for mmWave outdoor and indoor channels are remarkably similar to those used in today's wireless systems. However, at mmWave frequencies, diffraction becomes negligible, and reflection and large surface scattering become the primary propagation modes. Directional antennas and ray-tracing design methodologies become powerful engineering tools.

This chapter has illuminated the major fundamental issues surrounding mmWave radio propagation that are pertinent to creating mmWave wireless networks. We provided fundamental treatment of all key radio propagation and channel modeling areas for the design of wireless networks, for both fixed and mobile applications in indoor, outdoor, and vehicle-to-vehicle communications.

This chapter has provided a detailed compilation of real-world measurements and modeling techniques that give an early look at the radio channel properties of emerging mmWave systems. Weather effects, including rain and hail, as well as measurements and models for reflection, penetration, scattering, and path loss were provided. This chapter also described key channel statistics used by researchers and industry practitioners to develop wireless standards. We presented extensive mmWave propagation measurement data, and demonstrated many large- and small-scale channel models based on the body of existing measurements. These new measurements and models are well suited for international standards activities that will be needed for the development of future mmWave wireless networks, although much more work is needed in the field. We illustrated the IEEE 802.15.3c and IEEE 802.11ad channel models as particular examples, and we provided insights into how past modeling efforts, such as ray tracing and statistical modeling, may be applied to mmWave channels.

The field of radio propagation is complex and rich with research opportunities. As carrier frequencies and bandwidths increase, there will surely be a need for new models, measurements, and methodologies that can aid engineers in the development and deployment of wireless products.

Chapter 4

Antennas and Arrays for MmWave Applications

4.1 Introduction

The extremely short wavelengths of mmWave signals (e.g., 10.7 mm at 28 GHz, 5 mm at 60 GHz, and 789 μm at 380 GHz) offer enormous potential for mmWave antenna arrays that are adaptive, high gain, and inexpensive to fabricate and integrate in mass-produced consumer electronic products. There are both cost and performance advantages that result from extremely integrated and physically small antennas. From a cost perspective, mmWave antennas may be directly integrated with other portions of a transceiver and may be fabricated with either packaging or integrated circuit (IC) production technology. This is a stark departure from all existing wireless systems to date, which rely on coaxial cables, transmission lines, and printed circuit boards to connect antennas with the transmitter or receiver circuits in modern cellphones, laptops, and base stations.

The miniaturization caused by the smaller electrical wavelength now makes it possible to create entire wireless communication systems in one integrated circuit (IC) production process (also known as *circuit fabrication*, or *fab*), thereby eliminating costs associated with the interconnection cables and additional manufacturing steps that connect today's radio components together with many different processes. For example, rather than having to purchase a separate antenna for integration with a printed circuit board (PCB) that contains the rest of the transceiver, a mmWave on-chip antenna may be directly etched in on-chip metal during a complementary metal oxide semiconductor (CMOS) Back End of Line (BEOL) IC production process. Or, at slightly higher cost, the antenna may be fabricated in the packaging technology used to house the RF amplifier chip, or integrated in the printed circuit board used to house the transceiver. Both of these options will be less expensive than the use of a separate antenna with a separately packaged transceiver and will benefit further from lower ohmic losses due to the fact that less power is wasted when transferring mmWave signals between the antenna and the transceiver [HBLK14][LGPG09][RGAA09][RMGJ11][GJRM10][GAPR09].

In this text, we focus on emerging antennas that will likely be used in mobile and portable mmWave systems and devices of the future, as fixed antennas such as horn antennas or parabolic dishes are well known for conventional microwave and fixed mmWave wireless systems and are treated elsewhere in the literature. Our goal here is to introduce the reader to antenna topologies and fabrication methods appropriate for mmWave technologies. We also discuss various packaging technologies as they pertain to mmWave antennas that will be embedded in future cellular, personal/local area networking, and backhaul equipment. Proper characterization of mmWave antennas is challenging, due primarily to their unprecedented small size and implementation novelty. Before installing antennas in practical cellular or personal area networking systems or using integrated antennas for consumer or industrial connectivity equipment, antennas must be tested and understood in a laboratory setting. On-chip antennas, for example, may require the use of metal probing stations to excite the antenna in a laboratory. In-package antennas require precise coupling between the integrated circuit and the plastic package. Measurement gear, such as probing stations or custom test chips, is typically made of metal and introduce many obstacles that can interfere with pattern measurement by introducing multipath. Hence, accurate antenna patterns are difficult to ascertain in the laboratory, let alone for in-situ installations. An alternative to testing antennas with a probe station is to package the antenna with an active transmitter or receiver chip, or to place the antennas on an actual circuit board or enclosure, and to then use an anechoic chamber or outdoor antenna range for near-field or far-field patterns. This requires selection of a transmitter or receiver design, adding to testing cost. If the packaging process or enclosure is changed, all antenna measurements would have to be repeated due to the small wavelengths at mmWave frequencies.

Other challenges for mmWave antennas include design of the proper antenna pattern for the particular application and the proper design of passive feeding and/or active excitation elements such as baluns and hybrids. Even with adaptive arrays or multiple input multiple output (MIMO) systems, in which signal processing is used to alter the instantaneous antenna pattern, designers must know the efficiency and capabilities of antennas before installing them into actual systems and products. In this chapter, we discuss the challenges described above associated with mmWave antenna design and testing. We introduce the reader to both on-chip and in-package antennas, as well as their requirements and advantages. MmWave antennas are key to realizing the potential of mmWave systems such as 28 GHz, 60 GHz, and higher frequency transceivers, for either fixed (backhaul or fronthaul) or mobile/portable use. This chapter covers:

- review of certain mmWave antenna fundamentals, including array fundamentals;
- discussion of various antenna topologies that have been used for mmWave designs (including dipole, loop, Yagi-Uda, and traveling wave antennas such as Rhombic antennas);
- the on-chip antenna environment and associated challenges and solutions;
- in-package antenna environment;
- dielectric lens antennas;
- characterization methods for mmWave antennas.

In-package antennas, especially if fabricated using package technology and not simply placed inside the package, offer special challenges due to the relatively bulky size of antennas' elements and the limitations of integrated circuit packaging technology (e.g., the widths of metal vias and the height of metal layers above the ground plane). We describe various structures that have been used to improve the performance of mmWave antennas, including dielectric lenses and modern integrated lens antennas, and although recent advances in circuit board antennas have already been presented [HBLK14], we focus primarily on integrated on-chip and in-package antennas. We end the chapter with a discussion of characterization methods for mmWave antennas and describe the equipment that must be purchased to test mmWave antennas.

4.2 Fundamentals of On-Chip and In-Package MmWave Antennas

As discussed in Chapter 3, the short wavelengths at mmWave frequencies allow both the transmit and receive antennas or antenna arrays to be the size of many multiples of a wavelength and still easily fit within a package or on a chip. For example, at 60 GHz a quarter wave dipole is only 625 μm on a substrate with a relative permittivity of 4. A 100-element phased array, say a square 10×10 array of such dipoles, would have a maximum aperture length dimension of approximately $10 \times 625\,\mu m \times \sqrt{2} = 8.83$ mm $= \frac{3.53\lambda}{\sqrt{\epsilon_r}}$, where we have assumed a *relative permittivity* ϵ_r (where relative permittivity ϵ_r is also known as the *dielectric constant*) of the package substrate material equal to 4. At 2.4 GHz, 3.53 wavelengths would have required 0.22 m in the same material, or 24 times the length required at 60 GHz. The opportunity afforded by mmWave frequencies to integrate antenna arrays that are many multiples of a wavelength in a very small size is a key advantage. In fact, as should be clear from the preceding chapters, increased antenna gains that can be achieved in very small areas at mmWave frequencies are one of the keys to making a vast number of mmWave technologies feasible

Chapter 3 demonstrated how wireless systems in mmWave frequencies will most likely require beam steering in a very small form factor. Beam steering is possible due to the tight beamwidth that is achievable with *electrically large* (i.e., large compared to a wavelength in the particular material substrate) antenna arrays. This opens up the possibility of massive MIMO, improved link margin for cellular carriers, multi-Gbps personal area networks, and even hand-held radars (which may be useful, for example, to direct a user to a nearby object in indoor locations or other scenarios without the availability of GPS, for example, to find a car in an underground parking garage). From Chapter 3, Eqn. (3.3), the gain of an antenna or antenna array grows as the square of the electrical length D of the antenna or array—that is, gain grows by $\frac{D^2}{\lambda^2}$. Thus, if both the transmitter and receiver antenna sizes are grown, the path loss at mmWave frequencies may be easily compensated for since Eqn. (3.6) shows how antenna gain increases to the fourth power of antenna aperture length and free space path loss increases by the square of the transmitter-receiver separation distance in free space.

The beamwidth of an antenna or antenna array shrinks linearly with the increase in electrical size of the antenna or antenna array, with a good rule of thumb being [Bal05]

$$\text{Beamwidth} = \Theta \approx \frac{60°}{\left[\frac{D}{\lambda}\right]}. \qquad (4.1)$$

An implication of the different rates of growth and decay of antenna gain and beamwidth, respectively, is that for mmWave applications such as 60 GHz transceivers—and other low sub-terahertz devices—there will be two classes of devices: One class (for example, for personal area networking (PAN)) will be intended for short links and will use antennas that are only large enough to make short connections (by human standards), on the order of meters or tens of meters, and the other class of device (for example, for cellular/outdoor access) will be used in longer-range cellular, mobile, or backhaul systems. The antenna differences in these two classes are driven by application requirements. Personal area networking devices, for example, may not require small enough beam widths to require beam steering, and their design is likely to focus on very low cost, low power consumption, and extreme simplicity. The personal networking devices will likely operate with relatively simple baseband hardware and processing units, as the complexities of exotic beam steering will be avoided. Alternatively, mobile cellular, repeater/relay, or backhaul situations will be intended for longer-range connections at frequencies of 10 GHz and above, where these devices will enable network operation over distances up to 10-500 m, and also may be combined to use the spectrum simultaneously for outdoor backhaul and outdoor urban cellular mobile coverage. This second class of devices will require a substantial number of antennas or antenna elements, perhaps on the order of hundreds or even thousands, to meet link budget requirements. For example, early SiBEAM mmWave devices (now made by Silicon Image, which acquired SiBEAM in 2011) contain several dozen antenna elements for beam steering in a 60 GHz PAN (see Figs. 1.4 and 1.5). It is likely that future mmWave wireless long-distance devices, handsets, and infrastructure will employ 10 to 100 times as many antenna elements in the coming decade, and therefore these devices will have narrow and tunable antenna beamwidths that enable the transceiver to implement beam steering. Beam steering adds to the complexity of the physical layer protocols used to make over-the-air connections, as beam steering requires discovery and coordination between devices (see Chapters 7 and 8), as well as more sophisticated baseband processing hardware. Beam steering also forces the transceiver to implement additional RF or IF hardware including distribution circuits and possibly phase shifting circuits (assuming phase shifting is not performed at baseband)[Gho14]. At higher mmWave frequencies, sub-terahertz, and terahertz frequencies, all devices will eventually require the sophistication of spatial processing and a vast number of antenna elements, as those frequencies above 100 GHz will most likely require beam steering to meet link budget requirements and to exploit multipath and MIMO.

Despite their promise, mmWave antennas are associated with many challenges. High antenna efficiency (as high as possible) is critical and is not easy for on-chip or in-package antennas. The technical requirements of the two aforementioned classes of devices (i.e., long and short range) at mmWave frequencies has also added substantial confusion to the standards generating process—as technical standards such as IEEE 802.15.3c attempt to cover devices that individually and in networks operate according to very different principles and over varying distances. Fifth-generation (5G) cellular standards contemplated by 2020 will likely have to consider all aspects of indoor PAN, outdoor cellular, and backhaul for small cells [RRC14]. The rapidly emerging 60 GHz standards are discussed in Chapter 9.

Designing mmWave antennas and antenna arrays is also very challenging due to at least two factors: first, integrated mmWave antennas are in intimate contact with very complex environments (e.g., a stack of packaging material with other nearby metal objects or on the substrate of an integrated circuit), and they will generally not have the advantage of a separate radome to offer protection from the rest of the radio (although some packaging techniques provide some of the benefits of a radome). The complexity of the operating environment of mmWave antennas makes separating antenna design from transceiver design nearly impossible. An important implication of the requirement for an integrated design process is that a mmWave antenna, and especially a mmWave antenna array, may often decide the overall circuit IC, package, or circuit board layout (i.e., "floor plan") of a mmWave transceiver. This is especially true for applications that require isolation between antenna elements, for example, between receive and transmit elements. The importance of a floor plan to the design of any circuit board, integrated circuit, or packaged circuit should convince designers to begin antenna array design very early in the overall design process, as failure to lock down an integrated design can substantially delay development of other critical blocks, including active circuitry. A key to successful design of antenna array systems is to ensure that all transceiver circuitry fits on the substrate or in the package without being so far apart so as to cause *grating lobes* (undesired antenna pattern effects due to particular physical dimensions of the antenna).

The large electrical size of high gain, multi-element mmWave antennas also makes design and simulation of these antennas quite challenging. A popular electromagnetic simulation method, the *method of moments* (MoM), reveals that the amount of computer memory required to simulate a structure grows according to the level of discretization required to adequately represent the structure for acceptable simulation accuracy [Gib07, Chapter 3]. This is because the MoM works by breaking the structure into extremely small sections and finding a solution by enforcing boundary conditions over each small section. For example, a planar structure, such as an integrated antenna array, would require at least $\left[\frac{D}{\frac{\lambda}{10}}\right]^2$ sections, where D is the longest linear dimension of the array, to adequately simulate the radiation patterns of an antenna. This adds substantially to the circuit design and simulation time. Another popular method for solving computational electromagnetic problems, the *finite element method* (FEM), operates according to different principles but also requires the discretization of the antenna array or structure being simulated, and the number of elements also grows with the electrical size of the object [Dav10].

4.2.1 Antenna Fundamentals

Before proceeding further, we now review the desirable characteristics of an antenna, regardless of whether it is in an array or a single element. The most important consideration, especially in the context of mmWave antennas, is to attain high radiation efficiency. With traditional RF applications that use separate antennas, efficiency is often very high and thus may not be as important as gain, but efficiency for mmWave antennas may be challenging to attain. The efficiency of an antenna indicates the percentage of steady state power that is applied to the input of the antenna that radiates usefully. The rest of the input power is assumed to be lost in conduction currents in the antenna or the nearby environment (e.g., a chip or a package). Radiation efficiency is usually explained in terms of a radiation resistance and a loss resistance, as illustrated in Fig. 4.1. For an

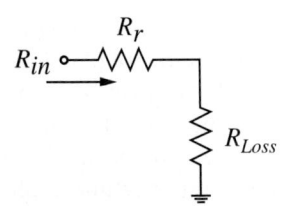

Figure 4.1 At resonance, an antenna may be modeled simply as a resistive circuit that includes both radiation and loss resistance.

input current I_o, the amount of power that is used to radiate is modeled by $I_0^2 R_r$, where R_r is the radiation resistance. The rest of the power is dissipated in R_{Loss}, resulting in an efficiency e_{ant} of:

$$e_{\text{ant}} = \frac{R_r}{R_r + R_{\text{Loss}}}. \tag{4.2}$$

As shown in Section 4.4, in-package mmWave antennas may achieve an efficiency of greater than 80% [SHNT05], whereas unsophisticated on-chip antennas typically have efficiencies closer to 10% [GAPR09]. Section 4.6 describes how on-chip antenna efficiencies can be increased to as high as 80% when fabricated out of more exotic materials than silicon, when lenses are used, or when antenna elements are elevated over a substrate with free-space carve outs.

The next most important characteristic of an antenna is its gain at the frequency of resonance. The gain of an antenna indicates how well it concentrates radiated power into a single beam. The antenna gain is related to the efficiency according to the antenna directivity, in addition to the fields radiated by the antenna:

$$G = \frac{|E \times H^*|(\theta, \phi)}{\frac{P_{\text{rad}}}{4\pi r^2}} = e_{\text{ant}} D \tag{4.3}$$

where G is the gain of the antenna, E and H are the radiated electric and magnetic fields in the far-field region, respectively, and θ and ϕ indicate that the gain of the antenna is in a certain direction (in spherical coordinates). P_{rad} is the power radiated by the antenna, r is the distance from the antenna to the observation point (in carrying out this calculation, r in the denominator will cancel with the distance dependence of E and H), E is antenna efficiency, and D is the directivity. As described in the introduction to this chapter, the easiest qualitative means for increasing gain is to use a larger array or antenna, but this is not always possible given other design constraints.

Computing fields in the far-field region of an antenna is fairly straightforward for an antenna in free space. In this case, the far-field is related to the spatial Fourier transform of the current on the antenna [Bal05][RWD94]:

$$E_{\text{ff}} \propto \int J_{\text{antenna}}(\vec{r}) e^{-j\vec{k}\cdot\vec{r}} dV_{\text{antenna}} \tag{4.4}$$

where J_{antenna} is the current on the antenna; $\vec{k}$ is the vector *wave number* that defines the radiation direction and relative wave propagation speed as described in Section 4.2.2,

4.2 Fundamentals of On-Chip and In-Package MmWave Antennas

$\vec{r} = \frac{X\hat{x}+Y\hat{y}+Z\hat{z}}{\sqrt{X^2+Y^2+Z^2}}$; X, Y, and Z are the coordinates within the antenna; and V is the volume of the antenna. The intuition that should be taken from (4.4) is derived from elementary knowledge of the Fourier transform as developed in most undergraduate science and engineering degrees: large antennas (current sources that occupy lots of space) will produce tighter beams than small antennas, which will radiate nearly isotropically when the small antenna dimension is much less than a quarter-wavelength. It should be noted that very large antennas may also produce large side lobes, so this intuition should only be carried so far in actual design.

MmWave on-chip and in-package antennas both typically exist on or in a material other than free space. These materials, including semiconductors such as doped Silicon, Gallium Arsenide, and Indium Phosphide, and dielectrics such as silicon dioxide or "low-k" (low relative permittivity, or low dielectric constant) dielectrics that are used in modern chip technologies to reduce coupling between metal structures, are characterized by a complex permittivity. There are many excellent texts on electromagnetics that discuss the constitutive parameter permittivity in great detail [RWD94][Poz05]. Permittivity dictates how much electric flux passes through a surface for a given electric field, and hence determines how densely an object can store electric energy. The permittivity $\epsilon = \epsilon_o \epsilon_r$ of a non-magnetic material also determines the speed of light in the material since $c = 1/\sqrt{(\mu\epsilon)}$. Thus, permittivity also governs the wavelength of an electromagnetic wave in the material. This is a key point, as the physical size of an antenna may be shrunk by $\sqrt{\epsilon_r}$ in linear dimension when compared with the physical size of an antenna in free space. We write the complex permittivity of a non-magnetic material ϵ as

$$\epsilon = \epsilon' + j\epsilon'' \tag{4.5}$$

where the complex portion ϵ'' of permittivity accounts a material's ability to interact with and remove energy from an electromagnetic wave. In chapter 3 we used the notation ϵ_r and ϵ_i for ϵ' and ϵ'', respectively, to denote real (i.e., non-lossy) and imaginary (i.e., lossy) portions of permittivity. The notation used in this chapter is more consistent with texts on electromagnetics and will therefore be used in this chapter. The ratio of ϵ'' to ϵ' determines another often used parameter for these materials called the *loss tangent* $\tan(\delta)$

$$\tan(\delta) = \frac{\epsilon''}{\epsilon'}. \tag{4.6}$$

[Poz05] shows that for rectangular resonators, the *relative bandwidth* of a resonance (assuming no additional metal loss) is determined by the loss tangent of the dielectric used to fill the resonator. The relative bandwidth of a signal is the ratio of its passband bandwidth to its center frequency. Passband bandwidth is related to another parameter called the *quality factor* Q, which roughly speaking is the inverse of the relative bandwidth (Q describes many things, and in the electromagnetic context often describes the ratio of energy stored to energy lost in one cycle of an electromagnetic wave)

$$\text{Relative Bandwidth} = \frac{B}{f_c} = \frac{1}{Q} \cong \tan(\delta), \tag{4.7}$$

where B is the passband bandwidth of a signal and f_c is the center, or carrier, frequency of the signal.

4.2.2 Fundamentals of Antenna Arrays

Antenna arrays are necessary to achieve acceptable link budgets for long-range applications (i.e., cellular or backhaul) and to exploit paths in a time-varying or spatially varying channel. There are many excellent antenna texts that cover antenna arrays (e.g., [Bal05]), so we only cover several fundamentals here. We refer the readers to the well-known textbooks that describe horn antennas, patch antennas, and other conventional antennas that may be used for backhaul, base station, and conventional antennas in handsets or mobile wireless consumer devices for microwave communications. We focus primarily on the futuristic needs of mmWave wireless that will demand tight integration to reduce cost; free up mechanical and form factor design considerations; and maximize power efficiency.

An antenna array is made by arranging a set of antennas periodically either in a one- or two-dimensional configuration. The phase difference between the excitations of different elements then determines the direction in which all the individual beams radiated by the antennas add constructively or destructively. Fig. 4.2 illustrates a basic array in which a set of antennas is arranged in a one-dimensional line with a separation between elements of d. By phasing the arrays such that each element experiences a progressive phase shift of $k_o d \cos\theta$ degrees, the outgoing phase fronts will add constructively at θ relative to the normal direction of the axis of the array. This is an approximation, but it is widely used and gives very good results in most situations.

Intuitively, this can be understood from Fig. 4.3, which illustrates phase fronts of each antenna propagating out from its respective antenna. Each antenna is "turned on" progressively later, resulting in phase fronts progressing downward in size for antennas A1, A2, A3, A4, and A5, respectively. The triangle on the right of the figure illustrates

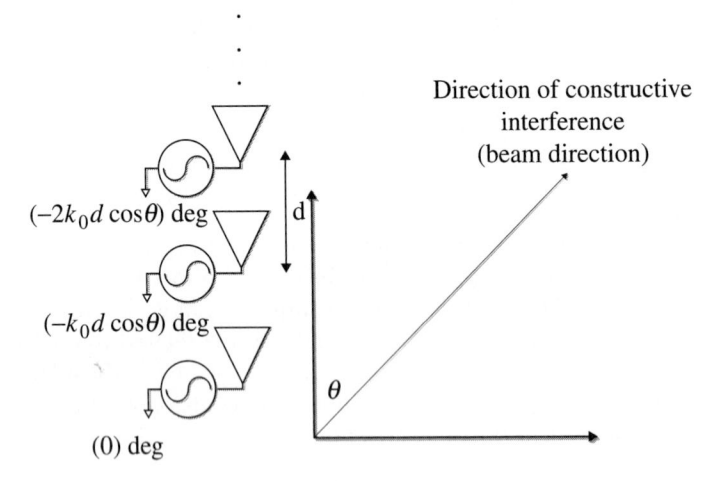

Figure 4.2 The phasing between different elements in the array will determine the direction of constructive interference.

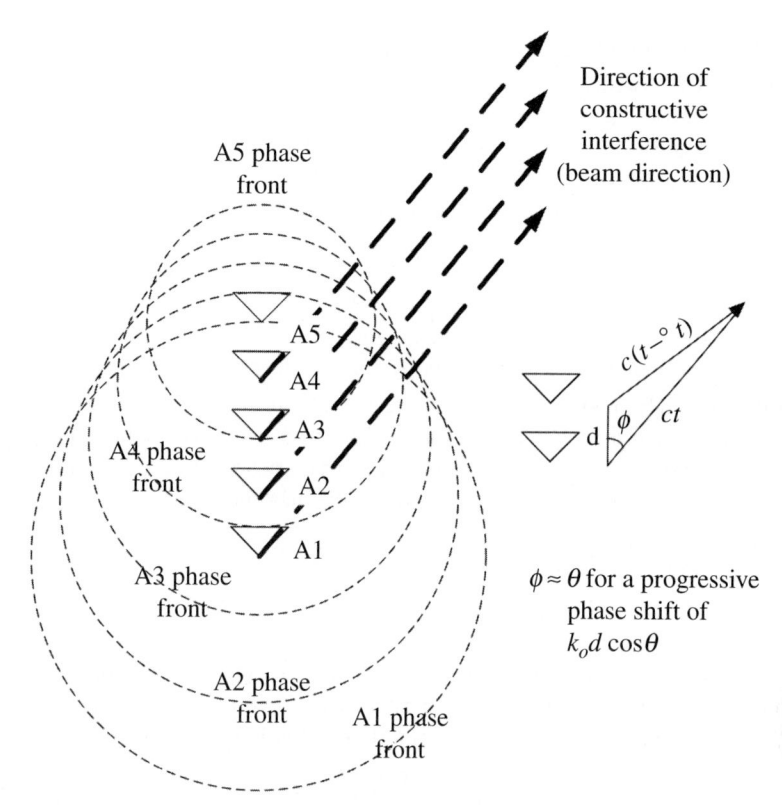

4.2 Fundamentals of On-Chip and In-Package MmWave Antennas

Figure 4.3 The progressive phase fronts are illustrated by progressively smaller circles for A1, A2, etc. The phase fronts progress downward in size due to the difference in time that each antenna "turns on," or equivalently due to the progressive phase shift of each antenna.

how, geometrically, the phase shift between each antenna results in phase fronts adding constructively at angle ϕ. By the law of cosines:

$$\left(\frac{d}{c}\right)^2 + t^2 - 2\left(\frac{d}{c}\right) t \cos\phi = (t - \Delta t)^2 \tag{4.8}$$

where c is the speed of light in the far-field region of the array. Solving for $\cos\phi$, we find

$$\cos\phi = \frac{\left[\left(\frac{d}{c}\right)^2 - \Delta t^2 + 2t\Delta t\right]}{2\left(\frac{d}{c}\right) t}. \tag{4.9}$$

By setting Δt equal to $\left(\frac{d}{c}\right) \cos(\theta)$ we find that

$$\cos\phi = \left(\frac{d}{2ct}\right) \sin^2\theta + \cos\theta. \tag{4.10}$$

In the far-field region in which this formulation is valid, the distance is at least $5\lambda_o$ away from the array (where λ_o is the operating wavelength). The spacing between the antennas is less than or equal to λ_o. Therefore, ct is at least $5\lambda_o$, indicating that the first term is very small and may be neglected. We see then that $\cos\phi = \cos\theta$. The time difference in excitation Δt corresponds to a phase shift of $2\pi f \Delta t = 2\pi f \left(\frac{d}{c}\right)\cos\theta = k_o d \cos\theta$, indicating that the phase shift prescribed in Fig. 4.2 does result in a beam direction in the θ direction. A question that may arise when designing an integrated array is whether to use the wave number for free space or for the on-chip or in-package environment when selecting element phase shifts. This geometric analysis indicates that it is the wave number of the medium in which the signal is propagating, free space in most cases, that should be used, and not the medium that performs the wave launching.

Mathematically, the result of arranging antennas in an array is to multiply the gain of a single element by a factor determined by the array geometry called the *array factor*, as explained by [Bal05]. At any given position of the far-field of the array, the fields of each antenna add with their respective phase shifts [Bal05]

$$E_{\text{ff}}^{\text{total}} = E_{A1} + E_{A2} + \ldots E_{A_N} \tag{4.11}$$

$$= E_{A1}\left(1 + e^{-jk_o d \cos\theta} + \ldots e^{-jk(N-1)k_o d \cos\theta}\right) \tag{4.12}$$

where $E_{\text{ff}}^{\text{total}}$ is the total far-field electric field and E_{A1}, E_{A2}, etc. are the individual antenna fields, assumed to be identical for this case (for non-uniform array elements this analysis does not hold). After some manipulation, this is found to equal [Bal05]:

$$e^{j\frac{(N-1)}{2}\psi}\left[\frac{\sin\left(\frac{N}{2}\psi\right)}{\sin\left(\frac{\psi}{2}\right)}\right] \tag{4.13}$$

where $\psi = k_o d \cos\theta$. The term in brackets is called the *array factor*. For broadside radiation in which $\theta = 90°$, the array factor converges to N, indicating that the power gain of the array in the broadside direction will be increased by N^2. For a two-dimensional array, a similar analysis indicates that there is an array factor for each dimension of the array. Therefore, for an $N \times M$ array with broadside radiation, we expect the gain to increase by $(NM)^2$.

The spacing between elements in an array is critically important because it determines the maximum angle of beam steering that the array can accomplish without grating lobes. A *grating lobe* is a large side lobe of the antenna array pattern in addition to the main lobe. Grating lobes result in lowered array directivity and gain and will destroy an array's ability to unambiguously determine the direction of arrival of a signal. In order to avoid grating lobes for a given maximum amount of beam steering θ_{max}, the spacing d between elements must be kept below

$$\frac{d}{\lambda_o} \leq \frac{1}{1 + \cos\theta_{\text{max}}}. \tag{4.14}$$

Element spacing greater than indicated by (4.14) will result in grating lobes.[1] This equation does not result from the antenna of a specific array. Rather, it has to do with

---

1. The authors thank Dr. Hao Ling of The University of Texas at Austin for pointing out this equation.

the fact that an array may be considered a sampled current source. The array factor is periodic in phase shift between elements, with a peak every 2π. By steering the beam, the window of this function shifts in the far field. The wider the element spacing, the wider the window, and if the window is too wide, then it may support two or more of the peaks in the array factor.

In addition to element spacing, the designer should take steps to ensure that the input impedance of all the elements in the array is nearly uniform. This is accomplished in part by preventing coupling between array elements that results in both non-uniform and time- and use-dependent input impedances for different elements in the array.

When we study arrays, it is useful to understand the concept of a *wave vector*, which describes how an electromagnetic wave propagates through space. The notion of a wave vector falls out from a solution to Maxwell's equations that have been decoupled to include only electric field or magnetic field. In a source-free medium (i.e., when no charges or currents are present), Maxwell's time-harmonic equations are written as

$$\nabla \times \vec{H} = j\omega\epsilon\vec{E} \qquad (4.15)$$

$$\nabla \times \vec{E} = -j\omega\mu\vec{H} \qquad (4.16)$$

$$\nabla \cdot \vec{E} = \frac{q}{\epsilon} = 0 \qquad (4.17)$$

$$\nabla \cdot \vec{H} = 0 \qquad (4.18)$$

where the first equation is Ampere's law (4.15), the second is Faraday's law (4.16), the third is Gauss' law (4.17), and the fourth (4.18) is the law that states that no magnetic charges are known to exist. These equations relate magnetic field H, electric field E, angular frequency $\omega = 2\pi f$, electric charge density q, and constitutive parameters permittivity ϵ and permeability μ. As is done in many texts [RWD94], we take the curl of (4.16) and insert (4.15) into the result to find the decoupled, non-linear homogenous Helmholtz equation that describes the propagation of the electric field (a similar procedure applies to the magnetic field)

$$\nabla \times \nabla \times \vec{E} = \nabla\left(\nabla \cdot \vec{E}\right) - \nabla^2 \vec{E} = -\nabla^2 \vec{E} = -j\omega\mu\nabla \times \vec{H}$$
$$-\nabla^2 \vec{E} = -j\omega\mu\left(j\omega\epsilon\vec{E}\right) \rightarrow \nabla^2 \vec{E} + \omega^2\mu\epsilon\vec{E} = 0 \qquad (4.19)$$

where the equation following the arrow is the homogenous Helmholtz equation for electric field E. As described in [Bal89, Chapter 3], the solution to (4.19) may be written:

$$\vec{E} = (E_x \hat{x} + E_y \hat{y} + E_z \hat{z}) e^{-jk_x x - jk_y y - jk_z z} \qquad (4.20)$$

where k_x, k_y, and k_z are constants with units m^{-1} (radians per meter) that describe how quickly the wave varies in the x, y, and z directions, respectively. For example, at a given instant in time if k_x is very large, then as we observe the wave at different points in the x direction it will appear to have many peaks and troughs. E_x, E_y, and E_z are the polarization vectors that describe how the electric field varies over time at a single point in space. By (4.17), the polarization vector $E_x \hat{x} + E_y \hat{y} + E_z \hat{z}$ must be orthogonal to the wave vector, which is given in terms of the constants k_x, k_y, and k_z: $k_x\hat{x} + k_y\hat{y} + k_z\hat{z}$. By plugging (4.20) into (4.19), we find:

$$k_x^2 + k_y^2 + k_z^2 = \omega^2 \mu\epsilon \qquad (4.21)$$

$$= k^2. \qquad (4.22)$$

From (4.22), it is clear that script $k = \omega\sqrt{(\mu\epsilon)}$, and since the speed of propagation script $c = 1/\sqrt{(\mu\epsilon)}$, it follows that script $k = 2\pi/\lambda = \omega_c = 2\pi f_c$.

This relationship in (4.21) and (4.22) is known as the *dispersion relationship*, and is basically a statement of how quickly energy may travel in the wave in different directions. It relates the wave number (the magnitude of the wave vector) k to the constants k_x, k_y, and k_z, and to the constitutive parameters of the medium of propagation. In general, the speed of propagation of energy of a wave is given by the group velocity v_g:

$$v_g = \frac{d\omega}{dk} = \frac{1}{\frac{dk}{d\omega}}. \qquad (4.23)$$

Now, for example, there may be a group velocity in the x direction given by $\frac{\delta\omega}{\delta k_x}$. Therefore, by taking the total derivative of the (4.22) we find that the group velocity in any given direction x, y, or z of the wave is related:

$$\frac{dk}{d\omega} = \frac{dk}{dk_x}\left(\frac{\delta k_x}{\delta\omega}\right) + \frac{dk}{dk_y}\left(\frac{\delta k_y}{\delta\omega}\right) + \frac{dk}{dk_z}\left(\frac{\delta k_z}{\delta\omega}\right) \rightarrow \frac{k}{v_g}$$

$$= \left(\frac{k_x}{v_{gx}} + \frac{k_y}{v_{gy}} + \frac{k_z}{v_{gz}}\right), \qquad (4.24)$$

where v_{gx} is the group velocity in the x direction, and so on.

The point of the above discussion is to show that a wave or signal may only carry energy in a given direction, and this direction is indicated by the wave vector. Therefore, when we say a signal arrives from a given direction at an array, we are also making a statement about its wave vector. As the wave vector is often a convenient way to describe the direction of travel, it is an important concept to understand. Fig. 4.4 illustrates the concept of a wave vector. The above equations are generally too complex to calculate, and circuit and antenna designers often rely on electromagnetic simulators to compute the fields and waves of array antennas.

4.3 The On-Chip Antenna Environment

MmWave antennas that are fabricated on a chip using integrated circuit production technology have been the focus of recent research, as they offer great potential production cost savings compared with packaged antennas if key challenges can be overcome to realize acceptable performance. The biggest difficulty faced by on-chip antennas is obtaining acceptable efficiency and gain, and much of the literature has focused on techniques to improve on-chip antenna gain. Before discussing these challenges, it is useful to review popular integrated circuit production technology. We will focus on complementary metal

4.3 The On-Chip Antenna Environment

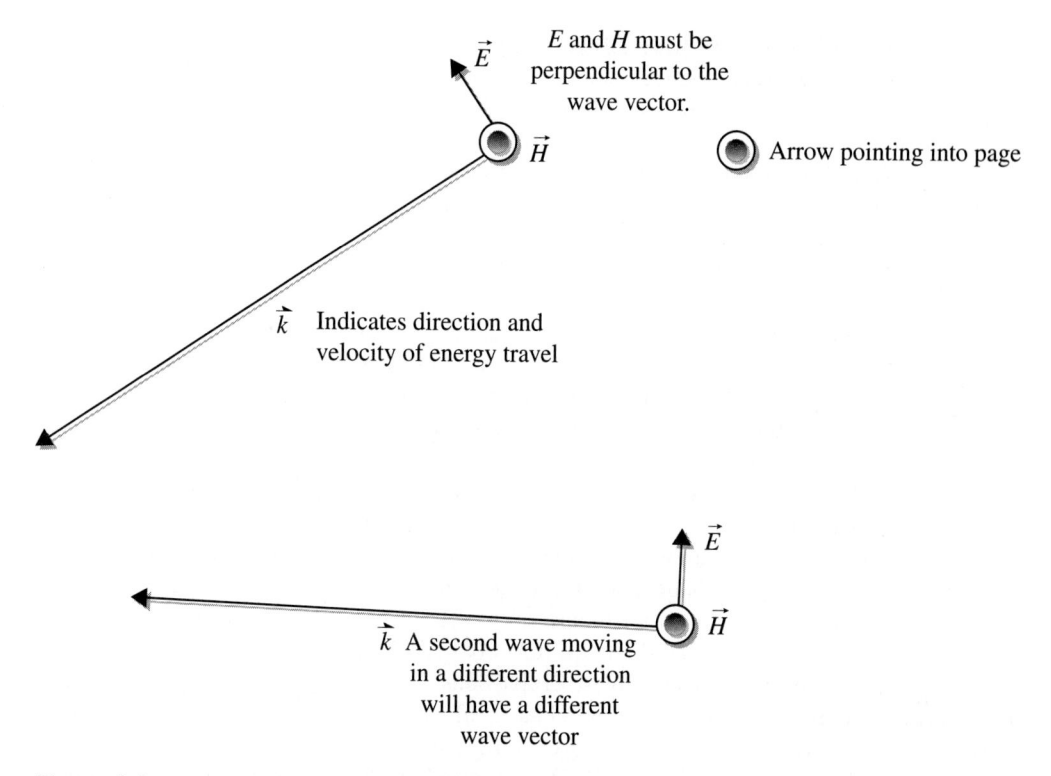

Figure 4.4 It is important to understand that the wave vector, which describes how an electromagnetic wave propagates through space, is related to the direction of propagation. This is important for arrays in which we may want to determine the direction of an incoming signal, or the direction of arrival (DOA).

oxide semiconductor (CMOS) technology, as it illustrates most of the challenges of the on-chip environment and also offers the lowest-cost means for creating on-chip antennas. Important points to take from this discussion include the production process requirements that impact antenna design and performance, including, for example, the requirement to used slotted metal. We will then use a simplified model of the substrate of an integrated circuit to describe four key challenges, and methods for overcoming these challenges: generation of substrate modes or surface waves, preferential antenna radiation into the substrate versus the surrounding environment, loss accrued by waves traveling through the substrate, and possible resonances of the substrate. Another key challenge that requires a more sophisticated approach discussed in this chapter is the effect of nearby metal structures, such as transmission lines. The designer should have an excellent knowledge of the placement of structures near the antenna, as coupling between structures can significantly impact the input impedance of an antenna and can detrimentally affect matching performance. We will introduce a possible method for reducing coupling between an on-chip antenna and other on-chip structures (including other antennas), including the use of periodic structures to interrupt surface waves.

It is worth noting that we have intentionally not covered emerging areas of mmWave antennas that show great promise but which use more exotic or cumbersome manufacturing processes or assembly techniques such as the use of circuit board edges for implementation of phased arrays, as demonstrated by Samsung [HLBK14]. As other promising examples not covered here, recent advances have been made in creating array antennas using *low-temperature co-fired ceramic* (LTCC) materials, and circuits with extensive cavity (free space) structures as well as integrated steerable lenses have been shown to offer impressive gains (over 30 dB) and impressive efficiencies (over 85%). LTCC is an integrated circuit technology that allows multi-layer circuits to be constructed by laminating single layers of thick filmed ceramic material, each with conductors such as copper, aluminum, or gold, as well as various passive components that may be embedded in each of the thick film layers. The individual circuit layers are stacked on top of each other, allowing three-dimensional structures and virtually unlimited signal paths in the circuitry, and the stack of films are then baked, or "fired" all at one time, at a relatively low temperature of 850°C. Although they offer exceptional performance in many cases, these structures are more difficult to manufacture in mass quantities using conventional circuit technologies employed in commercial foundries (see Chapter 5). Yet, many exciting and novel ideas have evolved for the creation of on-chip and on-substrate antennas that may find their way into future products, such as those presented in [GJBAS01][LSV08][ZSCWL08][EWA+11][KLN+11]. Also, Intel has recently demonstrated electronically steerable lens antennas using on-chip fabrication methods that may prove useful for wireless local area network/wireless personal area network (WLAN/WPAN) applications [AMM13] (see the description of lens antennas in Section 4.6). Many resources such as [LGPG09] and [TH13] provide up-to-date technical details regarding myriad promising mmWave antenna structures. Our purpose in this text is to focus on techniques and technologies that lend themselves to the lowest cost in mass quantities, as we attempt to treat the fundamentals that will allow researchers and engineers to learn the key concepts for all types of integrated on-chip and in-package antennas. We note that with advances in fabrication technologies, more exotic antennas such as those discussed above may eventually become cost effective and more efficient to use in mmWave communication systems. We now focus on the fundamentals of CMOS on-chip antennas, as this is the least expensive method for implementing on-chip antennas in mass quantities, and the fundamentals apply to other types of integrated circuits and materials, as well.

4.3.1 Complementary Metal Oxide Semiconductor Technology (CMOS)

CMOS chips are illustrative of the type of environment faced by most on-chip antennas. Fig. 4.5 illustrates a side view (cross-sectional view) of a typical CMOS chip. The top portion of the chip is occupied by a dielectric and contains the metal traces used to connect active circuitry or to construct an on-chip antenna. Semiconductor manufacturing processes are discussed in Chapter 5, where it is shown that the dimensions of the transistor gate length (such as 180 nm) are used to describe the technology process, or technology node, used in the CMOS chip foundry fabrication process. Certain modern

4.3 The On-Chip Antenna Environment

Thick metal trace (cu)

Metal trace (cu)

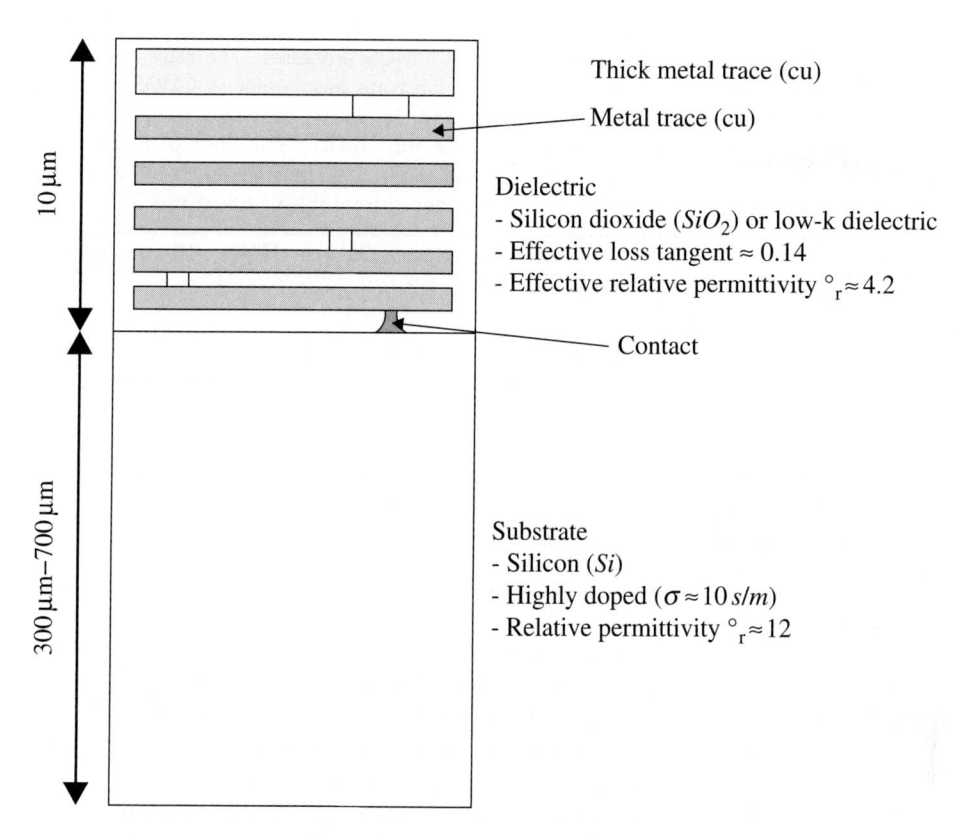

Dielectric
- Silicon dioxide (SiO_2) or low-k dielectric
- Effective loss tangent ≈ 0.14
- Effective relative permittivity $°_r$ ≈ 4.2

Contact

Substrate
- Silicon (Si)
- Highly doped (σ ≈ 10 s/m)
- Relative permittivity $°_r$ ≈ 12

10 μm

300 μm–700 μm

Figure 4.5 The cross section of a typical CMOS chip.

processes may use low-k dielectrics in place of silicon dioxide in order to reduce coupling between adjacent lines by reducing capacitance between lines [CLL+06]. The dielectric layer of the chip that contains the many layers of metal traces is often called the Back End of Line (BEOL) layer, is approximately 10 μm thick, and resides over a thicker layer of doped silicon that is roughly 300–700 μm thick. In older fabrication processes such as 180 nm CMOS, the metal traces were mostly aluminum [SKX+10]. In more modern processes, only the top metal layer is usually aluminum, and the other layers may be copper. Newer processes such as 45 nm (see Figure 4.5 and Table 4.1) may have as many as 8–12 metal layers, whereas 180 nm CMOS and older generations of CMOS have 6 or fewer. The top metal layer in most processes is thicker than lower metal layers (4 times thicker in 180 nm RF CMOS, 15 times thicker in 90 nm RF CMOS, and 10 times thicker in 45 nm digital CMOS [SKX+10]). In very mature (very old) technology nodes, this allowed the top metal layer to have a lower resistivity than the other layers. However, in more modern technology processes that have been developed for high speed digital circuits, despite the fact that the top layer is thicker than the bottom metal layers, the higher resistivity of the top metal layer causes the top layer to be more resistive in more recent process nodes. This makes analog antenna design more challenging in more modern process nodes. For example, aluminum has a higher resistivity than copper, so a

Table 4.1 A summary of the BEOL layer of recent CMOS processes. The table indicates that the distance from the top metal layer to the silicon substrate has shrunk as CMOS has evolved, and that there is more capacitance between metal and the silicon substrate in newer processes. Note the increasing resistivity and capacitance of the top BEOL metal layer of more advanced processes relative to older processes has increased the challenges associated with mmWave analog and integrated antenna design on the top layer. [reproduced from [SKX$^+$10] © IEEE].

Technology node (nm)	180 nm RF	90 nm RF	45 nm Digital
Top metal thickness	2 μm	4 μm	1 μm
Distance from top metal to silicon substrate	h	0.6h	0.6h
Metal thickness ratio (top metal/bottom metal)	4	15	10
Resistance ratio (top metal/bottom metal)	5	16	19
Normalized capacitance from top metal to silicon	C	1.1C	1.2C

thick top aluminum layer may be more resistive than a thinner lower copper layer. The reader is referred to [SKX$^+$10] for an excellent illustration of the evolution of CMOS BEOLs. Active devices exist at the interface between the dielectric BEOL layer and the thicker substrate, and are connected to metal using substrate contacts. In order to avoid a problem called *latch up* in digital circuits (in which "parasitic" structures such as an n region sandwiched between two p regions—a parasitic *Bipolar Junction Transistor* (BJT)—can turn on and damage the rest of the circuit) the substrate is highly doped (a typical value for conductivity of 180 nm CMOS is shown in Fig. 4.5) to prevent voltage buildup that leads to latch up. Other processes such as SiGe and III-V processes may offer better performance for RF operation than CMOS. This is because these processes usually have lower substrate doping levels, thus greatly reducing the loss experienced by passive devices constructed in the BEOL layer (see Chapter 5 for further details). Table 4.1, reproduced from [SKX$^+$10], summarizes the CMOS BEOL stacks for recent CMOS processes.

There are various requirements associated with the metal layers of most integrated circuit technologies. Four to keep in mind are: 1) Each layer in the process may have a minimum amount of metal to meet production rules. This is usually specified as a percentage of the metal layer and is required to meet what are known as *Chemical Mechanical Polishing* (CMP) rules. 2) In order to prevent damage to the chip during cutting or dicing operations (in which multiple chips are cut from a single large wafer), the chip often must have a ring of metal, sometimes called a *guard ring*, that encircles all portions of the chip. This ring is often connected to bond pads (in which case it is called a *pad ring*). This is important because the guard ring may have an impact on edge-mounted antennas. 3) Large pieces of metal usually require the addition of non-metal slots. Foundries require this due to the extremely high temperatures reached in production process. 4) There is a minimum size associated with metal structures, and this size is process dependent. Several of these requirements are illustrated in Fig. 4.6.

4.3 The On-Chip Antenna Environment

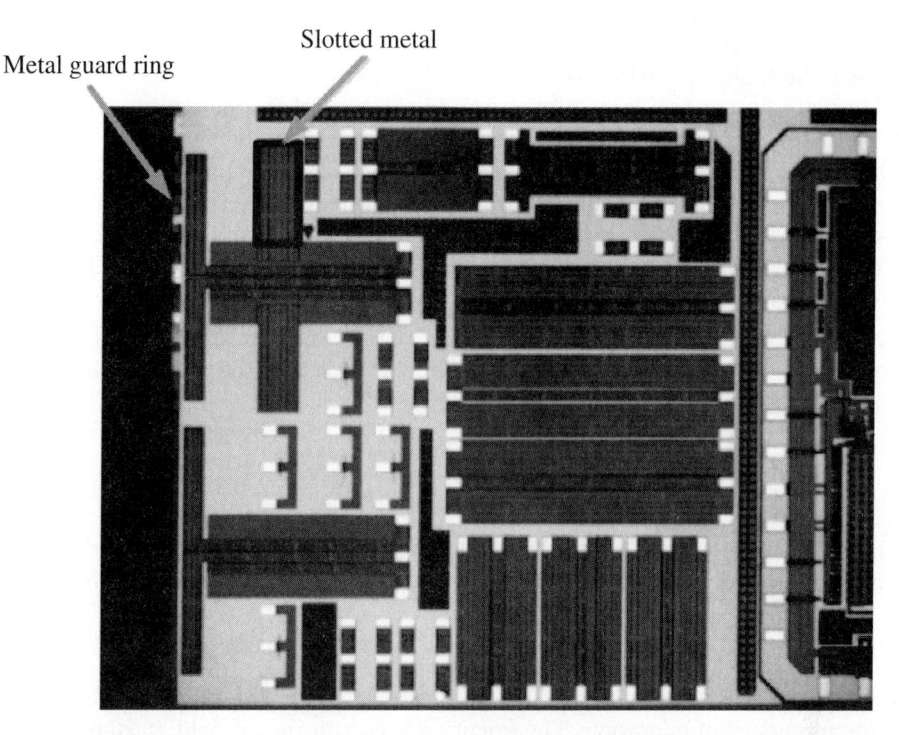

Figure 4.6 There are several considerations for on-chip antennas related to CMOS production rules: 1) All metal layers must meet a minimum fill requirement. This is reflected in the figure by the fact that there are no large portions of the chip left empty (the lighter-shaded portions of the figure). 2) A metal guard ring must often surround the chip to prevent damage during dicing. 3) Large areas of metal must be slotted to meet design rules. 4) Metal structures must meet a minimum size requirement, which in practice is usually satisfied by most designs.

Of these challenges, the requirement for a minimum metal density on each process layer in practice is not difficult to meet, even if antenna elements exist on only a single metal layer of the chip. This requirement should be especially easy to meet if an array is used. Also, in order to minimize conductive losses in the metal structure of an antenna, it is generally advisable to implement the antenna with all metal layers using interconnecting vias between layers (this may not be advisable if the resulting capacitance to the substrate is deemed by the designer to be too high to justify the decrease in conductive losses in the metal). If by design the antenna of interest is known to experience minimal conduction loss, it is advisable to place the antenna in the top metal layer as far from the substrate as possible to help avoid substrate loss.

Requirements 1, 2, and 4 should be taken into account when simulating an on-chip antenna using electromagnetic simulation software such as Ansoft's HFSS (see Chapter 5, e.g., Table 5.1). Slots are generally not included in simulations due to the fact that they greatly increase simulation time and memory requirements for analysis techniques such as method of moments and finite element method that are used by simulation programs. When drawing the substrate in a simulation tool, the designer should include a thin layer approximately 1 μm thick with a conductivity of 10 S/m at the top of the substrate to represent the highly doped portion of the substrate (this is because ions implanted into the substrate do not penetrate all the way to the base of the substrate). The remaining portion of the substrate may be modeled as bulk silicon [HFS08].

There are four key challenges posed by the on-chip environment, all of which derive from the substrate used to support an on-chip antenna [LKCY10][MVLP10][TOIS09]. First, the high permittivity of the silicon substrate below the antenna forces the antenna to "preferentially" radiate into the substrate instead of off of the chip. This is easily understood in terms of the Poynting vector of the far field of an antenna, which can be shown to be proportional $\epsilon_r^{\frac{3}{2}}$, where ϵ_r is the relative permittivity [RWD94] (i.e., the ratio of ϵ' to the permittivity of free space ϵ_0). Recall that the far-field radiation of an antenna, an aperture for example, is proportional to the wave number of the medium

$$(E_{\text{ff}}) \propto K \left(\frac{e^{-jkr}}{4\pi r} \right) \tag{4.25}$$

where E_{ff} is the far-field electric field, k is the wave number, and r is the distance from the antenna to the point of observation. From (4.22), the wave number is equal to:

$$k = 2\pi f \sqrt{\epsilon \mu} \tag{4.26}$$

where ϵ and μ are the permittivity and permeability of the point of space of interest, and f is the frequency of operation. The Poynting vector in the far field, if we recall that the magnetic field H is related to the electric field by wave impedance $\eta = \sqrt{\frac{\mu}{\epsilon}}$ by

$$H = \frac{j \left(\hat{k} \times E \right)}{\eta} \tag{4.27}$$

where $\hat{k}$ is the unit vector indicating the direction of propagation. The magnitude of the Poynting vector, which indicates the power density at a point in space, is then given by

$$|S| = |E \times H^*| \tag{4.28}$$

$$= \frac{|E|^2}{2\eta} = \frac{f^2 \epsilon \mu}{8\sqrt{\frac{\mu}{\epsilon}} r^2} \propto \epsilon^{\frac{3}{2}}. \tag{4.29}$$

Equation (4.29) indicates that the power density of radiation into the chip, barring the presence of shielding structures or other structures above the antenna, should be much higher in the substrate than in the region of the chip directly above the antenna. This "preferential" radiation would not be such a challenge if waves entering the substrate could exit without great attenuation so that they could still serve a useful purpose. But, the second key challenge of the on-chip environment, high substrate doping that serves to prevent latch up in digital circuits, causes radiation that enters the substrate to become greatly attenuated. The third key difficulty of the on-chip environment is due to the structure of the substrate, which may be qualitatively treated as a rectangular resonator. If we approximate the sides of the substrate as perfect electric conductors, then

4.3 The On-Chip Antenna Environment

we find that the lowest TM resonant frequency of the geometry illustrated in Fig. 4.7 is given by:

$$f_{\text{cutoff}} = \frac{c}{2\pi\sqrt{\epsilon_{r,si}}}\sqrt{\left(\frac{\pi}{a}\right)^2 + \left(\frac{\pi}{b}\right)^2}. \quad (4.30)$$

The fourth key challenge of on-chip mmWave antennas is due to the ease with which the chip substrate may support surface waves [AKR83]. A surface wave is a type of guided wave that travels along the axis of the substrate, as illustrated in Fig. 4.8. Designers of mmWave integrated antennas should seek to suppress these surface waves as much as possible as they reduce power radiated into the intended radiation field of the antenna and also increase coupling between the antenna and adjacent structures, including nearby antennas (e.g., in an array). [AKR83] showed that each surface wave mode has an effective substrate height that represents both the substrate thickness plus an apparent ray penetration depth above and below the antenna (when no ground plane is present immediately below the antenna). This effective height always appears in the denominator of the equation that predicts how much power will be coupled into each mode for a given frequency, dielectric constant, and physical substrate thickness, indicating that substrate thinning will reduce the amount of power coupled into substrate modes. They demonstrated that substrates thicker than one tenth of a free space wavelength will support substantial surface waves, greatly reducing antenna performance. [AKR83] also demonstrated that

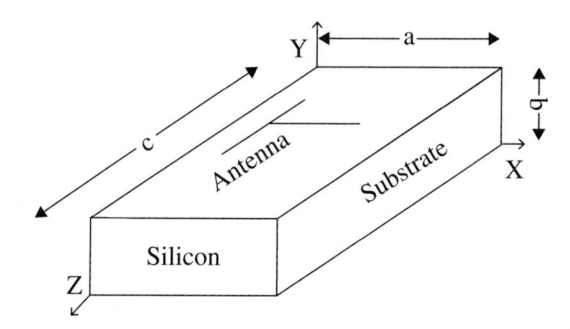

Figure 4.7 The geometry of the chip substrate indicates that the substrate may act as a very lossy resonator. Any waves that exist in the substrate will be highly attenuated.

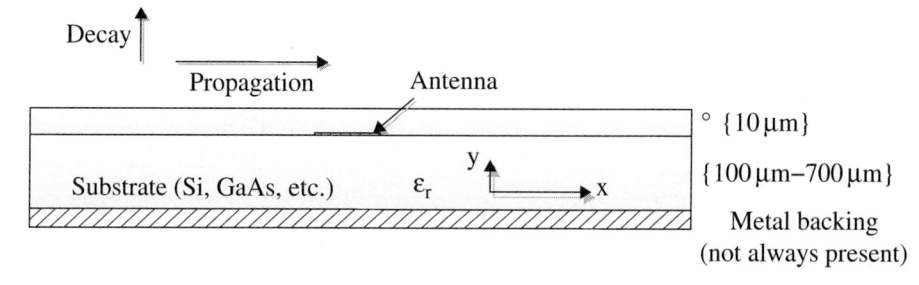

Figure 4.8 A guided wave propagates in the x direction and decays in the y direction. Numbers in the figure are representative only.

the ratio of power used to excite surface waves by a dipole to the power that would be radiated by a dipole in free space approaches $\sqrt{\epsilon_r}$ as the dielectric constant of the substrate increases (and approaches $\epsilon_r^{\frac{3}{2}}$ for a slot). Unfortunately, they also show that both slot and dipole antennas will radiate less into the air when on a substrate than when not integrated. The work of [AKR83] demonstrates that, if possible, lower permittivity substrates should be used for integrated antennas to avoid surface waves. [AKR83] also demonstrates that the efficiency of a microstrip dipole antenna is limited for a specific substrate thickness and antenna length, and for a relative permittivity of 12 cannot exceed 30%.

We will dwell slightly longer on the work presented by [AKR83] because it provides a very intuitive understanding of how surface waves contribute to detrimentally impact radiation efficiency. Fig. 4.9 is similar to Fig. 4.8, but the former illustrates a wave excited by the mmWave antenna. If the angle θ is greater than the critical angle $\theta_c = \sin^{-1}\frac{1}{\sqrt{\epsilon_{r,\text{substrate}}}}$, then the wave will be trapped as a surface wave. For a silicon substrate this crucial angle is quite small at 16.80°, showing that most of the waves excited by the antenna will be trapped in the substrate.

The multiple metal layers available in most integrated circuit BEOL processes (e.g., see Fig. 4.5) raises the possibility of implementing the antenna in a top metal layer while using the bottom layer to implement a shield to prevent formation of substrate modes. In 2006, [BGK+06b] studied this approach using the IE3D method of moments field solver. [BGK+06b] showed that this is a poor means of improving performance as it greatly reduces the input impedance of the antenna and the antenna's radiation efficiency resistance. This is to be expected based on intuition built from simple transmission line and capacitor models. As the distance between two metal plates decreases, capacitance increases, which reduces impedance as $Z = \sqrt{\frac{L}{C}}$. In Fig. 4.1, the lossy portion of resistance due to conduction loss will not be greatly changed by the proximity of a metal ground. Low radiation resistance in turn is also associated with lower efficiency and lower gain. [BGK+06b] found a radiation efficiency of only 5% for a separation between the antenna and ground plane of 15 μm. Work in [RHRC07] contemplates the use of metal lower layers just above the substrate to create frequency selective surfaces (FSSs) to increase

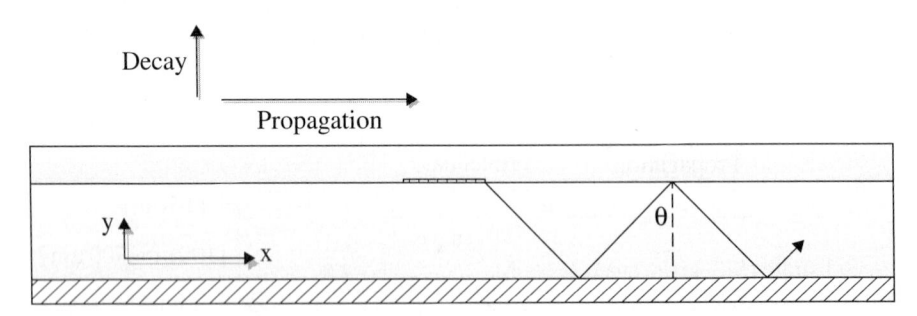

Figure 4.9 Once the angle θ passes the critical angle in the substrate, the wave generated by the antenna will remain trapped as a surface wave in the substrate, where it will then be highly damped by the high doping concentration of the substrate.

4.3 The On-Chip Antenna Environment

the radiation efficiency, which is a new idea that has gained substantial traction in the literature and among integrated chip developers in recent years.

All of the challenges to making on-chip antennas high-gain and efficient are lessened by thinning the bulk substrate. Fig. 4.10 produced by [KSK+09] summarizes this mitigation. The plot was generated for on-chip dipole antennas intended for inter-chip communication applications. The researchers performed an experiment in which they simulated the transmission between two equivalent antennas on two chips placed 5 mm apart. The plot shows the radiation efficiency of an on-chip antenna as a function of substrate thickness. ρ refers to the resistivity of $10 \, \Omega \cdot cm$ used in their analysis, d refers to the separation distance between chips, and L refers to the length of the dipole antennas used.

Many antenna topologies have been investigated for use on-chip for mmWave applications. Before discussing these, we present a circuit model that builds on the model in Fig. 4.1 that was developed by [ZL09] and is reproduced in Fig. 4.11. In the figure, R_r is the radiation resistance discussed in the introduction and accounts for the energy used by the antenna to produce useful radiation. L_d and C_d represent the inductance and capacitance of the antenna, respectively, and at the operating frequency $f = \frac{1}{\sqrt{L_d C_c}}$ these reactive components cancel out from the viewpoint of the outside world (they still exist within the antenna and store and exchange energy between each other at this frequency) and the antenna is said to resonate. R_{con} accounts for conduction loss in the metal of the antenna itself and is determined by the conductivity of the metal used to construct the antenna in addition to the magnetic field tangent to the antenna. R_{SUR} accounts for energy lost to surface wave generation along the top of the substrate on which the antenna resides. Surface waves are problematic not only because they drain energy from the antenna (reducing efficiency), but also because they may reduce isolation between two

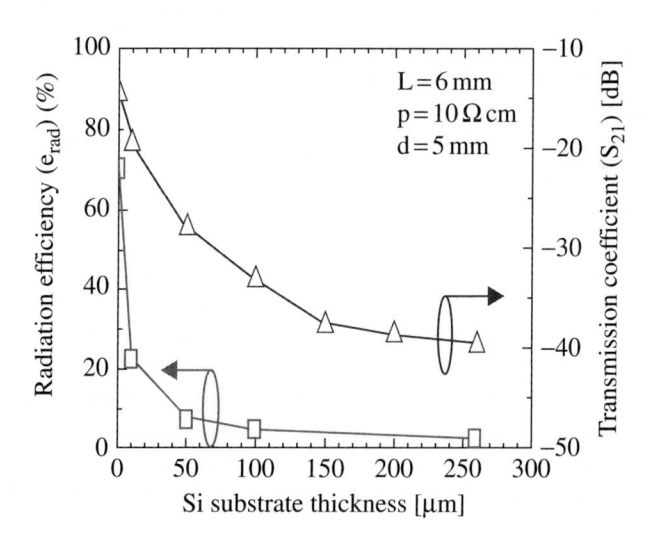

Figure 4.10 This figure indicates that the efficiency of on-chip antennas is reduced greatly by a thick substrate [from [KSK+09] © IEEE].

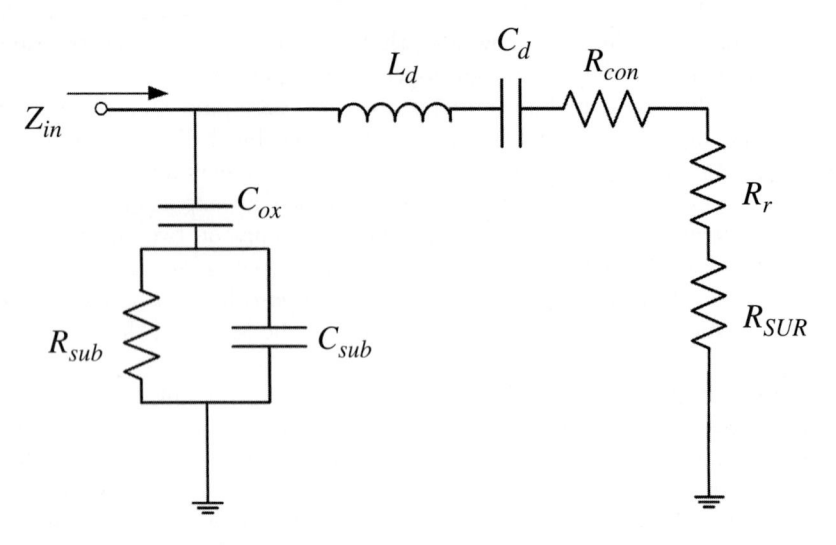

Figure 4.11 This figure illustrates many of the energy loss and storage mechanisms that may affect an on-chip antenna. The model is also very applicable to in-package antennas [reproduced from [ZL09] © IEEE].

closely spaced antennas, possibly leading to "active" impedances. An active impedance refers to a situation in which the input impedance of antenna is determined partly by the state (i.e., in use or not) of a nearby antenna. Mitigating surface waves is a major challenge for array design. C_{OX} accounts for the capacitance between the antenna and the IC substrate, and it will be lower for higher metal layers in Fig. 4.5 and lower in production processes that use "low-k" (i.e., low-permittivity) dielectrics between metal layers. R_{SUB} accounts for energy dissipated by conduction currents induced in the substrate, and it will be lower for higher doping concentrations. The energy lost to substrate currents will increase approximately by the term $\frac{V_{\text{in}}^2}{2R_{\text{SUB}}}$, indicating that a lower substrate resistance is not beneficial for antenna efficiency. The energy storage capacity of the substrate is represented by C_{SUB}.

A fifth challenge of on-chip antennas is illustrated in Fig. 4.11 by the loss due to conduction in the antenna R_{con}. This challenge will not be as significant as the other challenges facing on-chip antennas for small antennas, but it can be significant for antennas more than 1 millimeter in length. Should a design require very low conduction loss, then it is necessary to use a high-resistance substrate (e.g., a better insulator) with a resistivity of more than 100 $\Omega \cdot cm$ [LKCY10].[2] This is shown in Fig. 4.12, which illustrates the conduction loss in a transmission line in decibels per millimeter versus resistivity of the substrate. The loss on a transmission line and an antenna will not be equivalent in all cases due to the differences in how each structure supports current (and whether the antenna is a standing wave antenna or a traveling wave antenna like a Rhombic).

---

2. Certain foundries provide RF technology nodes that support higher resistivity substrates, with the express purpose of enabling efficient RF circuits.

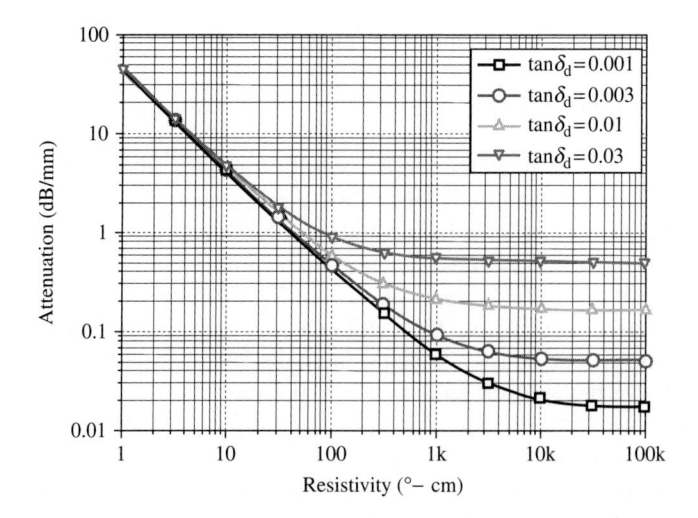

Figure 4.12 For low resistivity substrates, the loss due to currents carried by substrate dopants (i.e., conductive losses) is the major loss mechanism hurting performance [from [LKCY10] © IEEE].

Nevertheless, the plot indicates that high resistivity production processes such as Gallium Arsenide (GaAs) will give lower conduction loss because they generally require lower substrate doping and hence have higher substrate resistivity. Typical values for GaAs substrates are 10^7-10^8 $\Omega \cdot cm$ for the resistivity of GaAs substrates compared to 10 $\Omega \cdot cm$ for silicon processes such as CMOS [BGK+06b].

Modeling all of the loss mechanisms that account for the low efficiency of on-chip antennas requires consideration of the operating frequency relative to the substrate dimensions and permittivity to determine which substrate modes are excited. Knowledge of substrate conductivity is also needed to determine how much energy is lost to conduction currents. Recently, [OKF+05] used a very simple approach to model loss mechanisms that simply created a loss exponent α similar to path loss in an open air environment. This approach is not the most physical and should be applied primarily to inter-chip communications—that is, for applications in which the antenna need only communicate a signal to other structures on the same chip.

4.4 In-Package Antennas

An in-package integrated antenna is one that is manufactured with a packaging process. A typical package is composed of several layers of co-planar metal structures such as transmission lines, baluns, hybrids, and antennas. Electrical connections within the package are created between rows using either vias, which are vertical metal cylinders that run between different metal layers, or electromagnetic capacitive connections. Connection between the package and chip may be made using capacitive, ball-grid, or flip chip connections. From the view of circuit theory, a via resembles an inductor whereas a capacitive connection resembles a capacitor. An integrated circuit such as a 60 GHz mmWave radio is usually placed inside a small air cavity within the package. Small metal bumps or very thin bond wires connect the integrated circuit to the rest of the package. The metal structures of the package connect the chip to the rest of the world, generally for soldering onto a printed circuit board. Fig. 4.13 illustrates the structures of a typical package.

Packaged mmWave antennas have proven to be superior to on-chip antennas due to their higher achievable gain and efficiency. Many researchers have shown that packaged antennas can consistently attain gains as high as 10 dBi and efficiency greater than 80% [SZG+09][SHNT05][SNO08]. With the current level of development for on-chip antennas, most long- and short-range 60 GHz and other mmWave applications—any application for distances greater than roughly a meter—will need to use packaged antennas or carefully crafted antennas on circuit boards if the design calls for antenna integration. In-package antennas can achieve higher gain and efficiency because they are isolated from the lossy integrated circuit substrate that is responsible for the low efficiency of on-chip antennas. But, packaged antennas have a higher production cost than on-chip antennas. This is especially the case if the use of in-package antennas forces an integrated system such as a transceiver or radar to use a more expensive packaging process than would otherwise be required. For example, use of an in-package antenna might force a designer to use a packaging process with four metal layers instead of two to allow more design flexibility in the size of package via structures.

There are many types of popular packaging processes that have been studied for use in production of in-package antennas, including Teflon [SHNT05], low-temperature co-fired ceramic (LTCC) [SZC+08][ZL09][Zha09], fused silica [ZPDG07], and Liquid Crystal Polymer (LCP) [KLN+10]. When selecting a package technology, the most important considerations from the standpoint of an in-package antenna (besides cost, which is always the most important consideration for a mass-consumer marketplace) are the package's relative permittivity, minimum feature size, manufacturing accuracy, number of available metal layers and distance between metal layers, the types of interconnect technology compatible with the packaging process, and whether the packaging process would allow an air cavity between the antenna and packaging substrate.

The permittivity of a package technology is essential design knowledge because it determines the electrical size of the in-package antenna. The higher the relative permittivity of a packaging material, the larger its metal contents will appear electrically. Therefore, antennas manufactured with high permittivity processes will require less space

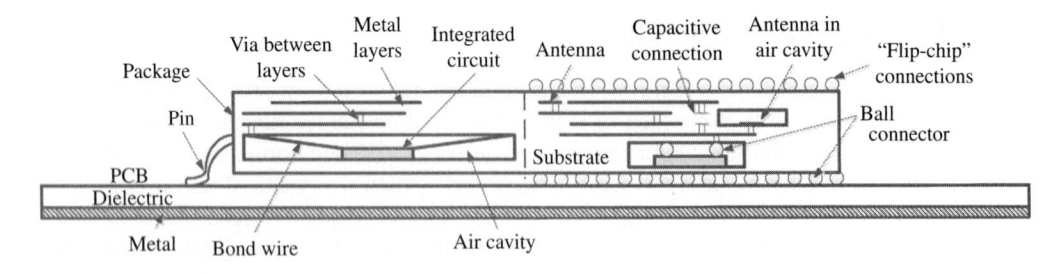

Figure 4.13 This figure illustrates various structures in a package (not drawn to scale). A given package is unlikely to have all of the structures illustrated.

to achieve the same gain as in competing package processes. But, high relative permittivities also exacerbate design problems due to inaccuracies in the production processes and will result in higher performance variability across devices unless production inaccuracies can be made extremely small (although often these variabilities may be compensated for by proper tunability features in the chip design). Another disadvantage of a high relative permittivity process is that it will result in a lower bandwidth for in-package antennas than a lower permittivity process, due to the fact that more energy will be stored in a high permittivity package than in a lower permittivity package. A common figure of merit for antennas, the "gain-bandwidth product" will generally be lower for an in-package antenna in a high permittivity process than in a lower permittivity process. The permittivities of common package technologies include 5.9-7.7 for LTCC [SZC+08][SNO08], 3.8 for fused silica [ZCB+06], 3.1 for LCP [KPLY05], and 2.2 for Teflon [SHNT05][ZS09a]. The loss tangent of the packaging material should be considered along with the permittivity, as the loss tangent will impact how much energy will be dissipated by electromagnetic fields in the package. The bandwidth of package resonances, which will adversely affect antenna performance, will also be determined by the loss tangent of the package (the relative bandwidth of a resonance increases approximately linearly with the loss tangent [Poz05]). Loss tangents of popular packaging technology include 0.0007 for Teflon [SHNT05], 0.002 for LTCC [SZC+08], 0.001 for fused silica [ZLG06], and 0.002-0.004 for LCP [ZL09].

The minimum feature size of metal structures, including vias, for a package technology is a major consideration when selecting a package material, and this will grow in importance as mmWave applications increase in frequency from 60 GHz to higher sub-terahertz and terahertz frequencies. The via length, in particular, will determine the minimum length of vertical structures in the package. The distance between metal layers in the packaging process is of major concern here, as this determines the length of via structures, and hence the inductance exhibited by each via. In general, as the frequency of operation increases, designers should consider using a process with a decreased interlayer thickness. The manufacturing precision, which determines how closely a manufactured chip will conform with design specifications, should be considered in addition to minimum feature size. In general, higher manufacturing precision will correlate with smaller feature size. The manufacturing precision of LTCC Fero A6, an example of a popular LTCC packaging technology, is 50 μm [SZC+08], greater than 50 μm for recent Teflon processes [SHNT05], and 10 μm for fused silica [ZLG06].

4.5 Antenna Topologies for MmWave Communications

Selection of an antenna topology is not trivial and depends on the intended application and the circuits that must interact with the integrated mmWave antenna. For example, polarization diversity may become an important aspect of mmWave communications, thereby dictating particularly well suited directional antenna structures, and 60 GHz devices intended for only short-distance links or sensing (under 1-2 m) may not require high gain but rather may place a premium on power efficiency and low cost. Longer links, including enterprise WiFi (using WiGig, IEEE 802.11ad, or Wireless HD) and backhaul or outdoor cellular networks, may need to emphasize pattern over antenna efficiency to achieve an acceptable element pattern for use in an array. The input impedance of an antenna may determine whether it is suitable for use with a particular driving circuit.

The dipole antenna is a simple antenna that provides a basis for much understanding of other antenna topologies. When fabricated on-chip or in-package, a dipole is often called a *planar dipole*. This type of antenna has been used or studied extensively for integrated applications [MHP+09][CDY+08][OKF+05][LGL+04][KO98] [FHO02][BGK+06b], and it has been used in conjunction with other radiating structures such as cylindrical resonators [BAFS08]. A planar dipole antenna is a type of broadside antenna [Reb92], indicating that its main beam is perpendicular to the axis of the antenna. Dipoles are easy to manufacture and may be easily interfaced with on-chip transmission lines such as coplanar transmission lines. A major disadvantage of this type of antenna for on-chip integrated applications is the ease with which it excites guided waves in the substrate [AKR83][Reb92], resulting in low efficiencies for on-chip antennas on unthinned substrates [GAPR09][Reb92], typically not substantially greater than 10%. An application for on-chip dipoles known as *inter-chip wireless interconnects* [KO98][OKF+05][LGL+04] may take advantage of substrate modes as it relies on communication between antennas on the same substrate rather than requiring radiation into free space. Researchers have suggested this approach as a means of replacing metal interconnects and for clock distribution across large integrated circuits [OKF+05][KO98].

A circuit model that was derived specifically for dipole antennas is shown in Fig. 4.14 and is useful for predicting the input impedance of a standard dipole. At mmWave frequencies, this model should be augmented to include effects of the integration environment, as shown in Fig. 4.11[HH97]. These include a capacitor to model coupling to the substrate for on-chip antennas, and a network model for the substrate itself that includes resistive and capacitive components to model substrate loss and energy storage. The model in Fig. 4.14 was developed to reproduce the input characteristics of a dipole antenna. These include a pole near resonance that results in a rapid increase in the real input impedance near resonance. In the simple model of Fig. 4.14, the capacitor C_0 is used to model the pole in the expression for the real input impedance, and L_0 is

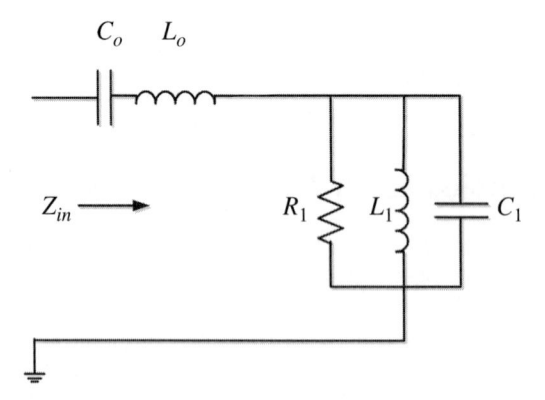

Figure 4.14 This simple model by [HH97] is useful for predicting the input impedance of a standard dipole. At millimeter-wave frequencies, this model should be augmented to include effects of the integration environment.

4.5 Antenna Topologies for MmWave Communications

used to offset the capacitor at other frequencies. The values of R_1, L_1, and C_1 are used to model higher-order resonances. A second parallel RLC network may also be added [HH97]. The reader should keep in mind that a resonant series LC circuit will appear as a short, whereas a parallel resonant LC circuit appears as an open. The form of the basic input impedance curve that results from Fig. 4.14 from [HH97] is reproduced in Fig. 4.15 (with the absolute numbers in the figure relevant only for a particular dipole antenna). A classic half-wavelength dipole in free space has an input impedance of $73 + j42.5\Omega$ [Bal05] at its resonant frequency.

The field of a half-wave dipole in free space is approximated by a "Donut" shape encircling the length of the dipole [Bal05]. The fields of integrated dipoles are different due to the presence of either the package or substrate. Of importance here is the potential of substrate modes excited by the dipole to radiate out the sides of the chip [BGK+06b], possibly resulting in a distorted pattern. [OKF+05] measured the patterns of on-chip dipole antennas by measuring the transmission between pairs of antennas fabricated on a chip at various angles. Their plots for linear and zigzag are shown in Fig. 4.16 for the pattern on the plane of the substrate. Measuring the pattern out of the plane of the substrate and in three dimensions has proven to be extremely challenging for most integrated antennas.

A dipole antenna is a balanced antenna and requires a differential feed (discussed further in Chapter 5). Therefore, single ended feed systems will require a balun for best performance for integrated dipoles. By using coplanar waveguide (CPW) lines (as shown below), it is possible to use a very simple balun structure [MHP+09][GJRM10][SRFT08] shown in Fig. 4.17. The lower metal strip used to connect the two ground ends is not

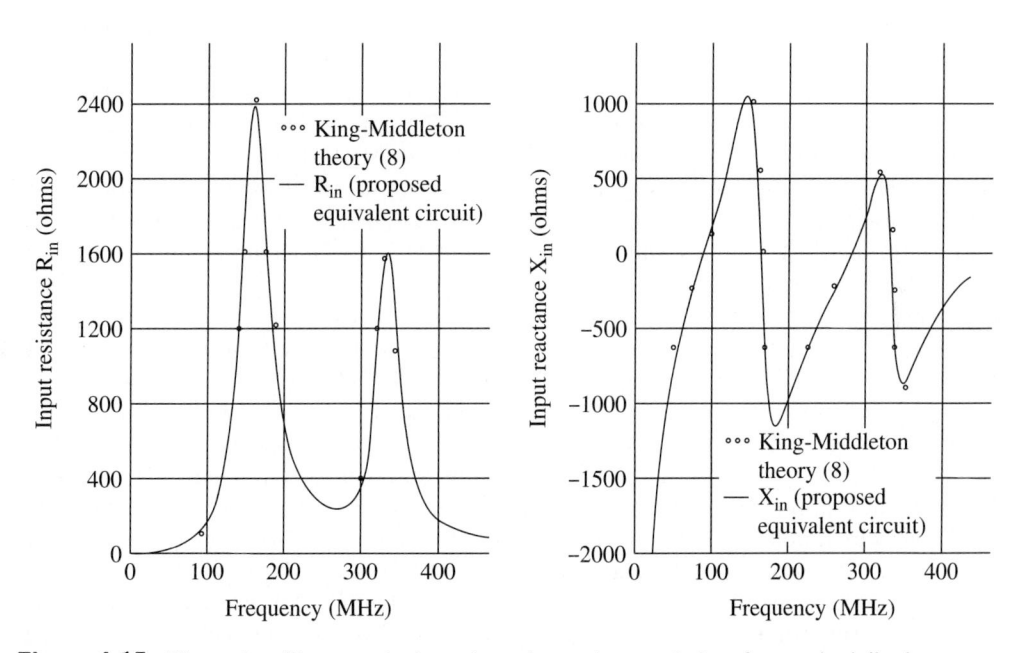

Figure 4.15 These plots illustrate the input impedance characteristics of a standard dipole antenna [from [HH97] © IEEE].

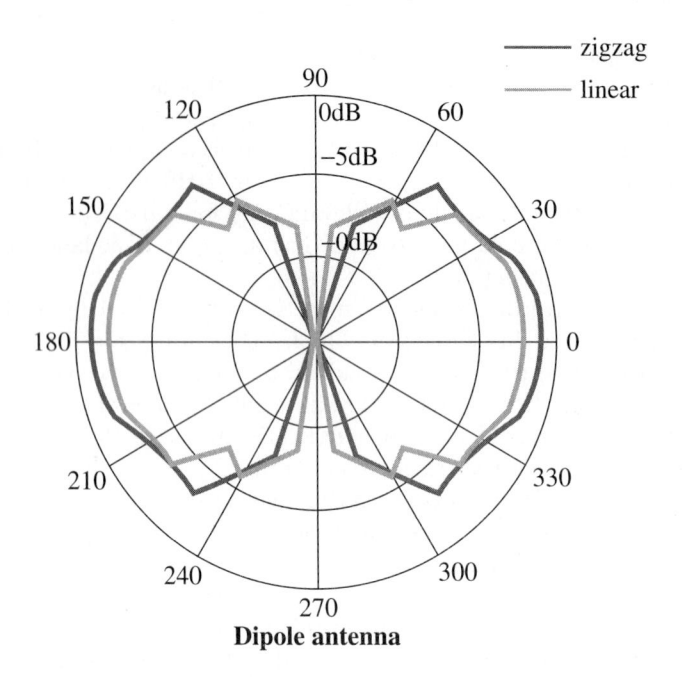

Figure 4.16 This plot shows the pattern of on-chip linear and zigzag antennas in the plane of the substrate [from [OKF+05] © IEEE].

always employed (e.g., in [MHP+09]). [MHP+09] presented a series of design curves for the return loss of this structure without the lower metal connecting strip, and these are reproduced in Fig. 4.18 based on a simulated dipole on a GaAs substrate that has relative permittivity of 12.9 and is 625 μm thick. The upper right curve shows that the resonant length of the dipole decreases with frequency [MHP+09]. The upper left curve indicates that the width of the dipole arms had nearly no effect on the resonant point of the dipole [MHP+09]. The two lower curves are for positive return loss (discussed in Chapter 5), for which a higher value indicates a better impedance match and transfer of power. The lower right curve is interesting and indicates that the dipole performed worse as its thickness in the vertical dimension increased, and this was probably due to the increased excitation of substrate modes [MHP+09]. The lower left curve indicates that the wider arms tended to give a better return loss [MHP+09].

Several variants to on-chip dipoles have been suggested, including linear, zigzag, and meander topologies. Zigzag topologies may offer slightly better efficiency compared to linear dipoles [CDY+08], but the difference is not substantial as seen in [OKF+05]. Fig. 4.19 illustrates these approaches. Zigzag antennas have also been shown to achieve higher gain than linear wire dipoles [FHO02]. [FHO02] used 30° angles to construct a zigzag antenna. When [CDY+08] implemented a zigzag design at 24 GHz on 0.13 μm CMOS, they used only the top three metal layers of their antenna to allow for 3.6 μm between the lowest metal of the antenna and the chip substrate (i.e., 3.6 μm of dielectric remained between the antenna and the substrate). By keeping the distance between the dipole and the substrate as high as possible, [CDY+08] traded off reduced coupling into substrate

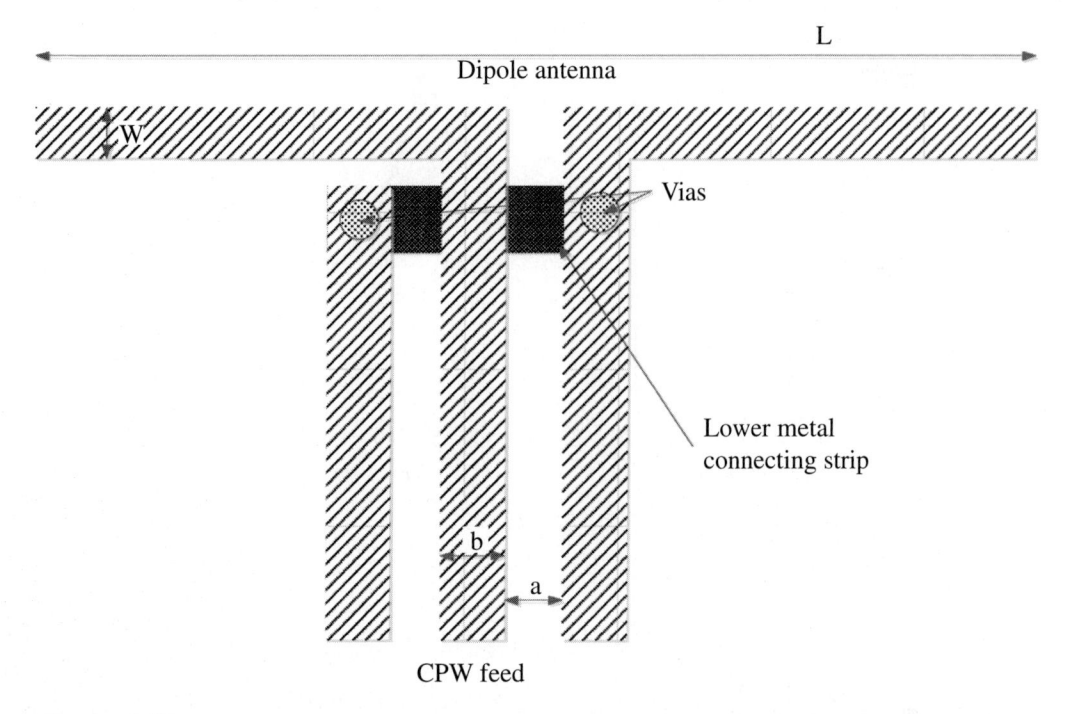

4.5 Antenna Topologies for MmWave Communications

Figure 4.17 A CPW line may be interfaced with a dipole antenna with this simple structure, as was used in [GAPR09]. The dipole length L is $\lambda/2$ for the substrate used.

guided waves for slightly higher conduction loss in the antenna, an effective choice given how easily substrate modes degrade performance [AKR83]. [CDY+08] attained an input impedance of 40-j100Ω and a gain of -10 dBi.

Slot antennas have also been used and studied for integrated antennas. Electromagnetically, a slot is the magnetic counterpart of a dipole, and consequently the analysis of each of the two is nearly identical to the other [Bal05]. The length of the slot should be $\frac{\lambda}{2}$, where λ is the operating wavelength. For integrated antennas, the effective wavelength (an average of the dielectric wavelength and the free space wavelength) should be used. The differences between integrated slots and dipoles are more pronounced, due largely to the differences in how these two topologies interact with the substrate or the package. Compared to the dipole, slot integrated antennas are slightly less susceptible to the challenges of substrate surface waves for thin substrates [Reb92][AKR83], but they suffer more degradation from substrate effects for thicker substrates [AKR83]. A slot antenna is implemented as a hole in the ground plane, as illustrated in Fig. 4.20. Compared to dipole and microstrip integrated antennas, slots have been claimed to have higher efficiency and be easier to match [Beh09], though this will depend on the nature of the substrate. Slots are also amenable to miniaturization and dual-mode techniques. [Beh09] presented miniaturization techniques for slot antennas based on loading an electrically small slot with inductive loads at its two ends (an electrically small antenna is defined as $kr < 1$, where k is the wave number for the medium of operation and r is the radius

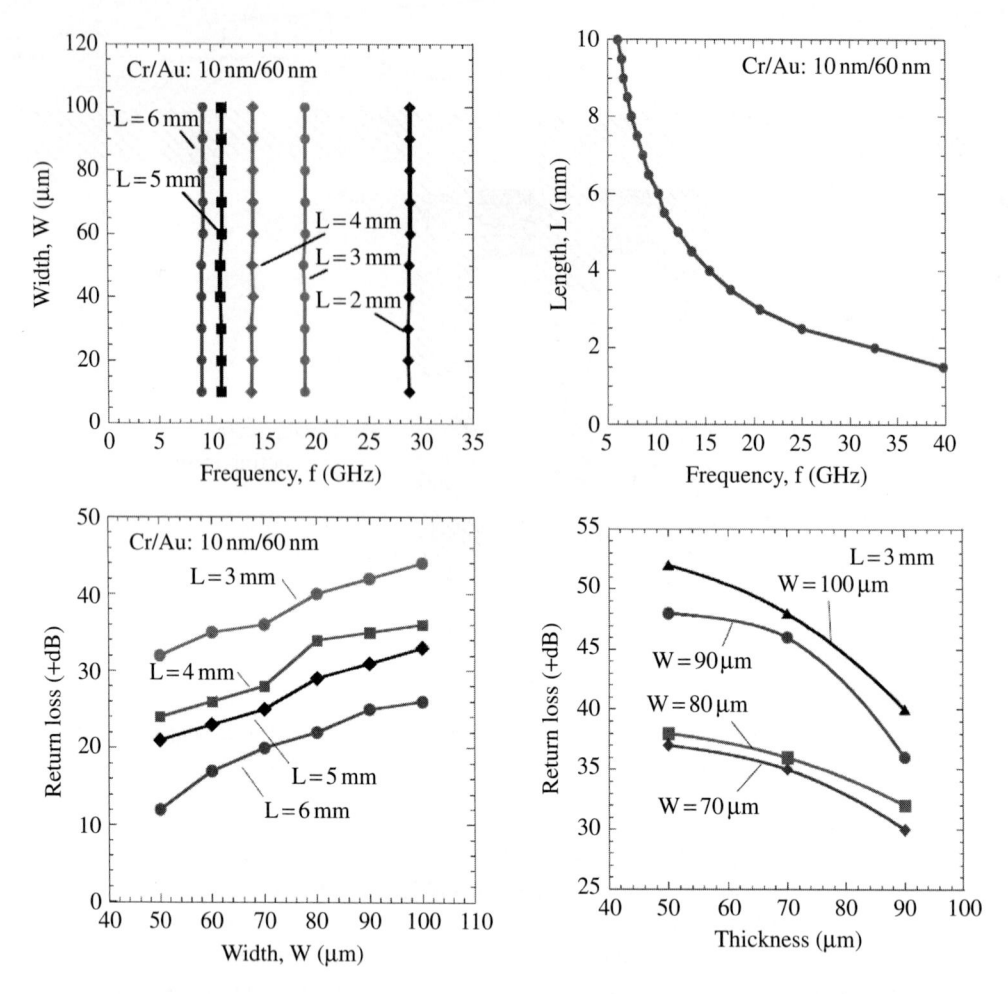

Figure 4.18 [MHP+09] presented these plots for the design of a planar dipole antenna on a 625 μm GaAs substrate with relative permittivity of 12.9 and 625 μm thick [reproduced from [MHP+09] © IEEE].

of a sphere that can completely enclose the antenna). The electrically thin slot on its own would resonate at a higher frequency, and the loads help to reduce this frequency down to the operating frequency. Fig. 4.21 illustrates this technique. [Beh09] used this technique to integrate a slot onto a CMOS substrate and achieved −10 dBi gain and an efficiency of 9%.

There are several special considerations to keep in mind when attempting to improve the performance of on-chip slots with lenses (a common approach to improve integrated antennas). Slot antennas individually are not suited for use with paraboloid lenses due to their non-uniform radiation and the large effect of the ground plane on a slot's pattern [KSM77]. [KSM77] showed that this topology can be effective in integrated receivers used

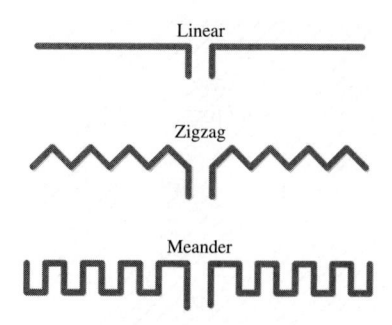

4.5 Antenna Topologies for MmWave Communications

Figure 4.19 Various approaches have been suggested and used for on-chip dipoles.

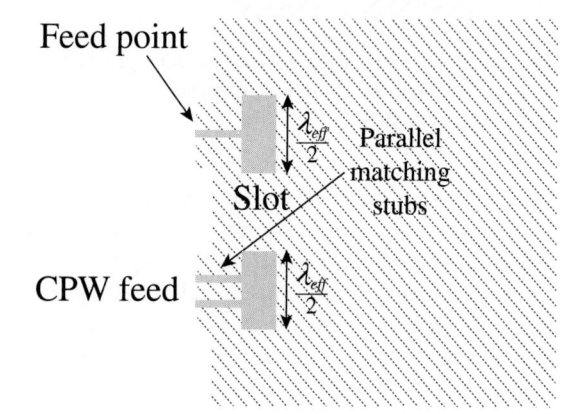

Figure 4.20 A basic slot antenna, implemented as a hole in a ground plane. The matching stubs shown were suggested by [Beh09].

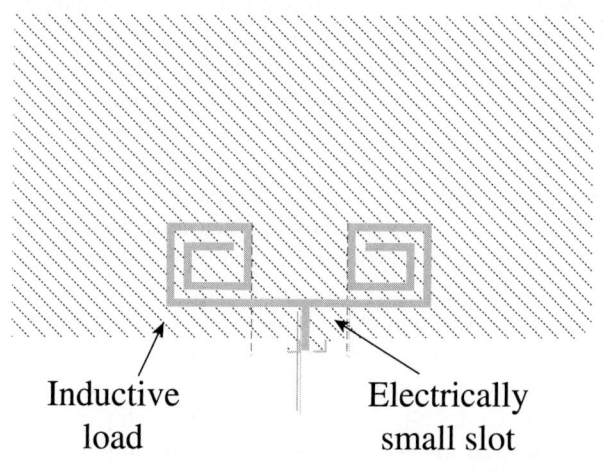

Figure 4.21 Slots may be miniaturized by loading their ends. [Beh09] used this technique, in which the two ends of a slot were loaded with inductive loads to lower its resonance frequency to the operating frequency.

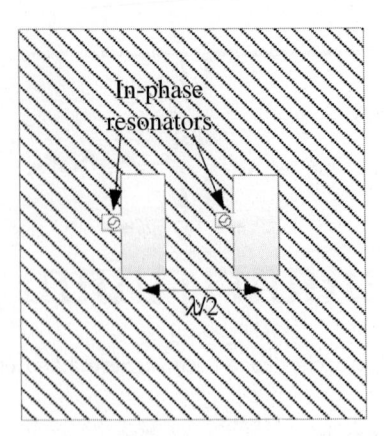

Figure 4.22 Two slot antennas separated by a quarter wavelength can work well as a means of reducing the sensitivity of a slot antenna to the effect of ground plane size and to provide uniform illumination suitable for use with a lens [KSM77].

with lenses when two slots are used together, are separated by a half wavelength, and are driven in-phase. This technique reduces the sensitivity of the pattern to the size of the ground plane and results in more uniform radiation. This technique is shown in Fig. 4.22.

Microstrip patch antennas may also be considered for integrated purposes. Electrically, microstrip patches are usually considered to be cavities that radiate out their sides, and they are often analyzed based with a resonant cavity approach. From a radiation perspective, patches may be considered to be a two-element array of magnetic current elements, as demonstrated using Huygens' principle in Fig. 4.23.[3] The advantages of microstrip antennas include their ability to conform to surfaces and relative ease to tune [Bal05]. Traditional patches suffer from low bandwidths and poor efficiencies [Bal05], though when integrated on-chip their efficiencies can be comparable to other on-chip antennas (e.g., [HW10] achieved a 15.87% efficiency without substrate thinning or a lens). Patch antennas have been used in mmWave systems. For example, the company SiBeam (now Silicon Image) recently presented a packaged 60 GHz transceiver that contained an array of patches in a low-temperature co-fired ceramic (LTCC) [EWA+11]. Among on-chip implementations without substrate processing, they have competitive gain (e.g., [HW10] achieved −10 dBi). In the event patch antennas are used for mmWave transceivers, their poor bandwidth may require high modulation efficiencies (in bits/hertz) to achieve very high data rates, indicating that these transceivers may also by necessity be higher power than other options. Low dynamic range systems that achieve high data rates through broad bandwidths with low spectral efficiency will probably not often use patch antennas. Should broad bandwidths be required for a patch, techniques such as the use of parasitic parallel metal strips may be used to increase the bandwidth of the patch [CGLS09]. An advantage of on-chip patches is that the area under the patch may be used for active circuitry, as was done in [HW10] (provided the patch's ground plane is not on the lowest metal layer of the chip [HW10]). Another

---

3. The authors thank Professor Hao Ling of The University of Texas at Austin for this insight.

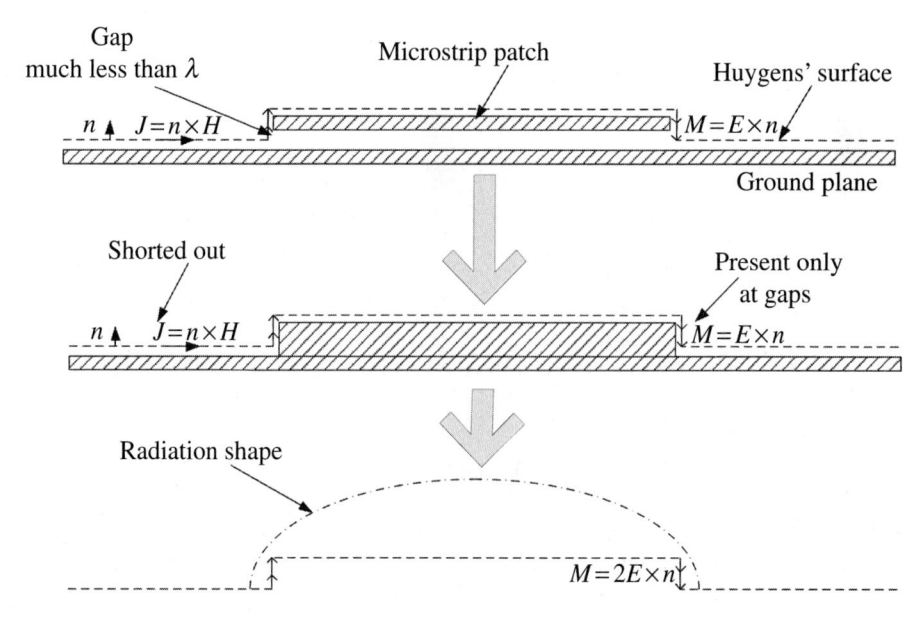

4.5 Antenna Topologies for MmWave Communications

Figure 4.23 A patch antenna may be considered an array of two magnetic elements at its ends, as shown by Huygens' principle. As the patch is lifted further from its ground plane, the magnetic elements become longer and hence more efficient as radiators.

disadvantage of patch antennas is their size compared to other integrated antennas—[YTKY+06] required 1.7×1.3 mm^2 to construct an on-chip patch for 60 GHz, [HW10] required 1.22 mm $\times$ 1.58 mm^2 for 60 GHz, and [SCS+08] required 200 μm $\times$ 200 μm for 410 GHz. This large size requires that on-chip patches have many slots to satisfy circuit design rules about the maximum size of a piece of metal without holes. If these slots are made much smaller than a wavelength and kept parallel to the length of the antenna [HW10], they should not greatly impact performance (e.g., [HW10] found that required ground-plane slots reduced efficiency by 3% in simulation).

Patches have been used for mmWave systems by [HW10][YTKY+06][SCS+08] [AL05][PNG+98][CGLS09][LDS+05]. There are many models that may be used to design patch antennas, including the transmission line model, which treats the patch as a resonant transmission line [Bal05]. The basic procedure to design the patch, as described by [Bal05] for the patch shown in Fig. 4.24, is first to specify a desired resonant frequency. Next, a set of design equations for the width and height based on the dielectric constant and the distance of the patch above the ground plane are used. We repeat these equations [Bal05] here in equations (4.31) through (4.34). In the equations, ϵ_{eff} is an effective permittivity based on how much electric field is contained between the top and bottom surfaces of the patch and the air. ϵ_o and μ_o are the constitutive parameters of free space, $\epsilon_{r,\text{dielectric}}$ is the permittivity of the dielectric, W is the width of the patch, L is its length, and ΔL is a length correction that accounts for the presence of fringing fields around the patch [Bal05]. When integrating on-chip, the height of the patch is critical to efficiency. This is seen in Fig. 4.25 by [HW10], which shows the simulated efficiency for a patch antenna designed using the transmission line model as a function of the height of the

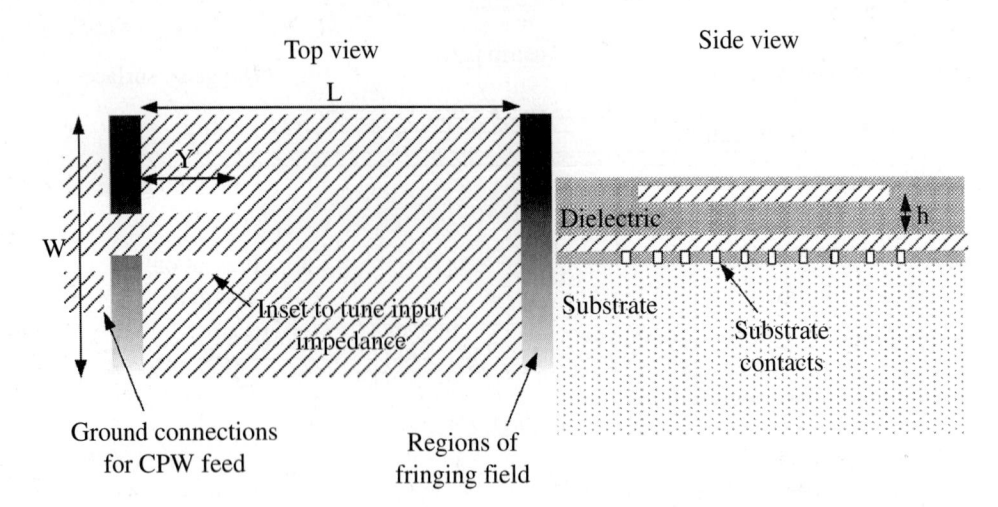

Figure 4.24 [Bal05] presents a set of design equations for patch antennas based on this geometry. The ground plane should be kept at the same potential as the substrate to help reduce coupling into the substrate [YTKY+06]. [Based on a figure in [Bal05].]

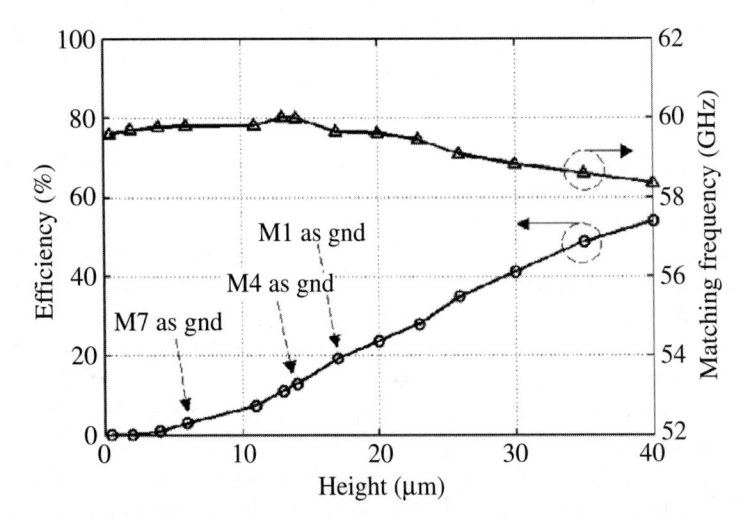

Figure 4.25 The height of the antenna above the substrate has a major impact on its efficiency [from [HW10] © IEEE].

patch above the ground plane in 0.13 µm CMOS. The plot also shows the effect of the height on the resonant frequency f_r

$$W = \frac{\sqrt{\epsilon_o \mu_o}}{2 f_r} \sqrt{\frac{2}{\epsilon_{r,\text{dielectric}} + 1}} \quad (4.31)$$

$$\epsilon_{\text{eff}} = \frac{\epsilon_{r,\text{dielectric}} + 1}{2} + \frac{\epsilon_{r,\text{dielectric}} - 1}{2} \sqrt{1 + \frac{12h}{W}} \quad (4.32)$$

4.5 Antenna Topologies for MmWave Communications

$$\Delta L = \frac{0.412h\left((\epsilon_{\text{eff}} + 0.3)\left(\frac{W}{h} + 0.264\right)\right)}{(\epsilon_{\text{eff}} - 0.258)\left(\frac{W}{h} + 0.8\right)} \qquad (4.33)$$

$$L = \frac{1}{2f_r\sqrt{\epsilon_{\text{eff}}}\sqrt{\mu_o \epsilon_o}} - 2\Delta L. \qquad (4.34)$$

The microstrip feed for the patch should be selected for the correct input impedance. [Bal05] presents the equations (4.35)-(4.37) that may be used before further refinement in simulation. In practice on-chip patches have achieved a range of input impedances. For example, [AL05]'s patch for 77 GHz had a very low input impedance near 5 ohms, whereas [SCS+08] achieved 50 ohms

$$G_1 = \frac{W}{120\lambda_o}\left[1 - \left(\frac{1}{24}\right)(k_o h)^2\right] \qquad (4.35)$$

$$R'_{\text{in}} = \frac{1}{2G_1} \qquad (4.36)$$

$$R_{\text{in}} = R'_{\text{in}}\cos^2\left(\frac{\pi}{L}Y\right). \qquad (4.37)$$

The ground plane for the patch should be kept at the same potential as the substrate if possible. This helps to reduce coupling into the substrate [YTKY+06]. One method to help ensure that the patch ground and substrate are at the same potential is to place substrate contacts between the patch ground and the substrate, as indicated in Fig. 4.24. Proper design of the ground plane will also help to ensure that the patch is isolated from nearby circuitry [SCS+08] by preventing coupling through the substrate.

The radiation pattern of a patch antenna is smooth and directed above the top metal layer of the patch. Directivities have varied for on-chip patch antennas, with [PNG+98] achieving 6.7-8.3 dB in a band near 60 GHz, [HW10] achieving 4.7 dB at 60.51 GHz, [CGLS09] achieving 6.34 dB at 60 GHz, and [SCS+08] achieving 5.15 dB at 410 GHz. Efficiencies for on-chip patches have generally been low but competitive with other on-chip mmWave antennas. [CGLS09] achieved 14% efficiency with the use of a patterned "artificial magnetic conductor" ground plane at 60 GHz, [SCS+08] achieved 22% at 410 GHz, [HW10] achieved 15.87%, and [PNG+98] achieved between 21% and 33% in a band near 60 GHz. Fig. 4.26, a combination of figures from [SCS+08] and [CGLS09], illustrates representative patterns of on-chip patch antennas.

Patch antennas have also been used in-package in addition to on-chip [LDS+05] (dual-polarized patch in LTCC or 60 GHz) [HRL10] (circularly polarized patch on glass package substrate for 60 GHz) [KLN+11] (16 element patch array in LTCC for 60 GHz) [LS08] (8 element patch on fused silica). In-package patches are generally higher in gain and efficiency than on-chip patches. For example, [HRL10] achieved 7.4 dBi gain with an efficiency of 60%, [KLN+11] achieved 4-6 dBi gain/element on LTCC, and [LS08] achieved 15 dBi gain for an 8 element patch array at 60 GHz on fused silica. A challenge in creating an in-package patch is to interface the patch with the packaged chip. There are various approaches that can be taken. For example, the left portion of Fig. 4.27 illustrates the use of a packaged via to excite the patch through a hole in the patch ground plane. The right portion of the figure is from [HRL10] and illustrates the use of a coupling connection from an on-chip CPW line to excite a patch antenna. A common approach

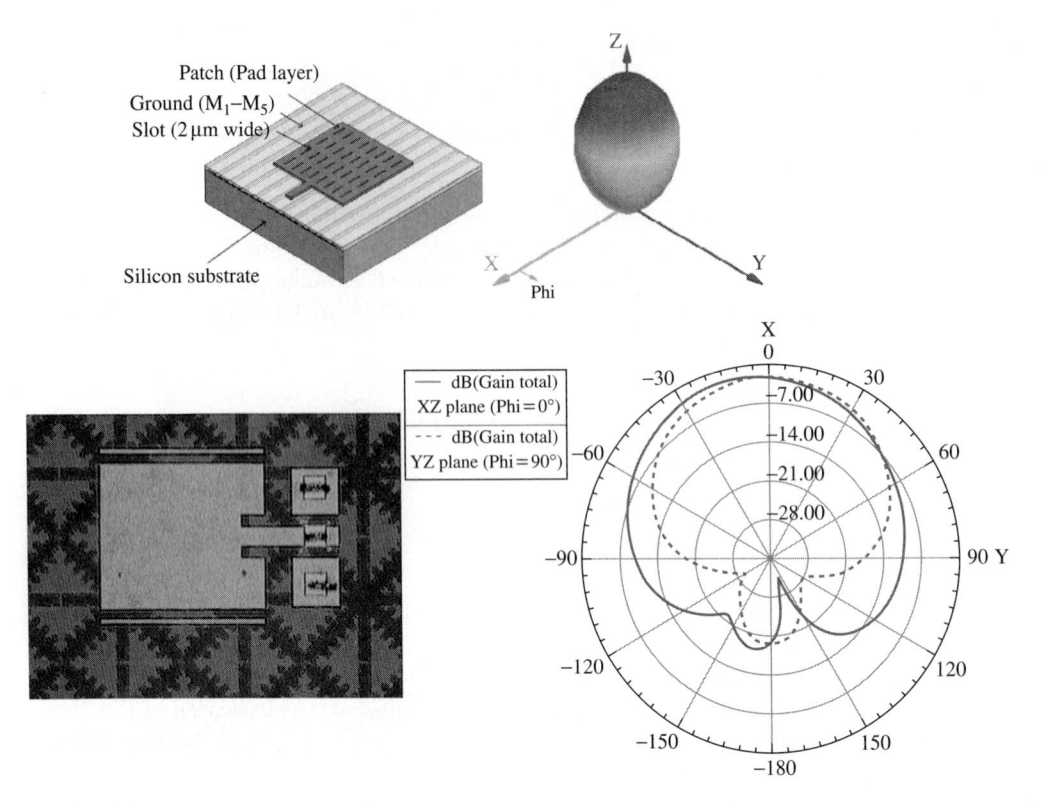

Figure 4.26 This figure shows that a patch antenna typically radiates above the top metal layer of the antenna. The top figure illustrates the metal slots that are usually required for on-chip patch antennas due to their large size. In the lower figure, [CGLS09] used two parallel metal strips on the edge of the patch to increase bandwidth. [This figure is a combination of figures from the literature ([SCS+08] above, [CGLS09] below) © IEEE.]

for improving the efficiency of packaged patch antennas is to include an air cavity below the patch [LS08][KLN+11], as illustrated in Fig. 4.28. This approach generally requires that the patch be fed from the side. The figure is based on a similar figure in [LS08].

Yagi-Uda antennas have also been studied for use in mmWave systems [HWHRC08][GAPR09][ZSG05][SZC+08][AR10]. As described by [Bal05], the Yagi-Uda antenna is actually an array consisting of a driven element, a reflector element, and one or more director elements. The driven element is usually a dipole antenna as in [GAPR09] and [HWHRC08]. Fig. 4.29 illustrates a basic Yagi-Uda antenna based on the description in [Bal05]. The basic principle behind the antenna is that the progressively shorter elements from reflector to driven element to directors results in a progressive phase shift of current on each element (in which voltage leads current in the reflector, current and voltage are in phase for the driven element, and in the directors current leads voltage) resulting in end-fire radiation. When on-chip or in-package, the multiple media through which waves may travel between elements (e.g., the top chip dielectric, the low substrate, and the air) complicate this view, usually requiring simulation-based optimization. For

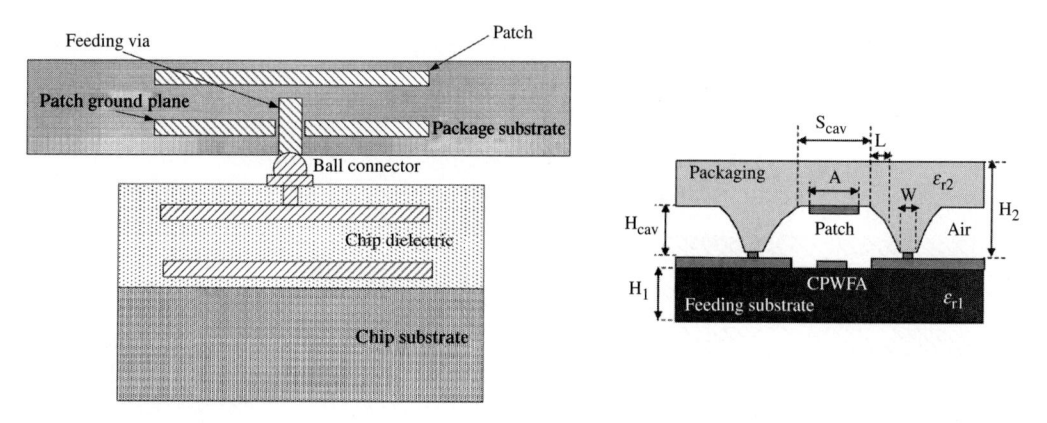

4.5 Antenna Topologies for MmWave Communications

Figure 4.27 There are various methods for feeding an in-package patch from a packaged chip. The ball connector (left) may, for example, be used in a flip chip connection. [KLN+11] found that this type of ball connector improves with a smaller radius of the ball and a smaller metal pad for the ball (represented here as a small rectangular piece of metal below the ball) [right portion from [HRL10] © IEEE].

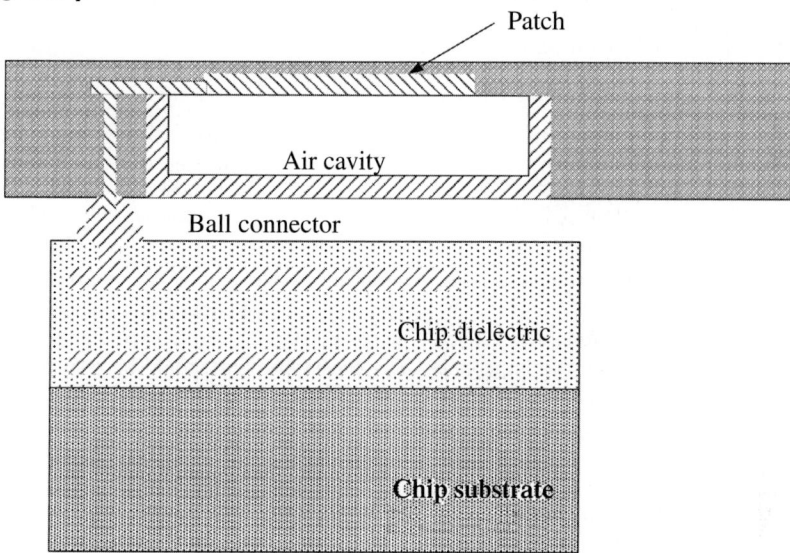

Figure 4.28 An air cavity may be hollowed out below the chip in some packaging processes. [LS08] and [KLN+11] used this technique to improve efficiency of in-package patch antennas.

example, [ZSG05] found that the distance between driven element and reflector should be greater than the distance between the driven element and the directors, while [GAPR09] found that $0.25\lambda_{\text{eff}}$ was the best spacing. An in-LTCC-package design by [SZC+08] found the best reflector length to be $0.25\ \lambda_g$, the length of the driver to be $0.56\ \lambda_g$, the length of the directors to be $0.32\ \lambda_g$, the distance between the reflector and driver to be $0.18\ \lambda_g$, and the distance between the director and driver to be $0.22\ \lambda_g$ (where λ_g is the guided wavelength and should be equal to the effective wavelength for the first mode). These

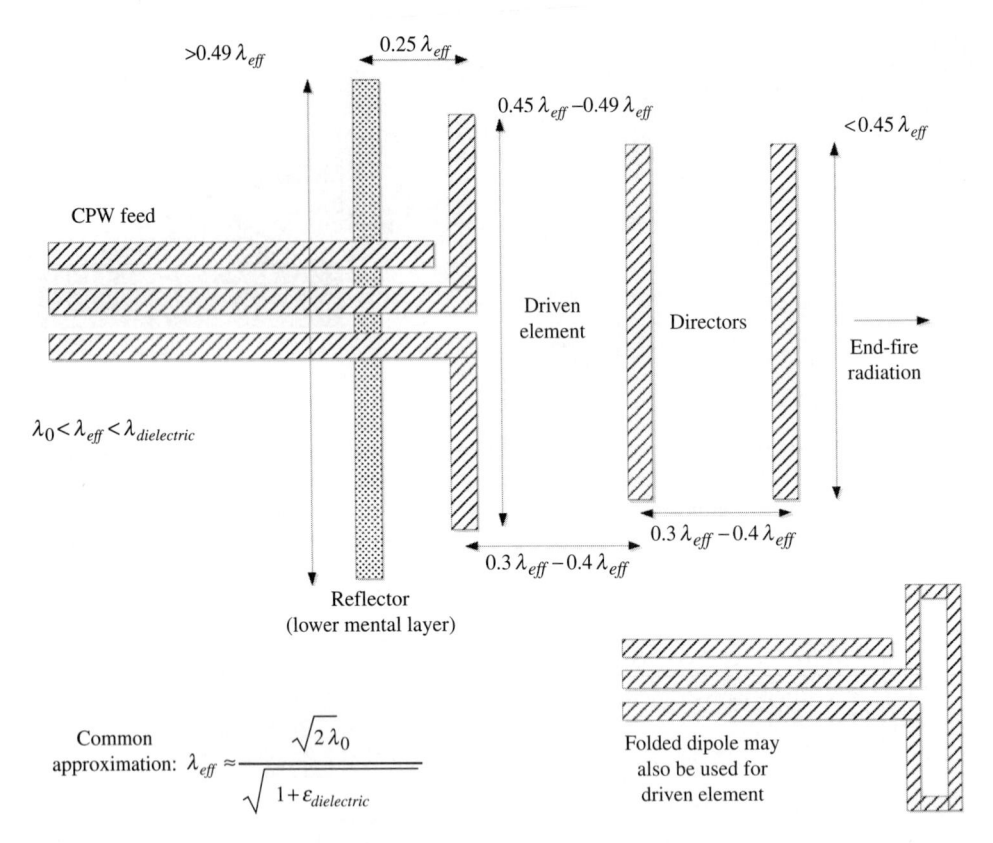

Figure 4.29 The Yagi-Uda antenna is actually an array of elements with one driven element and several parasitic elements [Bal05]. The lengths of the elements are chosen to create a traveling wave from the reflector to the directors. The reflector should be slightly longer than a resonant half-wave dipole so as to present an inductive impedance. The driven element should be resonant with a real input impedance, and the directors should be shorter than a resonant half-wave dipole to present a capacitive input impedance [Bal05].

variations underscore the importance of the integration environment in selecting final design dimensions. [GAPR09] used a clever simulation-based approach to find the effective wavelength λ_{eff}: They found the length of one-half of a driven element at resonance (i.e., when input impedance is purely real) and concluded this to be the effective quarter-wavelength.

Although a standard half-wave dipole may be used as a driven element, it may be beneficial to consider the use of a folded dipole [AR10]. This is because one of the traditional challenges of Yagi-Uda arrays is their low input impedance, and folded dipole may be used to increase input impedance. [AR10] used this approach to increase input impedance from 18 Ω to 153 Ω, which was beneficial given the impedance of their feeding line.

When designed correctly, the elements of the Yagi antenna should result in end-fire radiation (i.e., with maximum gain broadside to the driven dipole element, and with

4.6 Techniques to Improve Gain of On-Chip Antennas

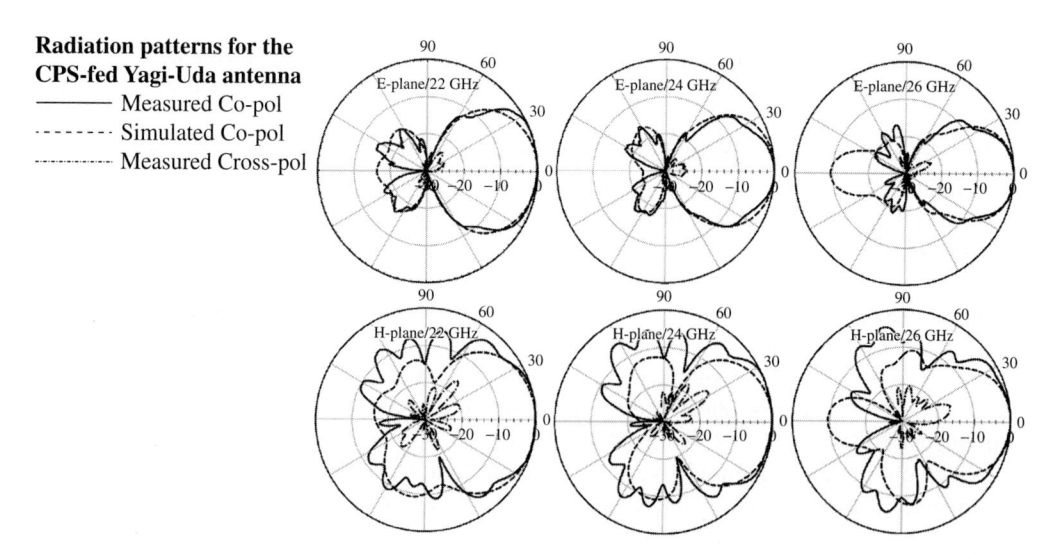

Radiation patterns for the CPS-fed Yagi-Uda antenna
—— Measured Co-pol
------- Simulated Co-pol
·········· Measured Cross-pol

Figure 4.30 This pattern shows the correct end-fire radiation (i.e., broadside to the driven element) for a Yagi-Uda array. If the elements are not correctly sized or placed, then non-end fire radiation may result [from [AR10] © IEEE].

maximum gain towards the direction of the smallest element). Fig. 4.30, taken from [AR10], demonstrates an end-fire (i.e., broadside to the dipole and toward the direction of the smallest element) Yagi-Uda pattern in a 24 GHz band for the array integrated in Teflon substrate.

[HWHRC08] used a Yagi antenna in 0.18 μm CMOS to achieve 10% efficiency and -10.6 dBi gain. [GAPR09] used 0.18 μm CMOS to achieve a simulated efficiency of 15.8% and a gain of -3.55 dBi. [ZSG05] achieved measured 5.6% efficiency and -12.5 dBi measured gain. [SZC[+]08] achieved a gain of 6 dBi and an efficiency of 93% for a Yagi-Uda in LTCC package for 60 GHz. [AR10] achieved 8-10 dBi gain and an efficiency better than 90% from 22 to 26 GHz for a Yaig-Uda array in Teflon.

4.6 Techniques to Improve Gain of On-Chip Antennas

Various techniques have been suggested as means of improving the gains of on-chip antennas. The discussion of the challenges facing on-chip antennas indicates that decreasing the substrate thickness or permittivity is the most obvious and effective means of improving gain and efficiency. Indeed, many authors cite this method [LKCY10][Reb92][KSK[+]09][AKR83]. Note that [AKR83] suggests keeping substrate thickness below $\frac{\lambda}{10}$ to achieve good gain and efficiency for on-chip antennas. Disadvantages of this approach are that it reduces mechanical stability of the substrate, possibly reducing manufacturing yield, and increases production cost and time. The very low cost of an integrated antenna, combined with the relatively high cost of the chip with which it is manufactured, results in virtually no loss of yield due to the antenna. Other techniques to improve gain and efficiency have been used. These include artificial magnetic conductors [CGLS09] and frequency selective screens [RHRC07] [CGLS09] [ZWS07], lenses [KR86] [Reb92] [Bab08] [BGK+06b] [FGR93] [LKCY10], substrate stacking [MAR10b] [Reb92], and arrangements of antennas to cancel substrate modes [Reb92] [RN88].

Considerable design freedom and performance enhancements can be obtained if stacked-substrates are used. For example, the distance between the antenna and substrate may be increased [MAR10b], or stacked antennas may be used to cancel substrate modes [Reb92].

The use of a "superstrate" above the main antenna substrate is one approach that has gained considerable attention [JA85][YA87]. [JA85] provided a detailed analysis of this technique based on transmission line theory under an assumption that radiated fields are approximated well by plane waves. If this is not the case (i.e., if $k_o R \gg 1$ does not hold [YA87]), then a plane wave decomposition would have to be performed before applying this analysis. Fig. 4.31 illustrates the basic approach, and Fig. 4.32 illustrates the transmission line analogy used to study the technique by [JA85]. The basic transmission line analogy provides easy intuition: If we assume that a "voltage" corresponding to an

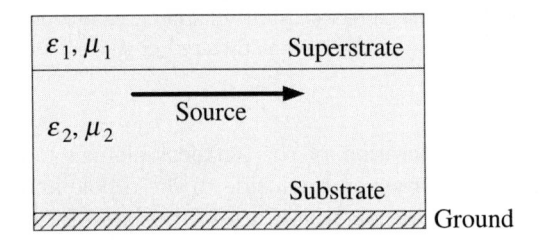

Figure 4.31 [JA85] provided a rigorous analysis of the use of a superstrate to improve gain. The source in the figure is the antenna.

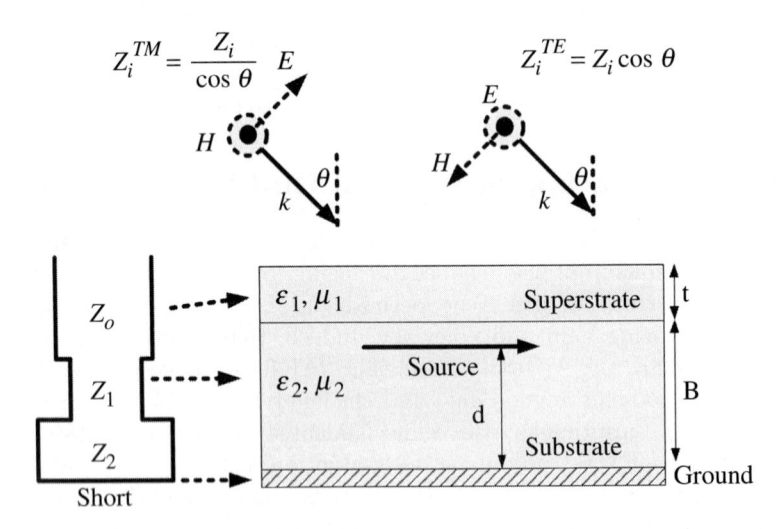

$$Z_i^{TM} = \frac{Z_i}{\cos \theta} \qquad Z_i^{TE} = Z_i \cos \theta$$

Figure 4.32 A transmission line analogy may be used to study the use of a superstrate to improve gain [JA85].

4.6 Techniques to Improve Gain of On-Chip Antennas

electric field exists at the position of the antenna, then if the lengths of the transmission lines are chosen correctly to induce a resonant condition, this voltage may be amplified by a factor of $\sqrt{\epsilon_1}$ at the top surface of the superstrate (this corresponds to a high gain case). We present the requirements derived by [JA85] here. If the substrate has a relative permittivity much greater than unity, the requirement for high gain is [JA85]:

$$\frac{n_2 B}{\lambda_0} = \frac{m}{2} \qquad (4.38)$$

$$\frac{n_2 d}{\lambda_0} = \frac{2n-1}{4} \qquad (4.39)$$

$$\frac{n_1 t}{\lambda_0} = \frac{2p-1}{4} \qquad (4.40)$$

where m, n, and p are positive integers, B, d, and t are thicknesses illustrated in Fig. 4.31, λ_0 is the wavelength of operation in free space, and n_1 and n_2 are the refractive indexes of the superstrate and substrate (where $n_i = \sqrt{\mu_{r,i}\epsilon_{r,i}}$), respectively. For example, on a substrate of silicon with permittivity 12.2, the substrate would have to be at least 715 μm thick, with the antenna positioned at 357 μm and a superstrate also of silicon to be 357 μm thick as well. This may not be possible for an on-chip antenna given foundry design rules, but the technique would also be applicable in a packaged environment where it may be more realistic. Provided these conditions are satisfied, the gain is approximated by [JA85]

$$G \approx \frac{8 n_2 B}{\lambda_0} \left(\frac{\epsilon_1}{n_2 \epsilon_2 \mu 1} \right). \qquad (4.41)$$

[CGLS09] presented the technique of using artificial magnetic conductors to improve antenna gain. This approach is similar to the frequency selective screen proposed by [RHRC07] and studied by [ZWS07]. A magnetic conductor is the dual of an electric conductor in which magnetic field must be perpendicular to the magnetic conductor just as electric field impinges on an electric conductor perpendicularly. Magnetic "conductors" are generally constructed by periodically spacing gaps or apertures in metal sheets. Magnetic conductors are often considered advantageous for purposes of shielding an on-chip antenna from the substrate because the image current of an electric current over a magnetic conductor is in-phase with the original current. Therefore, even though the shield is very close to the antenna, it will not result in a low input impedance nor "short out" the antenna as an electric shield would [BGK+06b]. This theory is incomplete, and [ZWS07] and others enriched it to include the effectiveness of the structure at suppressing surface waves that harm gain and efficiency. Huygens' principle, outlined in Fig. 4.33, indicates why such gaps can be modeled as magnetic conductors. [CGLS09] used a regular pattern in the lower metal layers of the BEOL metal layers of a 0.18 μm chip to create an artificial magnetic conductor to shield a patch antenna from the substrate. Fig. 4.34 from [CGLS09] shows the element used by [CGLS09]. This approach resulted in a gain of -2.2 dBi and an efficiency of 14% for the on-chip patch used by [CGLS09] at 60 GHz.

Artificial magnetic conductors and frequency selective screens are examples of more general patterned surfaces known as *high impedance surfaces* (HIS) and *electro-bandgap*

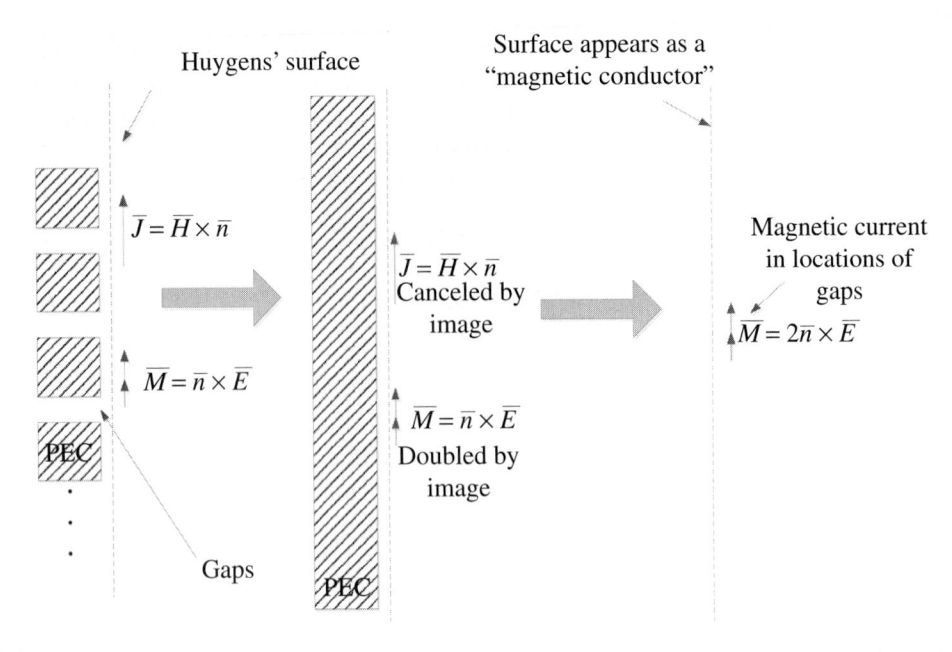

Figure 4.33 Huygens' principle reveals why certain periodic structures may be considered magnetic conductors.

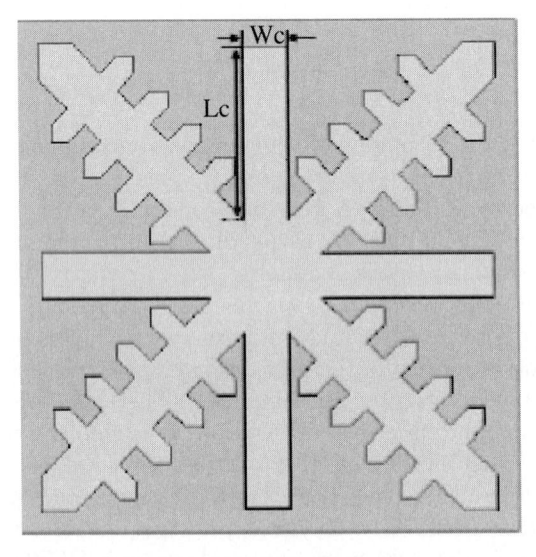

Figure 4.34 This element was cascaded periodically below an on-chip microstrip antenna to form an AMC to achieve a gain of −1.5 dBi [from [CGLS09] © IEEE].

surfaces. The approach is to create a regular pattern in metal below the antenna just as [CGLS09] did for their snowflake. Fig. 4.35 illustrates this approach, in which vertical vias have been added to add an inductive element to each unit cell (where a unit cell is an individual piece of the pattern, e.g., the snowflakes in [CGLS09]). There are many types

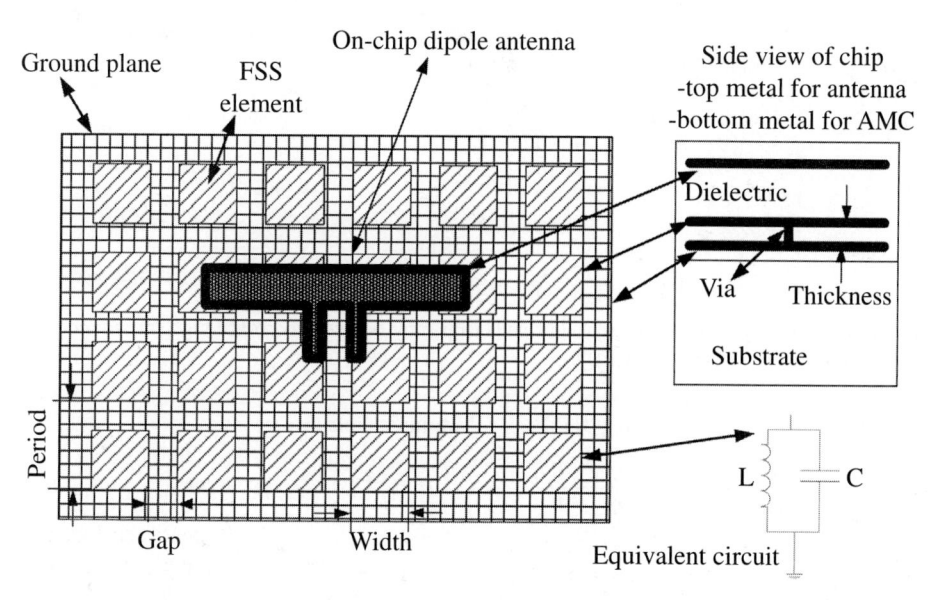

4.6 Techniques to Improve Gain of On-Chip Antennas

Figure 4.35 A metal grid or pattern can be used to construct an *artificial magnetic conductor*. [CGLS09] used this approach to achieve near 0 dBi gain for an on-chip antenna.

of patterns that may be used, including Peano curve elements, Hilbert curve elements, and mushroom elements [STLL05][SZB+99]. These surfaces are often also called *Sievenpiper surfaces* after D. Sievenpiper who worked with N. Alexopolous (partly responsible for determining ways to optimize substrate geometry [AKR83]). A circuit model can be created to represent a given unit cell. Fig. 4.36, based on a figure in [STLL05], summarizes several of these unit cells. When an HIS operates correctly it should not support or greatly dampen surface waves that adversely affect antenna performance (as explained in Section 4.3). Simulation can be used to determine whether the unit cell has been designed correctly. For example, there is a manual for designing meta-materials using Ansoft's HFSS that explains the proper way to simulate these structures in HFSS [HFS08] (available for free after registration with Ansoft). When functioning correctly, the reflection coefficient for electric field from these surfaces has zero phase. This can be understood by considering the reflection coefficient Γ from a surface for normal incidence (which is useful for understanding the principle) as illustrated in Fig. 4.37. When the second medium has a very high impedance, then the reflection coefficient is near 1. Unfortunately, these surfaces often operate on a resonance principle that prevents them from operating over all frequencies. Therefore, the surface needs to be redesigned for each operating frequency. Because the surface acts approximately as a "perfect magnetic conductor," it also shields the substrate, preventing the formation of substrate modes illustrated in Fig. 4.9. The fact that the electric field is not zero on the surface indicates that there exists an effective *magnetic* current on the surface. Also, the fact that this current takes time to build up (i.e., it does not exist for the first impinging wave) indicates that the surface may be characterized by a quality factor (i.e., energy must be stored in the surface to satisfy the boundary conditions). Therefore, we do not expect all high impedance surfaces to work equally well.

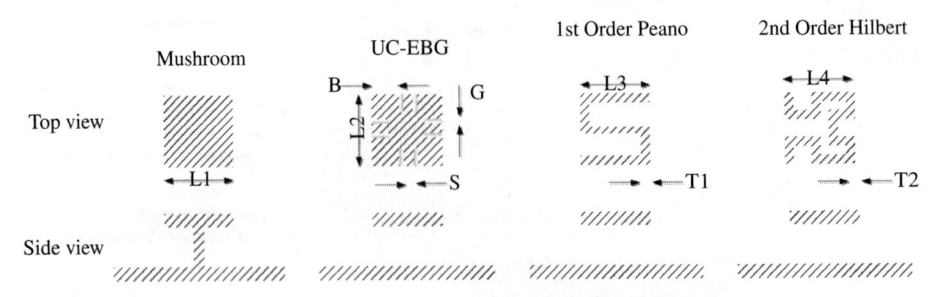

Figure 4.36 There are many types of unit cells that may be used to create high impedance surfaces [based on figure in [STLL05] © IEEE].

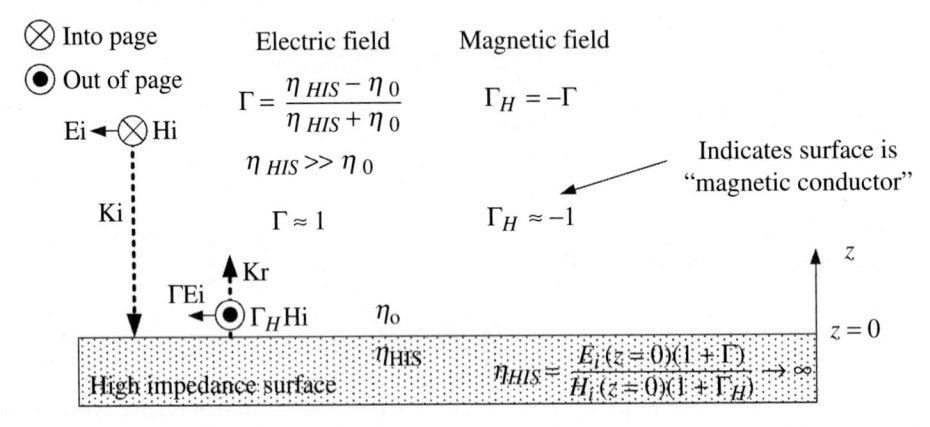

Figure 4.37 When the surface has a very high impedance, its reflection coefficient for the electric field is near 1. The reflection coefficient for the magnetic field is near -1, hence these surfaces are also often called *artificial magnetic conductors*. The figure also indicates that a high impedance on the surface results from the phases of the reflection coefficient on the surface. This shows why *high impedance surface* and *artificial magnetic conductor* are often interchanged as terms.

For lower frequency operation, the unit cells in Fig. 4.35 can be represented by an LC circuit with capacitance and inductance given by equations (4.42) and (4.43) [ZWS07]. L and C should be selected to give a resonance at the frequency of operation

$$C = \frac{\text{width} \times (\epsilon_{\text{air}} + \epsilon_{\text{sub}})}{\pi} \cosh^{-1}\left(\frac{\text{period}}{\text{gap}}\right) \qquad (4.42)$$

$$L = \mu \times \text{thickness}. \qquad (4.43)$$

The use of multiple phased elements to cancel out substrate modes [Reb92], [RN88] is an intriguing possibility for designs that use arrays and that do not face other system limitations that prevent the proper orientation of the elements to cancel substrate modes. This may be accomplished by properly spacing in-phase slot elements on the same substrate in broadside orientation [RN88]. [RN88] showed that this method can be used on a quarter-dielectric-wavelength thick substrate with a permittivity of 4 (e.g., a quartz substrate) to realize a slot efficiency of 70% when the elements are spaced slightly

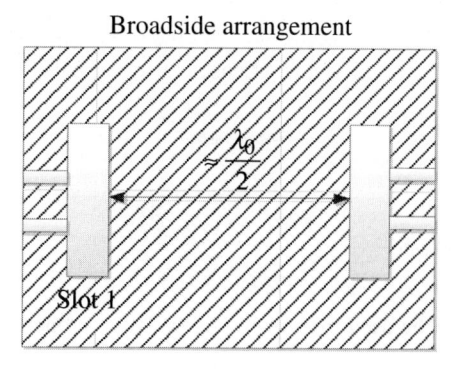

4.6 Techniques to Improve Gain of On-Chip Antennas

Figure 4.38 [RN88] provided an analysis of broadside slot elements to improve efficiency.

less than a half-free-space-wavelength apart. A broadside slot orientation is illustrated in Fig. 4.38. Although this method is not effective for dipoles on quarter-wave substrate, [RN88] showed that on thicker substrates that are $1.25 \times \lambda_d$, thick substrates with relative permittivity of 4 can also achieve near 70% efficiency. Unfortunately, this method will not work as well for higher permittivity substrates [RN88]. To determine the correct substrate thickness to use for a given pair of antenna elements, the first step is to know what substrate modes (TE or TM) will be excited in a broadside direction by an individual antenna for a given thickness. The operating point should be selected where broadside modes exist. Then, the elements should be spaced according to the half-wavelength of the broadside mode in the broadside direction.

4.6.1 Integrated Lens Antennas

Lenses have been used to improve the characteristics of planar on-chip antennas since at least the early 1990s [Reb92], when lenses were considered necessary for good performance. Lenses may be mounted on the top or bottom of the antenna substrate. Top mounted lenses increase radiation into free space, while bottom mounted lenses prevent substrate modes (and if back-side radiation is desirable may also be used to increase radiation into free space). Lenses on the back of an antenna should be at least $0.5\lambda_0$ in diameter for effective operation [Reb92]. Several types of lenses have been studied for use with antennas, including hyperhemispherical, elliptical, and hemispherical [Reb92]. A consideration when using lenses, especially if the antenna is to be used in an array, is the packing density of the lens-antenna system. The use of lenses was examined by [KR86], who used case studies of small spherical lenses to study the focusing properties of lenses as they become smaller. The basic idea behind a back mounted lens is to make the substrate appear infinite from the point of view of the antenna [Reb92], thereby making substrate modes impossible. The minimum size of the lens was found to be nearly proportional to the index of the lens by [KR86]. A basic top mounted lens is illustrated in Fig. 4.39. It is also possible to use a rectangular "lens" as shown in Fig. 4.40 [PW08]. This approach is fairly crude and may introduce problems due to incident-angle dependence on transmission through the lens. But, the rectangular lens does increase

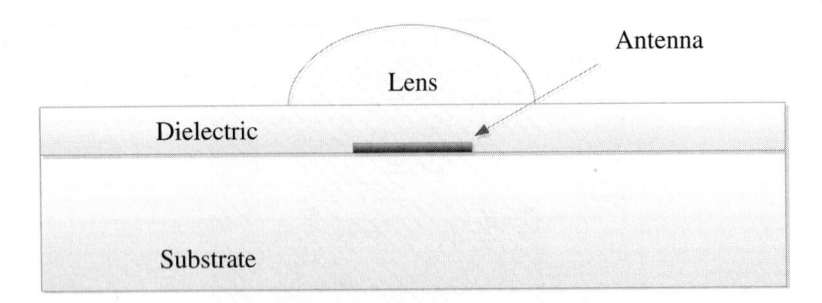

Figure 4.39 A lens can be used to improve the radiation properties of on-chip antennas.

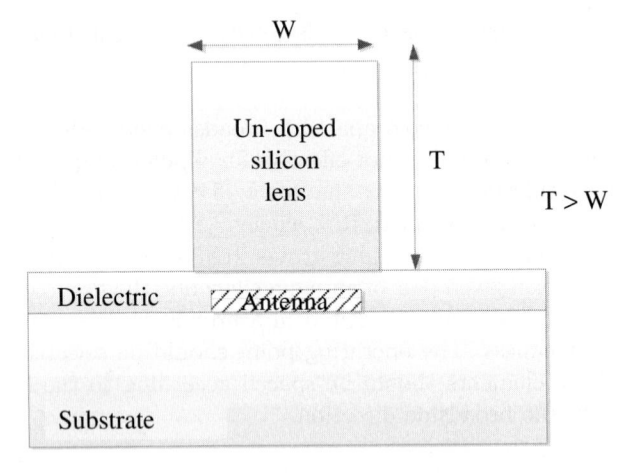

Figure 4.40 An undoped rectangular "lens" may be used to increase the gain of an on-chip antenna [PW08]. This is reminiscent of the use of stacked superstrates.

the dielectric constant of the medium above the antenna, and therefore we expect it to increase antenna gain by reducing the disparity in dielectric constant above and below the antenna. [PW08] used this technique to increase on-chip gain of an antenna for 28 GHz on a CMOS substrate by 8-13 dB. To be effective, [PW08] found that the lens must be taller than it is wide at its base (this leads [PW08] to hypothesize that the lens actually acts as a waveguide). A major advantage of lenses is that they effectively increase the effective area of the antenna from an electromagnetic perspective by $\epsilon_{r,\text{lens}}^2$ for higher gain [KR86] (an increase in physical area of the chip is also apparent, which may be a disadvantage for certain applications).

There are various types of lenses that have been considered for use with on-chip antennas, including hemispherical, hyperhemispherical, and ellipsoidal [FGR93][FGRR97]. These lenses are illustrated in Fig. 4.41. The best geometry of a hyperhemispherical lens is easily found by drawing an ellipse with one focal point at the antenna. The extension length L that results in best performance for this lens will cause the hyperhemispherical lens to approximate this arrangement, as indicated in Fig. 4.41. [FGRR97] created a set of very helpful contour plots for the hyperhemispherical lens that show how the

4.6 Techniques to Improve Gain of On-Chip Antennas

Hemispherical
(single focal point)

Elliptical
(two focal points)

Hyperhemispherical
(spherical with straight extension)

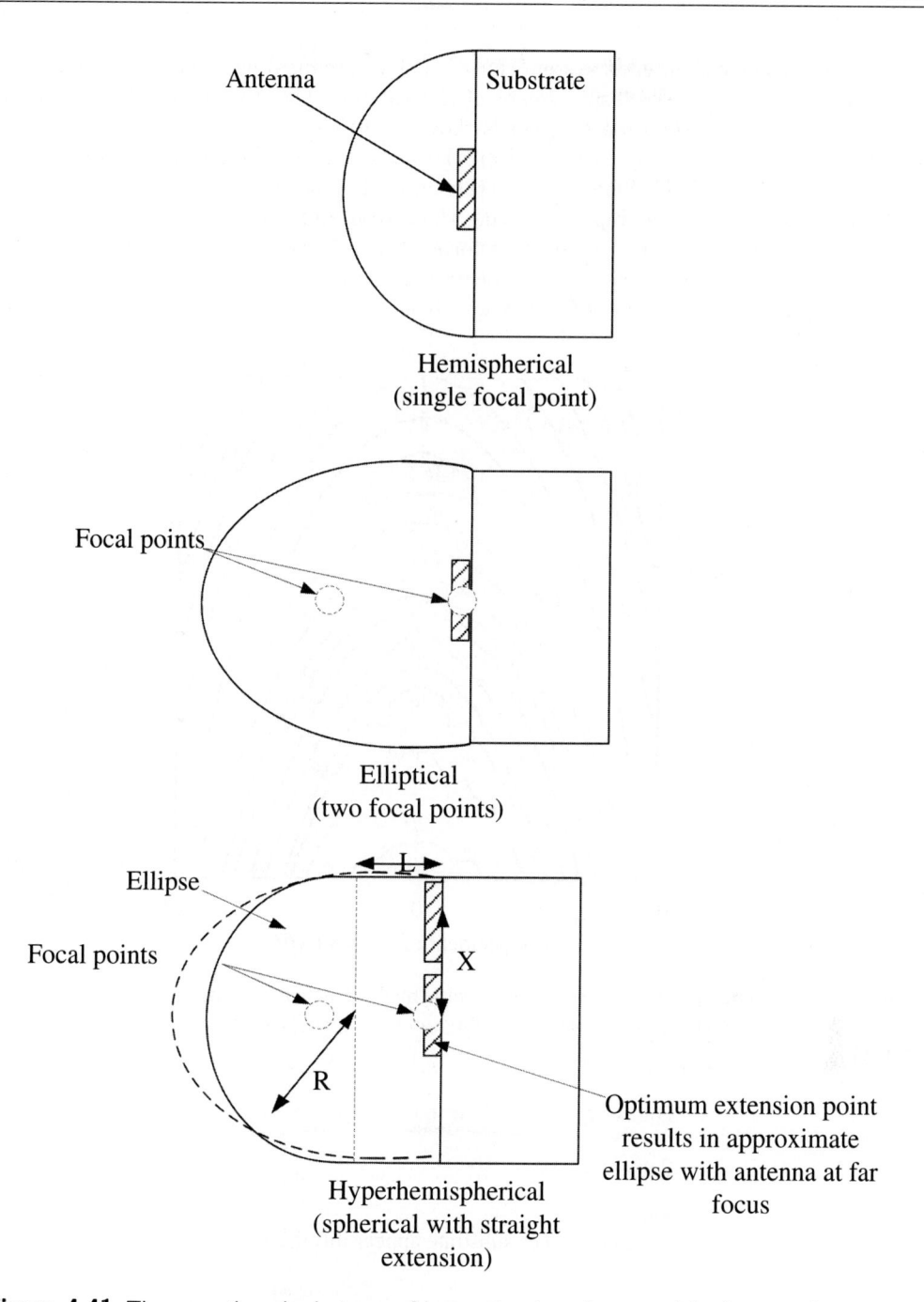

Optimum extension point results in approximate ellipse with antenna at far focus

Figure 4.41 There are three basic types of lenses that have been used for integrated antennas, including hemispherical, elliptical, and hyperhemispherical.

directivity of the lens+antenna system (for dual slot antennas) will change based on the lens geometry L/R and the displacement of the antenna from the center of the Lens X as shown in Fig. 4.41. This contour plot is shown in Fig. 4.42.

Lenses have also recently been used on the back-sides of chips [Bab08][BGK+06b], as illustrated in Fig. 4.43. The effect of the backside lens is to radiate any substrate modes out the back of the chip, preventing them from being trapped in the substrate, or preventing these substrate modes from forming at all. We may also consider the lens as a means of forcing the substrate to appear as an infinite dielectric half-space (i.e., the substrate appears to extend to infinity below dielectric top layer and antenna). Intel has

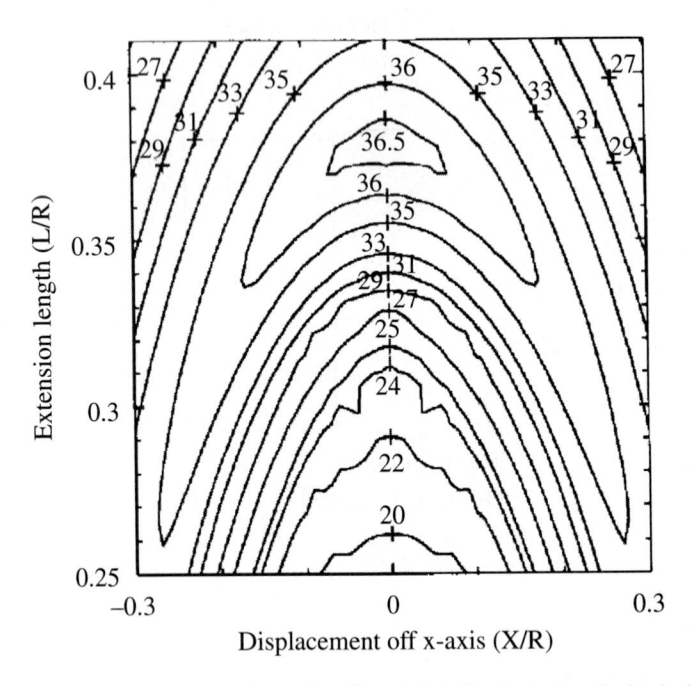

Figure 4.42 This plot demonstrates how the directivity of a hyperhemispherical lens + double slot antenna is impacted by the antenna's position and the geometry of the lens [from [FGRR97] © IEEE].

Figure 4.43 A back-mounted lens as used in [Bab08][BGK+06b] makes the substrate appear infinite, preventing the formation of substrate modes (i.e., an infinite half-space cannot be resonant) [RR97].

recently demonstrated electronically steerable lens antennas using on-chip fabrication methods that may prove useful for WLAN/WPAN applications [AMM13] and Nokia has recently demonstrated a lens for mmWave base station or mobile use.

4.7 Adaptive Antenna Arrays — Implementations for MmWave Communications

Integrated antenna arrays offer many advantages compared to single antenna systems, and they may be necessary to meet link budget requirements for 10 GHz (and higher) and device ranges much greater than just a few meters. In addition to improving antenna gain, an array can improve sensitivity by $10\log(n)$, where n is the number of elements in the array [GHH04]. As shown by [GHH04] and in more depth here with techniques from [Cou07], if a signal with power M is amplitude-modulated onto the outputs of n different antennas, then the overall signal power contained in the signal after the outputs are coherently added (which is the case for a signal arriving from the direction of the beam) is found by

$$P_{\text{sig}} = \text{Power}\left[M\cos(\omega t) + M\cos(\omega t) + \ldots + M\cos(\omega t)\right]$$
$$= \frac{1}{T}\int_0^T (nM\cos(\omega t))^2\, dt$$
$$P_{\text{sig}} = \frac{(nM)^2}{2}$$

where n is the number of elements, T is a period of the carrier, P_{sig} is the signal power, and ω is the frequency. As stated by [GHH04], any noise out of the each chain is uncorrelated with the noise from other chains. According to [Raz01] and [Cou07], the powers of n un-correlated random signals can be found

$$P_{\text{noise}} \geq N_1^2 + N_2^2 + N_3^2 + \ldots + N_n^2.$$

If the power in each noise signal is the same, then we may write this as

$$P_{\text{noise}} = nN^2. \tag{4.44}$$

Therefore, the signal-to-noise power is:

$$\text{SNR} = \frac{P_{\text{sig}}}{P_{\text{noise}}} = \frac{(nM)^2}{2} \times \frac{1}{nN^2} = \frac{n}{2}\left(\frac{M^2}{N^2}\right). \tag{4.45}$$

Therefore, a system with n elements has an advantage of n times the signal-to-noise (SNR) ratio compared to a single element system (where n is one). In addition to improving sensitivity by improving signal-to-noise ratio, an arrayed approach improves resiliency to strong undesired signal that may interfere with or block a desired signal, provided the undesired signal impinges on the array from a direction other than the beam direction. This is illustrated in Fig. 4.44 and can be considered a form of spatial filtering provided by an array. An implication of this spatial filtering is reduced linearity requirements for

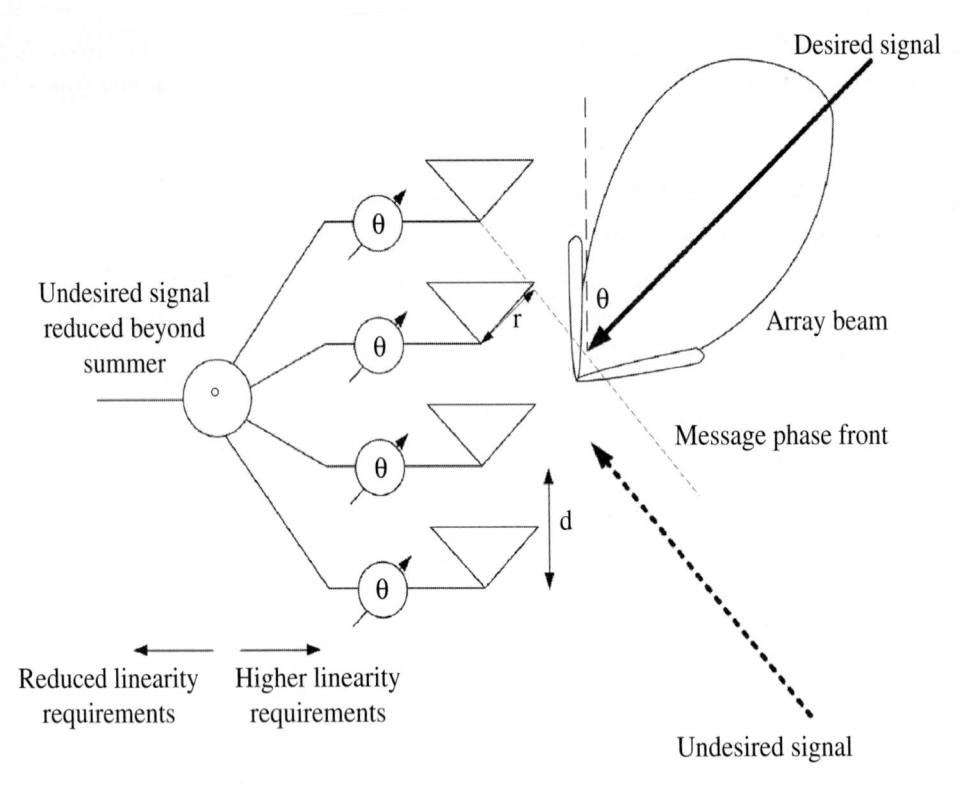

Figure 4.44 The effect of beamforming is to provide spatial filtering. The combined antenna, phase shifter, and summer system implements the phase shifting. After the last block in the spatial filter, the summer, components in the system will have reduced linearity requirements (i.e., they will not be required to handle signals as strong as components before the summer) [FNABS10].

components in the system after the summer (at which point undesired signals from directions other than the main beam have been filtered out) [FNABS10]. This filtering effect can be understood mathematically when we consider a desired signal arriving from angle θ_1 and an undesired signal arriving from angle ψ. The time required by the message phase front to travel the extra distance r (in the figure) between different elements will cause the signal arriving at the nth antenna from the top to have the form (using amplitude modulation for this example)

$$s_n(t) = m\left(t + \frac{nd}{c}\cos\theta\right)\cos\left(\omega\left(t + \frac{nd}{c}\cos\theta\right)\right)$$
$$+ u\left(t + \frac{nd}{c}\cos\psi\right)\cos\left(\omega\left(t + \frac{nd}{c}\cos\psi\right)\right)$$

where m is the desired message, d is the distance between elements, c is the speed of light, ω is the carrier frequency, and u is the undesired signal. We assume that the message signal bandwidth is small compared to the carrier frequency, so that $m\left(t + \frac{nd}{c}\cos\theta\right) \approx m(t)$.

4.7 Adaptive Antenna Arrays — Implementations for MmWave Communications

Then when a phase shift of $-\frac{nd}{c}\cos\theta$ is applied to each element, the summed signal is equal to

$$\sum s_n = nm(t) + 0 \qquad (4.46)$$

where the undesired signal was added incoherently and therefore summed to zero. If the message signal is not narrowband relative to the carrier frequency, as with an ultra-wideband signal, then a true-time delay element rather than a phase shifter must be used [RKNH06]. After the summation block, any signal not arriving from the beam direction need not be considered in first round designs (until sidelobe levels have been measured or simulated).

As explained in the introduction of this chapter, the general approach to building an array is to periodically array a set of antennas in one or two dimensions to achieve a smaller beamwidth (and higher gain) than any one single element can achieve alone. In order to achieve a wide coverage angle the beam must be steered, either mechanically or electronically. The first approach is not practical in an integrated environment, and therefore the second approach, relying upon electronic control to provide an adaptive beam, is much preferred.

4.7.1 Beam Steering for MmWave Adaptive Antenna Arrays

Electrical beam steering may be accomplished in many ways. Fig. 4.45 demonstrates several of these, in which a phase shifting element is visualized as a switched line. These methods include RF and IF/baseband analog phase shifting, providing differently phased local oscillator (LO) signals to mixers connected to different antennas, or at baseband using digital processing [FNABS10]. In both of these methods, a location of phase shift, in addition to the location of signal summation, must be selected. Signals from undesired directions will only be filtered out after the summation block. In the figure, the LO phase shifting is performed by switching the LO signal between various line lengths, but other approaches may be taken including slightly changing the frequencies provided by each LO, for example, with a voltage-controlled oscillator (VCO) or a phase-locked loop. There are various advantages and disadvantages to each approach, and we will discuss each. General goals include achieving phase shift in as little space as possible while still providing accurate phase-shift resolution, not degrading signal-to-noise ratio, and providing a sufficiently easy control mechanism. [FNABS10] provided a table to compare the requirements for different approaches that is reproduced in Table 4.2.

RF phase shifting has been used by [NFH07][RKNH06][KR07]. This approach generally requires less die space than others and also allows the summation block to be implemented before the mixer, reducing linearity requirements on the mixer due to elimination of signals outside the main beam after summation [FNABS10][SBB+08]. Disadvantages of this approach include difficulty in implementation of RF phase shifters at mmWave frequencies and the insertion loss of the phase shifter in the signal path.

Adaptive antennas (i.e., antennas with beam steering and possibly pattern shaping and nulling) may use either switched beam technology (also known as *code book* approaches), where a finite number of possible tap weights are used to selectively move the antenna radiation pattern between a number of preset patterns, or the array may

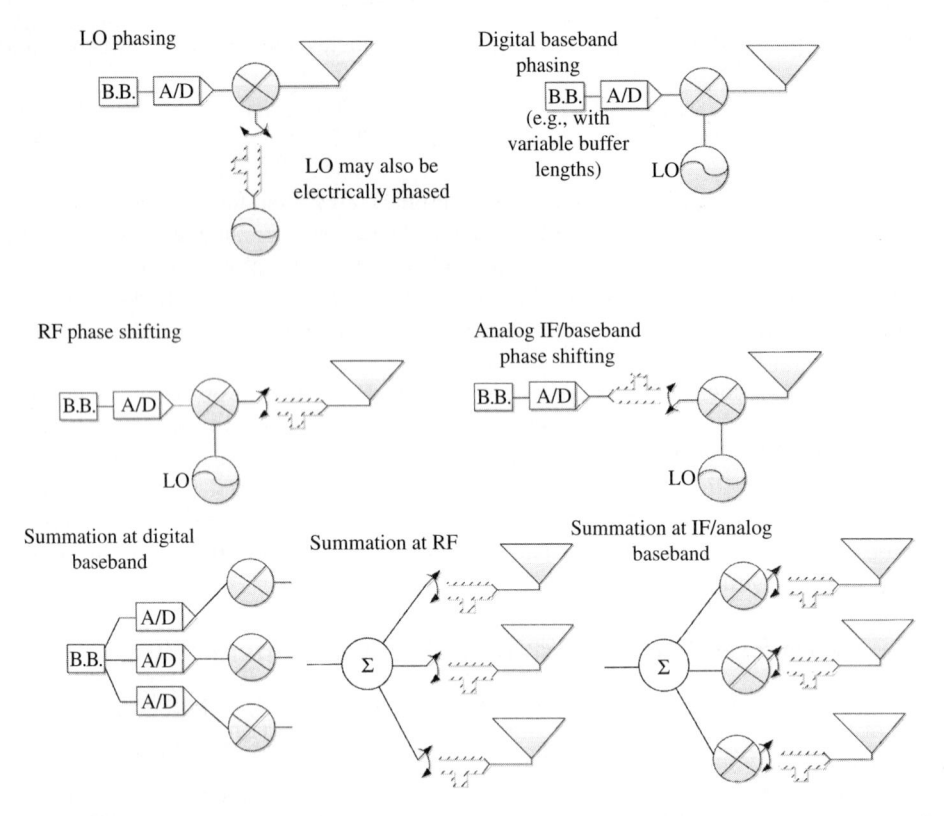

Figure 4.45 There are various approaches to electronically shifting the inputs to or outputs from different antennas.

Table 4.2 This table compares different approaches to perform phase shifting [reproduced from [FNABS10] © IEEE].

Architecture	Power Consumption	Chip Area/Cost	Design Challenge
RF Phase-shifting	Low	Low	Efficient front-end, phase shifter, beamforming algorithm
LO Phase-shifting	High	High	Linearity/LO distribution/Coupling
IF Phase-shifting	High	High	Linearity/LO distribution
Digital baseband Phase-shifting	High	High	Linearity/High dynamic range, fast ADC

be electrically steered over a continuous range of angles, and may also provide adaptive antenna nulling patterns to deliberately attenuate energy in certain directions while maximizing patterns in a desired direction. Well-known algorithms exist for implementing fully adaptive antennas [Muh96] and are discussed later in this chapter.

4.7 Adaptive Antenna Arrays — Implementations for MmWave Communications

All mmWave adaptive antennas will generally employ phase shifters to alter the electrical phase with minimum loss. There are multiple phase-shifter designs to consider, including analog phase rotators [BGK+06b], reflection-type phase shifters, loaded-line or switched line phase shifters, high-pass or low-pass phase shifters, tapped delay line phase shifters [RKNH06], and digital baseband phase shifters [PZ02]. A tapped delay line phase shifter is actually a true time delay element, and therefore should be used if the message signal has a bandwidth that is not narrow relative to the carrier frequency (i.e., when the assumption $m\left(t + \frac{nd}{c}\cos\theta\right) \approx m(t)$ does not hold). [RKNH06] presented this method for 60 GHz ultra-wideband radar, and Fig. 4.46 illustrates this approach. Variable gain amplifiers (VGAs) are used to compensate for the varying amount of gain for each length of delay line traversed by the signal on its way to the summation block. When a delay line approach is used, the delay through each step will be approximately equal to the group delay through each block τ_g (group delay is the derivative of the phase of a signal with respect to frequency). The figure gives an example of a simple amplifier with a load formed by a resistor R and capacitor C, and a transconductance g_m. V_{in} and V_{out} are the input and output voltages of each amplifier. If the capacitor is replaced by a *varactor*, a tuned capacitor, then the delay block may be implemented as a single tuned amplifier. Circuit details of phase shifters are explored in greater detail in Chapter 5.

While it may be necessary to use a true time delay element for very wideband systems, it will be adequate in many cases to use more standard phase shifter circuits. Most common phase shifters rely on a varactor [FMSNJ08], which will be described in greater detail in Chapter 5. At a high level, we must note that varactors generally have insertion

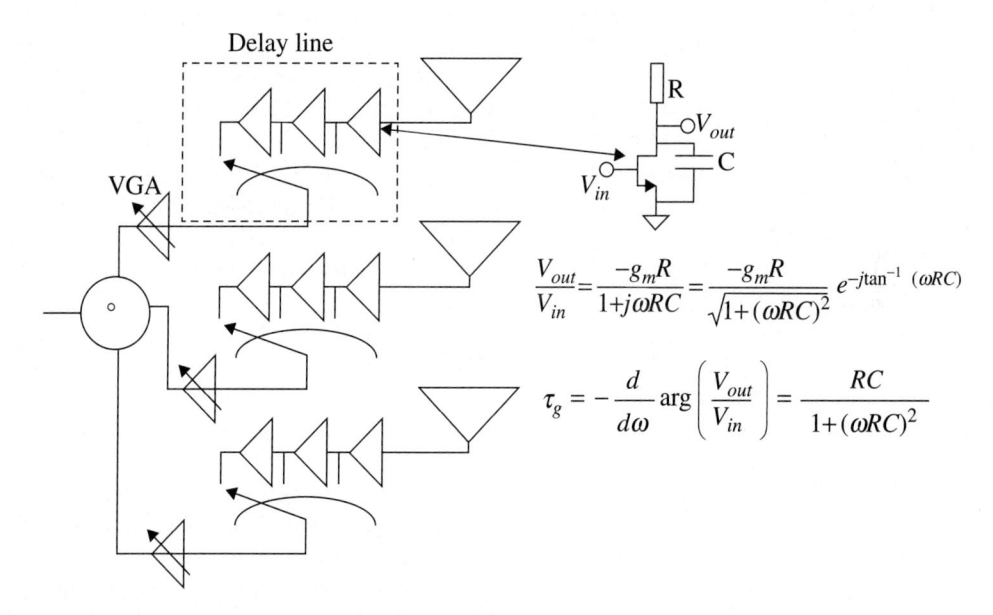

$$\frac{V_{out}}{V_{in}} = \frac{-g_m R}{1+j\omega RC} = \frac{-g_m R}{\sqrt{1+(\omega RC)^2}} e^{-j\tan^{-1}(\omega RC)}$$

$$\tau_g = -\frac{d}{d\omega}\arg\left(\frac{V_{out}}{V_{in}}\right) = \frac{RC}{1+(\omega RC)^2}$$

Figure 4.46 If the array is designed for very wideband signals, then a variable delay line may be used for phase shifting. The number of elements in the delay line grows proportionally to the bandwidth of the signal.

loss that varies as a function of the control voltage. There are ways to take advantage of this fact (e.g., [FMSNJ08]) to achieve higher array gain than would otherwise be impossible, but these are beyond the scope of this text.

IF or analog baseband phase shifting has been used by [KOH+09][WSJ09]. Like RF phase shifting, this technique relies on analog phase shifters in the signal path. Phase shifters may be more easily realized at the baseband or intermediate frequency (IF), but this technique will require all RF components and mixers to sustain all signals including signals arriving from undesired directions. This places increased linearity requirements on RF components. Phase shifters that may be used for this approach will be discussed in greater detail in Chapter 5.

LO phase shifting has been used by [GHH04], [HGKH05], [SBB+08], [BGK+06b], and [KL10]. An advantage of this method is that it removes the phase shifting element from the signal path, resulting in less susceptibility to variations through the phase shifter. The phase shifter may also be narrow band [KL10]. A disadvantage, especially if passive mixers that require high LO powers are used for high linearity, is that the phase shifter will reduce the amount of LO power arriving at the mixer unless an amplifier is used. Regardless of the type of mixer, loss through a distribution circuit or the need to produce differently phased LO signals from multiple VCOs may create problems from excessive power (due to possibly many VCOs), reduced space for other circuits (due to possibly many VCOs), and potentially large cross talk (due to the large number of lines needed). (Note that multiple phases may not require multiple VCOs, and even moderately clever students should consider ways to phase shift a single LO signal from a single VCO, for example, see [GHH04].) There are several methods to implement LO phase shifting, including cascading phase shifters following a bank of voltage-controlled oscillators (VCOs), distributing a single VCO signal to a signal distribution circuit and then to a bank of phase shifters, inputting the LO signal into an analog tapped-delay line, or implementing a VCO with multiple output taps of different phases [GHH04][HGKH05][SBB+08].

[GHH04] used LO phase shifting for a 24 GHz integrated CMOS array. Their implementation relied on 19 phased LO signals to be generated and distributed using a binary distribution tree. The 19 phases were generated with a ring-VCO with 19-output taps each of a different phase. When implementing this type of design, each amplifier in the VCO ring must either have the correct load to implement the desired phase shift, or must have its first pole frequency selected to give the desired phase shift. This is illustrated in Fig. 4.47, based on a similar figure in [HGKH05].

In addition to determining the method of phase shifting, the designer must also decide to use continuous or discrete phase shifting. Continuous phase shifting allows the main beam to be swept to any direction, whereas discrete phase shifting allows only a discrete set of beam directions. There are a number of advantages and disadvantages to each approach.

[GHH04], [HGKH05], [KOH+09], and [NFH07] used discrete phase shifting. This method requires a simpler beam-steering algorithm, often called a *codebook algorithm*, to determine which phase to distribute to each antenna. This may result in reduced baseband processing power and reduced time needed to sweep the beam. A disadvantage of this approach is that it will result in errors for signals that arrive from angles that are not angles to which the antenna array can be steered. If enough phases are included in

4.7 Adaptive Antenna Arrays — Implementations for MmWave Communications

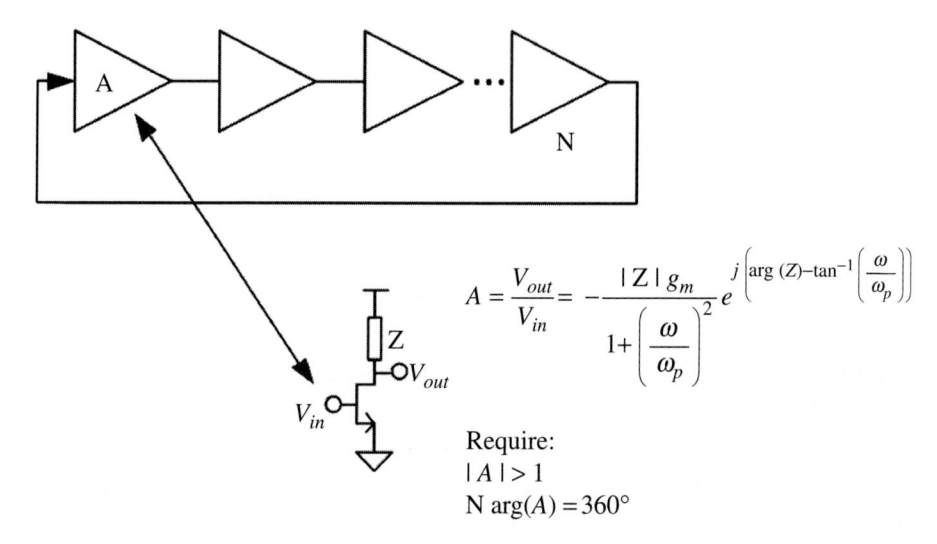

$$A = \frac{V_{out}}{V_{in}} = -\frac{|Z|g_m}{1+\left(\frac{\omega}{\omega_p}\right)^2} e^{j\left[\arg(Z) - \tan^{-1}\left(\frac{\omega}{\omega_p}\right)\right]}$$

Require:
$|A| > 1$
$N \arg(A) = 360°$

Figure 4.47 This figure explains the approach used by [GHH04][HGKH05] to implement an LO phase shifting method for an integrated phased array. N is the number of amplifiers, A is the gain of each amplifier, ω_p is the first pole frequency of each amplifier, ω is the oscillation frequency, and Z is the load on each amplifier.

QPSK constellation diagram

Signal arriving from exactly compensated angle

Signal arriving from un-compensated angle

Insufficient number of phases generated

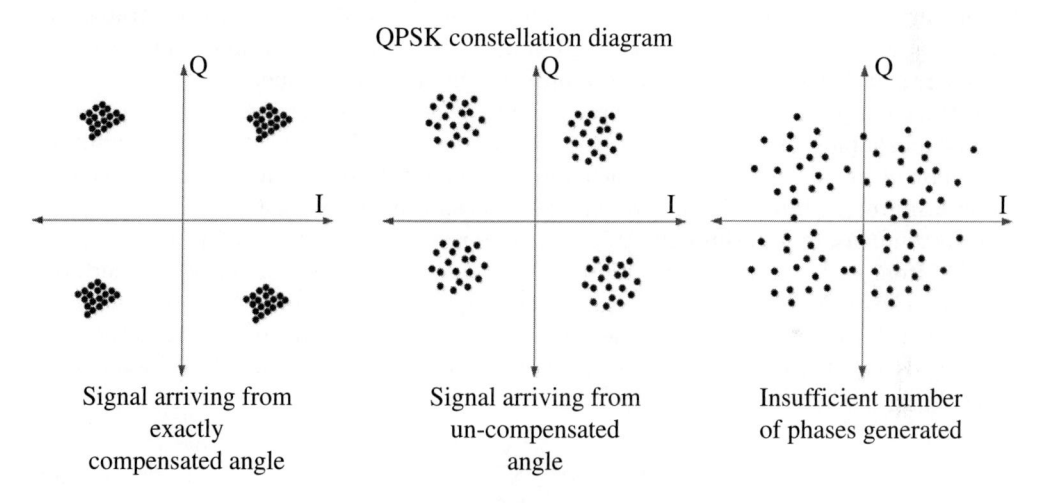

Figure 4.48 If an insufficient number of phases are generated, then constellation diagrams will be degraded, resulting in increased probability of error.

the discrete phase set, then these errors will be small and will be manifested as slight eye diagram or modulation constellation diagram degradation, or increased *error vector magnitude* (EVM), for signals not arriving from one of the exactly compensated directions [HGKH05]. Degradation of a constellation diagram for 4-QAM due to an insufficient number of phase-shifts is illustrated in Fig. 4.48.

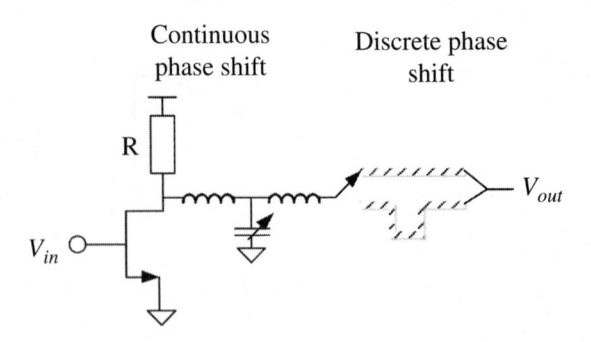

Figure 4.49 A combination of continuous and discrete phase shifting may be used to relax tuning range requirements for continuous phase shifters.

[SBB+08], [KL10], and [NFH07] used continuous phase shifting. This approach has the advantage of not increasing the probability of error for signals that do not arrive from a pre-determined set of directions, but it also requires a more complex phase-shifting algorithm. This approach may be difficult if a single continuous phase-shifter is used to shift from 0° to 180° with high linearity (due, e.g., to the non-linear nature of popular phase shifting elements such as varactors over wide tuning voltage ranges). [NFH07] used a combination of discrete and continuous phase shifting to relax the range of output phases of a continuous varactor-based phase shifter. Fig. 4.49 illustrates this concept in which a continuous phase-shift is provided by a varactor and common source amplifier, and discrete phase shifting is provided by a switched line.

Although a phase-shifter is adequate for simple beam-steering algorithms, more complex beam-shaping algorithms often rely on the ability to change both the magnitude and phase of signals arriving from each antenna. This may be accomplished by following a phase shifter with a variable-gain amplifier, as shown in Fig. 4.46. The VGA may be used to compensate loss through the phase shifter and to scale the magnitude of signals arriving from each antenna appropriately. Mathematically, this is described as performing a complex-weighted sum of the signals arriving from the antenna. To borrow notation from [CGY10], if we assume that the array has N antenna elements, and that the total signal arriving at the ith antenna element is $s_i(t)$, then the output of the summation block X_{total} in the array is given by

$$X_{\text{total}}(t) = \sum_{i=1}^{N} w_i S_i(t). \tag{4.47}$$

4.7.2 Antenna Array Beamforming Algorithms

Beamforming algorithms to steer the array beam and direction of arrival (DOA) algorithms to determine the direction of an incoming signal are both important for mmWave antenna arrays (and for some applications, they will be necessary). Beamforming results in reduced interference in a multiuser environment, and results in range extension [SR14] and lower delay spread [CGR+03]. Chapter 3 discussed signal propagation including delay spread. Beamforming and DOA estimation are both well explained in [CGY10],

4.7 Adaptive Antenna Arrays — Implementations for MmWave Communications

and we base our explanation on the understanding imparted by [CGY10]. DOA and beamforming can be broken down into a number of steps [CGY10]:

1. Determine the number of impinging wave fronts on the array. This is accomplished with algorithms such as the Akaike Information Criterion (AIC), a simple linear regression algorithm.

2. Determine the DOA of each wave front. This may be quite complex and require computationally intensive algorithms, but certain techniques based on a subspace of the signals impinging on the array may be used to quickly estimate DOA. DOA estimation algorithms include ESPRIT and MUSIC.

3. Reconstruct the signals arriving from each DOA. Use the reconstructed signals to determine the identity of the user responsible for generating each signal. Sequences of code embedded in the transmitted signal known as training sequences may be used for user identification.

4. Determine all DOAs attributed to a single user. This is especially important in the context of mmWave systems, which will often rely on reflected signals to establish a link. Fig. 4.50 illustrates the importance of this step, and Chapter 9 discusses in greater detail how mmWave standards such as IEEE 802.11ad, IEEE 802.15.3c, and WirelessHD take advantage of reflected signals. Chapter 3 illustrated how outdoor mmWave channels typically have several distinct angles of arrival for a given receiver location.

5. Track the user to ensure that the path selected for link establishment is not highly variable. This is important to ensure that the receive array beam does not incorrectly shift to detect a false signal. Steepest descent (SD) type algorithms are often used in this context.

6. Select the DOA with the strongest power to establish the link between the transmitter and the receiver. Apply phase shifts to maximize the signal arriving from this direction.

One of the key assumptions when implementing an array is that the received signal is narrowband relative to the carrier frequency. Consider a signal that is expressed as a complex phasor

$$S_i(t) = |m(t)| \, e^{j\omega t - j\angle m(t)} \tag{4.48}$$

where $m(t)$ is the message signal with magnitude $|m(t)|$ and phase $\angle m(t)$, and the carrier has frequency ω. If we assume that the signal emanates from the origin with wave vector (which describes the spatial frequency of the signal) $\vec{k} = k_x \hat{x} + k_y \hat{y} + k_z \hat{z}$, where $\hat{x}$, $\hat{y}$, and $\hat{z}$ are unit vectors in X, Y, and Z, and k_x, k_y, and k_z are wave numbers that describe how quickly the wave change in directions X, Y, and Z, then the signal will propagate from the origin according to

$$S_i(t) = |m(t)| \, e^{j\omega t - j\angle m(t) - j\vec{k}\cdot\vec{r}} \tag{4.49}$$

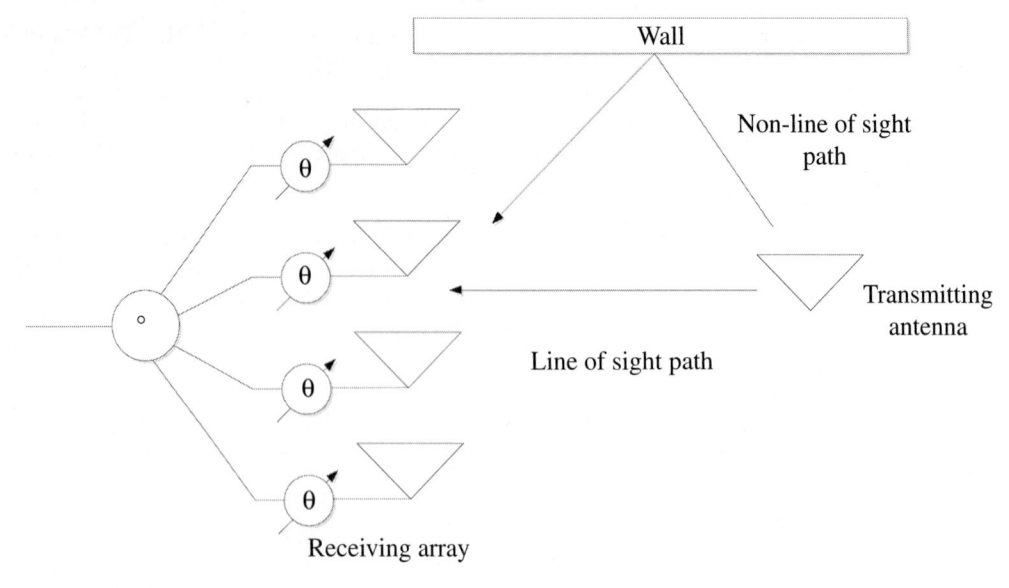

Figure 4.50 Many mmWave standards will rely on non-line of sight signal paths to establish a link between the transmitter and the receiver in the event a line of sight path is blocked or highly attenuated. It is important to determine the DOA of all sufficiently strong signals arriving from a single transmitting source.

where $\vec{r}$ is any vector from the origin. When the signal arrives at the array, it takes different amounts of time to reach each antenna element in the array. Therefore, the signal arriving at the mth element of the array may be written:

$$S_i^{(m)} = S_i^1 \left| m(t - \tau_m) \right| e^{j\omega(t-\tau_m) - j\angle m(t-\tau_m) - j\vec{k}\cdot(\vec{r_2} - \vec{r_1})}. \tag{4.50}$$

Because we assume that the signal is narrowband relative to the carrier *and* to the spatial frequency $\vec{k}$ (i.e., we assume that for a given time t that the signal exists on multiple phase-fronts in space) we may write

$$S_i^m = |m(t)| e^{j\omega(t-\tau_m) - j\vec{k}\cdot(\vec{r_2} - \vec{r_1})}. \tag{4.51}$$

Now, when we apply complex phase constants to each signal and take the summation, we have:

$$S_{\text{total}} = |m(t)| \sum e^{j\phi_i} e^{j\omega(t-\tau_m) - j\vec{k}\cdot(\vec{r_2} - \vec{r_1})} \tag{4.52}$$

where $e^{j\phi_i}$ is the phase shift applied to the signal arriving from each antenna element. Therefore, we set ϕ_i to cancel the phase shift between different elements:

$$\phi_i = -\omega(t - \tau_m) + j\vec{k}\cdot(\vec{r_2} - \vec{r_1}). \tag{4.53}$$

4.7 Adaptive Antenna Arrays — Implementations for MmWave Communications

This tells us that to compensate for the phase shift between elements, we must consider the difference in time between the arrivals of the signal at each element, and the difference in space represented by $\vec{r_2} - \vec{r_1}$. Now, we generally assume that the array is small enough for all elements to be illuminated at the same time so that signals from the *same instant in time* may be added, in which case there is no time difference between the signals at the elements in the array, only a difference (this may not be the case if the signal is especially wideband). Therefore, the phase shift to apply at the ith element is equal to

$$\phi_i = j\vec{k} \cdot (\vec{r_2} - \vec{r_1}). \tag{4.54}$$

This can seem confusing at first, especially given the introduction (see Figure 4.3) in which we argued from the perspective of a time difference. The difference between the arguments is that in the introduction we considered at what point in space the phase front emitted from each of the antennas would be equal (i.e., the same phase front). Here, because the receive antennas are fixed, the phase fronts arriving at a single point in time on the elements of the array are not equal. This is why (4.54) applies. Figs. 4.51 and 4.52 illustrate this concept.

In a real-world application, there are likely to be multiple signals impinging on a receiving array at any given time. If we assume that each of these signals has the same carrier frequency but a different direction of arrival, then each of the signals will also have a different wave vector associated with it. The total signal arriving at the ith antenna in the array will be given by

$$S_i = \sum_{\ell=1}^{N} m_\ell(t) e^{-j\vec{k_\ell} \cdot (\vec{r_i} - \vec{r_1})} \tag{4.55}$$

where $\vec{k_\ell}$ is the wave vector associated with the ℓth signal $m_\ell(t)$, $\vec{r_1}$ is the position of the phase-reference element in the array, $\vec{r_i}$ is the position of the ith antenna element, and S_i is the net signal arriving at the ith element. This is illustrated in Fig. 4.53.

The simplest method to determine DOAs of arriving signals is a so-called beamforming technique in which the array's main beam is swept to different angles, where for each angle the output power of the array is measured and recorded [CGY10]. The angles that correspond to peaks are then taken as the DOAs of impinging signals. Unfortunately, this method is often not very precise unless very large arrays are used [CGY10]. An example is illustrated in Fig. 4.54.

There are other, more sophisticated DOA estimation algorithms to consider when implementing a phased array, including the Bartlett, MUSIC, ESPRIT, and AP (Alternating Projection) algorithms [CGR+03][Muh96]. For the sake of brevity we will only consider ESPRIT and MUSIC. Most of these algorithms work by extracting DOA information from a covariance matrix of the signals arriving at the different elements of an array [CGY10]. This is a matrix that describes how highly or poorly correlated the signals are that arrive at the different antenna elements in the array [CGY10]. To compute this matrix, we need to collect multiple snapshots of the data coming out of the array.

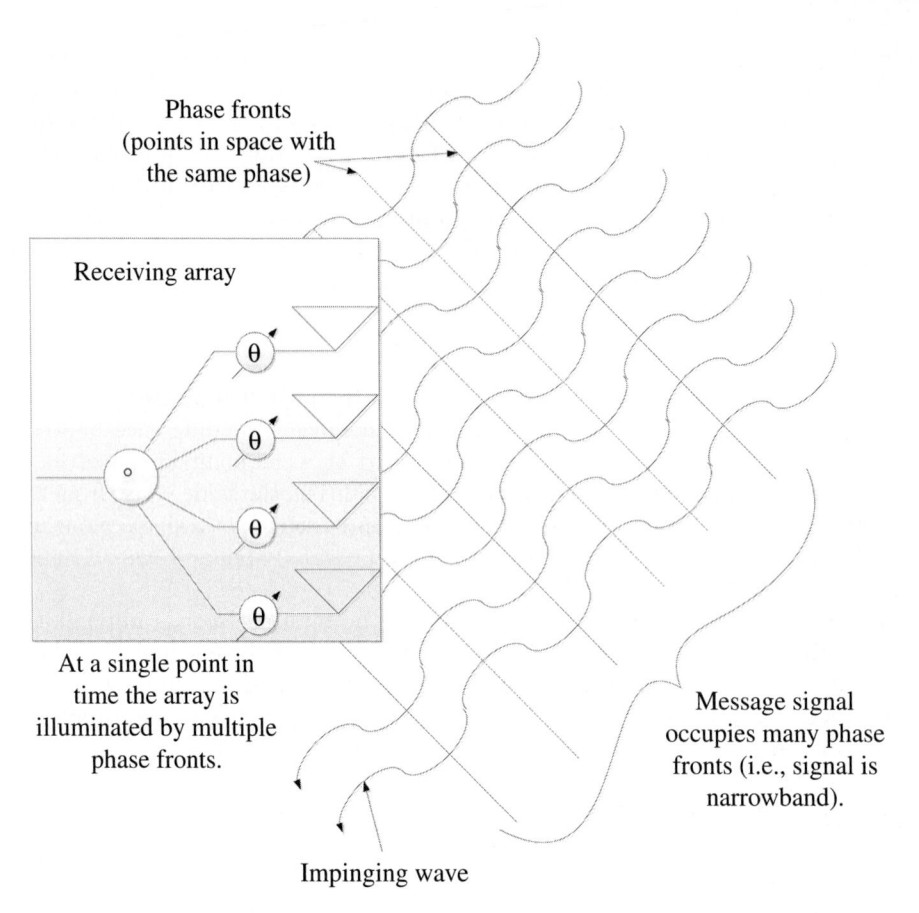

Figure 4.51 The array works in parallel and adds signals arriving at a single point in time at different points in space. The message is considered narrowband, which indicates that its bandwidth is small relative to the carrier frequency and that it occupies multiple phase fronts in the impinging wave.

If we have an array with M elements, and we take N snapshots of the output of each element S_i at equally spaced time intervals, then we may build up a matrix as

$$\mathbf{X} = \begin{bmatrix} S_1(t_1) & \ldots & S_1(t_N) \\ \ldots & \ldots & \ldots \\ S_M(t_1) & \ldots & S_M(t_N) \end{bmatrix}. \quad (4.56)$$

Note that $\mathbf{X}$ includes both impinging signals and noise arriving at the array. The covariance matrix may then be estimated according to [CGY10]:

$$\mathbf{R} = \left(\frac{1}{N}\right) \mathbf{X}^* \mathbf{X}. \quad (4.57)$$

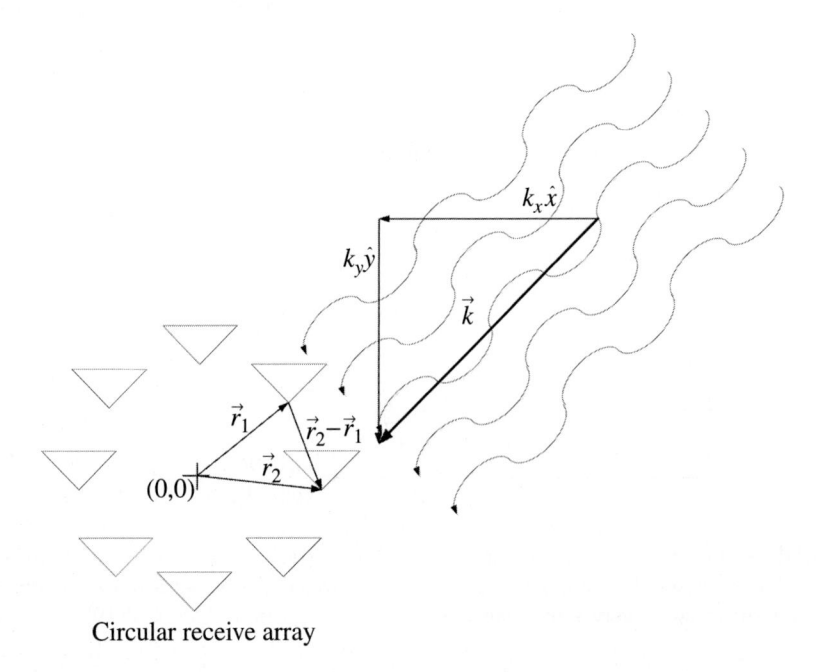

4.7 Adaptive Antenna Arrays — Implementations for MmWave Communications

Figure 4.52 The projection of the wave vector onto the vector between the two elements determines how their different spatial positions will result in a phase shift at a single point in time.

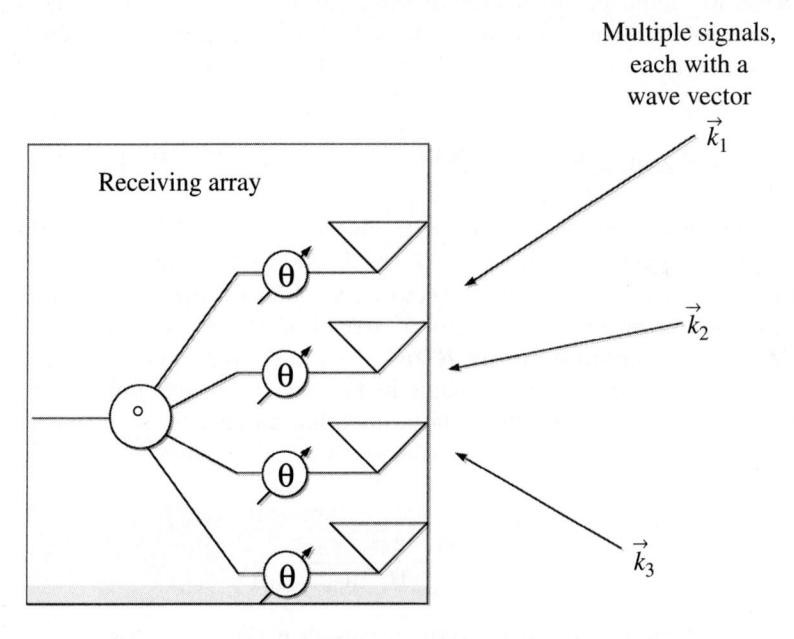

Figure 4.53 In a real-world application, many signals will impinge upon an array.

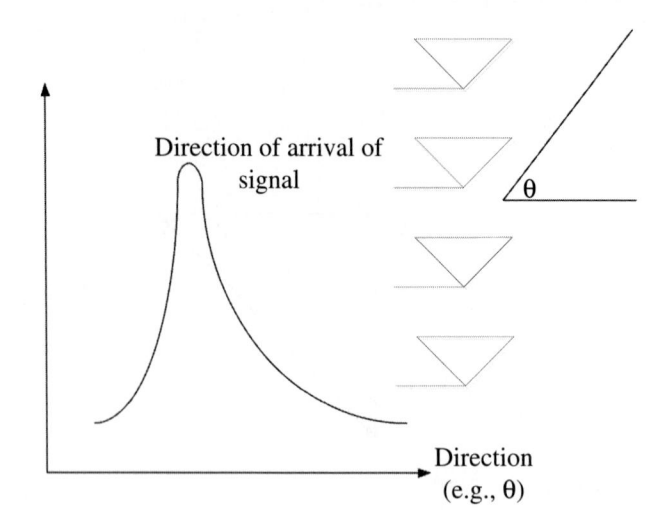

Figure 4.54 A plot of the power arriving from each direction may be used to estimate the DOA of different signals. A simple beam steering approach in which the received power is computed for different beam directions is very simple but does not have high resolution [CGY10].

[CGY10, Chapter 3] presents a number of equations that may be used to compute the weights to be applied to each signal according **R**. In addition to computing **R**, the directions that we would like to scan with the array must be selected. These directions correspond to a set of wave vectors as explained in the introduction. For each direction we wish to consider, we form a vector based on the wave vector associated with that direction

$$\mathbf{a}_{\text{direction}} = \begin{bmatrix} 1 & e^{-j\vec{k_l}\cdot(\vec{r_1}-\vec{r_2})} & \ldots & e^{-j\vec{k_l}\cdot(\vec{r_1}-\vec{r_M})} \end{bmatrix} \quad (4.58)$$

where $\vec{k_l}$ is the wave vector associated with the scanned direction, $\vec{r_1}$ is the position of the first phase-reference vector, and $\vec{r_M}$ is the position of the last element in the array. This vector is called a *steering* vector. Fig. 4.55 illustrates how the wave vector may be chosen based on the direction of arrival of a signal relative to the origin of an array. With the covariance matrix R and the phase shift vector $\mathbf{a}_{\text{direction}}$, we may find the weights to apply to each element in the array. For example, a method that gives better resolution than the simple beam steering method is the capon beam forming method, for which the weights are computed for each direction vector according to [CGY10]:

$$\mathbf{w} = \frac{\mathbf{R}^{-1}\mathbf{a}_{\text{direction}}}{(\mathbf{a}^*_{\text{direction}}\mathbf{R}^{-1}\mathbf{a}_{\text{direction}})}. \quad (4.59)$$

These weights may be used to plot power versus direction as in Fig. 4.54. This technique is more computationally intensive, as it requires matrix inversion, but has a better resolution [CGY10].

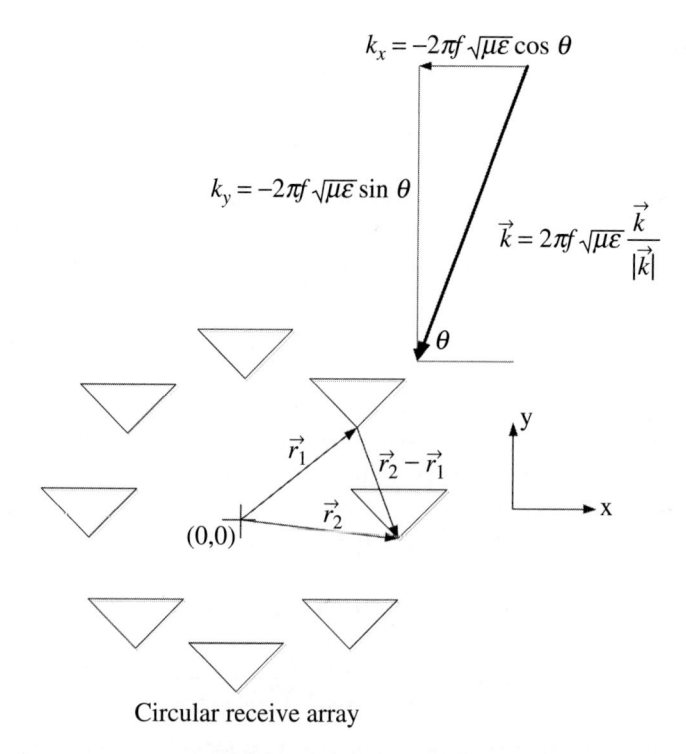

4.7 Adaptive Antenna Arrays — Implementations for MmWave Communications

$$k_x = -2\pi f \sqrt{\mu\varepsilon} \cos\theta$$

$$k_y = -2\pi f \sqrt{\mu\varepsilon} \sin\theta$$

$$\vec{k} = 2\pi f \sqrt{\mu\varepsilon} \frac{\vec{k}}{|\vec{k}|}$$

Circular receive array

Figure 4.55 The designer must specify a set of directions he or she would like to scan. Once the directions have been selected, a wave vector for each direction may be found as illustrated in the figure.

4.7.3 Specific Beamforming Algorithms — ESPRIT and MUSIC

Estimation of Signal Parameters via Rotational Invariance Techniques (ESPRIT) and Multiple Signal Classification (MUSIC) are types of subspace DOA estimation algorithms [LKS10][Muh96]. As summarized by [CGY10], a subspace based technique basically assumes that the column space of the covariance matrix may be spanned by two orthogonal subspaces: a signal subspace and a noise subspace. The column space is the set of all vectors that may be written as a linear sum of the columns of the covariance matrix. A subspace is, for example, the set of all vectors formed by only the first two columns of the covariance matrix. If two spaces are orthogonal, then any element for the first space is orthogonal to any element from the second space, and vice versa.

MUSIC is a type of spectral-based DOA estimation algorithm [LWY04] in which array weights are estimated using the array covariance matrix $\mathbf{R}$ in equation (4.57) [LKS10]. MUSIC assumes that the signals impinging on each element are not correlated or only poorly correlated, and that noise corrupts each signal but that the noise from two different antennas is uncorrelated [LKS10]. The gist of the approach is first to find the eigenvalues of the covariance matrix of the array. It can be shown then that the smallest eigenvalues will have the same value (or closely clustered values) and will correspond to the variance of the noise impinging on the array [CGY10]. Mathematically

$$|\mathbf{R} - \lambda \mathbf{I}| = 0 \tag{4.60}$$

where $\mathbf{R}$ is the covariance matrix, $\mathbf{I}$ is the identity matrix, and λ is an eigenvalue of $\mathbf{R}$. Equation (4.60) is solved for the values of λ. The smallest solution to (4.60), which is the characteristic polynomial of the covariance matrix, will occur more than once, that is, that the characteristic polynomial may be factored according to:

$$|\mathbf{R} - \lambda \mathbf{I}| = (\lambda - \rho_1)(\lambda - \rho_2)\ldots(\lambda - \sigma_n^2)(\lambda - \sigma_n^2)\ldots(\lambda - \sigma_n^2) \tag{4.61}$$

where σ_n^2 is the smallest eigenvalue, a multiple root of the characteristic equation, and equal to the noise variance. The larger eigenvalues are written as ρ_i. Eigenvectors $\mathbf{q}_i$ associated with the noise variance are then computed and used to form matrix $\mathbf{V}$ [CGY10]:

$$\mathbf{V} = \begin{bmatrix} \mathbf{q}_1 & \mathbf{q}_2 & \mathbf{q}_3 & \ldots \end{bmatrix}. \tag{4.62}$$

Then, the set of steering vectors of the form (4.58) are inserted into [LKS10]:

$$P_{\text{direction}} = \frac{1}{\mathbf{a}_{\text{direction}} \mathbf{V} \mathbf{V}^* \mathbf{a}_{\text{direction}}}. \tag{4.63}$$

The directions whose steering vectors correspond to a local maximum in MUSIC are the directions of arrival of the various signals.

ESPRIT is a type of parametric subspace-based DOA estimation algorithm that can be computationally intensive as it requires one or more steps of Eigenvalue decomposition (EVD) of a high dimensional matrix [LWY04]. But, it will generally be less computationally and memory intensive than MUSIC [LKS10]. The performance of ESPRIT degrades with decreasing SNR [LWY04]. There are two popular classes of ESPRIT algorithms: temporal-ESPRIT (T-ESPRIT) for determining DOA [Muh96] and Spatial-ESPRIT (S-ESPRIT), which is actually a propagation delay algorithm [LWY04].

The basic idea of ESPRIT is to first break the array into smaller sub-arrays and then to use the phase shift between the two arrays to determine the DOA of signals impinging on the array [LKS10]. The sub-arrays must be identical except for a translational change in position but not orientation (i.e., the sub-arrays are all oriented in the same direction) [LKS10]. If we have d signals impinging on the array each with an associated steering vector of the form (4.58), then we may write the total signal arriving at the aggregate array as [LKS10]

$$\mathbf{X}(t) = \begin{bmatrix} \mathbf{a}_1 & \ldots & \mathbf{a}_2 \end{bmatrix} \begin{bmatrix} m_1(t) \\ \ldots \\ m_d(t) \end{bmatrix} + \sigma_n^2 I = \mathbf{AM} + \sigma_n^2 I \tag{4.64}$$

where m_I is the ith message signal impinging upon the array with associated steering vector (determined by the direction of arrival of m_i), and σ_n^2 is the noise variance. $\mathbf{A}$ is a steering vector matrix, **M** is the vector of signals impinging on the array. For two sub-arrays, we may write [LKS10]:

$$X_1(t) = A_1 M + \sigma_n^2 I \tag{4.65}$$

$$X_2(t) = A_1 \Psi M + \sigma_n^2 I \tag{4.66}$$

where $\mathbf{A}_1$ is smaller than $\mathbf{A}$ to account for the reduced number of elements in a sub-array compared to the entire array. Ψ is a diagonal unitary matrix that accounts for the phase shift between the sub-arrays due to their different positions in space. The diagonal values λ_i of Ψ may then be used to estimate the DOAs (where we assume that the array is linear) [LKS10]:

$$\theta = \sin^{-1}\left(\frac{\arg(\lambda_i)}{\omega d\sqrt{\mu\epsilon}}\right) \tag{4.67}$$

where d is the spacing between elements in the array, ω the angular frequency, and μ and ϵ are the constitutive parameters of free space.

Many students and researchers have studied ESPRIT and MUSIC in depth since their origins, such as in [Ron96] and [Muh96], to compare their accuracy, computational burdens, and resilience to closely spaced signals of arrival. A few texts exist on the subject [LR99b][SM05], and a compendium of key research papers on smart antennas was published by the IEEE in 1998 [Rap98].

4.7.4 Case Studies of Adaptive Arrays for MmWave Communications

It is worth considering some of the reported results for integrated phased array antennas for mmWave communications. In 2007, [Emr07] presented a thesis on the design of on-chip antenna arrays. He presented a detailed system-level analysis of the requirements of such an array, including effects of multipath on mmWave systems. Two primary metrics used to perform this analysis were IC noise figure and transmit power. He found that for an acceptable SNR for 1.25 Gbps over 10 m, systems would require transmit and receive antenna gains of at least 25 dBi in NLOS conditions.

In 2006, [BBKH06] presented a SiGe 60 GHz transceiver array based on a scalable, 2×2 element design. An injection locking scheme was used to phase lock the different VCOs of the array together. This scheme achieved a 200 MHz locking range, while a 1×4 array had a 60 MHz locking range. The design was intended for easy "aggregation" of multiple chips for increased array size. This decision influenced the selection of sub-harmonic frequency. Buckwalter et al. [BBKH06] used a subharmonic for inter-chip coupling, as this would keep the interconnects electrically shorter than if the first harmonic had been used, and a subharmonic did not exceed the cutoff frequency of wire interconnects (20 GHz). The desired electrical length of each interconnect was half-wavelength.

Buckwalter et al. specifically designed their distribution system to minimize the amount of energy absorbed by their signal transmission line interconnects [BBKH06]. The approach relied on "severed" ground shields. By avoiding a continuous ground shield, they reduced the amount of energy absorbed by the substrate.

In 2004, Guan et al. [GHH04] presented a fully integrated eight-element 24 GHz array in SiGe BiCMOS. Their design was based on a heterodyne architecture with low intermediate frequency (IF) of 4.8 GHz, indicating that the design needed to place extra emphasis on image-response control. They relied on a 16-phase ring oscillator, indicating that the design's layout was probably very challenging. Such a design would benefit from a process with as many metal layers as possible. The design permitted independent phase distribution to each path. In practice such a general approach may not always be called for, especially if, for the intended design, it is acceptable to keep the array co-phasal (i.e., to have constant phase shift between elements).

In 2010, Lee et al. presented a beamforming lens antenna on a high-resistivity silicon substrate for 60 GHz applications [LKCY10]. The technique achieved up to 70% radiation efficiency and was able to steer the main beam to $-29.3°$, $-15.1°$, $0.2°$, $15.2°$, and $29.5°$, making it useful for switched-beam designs.

4.8 Characterization of On-Chip Antenna Performance

There are two general methods in the literature for testing on-chip antennas: measurement of packaged chips or probe-station measurement [PA09]. Both of these methods are in the early stages of their development, and there exists substantial opportunity determining accurate and reliable on-chip characterization techniques [PCYY10]. These techniques are challenging to develop because of the low gain characteristics of most on-chip antennas [PCYY10], and because of the difficulty of creating re-creatable standardized measurement environments with no or little multipath.

Various techniques exist for probe-station characterization, including scanning near-field measurements with a probe combined with near-to-far field transformation, and excitation with a separate transmitter tuned with a signal generator [PA09]. Probe-station characterization has been performed with both known test antennas and two identical unknown antennas[PCYY10][PA09]. For compete accuracy, a known antenna should be used.

After a test setup is completed and measurements are taken, it is possible to measure the gain of the antenna with a link budget approach:

$$G_{\text{AUT}} = 20\log(S_{21}) - 20\log\left(\frac{d}{4\pi f R}\right) - 10\log\left(1 - |S_{11}|^2\right)$$
$$- 20\log\left(\rho_{\text{aut}}\rho_{RX}\right) - 10\log\left(G_{RX}\left|1 - |s_{22}|^2\right|\right) \quad (4.68)$$

where S_{ii} represent the s-paramter measurements, R is the distance between transmit and receive antenna, G_{RX} is the gain of the receive antenna, which is assumed to be known, c is the speed of light, and $\overrightarrow{\rho_{\text{AUT}}} \cdot \overrightarrow{\rho_{RX}}$ is the polarization mismatch of the two antennas. If the polarization mismatch is not known, then an assumption must be made.

It is often necessary to de-embed feeding structures to accurately characterize on-chip antennas. Proper calibration of a probe setup can be used to shift the measurement reference plane past feeding structures for accurate measurement [AR08]. For example, Alhalabi et al. [AR08] used a Through Line Reflect (TRL) calibration to remove the effect of a CPW-to-microstirp transition in their angled dipole design. Gutierrez used a large horn antenna on a probe station to determine gain [GJ13].

4.8.1 Case Studies of MmWave On-Chip Antenna Characterization

Once an on-chip antenna is fabricated, the designer must be able to reliably measure the antenna's performance to validate the fruits of his or her efforts. This is often quite difficult to do at mmWave frequencies and requires great care, since at such small wavelengths, the high measurement cable loss, high degree of variability when probing the chip, and tiny movements by a probe or a cable can combine to create very large signal swings (>10 dB) when making antenna measurements on a probe station. Great care and repeated calibrations with systematic methods must be used to make reliable and repeatable antenna measurements at such high frequencies. We now review some recent work that shows methods to accurately measure the performance of on-chip and in-package antennas.

In 2009, Payandehjoo et al. [PA09] presented a technique for the characterization of on-chip antennas that used two identical on-chip dipole antennas operating at 55 GHz. Each antenna chip was connected to a probe that was in turn connected to a vector network analyzer. The transmitting antenna chip was held stationary, and several copies of the receiving antenna chip were pasted to a glass sheet at various positions (pasting was used to improve repeatability). The s-parameters were measured for each receive antenna. Their measured results agreed approximately with simulation based on a free space propagation assumption. They did not mention use of absorbing material to mitigate multipath, and the measurements may have been closer to simulation if this had been used.

In 2010, Park et al. [PCYY10] presented a technique for characterizing on-chip antennas. This relied on a known high-gain off-chip antenna to test the on-chip antenna. Before characterizing the on-chip antenna, the known antenna was calibrated to the vector network analyzer. This should lead to improved accuracy over a method that uses two identical unknown antennas, which would be impossible to calibrate to the vector network analyzer.

In 2005, Seki et al. [SHNT05] presented a characterization technique for antennas in-package based on a modified RF probe station. Fig. 4.56 illustrates their approach. The close agreement between simulated and measured antenna performance with this technique indicates its reliability for characterizing 60 GHz on-chip or in-package antennas.

In 2006, Zwick et al. [ZLG06] used a probe-based technique for the characterization of flip-chip folded dipoles on a fused-silica substrate and superstrate. The measurement setup consisted of a WR15 standard gain horn antenna for a reference antenna, WR15 connector cables, and modified probe station with specialized sample holder, all inside a 1.4 m × 1.2 m × 1.2 m anechoic chamber [ZBP+04]. Fig. 4.57 illustrates their test setup. In order to use a one-probe setup, two calibrations were needed: a calibration for the standard horn gain in which the probe and device under test (DUT) are replaced by a known horn antenna, and an short open line (SOL) calibration with the probe to allow for measuring S11 with only a single probe. The minimum measured gain of the probe setup ranges from −40 dBi to −30 dBi. Zwick et al. found that an unloaded probe can have a maximum gain of −15 dBi in the V-band [ZBP+04], though when loaded, the

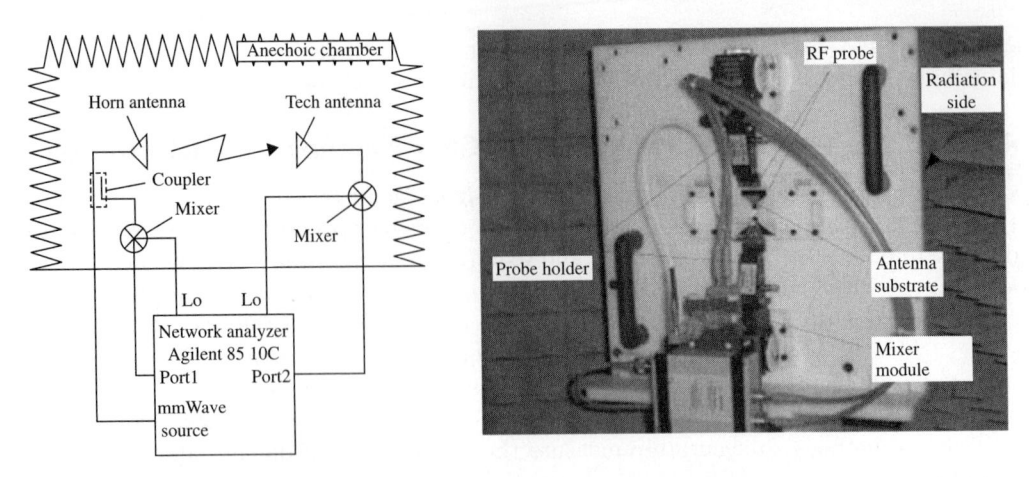

Figure 4.56 Seki et al. [SHNT05] proposed a modified RF probe station in an anechoic chamber to characterize their in-package patch antenna array [from [SHNT05] © IEEE].

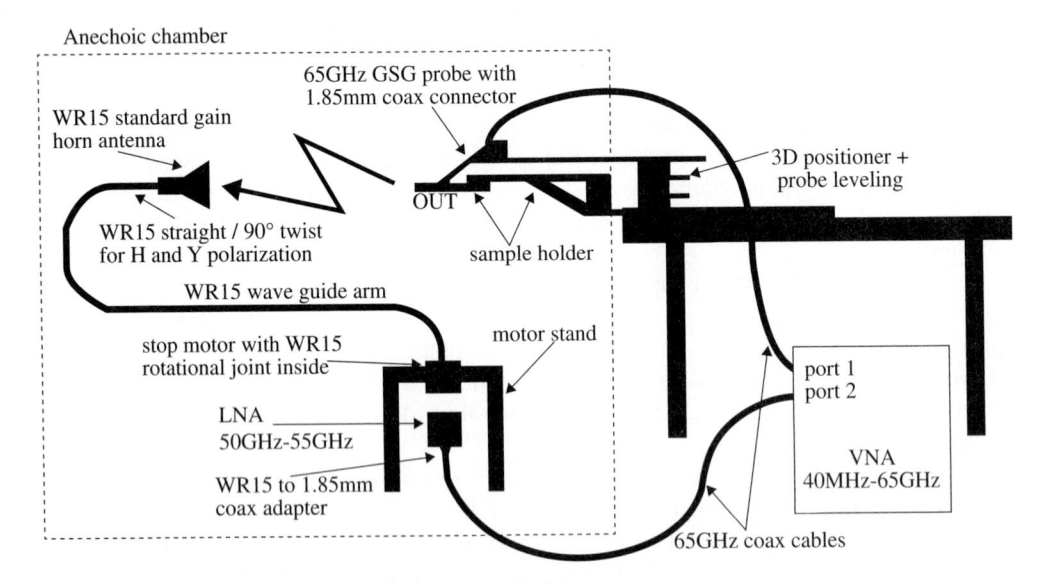

Figure 4.57 Zwick et al. developed a specialized probe-based measurement setup for characterization of on-chip antennas [from [ZLG06] © IEEE].

probe was found to impact measurements much less than anticipated from an unloaded measurement.

In 2008, Alhalabi et al. used a zero-bias Schottky diode detector and lock-in amplifier to measure the patterns of their angled dipoles on Teflon substrate [AR08].

In 2009, Chen et al. used a three-antenna test procedure to measure the gain of their on-chip dipole antenna [CCC09]. The three antennas used for their test were the on-chip dipole antenna, a Flann lens horn antenna, and a QuinStar standard horn

4.8 Characterization of On-Chip Antenna Performance

antenna. The three-antenna technique is useful for accurately measuring antenna gains because it eliminates assumptions about the antenna under test needed in simpler techniques.

4.8.2 Improving Probe Station Characterizations of On-Chip or In-Package Antennas

Although it is common to use a probe station to characterize on-chip antennas, these measurements face two major challenges that may reduce the accuracy of these measurements: radiation from probes used to excite on-chip antennas and scattering from the probe-station environment. Fig. 4.58 illustrates a probe-station environment and highlights these challenges.

When two identical integrated mmWave antennas (e.g., on-chip 60 GHz antennas) are used in the probe station environment, then in total there are four radiators present on the probe station: the two probes and the two integrated antennas. For purpose of explanation, let us assume that one antenna is on the right and the other on the left. In order to remove the effects of the probe radiation, a simple approach is to first measure four sets of s-parameters. (The reader is referred to Chapter 5 for a review of s-parameters.) The first set of measurements are measured with both antennas excited by a probe. This set is denoted (P,A)(P,A). The second and third set of s-parameters are measured with only one of the two antennas contacted by a probe. The second set is denoted (P,A)(P) to indicate that only left antenna is excited, while the third is denoted (P)(P,A) to indicate that only the right antenna is excited. The antenna that is not excited remains on the probe station, with its probe raised slightly above the antenna's feeding point. The fourth set of measurements is taken with neither of the antennas excited and with both probes raised slightly above the two antenna feed points, denoted (P)(P). The S_{21} components from each set are used to find the S_{21} components of only the two antennas according to:

$$(P, A)(P, A) = (P)(P) + (A)(A) + (P)(P, A) + (P, A)(P) \quad (4.69)$$
$$(A)(A) = (P, A)(P, A) - (P)(P) + (P)(P, A) + (P, A)(P). \quad (4.70)$$

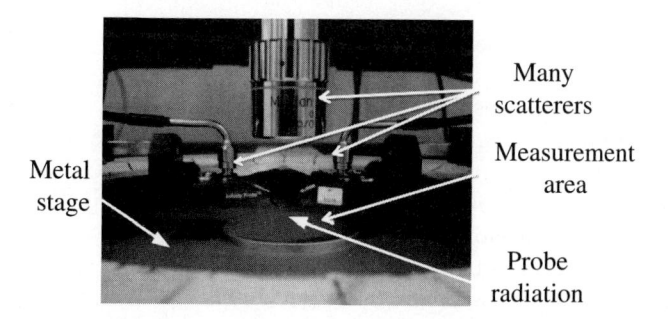

Figure 4.58 A probe station is often used to characterize mmWave antennas. These measurements may be inaccurate due to radiation from probes and scattered fields from the many surrounding metal objects [from [MBDGR11] © IEEE].

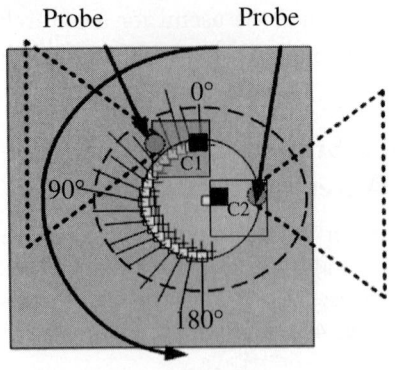

Figure 4.59 The two antennas were swept in angle across each other. The chips on which the antennas were fabricated are represented by squares, and the antennas are represented by smaller black boxes [from [MBDGR11] © IEEE].

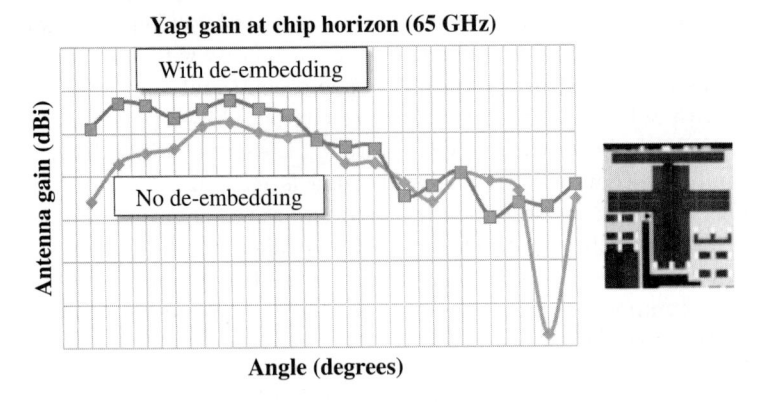

Figure 4.60 The de-embedding method indicates that the on-chip Yagi pattern was distorted by the presence of other nearby metal structures [from [MBDGR11] © IEEE].

This phenomenological approach was used by [MBDGR11] to measure an on-chip Yagi antenna for 60 GHz. Two Yagi antennas were placed on chip gel carrier packs (small plastic boxes with a silicone cushion on the interior of the box) and placed 6 mm apart, which satisfied the far-field requirement based on the 1 mm size of the antennas. One of the Yagi antennas was then swept in an arc in 10° increments, and the four sets of s-parameters were taken at each point on the arc. Fig. 4.59 from [MBDGR11] illustrates the measurement. Antenna gain patterns were then computed with the s-parameters found using Eqns. (4.70) and (4.68). The results of the campaign are shown in Fig. 4.60. Notably, the pattern seems to be pulled to on-side for the antenna. Simulations confirmed that this was due to the non-symmetric arrangement of on-chip metal structures that surrounded the on-chip Yagi. Simulations confirming this pattern pulling are shown in Fig. 4.61. For the simulations, an HFSS model of the on-chip Yagi was surrounded by increasingly numerous metal structures to represent the other devices on the same chip.

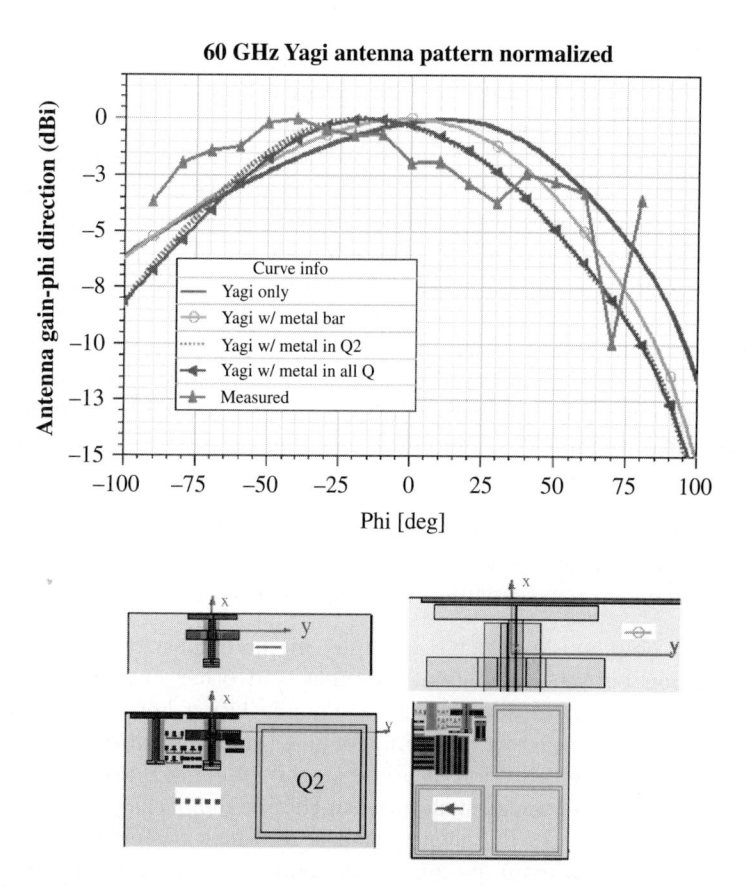

Figure 4.61 Simulations confirmed measurements that indicated the distortion of the antenna pattern was caused by surrounding metal structures. This indicates that isolation between integrated antennas and other nearby structures on the chip or in the package is key to successful design [from [MBDGR11] © IEEE].

This pattern distortion clearly indicates the need for high isolation between an on-chip (or in-package) antenna and other objects to ensure an undistorted pattern.

4.9 Chapter Summary

Integrated mmWave antennas will become increasingly important to consumer technology and other applications as demand for bandwidth from portable devices increases. Examples of where these systems will be useful abound. As media production continues to advance to bring the world future generations of 3-D movies that are based on true three-dimensional images, consumers will require ever greater bandwidth to access these media products from wireless devices like smartphones and tablets. Future mobile networks will carry tens of gigabits per second transmissions using adaptive arrays smaller than a human fingernail. Cancer detection may soon be accomplished by a doctor sitting at the bedside of a patient with a small terahertz device with an integrated antenna array for imaging tumors below the skin. Toy companies will be able to use small mmWave transceivers embedded in products to create elaborate Lego toy cities with fun and engaging screen technology and synchronized toy movement. Drivers will soon ubiquitously use mmWave radars with mmWave antennas to detect nearby vehicles and icy conditions on the road. Small integrated terahertz transceivers with integrated antennas will soon be capable of spectroscopy, enabling a vast number of devices and applications ranging from improved gas leakage detection technology to wire insulation failure detection for the power utility business. Creative readers will have little trouble in devising and imagining more applications, some of which were hinted at in Chapter 1.

Before widespread integration and proliferation of reliable integrated on-chip and in-package antennas are possible, many challenges associated with the design and characterization of integrated mmWave antennas must be overcome. This is a current, active area of research, as the following challenges must be addressed:

1. For on-chip antennas, methods must be developed to enable high efficiency directive radiation. Approaches such as frequency selective screens and electro-bandgap materials may prove successful in enabling low-loss radiation.

2. For both on-chip and in-package antennas, methods must be developed to achieve high isolation between individual antennas and other objects and antennas on the same chip or in the same package. Poor isolation between different antenna elements can result in active input impedances (an input impedance for an antenna that depends on the radiation of other nearby antennas). Poor isolation between an antenna and other metal structures on-chip or the package itself for in-package antennas may result in distorted patterns in which a single beam is split into multiple beams, or a main antenna beam is pulled in an unintended direction.

3. For integrated millimeter arrays (both on-chip and in-package), work is needed to develop low complexity and effective beam steering and direction of arrival detection algorithms that are low-latency and require minimal baseband processing.

4. For in-package antennas, antenna topologies must be developed that are not overly affected by the manufacturing imprecision that characterizes package fabrication. Although a difference of 10 μm between a designed dimension and a manufactured dimension may seem small, such inaccuracies may detune antennas or such structures as frequency selective screens.

5. For all highly integrated mmWave antennas, work is needed on characterization methods that produce reproducible and accurate three-dimensional pattern measurements. This may depend on development of specialized anechoic chambers.

Chapter 5

MmWave RF and Analog Devices and Circuits

5.1 Introduction

This chapter presents recent developments and design approaches in analog devices and circuitry that will enable incredible bandwidths and multi-gigabit per second (Gbps) data rates for mobile devices through the use of mmWave wireless carrier frequencies. We highlight modern analog design techniques and practical specifications for mmWave devices, circuits, and components within the last decade.

First, we present the basic concepts for mmWave transistors, devices, and systems for modern applications, as the transistor forms the basis of all the circuit blocks in a wireless transceiver and baseband processor. To characterize transistors and circuits, we introduce the various measurement parameters used in industry and research to quantify the behavior of transistors and circuits, and we describe the fundamentals of passive devices, such as transmission lines, inductors, and capacitors, as they are implemented in integrated circuits. We show how the fabrication of integrated circuits in a commercial foundry requires proper modeling and careful layout of both active transistors and passive devices. We then discuss analog mmWave components that are critical for creating mmWave wireless communication systems and subsystems, including the basic circuit blocks used in all wireless transceivers. After reviewing the signal sensitivities and link budget expectations for mmWave radios, we then present the fundamentals and modern specifications of power amplifiers (PAs) and low noise amplifiers (LNAs). Following amplifiers, we discuss mixers, voltage-controlled oscillators (VCOs), and phase-locked loops (PLLs). Together, these elements form all of the critical analog circuit blocks needed to construct a mmWave radio.

Then, a fundamental, and easy to use theory for power consumption, called the *consumption factor theory*, is presented. The consumption factor theory allows engineers to compare the power consumed by a cascade of individual circuit components, thereby allowing different transmitter and receiver architectures to be explored and compared.

Examples are presented that allow communications and circuits engineers to compare the power efficiencies of different circuits, cascaded systems, or architectures. This theory also is shown to support the analysis of the power efficiency of propagation channels, and the theory may be used for determining the most power-efficient system and circuit architectures.

5.2 Basic Concepts for MmWave Transistors and Devices

The most basic and lowest-cost material used to construct CMOS semiconductor circuits is silicon (Si). On the periodic table, silicon occupies the column reserved for elements with four valence electrons (electrons that occupy the highest energy states and are furthest from the nucleus of the atom). Valence electrons are generally the electrons that participate in chemical reactions. The simple "rule of eight" specifies that an atom is least reactive and most stable (i.e., will not give, take, or share electrons with other atoms) when it has eight valence electrons.

If many million silicon atoms (i.e., 10^{23} or more of them) are placed together, they begin to share electrons between each other, and reach a stable value of eight valence electrons per atom. Each silicon atom achieves this stability by forming a covalent bond (i.e., a bond in which electrons are shared, not given or taken) with four of its nearest neighbors. For all silicon atoms to do this, the atoms must arrange themselves into a regular lattice, as illustrated in Fig. 5.1. Similar figures have appeared in many textbooks, for example, [SB05].

If a pure lattice of silicon is doped (implanted) with a sufficiently large (but not too large) number of atoms with only three valence electrons (e.g., boron), then the hole where the electron should be, in effect, becomes a positive charge that may move throughout the lattice. If pure silicon is doped with a sufficiently large number of atoms with five valence electrons (for example, arsenic), then the extra electron in the lattice acts as a negative charge carrier. These two forms of doped silicon are referred to as P-type and N-type silicon, respectively. CMOS devices are constructed by interfacing different sections and layers of P- and N-type silicon and creating physical gaps through which electrons may flow between the different sections of the semiconductor.

The two elementary non-linear devices used to create mmWave circuits are the diode and the transistor. These devices are critical for creating electrical switches, or *gating circuits*. The diode is a well-known device that rectifies an AC signal and has a diode turn-on voltage (a threshold voltage) at which point the diode acts as a conductor. When the voltage differential across the diode is below the diode's threshold voltage, the diode appears as an open circuit. A CMOS transistor has three terminals: the source, the gate, and the drain,[1] as discussed in Section 5.5 and subsequent sections. When implemented in CMOS, gates (switches) are formed by exploiting the effects of the electromagnetic fields between tiny wells of doped material that form the drain and the source of the transistor while a voltage is applied to the gate of the transistor. The gate is ever so slightly elevated by a tiny insulating layer of silicon dioxide and is formed by a tiny

---

1. A fourth connection, the body, is usually not used, except when it is desirable to impact certain device parameters such as the the threshold voltage.

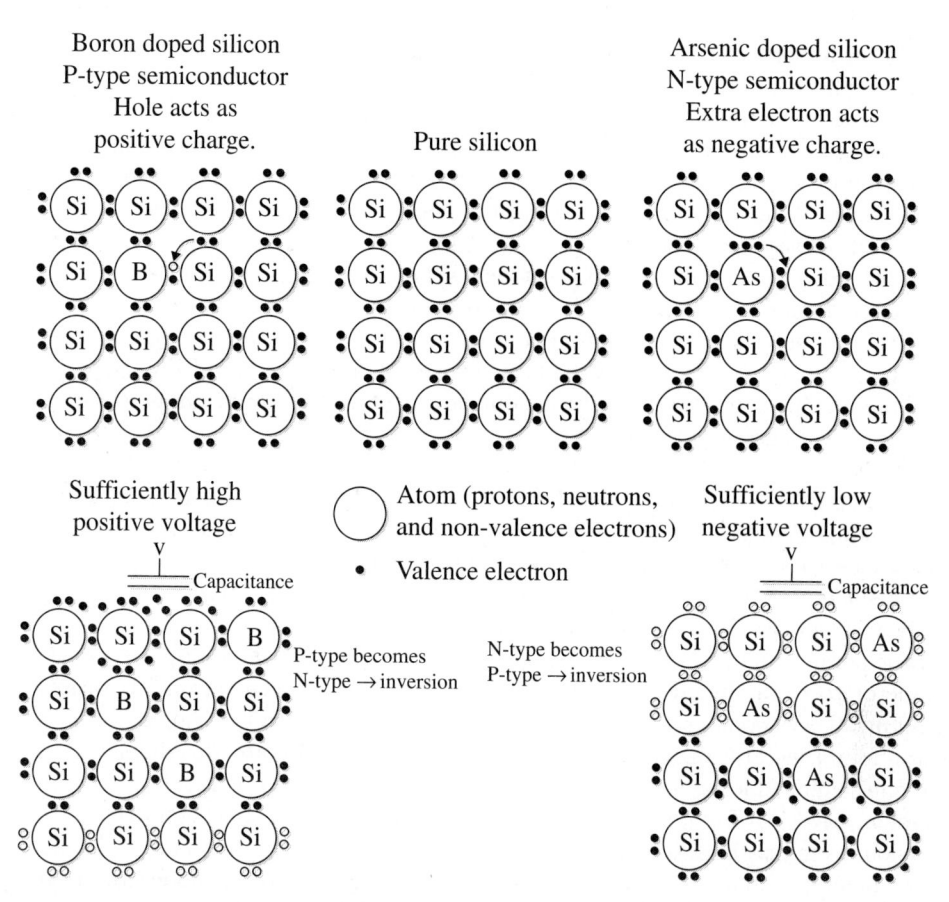

5.2 Basic Concepts for MmWave Transistors and Devices

Figure 5.1 Pure silicon is a semiconductor. When doped with elements such as boron (which has only three valence electrons), silicon becomes a P-type semiconductor. When doped with elements like arsenic (which has five valence electrons), silicon becomes an N-type semiconductor. A sufficiently high voltage applied to a P-type semiconductor may create an inverted region, in which there are extra electrons so that the inverted region becomes N-type. A sufficiently low negative voltage applied to an N-type semiconductor may create an inverted region in which very few valence electrons are present, so the inverted region becomes P-type.

metal layer or conducting polysilicon layer. See Fig. 5.11 and Figs. 5.15 and 5.16. This metal or polysilicon layer has a length called the *gate length,* and it is equal to the length of the channel between the drain and source. Depending on the applied gate signal voltage, the fields between the source and drain are varied, and this causes electrons to flow in various directions. It is for this reason that a CMOS transistor is often called a *field-effect transistor* (FET). Because of the metal oxide semiconductor (MOS) materials used to create transistors in CMOS, CMOS transistors are typically called *metal oxide semiconductor field-effect transistors* (MOSFETs).

It is important to note that all of these layers of a CMOS transistor (e.g., MOSFET), as well as all of the doped wells, are fabricated in popular (yet complex) automated manufacturing processes in a semiconductor foundry. The "magic" of all analog and digital circuits that compose all of the electronic devices that we enjoy today, at all possible operating frequencies, is provided by microscopic chunks of doped silicon that are deposited into bulk silicon. The operation of these circuits is based on the proper application of voltages that cause electrons to flow within and between the tiny wells of doped silicon. When an appropriate voltage is placed on the gate of a transistor, a voltage differential between the gate and the source is formed, where the gate controls the current that flows between the source and the drain. When the proper voltages are applied, and the proper doping and geometrical structures are created with the right doping levels, electrons flow into and out of different materials within the semiconductor, causing a switching action to occur. This is the basic concept behind the MOSFET. Depending on the depletion (i.e., the flow of electrons out of a region of the semiconductor) caused by the various voltages applied to the gate, a MOSFET can have a hard or soft switching mechanism.

It is possible to form *inversion layers* in doped silicon by applying sufficiently high (magnitude) voltages. For example, if we apply a sufficiently high voltage through a capacitance to a P-type piece of silicon, then a portion of the silicon may invert into N-type silicon. The ability to create inversion layers is key to MOSFET operation, as this is what allows the MOSFET to operate as a basic switch.

In general, a sufficiently high voltage (in magnitude) must be applied to the semiconductor to result in inversion. This voltage is related to the threshold voltage V_t of a MOSFET. Without getting into the physics involved, circuit designers and semiconductor foundries are able to empirically determine appropriate threshold voltages for different semiconductor materials and applications. There are generally three kinds of inversion: *weak*, *medium/moderate*, and *strong* [CH02]. Weak inversion occurs when an inversion layer just begins to form, and occurs for applied voltages across the semiconductor junctions that are slightly less than the threshold voltage. Moderate inversion is when the inversion layer is fully formed and the applied voltage is equal to the threshold voltage. Strong inversion is when the applied voltage is greater than the threshold voltage [SN07][CH02].

In later sections of this chapter, we provide extensive details on the operation of MOSFETs, including popular models that are used to mathematically represent the behavior of the MOSFET for use in circuit design, and mmWave circuit design in particular. For now, though, it is sufficient to realize that CMOS devices are the world's most popular and least expensive semiconductor devices. By configuring dozens of MOSFETs in parallel, it is possible to handle higher currents, and by using MOSFETs in circuit designs, it is possible to make complex circuits for use in mmWave transceivers. Often, MOSFETs are used in mixed-signal designs, in which both digital switches and analog amplification are provided in the same CMOS circuit (by using different voltages to obtain different diffusion levels in the MOSFETs, depending on the portion of the circuit that is designed to be digital or analog). Later in this chapter, we present many of the key analog and RF components of communication systems, and we show how the MOSFET transistor models help to create effective circuit designs. To properly analyze RF and analog circuits and devices, it is important to move up from the device level of transistors and define some fundamental circuit analysis techniques that are universally used to characterize and measure analog and RF circuits.

5.3 S-Parameters, Z-Parameters, Y-Parameters, and ABCD-Parameters

A basic understanding of S-, Z-, Y-, and ABCD-parameters is needed to design and test mmWave circuits and devices. Consider a mmWave device or structure that has N ports, as illustrated in Fig. 5.2. We use S-, Z-, Y-, or ABCD-parameters to relate quantities of voltage and current, as well as voltage standing wave ratio (VSWR, or simply SWR), gain, and coupling of signals at each port. [Poz05] is an excellent text that discusses these parameters at length, and we only briefly discuss them here.

S-parameters, also known as *scattering parameters*, relate incoming and scattered (e.g., reflected) voltage waves at each port and are most frequently used to characterize devices where the signals being measured have wavelengths that are smaller than the physical structures under test, such as in coaxial cables, waveguides, and circuits. S-parameters are particularly useful for mmWave circuit design and characterization because they are based on a known reference impedance, usually 50 ohms, and they do not rely upon the ability to create perfect short circuits and open circuits during measurement (which are difficult to implement at mmWave frequencies due to the fact that all leads and junctions tend to have some inductance or capacitance). An engineer uses S-parameters (which may be easily related to the other parameters mentioned above) to rapidly determine if a structure has gain or loss, if it is well matched for maximum power transfer, if it is a stable circuit, or if it behaves as a capacitive or inductive structure. For example, S_{11} indicates the voltage that is reflected or scattered out of a structure at port 1, divided by the voltage applied to the input of port 1, thus giving the input reflection coefficient of port 1 of the structure. S_{11} is related directly to voltage standing wave ratio, or VSWR (e.g., a smaller value of S_{11} implies that smaller reflected voltage occurs for a particular applied voltage, thereby indicating a much lower SWR and better match), whereas S_{21} indicates the voltage out of port 2 divided by the voltage applied to the input port 1, thereby indicating the forward *voltage gain* of the device from port

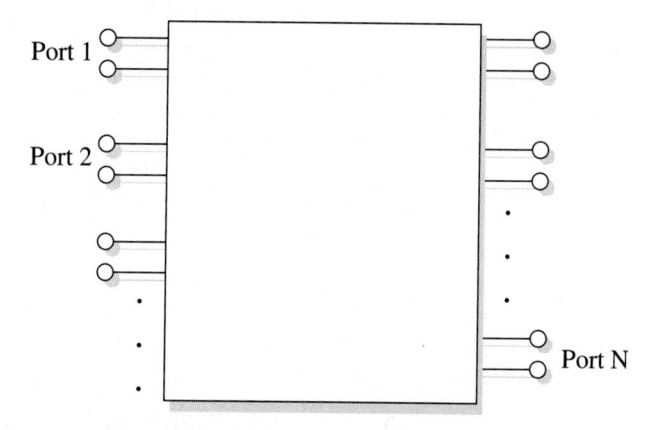

Figure 5.2 S-, Z-, Y-, and ABCD-parameters related voltage and current quantities at the ports of a structure. ABCD-parameters are useful for systems with many cascaded structures and are typically only applied to two-port structures.

1 to port 2. In other words, S_{21} represents the voltage produced by the device at port 2 compared to the voltage applied to port 1. Z-parameters relate the voltages at each port to the currents at each port, thus providing insights into the impedance of each port of the device or structure.

Y-parameters relate the currents at each port to the voltages at each port, thus indicating the admittance (the inverse of impedance) at each of the ports of the device. The ABCD-parameters relate the currents and voltages at one port to the currents and voltages at a second port and are typically used only for two-port structures. Generally, all of these different parameters are represented in matrix form to provide a means for engineers to characterize the circuit behavior of all of the ports of a device or structure, without having to know the particular circuit details of what is inside the device or structure.

The Z-parameters relate the voltage at a given port to the currents at all the other ports, and are denoted

$$\begin{bmatrix} V_1 \\ \cdots \\ V_N \end{bmatrix} = \begin{bmatrix} Z_{11} & \cdots & Z_{1N} \\ \vdots & \ddots & \vdots \\ Z_{N1} & \cdots & Z_{NN} \end{bmatrix} \begin{bmatrix} I_1 \\ \cdots \\ I_N \end{bmatrix} \quad (5.1)$$

where V_i is the voltage at the ith port and I_k is the current at the kth port. To determine Z_{ik}, apply a current to the kth port and leave all other ports open and measure the voltage at the ith port.

The Y-parameters relate the current at a given port to the voltages at all other ports, and are denoted

$$\begin{bmatrix} I_1 \\ \vdots \\ I_N \end{bmatrix} = \begin{bmatrix} Y_{11} & \cdots & Y_{1N} \\ \vdots & \ddots & \vdots \\ Y_{N1} & \cdots & Y_{NN} \end{bmatrix} \begin{bmatrix} V_1 \\ \vdots \\ V_N \end{bmatrix}. \quad (5.2)$$

To determine Y_{ik}, apply a voltage to the kth port and ground all other ports, and measure the current at the ith. The Y-parameters for a simple pi-network (named as such because the circuit structure looks like the Greek letter "pi") are shown in Fig. 5.3.

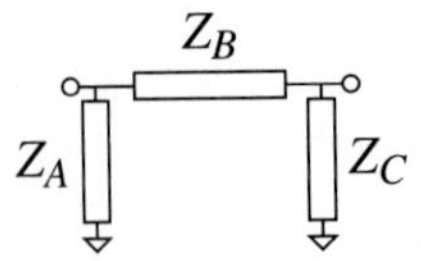

$$Y_\Pi = \begin{bmatrix} (Z_A+Z_B)/(Z_A Z_B) & 1/Z_B \\ 1/Z_B & (Z_C+Z_B)/(Z_C Z_B) \end{bmatrix}$$

Figure 5.3 The Y-parameters for a simple pi-network.

5.3 S-Parameters, Z-Parameters, Y-Parameters, and ABCD-Parameters

The ABCD-parameters relate the current and voltage at one port of a two-port structure to the voltage and current at the second port, and are denoted

$$\begin{bmatrix} V_1 \\ I_1 \end{bmatrix} = \begin{bmatrix} A & B \\ C & D \end{bmatrix} \begin{bmatrix} V_2 \\ -I_2 \end{bmatrix}. \tag{5.3}$$

To determine the ABCD-parameters, a voltage source must be applied at the input port (starting from the left, as an input signal), while measuring the current and voltage at the output port to the right. A cascade of devices or circuits may be characterized by simply continuing the process of successively applying input voltage sources to subsequent stages, and working from left to right. The advantage of ABCD-parameters is that they may be multiplied to represent a cascaded structure, such as a pi-network, as illustrated in Fig. 5.4.

The S-parameters of a structure are denoted

$$\begin{bmatrix} V_1^- \\ \vdots \\ V_N^- \end{bmatrix} = \begin{bmatrix} S_{11} & \cdots & S_{1N} \\ \vdots & \ddots & \vdots \\ S_{N1} & \cdots & S_{NN} \end{bmatrix} \begin{bmatrix} V_1^+ \\ \vdots \\ V_N^+ \end{bmatrix} \tag{5.4}$$

where V_i^- is the wave that is outgoing of the ith port as a result of all the incoming waves to all the other ports, denoted V_k^+ for the kth port.

Fortunately, simple conversion formulas exist between S-, Y-, Z-, and ABCD-parameters [Poz05, Chapter 4]. Often, it is easier with a hand calculation to determine the Z- or Y-parameters, and then to convert to S- or ABCD-parameters. Fig. 5.5 shows the Z- parameters for T-networks and pi-networks, which are two typical formulations for mmWave networks. A common approximation in mmWave context is to relate S_{11} to Z_{11} according to

$$S_{11} \approx \frac{1 + Z_{11}/Z_0}{1 - Z_{11}/Z_0}. \tag{5.5}$$

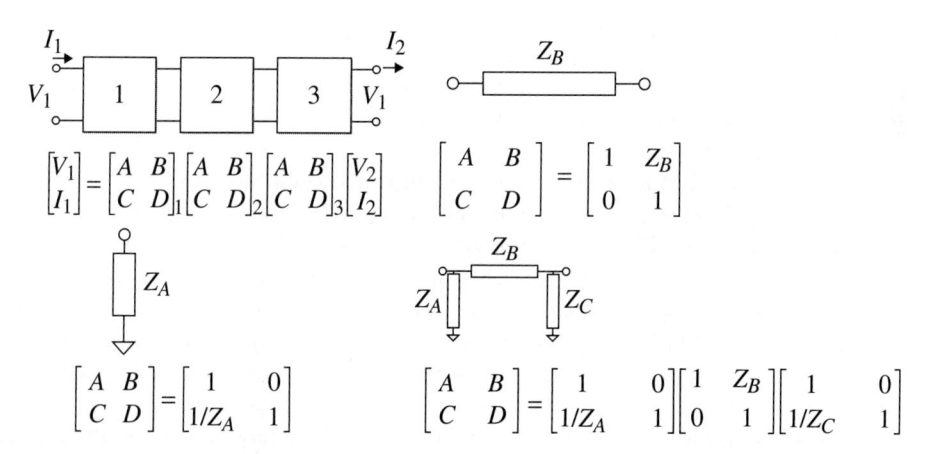

Figure 5.4 ABCD-parameters have the advantage of allowing cascaded structures to be modeled through multiplication.

$$Z_\Pi = \frac{1}{Z_A+Z_B+Z_C}\begin{bmatrix} Z_A(Z_B+Z_C) & Z_AZ_C \\ Z_AZ_C & Z_C(Z_B+Z_A) \end{bmatrix} \quad Z_T = \begin{bmatrix} Z_1+Z_2 & Z_2 \\ Z_2 & Z_3+Z_2 \end{bmatrix}$$

$$Z_1 = \frac{Z_A Z_B}{Z_A+Z_B+Z_C}$$

$$Z_2 = \frac{Z_A Z_C}{Z_A+Z_B+Z_C}$$

$$Z_3 = \frac{Z_C Z_B}{Z_A+Z_B+Z_C}$$

Figure 5.5 Once Z-parameters are found, they can be easily converted to Y-, S-, or ABCD-parameters (although ABCD are just used for two-port networks). Shown here are pi- and T-networks. Pi- and T-networks can also be interchanged, as shown here.

When working with S-parameters, two common terms are *return loss* and *insertion loss*. Return loss describes how much power is reflected back by a circuit at a particular port. The lower the return loss, the more energy the circuit accepts (and the less it reflects) from the source that feeds it. Therefore, most designs should strive for the smallest return loss possible, close to 0 in absolute terms, or as low as possible in decibel values (e.g., −13 dB or smaller is often desired for good matching, and −20 dB or smaller is even more desirable). The return loss is the decibel representation of the voltage ratios characterized by S_{11}:

$$\text{Return Loss [dB]} = 20\log_{10}(S_{11}). \tag{5.6}$$

Coaxial cable and waveguide transitions, most circuit connections, and amplifiers such as power amplifiers or low noise amplifiers (LNAs) should have return losses that are negative in decibels (e.g., very little reflected or scattered energy). However, unlike most other circuits, oscillators such as voltage-controlled oscillators (VCOs) must have a positive return loss. This is because a value for S_{11} greater than 1 (or greater than 0 dB) indicates instability, or oscillation, at the ith port of a device. If an amplifier or other circuit component has a positive return loss, then it is said to be unstable, and it will act as an oscillator, not an amplifier. The stability of amplifiers may be heavily dependent on the biasing conditions of the amplifier, but under certain conditions an amplifier can be made unconditionally stable, so that its return loss is always less than 0 dB for any possible bias condition.

The insertion loss indicates how much energy is lost (or how much energy is gained if insertion loss is negative in decibel values) as it traverses through the device under test (DUT). It is simply the dB representation of the voltage ratios characterized by S_{21}. To avoid power loss through a device, passive devices such as transmission lines or waveguides should have an insertion loss as small as possible, for example, as close to 0 dB as possible. Note that passive, lossy devices always have an insertion loss that is greater than 0 decibels, since the voltage out of the output port 2 will always be less than that applied to the input port 1. However, an active device, such as an amplifier, may have an insertion loss that is less than 0 dB, as it may provide gain. The insertion loss is defined as

$$\text{Insertion Loss [dB]} = -20 \log_{10}(S_{21}). \tag{5.7}$$

Isolation is another term that appears in design of amplifiers or mixers. Isolation describes how well a circuit prevents energy at one port from arriving at another port. In general, amplifiers and mixers should have good isolation (i.e., very low voltage transfer) between input ports and from output ports back to input ports. We may define the backward isolation, or reverse isolation, to describe how well the output port of a circuit or structure is isolated from the input of that circuit in a two-port network, such as an amplifier. High isolation helps ensure that if there are impedance matching problems (e.g., standing waves) at the output of an amplifier, the input signal will not be affected (e.g., the input port will not see any of the reflections occurring at the output, which could then be further reflected at the input and reradiated back through the amplifier). Isolation from the output back to the input of a circuit is simply the decibel representation of the voltage ratios expressed by S_{12}

$$\text{Isolation [dB]} = 20 \log_{10}(S_{12}). \tag{5.8}$$

When S_{12} is near zero (i.e., isolation is very good, a very large negative value in decibels), the circuit protects the output from the input by not allowing much energy at the output to leak back to the input. Poor isolation can result in such problems as unintended receiver radiation, for example, where a received signal can leak "backwards" through a low noise amplifier (LNA) of a receiver, back up to the antenna, and be reradiated, causing unintended interference and problems to other receivers in the vicinity.

5.4 Simulation, Layout, and CMOS Production of MmWave Circuits

The creation of mmWave integrated circuits requires a number of steps that are often iterated in tandem. After the initial theoretical design of a circuit, two critical and intertwined steps are simulation and layout. The complex environments of most production processes, which include multiple layers of dielectrics and metal layers, make it prohibitively difficult in most cases to rely solely on theory to design a circuit. Rather, theories about circuit operation offer springboards into design refinement through simulation.

There are several circuit simulation tools that should be considered when designing an analog mmWave circuit. These include both SPICE circuit simulators and electromagnetic simulators. Table 5.1, which is based on a table in [RGAA09], summarizes several popular simulation tools and their uses in circuit design. The electromagnetic structure simulators are designed to solve Maxwell's equations on a user-defined metal or dielectric geometry for a user-defined excitation (e.g., a voltage source, current, or

Table 5.1 There are many software tools that should be considered when designing mmWave circuits [modified from [RGAA09] © IEEE].

Software	Company	Use	Electromagnetic Simulation Type
Ensemble (Designer)	Ansoft	Electromagnetic structure simulation	Moment method
IE3D	Zeland	Electromagnetic structure simulation	Moment method
Momentum	HP	Electromagnetic Structure simulation	Moment method
EM	Sonnet	Electromagnetic Structure simulation	Moment method
PiCasso	EMAG	Electromagnetic structure simulation	Moment method/ genetic algorithm
FEKO	EMSS	Electromagnetic structure simulation	Moment method
PCAAD	Antenna Design Associates, Inc	Patch antenna design	Cavity model
Micropatch	Microstrip Designs, Inc.	Electromagnetic structure simulation	Segmentation
Microwave Studio (MAFIA)	CST	Electromagnetic structure simulation	Finite difference Time domain
Fidelity	Zeland	Electromagnetic structure simulation	Finite difference Time domain
HFSS	ANSYS/Ansoft	Electromagnetic structure simulation	Finite element
Cadence Virtuoso Analog Design Environment	Cadence	Circuit simulation Circuit layout Layout extraction	N/A
Advanced Design System (ADS)	Agilent	Circuit simulation	N/A
PeakView	Lorentz Solution	Electromagnetic structure simulation	Moment method

impinging wave). There are three primary means of simulating electromagnetic structures: method of moments (aka moment method), finite element method (FEM), and finite difference time domain (FDTD) [RGAA09]. There are also hybrids of these techniques that often combine the advantages of two or more different techniques. Of these simulation approaches, FEM works well for electrically small three-dimensional structures that contain dielectrics and metals. Method of moments is superior for large, mostly planar structures (up to many tens of cubic wavelengths or even larger) that are composed mostly of metal. FDTD is implemented to actually simulate waves in time, and it may not be the most desirable for frequency domain information, but offer the advantage of providing temporal information that may be useful for some designs.

As described by [RGAA09], design of electromagnetic structures requires knowledge of the characteristics of the materials used in the device, in addition to their physical placement in the device. Typically, simulators will allow a designer to draw a three-dimensional or two-dimensional model of the structure and the location of all excitations such as voltage sources. Unfortunately, the complex permittivity of most materials used in semiconductor production is not provided by most foundries at this time, requiring designers to guess or measure this vital parameter [RGAA09]. The permittivity may not be greatly different compared with low frequencies, but if there are loss mechanisms (e.g., due to the presence of dopants in a substrate) that are not accounted for in low frequency models, then this parameter will likely have a different value at mmWave frequencies than at lower RF frequencies.

CMOS and other foundry technology production processes (e.g., Silicon Germanium [SiGe]), have metal layers that are added during the production steps known as the Back End of Line (BEOL). These metal lines, of which there may be as many as seven or more, are embedded in a dielectric. Traditionally, silicon dioxide was used for this purpose, but increasingly low-k (i.e., low relative permittivity) materials are used as the dielectric that contains metal layers. This is because low-k dielectrics offer lower coupling between the metal layers and the substrate, and between individual metal layers. As shown in Chapter 4 (see Fig. 4.5), the cross section of a typical CMOS chip has a bulk substrate with a small number of closely spaced metal layers above the substrate. In Fig. 4.5, the metal lines and (possibly low-k) dielectric are supported by a much thicker doped semiconductor called the *substrate* (made of silicon for CMOS processes). Active devices such as transistors reside at the very top of the silicon substrate, and the first micro meter of the substrate has the highest conductivity and concentration of dopants. The metal layers in the upper dielectric are connected together by *vias*. The typical feature (once the process has been optimized for analog operation) of a thick top metal layer is also illustrated. This thick layer should be used for long metal pieces (aka traces), such as those used in a long transmission line.

For commercial mass-market mmWave circuit designs at or above 20 GHz, at the time of this writing, 130 nm, 90 nm, 65 nm, and 40 nm production processes (aka nodes) are best suited to achieve design specifications, with 65 nm offering a good tradeoff between speed and power handling ability. Process nodes below 65 nm offer greater speeds for mmWave applications due to smaller gate lengths, but they have more difficulty handling greater power levels than nodes at 65 nm and above. Many commercial foundries offer special RF-processing, including thicker conducting lines and special intellectual-property (IP) for improved power efficiency and yield. Silicon on Insulator (SOI) processes may also be considered. An SOI process allows for separate doping of the substrate and the channel regions of a transistor. This allows the substrate to be manufactured with a lower loss tangent than that of "regular" CMOS processes, in which the transistor channel and substrate are connected. Passive components such as antennas and transmission lines may therefore be made with much lower loss in SOI than in standard CMOS processes.

IC probe pad cross section

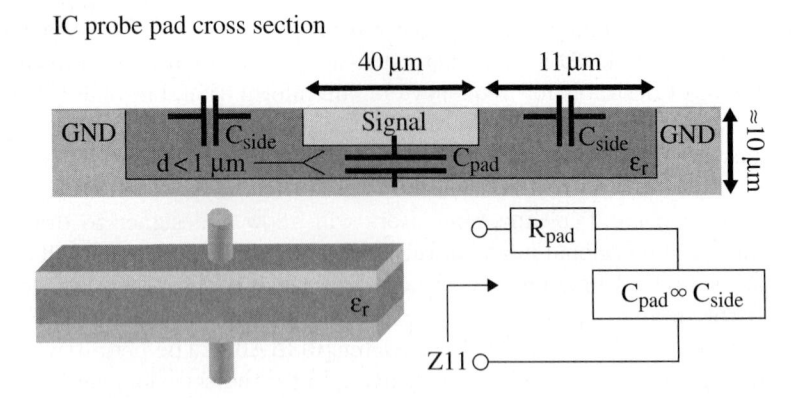

Figure 5.6 A simple pad implemented for a co-planar waveguide (CPW) probe. CPW transmission lines are discussed in the section on transmission lines (Section 5.7). A measurement of the pad's capacitance can be used to estimate effective relative permittivity of the substrate and dielectric that supports metallic BEOL structures [from [GJRM10] © IEEE].

Measurement of the characteristics of the substrate and dielectric that support BEOL structures is a major concern for mmWave frequencies. The two most important factors to measure are the permittivity and loss tangent. The permittivity of the substrate will partially determine the electrical length of metallic structures in the dielectric, though the dielectric will have a larger impact on the electrical length of metal traces than the substrate. In practice, it is typical to measure an effective permittivity and loss tangent that accounts for the impact of both the substrate and the dielectric that contains the BEOL structures. There are several methods to accomplish this.

A very simple method relies on measuring the capacitance of a small "pad," as illustrated in Fig. 5.6 from [GJRM10]. In the figure, the pad is implemented for a ground-signal-ground (GSG) probe, and is intended to interface with a co-planar waveguide (CPW) transmission line. CPW lines are described in Section 5.7. By measuring the S_{11} of the pad, it is possible to estimate Z_{11} according to [GJRM10]

$$Z_{11} = Z_0 \frac{S_{11}+1}{S_{11}-1} = R_{\text{pad}} + \frac{1}{j\omega C_{\text{pad}}}. \tag{5.9}$$

Here, Z_0 is the impedance of the transmission line, and the imaginary part of Z_{11} can be used to estimate the capacitance of the pad. The geometry of the pad indicates that the capacitance should be approximated well by the equation for a parallel-plate capacitor

$$C = \frac{\epsilon_0 \epsilon_{r,\text{eff}} A}{d} \tag{5.10}$$

where A is the area of the pad and d is the distance between the pad and the ground layer or substrate. [GJRM10] uses this method for a pad with $A = 40\ \mu\text{m} \times 60\ \mu\text{m}$ and $d = 1\ \mu\text{m}$ to estimate the effective permittivity to be 4.32 for 180 nm CMOS.

A more advanced technique based on measuring the S-parameters of a transmission line may be used to estimate both relative permittivity and loss tangent. First, the

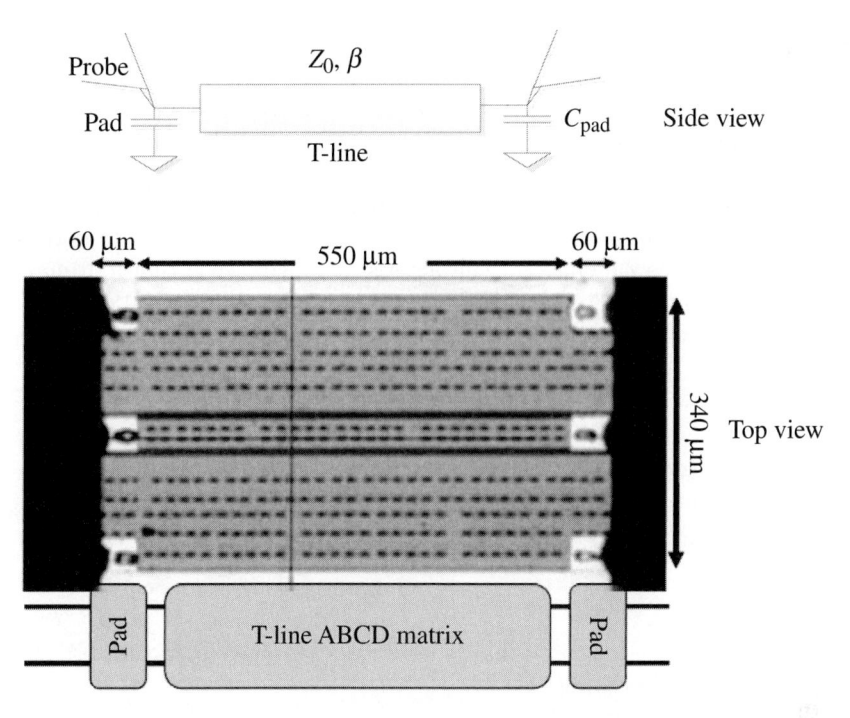

5.4 Simulation, Layout, and CMOS Production of MmWave Circuits

Figure 5.7 The S-parameters of a transmission line can be used to determine the effective relative permittivity and loss tangent of a CMOS process. The effects of the probe pads must be de-embedded for this measurement to be accurate [from [GJRM10] © IEEE].

S-parameters of a transmission line must be measured, and the effects of the pads used to interface a probe with the transmission line (e.g., a probe from a network analyzer) must be de-embedded, as illustrated in Fig. 5.7, which is modified from [GJRM10]. To de-embed the capacitors, note from Fig. 5.4 that the ABCD matrix of the line plus capacitors is given by

$$\begin{bmatrix} A & B \\ C & D \end{bmatrix}_{\text{measure}} = \begin{bmatrix} 1 & 0 \\ j\omega C_{\text{pad}} & 1 \end{bmatrix} \begin{bmatrix} A & B \\ C & D \end{bmatrix}_{\text{line}} \begin{bmatrix} 1 & 0 \\ j\omega C_{\text{pad}} & 1 \end{bmatrix} \quad (5.11)$$

so that the ABCD matrix of the line may be found according to

$$\begin{bmatrix} A & B \\ C & D \end{bmatrix}_{\text{line}} = \begin{bmatrix} 1 & 0 \\ j\omega C_{\text{pad}} & 1 \end{bmatrix}^{-1} \begin{bmatrix} A & B \\ C & D \end{bmatrix}_{\text{measure}} \begin{bmatrix} 1 & 0 \\ j\omega C_{\text{pad}} & 1 \end{bmatrix}^{-1}. \quad (5.12)$$

The ABCD matrix of a transmission line, as described in the section on transmission lines (Section 5.7), is given by [GJRM10]

$$\begin{bmatrix} A & B \\ C & D \end{bmatrix}_{\text{line}} = \begin{bmatrix} \cosh(\gamma\ell) & Z_0 \sinh(\gamma\ell) \\ \sinh(\gamma\ell)/Z_0 & \cosh(\gamma\ell) \end{bmatrix} \quad (5.13)$$

where γ is the complex propagation constant of the line and is given by $\alpha + j\beta$, where α describes how a wave is phase shifted as it propagates down the line and β is a damping factor. To de-embed the pad, it is necessary to have a pad structure that is not connected to a transmission line available for measurement with a network analyzer. The measured S-parameters of the pad may be converted to ABCD-parameters for use in (5.12). The A-parameter of the line may be used to determine the propagation factor of the line. The propagation factor may then be used along with the C-parameter to determine the characteristic impedance Z_0. Once the characteristic impedance is known, it may be used to compute the effective loss tangent and characteristic impedance [GJRM10]

$$Z_0 = \frac{\frac{\eta_0}{2\sqrt{\epsilon_{\text{eff}}}}}{\left(\frac{K(k)}{K(k')}\right) + \left(\frac{K(k_1)}{K(k'_1)}\right)} \tag{5.14}$$

where $K(k)$, $K(k')$, $K(k_1)$, and $K(k'_1)$ are functions of the geometry of the line described in [Wad91], η_0 is the characteristic impedance of free space (377 ohms), and ϵ_{eff} is the complex effective permittivity

$$\epsilon_{\text{eff}} = \epsilon' + j\epsilon'' \tag{5.15}$$

where ϵ' is the relative permittivity that is used to determine electrical lengths. The loss tangent may be found according to

$$\tan \delta = \epsilon''/\epsilon'. \tag{5.16}$$

This process was used successfully for transmission lines manufactured in 180 nm CMOS to measure the effective permittivity of the 180 nm CMOS substrate, as shown in Figs. 5.8

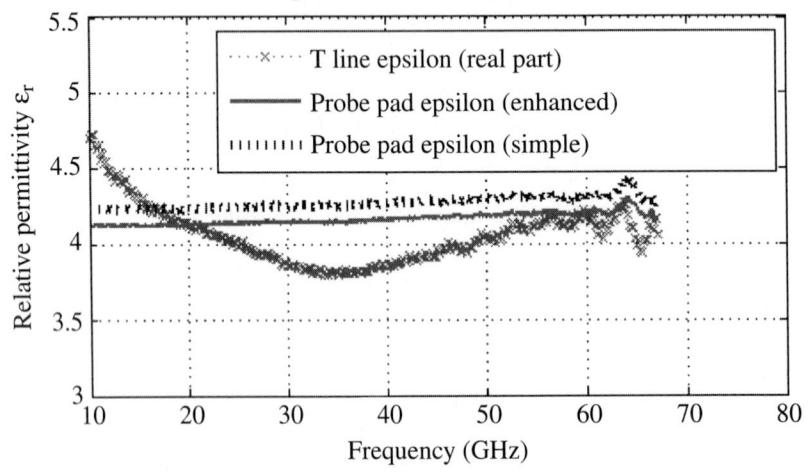

Figure 5.8 The effective relative permittivity may be measured in a number of ways and is a vital parameter for the design of passive structures [from [GJRM10] © IEEE].

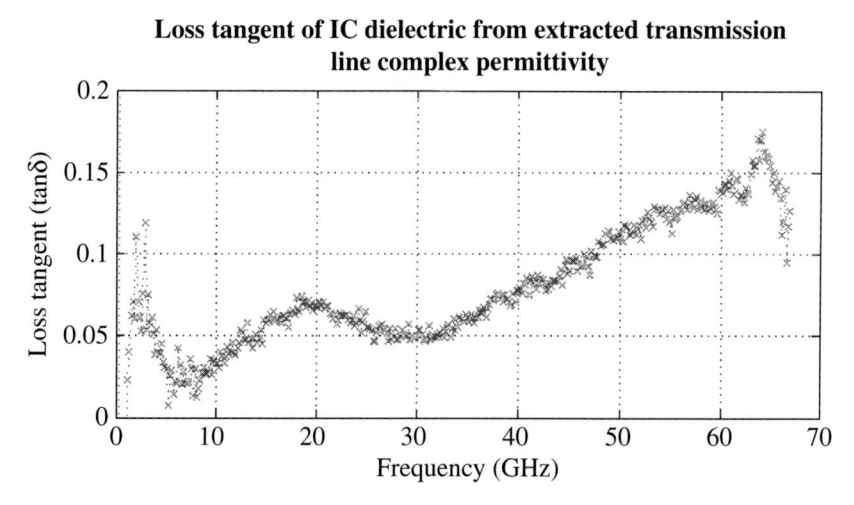

Figure 5.9 The effective loss tangent is a vital parameter to predict loss of passive structures [from [GJRM10] © IEEE].

and 5.9 from [GJRM10]. Fig. 5.8 also compares this method to results obtained from the simple capacitor based model and an enhanced capacitor method described in [GJRM10].

5.5 Transistors and Transistor Models

There are many classic texts that discuss transistor physics, including [Raz01], [Yng91], and more recent texts such as [NH08]. We only briefly treat the basic concepts of transistor modeling before focusing on recent research developments. Because of the low cost and massive scaling for mmWave devices, the treatment here focuses on metal oxide semiconductor field-effect transistors (MOSFETs), although readers may consider [Yng91] for more information on Bipolar Junction Transistors (BJTs), and [Lee04a] for such devices as Heterojunction Bipolar Transistors (HBTs) and High Electron Mobility Transistors (HEMTs). The physics governing all of these devices differs significantly, as do the descriptive mathematics. Nevertheless, much intuition may be utilized across different transistor devices.

At a basic level, a MOSFET is a voltage controlled switch with four connections, as shown in Fig. 5.10. The four connections are the gate, drain, source, and body. For a common source connection (i.e., grounded source of a NMOS, or source tied to supply voltage of a PMOS), a sufficiently high (low) voltage applied to the gate of an N-type MOSFET (P-type MOSFET) will fill the channel with electrons (holes, positive charge carriers). A positive (negative) voltage applied to the drain will then result in the electrons (holes) flowing up (down) to the drain. The minimum (maximum) gate voltage required (allowed) to fill the channel with electrons (holes) is called the *threshold voltage* V_T. The amount by which the gate voltage exceeds the *threshold voltage* is called the *overdrive voltage*.

Once the channel is filled with charge carriers (electrons in an NMOS, holes in a PMOS), there are several regions of operation. For low voltages applied to the drain of an NMOS, the transistor will operate in linear or triode region, where the current

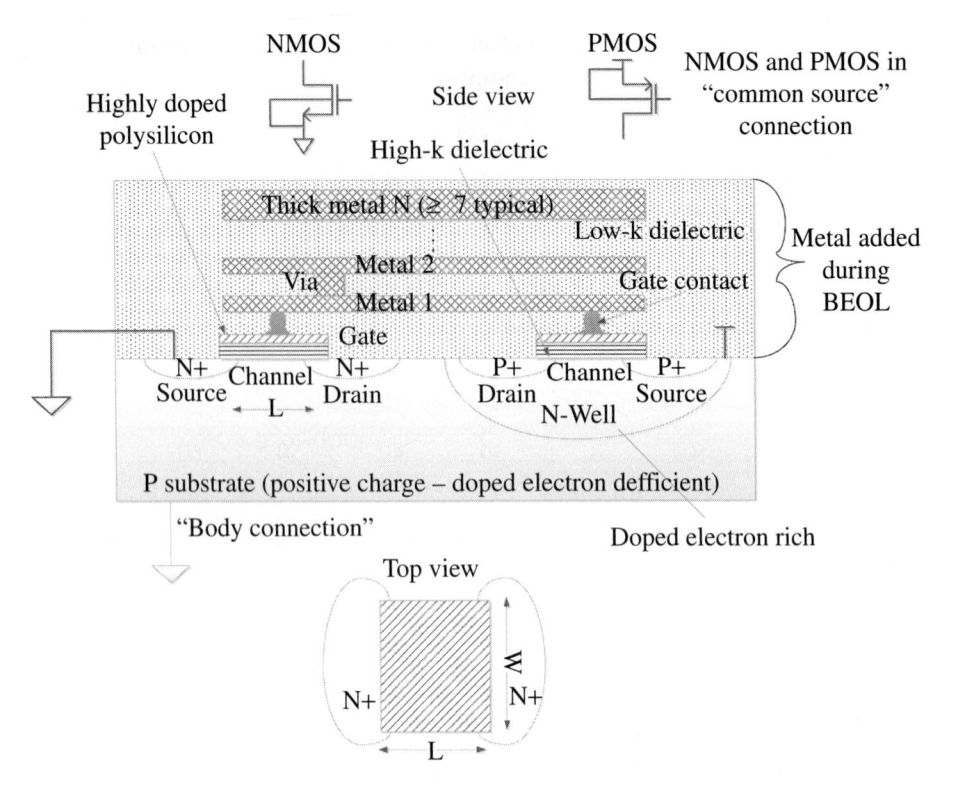

Figure 5.10 A basic MOSFET is a voltage-controlled switch. When a high voltage is applied to the gate of an N-type metal oxide semiconductor (NMOS) or a low voltage to the gate of a P-type metal oxide semiconductor (PMOS), the channel will be able to conduct electrons or holes, respectively. The length and width of the channel are indicated by L and W. Transistor scaling has resulted in newer generations of devices with shorter manufactureable lengths.

though the drain is related to the gate to source voltage V_{gs} and the drain to source voltage V_{ds} by

$$I_D = k_n \left(\frac{W}{L}\right) \left((V_{gs} - V_t)V_{ds} - \frac{V_{ds}^2}{2}\right) \tag{5.17}$$

where I_D is the current flowing through the drain to the source. In this equation, k_n is a parameter that indicates how easily charge flows through the channel in addition to the amount of capacitance between the gate and the channel. W is the width of the channel, and L is its length. More accurately, a gate-source voltage greater than V_t results in "inversion" of the channel. The channel (on an NMOS) is said to be inverted because, although the channel in its resting state (i.e., no gate voltage) is a p-type semi-conductor (i.e., it is filled with positive charge carriers), the high gate voltage results in the channel being filled with electrons. There are technically two regions of inversion: weak/moderate inversion to strong inversion. Weak or moderate inversion occurs when the gate voltage is high enough to fill the channel with enough electrons to equal or

5.5 Transistors and Transistor Models

barely exceed the number of holes that would usually exist in the channel. Strong inversion occurs when the gate voltage is high enough (greater than V_t) to result in the number of electrons being equal to at least twice the number of holes that would normally occupy the channel. When a voltage higher than the overdrive voltage $V_{gs} - V_t$ is applied to the drain, the device enters the saturation region of operation, in which the drain current is related to the gate voltage according to the well-known square law [Raz01]

$$I_D = \frac{1}{2} k_n \left(\frac{W}{L}\right)(V_{gs} - V_t)^2. \tag{5.18}$$

In the saturation region, the region of the channel closest to the drain becomes "pinched off" or depleted of charge. This is because the drain voltage is sufficiently high (in an NMOS) to cause the electrons to be swept out of the channel into the drain before an effective electron charge density can be established in the region of the channel closest to the drain.

Fig. 5.11 summarizes the basic regions of operation for a MOSFET just described. When referring to the operation of a device, it is common to say that the device is *biased* in a particular region of operation. The simple model described here shows that the way in which a circuit or device is biased has a major effect on how it operates, and certain device effects may only occur for certain bias conditions.

When a transistor operates in the saturation region, it is common to linearize the square law about a gate voltage operating point, so that the drain current may be related linearly to the gate voltage. As is done is such texts as [Raz01], if one considers the slope of (5.18) with respect to V_{gs}, it is possible to obtain a parameter known as the *transconductance* of the device, denoted by g_m where

$$g_m = k_n \left(\frac{W}{L}\right)(V_{gs} - V_t). \tag{5.19}$$

Since

$$\frac{1}{2} g_m (V_{gs} - V_t) = I_D \tag{5.20}$$

it follows that

$$g_m = \frac{2 I_D}{V_{gs} - V_t}. \tag{5.21}$$

Equations (5.19) and (5.21) relate transconductance to device aspect ratio (based on the physical size of the transistor junction in two dimensions), W/L overdrive voltage $V_{gs} - V_t$, and drain current I_D. Once the transconductance is defined, it is often placed in a simple circuit model with a voltage dependent current source to model the operation of the transistor in the saturation region, and for inclusion in circuit analyses. Such a simple model is shown in Fig. 5.12. The model also includes the output resistance of the device $r_0 = \frac{\delta I_D}{\delta V_{ds}}$, and its gate resistance R_g. C_{gs} is the capacitance between the gate and the source, and is responsible for the fact that a voltage at the gate can induce charge in the channel of the device. The gate resistance and gate-source capacitance together add

NMOS regions

$V_{gs} < V_T$, V_{ds}

Gate — Source, Polysilicon, Drain
Channel
e^-

"Cutoff"
Charges are not in channel; no current from drain to source.

$I_D = 0$

$V_{gs} \geq V_T$, $V_{ds} < V_{gs} - V_T$

Polysilicon

"Linear"
Charges fill channel; channel is "inverted."

$$I_D = k_n \left(\frac{W}{L}\right)\left((V_{gs} - V_T)V_{ds} - \frac{1}{2}V_{ds}^2\right)$$

$V_{gs} \geq V_T$, $V_{ds} \geq V_{gs} - V_T$

Polysilicon
Depleted region

"Saturation"
Charges partially fill channel; high drain voltage results in "depleted" region.

$$I_D = \frac{1}{2} k_n \left(\frac{W}{L}\right)(V_{gs} - V_T)^2$$

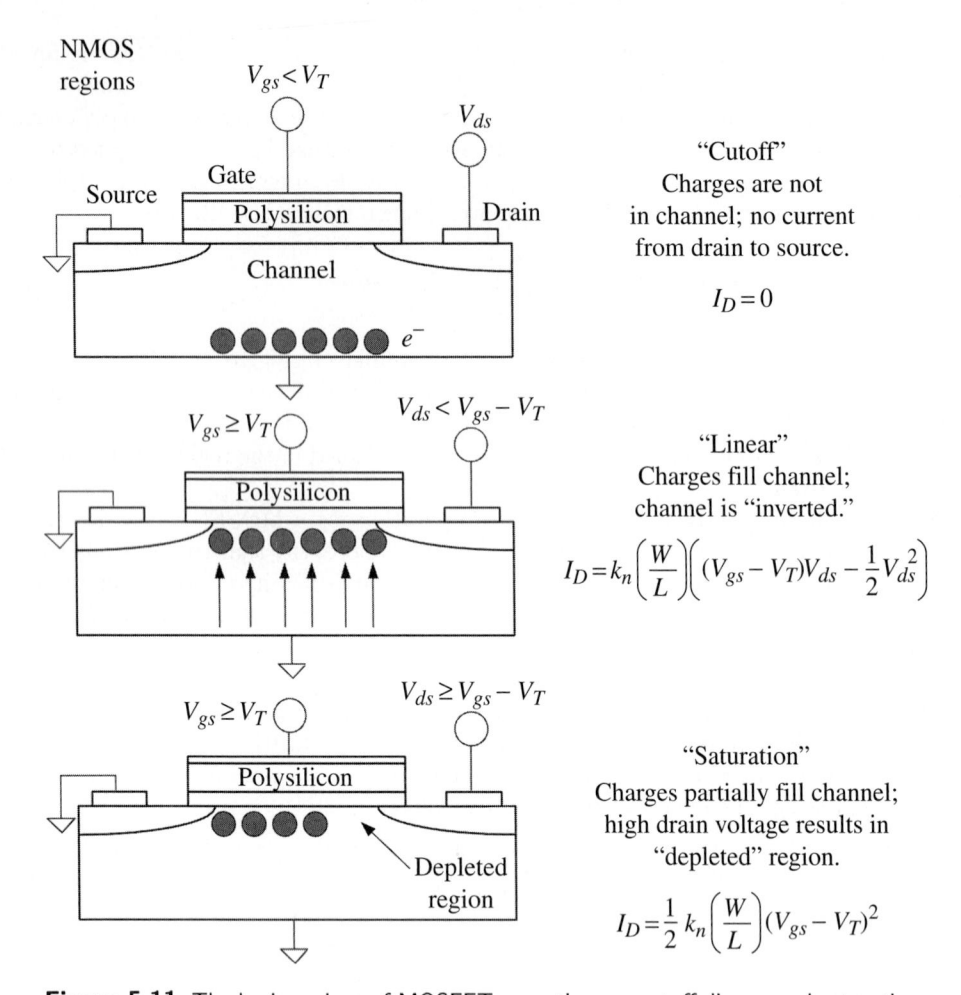

Figure 5.11 The basic regions of MOSFET operation are cutoff, linear, and saturation.

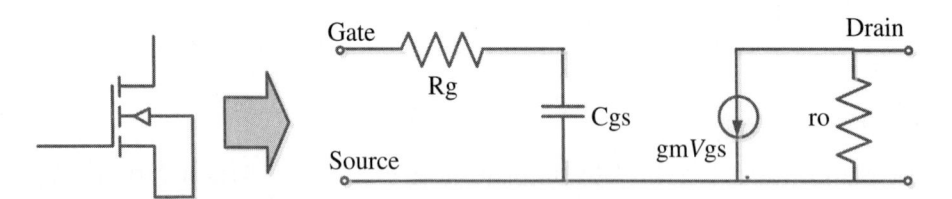

Gate — R_g — C_{gs} — $g_m V_{gs}$ — r_o — Drain
Source

Figure 5.12 A simple model for the operation of an NMOS MOSFET in the saturation region.

an RC time constant that describes how quickly a voltage applied at the gate can induce charge in the channel.

By linearizing the output current in terms of the input voltage through transconductance, it is implied that the input voltage is not too large, such that the transistor operates in the "small-scale" or "small-signal" regime, when transconductance is a valid

5.5 Transistors and Transistor Models

and useful parameter. When a large input signal is applied, the transistor is said to operate in the "large-scale" or "large-signal" regime.

Equation (5.21), when rewritten, provides one of the simplest figures of merit for a transistor — the ratio of transconductance to drain current, which is an indication of the efficiency of the device [Enz02]. According to (5.21), this ratio falls proportionally to the inverse of the overdrive voltage. According to the EKV model, which will be discussed shortly (named for the initials of its inventors C. C. Enz, F. Krummenacher, and E. A. Vittoz [Enz08]), this ratio decreases as $1/\sqrt{I_D}$, where I_D in this case refers to the DC drain bias current [Enz02]. This indicates that, although increasing current can increase g_m, it also may result in a less efficient device. In proper low power designs (and in mmWave applications, anything that can reduce power is an advantage), efficient devices will have a low overdrive voltage and well-chosen bias currents.

MOSFET transistors are manufactured most commonly with complementary metal oxide semiconductor (CMOS) production processes. CMOS processes have many advantages, but perhaps most importantly, they are relatively cheap compared to other processes and are excellent for use in systems that require digital components. Most digital circuits, such as computer processors, are manufactured with CMOS processes, and a CMOS production process is usually optimized for digital circuitry before it is enhanced with special features required for analog circuitry such as amplifiers and mixers. These enhancements often involve the introduction of Back End of Line (BEOL) features. The BEOL encompasses the steps in the production process in which metal connections are added to the circuit, connecting transistors together. For example, most analog circuits benefit from the availability of thick metal layers as far from the substrate as possible, and these layers are usually made available after the process has been used in purely digital circuits for several months to more than a year.

In Fig. 5.10, the adjectives "high-k" and "low-k" refer to the dielectric constant (i.e., relative permittivity ϵ_r) of the dielectrics used in different portions of the chip. The dielectric below the gate often has a high relative permittivity to give the gate as much control over the channel as possible and to prevent the drain from competing with the gate for control over the channel (i.e., preventing the drain voltage from determining whether charges are available in the channel for conduction. The drain should only determine whether the charges in the channel flow as a current, while the gate voltage should determine the presence or lack of charge in the channel). The dielectric surrounding the metal layers used to actually connect different transistors together are often embedded in "low-k" or low relative permittivity dielectric to reduce capacitance between different metal layers (high capacitance can lead to such effects as *cross talk*, in which the voltage on one layer may determine the voltage on another layer) and to reduce capacitance between metal layers and the substrate (a high capacitance between metal and the substrate may result in substrate currents, which are very lossy and hurt circuit performance). Low-k dielectrics also help reduce capacitances on metal lines in a chip, reducing the power needed to drive signals through these lines at a given frequency.

As discussed earlier, the channel length L available to a circuit designer has been reduced at nearly periodic intervals over the last several decades, due to Moore's law. The gate length has shrunken remarkably and now can be made as small as 10-20 nm, thereby increasing the operating frequency of transistors up to several hundreds of gigahertz, while reducing the required voltage levels to on the order of 1 or 2 volts for substantial amplification. Modern MOSFETs owe their ability to operate for mmWave circuits to Moore's law, as this shrinking has made MOSFETs fast enough (i.e., capable of modulating the current in the channel in response to a gate voltage in a short enough time) to keep up with the speeds required for mmWave operation. But, this shrinking has also reduced the accuracy of basic models that describe MOSFET behavior. The shrinking gate length of CMOS processes is usually described in terms of gate length, and mmWave devices have typically relied on processes with gate lengths of 130 nm [LWL+09][DENB05], 90 nm [HBAN07b][HKN07a][BSS+10][KLP+10][SBB+08], or 65 nm [VKKH08][LKN+09][LGL+10]. Older processes such as 250 nm CMOS have been used on occasion to demonstrate mmWave operation [Wan01], though these older processes (also known as older "nodes") are generally not adequate to achieve required gain to meet link budge requirements at mmWave frequencies. Nodes at or below 45 nm are ideal for mmWave operation [SSM+10], but there is a limit to how small the gate length can be before power handling limitations become more of an impediment for analog operation than speed. The square law is the first device model of a MOSFET most students encounter when learning about transistors. Although it is still excellent for use in developing intuition and for making basic design decisions or estimates, it fails to predict device performance precisely for modern CMOS process nodes, such as 65 nm or smaller CMOS.

The price of greater device complexity and the resulting device model complexity needed to achieve mmWave operation is well worth the cost. To understand this quantitatively, consider two fundamental measures of how quickly a transistor may operate: its *transit frequency* f_T and its *maximum frequency of operation* $f_{\max}$. f_T is defined as the operating frequency for which a transistor's intrinsic current gain falls to zero, and $f_{\max}$ is defined as the frequency at which the power gain (the product of voltage gain and current gain) falls to zero. As shown in Fig. 1.6, the transit frequencies of modern MOSFETs reach well above 100 GHz. The simple MOSFET model of Fig. 5.12 estimates the transit frequency of a common-source MOSFET and its $f_{\max}$, as shown in Fig. 5.13.

By setting the current gain equal to 1 and solving for frequency, one easily finds the transit frequency of the device to be equal to

$$f_T = \frac{g_m}{2\pi C_{gs}}. \tag{5.22}$$

By setting the power gain equal to 1 and solving for frequency, the maximum frequency of operation, $f_{\max}$, is found to be equal to

$$f_{\max} = \frac{g_m}{2\pi C_{gs}} \sqrt{\frac{r_0}{R_g}}. \tag{5.23}$$

Equation (5.22) is the most commonly used form of the transit frequency of the device, and in general, most publications focus on the transit frequency. It is important to note that the power gain of the device is proportional to the ratio of the transit frequency to the frequency of operation squared:

$$|A_s| = \left(\frac{f_T}{f}\right)^2 \frac{r_0 C_{gs}}{\sqrt{\left(\frac{1}{2\pi f}\right)^2 + (R_g C_{gs})^2}} \tag{5.24}$$

where A_s is the power gain of the device, and f is the frequency of operation.

Find current gain: I_{in}, R_g, C_{gs}, $+V_{gs}$, $g_m V_{gs}$, r_o, I_{out}

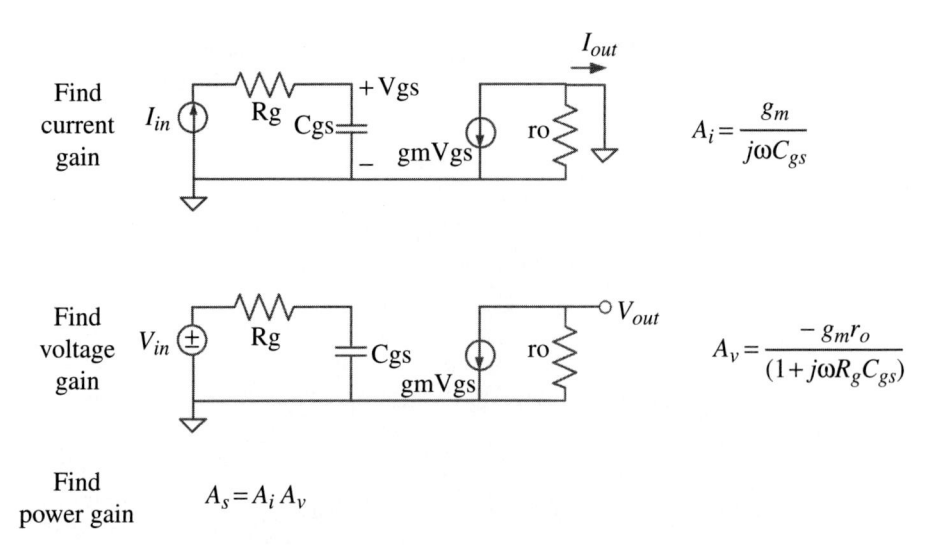

$$A_i = \frac{g_m}{j\omega C_{gs}}$$

Find voltage gain: V_{in}, R_g, C_{gs}, $g_m V_{gs}$, r_o, V_{out}

$$A_v = \frac{-g_m r_o}{(1 + j\omega R_g C_{gs})}$$

Find power gain

$$A_s = A_i A_v$$

Figure 5.13 An analysis of the simple model of Fig. 5.12 allows us to compute the current, voltage, and power gains, A_i, A_v, and A_s, respectively.

One of the major advantages to the use of CMOS over other technologies is its portability across technology nodes, which is made possible by constant field scaling rules [YGT+07]. These scaling rules have resulted in peak f_T, $f_{\max}$, and optimum noise figure current densities of 0.3-0.35 mA/μm, 0.2 mA/μm, and 0.15 mA/μm for a cascode nMOSFET, respectively [YGT+07][DLB+05]. The relative proximity of these densities, and their insensitivity to threshold voltage and gate length, are huge advantages for analog design [YGT+07].

In contrast to CMOS, BJTs and HBTs do not offer the same level of technology-node independent current density performance optimization [DLB+05]. Also, bipolar designs will usually require high supply voltages (which can be an advantage in certain cases, for example, for dynamic range or high output power) [DLB+05]. BJTs offer higher transit frequencies than MOSFETs, but are not nearly as easy to integrate with digital circuits, such as required for a mmWave system-on-chip (SoC) or for integrating many analog components onto a single chip [DLB+05].

5.6 More Advanced Models for MmWave Transistors

Fig. 5.14 illustrates a basic transistor. There are many excellent texts that discuss layout techniques for transistors and circuits (e.g., [SS01], which is also a very enjoyable read), and Fig. 5.14 is only meant to illustrate that most radio frequency (RF) mmWave devices have multiple fingers. A "finger" is a portion of the device width W, and the device's overall width is reached by adding together multiple fingers (i.e., many individual transistors are built in parallel across the junction to form a single transistor), with more fingers in parallel offering smaller resistance, and thus greater current-carrying capacity.

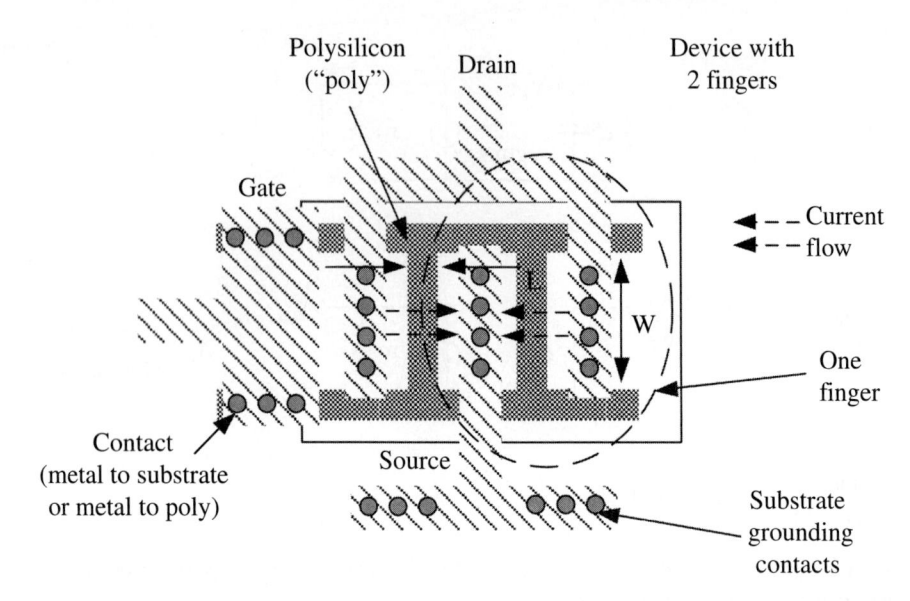

Figure 5.14 Most devices used in mmWave circuits will have multiple fingers. Generally, a designer will select a best design for a single finger, and then vary a transistor's width by varying the number of fingers in the device.

The fact that most RF devices have multiple fingers has an important impact on the model used to describe a MOSFET.

More accurate device models must accurately reflect the effects of all physical phenomena important to the device's operating regime. The layout of most RF mmWave transistors, which are multi-finger with large aspect ratios (W/L) and minimum gate lengths, requires models to give accurate values for gate resistance and short-channel effects [SSDV+08][Enz02]. Operation closer to the f_T of devices requires that device models transition away from quasi-static models in which all changes in the channel current are assumed to follow instantly a change in gate voltage [Yos07], and that DC non-linear characteristics can account for non-linear effects at higher frequencies (i.e., that a DC sweep is adequate to capture non-linear performance at higher frequencies) [EDNB04]. The models must also accurately capture the non-linear, large-signal characteristics of devices, especially for use in such non-linear components as power amplifiers and voltage-controlled oscillators (VCOs) [EDNB04][Poz05, Chapter 11]. Also important are the parasitic capacitances that hinder operation, in addition to the finite time required for the channel to respond to changes in drain or gate voltage (also known as *non-quasi static effects*). A substrate network may also be added to account for the fact that substantial energy may be dissipated in the substrate of the device, and the voltages on the four nodes of the device (source, drain, gate, and body) may induce charge on the other nodes through capacitances in the substrate. An accurate DC model that captures effects in the region of operation (either strong inversion, weak inversion, etc.) is important to predict such performance metrics as power requirements [Enz02]. We will now briefly consider gate resistance, short-channel effects, operation close to

f_T, large-signal operation, parasitic capacitances, non-quasi static effects, and substrate networks.

The layout of a device has a critical impact on the gate resistance, which (5.23) indicates is an important parameter as a large gate resistance severely limits the power gain that can be achieved with a device. The gate resistance is given approximately by [SSDV+08][Lit01]

$$R_g = \frac{1}{N_f}\left(\frac{\frac{\rho_{\text{sil}_{\text{sq}}}}{3N_{\text{gate con}}^2}W_f}{L} + \frac{(\rho_{\text{con}} + \rho_{\text{poly}})}{W_f L}\right) \qquad (5.25)$$

where N_f is the number of fingers, W_f is the width of each finger, L is the gate length, $N_{\text{gate con}}$ is the number of gate contacts (as indicated in Fig. 5.14), $\rho_{\text{sil}_{\text{sq}}}$ is the sheet resistance of the silicide (approximately 4 Ω/sq [Lit01]), which is presumed to coat the gate polysilicon, ρ_{con} is the contact resistance between the silicide and the polysilicon (approximately 7.5e-5 $\Omega \cdot m^2$), and ρ_{poly} is the vertical polysilicon resistance (i.e., the resistance seen by a charge as it moves vertically down through the gate, $\rho_{\text{poly}} \approx R_{\text{sh}_{\text{poly}}} \cdot \min(\delta^2, t_{\text{poly}}^2)$, where $R_{\text{sh}_{\text{poly}}} \approx 150\frac{\Omega}{sq}$, $\delta = \sqrt{\frac{1}{\pi f \mu_0 \sigma_{\text{poly}}}}$ where $\sigma_{\text{poly}} \approx 10^4 \Omega^{-1}/cm$ [SN07] is the skin depth of the polysilicon, and t_{poly} is the thickness of the polysilicon). These concepts are illustrated in Fig. 5.15. The most important lesson from this discussion of gate resistance is that the gate resistance plays a major role in degrading the accuracy of such simple models as the square law.

One common modeling technique to improve the model of mmWave devices is to add gate resistance in the simulation model. This is an easy way to modify the traditional MOSFET transistor model to account for high frequency effects. The addition of a gate resistance, in series with the well-known C_{gs}, adds an RC time constant that practically accounts for delays and other effects [EDNB04][WSC+10].

Gate resistance is one of the most important parameters to minimize for high frequency designs in which high $f_{\max}$ is desired [HBAN07b]. In addition to the gate resistance, there are also resistances associated with each of the contacts, including resistances to the gate and source, and these often are the dominant resistances in the extrinsic portion of the device (contrast an intrinsic resistance, such as the channel resistance, associated with the portions of the device that are essential to its operation) [Enz02]. Reducing the contact resistance is desirable at mmWave frequencies (and at any other frequency). The parasitic extrinsic gate resistance should be engineered by correctly selecting the number of fingers and gate contacts [SSDV+08][Enz02] to obtain the resistance needed to match the gate to the driving circuit. Unfortunately, parasitic resistance values will increase with device scaling (i.e., with new process generations) to some extent regardless of operating frequency due to thinner line widths, closer spacing between devices, and reduced distance between interconnects/signal lines [JGA+05]. Fortunately, once a new process has been optimized for analog use (after initial deployment for digital use), some of these resistances will be reduced by enhancements in the Back End Of Line (BEOL, the production steps in which metal traces are added) added for analog development.

Poor device layout can have a major impact on device performance by increasing parasitic resistances and capacitances in addition to the gate resistance. Such effects

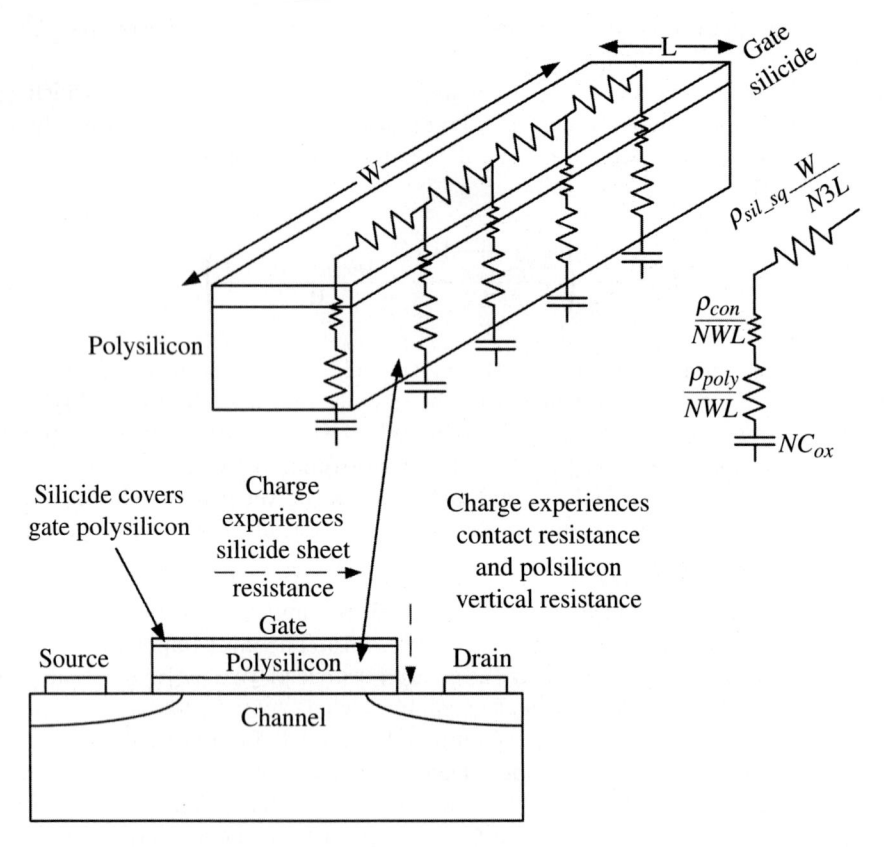

Figure 5.15 The gate resistance depends critically on the device layout, and in particular on the number of fingers. It also depends on the nature of the materials used to construct the gate, including the silicide resistance, silicide-polysilicon contact resistance, and vertical polysilicon resistance. These topics are explored in [Lit01].

could often be ignored at lower frequencies due to the electrical size of components, but at mmWave frequencies, the idiosyncrasies of a layout, including improper interconnect placement, can have a major effect and must be included in a model [HBAN07b]. The best way to do this is to use a software-based parasitic extraction tool (provided by such companies as Cadence) to estimate the values of parasitic resistances and capacitances caused by a particular layout. After a test device is manufactured, a more refined model can then be generated through careful measurement and comparison with the initial parasitic model. An implication of the importance of layout at mmWave frequencies is that layout effects are not only a possible cause of failure, but also a tool in the hands of a circuit designer, unlike at lower frequencies in which the designer could generally ignore these effects [HBAN07b]. Incorporation of these effects into a model can also improve the designer's physical intuition and understanding of a process. For example, the model developed by Heydari et al. in 2007 [HBAN07b] for a common source MOSFET very neatly informs the designer of the effects of the layout of the devices. Note that for

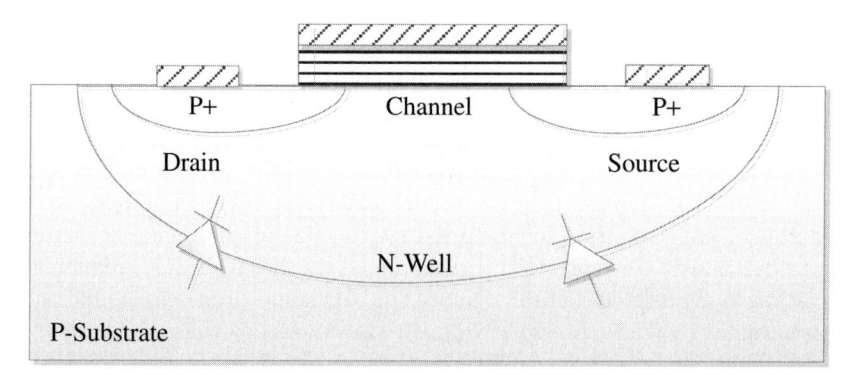

5.6 More Advanced Models for MmWave Transistors

Figure 5.16 There are many parasitic diodes that exist in a device, and substrate contacts are usually used to prevent these from becoming destructive. Unfortunately, the substrate contacts used for this purpose can increase parasitic capacitances to the substrate.

oscillators it is very important to accurately account for parasitics because parasitics will substantially impact the frequencies at which an oscillator can operate.

Another source of parasitic resistances comes from the substrate contacts that must be used to ground the substrate near the device. Substrate contacts for devices are necessary, for example, to prevent parasitic diodes, as illustrated in Fig. 5.16 for a PMOS, from becoming forward biased and thus destructive [SS01]. A *well tie*, also known as a *body tie*, or *substrate tie*, is a heavily doped (n+ or p+) portion of silicon sitting in a larger well or substrate of the same type conductivity, but doped at a lower level (e.g., p+ in p-well, or n+ in an n-well). The purpose of the well tie is to supply a solid, well-defined voltage to the well or substrate body, or to make a metal interconnection to the semiconductor. In physical verification of transistors, the term *well tie* is used to describe an (artificial) via connecting active (diffusion region) with the "well" ("substrate"). The well tie provides Ohmic contact between metal interconnect lines and lower doped semiconductor. A physical contact area between the well tie and the metal contact (going from the lowest metal layer to silicon) creates an Ohmic contact, so that the voltage in metal is equal to the voltage in semiconductor. With increased substrate ties, parasitic capacitances will increase [JGA+05], indicating that a compromise must be reached between device safety and speed.

In general, symmetrical layouts are always best for matching and performance [SS01]. The increased importance of parasitic components at mmWave frequencies and above also makes modular layouts advantageous, as the parasitics of each module can be formulated and minimized in advance before inclusion in the overall circuit [HBAN07b][JGA+05]. For example, Heydari et al. were able to use such an approach to achieve an $f_{\max}$ near 300 GHz in a 90 nm CMOS process [HBAN07b]. In the event that vias are used to contact any portion of the device, this also facilitates selecting the correct number of vias per modular unit.

Short-channel effects refer to the changes in device parameters that occur for devices with very short gate lengths ("deeply submicron," meaning gate lengths much shorter than $1\mu m$). Two of the chief effects of very short gate lengths are reduction in the threshold voltage V_t, also known as the *short channel effect*, so that the device is more difficult to "turn off" (as even small gate voltages may result in charge in the channel), and poor isolation between the drain voltage and the channel (i.e., when the drain voltage, rather than the gate voltage, determines whether charge is present in the channel) [SN07]. This phenomenon is often called *Drain Induced Barrier Lowering* (DIBL), and process development often focuses on achieving the correct doping profile (i.e., doping concentration level) for the transistor channel to prevent such effects as DIBL. The basic approach is to increase doping concentration levels to reduce the size of the region of channel semiconductor whose charge concentration is determined by the drain (i.e., reducing the size of the depletion layer near the drain). Note that, for some processes, reducing channel length may actually increase threshold voltage, also known as the *reverse short channel effect*.

Device operation close to the transit frequency is important because the transit frequency defines the regions of operation over which a MOSFET can achieve meaningful gain and often indicates what kind of parasitic components (i.e., unwanted capacitances) act on the device.

Large-scale operation refers to the fact that a linear model, based on such parameters as transconductance, break down when the input signal may vary greatly in amplitude. The basic transconductance model for transistors operating between small-signal and large-signal regimes assumes that there are only three basic modes of operation: linear (aka triode), saturation, and cutoff (when the gate voltage is insufficient to fill the channel with charge). The simple model also assumes that the mode of operation depends only on the gate-to-source voltage relative to the threshold voltage V_t, and the drain-to-source voltage relative to $V_{gs} - V_t$. The two most important capacitances that must be included explicitly in the transconductance model are gate-source capacitance and gate-drain capacitance [EDNB04][HKN07a]. Certain older models did not explicitly account for these capacitances as they did not allow for the addition of poles in the transfer function of the device due to these capacitances, but only accounted for gain reduction caused by these capacitances.

Along with the gate-source capacitance, the gate-drain capacitance should be kept low by keeping portions of the gate electrode not over an active device to a minimum [Enz02]. The gate-source capacitance increases with the number of fingers [JGA+05], while the gate resistance decreases, so a compromise should be reached for desired gate resistance and gate-source capacitance (increasing C_{gs} directly reduces transit frequency, while increasing gate resistance decreases maximum frequency of oscillation). The gate-drain capacitances is also more important at mmWave frequencies than at other frequencies [HKN07a] and may be minimized, for example, through resonance techniques (i.e., use of inductors).

The substrate network accounts for loss caused by currents excited in the substrate through capacitive coupling with the device. Enz's extension of the EKV model for RF operation included a simple network of three resistors in a pi-network [Enz02]. Emami et al. also used a three resistor model arranged in a T-network [EDNB04]. Certain authors chose to forego a complex network in favor of a single or double resistor network to represent loss in the substrate [WSC+10][HKN07a]. Any resistance between the source and bulk (i.e., substrate) will scale inversely with finger width and number of contacts [Enz02].

Non-quasi static effects are due to the finite time required for a voltage or current at one node in a circuit model to result in a change at another point in the model. The gate resistance may be adequate to account for non-quasi static effects as it adds an RC-time constant to how quickly the channel may respond to a change in the gate voltage. Certain authors will more explicitly account for non-instantaneous response of the channel by adding a phase delay factor $e^{-j\omega\tau}$ to their expression for channel transconductance [WSC+10].

More accurate and advanced models such as the EKV model and BSIM model are generally used today, as they are far superior to the simple square-law transistor model for the case of mmWave designs that require the use of deeply scaled technologies operating at very high frequencies with low supply voltages. In general, the EKV model is more accurate than the BSIM model. Most models developed for RF and mmWave devices contain an intrinsic portion that accounts for the operation of the device and an extrinsic portion that accounts for the effect of external components and interconnects on the device (e.g., inductors and resistors in series used to model connections to the outside world [EDNB04]) [JGA+05]. For example, C. Enz developed such a model based on the EKV model for RF design [Enz02]. Fig. 5.17 illustrates some effects that must be included. The best model will most accurately account for all of the effects described in this section, will accurately predict the current flowing through each of the terminals of the device (gate, body, source, and drain), and will provide a limited set of parameters that a designer may use to design her or his circuit.

Fig. 5.17 indicates that noise sources should also be included in more advanced models. These are especially useful for calculating signal-to-noise ratios (SNRs) out of the device. There are many sources of noise in RF MOSFETs, including channel thermal noise, flicker noise, gate-induced noise and gate-resistance noise, and noise due to contact and substrate resistances, of which the channel thermal noise is the most important [Enz02][HKN07a]. Gate thermal noise is also very significant [WSC+10], though less so than channel thermal noise [WSC+10]. A figure of merit that is useful in comparing noise performance of different devices is the power spectral density of channel noise [WSC+10]. The channel noise can be approximated as [WSC+10]

$$S_{id} = \frac{4k_B T I_D}{V_{D,\text{sat}}} \qquad (5.26)$$

where I_D is the drain current, k_B is Boltzman's constant, T is ambient temperature, and $V_{D,\text{sat}}$ is the drain saturation voltage [WSC+10].

In 2010, Wang et al. [WSC+10] studied the noise performance of 65 nm CMOS for mmWave applications. They obtained values for minimum noise figure and equivalent noise resistance as well as values for the power spectral density of drain, gate, and imaginary cross correlation between gate and drain current noise sources [WSC+10]. They found that drain current noise power is independent of frequency and increases with a shrinking gate length, with increasing drain current, and inversely with decreasing drain saturation voltage [WSC+10]. Jagannathan et al. also found that channel noise increases with decreased gate length [JGA+05]. They found that gate noise power increases as the square of frequency and decreases with shrinking gate length [WSC+10]. The imaginary cross correlation between gate and drain noise increased proportionally to frequency

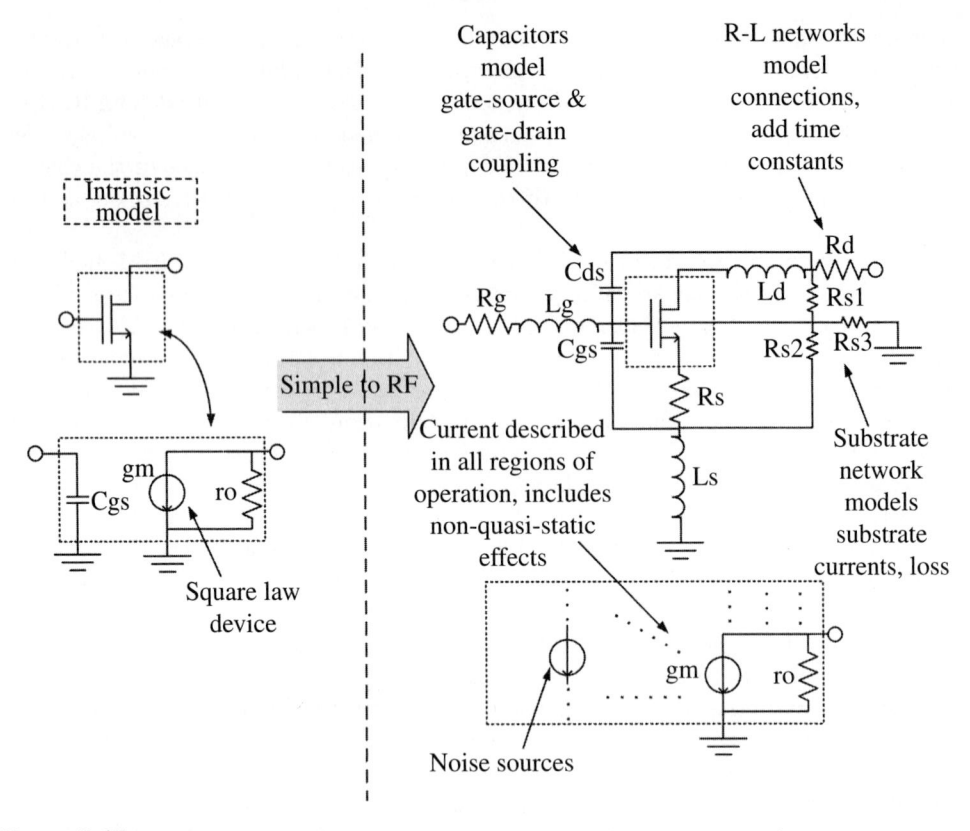

Figure 5.17 Depiction of key effects that must be considered in a more complex yet more accurate mmWave device model, based on the work of [Enz02].

[WSC+10]. In 2008, Varonen et al. [VKKH08] presented noise data for a W/L = 90/0.07 V-band transistor. They measured the transistor using a co-planar open and shorted waveguides. They found a maximum stable gain for the transistor between 9.75 and 8.25 dB from 50 to 75 GHz for a current draw of 18 mA. The equivalent noise resistance of the resistor was between 20 and 30 ohms, and the noise figure ranged between roughly 2 and 4 dB over the same frequency range.

5.6.1 BSIM Model

The various versions of the BSIM model (BSIM 1-4) [Kay08, Chapter 2] are perhaps the most well-known models and are more advanced than the basic square law, threshold-voltage model presented above. The BSIM model takes into account some of the effects just described, but in general is not as accurate as more advanced models such as the EKV model. The BSIM3 model has been used as an industry standard for lower frequency analog designs [Kay08, Chapter 2]. The BSIM3 model is a threshold-voltage based model (just as the simple model described above is) that accounts for such effects as non-uniform doping effects, charge sharing and DIBL, the reverse short channel effect, the normal narrow width effect, the reverse narrow width effect, the body effect, subthreshold conduction, field dependent mobility, velocity saturation, channel length modulation, impact ionization substrate current, gate-induced drain leakage, polysilicon gate depletion, inversion layer quantization effect, velocity overshoot, and the self-heating effect [CH02]. Discussing all of these is beyond the scope of this text but is foundational to the characterization and modeling of mmWave transistors.

Although the BSIM model accurately accounts for many of the mal-effects that result from transistor scaling, it is hampered by the fact that such concepts as the threshold voltage V_t are not physically based (i.e., it is not an intrinsic parameter of the device), but are simply empirically descriptive (e.g., like saying the sun rises every day without explaining why). As the BSIM models have progressed to account for ever more effects caused by device scaling and high frequency operation, they have had to add many fitting terms and parameters to maintain their accuracy. The BSIM1 and BSIM2 models required 60 and 90 DC fitting parameters, respectively [Kay08]. The BSIM3 model is more physically descriptive (based in physics) as opposed to empirically descriptive, but it still requires approximately 40 fitting parameters [Kay08].

One important feature of the BSIM3 model that is not attained by the BSIM1 or BSIM2 models is that it has a single expression for the drain current that is applicable over all regions of operation [Kay08][CH02]. This is in contrast to the simple model presented earlier, in which the user must know, a priori, the region of operation in order to apply the correct equation for the drain current. The difficulty with such a piece-wise approach is that it can result in discontinuities and points of non-differentiability [Kay08].

5.6.2 MmWave Transistor Model Evolution — EKV Model

The requirements of RF-modeling and the need for improved device performance prediction have resulted in substantial evolution of small- and large-signal models in recent years. As early as 2004, much effort was focused on modifying the BSIM3 model to account for non-quasi static effects [EDNB04]. These approaches focused on either look-up table approaches or physical approaches in which circuit elements were mapped to physical phenomena [EDNB04]. Later iterations of the BSIM model, including BSIM4, still did not accurately capture all the high frequency effects or operation for MOSFETs in weak-inversion [Yos07] (i.e., a gate-source voltage slightly less than the threshold voltage). Incorporation of such theories as charge-based modeling with transcapacitances (EKV model) and channel surface potential have since displaced some of these models [EV06][SSDV+08][Yos07]. Charge-based modeling refers to the way in which the EKV model explicitly accounts for charge in the device. This is in contrast to the empirical BSIM models and the simple model first described, in which charge formed the motivation for our description but was not explicitly accounted for (rather, the simple model subsumes the presence or lack of charge in the channel into the empirical parameter V_t). A transcapacitance is a voltage-dependent charge source (i.e., in which a voltage at one point of a circuit model results in a charge at another point in the model). A channel surface approach refers to how the EKV model explicitly accounts for and depends on the surface potential of the channel, which is the voltage at the surface just below the gate dielectric in Fig. 5.10. These newer models more accurately account for high frequency, non-quasi static, weak-inversion, conductance, and small-size effects [Yos07]. Unfortunately, the frequency at which non-quasi-static effects become important is bias-dependent and increases as the square of the drain current in forward strong saturation [Enz02] (forward saturation refers to the fact that the device is forward biased, that is, a sufficiently great gate bias is applied to result in strong inversion).

The shrinking of gate lengths with new technology nodes has resulted in a dramatic reduction of power supply voltages for newer CMOS technology generations (e.g., for 180 nm CMOS, a common supply voltage was 1.8 V; and for 65 nm CMOS, a more common figure is 1 V [Lee10] to avoid breakdown). However, in addition to avoiding breakdown of the transistor circuits, the lowered supply voltage has also led to increased operation in the weak inversion region of MOSFETs [Yos07]. New models that may displace or augment the BSIM models, including the EKV model, better predict behavior in this region.

The EKV model is based on relationships that govern the amount of free charge in a MOSFET model as opposed to simply specifying basic regions of operation (e.g., below threshold, above threshold). It more accurately captures intrinsic high frequency, non-quasi static, and low-supply voltage effects than the BSIM models [Yos07][Enz08]. It has the advantage of being grounded in physics, so it gives insight to the designer [Enz02]. In addition, it results in a coherent description from DC to RF [Enz02]. It is similar in some respect to the PSP model in that it accounts for inversion charge in terms of the surface potential of the channel [Enz08]. It also more accurately models device noise and dynamic behavior through improved representation of source-drain symmetry [Enz08].

The reader should note that modeling MOSFETs is a very deep field; we have only barely scratched its surface. The reader is referred to any number of excellent texts for a more thorough discussion of modeling MOSFET devices [SN07][Kay08][CH02]. We now move on to discuss passive devices, as they are also critical to mmWave circuits.

5.7 Introduction to Transmission Lines and Passives

Transmission lines and passives are as fundamental to the operation of an integrated circuit as the transistors used to construct the amplifiers, mixers, oscillators, and other analog circuit blocks in a mmWave device. Generally, the key to a good circuit design is to implement passive devices for lowest loss and best impedance matching. The *quality factor* (Q) of a circuit, often defined as the operating center carrier frequency divided by the 3-dB passband region or bandwidth of the circuit, will directly determine its loss characteristics, and in general high Q values should be used for lowest loss (another definition of Q relates lost to stored energy, and is given subsequently). It is worth noting that higher Q values will also result in circuits that take longer to start up, so the highest Q may not be advantageous if a very fast startup time is desired in a circuit. The quality factor Q also determines bandwidth of passive filters, loss in matching circuits, and phase stability of oscillators [O98][BBKH06]. Passive components are important in the context of interconnections and matching between circuit blocks, for phase or signal distribution systems, and in phase-locking systems used for phased arrays. The values of

5.7 Introduction to Transmission Lines and Passives

Q for individual components and complete circuits determine the phase difference and locking range of coupled oscillators. The passive components and their Q values are also the primary determinants of the phase noise of many systems.

Accurate models that account for high frequency effects are essential to the success of RF or mmWave circuits [Yos07]. In the short run, researchers and designers who rely on CMOS or SiGe technologies will need to create their own models for passive devices such as transmission lines and inductors, as most foundries currently do not supply models for such structures beyond 20 GHz [BKKL08]. This is rapidly improving, however, as new low-power, low-loss RF processes are being developed for 60 GHz products. Commercial success of silicon mmWave devices will continue to remove this need in the near future. A general approach for model creation is to fit a mathematical and circuit description to a set of measured S-parameters (or converted S-parameters). Most resulting models of transmission lines and inductors take the form of pi-, double-pi-, or t-circuits as shown in Figs. 5.4 and 5.5 [BKKL08][CGH+03][HKN07b][HLJ+06]. The shunt branch of these models usually accounts for substrate effects [BKKL08], and the series branches account for conductor effects. When developing a model, the choice arises between basing the model on physical phenomena or on observed data (an empirical model). The latter may be more accurate but may also provide little physical insight. Also, non-physical empirical models can yield behavior that is very non-physical when they are extrapolated beyond their intended use range, for example, by predicting negative inductances [BKKL08]. Whenever possible, a model should be as scalable with frequency and dimensions as possible. This often favors the use of per-unit-length parameters [BKKL08].

When constructing a model for a passive device, it is essential that the model have the properties of stability, causality, and passivity [TGTN+07]. Poor measurement techniques or model fitting can both account for violations of these conditions. Subtle problems can arise that make a model inapplicable to a physical system due to non-causalities if measurements fail to satisfy these requirements. If a detailed analysis of the mathematical properties of a model is not possible (e.g., due to time constraints), then at a minimum all circuit components used in a model should be kept passive, which is enough to ensure causality and stability [TGTN+07].

An important consideration for model creation is to include sufficient de-embedding structures in test-chip layouts, to prevent models of passive devices from being corrupted for example, by pad capacitances [RGAA09]. De-embedding structures allow the circuit designer to indirectly measure S-parameters of needed circuit interconnect items, such as bond pads or feedlines, so that the impact of such connecting structures may be factored out and "de-embedded" in circuit simulation or measurement, allowing one to ascertain the true characteristics of a device or circuit of interest. Some research has focused on developing de-embedding techniques that allow the measured parameters of one structure to be used for de-embedding other structures [BKKL08].

For inductive components, it is possible to use both transmission lines and passive spiral inductors. An important consideration when selecting which to use is the space required to implement a given solution. Inductors may be designed to offer space savings over transmission lines [YGT+07][DLB+05]. But, transmission lines offer more freedom in routing than inductors [PMR+08], which makes the choice between inductors and transmission lines a non-trivial one. Highly integrated devices that require extensive matching networks may want to approach selection from the standpoint of a tiling problem, in which the goal is to make best utilization of space on the chip.

Another important consideration for any passive device is its self-resonant frequency — the frequency at which the device naturally resonates. Fig. 5.18 illustrates the concept of self-resonant frequency. On a Smith chart of the device's S-parameters, this frequency would correspond to the point at which S11 is purely real — or equivalently when the device appears to be a resistor. This is an important point, especially for mmWave devices (and devices operating at higher frequencies), due to the wide range of frequencies present in a single device (e.g., DC for power, RF for output or input, IF or LO at other ports). Coupling capacitors used for power supply purposes, for example, may have self-resonant frequencies below the RF frequency [HBAN07b], so that at the RF frequency, the coupling capacitors will appear to be inductive. A smart designer takes into account this effect and uses it to his or her advantage for RF performance. Fig. 5.18 emphasizes this point. One very important implication drawn from this figure is that device models must be optimized over a very wide frequency range (i.e., from DC to daylight), so that proper operation of the circuit design may be determined over all possible frequencies.

When performing simulations of passive components, it is important to consider loss effects that result from the finite conductivity of metals used. More generally, in addition to topology, the material used to construct passive elements such as transmission lines can have a major impact on quality factor and insertion loss and other performance metrics [Emr07]. Simulations, for example, in such programs as HFSS, should not be conducted at the level of perfect electric conductor analysis, as the use of real metals can increase insertion loss by as much as 15 dB versus idealized *perfect electrical conductors* (PECs) [Emr07]. Most modern production processes offer multiple metal layers in their BEOL

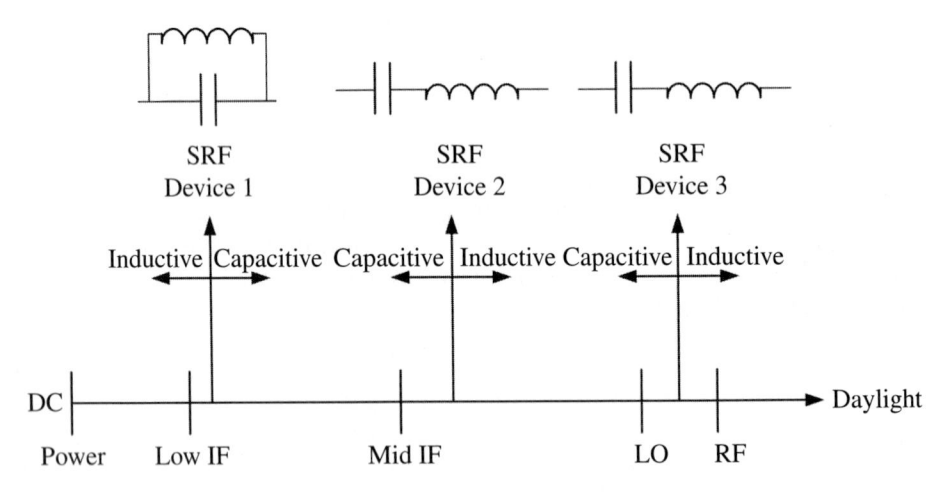

Figure 5.18 As carrier frequencies increase, the range of frequencies that may be encountered in any one device will also increase. This implies that designers will have to consider the frequency-dependent nature of all passive devices, especially where the device is operating relative to its self-resonant frequency.

5.7 Introduction to Transmission Lines and Passives

(e.g., 180 nm TSMC CMOS offers 6 metal layers). The metal layer selected to perform a particular function should be decided by the characteristics of each layer. If thick metal layers or higher conductivity layers are offered (e.g., in STM BiCMOS9 [TDS+09]), then these should be used when low-loss is required (e.g., To et al.'s use of thick top metal for matching in LNA design [TWS+07]). For example, STM's BiCMOS9 130 nm process offers three thick top metal layers that offer very low attenuation constants of 0.5 dB/mm for 50 Ω lines, nearly as good as on a printed circuit board [TDS+09]. High metal layers should be used when a lower capacitance or higher impedance is required [LKBB09].

As mentioned earlier, the quality factor, Q, is an important parameter and was also considered in Chapter 4 in the context of antennas. For passive components such as transmission lines and inductors, the quality factor is defined as the ratio of energy stored to energy lost in the device in one cycle of the carrier frequency. For passive components, it is common to model devices using lumped components including ideal capacitors, inductors, and resistors, even though such components are often distributed in chip design. For a given lumped model of a device, it is possible to calculate the quality factor. Two possible lumped models are a series reactance X and resistance R, or a parallel combination of reactance X and resistance R, as illustrated in Fig. 5.19. For these two simple models, it is very easy to derive the quality factor:

$$Q_{\text{series}} = \frac{P_{\text{stored}}}{P_{\text{lost}}} = \frac{\{\frac{1}{2}I^2 X\}}{\{\frac{1}{2}I^2 R\}} = \frac{X}{R} \tag{5.27}$$

$$Q_{\text{parallel}} = \frac{P_{\text{stored}}}{P_{\text{lost}}} = \frac{\{\frac{V^2}{2X}\}}{\{\frac{V^2}{2R}\}} = \frac{R}{X} \tag{5.28}$$

where I is the current through the series circuit, and V is the voltage across the parallel circuit.

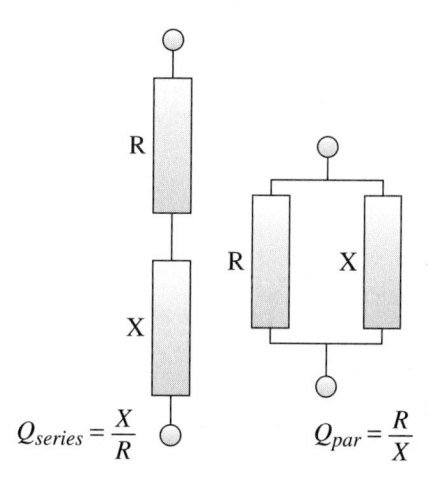

Figure 5.19 Series and parallel resonant circuits and their Qs.

5.7.1 Transmission Lines

Transmission lines are necessary to carry signals and to act as reactive components in resonators, and they can serve as transformers [MN08]. Though larger than inductors, they are easier and faster to design and so can improve first-pass design success rate of inductive reactive components [TWS+07]. One of the most important factors in transmission line design is the quality factor of the line [HKN07b].

To understand transmission lines, it is useful to understand that every transmission line is characterized by two key parameters: its *characteristic impedance* and its *propagation constant*. The characteristic impedance, Z_0, of a transmission line determines the ratio of voltage to current on the line, and is an important parameter for determining whether the line is impedance matched to other circuits. The propagation constant, α, determines the loss on the line and also how long the line needs to be to achieve a given input impedance. These two key parameters are functions of the semiconductor properties and the physical size of the transmission line itself.

A basic transmission line is illustrated in Fig. 5.20, and its properties arise as a result of the nature of the infinitesimal cells that are cascaded together to form the line. In Fig. 5.20, L and C are the inductance and capacitance of the line, respectively. The basic equations that relate the voltage and current on the line to the position on the line ℓ, and the propagation constant and characteristic impedance to the inductance and capacitance are

$$I(\ell) = \frac{V^+}{Z_0} e^{-j\omega\sqrt{LC}\ell} - \frac{V^-}{Z_0} e^{j\omega\sqrt{LC}\ell} \tag{5.29}$$

$$V(\ell) = V^+ e^{-j\omega\sqrt{LC}\ell} + V^- e^{j\omega\sqrt{LC}\ell} \tag{5.30}$$

$$Z_0 = \sqrt{\frac{L}{C}} \tag{5.31}$$

$$\alpha = \omega\sqrt{LC} \tag{5.32}$$

where V^+ and V^- are two parameters that determine the values of the forward and backward propagating voltage waves, respectively. Two more relations that prove useful

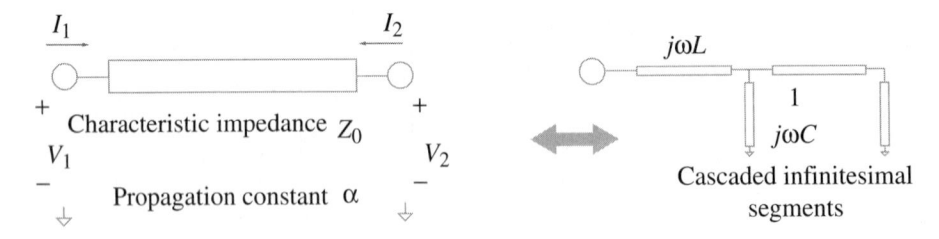

Figure 5.20 A transmission line has two basic parameters: its characteristic impedance and its propagation constant.

are the input impedance seen looking into a line for a given load and the matrix relating the voltages and currents and the two ends of the line [Poz05]

$$Z_{\text{in}} = \frac{Z_0(Z_L + jZ_0 \tan(\alpha\ell))}{Z_0 + jZ_l \tan(\alpha\ell)} \quad (5.33)$$

$$\begin{bmatrix} V_1 \\ V_2 \end{bmatrix} = \begin{bmatrix} -jZ_0 \cos(\alpha\ell) & Z_0 \tan(\alpha\ell) \\ Z_0 \tan(\alpha\ell) & -jZ_0 \cos(\alpha\ell) \end{bmatrix} \begin{bmatrix} I_1 \\ I_2 \end{bmatrix}. \quad (5.34)$$

The ABCD-parameters for a transmission line are given by [GJRM10]

$$\begin{bmatrix} V_1 \\ I_1 \end{bmatrix} = \begin{bmatrix} \cosh(\alpha\ell) & Z_0 \sinh(\alpha\ell) \\ \sinh(\alpha\ell)/Z_0 & \cosh(\alpha\ell) \end{bmatrix} \begin{bmatrix} V_2 \\ -I_2 \end{bmatrix}. \quad (5.35)$$

In general, the topology of a transmission line can have a large impact on its characteristics and performance. For example, some authors have experimented with tapered designs in order to obtain improved transmission line quality factor for use in resonators [MN08]. Two common topologies with standardized design formulas are microstrip and co-planar waveguide (CPW) topologies as illustrated in Fig. 5.21 from [RMGJ11]. In general, CPW transmission lines offer higher inductive quality factors than microstrip lines and are thus often preferable in cases in which high inductive quality factors are needed (e.g., in amplifier design) [DENB05]. Due to their popularity in mmWave 60 GHz designs, the CPW lines are presented here.

There are many considerations when constructing transmission lines, and one of the most important is the ground current return path. Transmission lines are advantageous as their ground return paths are explicitly defined. The ground return path should be engineered to minimize the extent to which ground return currents induce lossy currents in the substrate — for example, by keeping the ground as far from the substrate as possible. For example, when Pfeiffer et al. designed transmission lines for their 100-120 GHz Schottky diode upconverter, they specifically considered the ground return paths in all their transmission lines [PMR+08].

Transmission line topologies for mmWave devices vary substantially — for example, there are various "flavors" of co-planar transmission lines. Two popular co-planar designs

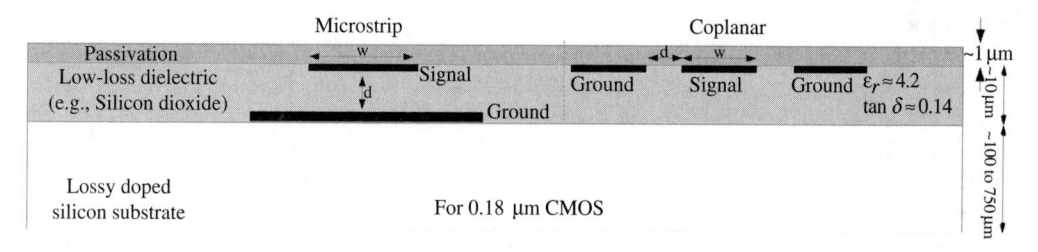

Figure 5.21 Two common transmission line topologies are microstrip and co-planar. An advantage of the co-planar design is that it enables contacts to ground at the same level as signal contacts [from [RMGJ11] © IEEE].

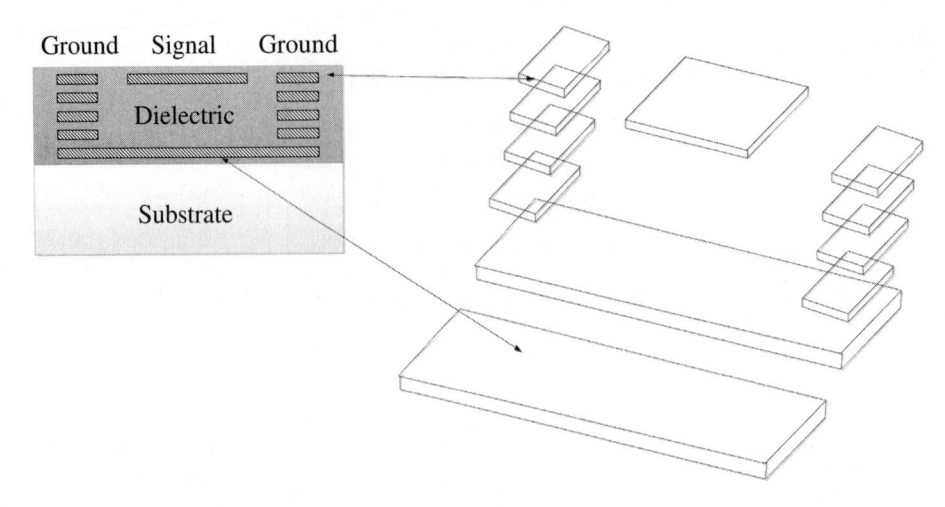

Figure 5.22 Shielded transmission lines prevent currents from being induced in the substrate.

are shielded and non-shielded lines [VKKH08]. Shielded lines, illustrated in Fig. 5.22, have an array of strips below the line between the line and the substrate (on a lower metal layer, e.g., metal layer 1 of a CMOS process) and can offer quality factors three times higher than unshielded lines [VKKH08], but they suffer from greater variability in performance [VKKH08]. Also shown in Fig. 5.22 is the use of multiple metal layers to form the ground traces for the CPW. These traces help to keep electric and magnetic fields from extending beyond the borders of the transmission line.

Microstrip lines have received less attention by researchers for mmWave CMOS design, but they remain a design option and have been studied. For example, Brinkhoff et al. studied the frequency behavior of microstrip transmission lines, and found several effects [BKKL08]: the capacitance of the line to ground depends very little on frequency and increases nearly linearly with signal line width. Inductance between the two ends of the line depends on frequency only near DC, past which it shows a decreasing trend with increasing line width. Conductance to ground increases nearly linearly with frequency and line width. Resistance increases with frequency and decreases with line width [BKKL08].

As with all passives, the circuit model used to study transmission lines should accurately capture the energy storage and loss mechanisms present in the line. Loss mechanisms important in transmission line design include substrate capacitive loss and conductive loss, and for very short lines, eddy current losses [HKN07b]. A very simple model developed in 1993 by W. Heinrich is shown in Fig. 5.23 [Hei93]. Although simple in structure, the model is fairly adaptable and may be modified to account for a variety of loss mechanisms present, including the skin effect and the thickness of each conductor [Hei93], although it is not clear whether this approach could be used for CPW transmission lines designed with modern BEOL processes. What is interesting about the Heinrich model is that if it is not practical or possible to explicitly calculate all the behaviors of components from physical principles, then this model can be used in the context of simply assigning the line a characteristic impedance and complex propagation constant based on measurement data [HBAN07b][Poz05]. In

5.7 Introduction to Transmission Lines and Passives

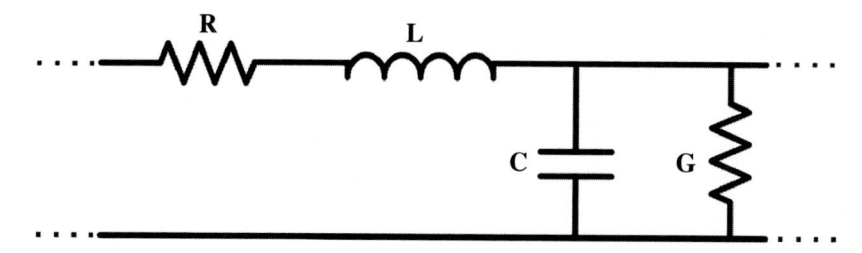

Figure 5.23 A very simple CPW model. The model developed by W. Heinrich used this structure [Hei93].

general, the more effects accounted for, the more involved the analysis required must be. The various line geometries that determine each circuit parameter have been well characterized and explained in the literature [Hei93].

It is useful to consider a few case studies of model creation for transmission lines. In 2007, Hasani et al. presented a T-model that accounts for conductor loss, eddy current loss, and substrate loss [HKN07b]. They accounted for eddy current loss by the simple inclusion of a resistor in parallel with the inductor used to represent line inductance [HKN07b]. In 2008, Brinkhoff et al. presented a model for microstrip transmission lines [BKKL08]. Their model was designed to accurately capture capacitance, inductance, conductance, and resistance behavior. They chose to develop an empirical model to make fitting to measured data easier than it would be with a physically based model. The structure of their model, a set of cascaded shunt sections followed by a series of cascaded series sections followed by a second series of cascaded shunt sections, allowed for easy extension to very high frequencies on the order of 100 GHz.

The basic design of transmission lines (microstrip or CPW) begins by selecting the desired impedance for the line to offer the best impedance match to the circuits to which the line interfaces, and the lowest loss along the line. In equation (5.32), the propagation factor is purely real, but if it were to have an imaginary component then the line would be lossy. To minimize loss, techniques such as shielding from the substrate should be used to keep currents induced in the substrate to a minimum, and the thickest metal layers available should be used to carry a signal, provided these are not too close the substrate (if too close, then there is an increased chance that lossy currents will be induced in the substrate). To select the correct characteristic impedance Z_0, the best approach is generally to design for a conjugate impedance match, in which [Poz05]

$$Z_s^* = Z_0 \frac{Z_L + jZ_0 \tan \alpha \ell}{Z_0 + jZ_L \tan \alpha \ell} \qquad (5.36)$$

where Z_s is the source impedance and Z_L is the load impedance. This is illustrated in Fig. 5.24. This condition ensures that maximum power is delivered to the load. Once a desired characteristic impedance is chosen, the dimensions of the lines can be selected and refined through simulation based on equations presented in such texts as [Poz05] and [Sim01]. These equations are often given in terms of elliptic functions or other complex formulations that may or may not be applicable to the CPWs that may be implemented in a particular technology. Therefore, it is often best to refine the capacitance based on

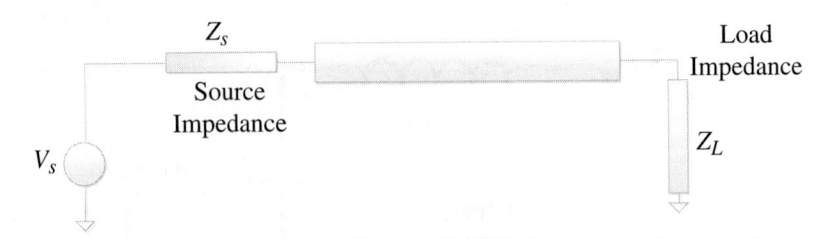

Figure 5.24 How a voltage source drives a load through a transmission line, with maximum power transfer occurring when Z_L is the conjugate of Z_s. Note that frequency effects cause both Z_s and Z_L to change with frequency.

simulations with such software as HFSS or PeakView. A general "rule of thumb," based on the understanding that as capacitance increases the impedance of a structure decreases as $\sqrt{\frac{L}{C}}$ (where L is inductance and C is capacitance), indicates that as the distance between the signal line and the ground lines increases, or as the distance to the substrate increases, the characteristic impedance of the line increases. Also, as the width of the main signal line increases, the capacitance increases so that the impedance decreases. It may not be practical to design each line according to equation (5.36), especially for large designs. For short traces it may be adequate to use a line that was optimized for a selected impedance, for example, 50 ohms, and lowest loss, without concern for achieving a conjugate match.

In theory, CPW transmission lines should be quasi-transverse electromagnetic (TEM) (e.g., the traveling wave propagates in both the substrate, and the air above the substrate, but because of the tiny geometry of semiconductors, the E and H fields are perpendicular for all intents and purposes). Thus, CPW lines should have a very wide bandwidth. In the event precise frequency tuning is needed, certain techniques such as the addition of a ladder reflector can facilitate frequency tuning [TWS+07].

Quasi-TEM indicates that the characteristics of the transmission line may be determined by analyzing a small cut of the line that is small relative to the wavelength of operation. Since the structure is small, over a small interval of time the fields over it are nearly constant, and therefore electro-static and magneto-static equations may be used to analyze the line. The primary means of tuning the characteristic impedance of a CPW transmission line is through control of the center conductor width and spacing between the center conductor and ground traces.

The quality of the "ground" designed on a chip is extremely important to properly provide transmission lines with minimal loss. Good grounding also avoids detrimental noise and distortions. Generally, all of the analog grounds (i.e., grounds used in analog portions of a chip, as opposed to digital portions of the chip) should be at the same voltage. This impacts CPW design by encouraging designers to merge the grounds traces of proximate transmission lines into a single plane [PMR+08]. Fig. 5.25 illustrates this concept. Fig. 5.26, illustrates the use of a common ground plane [MTH+08]. In Fig. 5.26, CMP refers to *Chemical Mechanical Polishing* (CMP), a step in the production of metal layers in a CMOS process in which metal layers are polished with chemicals. For these steps to be possible, no piece of contiguous metal may be too large (the largest size

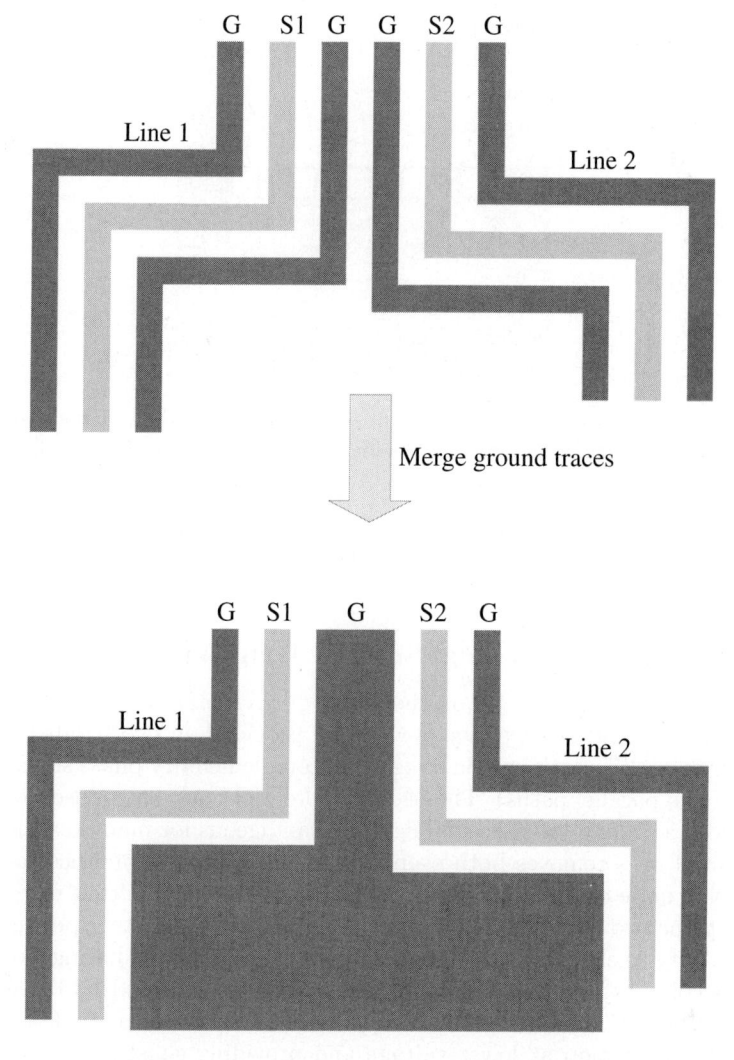

5.7 Introduction to Transmission Lines and Passives

Figure 5.25 Note how grounds may be merged for adjacent lines.

depends on the production process), and therefore metal holes must occasionally be added.

Another approach worth investigating is the use of dual ground planes (e.g., a digital ground and an analog ground, or two analog grounds) [AKD+07], for cases in which two sub-systems of the circuit share few components and may operate using different supplies. Another case in which separate ground planes may be needed is when one sub-system requires a nearly "noise-free" ground plane, while another sub-system present in the system has lots of very fast switching or other activity that may result in noisy signals being carried by the ground plane. The best approach for grounding is often based on chip layout (e.g., the footprint) and simulation results. In all cases, closed ground-plane

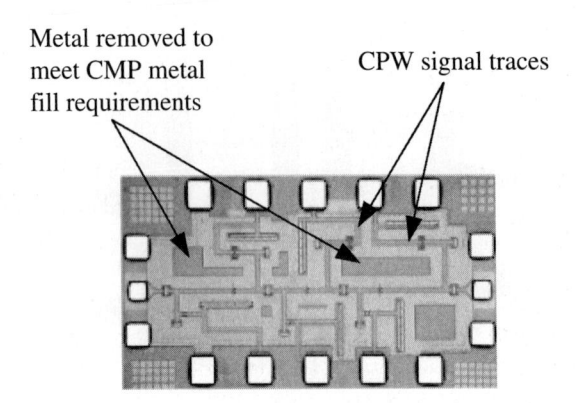

Figure 5.26 A common ground plane is evident in this layout. Portions of the metal have been removed in order to meet metal fill requirements [based on a figure from [MTH+08] © IEEE].

loops (e.g., to meet CMP requirements) should not have perimeters that are multiples of a half wavelength, to avoid formation of ground currents around these holes that could radiate.

5.7.2   Differential versus Single-Ended Transmission Lines

Modern circuit design relies greatly on the use of differential circuits, where the outputs of amplifiers, mixers, and other active devices are provided on two parallel transmission lines being driven in differential mode (where the complementary phase signal is provided on each of the two parallel paths). The use of differential lines, as opposed with a single-ended line above a ground reference, is highly advantageous for many reasons. First, the use of differential lines removes both even mode and odd mode variances or distortion, thereby reducing noise compared with single-ended circuit designs that rely on a ground plane. In effect, the two differential lines cancel out the noise or voltage perturbations and also offer immunity to electromagnetic interference that may be radiating from elsewhere on the chip. Second, the use of differential lines allows each circuit that drives one of the lines to run at half of the voltage as compared to the case of a single-ended circuit, thereby allowing operation at lower voltage and providing substantial headroom from the device breakdown voltage. CPW transmission lines are particularly well suited for differential circuits.

Often times, circuits are designed using differential lines (such lines are called *balanced lines*) and are converted to single-ended (*unbalanced*) lines at the high power amplifier stage at the edge of the chip. The device that converts balanced lines to unbalanced lines is called a *balun* and is a special type of transformer (discussed later in this chapter).

5.7.3   Inductors

Inductors may be made by using various lengths of shorted transmission lines, especially at mmWave frequencies. On-chip spiral inductors, however, are another way to fabricate inductive components. Inductors that are spiral in nature, and which utilize several layers of the chip to build a vertical overlay of coils, may be designed to offer space-savings over stub transmission lines [YGT+07][DLB+05][CO06]. However, transmission lines offer more freedom in routing than do spiral inductors, which makes the choice between inductors and transmission lines non-trivial, especially as frequencies increase and transmission lines become shorter.

Highly integrated devices that require extensive matching networks may wish to approach the selection of inductors from the standpoint of a tiling or packing problem, in which the goal is to make best utilization of space on the chip (i.e., "pack" as many components as possible in a given area). The primary design considerations when designing an inductor are line width, spacing between lines, number of turns, and whether to use stacked or planar topologies. There are tradeoffs involved in the selection of each parameter. The overall design should seek to minimize space and loss while maintaining a self-resonant frequency as high as possible and obtaining the desired value of inductance.

Electromagnetically, an inductor is a component designed to store magnetic energy. Fig. 5.27 illustrates a basic uni-spiral inductor. The inductance L of the spiral is defined as

$$L = \frac{\Phi}{I} \tag{5.37}$$

where Φ is the magnetic flux "linked" by the inductor, and I is the current that flows on the coils of the inductor. Based on (5.37), we may estimate the inductance using Ampere's law

$$\oint H \cdot d\ell = I \rightarrow H = \frac{I}{2\pi r} \tag{5.38}$$

$$\begin{aligned} \Phi &= \int \mu_{\text{eff}} H \cdot dS = \int_0^R \mu_{\text{eff}} \left(\frac{I}{2\pi r}\right) 2\pi r dr \\ &= \mu_{\text{eff}} R I \rightarrow L = \mu_{\text{eff}} R \end{aligned} \tag{5.39}$$

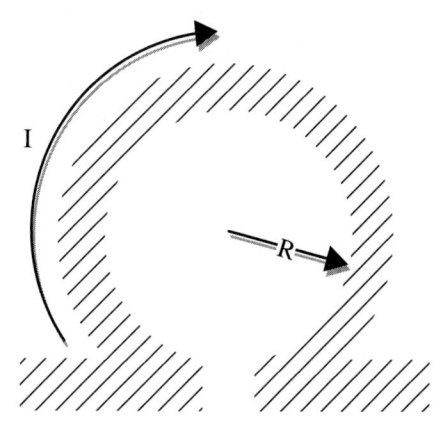

Figure 5.27 The inductance of the inductor increases with the radius, and sometimes with the number of links or turns in multiple metal layers.

where μ_{eff} is the effective permeability of the medium, S is the surface area enclosed by the inductor, and R is the radius of the inductor. If the inductor were composed of multiple turns, then the inductance would be multiplied by the number of turns squared, N^2. The reason N is squared is that the flux from one turn links or passes through all $N-1$ other turns and itself, and this is true for all N turns. Therefore, it can be seen that the inductance increases with the radius of the loop. The value of inductance required for an application should decrease with increasing frequency of operation, which is fortunate, given that the self-resonance frequency of the inductor is approximated by one divided by the radius (and the inductor should only be used at a frequency well below its self-resonant frequency).

Most inductor designs use multiple loops to increase the inductance proportionally by the number of loops squared. The number of loops should not be made so large that the loops extend all the way to the center of the inductor. Such small inner loops would be ineffective at storing flux because their inner radii are so small. Therefore, extremely small inner loops contribute only to loss.

The quality factor is an important factor for characterizing an inductor [O98]. Yao et al. found that the best quality factor results were for spiral inductors (which occupy more than one metal layer) versus single-layer inductors [YGT+07]. In general, the lowest substrate loss will be obtained if the footprint of the inductor is kept as small as possible by using small line widths and total area [DLB+05]. Past a point, reducing line widths may increase loss [LKBB09], due to increased conduction loss — the width that gives minimum loss should be found in simulation before taping out [CDO07]. Also, extremely thin lines may present difficulties due to electromigration [DLB+05]. Using stacked inductors instead of planar designs can be used to minimize the area required for a given inductance [DLB+05] and can improve the coupling between lines [CTYYLJ07][LHCC07] used to create an inductor.

The self-resonant frequency (SRF) of an inductor used in a communication circuit should be significantly higher than the operating frequency. Although at first blush it may appear attractive to operate an inductor near its self-resonant frequency so as to obtain a high value of impedance, this is unwise because phase noise (see Section 5.11.4.2) could move the inductor past SRF, causing it to act as a very large capacitor instead of an inductor. The self-resonant frequency arises from the fact that the inductor may store electric energy in addition to magnetic energy between its two terminals. A simple model for an inductor is shown in Fig. 5.28. This model captures the self-resonant frequency,

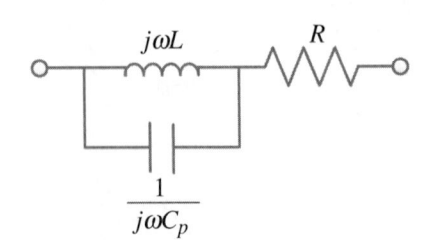

Figure 5.28 A simple model for an inductor.

5.7 Introduction to Transmission Lines and Passives

as it indicates that the impedance between the two nodes is given by

$$Z = \frac{j\omega L}{1 - \omega^2 LC} \quad (5.40)$$

where $\frac{1}{\sqrt{LC}}$ is the self-resonant frequency (SRF) of the inductor, so that at frequencies much below the SRF the impedance is approximately that of an ideal inductor, while at frequencies higher than the SRF the impedance becomes that of the capacitor: $\frac{1}{j\omega C}$.

Shields around the inductor can be used to improve the quality factor of spiral inductors [NH08, Chapter 3], for example, by reducing losses to eddy currents in the substrate [CGH+03]. A shield must be carefully designed in simulation if it is to aid performance. For example, a ground shield below the inductor can actually result in lower quality factor [BKKL08]. A shield under the inductor can also lower the self-resonant frequency [CGH+03]. It is also possible to implement a ground shield using the polysilicon layer of a process, as was done by Cao et al. for the design of a circular loop inductor [CO06]. Elements of a ground shield (e.g., metal strips) that are under the inductor should be placed perpendicularly to the inductor at their intersection points to reduce losses to eddy currents generated in the ground shield. Fig. 5.29 illustrates this point.

The self-resonant frequency, effective inductance, and the quality factor are usually the three parameters of most interest when measuring an inductor and can be estimated by measurements of the network parameters of an inductor (e.g., the inductor's S- or Z-parameters). The low frequency inductance is found by taking the ratio of the negative

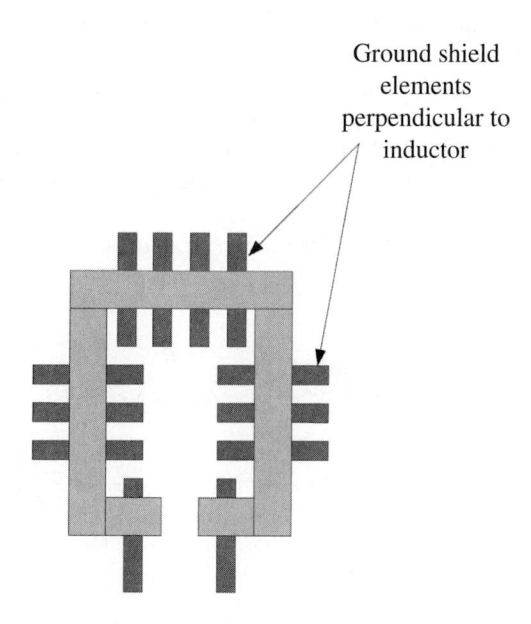

Figure 5.29 Ground shield elements should be perpendicular to the inductor.

of the imaginary portion of Z_{12} to angular frequency [DLB+05]. From Fig. 5.28 we may estimate the quality factor of a series network according to:

$$Y_{11} = \frac{1}{j\omega L + R} \quad (5.41)$$

$$Q_{series} = \frac{X}{R} = \frac{Im\left(\frac{1}{Y_{11}}\right)}{Re\left(\frac{1}{Y_{11}}\right)}$$

where Y_{11} is the first Y-parameter of the measured structure, Q is the quality factor, and X and R are the imaginary and real parts of the inductive component (assuming the measurement is well below the SRF). This may give inaccurately low results at high frequencies close to the self-resonant frequency and in cases in which the capacitive quality factor of the inductor is reasonably high [O98]. This is because near the SRF the inductor begins to look more capacitive, so that above the SRF the Q of the series network would become $\frac{1}{R\omega C}$. K. O [O98] has proposed an alternate means of measuring the quality factor of on-chip inductors through the numerical addition of a parallel capacitor.

An accurate model for on-chip inductors is vital in design and performance prediction. Several authors have presented models for on-chip inductors, including [DLB+05], [HLJ+06], and [CGH+03]. As with any model, inductor circuit models should accurately capture the loss mechanisms that impact the operation of the device in the desired frequency ranges. Successful extraction of model parameters and their use in the model to predict device performance indicate that the model accurately captures the energy loss and storage mechanisms present. In addition to modeling energy phenomenon, the model should provide formulas that convert a layout to model parameters [CGH+03]. The topologies of two popular models include the pi- and the double pi-models [CGH+03], although double pi-models can be more accurate as they are superior for capturing effects between closely spaced lines [CGH+03]. A simple pi-model is shown in Fig. 5.30. At low frequencies, the two shunt capacitors have a very high impedance and the model reverts to Fig. 5.28.

To fit measurements to the simple PI-model, first measure the S-parameters of the inductor and convert these to Y-parameters. At frequencies below the SRF, one may

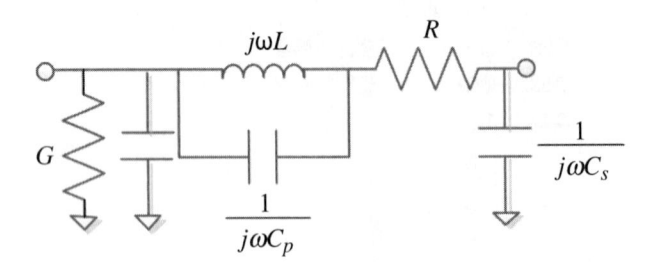

Figure 5.30 PI-models are very simple and popular models for inductors.

estimate L, R, and C_s using the Y-parameters. First, note that from Fig. 5.3 the shunt capacitors give Y_{12} and Y_{21} directly

$$Y_{12} = \frac{1}{j\omega L + R} \rightarrow L = -\frac{1}{Im(Y_{12})\omega}, \quad R = \frac{1}{Re(Y_{12})}. \quad (5.42)$$

From Fig. 5.3, Y_{11} and Y_{22} are identical for this network. After some algebra,

$$C_s = \frac{Im(Y_{11} - Y_{12})}{\omega}, \quad G = Re(Y_{11} - Y_{12}). \quad (5.43)$$

To determine C_p, determine the SRF as the first frequency at which the input impedance becomes real (i.e., from measurement of S_{11}, find the frequency at which S_{11} is purely real). C_p may then be estimated according to:

$$C_p = \frac{1}{4\pi^2 \text{SRF}^2}. \quad (5.44)$$

This is a simple model and does not capture all of the effects that may impact very high frequency mmWave inductors. Three major energy loss mechanisms that must be accounted for in an inductor model include conductive loss through the substrate, eddy current substrate loss caused by inductive coupling between the substrate and metal traces, and conductive and eddy current losses in the metal traces [CGH+03]. In addition to loss mechanisms, it is important to accurately capture frequency-dependent impedance of metal caused by different modes of carrying current at different frequencies [BKKL08]. At low frequencies a strip of metal may be modeled as a series inductor and resistor. As frequency increases, AC current crowds to the outside of the conductor due to the skin effect, requiring the inclusion of inductor-resistor branches in series with the original resistor. As frequency climbs higher, the proximity effect due to the mutual inductance between close lines must be modeled with an additional series mutual-inductor [CGH+03]. Such a simple model is illustrated in Fig. 5.31.

To help illustrate the challenges in modeling inductors, consider a few "case studies." In 2005, Dickson et al. [DLB+05] presented a circuit model for on-chip spiral inductors and techniques for extracting the model's parameters. In 2003, Cao et al. [CGH+03] presented a frequency-independent double pi-model for on-chip spiral inductors, in part to address concerns with single pi-models. The concerns include lack of frequency dependence of device parameters to account for current crowding due to the skin and proximity effects, inadequate representation of distributed nature of large inductors including line coupling, and difficult applicability to broad-band design [CGH+03]. In their model, Cao et al. explicitly account for the skin effect of transistors by adding parallel R-L branches to series resistors. In 2006, the same research group [CO06] presented a circular inductor with ground shield fabricated on the polysilicon layer. This may be advantageous if the design rules for the polysilicon layer permit a more flexible and successful design of the ground shield. At 60 GHz for a circular inductor with a diameter of 89.6 μm, this design achieved a Q of 35 and an inductance of 200 pH [CO06].

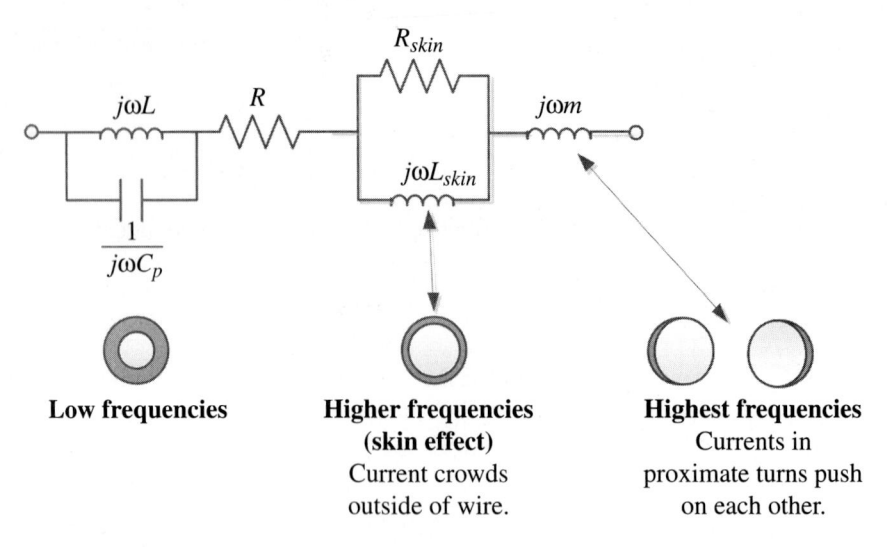

Figure 5.31 The series component of the inductor may be modified to account for higher order effects, including the skin and proximity effects.

5.7.4 Parasitic Inductances from Bond Wire Packaging

In addition to inductors designed intentionally within the integrated circuit, a mmWave chip can be adversely affected (or occasionally helped) by inductances that are formed by packaging or interconnections to the chip, most commonly from bond wires. A straight piece of wire has an inductance simply as a result of its current carrying capability. The specific inductance of a bond wire will vary by packaging, but an estimate of 1 nH/mm is a typical value. A simple formula that can be used to estimate the inductance in Henries/meter of a bond wire is:

$$L\left[\frac{H}{m}\right] = \frac{\mu_0}{2\pi} \ln\left(\frac{x}{r}\right) \tag{5.45}$$

where r is the radius of the bond wire, and x is the distance of the bond wire above the ground plane of the package.

5.7.5 Transformers

Transformers are one of the key passive blocks for the design of integrated circuits, as they are capable of converting between two AC voltages. The essence of a transformer is to use current in one branch of a circuit to induce a voltage in another branch. The amount of magnetic flux captured by the second branch is determined by the branch's geometry and controls the induced voltage. Fig. 5.32 illustrates this concept, in which a current in one loop induces a voltage in the second loop.

By Ampere's law, the magnetic field induced by the current I in the first loop of Fig. 5.32 is given approximately by

$$\oint H_1 \cdot d\ell = I \quad \rightarrow \quad H_1 = \frac{I}{2\pi r}. \tag{5.46}$$

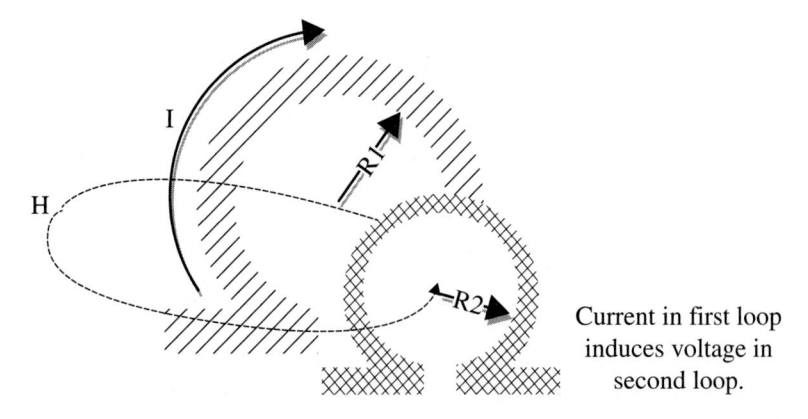

Figure 5.32 A transformer works by capturing flux generated in one branch of a circuit in another branch of the circuit.

Current in first loop induces voltage in second loop.

Now assume that the second loop is close enough that the field has not decayed substantially before reaching the second loop (e.g., the loops may actually be concentric). By Faraday's law, the voltage in the first and second loops can be found according to

$$V_1 = -\int E \cdot d\ell = \mu j\omega \int H \cdot dS = \mu j\omega \int_0^{R_1} \frac{I}{2\pi r} 2\pi r \, dr = j\omega \mu I R_1 \quad (5.47)$$

$$V_2 = -\int E \cdot d\ell = \mu j\omega \int H \cdot dS = \mu j\omega \int_0^{R_2} \frac{I}{2\pi r} 2\pi r \, dr = j\omega \mu I R_2 \quad (5.48)$$

$$j\omega \mu I = \frac{V_1}{R_1} \rightarrow V_2 = V_1 \left(\frac{R_2}{R_1}\right). \quad (5.49)$$

We see that the AC voltage in the first loop is transformed by the ratio of the loop radii to an AC voltage in the second loop. If the first loop had N turns, while the second loop had M turns, the induced voltage V_2 would be

$$V_2 = V_1 \left(\frac{M_2 R_2}{N_2 R_1}\right). \quad (5.50)$$

It is customary to call the first loop of the transformer the *primary* loop or branch, and the second loop or branch the *secondary* loop or branch. From a circuit perspective, as shown in Fig. 5.33 we define what is called a *mutual inductance*, m, to describe how a current in the first loop induces a voltage in the second loop:

$$V_2 = j\omega m I_1. \quad (5.51)$$

In reality, not all of the flux through the primary loop will link to the secondary branch. For the simple loop example, Fig. 5.32 shows that the secondary loop has a smaller radius than the first loop, indicating that only $\frac{R_s}{R_p}$ of the flux in the first loop links to the secondary loop. Assuming that the loops are co-centric, all of the flux in

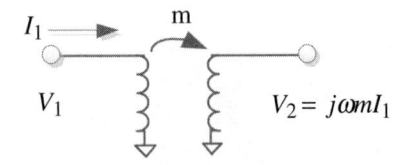

Figure 5.33 The mutual inductance between the two inductive elements determines the voltage induced in the second branch by the current in the first branch. We assume here that the second branch has no self-inductance.

the secondary loop links to the primary loop (note that if the loops were not co-centric, not all the flux in the secondary loop would link to the primary loop). To account for the fact that not all the flux in one loop reaches the other, it is customary to define a coupling coefficient k_m that describes the percentage of flux in one loop that links to the other loop

$$k_m = \sqrt{\frac{\Phi_{ps}\Phi_{sp}}{\Phi_s \Phi_p}} \qquad (5.52)$$

where Φ_{ps} is the flux in the primary loop due to current in the secondary loop, and Φ_{sp} is the flux in the secondary loop due to current in the primary loop. Φ_p is the total flux in the primary loop, and Φ_s is the total flux in the secondary loop. $\Phi_{ps} = m \times I_s$, and $\Phi_{sp} = m \times I_p$, while $\Phi_p = L_p \times I_p$ and $\Phi_s = L_s \times I_s$, so that

$$k_m = \frac{m}{\sqrt{L_s L_p}}. \qquad (5.53)$$

If all of the flux in one loop links to the secondary loop and vice versa, then the coupling factor is equal to 1.

The similarity of the mathematics and structure used to describe a transformer should convince us that transformers and inductors are closely related, and the design of one is not far removed from the design of the other. Transformers have many uses, including power combining, single-ended to differential conversion (and vice versa), voltage conversion, and providing ESD protection [Nik10][LKBB09]. To understand their use for differential to single-ended conversion (e.g., using a transformer as a balun), consider a main loop that links flux to two sub-loops, where the leads of the two sub-loops are inverted relative to one another, as shown in Fig. 5.34. If both sub-loops have the same radius, then their voltages will be equal in magnitude but 180 degrees out-of-phase.

When designing a transformer, it is desirable to obtain high Q and coupling between secondary and primary windings [LKBB09]. The Qs of both the primary and secondary branch should also be high. It is also desirable, as with an inductor, to have a self-resonant frequency (SFR) much higher than the operating frequency. The types of turns or loops in the transformer have a major impact on both loss and SRF. Leitre et al. [LKBB09] found that single-turn transformers offer lower loss and higher resonant frequency than multi-turn transformers. They also found that octagonal turns offer higher quality factor than square turns.

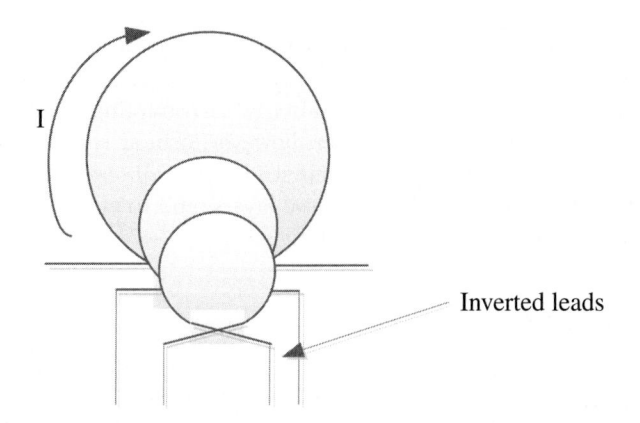

5.7 Introduction to Transmission Lines and Passives

Figure 5.34 A transformer can be used for single-ended to differential conversion in a balun.

Measurements of the Z-parameters of a transformer can be used to determine inductances, quality factors, and magnetic coupling between primary and secondary arms [LKBB09]

$$L_p = \frac{Im(Z_{11})}{\omega} \quad (5.54)$$

$$L_s = \frac{Im(Z_{22})}{\omega} \quad (5.55)$$

$$Q_p = \frac{Im(Z_{11})}{Re(Z_{11})} \quad (5.56)$$

$$Q_s = \frac{Im(Z_{22})}{Re(Z_{22})} \quad (5.57)$$

$$k_m = \sqrt{\frac{Im(Z_{12})Im(Z_{21})}{Im(Z_{11})Im(Z_{22})}} \quad (5.58)$$

where L_p is the inductance of the primary loop, L_s is the inductance of the secondary loop, Q_p is the quality factor of the primary loop, Q_s is the quality factor of the secondary loop, and k_m is the coupling factor. Note that because the inductor is a passive device, $Z_{12} = Z_{21}$. These equations are understandable when we write the currents in each branch as

$$V_1 = j\omega L_1 i_1 + R_1 i_1 + j\omega m i_2 \quad (5.59)$$
$$V_1 = j\omega L_1 i_1 + R_1 i_1 + j\omega m i_2 \quad (5.60)$$

where:

$$Z_{11} = \left.\frac{V_1}{I_1}\right|_{I_2=0} \quad Z_{22} = \left.\frac{V_2}{I_2}\right|_{I_1=0} \quad Z_{12} = \left.\frac{V_1}{I_2}\right|_{I_1=0} \quad (5.61)$$

so that:

$$Z_{11} = j\omega L_1 + R_1, \quad Z_{22} = j\omega L_2 + R_2, \quad Z_{12} = j\omega m. \quad (5.62)$$

In 2009, Leite et al. [LKBB09] studied the performance of various geometries of spiral inductors in 65 nm CMOS. They found that the diameter of the transformer primarily impacts its resonant frequency but not its quality factor, coupling coefficient, or insertion loss. The trace width used in the transformer, however, primarily impacted the insertion loss, with an intermediate width (8 μm in their study) providing the lowest insertion loss. The process used by [LKBB09] provided 7 metal layers, and to obtain lowest attenuation, they used the two top thick layers in their design.

5.7.6 Interconnects

Interconnects add parasitic capacitances, resistances, and inductances to a circuit and can have a major negative impact on device or circuit operation [EDNB04]. Simulations and experimental tape-outs are always required to understand the impact of interconnects at new, higher frequencies or when using new, more advanced technology nodes, and digital tuning (e.g., designing digital switches to add or couple different filters at an interconnect point) is a good approach for ameliorating problems due to interconnections after chip fabrication. Steps taken to reduce parasitic resistances (e.g., through an increased number of gate fingers for reduced gate resistance [SSDV$^+$08]) will also often lead to increased capacitances (e.g., through increased overlap capacitance from gate to source [JGA$^+$05]). The RC product of a structure's parasitic capacitances, together with knowledge of design goals (e.g., for matching and transit frequency) are important to understand.

Accurate models of interconnects such as bond wires and transmission lines are very important in the analysis of high frequency structures. All on-chip interconnects, especially at mmWave frequencies, should be treated as transmission lines and designed as such [TDS$^+$09]. While the purpose of interconnections is to simply connect devices to each other on the integrated circuit, the high frequency consequences of interconnects can be difficult to analyze. With today's modern simulation tools, it is possible to forego hand calculation, but hand calculation can often reveal details about a structure that would otherwise be missed. Caverly, for example, used a simple conformal mapping to determine an effective dielectric constant of a wire air dielectric interface [Cav86].

The parasitic capacitances and inductances associated with interconnects can greatly impact or even dominate the performance of a device [EDNB04]. When designing devices for mmWave frequencies, the effects of parasitic components should be considered in all levels of the design, including at the schematic phase, as these will decide such things as matching structures and possibly active device topologies [TDS$^+$09].

5.8 Basic Transistor Configurations

Before progressing further, it will be helpful to summarize the basic configurations of transistors used in a mmWave radio, or any radio. The most common configurations are: common-source, common-gate, and common-drain or source follower. These basic configurations are indicated in Fig. 5.35.

The *cascode* design is very common in integrated circuits (not be confused with a *cascade*). While a cascade transistor circuit simply has the output of one transistor driving the input of a second transistor, and so on until sufficient gain is obtained,

5.8 Basic Transistor Configurations

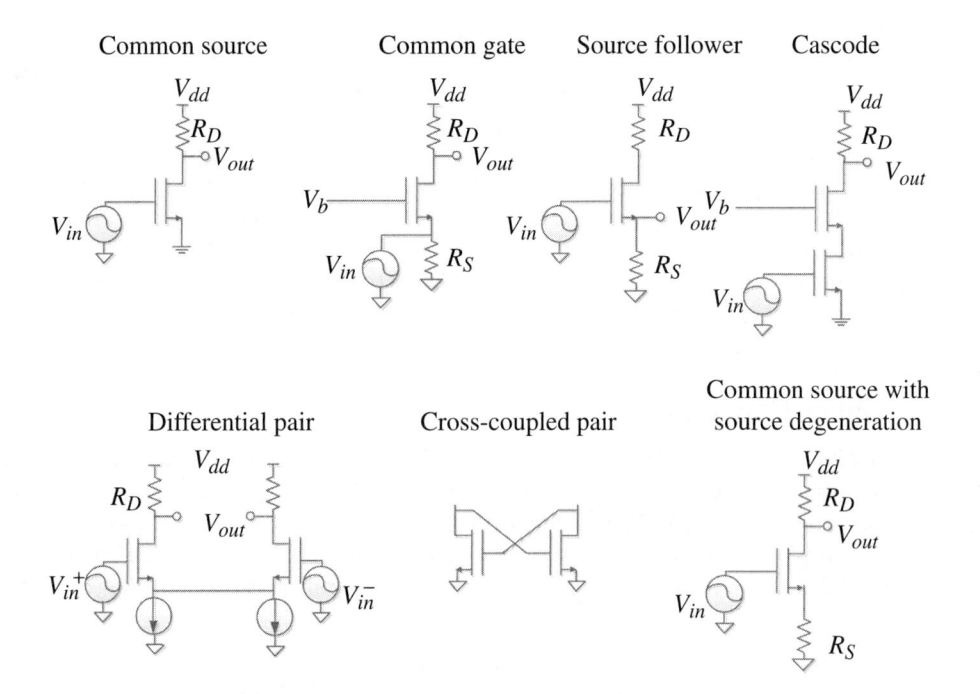

Figure 5.35 The basic configurations for transistors.

the cascode consists of two transistors and obtains greater gain and less noise. In the cascode, the bottom transistor whose gate is connected to the input is in a common-source configuration because its source is grounded. The second transistor is in a common-gate configuration because its gate is connected to a DC voltage.

Consider each NMOS as having an output resistance of r_0. For the common-source configuration, if the resistor R_D is replaced with a parallel RLC load, then the input-output voltage relationship is:

$$V_0 = -\frac{j\omega L g_m V_{in}}{1 + \frac{j\omega L}{R_D || r_0} - \omega^2 LC} + V_{dd}\left(1 + j\omega\left(\frac{L}{R_D}\right) - \omega^2 LC\right) / \left(1 + j\omega L/(R_D||r_0) - \omega^2 LC\right) \quad (5.63)$$

where the '||' notation indicates the parallel notation for two resistors, where the overall resistance is the product of the two resistors divided by their sum. Deriving an input-output relationship such as (5.63) is an important step in design because it allows the designer to determine what parameters affect important figures of merit such as bandwidth and gain.

5.8.1 Conjugate Matching

Matching for maximum power transfer from a source to a load is important in many areas of circuit design. To transfer the maximum power from a source to a load, the source impedance should be set equal to the complex conjugate of the load impedance

$$Z_s = Z_L^*. \tag{5.64}$$

Fig. 5.36 illustrates this concept. If a device is a two-port (i.e., it has an input and an output), then it should be matched at both ports for maximum power transfer. This can be complicated for some devices, such as common-source amplifiers, as the input impedance and output impedance may be difficult to tune independently. A cascade design is popular for amplifiers because it allows separate matching of its input and output impedance.

5.8.2 Miller Capacitance

Miller capacitance refers to the gate-drain capacitance of a transistor, and this capacitance has a negative impact on the gain of the circuit. Consider the common-source amplifier shown in Fig. 5.37. Without considering the gate-drain capacitance C_{gd}, the gain from the input to the output is given by $g_m R_L$, where g_m is the transistor's transconductance and R_L is the load resistance (we assume the output resistance of the transistor

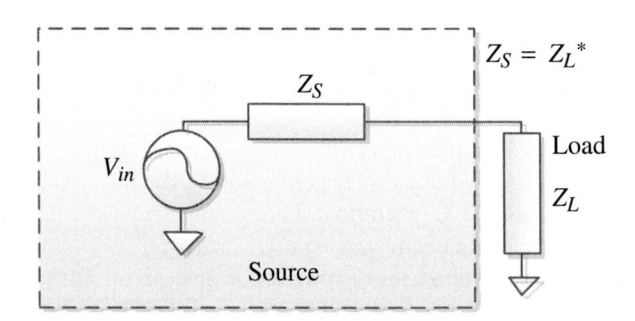

Figure 5.36 Basic circuit showing a load Z_L connected to a voltage source having an output impedance of Z_s.

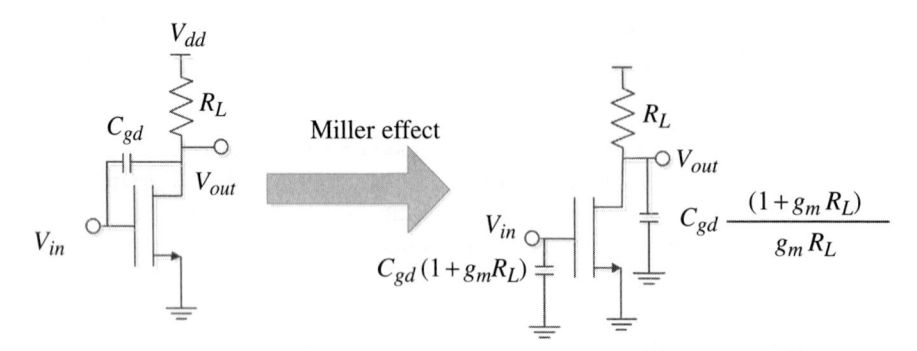

Figure 5.37 The gate-drain capacitance of a common-source amplifier can cause the input pole to be at a lower frequency than the output pole.

to be infinite). The Miller theorem states that the impedance represented by the capacitor may be split into two separate impedances, one from the input to ground and one from the output to ground, as shown in Fig. 5.37 [Raz01, Chapter 6]:

$$Z_{\text{out}} = \frac{Z}{1 - A^{-1}}, \qquad Z_{\text{in}} = \frac{Z}{1 - A} \tag{5.65}$$

$$Z = \frac{1}{j\omega C_{gd}}, \qquad A = -g_m R_L \tag{5.66}$$

where A is the gain of the amplifier and Z is the impedance of the gate-drain capacitance. The problem caused by the Miller effect on the gate-drain capacitance is that it can make the input pole frequency lower than the output pole frequency. This is bad because the load resistance may be kept high to realize high gain, which usually necessitates a fairly low output pole frequency. If the input pole has a lower frequency, then the gain of the amplifier would have to be reduced to maintain stability.

5.8.3 Poles and Feedback

It is common to consider poles and zeros when discussing circuit design in a mmWave context or in a more general analog circuit design context. A pole is a factor such as $(1 - 2jA\pi f)$ that appears in the denominator of a transfer function or gain expression. In this expression, f is frequency, and when $f = \frac{1}{2A\pi}$ the expression reaches its maximum, and this is said to the be frequency of the pole. Past this frequency, the expression diminishes as the inverse of f. Generally, the output pole of an amplifier should be the dominant pole of the amplifier, indicating that its frequency is the lowest of any pole for the amplifier. For example, consider the common-source amplifier in Fig. 5.38. The gain, A, of this amplifier is found to be

$$V_{\text{out}} = -\frac{g_m R_L}{1 + j\omega C_L R_L} V_{\text{in}} \tag{5.67}$$

$$A = -\frac{g_m R_L}{1 + j2\pi f C_L R_L} \tag{5.68}$$

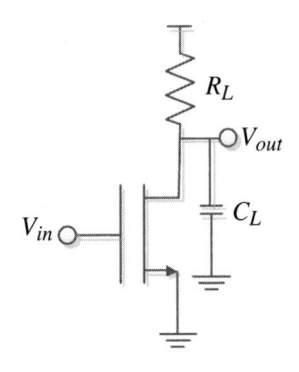

Figure 5.38 The output pole of the common source amplifier should be the dominant pole.

where ω is the angular frequency. This amplifier has an output pole that occurs at $\frac{1}{2\pi R_L C_L}$, and this should be the dominant pole. If non-zero gate resistance is added, as shown in Fig. 5.39, then the gain becomes

$$A = -\frac{g_m R_L}{(1+j\omega C_{gs}R_g)(1+j\omega C_L R_L)} \tag{5.69}$$

where C_{gs} is the gate-source capacitance of the amplifier. In order for the output pole to remain the dominant pole, we need $R_L C_L > R_g C_{gs}$.

Feedback is another important topic and a vital part of circuit design. There are many excellent texts that discuss feedback in detail (e.g., [Raz01, Chapter 8]). As is shown in many introductory texts, the left portion of Fig. 5.40 illustrates basic feedback. Feedback is very common in integrated circuits, as illustrated by the example on the right side of Fig. 5.40. Feedback has many wonderful properties, but it can be dangerous if ever $A\beta = 1$, as this causes the amplifier to become unstable and to oscillate rather than to amplify. In general, the stability of an amplifier is characterized or measured by its *phase margin*. The phase margin of an amplifier is defined as the amount by which the phase of the gain stage output is greater than 180 degrees (as compared to the input) at the unity-gain frequency (e.g., the frequency at which the gain has fallen to 0 dB, or 1 in absolute terms).

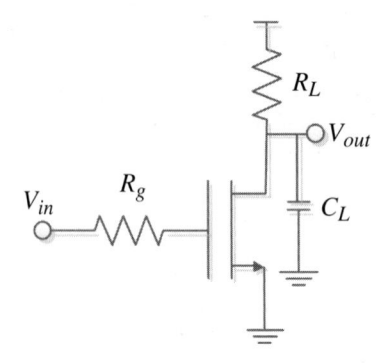

Figure 5.39 A common source amplifier with gate resistance to adjust dominant poles.

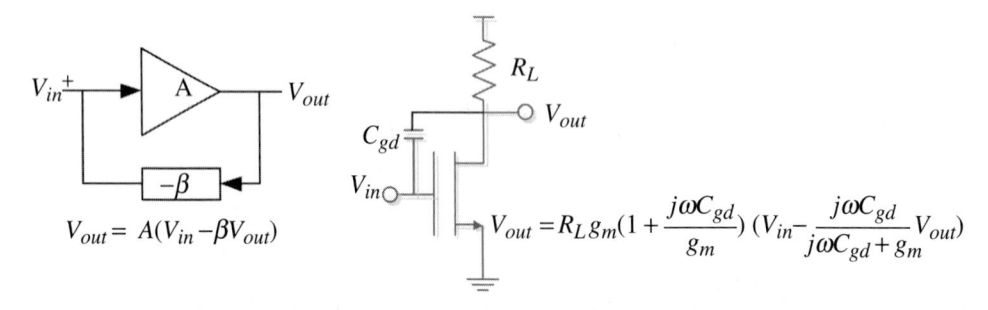

Figure 5.40 Feedback is very common in differential circuits.

5.8 Basic Transistor Configurations

When discussing feedback, two important terms are closed-loop gain and open-loop gain. The closed-loop gain of the circuit on the left in Fig. 5.40 is given by

$$G_{\text{closed}} = \frac{A}{1-A\beta}. \quad (5.70)$$

The open-loop gain is found by breaking the loop and finding the gain from the beginning of the loop to the end of the broken loop. In the left circuit of Fig. 5.40, the open-loop gain is

$$G_{\text{open}} = A\beta. \quad (5.71)$$

If the open-loop gain of a circuit is ever greater than one with zero phase, then the closed-loop circuit will oscillate.

5.8.4 Frequency Tuning

The amplifiers discussed in this chapter are all intended to be tuned to a specific frequency. Tuning a circuit means that it has been optimized for best performance at a selected frequency or frequency band. Inductors and capacitors are essential components for tuning amplifiers and other circuits. This is because, at the resonant frequency for a network of inductors and capacitors, the proper design will ensure that the impedance of the network will be real at the frequency for which the network is intended to operate, for example, as the load of an amplifier. For example, consider the simple amplifier shown in Fig. 5.41. We may find the Z-parameters of this circuit:

$$Z_{11} = \frac{1}{j\omega C_{\text{in}}} + j\omega L_s + g_m \left(\frac{L_s}{C_{\text{in}}}\right) \quad (5.72)$$

$$Z_{22} = \frac{j\omega L_L}{1 + \frac{j\omega L_L}{R_L} - \omega^2 L_L C_L} \quad (5.73)$$

$$Z_{12} = 0 \quad (5.74)$$

$$Z_{21} = \frac{g_m \left(\frac{L_L}{C_L}\right)}{1 + \frac{j\omega L_L}{R_L} - \omega^2 L_L C_L}. \quad (5.75)$$

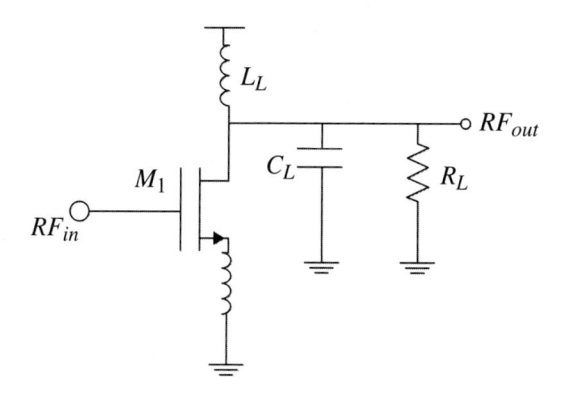

Figure 5.41 Amplifiers for mmWave frequencies are tuned circuits.

By converting these parameters into S-parameters, we find:

$$S_{11} = \frac{Z_{11} - Z_0}{Z_{11} + Z_0} = \frac{(1 - \omega^2 L_s C_{in}) + j\omega(L_s g_m - C_{in} Z_0)}{(1 - \omega^2 L_s C_{in}) + j\omega(L_s g_m + C_{in} Z_0)} \tag{5.76}$$

where

$$\Re(S_{11}) = \frac{(1 - \omega^2 L_s C_{in})^2 + \omega^2(L_s^2 g_m^2 - C_{in}^2 Z_0^2)}{(1 - \omega^2 L_s C_{in})^2 + \omega^2(L_s g_m + C_{in} Z_0)^2} \tag{5.76a}$$

$$\Im(S_{11}) = \frac{-2j\omega C_{in} Z_0 (1 - \omega^2 L_s C_{in})}{(1 - \omega^2 L_s C_{in})^2 + \omega^2(L_s g_m + C_{in} Z_0)^2} \tag{5.76b}$$

$$S_{21} = \frac{2 Z_0 Z_{21}}{(Z_{11} + Z_0)(Z_{22} + Z_0)} \tag{5.76c}$$

$$S_{21} = \left(\frac{2 Z_0 g_m L_L}{C_L}\right) \Bigg/ \Bigg(\left[\frac{Z_0 g_m L_s}{C_{in}}(1 - \omega^2 L_L C_L)\right.$$
$$-\omega L_L \left(1 + \frac{Z_0}{R_L}\right)\left(\omega L_s - \frac{1}{\omega C_{in}}\right)\Bigg]$$
$$+ j\left[Z_0\left(\omega L_s - \frac{1}{\omega C_{in}}\right)(1 - \omega^2 L_L C_L)\right.$$
$$\left.+ \frac{g_m L_s}{C_{in}} L_L \omega \left(1 + \frac{Z_0}{R_L}\right)\right]\Bigg). \tag{5.77}$$

Figs. 5.42 and 5.43 show the magnitudes of these functions for $g_m = 5$ mS, $C_{in} = 37$ fF, $L_s = 0.187$ nH, $C_L = 7$ fF, $L_L = 6.32$ nH, $Z_0 = 50\,\Omega$, and $R_L = 50\,\Omega$. We see that the value of S_{11} falls to near -10 dB near 60 GHz, indicating that the amplifier accepts energy from an assumed 50 ohm line, and that S_{21} peaks near 12 dB at 60 GHz, indicating that the amplifier provides gain.

5.9 Sensitivity and Link Budget Analysis for MmWave Radios

Before discussing the individual blocks in a mmWave radio, we take a large-scale view to familiarize the reader with the overall radio system. Fig. 5.44 shows the major blocks in a basic radio. Antennas were discussed in Chapter 4. From the illustration, we see that the major blocks for any radio are the power amplifier (PA), low noise amplifier (LNA), mixer, and voltage-controlled oscillator (VCO). Designs that must meet a spectral mask requirement or perform complex modulation will use a frequency synthesizer such as a phase-locked loop (PLL) in place of a simple VCO, and we briefly discuss PLLs in the sections that follow. Table 1 in [RMGJ11] gives examples of several 60 GHz and mmWave radios and their performance. This section includes an updated and expanded discussion that first appeared in [RMGJ11], written by two of that article's authors.

For a mmWave receiver to operate properly, the radio must have an adequate signal-to-noise ratio (SNR) to support the modulation scheme used by the radio. Modulation

5.9 Sensitivity and Link Budget Analysis for MmWave Radios

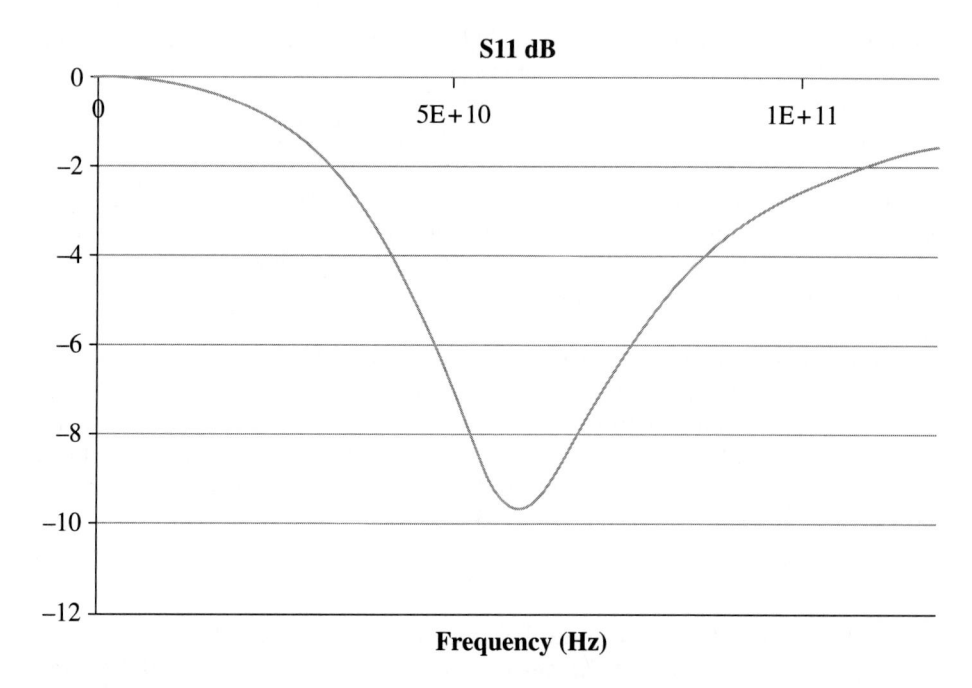

Figure 5.42 The amplifier accepts energy in only a certain range of frequencies.

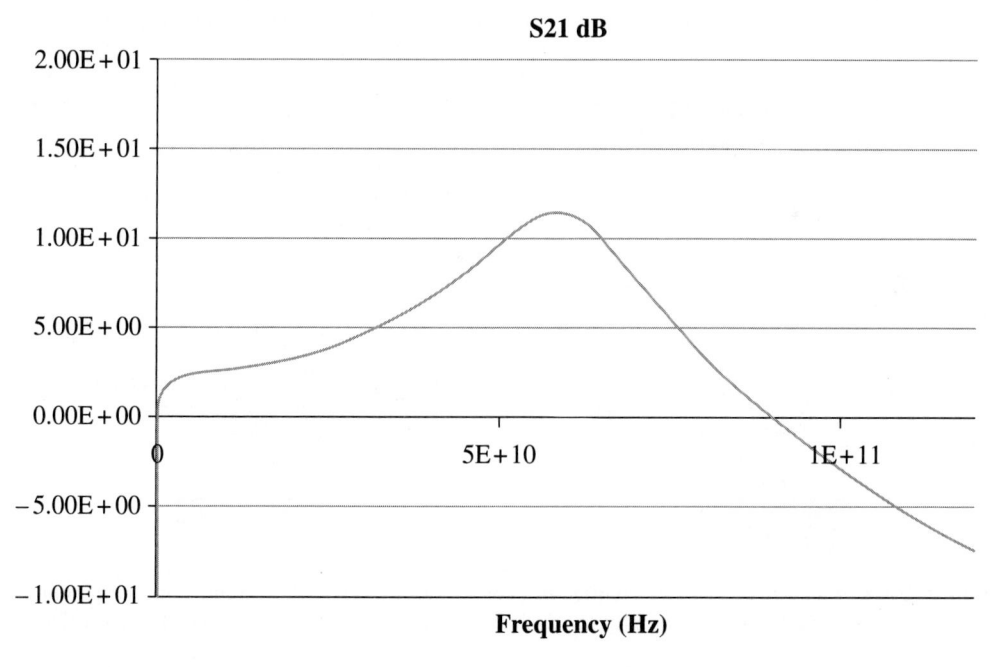

Figure 5.43 The value of S_{21} gain is only high in a certain band of frequencies.

was discussed in Chapter 2 and is discussed in Chapter 7. In addition to adequate SNR, the signal power must not be so strong that it saturates or compresses the receiver. Link budget calculations help to determine the required SNR, while studies of system linearity and use of Automatic Gain Control (AGC) circuits determine when the received power is too high. Maintaining linearity becomes very important when the transmitter and receiver are in close proximity, and maintaining adequate SNR is more challenging when the transmitter and receiver are far apart. In addition to a link budget to determine received power and linearity studies, it is often helpful to study the required noise performance of individual blocks. We now discuss link budgets, linearity, and noise figure.

A link budget helps to understand power and equipment the requirements of a mmWave radio. A link budget describes how much power from the transmitter is received by the receiver, and the signal-to-noise ratio (SNR) at the receiver. [RMGJ11], [TAY+09], [YSH07], [BNVT+06], [YC07], [XKR02], and [AR04] have presented link budgets for 60 GHz or mmWave radios. See Fig. 5.44 for a direct conversion architecture. We first find the power delivered to the receiver [Rap02][RMGJ11]

$$P_{RX}[\text{dB}] = P_T[\text{dB}] - PL_d[\text{dB}] + G_T[\text{dB}] + G_R[\text{dB}] \qquad (5.78)$$

where terms in (5.78) are all in decibels and P_T is the transmitted power, PL_d is the channel path loss for omnidirectional antennas at the transmitter and receiver for the transmitter receiver separation distance d, G_T is the transmit antenna gain, and G_R is the receive antenna gain [RMGJ11]. Equation (5.78) indicates that the transmitted power fundamentally determines the received power. The output power of the power amplifier (PA) used in the system is the key determinant of the transmitted power and is one of the most important specifications for a power amplifier. The noise power in decibel units at the output of the receiver's low noise amplifier (LNA) is given by [Rap02][RMGJ11]

$$P_{\text{noise}}[\text{dB}] = 10\log_{10}(kT_{\text{syst}}BNF_{RX}G_R) \qquad (5.79)$$

$$P_{\text{noise}}[\text{dBm}] = -174 \text{ dBm} + 10\log_{10}(B) + 10\log_{10}(NF_{RX}) + 10\log_{10}(G_R) \qquad (5.80)$$

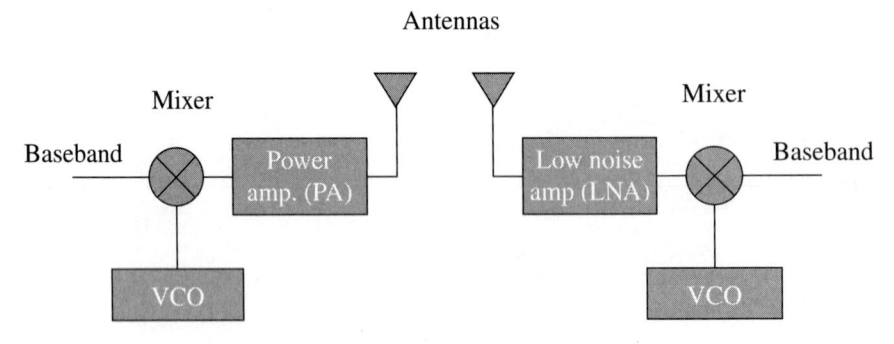

Figure 5.44 A direct conversion architecture for a transmitter and receiver. This is a popular architecture for today's cellphones. In many designs, the VCO is part of a phase-locked loop (PLL).

where $10\log(kT_{\text{syst}})$ is equal to -174 dBm for a system temperature of $17°$ Celsius. NF_{RX} is the noise factor (see Section 5.10.2) of the receiver LNA, and B is the bandwidth of the signal (also see Section 5.11.2 regarding noise factor and noise figure[2]). This equation indicates the importance of the noise figure of the LNA, which we will discuss in Section 5.11.2. A 1 m link with a free space path loss exponent of 2 (see Chapter 3 for a discussion of propagation and path loss exponent) has a path loss of 68 dB at 60 GHz. If we use antennas with 0 dB gain, and assume a transmit power of 10 dBm and an LNA gain of 15 dB, it is seen that the received power [RMGJ11]

$$P_{RX}[\text{dBm}] = 10 \text{ dBm} - 68 \text{ dB} + 0 + 0 = -58 \text{ dBm}. \tag{5.81}$$

A typical noise figure for a mmWave LNA is 6 dB. Assuming a 1.25 GHz bandwidth channel, we can use (5.78) to find the noise power at the receiver front end (note that IEEE 802.15.3c uses channels with 2.16 GHz bandwidth [RMGJ11]):

$$P_{\text{noise}}[\text{dBm}] = -174 \text{ dBm/Hz} + 10\log(1.25 \text{ GHz}) + 6 \text{ dB} = -77 \text{ dBm}. \tag{5.82}$$

We therefore find a signal-to-noise ratio (SNR) in this example of -43 dBm $+ 62$ dBm $= 19$ dB. Chapter 7 discusses the various SNR requirements for modulation schemes used for 60 GHz systems. Any modulation scheme (ranging from simple amplitude modulation to orthogonal frequency division multiplexing [OFDM]) has a minimum SNR required to achieve a sufficiently error free reception, or a reception that has a low enough bit error rate (BER) for the transmitted signal to be meaningfully recovered. The amount by which the SNR exceeds the minimum required SNR for the modulation scheme used is called the Link Margin.

5.10 Important Metrics for Analog MmWave Devices

5.10.1 Non-Linear Intercept Points

Radios are inherently non-linear. Our discussion of basic transistor operation indicates that the fundamental blocks — transistors — of radios are non-linear. An understanding of non-linearity is especially important in the study of power amplifiers (PAs), as the power amplifier must typically handle the largest signal swings unless the receiver is very close to the transmitter, in which case the low noise amplifier (LNA) may also experience high input powers. For a non-linear system, we may assume the output (assumed to be a voltage) may be represented through a geometric series of the input:

$$V_0 = a_0 + a_1 V_{in} + a_2 V_{in}^2 + a_3 V_{in}^3 + .. + a_n V_{in}^n. \tag{5.83}$$

---

2. As shown in Section 5.11.2, noise factor (NF) is the ratio of noise coming out of a device compared with the noise applied to the input of the device, and noise figure (F) is simply the NF given in decibels.

To find the values of these coefficients, we need a model of our device. For example, for a simple common-source transistor with a real load resistance as shown in Fig. 5.45, we may use the square law to write

$$I_d = \frac{1}{2} k_n \left(\frac{W}{L}\right) (V_{in,DC} - V_t + v_{in,RF})^2 \rightarrow = \frac{1}{2} k_n \left(\frac{W}{L}\right)$$
$$\cdot (V_{GST}^2 + V_{GST} v_{in,RF} + v_{in,RF}^2) \tag{5.84}$$

$$V_{GST} = (V_{in,DC} - V_t) \tag{5.85}$$

$$\frac{V_{dd} - V_0}{R_L} = I_d \tag{5.86}$$

$$V_0 = V_{dd} - \frac{R_L}{2} k_n \left(\frac{W}{L}\right) V_{GST}^2 + R_L k_n \left(\frac{W}{L}\right) V_{GST} v_{in,RF}$$
$$+ \frac{R_L}{2} k_n \left(\frac{W}{L}\right) v_{in,RF}^2 \tag{5.87}$$

$$a_0 = V_{dd} - \frac{R_L}{2} k_n \left(\frac{W}{L}\right) V_{GST}^2 \tag{5.88}$$

$$a_1 = R_L k_n \left(\frac{W}{L}\right) V_{GST} \tag{5.89}$$

$$a_2 = \frac{R_L}{2} k_n \left(\frac{W}{L}\right). \tag{5.90}$$

The input to any block may be considered to be a modulated sinusoid, where the signal envelope (i.e., the magnitude of the sinusoid) changes much more slowly than the carrier frequency (i.e., the envelope is narrowband relative to the carrier's frequency):

$$V_{in} = A(t) \sin(\omega t + \phi). \tag{5.91}$$

Because $\frac{dA(t)}{dt}$ is much less than $\frac{d\sin(\omega t)}{dt}$, A may be assumed to be a constant:

$$V_{in} = A \sin(\omega t + \phi). \tag{5.92}$$

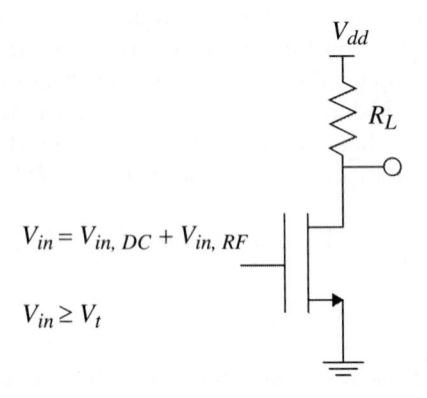

Figure 5.45 A simple power amplifier illustrates the concept of power added efficiency.

5.10 Important Metrics for Analog MmWave Devices

Recall from basic trigonometry that $\sin(x)^2 = 0.5(1 - \cos(2x))$, and $\sin(x)^3 = 0.25(3\sin(x) - \sin(3x))$. If only the first three terms of (5.83) are used, then the output may be written in terms of the input as

$$V_0 = \frac{a_2}{2}A^2 + \left[a_1 A + \frac{3}{4}a_3 A^3\right]\sin(\omega t) + \frac{a_2}{2}e^{\frac{j\pi}{2}}A^2 \sin(2\omega t) - \frac{a_3}{4}A^3 \sin(3\omega t). \quad (5.93)$$

In (5.93), the coefficient of $\sin(\omega t)$ is the in-channel component. In addition to the in-band component, there exists substantial energy at baseband, and the second and third harmonic. (5.93) allows us to define several parameters that are of interest for radio design. First, the coefficient a_3 is often negative, indicating that as the amplitude A increases, a point is reached at which the voltage in the fundamental harmonic (i.e., the coefficient of $\sin(\omega t)$) will decrease. A common measure of linearity performance is the *input-referred 1-dB compression point*, or IP1dB, or simply P_{1dB}. This is the input amplitude A at which the output is 1 dB (on a $20\log_{10}$ scale) below what the output would be if a_3 were zero:

$$\frac{a_1 A + \frac{3}{4}a_3 A^3}{a_1 A} = 0.891 \rightarrow A = \sqrt{\frac{4}{3}\left|\frac{a_1}{a_3}\right|} \times 0.109. \quad (5.94)$$

It is also common to refer to the output signal magnitude at which the output at the fundamental frequency is 1 dB below $a_1 A$, and this is called the *output 1 dB compression point*, OP1dB. Fig. 5.46 illustrates the 1 dB compression point. If an infinite number of terms of (5.83) were included in the analysis to accurately model the output power over all input signal ranges, then one would find that at a certain point the output power compresses to the point that it stops increasing with increased input power. The output power at which this occurs is called the *saturated output power*. When the amplifier operates near this saturated output power, it is said to operate in the saturated power regime. When the amplifier operates below its saturated output power, it is said that the amplifier is operating under "back-off." The amount of back-off, usually specified in decibels, is the amount by which the input/output power is below the input/output power for the device when it is saturated.

Another common test is the *two-tone test*, in which two tones are simultaneously applied to the device or the radio, one at ω_1 and a second at ω_2. This gives rise to intermodulation, in which the mixing of the two tones can result in jamming of the desired signal. From (5.93), it can be seen that the third harmonic for a single tone test rises at thrice the rate of the fundamental harmonic. For a two-tone test with input $A_1 \sin(\omega_1 t) + A_2 \sin(\omega_2 t)$, the output is

$$a_1[A_1 \sin(\omega_1 t) + A_2 \sin(\omega_2 t)]$$
$$+ a_2\left[\frac{A_1}{2} + \frac{A_2}{2} - \frac{A_1}{2}\cos(2\omega_1 t) - \frac{A_2}{2}\cos(2\omega_2 t)\right.$$
$$\left. + A_1 A_2 \left\{\cos(\omega_1 - \omega_2)t - \cos(\omega_1 + \omega_2)t\right\}\right]$$

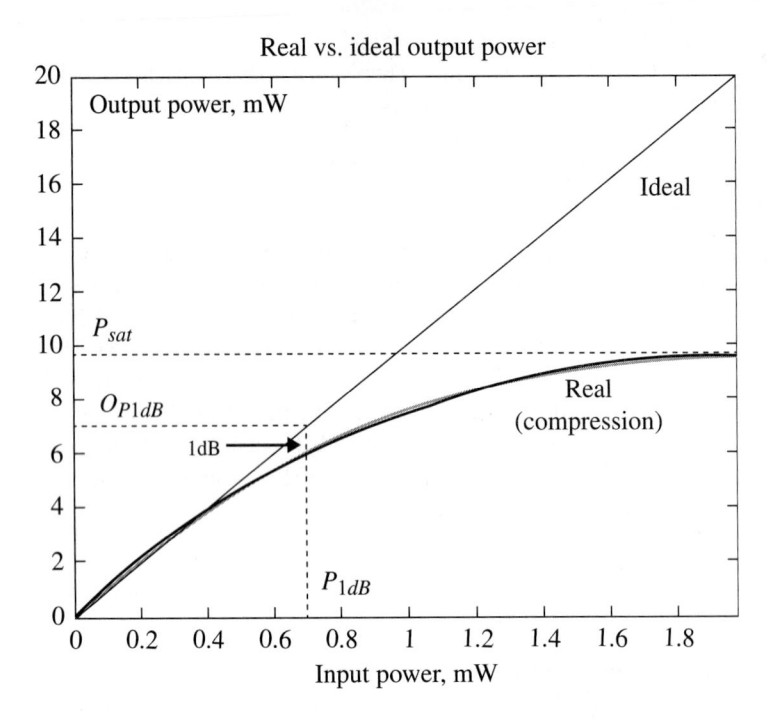

Figure 5.46 The non-linearity of most devices results in the compression of the output power of the fundamental harmonic.

$$+ a_3 \left[\frac{A_1}{4}^3 \{3\sin(\omega_1 t) - \sin(3\omega_1 t)\} \right. \tag{5.95}$$
$$+ 3A_1^2 A_2 \left\{ \frac{1}{2}\sin(\omega_2 t) + \frac{1}{4}\sin(\omega_2 + 2\omega_1)t + \frac{1}{4}\sin(\omega_2 - 2\omega_1)t \right\}$$
$$+ 3A_2^2 A_1 \left\{ \frac{1}{2}\sin(\omega_1 t) + \frac{1}{4}\sin(\omega_1 + 2\omega_2)t + \frac{1}{4}\sin(\omega_1 - 2\omega_2)t \right\}$$
$$\left. + A_2^3 \left\{ \frac{3}{4}\sin(\omega_1 t) - \frac{1}{4}\sin(3\omega_2 t) \right\} \right].$$

It is customary to assume that $A_1 = A_2$ and that $2\omega_1 - \omega_2 \approx \omega_1$, where the desired signal is at ω_1.

Fig. 5.47 illustrates in a log scale the different non-linearities that result from equation (5.83). The y-intercept of the fundamental frequency is the gain of the system. Once the linearity parameters of a single block such as IIP3 and P1dB have been defined, it is often necessary to determine the linearity of a cascaded set of components or circuit blocks. For example, in some mmWave systems it may be necessary to cascade multiple amplifiers to obtain enough gain for link budget requirements. A multi-amplifier approach can also be beneficial for improved power efficiency. The types of amplifiers used in such a cascade will

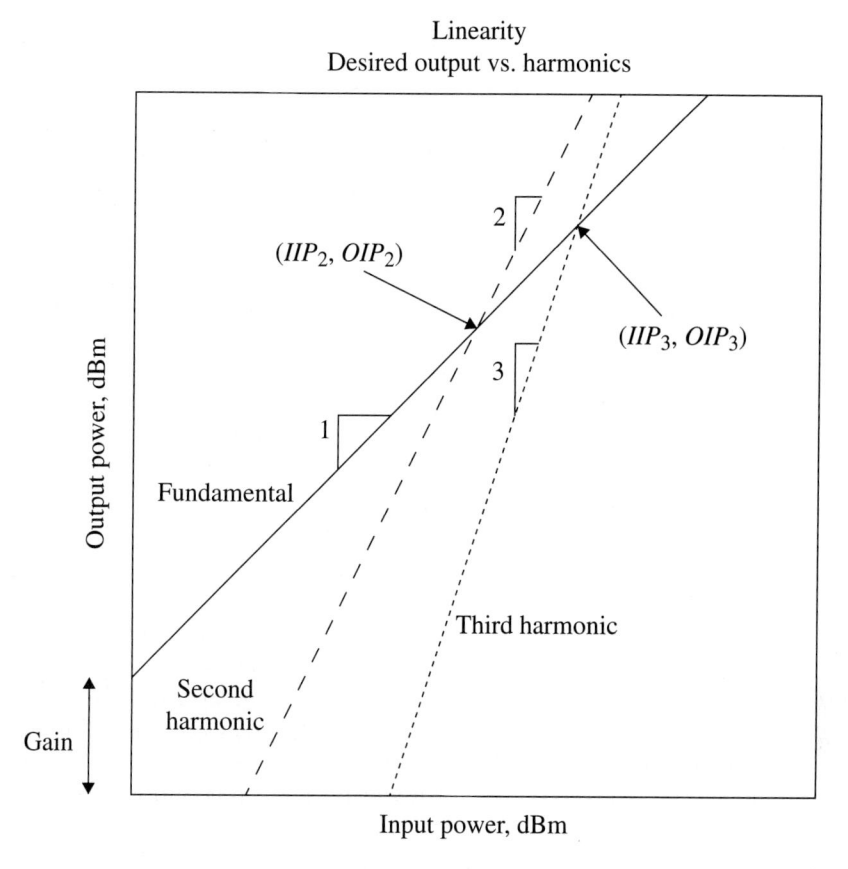

5.10 Important Metrics for Analog MmWave Devices

Figure 5.47 The non-linear nature of most signal blocks results in the various measures of non-linearity, including IP1dB, IIP3, and IP2. It is important to understand that the power in the nth harmonic grows n times as quickly as the power in the fundamental.

have a major impact on linearity. A drawback of the Class AB and Class E amplifiers, which will be described in the next section, is that they are not as linear as Class A amplifiers [Poz05]. However, non-linear amplifiers are often more efficient than their linear counterparts. If a modulation scheme requires a high degree of linearity or high peak-to-average-power ratio (PAPR), then these more efficient amplification techniques may not be feasible.

If a cascaded design is used for increased gain or output power, it is necessary to size each stage such that the last stage compresses first to prevent the dynamic range and output current of the last stage from being wasted [BKPL09]. This means that the last stage should be the first to experience an input power that would result in the last stage's compression (i.e., an input power above the last stage's P1dB). The last stage of the amplifier will dominate the linearity, as reflected in the expressions for the third order voltage input intercept point for an arbitrary number of cascaded devices and the output power third order intercept point for a design with three cascaded stages

[Lee04b, Chapter 13][YGT+07] (illustrated in Fig. 5.48):

$$\frac{1}{IIV3_{tot}^2} = \sum_{j=1}^{n} \left\{ \frac{1}{IIV3_j^2} \prod_{i=1}^{j} A_{vi}^2 \right\} \quad (5.96)$$

$$\frac{1}{OP1dB_{cascade}} = \frac{1}{OP1dB_3} + \frac{1}{OP1dB_2 \times G_3} + \frac{1}{OP1dB_1 \times G_2 \times G_3} \quad (5.97)$$

where $OP1dB_i$, G_i, $IIV3_i$, and A_{vi} are the output power third order intercept, power gain, input voltage third order intercept, and voltage gain of the ith stage.

5.10.2 Noise Figure and Noise Factor

The noise performance of a system, device, or cascade of circuits is also very important in determining the signal-to-noise ratio (SNR). The noise performance is specified in terms of a *noise figure* F, which indicates the amount of additional noise contributed by a device or system compared with the amount of noise applied to the input of said device or system. In other words, the noise figure is the amount by which a block, such as a low noise amplifier (LNA), will degrade the signal-to-noise ratio in decibels, as illustrated in Fig. 5.49. The noise figure is also sometimes given in absolute (linear) terms, not decibels, in which case it is called the *noise factor* (*NF*). It is common for engineers to only use the term *noise figure*, since it is understood that NF and F mean the same thing, with noise figure (F) typically represented in decibels. However, the term *noise figure* is commonly used in place of noise factor when using absolute (linear) values, as we show in (5.98). Under matched conditions (i.e., conditions in which the output impedance of a block is

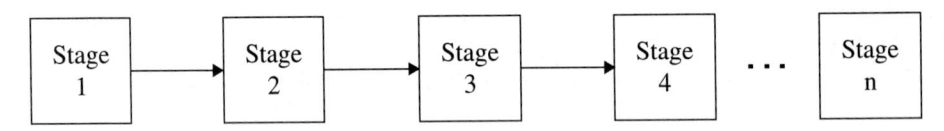

Figure 5.48 A cascaded stage of circuits used to create gain stages, or to implement proper voltage or current levels.

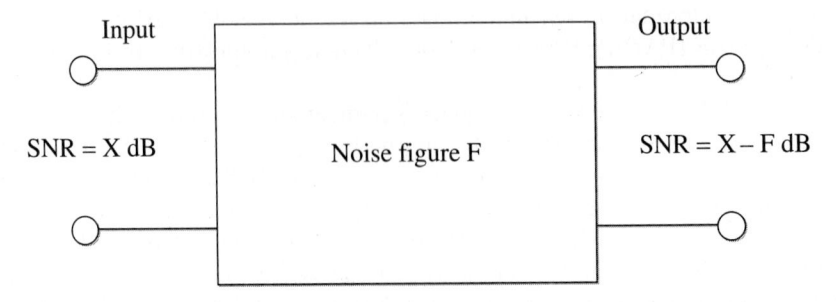

Figure 5.49 The noise figure indicates the amount of excess noise contributed by a device and is a measure of how SNR is degraded by additive self-noise of a device [Rap02].

conjugately matched to the input impedance of the block that follows it), the following equation applies for the noise figure of a cascaded system [Cou07][Rap02]:

$$F_{\text{sys}} = F_1 + \frac{F_2 - 1}{G_1} + \frac{F_3 - 1}{G_1 G_2} + \cdots \qquad (5.98)$$

where F_1 is the noise figure of the input block, and F_i is the noise figure of the ith block in the cascade. G_1 is the power gain of the first input block, and G_i is the power gain of the ith block in the cascade. F_{sys} is the noise figure of the overall cascaded system, where $F \geq 1$, and $F = 1$ (e.g., a noise figure of 0 dB) is an ideal noiseless system such that there is no additional noise temperature contribution from the noiseless device (see (5.79) and [Cou07][Rap02]).

5.11 Analog MmWave Components

This section describes the specifications and key fundamentals of analog building blocks that make up the amplifiers, mixers, oscillators, and other analog components required in any mmWave communication system.

5.11.1 Power Amplifiers

A power amplifier (PA) is usually the last active block in a transmitter before the antenna. The goal of a power amplifier is to provide as much output power as possible while maintaining high efficiency and sufficient linearity. Unlike low noise amplifiers (LNAs), the noise performance of a power amplifier is not critically important because it is preceded by several stages of gain, and because it is in the transmitter, not the receiver. Table 5.2 lists several important attributes of state-of-the-art mmWave PAs and LNAs from the literature. To understand the requirements for power amplifiers for mmWave devices, it is useful to examine several examples of power amplifiers, as shown in Table 5.3 from [RMGJ11].

It can be seen from Table 5.3 that the most important criteria by which to judge a power amplifier are its gain, output power (i.e., saturated output power), Power Added Efficiency (PAE), and consumed power. The output power is important, as seen from the link budget calculations of (5.78), because the received power at the receiver is directly proportional to the output power of the power amplifier. The gain is important if the blocks that precede the PA do not have sufficient output powers to ensure an adequate link budget. The linearity of the power amplifier is also important as it follows all of the gain stages that precede it, so it may need to handle large input signals. Metrics of power compression such as P_{1dB} and intermodulation distortion such as IIP3 are therefore important also. The best approach for designing a power amplifier is dependent on the chosen topology, which in turn depends on the intended application. A useful figure of merit FoM_{PA} that can be used to vet different designs is defined in the International Technology Roadmap for Semiconductors (ITRS) as [YGT+07]

$$FoM_{PA} = P_{\text{out}}\ G\ \text{PAE}\ f^2 \qquad (5.99)$$

where P_{out} is the output power, G is the gain, PAE is the *power added efficiency*, and f is the design frequency, which is included to account for the fact that power gain of a transistor falls as $(\frac{f_T}{f})^2$ [Poz05].

Table 5.2 Key attributes and comparisons of mmWave power amplifiers and low noise amplifiers.

Reference, Year	TX	RX	Output Power	Gain	Noise Figure	IIP_3^1, $IP_{1\,dB}$, etc.	Bandwidth	Power Cons.	Size	Process
[MID+00], 2000	x	x	PA: 12 dBm, Total TX: 10 dBm	PA: 12 dB, LNA: 18 dB, Complete RX: 10 dB	LNA: 5 dB	N/A	Complete TX/RX: 59-60 GHz	N/A	$\sim 2 \times 1$ cm^2	Separate III-V MMICs on LTC
[OMI+02], 2002	x	x	PA: 14 dBm, Total TX: >10.6 dBm	PA: 12 dB, LNA: 18 dB	N/A	N/A	LNA: 59-64 GHz, TX: 1.74 GHz	N/A	TX + RX: 82 mm × 53 × 7 mm^2	Separate III-V MMICs
[BFE+04], 2004		x	10 dBm to antenna	LNA: 33 dB	Total TX: 6.5 dB	N/A	Total RX: 5 GHz	N/A	N/A	Separate III-V MMICs
[FRP+05], 2005	x	x	PA: 16.2 dBm, VCO: −8 dBm	LNA: 14.7 dB, RX w/o LNA: 18.6 dB,	LNA: 4.5 dB, RX w/o LNA: 14.8 dB	RX w/o LNA IP1dB: −17 dBm, LNA P1dB: −20 dBm	VCO: 65.8-67.9 GHz	LNA: 10.8 mW RX w/o LNA: 302 mW, PA: 270 mW	LNA: 0.9 × 0.6 mm^2, RX w/o LNA: 1.9 × 1.65 mm^2, PA: 2.1 × 0.8 mm^2	Separate 0.12 μm SiGe Bipolar MMICs
[GKZ+05], 2005	x	x	Total TX: 3.71.5 dBm	Total RX: 7.11.5 dB, Total TX: 5.2 dB	Total RX: 10.5 dB	RX IIP$_3$: −11 dBm, −19 IP1dB, TX: OP1dB: 0 dBm	TX: 54-61 GHz, RX: 59.5-64.5 GHz	Total TX: 820 mW, Total RX: 990 mW	Total RX: 5.7 × 5.0 mm^2, Total TX: 5.0 × 3.5 mm^2	Separate GaAs pHEMT MMICs

Table 5.2 (continued)

[Raz06], 2006	×	N/A	RX: 28 dB Voltage Gain	Total RX: 12.5 dB	Total RX IP1dB: −22.5 dBm	57-64 GHz	Total RX: 9 mW	.3 × .4 mm² excluding pads	Single 0.13 μm CMOS chip
[ACV06], 2006	×	N/A	Total RX: 16 dB	RX: <7 dB	Total RX P1dB: −21 dBm	58.5-60.5 GHz	60 mW	0.6 × 0.475 mm²	Single 90 nm CMOS chip
[SHW+06], 2006	×	N/A	LNA: 18 dB, Mixer; 10.8 dB, Total RX: 28 dB	LNA: 6.8 dB, Mixer: 14 dB (simulated)	Total RX OP1dB: −1.6 dBm	57-64 GHz	LNA: 66 mW, Mixer: 21 mW	0.8 mm²	Single SiGe:C BiCMOS chip
[RFP+06], 2006	×	Total TX: 15-17 dBm	RX: 38-40 dB, TX: 34-37 dB	Total RX: 6 dB	Total RX IIP₃: −30 dBm, Total TX OP1dB: 10-12 dBm	55-64 GHz, 1.5 GHz VCO tuning	RX: 500 mW TX: 800 mW	Total RX: 3.4 × 1.7 mm², Total TX: 4 × 1.6 mm²	Single 0.13 μm SiGe BiCMOS chip
[MFO+07], 2007	×		LNA: 13.7 dB	LNA: 7.8 dB (simulated)	N/A	VCO: 61.2-64.4 GHz, PLL: 1.7 GHz	LNA: 39 mW, Total RX: 144 mW	2.4 × 1.1 mm² without pads	Single 90 nm CMOS chip
[DSS+09], 2009	×	PA: 2.1 dBm, Total TX: 5.7 dBm	PA: 17 dB, Total TX: 8.6 dB	N/A	PA OP1dB: 2.1 dBm, Total TX OP1dB: 1.5 dBm	VCO: 48.5-55 GHz, Total TX: 57-65 GHz	PA: 44 mW, Total TX: 76 mW	1.4 × 1.5 mm²	Single 90 nm CMOS chip

continued on next page

Table 5.2 (continued)

Reference, Year	TX	RX	Output Power	Gain	Noise Figure	IIP_3^1, IP_{1dB}, etc.	Bandwidth	Power Cons.	Size	Process
[DSS+09], 2009	x		PA: 8.4 dBm, Total TX: 8.6 dBm	PA: 17 dB, Total TX: 12.4 dB	N/A	PA OP1dB: 8.4 dBm, Total TX OP1dB: 4.1 dBm	VCO: 53.4-55.7 GHz, Total TX: 57-65 GHz	PA: 54 mW, Total TX: 112 mW	1.3×1.5 mm^2	Single 90 nm CMOS chip
[PR09], 2009		x	Total TX: −7.2 dBm	Total RX: 19.8-22 dB	Total RX: 5.7-7.1 dB	Total RX IP1dB: −27.5 dBm, Total TX OP1dB: −8.6 dBm	N/A	Total RX: 36 mW, Total TX: 78 mW	Total RX: $0.5 \times .37$ mm^2 active area, Total TX: 0.495×0.425 mm^2 active area	Single 90 nm CMOS chips (separate TX & RX)
[TAY+09], 2009	x	x	Total TX: −0.7 dBm	Total RX: 8.9 dB	Total RX: 5.8 dB	Total RX IP1dB: −22 dBm	Total TX: 50-66 GHz	Total TX + RX: 232 mW	Total TX+RX: 1.28×0.81 mm^2	Single 65 nm CMOS chip (combined TX and RX)

[1] IIP3 is the input referred third order compression point, and is the extrapolated input power at which the third order harmonic overtakes the first order harmonic in the output.

$IP_{1\,dB}$ is the input referred 1 dB compression point, and is the input power at which the output power is 1 dB below expected output power based on extrapolation of output power with input power at low input powers.

5.11 Analog MmWave Components

Table 5.3 Several examples of power amplifiers [from [RMGJ11] © IEEE].

Reference	Topology	Gain and Output Power (measurement frequency)	PAE, Power Consumption, Supply Voltage
[YGY+06]	3 single ended CS (Common Source) Class A[1] in 90 nm CMOS	5.2 dB gain, 9.3 dBm (61 GHz)	7% PAE, 39.75 mW from 1.5 V
[FRP+05]	2-stage AC balanced CE (Common Emitter) in 0.12 μm SiGe HBT	9 dB, 10 dBm (61.5 GHz)	143 mW from 1.1 V
[PG07a]	Single-stage push-pull with microstrip and differential cascode in 0.13 μm SiGe BiCMOS	18 dB, 13.1 dBm	12.7% PAE, 248 mW from 4V
[HBAN07a]	Cascaded 2-stage, CS to CS in 90 nm CMOS	9.8 dB, 6.7 dBm (56 GHz)	20% PAE
[LLC09]	Transformer-coupled 3-stage cascade in 90 nm CMOS	15 dB, 12.2 dBm at 61 GHz	84 mW from 1.2 V
[CRN09]	2-stage transformer-coupled cascode to CS, with differential-to-single ended conversion through transformer in 90 nm CMOS	5.6 dB, 12.3 dBm	8.8% PAE, 1 V
[CRN09]	3-stage transformer coupled cascode to CS to CS in 90 nm CMOS	13.8 dB, 11.0 dBm	14.6% PAE, 1 V supply
[AKBP08]	Cascaded CS to CS in 65 nm CMOS	7.6 dB, 8.9 dBm	PAE <11%, 64.8 mW from 0.9 V
[WSE08]	Doherty in 0.13 μm CMOS	13.5 dB, 7.8 dBm	3.0% PAE, 1.6 V
[DSS+08]	3-stage cascode to CS in 90 nm CMOS	17 dB, 8.4 dBm	5.8% PAE, 54 mW

[1] Class A indicates that the amplifier is highly linear and operates as a current source over the entire period.

In 2009, Zhang et al. [ZL09] published Table 5.4 summarizing the transmit power regulations for 60 GHz in 7 different regulatory regions. Table 5.4 indicates the levels of output power that should be achieved by single power amplifiers or amplifiers used in an array. For example, in the USA a mmWave array may use up to 500 mW of transmit power, and the power amplifiers of the device must supply this power. Table 5.3

Table 5.4 Transmission power regulations for 60 GHz band [from [ZL09] © IEEE].

Countries	Frequency Band (GHz)	Maximum TX Power (mW)	Maximum Antenna Gain (dBi)
Japan	59–66	10	47
USA	57–64	500	Not Specified (NS)
Canada	57–64	500	NS
Australia	59.4–62.9	10	NS
Europe	57–66	20	37
China	57–66	10	NS
Korea	57–64	10	NS

indicates that 10 dBm or 10 mW is a typical output power for mmWave power amplifiers. One approach to improve output power is to use power combining, which becomes very important at mmWave frequencies where transistor power handling capability is lower. This approach can also be used to ease antenna gain requirements provided regulatory restrictions are met [Emr07]. A key challenge to power combining is achieving a matching structure with a wide enough bandwidth that can support more than just a few elements [Emr07]. [Emr07] reported that typical schemes limit the number of elements combined with any power combining system to under 10.

The efficiency of a power amplifier is described by the percentage of its consumed power that drives its load. This is often explained in terms of the power added efficiency (PAE), which is defined as

$$\text{PAE} = \frac{P_{RFout} - P_{RFin}}{P_{DC}} \tag{5.100}$$

where P_{RFout} is the output power at the operating frequency, P_{RFin} is the input power at the operating frequency, and P_{DC} is the power consumed by the device. It is also common to refer to the efficiency η, which is simply the ratio of the output RF power to the DC power

$$\eta = \frac{P_{RFout}}{P_{DC}}. \tag{5.101}$$

Take as an example a simple common-source amplifier (*common-source* indicates that the source of the device is connected to ground), as shown previously in Fig. 5.45.

Now assume that the DC input voltage $V_{in,DC}$ is greater than the threshold voltage of the NMOS transistor. For an input impedance equal to the gate resistance followed by the gate to source capacitance, the input RF power is equal to

$$P_{RF,in} = \frac{V_{in,RF}^2}{2} \left(\frac{j\omega C_{gs}}{1 + j\omega C_{gs} R_g} \right) \approx \frac{V_{in,RF}^2}{2} j\omega C_{gs}. \tag{5.102}$$

5.11 Analog MmWave Components

Assuming square-law operation, the drain current I_D is given by

$$I_D = \frac{1}{2}k_n \left(\frac{W}{L}\right)\left[(V_{inDC} - V_t)^2 + v_{in,RF}(V_{inDC} - V_t) + v_{in,RF}^2\right]$$
$$\approx \frac{1}{2}k_n \left(\frac{W}{L}\right)\left[(V_{inDC} - V_t)^2 + v_{in,RF}(V_{inDC} - V_t)\right]. \qquad (5.103)$$

Therefore, the output RF power is given by

$$P_{RF,out} = \frac{1}{4}k_n \left(\frac{W}{L}\right)\left[v_{in,RF}(V_{inDC} - V_t)\right]R_L \qquad (5.104)$$

and the DC power (approximately the "consumed power") is given by

$$P_{DC} = \frac{1}{2}k_n \left(\frac{W}{L}\right)\left[(V_{inDC} - V_t)^2\right]R_L. \qquad (5.105)$$

The PAE is then found to be

$$\text{PAE} = 50\% - v_{in,RF}^2 \left(\frac{j\omega C_{gs}}{R_L}\right)\left[\frac{1}{k_n\left(\frac{W}{L}\right)(V_{inDC} - V_T)^2}\right]. \qquad (5.106)$$

As the aspect ratio W/L of the device becomes large, much greater than 10, the PAE approaches 50%. This type of amplifier is technically called a Class A amplifier, which indicates that the input voltage is always high enough to cause the drain current to flow. Note that by using the square law, this analysis is limited to low to moderate output powers. For high output powers, the drain-source voltage may not be high enough for the transistor to operate in the square-law saturation regime. The amplifier is still Class A so long as its gate-source voltage never falls below the threshold voltage.

The voltage gain of the example amplifier is given by

$$A_v = \frac{\frac{1}{2}k_n\left(\frac{W}{L}\right)[v_{in,RF}(V_{inDC} - V_t)]R_L}{v_{in,RF}} = \frac{1}{2}k_n\left(\frac{W}{L}\right)(V_{inDC} - V_t)R_L. \qquad (5.107)$$

The gain of a power amplifier may be important in link budget calculations if previous stages of the power amplifier have low output powers. Note that the power gain per transistor is proportional to $20\log\left(\frac{f_T}{f}\right)$ [Poz05][NH08], and because the transit frequency f_T has increased less quickly with the advance of technology and new process generations than the operating frequency in the transition to mmWave regimes, each transistor is capable of less gain than at lower frequencies (e.g., a 90 nm process has an f_T of 120 GHz, so that at 60 GHz a 90 nm CMOS transistor can produce approximately 6 dB power gain, whereas at 5 GHz it is capable of 27 dB power gain). This will often necessitate multi-stage power amplifier designs [YGT+07].

There are several *classes* of power amplifiers: Classes A, B, AB, C, D, E, and F. Classes A, B, AB, and C are types of linear amplifiers, in the sense that their output power is proportional to their input power. Classes D, E, and F are types of non-linear amplifiers in which the transistor is operated as a switch.

The distinction between Class A, B, AB, and C amplifiers is their different *conduction angles*, denoted θ [Kaz08]. The conduction angle θ refers to the percentage of a half period of the input sinusoid over which the transistor conducts current. For Class A amplifiers as described previously, drain current is conducted over 100% of the period, so the conduction angle θ is 180° over the half period, and 2θ is 360° over the whole period. For Class B amplifiers, current is conducted over exactly 50% of the period of the input sinusoid, so the conduction angle 2θ is 180° for the entire period. For Class AB amplifiers, the conduction angle for the entire period is between 180° and 360°, indicating that current is conducted between 50% and 100% of the period. For Class C amplifiers, the drain current is non-zero over less than half of the input period, so the conduction angle for the entire period 2θ is less than 180°. The bias point of the transistor determines the class of the amplifier. Specifically, Class A amplifiers must have their gate-source voltage (sum of DC and RF voltage) greater than the threshold voltage at all times [Kaz08]. Class B amplifiers are biased such that their input DC voltage is equal to the threshold voltage [Kaz08]. Class C amplifiers have their inputs biased at less than the threshold voltage [Kaz08]. Class AB amplifiers are biased with their DC input voltages above the threshold voltage, but such that with the addition of the RF input voltage, the gate-source voltage falls below V_t over part of the input period [Kaz08]. Fig. 5.50 illustrates the bias conditions for these classes of amplifiers.

The conduction angle determines the efficiencies of the linear amplifiers. To see this, first note that we may represent the conduction current of each class of amplifier as our drain current expression multiplied by a square wave $S(t)$ with an appropriate duty cycle d

$$S(t) = \Pi_p(t) \star \sum \delta(t - nT) \tag{5.108}$$

where $\star$ represents convolution, δ is a Dirac delta function, T is the period of the waveform and is equal to the inverse of the operating frequency f, and Π_p is a square pulse with support of p (i.e., is non-zero from $-p/2$ to $p/2$). ω is the angular operating frequency equal to $2\pi f$. The duty-cycle of the square wave $S(t)$ is $d = p/T$. The conduction angle θ is equal to $180° \times d$ in degrees, or πd in radians.

For simplicity, when analyzing the efficiency of power amplifiers, we usually write the drain current and output voltage as [Kaz08]

$$I_d = I_m \cos(\omega t) S(t) \tag{5.109}$$
$$V_d = V_I - V_{RF} \cos(\omega t) \tag{5.110}$$

where V_I is the DC drain voltage and V_d is the magnitude of the RF drain voltage. Note that the square law has not been invoked to represent the drain current. This is because the square law implies operation in the saturation region of the device, which would limit this analysis unnecessarily. The nth Fourier series coefficient of the current waveform is

$$a_{n,n>0} = 2f \int_{-p/2}^{p/2} f(t) \cos(n 2\pi f t) dt,$$

$$b_{n,n>0} = -2f \int_{-p/2}^{p/2} f(t) \sin(n 2\pi f t) dt$$

5.11 Analog MmWave Components

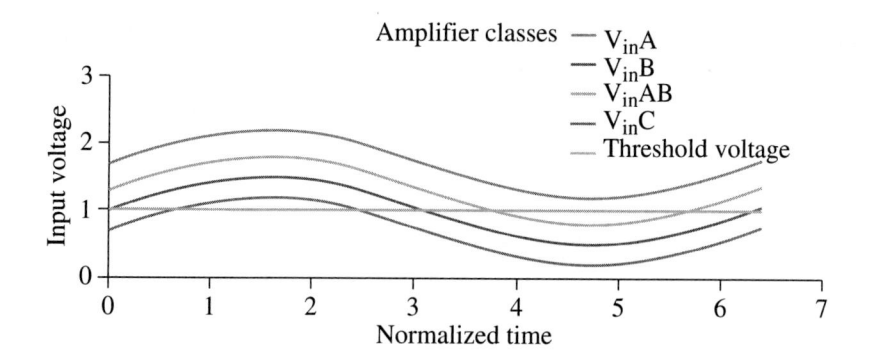

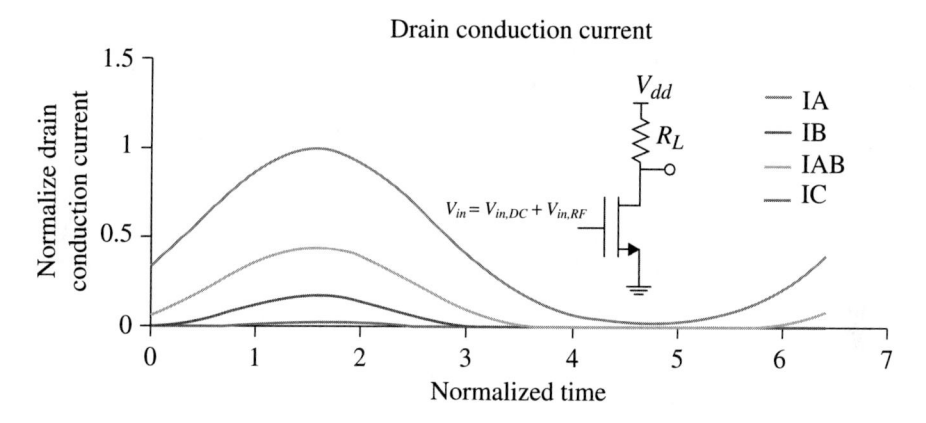

Figure 5.50 The bias point of the amplifier determines the amplifier's class. Class A amplifiers conduct current over the entire period. Class B amplifiers conduct over half the period, Class C conduct over less than half the period, and Class AB conduct over more than half the period, but less than the entire period.

$$a_0 = \frac{1}{\pi}\sin(\pi d),$$
$$a_{n,n>0} = \frac{1}{\pi}\left[\frac{\sin((n-1)\pi d)}{n-1} + \frac{\sin((n+1)\pi d)}{n+1}\right], \quad b_n = 0. \quad (5.111)$$

Using the Fourier series of the square wave coefficients, the drain current may be written as

$$I_d = \frac{1}{\pi}\sin(\pi d)I_m + \frac{1}{\pi}\left[\pi d + \frac{\sin(2\pi d)}{2}\right]I_m\cos(\omega t) + \ldots \quad (5.112)$$

See that the DC current is related to the magnitude of the RF conduction current over the conduction period

$$I_{DC} = \frac{I_m}{\pi}\sin(\pi d) \quad (5.113)$$

and the current at the fundamental frequency is found to be

$$I_{RF} = \frac{I_m}{\pi}\left[\pi d + \frac{\sin(2\pi d)}{2}\right]. \tag{5.114}$$

The DC power is therefore given by

$$P_{DC} = V_I I_{DC} = \frac{V_I I_m}{\pi}\sin(\pi d). \tag{5.115}$$

The RF power is then

$$P_{RF} = \frac{V_{RF} I_m}{2\pi}\left[\pi d + \frac{\sin(2\pi d)}{2}\right]. \tag{5.116}$$

By taking the ratio of P_{RF} and P_{DC} we can find the efficiency:

$$\eta = \frac{\frac{V_{RF}}{V_I}[2\pi d + \sin(2\pi d)]}{4\sin(\pi d)} = \frac{\frac{V_{RF}}{V_I}[2\theta + \sin(2\theta)]}{4\sin(\theta)} \tag{5.117}$$

where θ is the conduction angle. As previously stated, the conduction angle for a Class B amplifier is $90° = \pi/2$. Therefore, the efficiency (not the PAE, as RF input power is not considered here) of a Class B amplifier is given by

$$\eta_{\text{CLASSB}} = \frac{\frac{\pi}{4}V_{RF}}{V_I} \tag{5.118}$$

with a maximum value of $\pi/4$ for the highest possible RF voltage, which is equal to V_I. The tradeoff between efficiency and output power is indicated by (5.117) and (5.116): as the conduction angle decreases, the efficiency increases, but the output power decreases. In addition to trading efficiency and output power, these designs trade linearity and efficiency. In general, highly linear power amplifiers will have lower efficiencies. When efficiency is the design goal, then non-linear amplifiers, where the power device may spend considerable time in saturation, are preferable [FPdC+04]. If efficiency and linearity together are desired, such as for applications intended to operate a long time between battery recharging, it may be necessary to consider approaches such as Doherty (explained below), envelope elimination and restoration, outphasing, digital polar modulation, and transformer-based power combining [ALK+09]. At lower frequencies another approach has been voltage supply regulation with DC-DC converters, though these have found limited use in mmWave systems. Transformer-based approaches generally give the highest efficiency in circuit-board implementations, so they are worth investigating for on-chip mmWave applications.

The Doherty design, shown in Fig. 5.51, has traditionally been used in cellular base stations to improve efficiency under back-off conditions as high as 6 dB. Three-stage structures are also possible for greater improvements in efficiency [ALK+09]. Traditionally, this architecture has been difficult to implement on-chip or in handsets due to the large size of the transmission lines required [ALK+09], but at mmWave frequencies this difficulty is largely resolved as evidenced by 60 GHz on-chip implementations [WSE08]. The basic theory of its operation is to combine a Class A main amplifier with Class B

5.11 Analog MmWave Components 333

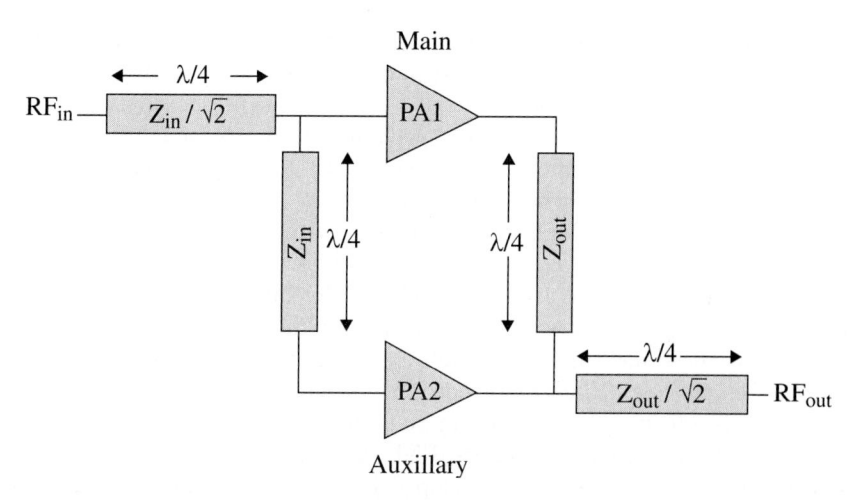

Figure 5.51 The Doherty amplifier uses an auxiliary power amplifier in conjunction with a primary power amplifier. The clever power combining technique provides additional power efficiency improvements for modulations that require some back-off. Greater power efficiency is achieved by allowing the primary power amplifier to operate at an efficient operating point for the back-off power level. When more power is needed, the auxiliary amplifier contributes power to the primary amplifier to meet instantaneous needs, without compromising linearity of the amplified signal. Both amplifiers may be set to operate at more efficient operating points for greater overall power efficiency in the amplifier circuit [from [WSE08] © IEEE].

or Class AB auxiliary amplifier. In their design, Wicks et al. used a five-stage cascade of cascodes for both the main and auxiliary amplifier to increase gain of each.

When working with modern power amplifiers, the terms *back-off* and *peak-to-average-power ratio* are commonly used. Chapter 7 has more detail on peak-to-average-power ratios (PAPRs). For power amplifiers, this refers to the ratio of the peak output power to the average output power that the amplifier must produce for the modulation scheme used. A high PAPR forces the amplifier to operate in the "back-off" region, where it is more linear but less efficient.

Class D, E, and F amplifiers can be made much more efficient than Class A, B, AB, or C power amplifiers. In principle, Class D, E, and F amplifiers are switching amplifiers that are similar to DC-DC converters. Their precise operation is beyond the scope of this book because these amplifiers have not been used as frequently in mmWave applications.

Because power amplifiers must often provide high output powers, the designer may need to consider their reliability because high output powers will stress the transistors in the amplifier and may lead to breakdown. Common ways in which a device can break include hot-carrier degradation, dielectric breakdown, and negative bias temperature instability [MTH+08]. Use of CMOS for power amplifier design results in increased susceptibility breakdown effects/degradation effects compared with amplifiers constructed in compound semiconductor processes (e.g., GaAs) [MTH+08]. When a design requires more gain, it is sometimes attractive to increase the supply voltage, as this will increase device gain and output power. Maruhashi et al. found that this resulted in a decreased expected lifetime by a factor of approximately 1000 ($> 10^8$ hours to $> 10^5$ hours for higher supply voltage) [MTH+08].

After considering required output power, gain, linearity, and efficiency to determine the targeted power amplifier topology, a load-pull analysis may be used to more rigorously design the power amplifier. Load-pull analysis is a technique for developing a large-signal model of a power amplifier that takes into account the source and load impedance on the amplifier [Ito00]. This technique can be especially useful for achieving high power densities and power added efficiencies. For example, in 2008 Ferndahl et al. used load-pull simulations with 40 nm CMOS for a transistor with a gate width of 192 μm (W/L of nearly 5000) to achieve a PAE of 33%, an output power of 11.7 dBm per transistor, and a power density of 100 mW/mm at 35 GHz [FNP+08]. The basic principle of the load-pull technique is to measure a transistor or amplifier with varying load and source impedance. From these measurements, plots of constant output power, gain, linearity, and power added efficiency can be plotted on a Smith chart to facilitate future designs [dMKL+09]. Load-pull characterization can be performed by varying the load and source impedance with Maury impedance tuners [FNP+08].

Recently, Buckwalter, et al. have developed high-efficiency mmWave amplifiers in a 45 nm Silicon on Insulator (SOI) process using stacks of FET transistors with floating bias voltages to avoid device breakdown. Their results show promising efficiencies and reasonably large power outputs from 45 to 90 GHz [DHG+13] [AJA+14].

5.11.2 Low Noise Amplifiers

Unlike for power amplifiers, the noise performance of low noise amplifiers (LNAs) is their most important figure of merit. The low noise amplifier is typically the first active block in the receiver chain following the receive antenna. While gain is important for power amplifiers, it is very important for low noise amplifiers. The role of low noise amplifier is to boost the received signal strength without adding substantial noise. As seen from equation (5.98), by keeping the noise figure of the low noise amplifier low and its gain high, the noise performance of the entire receiver chain is improved. The International Roadmap of Semiconductors (ITRS) provides a useful figure of merit for low noise amplifiers:

$$FoM_{\text{LNA}} = \frac{G \times IIP3 \times f}{(F-1) \times P} = \frac{OIP3 \times f}{(F-1) \times P} \qquad (5.119)$$

where F is the noise factor, G is the gain, f is the frequency, and $OIP3$ is the output third order intercept gain compression point. At mmWave frequencies, successful design becomes more challenging for several reasons. First, there is the need for more stages to achieve needed gain and the greater variations in device parameters for scaled devices [NH08]. Also, at mmWave frequencies, all metal can be considered a distributed passive device capable of impacting device operation, adding noise, and complicating design [STD+09]. Table 5.5, adapted from [RMGJ11], gives examples of specifications of mmWave 60 GHz LNAs.

Both transistors and resistors in an LNA will contribute noise. The basic noise source for a resistor is thermal noise [Pul10]. The basic noise sources for a transistor include thermal noise, shot noise, and flicker noise [Pul10]. Thermal noise is related to the random motion of electrons moving through a resistance [Pul10]. For a resistor, we may model thermal noise as either a series voltage source or as a parallel current source, as shown in

Table 5.5 Examples of mmWave LNAs from the literature [adapted from [RMGJ11]].

Reference	Gain	Noise Factor	Topology	Linearity IIP3/P1dB	Power, Voltage	Process
[YGY+06]	14.6 dB	4.5 dB, simulated	2-stage cascode	−6.8 dBm at 58 GHz IIP3, −0.5 dBm at 58 GHz OP1dB	24 mW from 1.5 V	90 nm CMOS
[LLW06]	20 dB from 51-57.5 GHz	8 dB from 50-57 GHz	3-stage cascode	−12 dBm at 56 GHz IIP3, −22 dBm IP1dB	79 mW from 2.4 V	0.13 μm CMOS
[Raz06]	13 dB	4.6 dB simulated	2-stage CG to CS	N/A	N/A	0.13 μm CMOS
[FRP+05]	14.7 dB at 61.5 GHz	4.5 dB at 61.5 GHz	2-stage single ended CB (Common Base) to degenerated cascode	−8.5 dBm IIP3, −20 dBm IP1dB	10.8 mW from 1.8 V	0.12 μm SiGe BJT
[Flo04]	15 dB	4.5 dB	unbalanced CB to degenerated cascode	−9 dBm IIP3, −20 dBm IP1dB	10.8 mW from 1.8 V	0.12 μm SiGe BJT
[AKD+07]	14.5 dB at 59 GHz	4.1 dB	cascode single stage	−2 dBm IIP3, +1.5 dBm OP1dB	8.1 mW from 1.8 V	0.12 μm SiGe BiCMOS

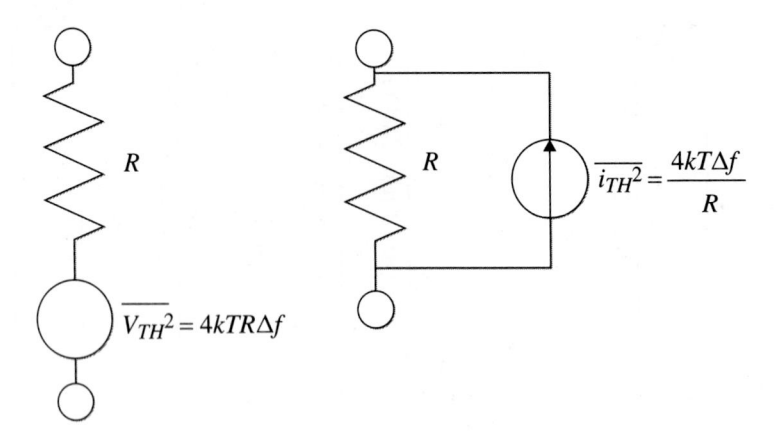

Figure 5.52 Thermal noise is the type of noise that results from resistive elements such as resistors.

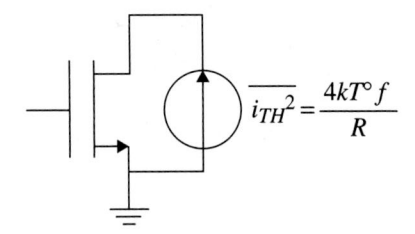

Figure 5.53 Thermal noise in a transistor.

Fig. 5.52. For a transistor, thermal noise is typically modeled as a noise current in parallel with the drain current, as shown in Fig. 5.53. In the figures, k is Boltzmann's constant, R is the resistance, T is the absolute temperature, γ is a factor that is usually equal to 2/3 for long-channel devices and is closer to 1/3 for short-channel devices, and Δf is the bandwidth of operation. In Figs. 5.54 and 5.55, the direction of current is purely arbitrary and may be changed for analysis, so long as it is kept constant throughout analysis. We write the thermal noise of a resistor as

$$\overline{V_{TH}^2} = 4kTR\Delta f \tag{5.120}$$
$$\overline{I_{TH}^2} = 4kT\gamma g_m \Delta f. \tag{5.121}$$

As indicated by (5.120) and (5.121), thermal noise is "flat" across frequencies, indicating that the amount of noise power grows linearly with the bandwidth of the system.

Shot noise has to do with the flow of charge through a potential barrier [Pul10]. Like thermal noise, shot noise is flat across frequencies. The shot noise current of a transistor is modeled as a parallel current source identical to the model for thermal noise current in Fig. 5.53. The value of shot noise is related to the DC current through the drain

$$\overline{I_{shot}^2} = 2qI_{DC}\Delta f \tag{5.122}$$

where q is the charge of an electron and I_{DC} is the DC drain current.

5.11 Analog MmWave Components

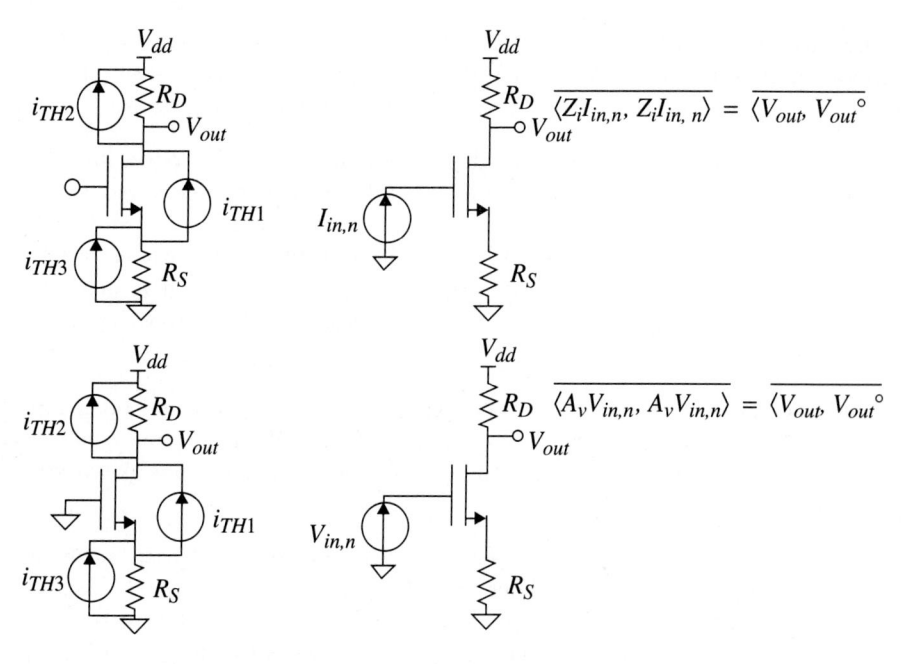

Figure 5.54 Solve for $V_{in,n}$ and $I_{in,n}$ in noise modeling.

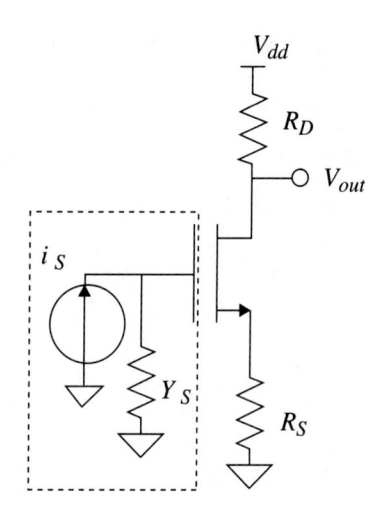

Figure 5.55 The source driving the circuit contributes its own noise.

Flicker noise differs from shot noise and thermal noise because it is not flat across frequency, but rather is proportional to $\frac{1}{f^n}$. The flicker noise is usually written in terms of a process-dependent parameter K [Lee04a]

$$\overline{I_{\text{flicker}}^2} = \frac{K}{f^n}\Delta f. \tag{5.123}$$

The parameter n is typically taken as unity [Lee04a].

As indicated by (5.119), the noise factor (or noise figure, equivalently) is one of the most important figures of merit of an LNA. [Raz01, Chapter 7] is an excellent introduction to noise modeling. To find the noise figure, we perform the following steps:

1. Place each noise source in the circuit individually. For each source, find the output voltage for both an open-circuited input and a closed-circuited input. Use superposition to write: $V_{n,out}^{open} = V_{out,source1} + V_{out,source2} + \cdots$ and $V_{n,out}^{short} = V_{out,source1} + V_{out,source2} + \cdots$. $V_{n,out}^{open}$ is the output voltage when the input is open circuited, and $V_{n,out}^{short}$ is the output voltage when the input is shorted. Each source (e.g., thermal noise source of each transistor) must be considered for both the open and shorted input case.

2. Find the total output noise power for each case — shorted and open input — by taking the auto-correlation of $V_{n,out}^{open}$ and $V_{n,out}^{short}$. We may often consider each individual source un-correlated with other sources, but this is not always true and requires careful analysis.

3. Divide the output noise voltage power for the shorted-input case by the square of the voltage gain of the amplifier to find $V_{n,in}$, the input referred noise voltage. Divide the output noise voltage power for the open input case by the square of the circuit's transimpedance (i.e., the relationship between the output voltage and an input current) to find $I_{n,in}$, the input referenced noise current.

For example, for the very simple source-degenerated single-ended amplifier shown in Fig. 5.54, we have three noise currents. If we open the input and find the output voltage from this device, we find:

$$V_{out}^{open} = -R_D(i_{th1} + i_{th2}) \rightarrow \overline{V_{out}^{open^2}} = R_D^2\left(\overline{i_{th1}^2} + \overline{i_{th2}^2}\right). \tag{5.124}$$

The transimpedance Z_s of the device is

$$V_{out} = Z_i I_{in} = -\frac{R_D g_m}{j\omega C_{gs} I_{in}}. \tag{5.125}$$

Therefore,

$$\left(\frac{R_D g_m}{\omega C_{gs}}\right)^2 \overline{I_{n,in}^2} = R_D^2\left(\overline{i_{th1}^2} + \overline{i_{th2}^2}\right) = R_D^2\left(4kT\gamma g_m + \frac{4kt}{R_d}\right)\Delta f \tag{5.126}$$

$$\overline{I_{n,in}^2} = \left(\frac{\omega C_{gs}}{g_m}\right)^2 4kT\left(\gamma g_m + \frac{1}{R_D}\right) \tag{5.127}$$

where C_{gs} is the gate-source capacitance of the transistor, and g_m is the transconductance of the transistor. If we short the input and find the output voltage,

$$V_{out}^{short} = -\frac{R_D}{2}\frac{\left(1 + \frac{j\omega}{2}C_{in}\left(2R_s + \frac{1}{g_m}\right)\right)}{1 + j\omega C_{in}\left(\frac{1}{g_m} + R_s\right)}i_{th1}$$

$$- R_D i_{th2} - \frac{R_D(1 + j\omega R_s C_{in})}{1 + j\omega C_{in}\left(\frac{1}{g_m} + R_s\right)}i_{th3} \tag{5.128}$$

5.11 Analog MmWave Components

$$\overline{V_{out}^{short^2}} = \frac{R_D^2}{4} \frac{\left(1 + \frac{\omega^2}{4} C_{in}^2 \left(2R_s + \frac{1}{g_m}\right)^2\right)}{1 + \omega^2 C_{in}^2 \left(\frac{1}{g_m} + R_s\right)^2} \overline{i_{th1}^2} \quad (5.129)$$
$$+ R_D^2 \overline{i_{th2}^2} + \frac{R_D^2 \left(1 + \omega^2 R_s^2 C_{in}^2\right)}{1 + \omega^2 C_{in}^2 \left(\frac{1}{g_m} + R_s\right)^2} \overline{i_{th3}^2}.$$

Dividing (5.129) by the voltage gain of the device we find the input referred voltage. Due to the dependence of $V_{in,n}$ and $I_{in,n}$ on i_{th1} and i_{th2}, $V_{in,n}$ and $I_{in,n}$ are correlated. Therefore, it is typical to separate $I_{in,n}$ into a term that is correlated with $V_{in,n}$ and a term that is uncorrelated with $V_{n,in}$ [Lee04a]:

$$I_{in,n} = I_c + I_u \quad (5.130)$$

where I_c is correlated with $V_{n,in}$ and I_u is not. The relationship between I_c and $V_{in,n}$ is specified in terms of a correlation admittance Y_c [Lee04a]

$$I_c = Y_c V_{n,in}. \quad (5.131)$$

Next, assume that the circuit is driven by a noise source with admittance Y_s, as shown in Fig. 5.55 [Lee04a]. With $V_{in,n}$, I_c, I_u, Y_c, and I_s the noise factor F may be found [Lee04a]:

$$F = 1 + \frac{\left(\overline{I_{in,n}^2} + |Y_c + Y_s|^2 \overline{V_{in,n}^2}\right)}{\overline{I_s^2}}. \quad (5.132)$$

[Lee04a] and Chapter 11 have a more advanced presentation of noise-figure calculations. [Lee04a] describes that there is an optimal source impedance/admittance that should be used for the optimal noise figure. Unfortunately, in general, this optimal noise impedance is not always equal to the complex conjugate of the input impedance of the amplifier. The difficulty comes from the fact that the source impedance should be set to the complex conjugate for maximum power transfer from the source to the amplifier. A common narrowband technique to overcome this problem is inductive degeneration, as illustrated in Fig. 5.56. The figure also illustrates a cascode design, which is commonly used to improve isolation between the input and output (i.e., to prevent the output from feeding back to the input, where it may reradiate through the antenna, jamming the receiver). To see why inductive degeneration works, consider the simple amplifier shown in Fig. 5.57, in which we consider the load resistance to be noiseless.

After some algebra, we find that the input referred noise voltage is

$$V_{n,in} = j\omega L_s i_{th} \quad (5.133)$$

and the input referred noise current is

$$i_n = \frac{j\omega C_{in}}{g_m} i_{th}. \quad (5.134)$$

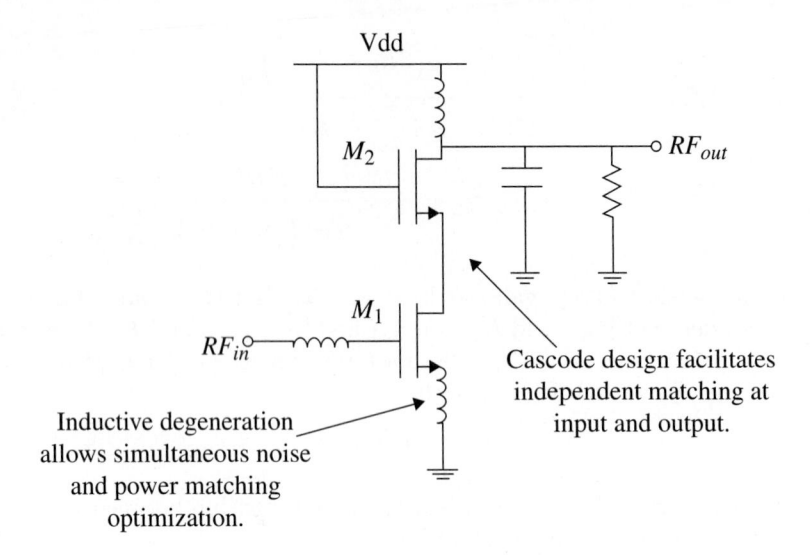

Figure 5.56 Inductive degeneration is a common narrowband technique for simultaneous noise and power matching.

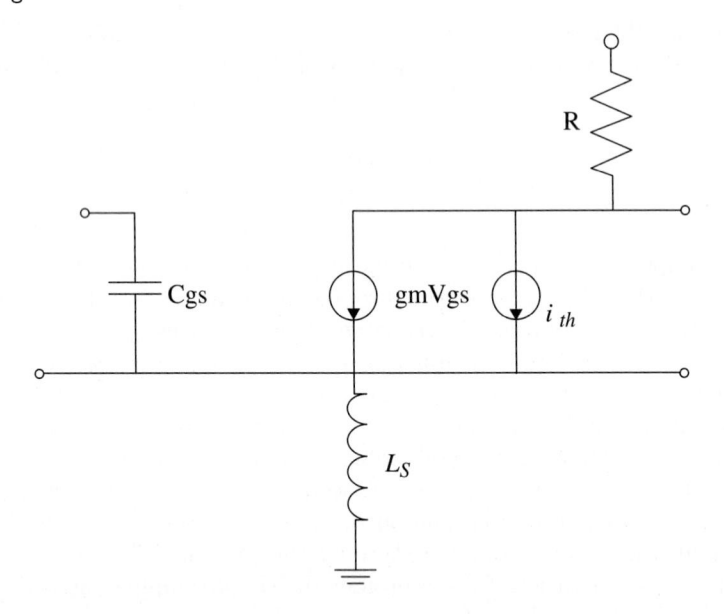

Labels in figure: Vdd; M_2; M_1; RF_{in}; RF_{out}; Cascode design facilitates independent matching at input and output. Inductive degeneration allows simultaneous noise and power matching optimization.

Figure 5.57 Inductive degeneration for a very simple LNA.

Labels in figure: C_{gs}; $g_m V_{gs}$; i_{th}; R; L_s

According to [Lee04a], we may use (5.133) and (5.134) to find the optimal source admittance for lowest noise figure

$$Y_{\text{opt,source}} = \frac{C_{gs}}{g_m L_s}. \tag{5.135}$$

(5.135) indicates that the optimal source impedance from a noise perspective may be set by selecting the source inductance. The advantage of this becomes clear when we find the input impedance of the multiplier

$$Z_{in} = j\omega L_s + \frac{1}{j\omega C_{gs}} + \frac{g_m L_s}{C_{gs}}. \tag{5.136}$$

(5.135) and (5.136) indicate that the optimal noise source admittance determines the real part of the input impedance. Therefore, we may set $Y_{\text{opt,source}}$ using the source inductor so that the optimal input admittance is equal to the real output admittance of the source. We may then add a gate inductor to cancel the imaginary portion of the input impedance. We see then that the source inductor has allowed simultaneous noise and power matching.

Fig. 5.56 illustrates a cascode design, which along with inductive degeneration is perhaps the most common design choice for mmWave LNAs. This design provides isolation between the output and input stages (i.e., little energy is leaked from the output back to the input, and S_{12} is near zero) and facilitates separate matching of the input and output [Lee04b][TWS+07]. This approach is very appropriate for high gain and low noise figure when the ratio of operating frequency to transit frequency is small (e.g., 0.2) [GHH04] (though it does not offer lower noise figure than all other designs [TWS+07]). The inductively degenerated cascode can also be made unconditionally stable, which results in simple matching networks [HBAN07b]. It also has the advantage of Miller-capacitance immunity [AKD+07]. A drawback to the inductively degenerated cascode approach is that the pole between the common source and common gate transistors is poorly accessible for resonant compensation, though an inductor between the two may work as a compensating structure [YGT+07][Raz06]. This pole is due in part to the parasitic source drain capacitance of the common source transistor and results in lower gain and higher noise figure than for the cascode topology at lower frequencies [Raz06]. This design may also be easily affected by package parasitics due to the small inductive degeneration it often requires [Raz06]. Low degeneration of the cascode device near f_T can also result in higher noise figure relative to other topologies [HBAN07b]. An alternative approach that avoids the difficulties of a cascode is a common-gate design [Raz06]. This second approach is very useful in cases when the operating frequency is comparable to the transit frequency of the technology used [GHH04].

There are many factors that can contribute to a high noise figure, including poorly designed current return paths and a circuit layout that does not minimize parasitics. The term *current return path* refers to the route that current follows as it flows back toward the source of a transmission line connected to a circuit load. In any transmission line, current flows down the signal conductor of the line and returns in equal amount, as illustrated in Fig. 5.58. If a conductor is not explicitly placed for the return current, then it will flow through the substrate, and this can generate extra noise. Alvarado et al. have made very good use of current return path engineering to minimize noise figure [AKD+07]. An advantage of a transmission line as opposed to a trace of metal for the input of an LNA, or any circuit, is that a transmission line explicitly defines a metal ground that is used by current as the return path.

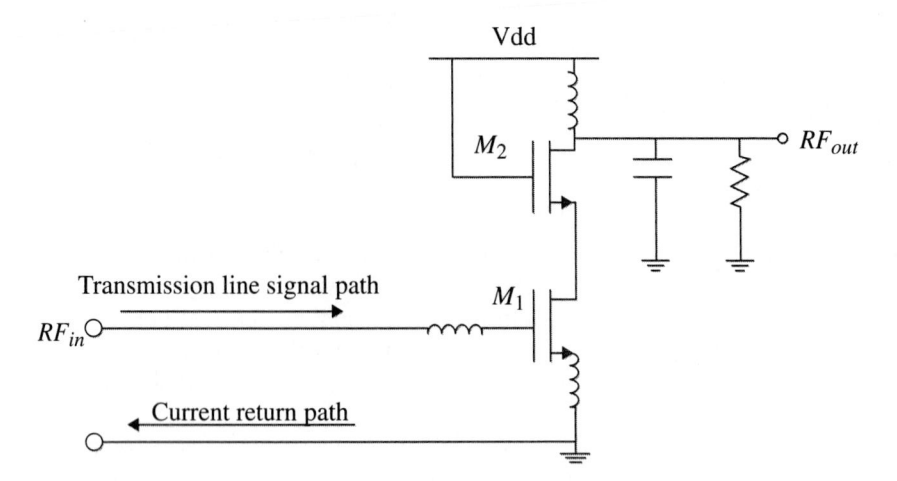

Figure 5.58 Current will return along the ground current return path.

Parasitics are circuit components that are not intended in the design but exist nevertheless due to the physical layout of a device. For example, an extra-long metal trace to the gate of a transistor will have the effect of an added inductance. When laying out a design, it is essential to consider parasitic capacitances and resistances with a parasitic extraction tool [STD+09][JGA+05]. For example, the circuit tool cadence can be used to extract values of parasitic components that can be used in simulation to determine the performance of the laid-out circuit. Inattention to parasitics can cause a significant downshift in resonant frequency of the LNA [STD+09] and increase in noise figure through higher parasitic values. Gate parasitic resistance is generally kept close to the optimum value by properly selecting the number of fingers (most LNAs have a large number of fingers per MOSFET) [JGA+05]. In addition to device parasitic capacitance and resistance, it is necessary to consider parasitic capacitances on metal interconnects between devices. After parasitic extraction, it is possible to design successful matching networks [STD+09].

5.11.3 Mixers

Mixers are non-linear or time-variant circuits that are responsible for downconverting or upconverting a signal to a new carrier frequency so that it may be transmitted or used by lower-frequency components of a receiver. Table 5.6, adapted from [RMGJ11], summarizes several mmWave 60 GHz mixers that have been presented in the literature. The basic action of a mixer is summarized in Fig. 5.59, which shows a downconversion mixer. An upconversion form of this mixer would exchange the RF and IF ports. In the figure, the input RF frequency is "mixed" with the input LO frequency, and the result is an IF frequency equal to the sum or difference of the RF and LO frequencies.

There are three basic criteria by which to judge a mixer: its conversion gain, linearity, and isolation. *Conversion* refers to the ratio of the power of the input signal to the power of the output signal of the mixer. A positive conversion gain indicates that the output

Table 5.6 Millimeter wave mixers [adapted from [RMGJ11]].

Reference	Topology and Type	Conversion Gain/Loss	Process	RF, IF Frequency	Tested LO Frequency and Power	RF-LO Isolation & Linearity[1]
[ZSS08]	Gilbert-cell upconversion double-balanced, active	< 2 dB (Gain)	0.13 μm CMOS	57-65 GHz RF, baseband IF	60 GHz at 0 dBm	−37 dB isolation, −5.6 dBm OP1dB
[ZSS07]	Gilbert-cell downconversion double-balanced, active	< 2 dB (Gain)	0.13 μm CMOS	57-64 GHz RF, baseband IF	60 GHz at 0 dBm	−36 dB isolation, −8 dBm IIP3
[EDNB05]	quadrature single-balanced single-gate, active	< −2 dB (Loss)	0.13 μm CMOS	51-63 GHz RF, 2 GHz IF	58 GHz at 0 dBm	−15 dB isolation, −3.5 dBm IP1dB
[CR01c]	unipolar subharmonic antiparallel diode pair, single-ended, passive	−13.2 dB (Loss)	GaAs MSAG5 process	58.5-60.5 GHz RF, 1.5-2.5 GHz IF	14-14.5 GHz at 3-4 dBm	−17 dB isolation
[BSS+10]	4th subharmonic antiparallel diode pair, single-ended, passive	−17 dB (Loss)	GaAs on liquid-crystal polymer	60 GHz RF, DC 1.25 GHz IF	16 GHz at 8 dBm	−30 dB isolation, −2 dBm IIP3

continued on next page

Table 5.6 (continued)

Reference	Topology and Type	Conversion Gain/Loss	Process	RF, IF Frequency	Tested LO Frequency and Power	RF-LO Isolation & Linearity[1]
[TH07]	2nd subharmonic Gilbert cell, double-balanced, active	upconversion: −6 dB (Loss), downconversion: −7.5 dB (Loss)	0.13 μm CMOS	35-65 GHz RF, baseband IF	20-32.5 GHz at 7-8 dBm	−45 dB isolation, −5 dBm IP1dB
[SQI01]	push-pull dielectric resonator, double-balanced, passive	> −15 dB (Loss)	Fujitsu FHR02X K-band pHEMT	60-61.5 GHz RF, 1 GHz IF	self oscillating	isolation through integrated Yagi antenna
[Rey04]	single-balanced Gilbert cell, active	> 9 dB (Gain)	0.12 μm SiGe BJT	57-64 GHz RF, 8.3-9.1 GHz IF	52 GHz at −3 dBm	−26 to −30 dB isolation, −7 dBm IP1dB (includes buffer)
[MGFZ06]	single-ended resistive mixer, passive	−11.6 dB (Loss)	90 nm CMOS	57-63 GHz RF, 2 GHz IF	60 GHz at 4 dBm	6 dBm P1dB, 16.5 dBm IIP3
[VKR+05]	single-balanced resistive mixer, passive	−11.5 dB (Loss)	0.25 μm pHEMT	57-67 GHz RF, 5.3 GHz IF	56 GHz at 8 dBm	34 dB isolation,
[VKR+05]	single-balanced image reject mixer, passive	−13 to −16 dB (Loss)	0.25 μm pHEMT	57-66 GHz RF, 5.3 GHz IF	57 GHz at 8 dBm	−36 dB isolation, −13 dBm OP1dB, 4 dBm IP3

[1] Most authors report either P1dB or IIP3. The application determines which is most appropriate.

5.11 Analog MmWave Components

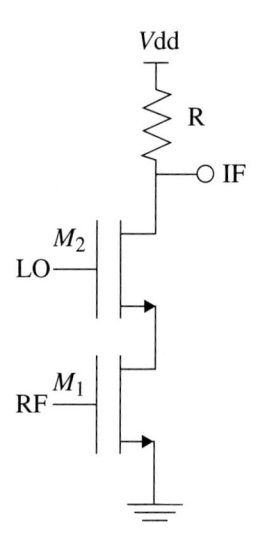

Figure 5.59 The mixer produces a signal at the IF frequency based on the mixing on the RF and LO signals.

signal has more power than the input signal. *Linearity* refers to the extent to which the output power is proportional to the input power. As with an amplifier, a mixer will provide power at the intended output frequency in addition to harmonics of the intended output frequency, and a very non-linear mixer will provide substantial power to the harmonics. *Isolation* refers to the ability of the mixer to shield the output from the input RF or local oscillator (LO) frequency and to provide only the IF (intermediate frequency) at its output.

To understand conversion gain, consider the downconversion mixer in Fig. 5.59. There are two ways we could operate this mixer: with a high LO signal voltage that will cause the cascode transistor to act as a switch, or a low LO signal voltage that will simply modulate the transconductance of the cascode transistor. Consider first the latter case. For accuracy, we must consider the output impedance of the common-source input transistor, which is the resistance that we would see if we looked into the drain of the common-source device. This resistance is written as $\frac{1}{g_{ds}}$, where g_{ds} is a conductance. When current produced by the transconductance of the common-source device exits the drain of the common-source device, it may flow into the source of the cascode device or be shunted back through the output resistance of the common-source device. The ratio that describes the amount of current that usefully flows up through the cascode source is given by

$$\beta = \frac{g_m^{cas}}{g_m^{cas} + g_{ds}} \quad (5.137)$$

where g_m^{cas} is the transconductance of the cascode device, and g_{ds} is the inverse of the output resistance of the common-source device. The transfer function of the mixer may be written for the low LO magnitude regime:

$$V_{\text{out}} = -g_m R \beta V_{RF} = -g_m R \left(\frac{g_m^{\text{cas}}}{g_m^{\text{cas}} + g_{ds}} \right) V_{RF} = -\frac{g_m R \left(\frac{g_m^{\text{cas}}}{g_{ds}} \right)}{1 + \frac{g_m^{\text{cas}}}{g_{ds}}} V_{RF}. \qquad (5.138)$$

We now make use of the Taylor series expansion of the inverse of $1 + x$:

$$V_{\text{out}} = -g_m R \left(\frac{g_m^{\text{cas}}}{g_{ds}} \right) \left[1 - \left(\frac{g_m^{\text{cas}}}{g_{ds}} \right) + \left(\frac{g_m^{\text{cas}}}{g_{ds}} \right)^2 - \ldots \right] V_{RF}. \qquad (5.139)$$

We may write the input RF signal and the cascode transconductance as

$$V_{RF} = A \cos(\omega_{RF} t) \qquad (5.140)$$
$$g_m^{\text{cas}} = g_{mO}^{CAS} \cos(\omega_{LO} t) \qquad (5.141)$$

so that the output signal becomes

$$V_{\text{out}} = -A \frac{g_m}{g_{ds}} R g_{mO}^{CAS} \cos(\omega_{LO} t) \cos(\omega_{RF} t)$$
$$\cdot \left[1 - \left(\frac{g_m^{CAS}}{g_{ds}} \right) \cos(\omega_{LO} t) + \ldots \right] \qquad (5.142)$$

$$V_{\text{out}} = -\frac{A}{2} \frac{g_m}{g_{ds}} R g_{mO}^{CAS} \left(\cos(\omega_{LO} t + \omega_{RF} t) + \cos(\omega_{LO} t - \omega_{RF} t) \right)$$
$$\cdot \left[1 - \left(\frac{g_m^{CAS}}{g_{ds}} \right) \cos(\omega_{LO} t) + \ldots \right].$$

At the IF frequency $f_{IF} = f_{RF} - f_{LO}$, we find the signal to be

$$V_{IF} = -\frac{A}{2} \left(\frac{g_m}{g_{ds}} \right) R g_{m0}^{CAS} \cos(\omega_{IF} t). \qquad (5.143)$$

The conversion gain is found by taking the ratio of the power at the IF frequency to the power at the RF frequency [Poz05, Chapter 12]

$$G_c = \frac{P_{IF}}{P_{RF}} \qquad (5.144)$$

where P_{IF} is the power available at the IF frequency and P_{RF} is the power available at the RF frequency. The RF voltage drops across the gate-source capacitance of the common-source transistor, while the IF voltage drops across the output resistor:

$$P_{RF} = \omega_{RF} C_{gs} \frac{A^2}{2} \qquad (5.145)$$

$$P_{IF} = \frac{A^2}{8} \left(\frac{g_m}{g_{ds}} \right)^2 R g_{m0}^{\text{cas}^2}.$$

5.11 Analog MmWave Components

The conversion gain is therefore

$$G_c = \frac{Rg_{m0}^{CAS^2}}{4\omega_{RF}C_{gs}}\left(\frac{g_m}{g_{ds}}\right)^2. \qquad (5.146)$$

This analysis only holds for low-magnitude RF signals and low-magnitude LO signals. At higher LO power, the beta of the cascode may be treated as a square wave, with duty cycle d based on the LO voltage signal and bias point, as illustrated in Fig. 5.60. The beta factor may be written as a Fourier series

$$\beta = a_0 + a_1\cos(\omega_{LO}) + b_1\sin(\omega_{LO}t) + \ldots$$

$$a_0 = \frac{1}{\pi}\sin(\pi d),\, a_{n,n>0} = \frac{1}{\pi}\left[\frac{\sin((n-1)\pi d)}{n-1} + \frac{\sin((n+1)\pi d)}{n+1}\right], b_n = 0$$

$$\beta = \frac{1}{\pi}\sin(\pi d) + \frac{1}{\pi}\left[\pi d + \frac{\sin(2\pi d)}{2}\right]\cos(\omega_{LO}t) + \ldots. \qquad (5.147)$$

The IF output is therefore

$$V_{IF} = -\frac{A}{2\pi}g_m R\left[\pi d + \frac{\sin(2\pi d)}{2}\right]\cos(\omega_{IF}t). \qquad (5.148)$$

Therefore, under high-LO signal levels, the conversion gain is

$$G_C = \frac{Rg_m^2}{4\pi^2\omega_{RF}C_{gs}}\left[\pi d + \frac{\sin(2\pi d)}{2}\right]^2. \qquad (5.149)$$

To understand isolation, first realize that up until to now we have assumed that the output resistance of the cascode transistor is infinite (we have implicitly assumed this for ease of analysis), while in reality the output resistance is finite. Poor isolation indicates that a signal at the IF port will cause a signal at the RF port in the simple downconversion mixer. Consider the simple-small signal model shown in Fig. 5.60, in which we have added a degeneration resistor R_s for sake of analysis. It is easily shown that a voltage signal applied at the output port will appear at the input port as

$$V_{in} = \left(\frac{V_0}{1 + \frac{R_0}{R_s}}\right)\left[\frac{R_0 + R_s}{R_0^{CAS} + R_0 + R_s + g_m^{cas}R_0^{CAS}(R_0 + R_s)}\right]. \qquad (5.150)$$

This is a simple example that only applies to low-magnitude signals, but it serves to illustrate that a signal at the output port will appear at the input port if the cascode transistor has finite output resistance.

To understand linearity, consider the possibility that the input RF amplitude is strong enough to modulate the output resistance of the input transistor. In this case, for very high RF powers, the common-source transistor may enter the linear region, at which point its g_{ds} may rapidly decrease. This is because in the linear region of operation the transistor has a lower drain current than in the saturation region, and the output resistance of a transistor is generally inversely proportional to its drain current [Raz01].

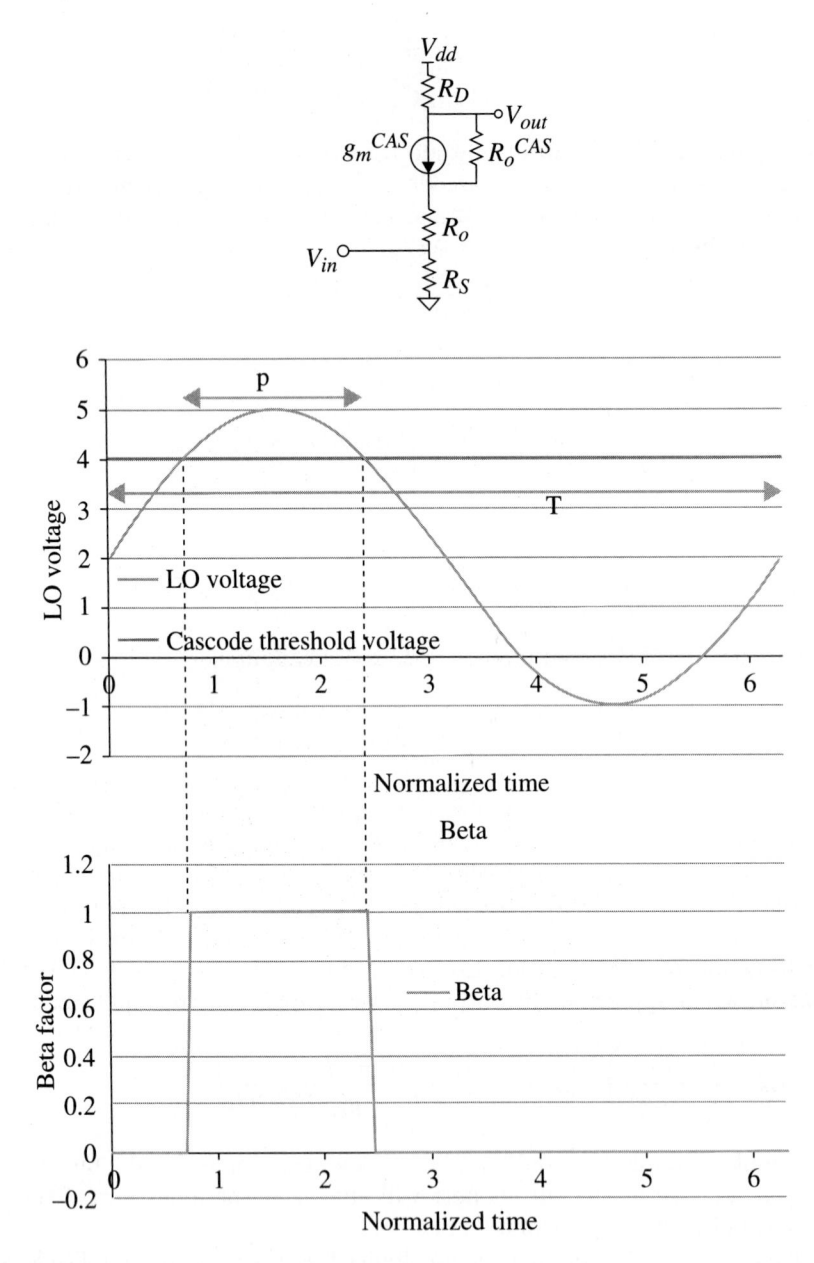

Figure 5.60 Based on the nature of the LO signal magnitude and bias point, we may treat the transconductance of the cascode as a square wave that switches on and off. This figure shows how the gain of the switching mixer is gated by the LO voltage to create the switching effect of the mixer. This approach is used in double balanced mixers, such as Gilbert cells.

5.11 Analog MmWave Components

At mmWave frequencies it may be difficult to generate large LO signal powers to operate a mixer. This is important because the LO signal level has a large effect on the operation of the mixer; for example, conversion gain generally falls quickly with a reduction in LO power [NH08].

There are two basic classes of mixers: active and passive. Within these two classes, there are three basic sub-classes: *single-ended*, *single-balanced*, and *fully balanced* (also called *double-balanced*). Active mixers and passive mixers differ in their conversion gain: whereas active mixers provide a positive conversion gain, passive mixers provide a loss. The difference between single-ended, single-balanced, and fully balanced has to do with linearity and frequency suppression characteristics as well as what frequencies appear at the output of the mixer. Single-ended mixers will have a signal at the LO, RF, and IF frequencies that appears at the output. Single-balanced mixers will have only two of these three frequencies at the output, and double-balanced mixers have superior linearity, port isolation, and spurious frequency suppression characteristics and have only one frequency (the RF or IF, depending on whether the mixer is an upconversion or downconversion mixer) at the output. Fig. 5.61 gives examples of each of these three types of mixers.

To see how isolation improves when going from single-balanced to double-balanced designs, consider the output of the single-balanced mixer shown in Fig. 5.61. The RF current generated at the base current source is split evenly between the two outputs, as it sees the same resistance looking into the sources of the two transistors whose gates are connected to the LO signal. The half-current that splits to the right generates a voltage $-I_{RF}R/2$, and the current that splits to the right generates the same voltage. Therefore, at the output, which is taken differentially, these voltages cancel so that the RF frequency signal does not appear at the output. Double-balanced mixers use differential inputs for the IF, LO, and RF signals, providing improved isolation and enhanced rejection of unwanted signals at the RF output.

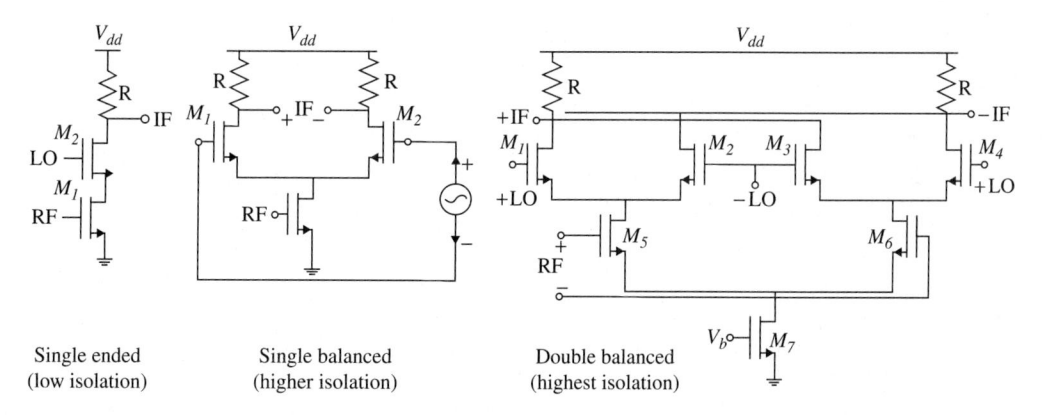

Figure 5.61 Single-ended mixers have the lowest isolation. In general, as a port is made to be differential, it will be isolated from the output. For example, in a double-balanced mixer, both the RF and LO ports are differential.

5.11.4 Voltage-Controlled Oscillators (VCOs)

Table 5.7, as adapted from [RMGJ11], gives examples of VCOs that have been developed for mmWave applications.

5.11.4.1 Basic VCO Design and LC Tank VCOs

Voltage-controlled oscillators (VCOs) are responsible for generating the local oscillator frequencies that are used by mixers to perform frequency conversion. The output frequency of a VCO is controlled by an input voltage to the VCO. Unlike an amplifier, a VCO, or any oscillator, is fundamentally unstable, which is why it oscillates. To become unstable, an oscillator must contain a feedback loop with sufficient loop gain and delay for oscillation amplitude to accumulate. As is explained in many texts such as [Poz05, Chapter 12], a basic feedback circuit such as that shown in Fig. 5.40 may become unstable

$$V_0 = \frac{A}{1 + A\beta(\omega)} V_{in}. \tag{5.151}$$

If $1 + A\beta(\omega) = 0$, then the output will oscillate provided there is an initial input. To oscillate, the open-loop gain of the circuit must be greater than 1, and it must have a phase shift of 0 degrees [Raz01, Chapter 14].

As a simple example, consider the circuit shown in Fig. 5.62. The open-loop voltage-gain function is

$$A_v^{\text{open}} = g_m R_L^2 \left[\frac{j\omega C_s}{1 - \omega^2 L_s C_s + j\omega \left(\frac{C_s}{g_m} + L_s g_m \right)} \right]. \tag{5.152}$$

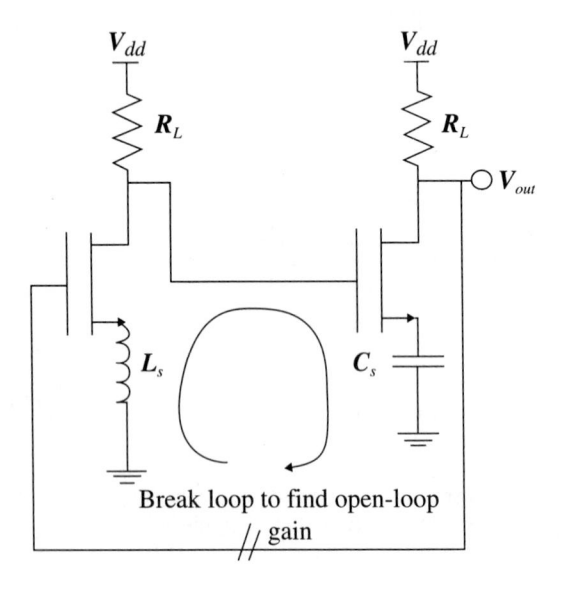

Figure 5.62 A simple wideband LC oscillator.

Table 5.7 Examples of oscillators used for mmWave applications.

Reference	Frequency Range	Topology & Process	Phase Noise	Output Power	Power Consumption, Voltage
[Flo04]	52–53.955 GHz	differential colpitts in 0.12 μm SiGe	−100 dBc/Hz at 1 MHz offset	−8 dBm	25 mW from 2.5 V
[Flo04]	65.8–67.9 GHz	differential colpitts in 0.12 μm SiGe	−98 to −104 dBc/Hz at 1 MHz offset	−8 dBm	24 mW from 3.0 V
[LSE08]	64–70 GHz	fundamental 30 GHz with push-push buffers in 0.13 μm CMOS, single-ended output	−90.7 dBc/Hz at 1 MHz offset	−10 dBm	1.5 V and 1.0 V supply tested
[YCWL08]	53.1–61.3 GHz	variable inductor LC-tank VCO in 90 nm CMOS, single-ended output	−118.75 dBc/Hz at 10 MHz offset	−6.6 dBm	8.7 mW from 0.7 V
[CCC+08]	66.7–69.8 GHz	intrinsic tuning LC tank VCO in 0.13 μm CMOS, differential output	−98 dBc/Hz at 1 MHz offset, −115.2 dBc/Hz at 10 MHz offset	> −24.8 dBm	4.32 mW from 0.6 V
[BDS+08]	55.5–61.5 GHz	cross-coupled pair with LC tank (shielded slow-wave inductor, MOSCAP varactors) in 0.13 μm CMOS	> −90 dBc/Hz at 1 MHz offset	−15 dBm	3.9 mW from 1 V

continued on next page

Table 5.7 (continued)

Reference	Frequency Range	Topology & Process	Phase Noise	Output Power	Power Consumption, Voltage
[BDS+08]	59-65.8 GHz	cross-coupled pair with LC Tank (shielded slow-wave inductor, MOSCAP varactors) in 0.13 μm CMOS	> -90 dBc/Hz at 1 MHz offset	-15 dBm	3.9 mW from 1 V
[CLH+09]	0.1-65.8 GHz	ring-based triple push in 90 nm CMOS	-99.4 to -78 dBc/Hz at 1 MHz offset, -107.8 to -94.6 dBc/Hz at 10 MHz offset, measured up to 47.4 GHz	-27 to -7.5 dBm	1.2 to 26.4 mW from 1.2 V
[LRS09]	53.2-58.4 GHz	inductive division LC tank in 90 nm CMOS	-91 dBc/Hz at 1 MHz offset at 58.4 GHz operating frequency	-14 dBm	8.1 mW from 0.7 V supply, 1.2 mW from 0.43 V supply
[Wan01]	49-51.1 GHz	LC-resonator with cross-coupled NMOS in 0.25 μm CMOS	-99 dBc/Hz	-11 dBm at 49.4 GHz	4 mW from 1.3 V
[CTC+05]	52-52.6 GHz	push-push with thin film microstrip lines in 0.18 μm CMOS	-97 dBc/Hz at 1 MHz offset at 53 GHz	-16 dBm	27.3 mW from 2.1 V supply

5.11 Analog MmWave Components

At the resonant frequency $\frac{1}{\sqrt{LC}}$, the transfer function becomes

$$A_v^{\text{open}} = \frac{g_m^2 R_L^2}{1 + g_m^2 \left(\frac{L_s}{C_s}\right)}. \tag{5.153}$$

Therefore, the open-loop gain has a phase shift of 0 degrees, and provided that

$$g_m \geq \frac{1}{\sqrt{R_L^2 - \frac{L_s}{C_s}}}. \tag{5.154}$$

The open-loop gain will be greater than or equal to 1, indicating that the circuit will oscillate. Fig. 5.63 illustrates the open-loop gain and phase for $g_m = 5mS$, $R_L = 500\,\Omega$,

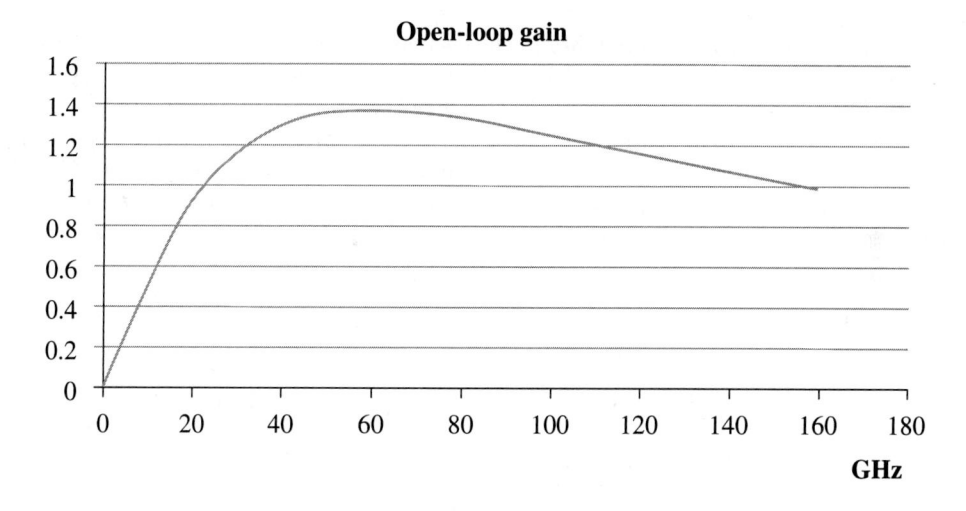

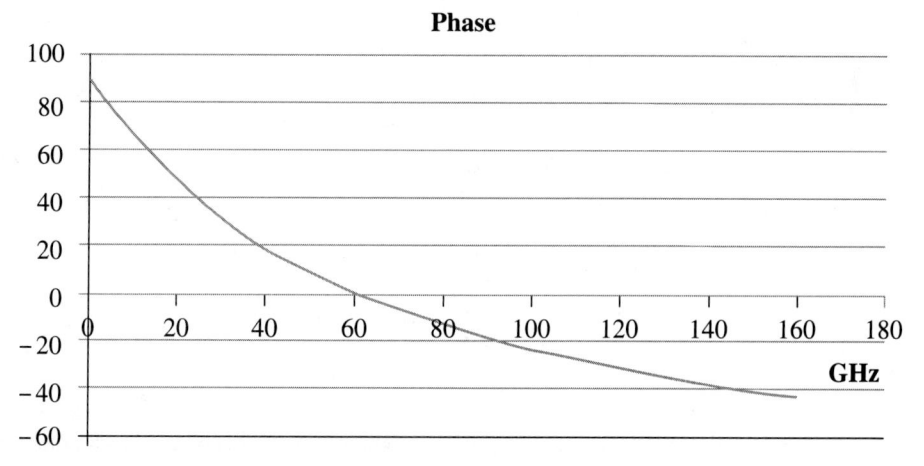

Figure 5.63 The open-loop gain (top) and phase shift (bottom) of the oscillator in Fig. 5.62.

$C_s = 7$ fF, and $L_s = 1$ nH. We see that the open-loop gain is greater than 1 at 60 GHz, and that the open-loop phase is 0 at 60 GHz.

The oscillator in Fig. 5.62 is not the most common forms and is mostly intended to illustrate oscillator analysis. One of the most common form of oscillator is the LC tank cross-coupled oscillator, shown in Fig. 5.64. [Raz01, Chapter 14] presents an analysis of this oscillator that is very intuitive and recommended. The oscillator provides a differential output signal. The open-loop gain of the oscillator is easily found and is illustrated in Fig. 5.65 for $R_p = 5$ kΩ, $g_m = 5$ mS, $C_p = 7$ fF, and $L_p = 1$ nH:

$$A_v^{\text{open}} = \frac{g_m^2 \omega^2 L_p^2}{\left[(1 - \omega^2 L_p C_p)^2 - \omega^2 \left(\frac{L_p}{R_p}\right) + \frac{2j\omega L_p}{R_p}(1 - \omega^2 L_p C_p)\right]}. \quad (5.155)$$

The resonant frequency of L and C in the load circuit of the LC oscillator determines the frequency of oscillation of the circuit.

Most applications call for tunability in the oscillation frequency. To achieve tunability, it is necessary to change the value of L, or C in the simple LC tank oscillator shown in Fig. 5.64. Although it is possible to tune the value of L with certain techniques [YCWL08], in general the value of capacitance is tuned by using a varactor to provide capacitance instead of a simple capacitor. A *varactor* is a diode or specially connected MOSFET to form what is known as a metal oxide semiconductor capacitor (MOSCAP). These simple varactor types are illustrated in Fig. 5.66.

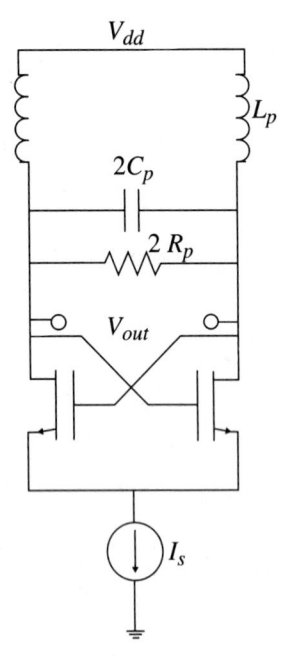

Figure 5.64 The LC tank VCO is perhaps the most common form of oscillator. This topology can be difficult to use if there is insufficient loop gain to achieve oscillation.

5.11 Analog MmWave Components

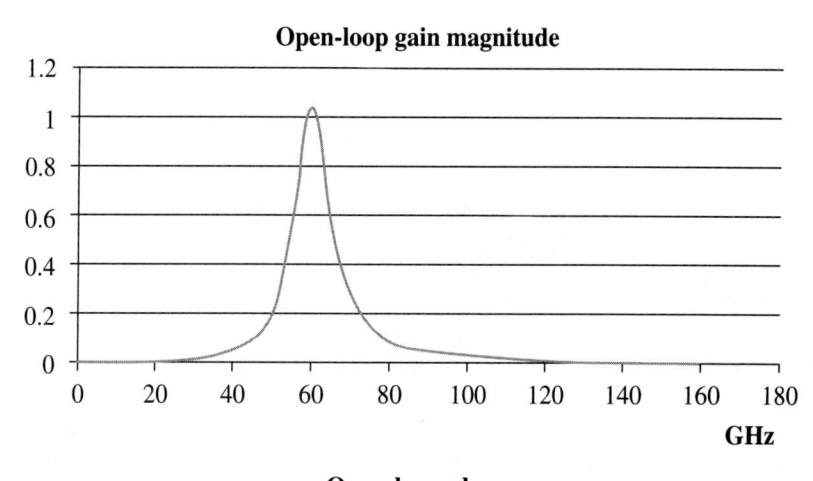

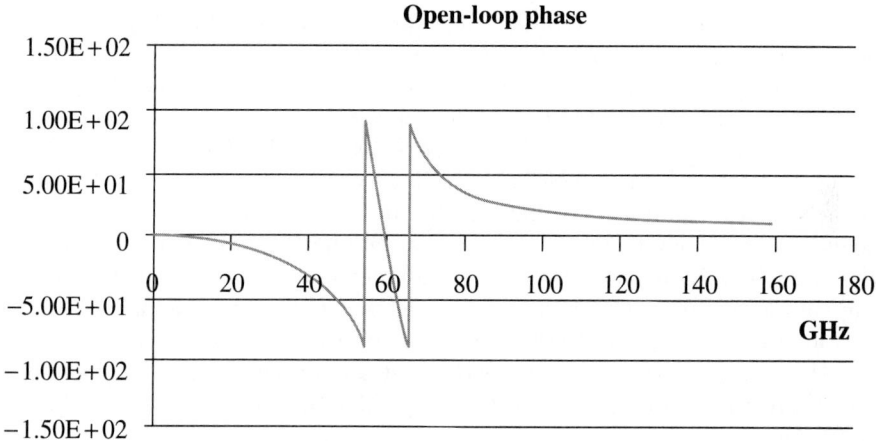

Figure 5.65 The gain and phase of a simple LC tank oscillator.

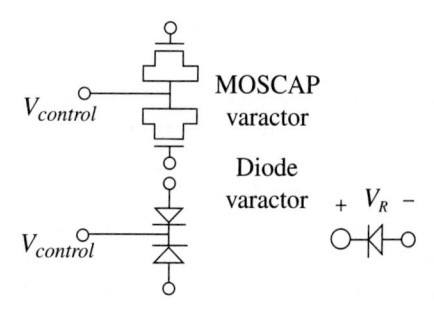

Figure 5.66 MOSFETs with their source and drain connected together to form a MOSCAP may serve as a varactor. Reverse biased diodes [Raz01, Chapter 14] may also serve as a varactor.

The physics of a MOSCAP are explained in [SN07]. [Raz01, Chapter 14] gives the following simple equation for the capacitance of a diode-varactor:

$$C_{var} = \frac{C_0}{\left(1 + \frac{V_R}{\phi_B}\right)^m} \tag{5.156}$$

where m is usually between 0.3 and 0.4, C_0 is the capacitance at $V_R = 0$, V_R is the reverse bias on the varactor diode, and ϕ_b is the built-in-potential drop of the diode [Raz01]. The tuning range of an LC tank oscillator is determined by the range in values that may be taken by the varactor, and by the range of frequencies over which the cross-coupled transistors can provide a negative resistance to drive the LC tank.

In 2008, Xu et al. proposed a varactor design using thick-gate-oxide sub-design rule channel length MOS structures that obtained quality factors as high as 100 and tuning ranges of 1:1.6 at 24 GHz [XK08]. They demonstrated that a thick gate dielectric provides a higher quality factor than varactors with thin gate oxides (approximately a 3× improvement). Unfortunately, as should be expected, the higher quality factor resulted in a lower tuning range for this varactor design (1.6 for thick oxide vs. 2.8 for thin oxide designs) [XK08], so this approach may not be applicable to the most modern of CMOS processes. The use of sub-design rule channel lengths is not problematic because the source and drain of a MOSCAP are connected together [XK08]. In 2006, Cao et al. [CO06] studied varactor design for a 60 GHz VCO in 130 nm CMOS. Their interest was in the MOSFET gate length that resulted in the best combination of tuning range and quality factor. They found that larger than minimum gate lengths of 0.18 μm and 0.24 μm at 24 GHz offered the best combination of tuning range and quality factor. The 0.18 μm gate length provided lower tuning range but improved phase noise (see Section 5.11.4.2) for the resulting VCO versus the 0.24 μm length varactor [CO06]. A 0.12 μm gate length varactor, however, provided the best quality factor of approximately 25-35 over the entire tuning range (-1.5 V to 1.5 V, 24 GHz) and thus had the best phase noise performance.

5.11.4.2 Phase Noise

So far, we have implicitly assumed that the output frequency of an oscillator is spectrally clean, in the sense that it contains energy at only one frequency — the frequency at which the loop gain is greater than unity and has a zero-degree phase shift, as illustrated in Fig. 5.67. This is not true for real circuits. Fig. 5.67 also illustrates that the spectral content of an oscillator's output actually contains significant energy at frequencies near the center frequency of oscillation. This phenomenon is called *phase noise*, and low phase noise is a major design goal for any oscillator.

Phase noise is measured in decibels below the carrier per-hertz (dBc/Hz), and it can be calculated or measured by determining the amount of power in a 1 Hz bandwidth at a given offset from the carrier relative to the carrier power. Figs. 5.68 through 5.73 illustrate this concept, and (5.157) encapsulates the definition [HL98].

$$\mathcal{L}\{\Delta\omega\} = 10 \log \left\{ \frac{P_{sideband}(\omega_0 + \Delta\omega, 1Hz)}{P_{carrier}} \right\} [\text{dBc/Hz}] \tag{5.157}$$

5.11 Analog MmWave Components

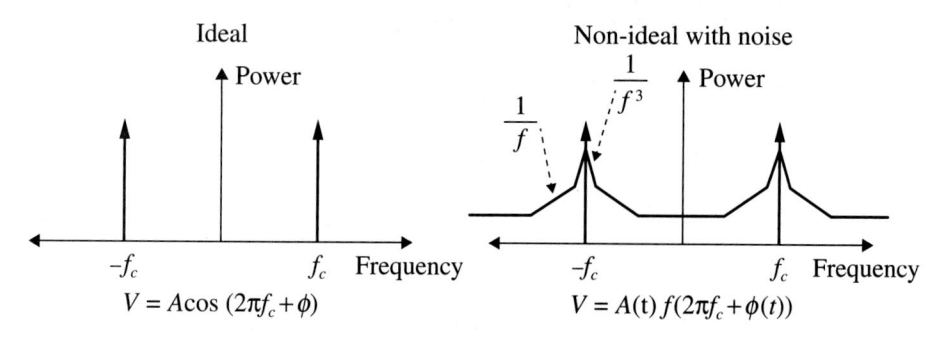

Figure 5.67 The output of an oscillator actually contains energy at frequencies other than the intended center frequency. This energy is known as *phase noise*.

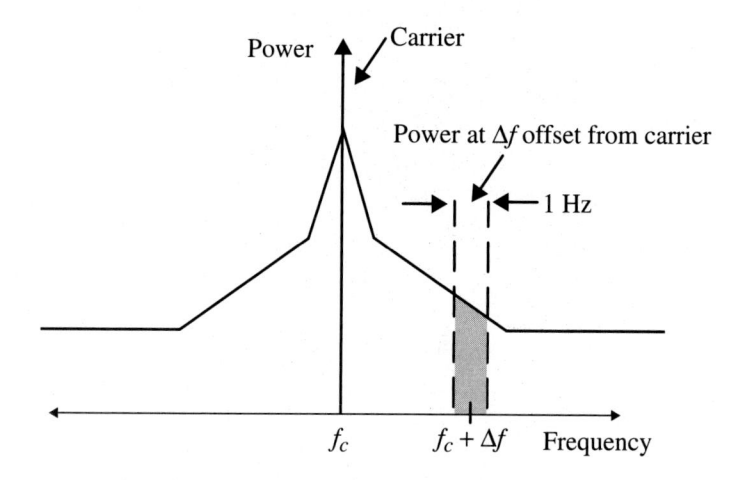

Figure 5.68 Phase noise is measured as the decibel (10log[...]) ratio of power in the carrier to the power in a 1 Hz interval at some offset from the carrier. The offset frequency is generally 1 or 10 MHz for mmWave VCOs.

where $\mathcal{L}$ is the phase noise at $\Delta\omega$ away from the center or carrier frequency, ω_o, and $P_{sideband}$ is the power in the 1 Hz-wide sideband $\Delta\omega$ away from ω_0. $P_{carrier}$ is the power of the center frequency. In general, equation (5.157) includes effects of both phase and amplitude noise, but for ease of understanding we treat it as referring to phase noise only, and we assume that amplitude limiting devices such as limiters in the circuit control the effect of amplitude noise [HL98]. If amplitude noise were present, it would be evidenced by a non-constant power in the carrier frequency. Fig. 5.72 illustrates how a circuit can dampen the amplitude noise while the impact of phase noise persists, as can be seen by the continuing phase error after the amplitude noise event.

The phase-noise characteristic generally falls as $\frac{1}{f^n}$ away from the carrier. Very close to the carrier, the dependence is $\frac{1}{f^3}$, and eventually the spectrum follows a $\frac{1}{f}$ dependence. Phase noise is problematic in that it can result in undesired signals being mixed down to baseband or the IF frequency. This is illustrated in Fig. 5.69.

Phase noise results from the noise sources that exist in the circuit. For example, we learned in Section 5.10.2 that all transistors naturally have noise currents through their

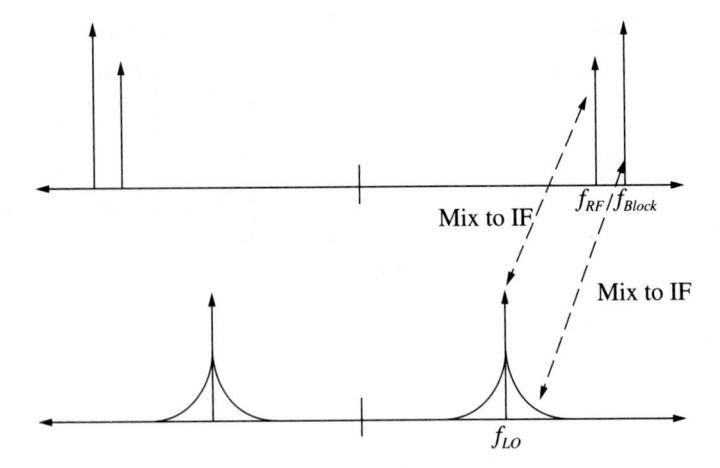

Figure 5.69 Phase noise is problematic because it can result in undesired signals mixing down to the IF frequency or to baseband.

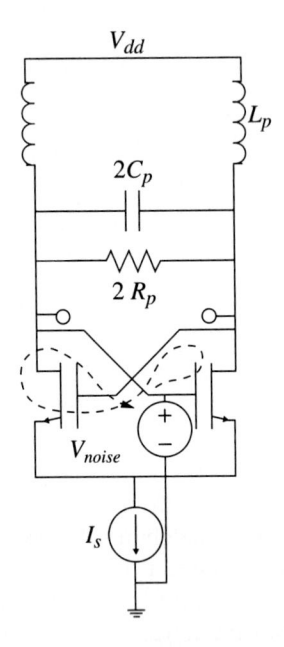

Figure 5.70 The noise in the oscillator will appear at the output as phase noise.

drains that have energy from DC to daylight. In the simple LC oscillator, consider a spectrally white noise voltage that is injected into the gate of one of the devices. This noise goes through the loop an infinite number of times as illustrated in Fig. 5.70, so that the output due to this noise voltage may be written

$$V_{\text{noise}}^{\text{out}} = V_{\text{noise}} + A_v(\omega)V_{\text{noise}} + A_v^2(\omega)V_{\text{noise}} + \ldots = \frac{V_{\text{noise}}}{1 - A_v(\Omega)} \quad (5.158)$$

5.11 Analog MmWave Components

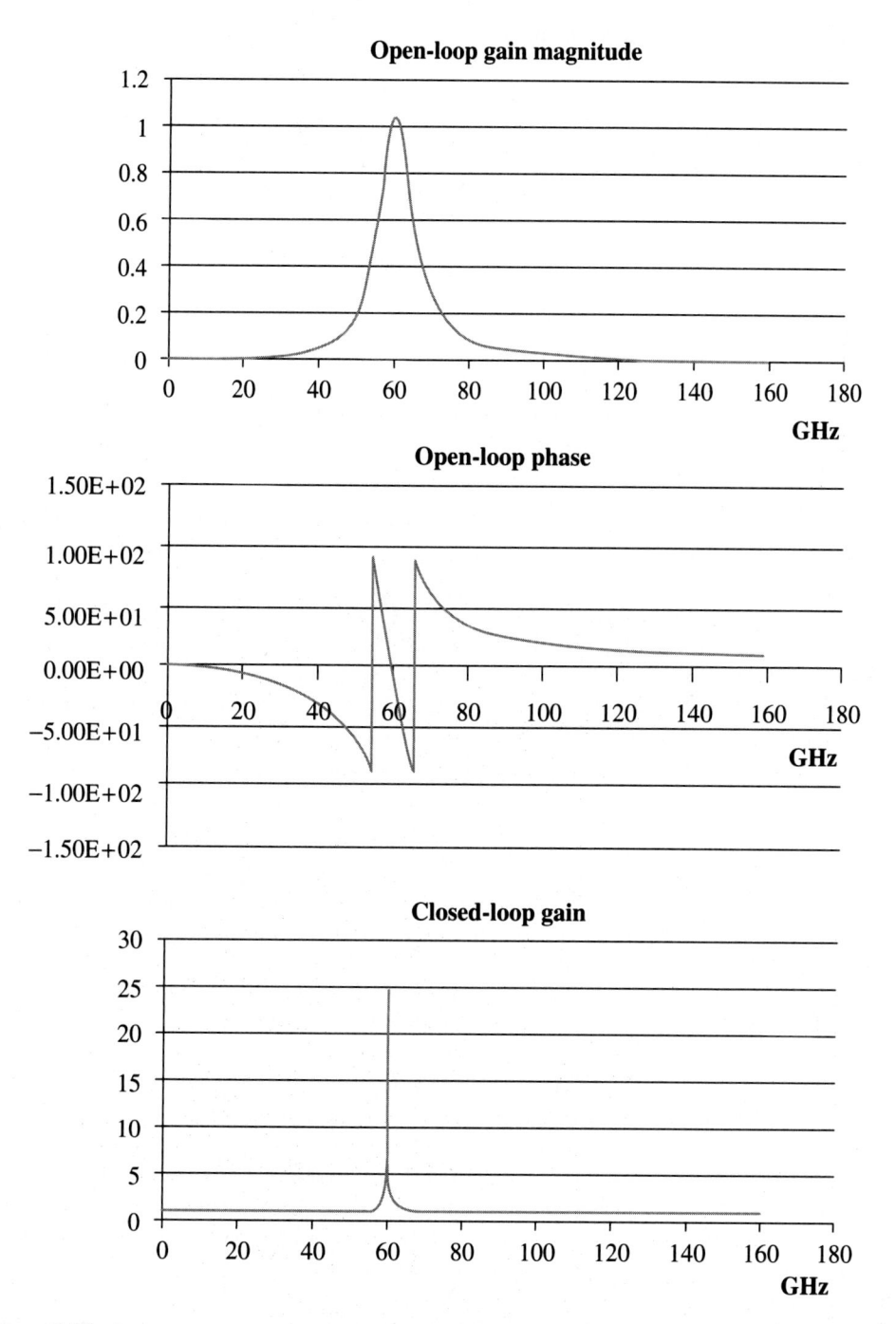

Figure 5.71 For real-world oscillator circuits, the output spectrum will be polluted by power at frequencies other than the intended operating frequency.

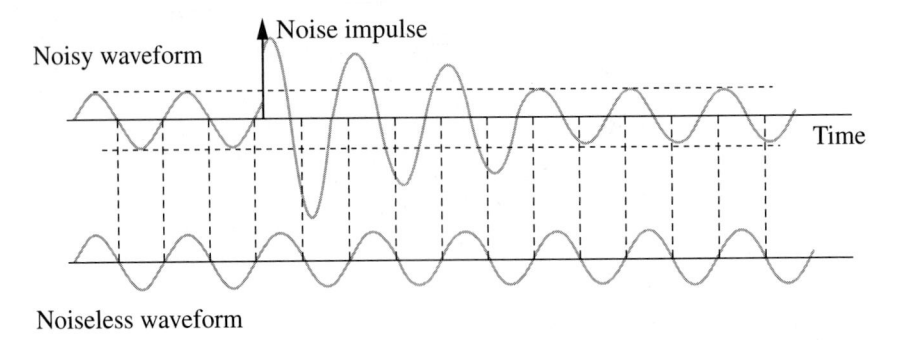

Figure 5.72 Noise events, such as the noise impulse represented here, will affect both the amplitude and phase of a circuit. In general, the amplitude impulse response of the circuit will act to remove amplitude noise over time. But phase noise persists, as is evident when we compare the phase of the noisy waveform to the noiseless waveform.

where the last equality is true, provided that the gain is less than unity. As the loop gain approaches unity and zero-phase, the oscillator begins to oscillate. At frequencies for which this is not true, there is still spectral energy at the output. For example, if we take the simple LC oscillator with $R = 204\ \Omega$, $L_p = 1$ nH, and $C_p = 7.036 fF$, we see that the closed-loop transfer function (5.158) is not zero at frequencies outside the intended oscillation frequency.

The quality factor of the LC tank circuit used in an LC oscillator has a major impact on the phase noise of the oscillator. In general, the higher the quality factor of the tank, the lower the phase noise. The quality factor for a parallel RLC circuit may be written

$$Q_{\text{tank}} = \frac{R_p}{\omega L_p}\left(1 - \omega^2 L_p C_p\right). \tag{5.159}$$

For example, in Fig. 5.73, we see that the spectrum of the output for a value of L_p of 0.1 nH is much more pure than the output of LC tank oscillator whose inductor has a value of 1 nH. But, there is a limit to how much we can or want to reduce the inductance to improve the quality factor of the circuit. In reality, the presence of parasitic resistances in the inductor and capacitor/varactor limits the Q of the tank used in an LC oscillator.

The varactor used in an LC tank VCO determines the phase noise of an LC tank oscillator to a large degree [OKLR09][XK08]. At very high frequencies, for example, mmWave regimes, the quality factor of the varactor can be a larger impediment to a high-Q tank circuit than the inductor [OKLR09][CO06]. Unfortunately, the quality factor of a MOSCAP varactor decreases as the inverse of frequency (unlike inductors, whose Q increases with frequency [XK08]) and will also decrease with increased tuning range [OKLR09]. The extrinsic gate resistance and the intrinsic resistance of the MOSCAP varactor should be minimized to obtain a high Q.

Supply noise rejection circuits or supply regulators can also help reduce phase noise caused by supply voltage variations [CDO07]. Also, inductive degeneration in LC tank VCOs constructed with BJTs can reduce phase noise by 3-4 dB [DLB+05].

At mmWave frequencies, achieving high output powers from oscillators is a major challenge. The output power of an oscillator is difficult to determine theoretically, as

5.11 Analog MmWave Components

Output spectrum Q comparison

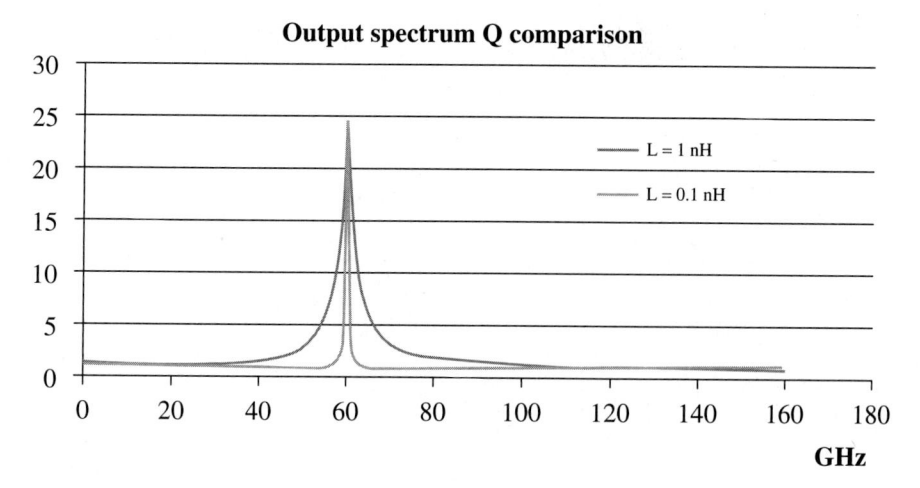

Figure 5.73 The output spectrum of an LC oscillator becomes more spectrally pure as the quality factor of the tank increases.

the oscillator is assumed to have infinite closed-loop gain at the oscillation frequency. In general, it is desirable to have as high an output power as possible, as the mixers that are driven by the oscillator will have better conversion gain if the LO signal from the oscillator is higher power.

5.11.4.3 Subharmonic Oscillator

Subharmonic VCOs are advantageous in cases where active devices are not capable of providing enough power at the fundamental frequency to give an open-loop gain greater than 1.

The output of a subharmonic should be taken from a common-mode node (a ground point for the fundamental). The common mode node that should be selected for the output point should have the lowest parasitic capacitance to ground of the available common-mode nodes [CDO07]. If possible, the parasitic capacitance to ground at the selected output point should be resonated out to provide improved output voltage dynamic range [CDO07].

To understand a basic subharmonic amplifier, consider the closed-loop LC amplifier shown in Fig. 5.74. Fig. 5.75 shows a more conventional illustration of this VCO and gives some basic notes on its operation. There are four paths that may be taken from the output around the loop. The open-loop transfer function may be found by summing the open-loop transfer functions from each of these paths:

$$A(\omega) = \frac{\frac{2g_m(j\omega L_p)}{1-\omega^2 L_p C_p + \frac{j\omega L_p}{R_p}}\left(1+\frac{j\omega C_{in}}{g_m}\right)}{2+\frac{j\omega C_{in}}{g_m}} \left(1-\omega^2 L_p C_p + j\omega\left(\frac{L_p}{R_p}-\frac{L_p g_m}{2}\right)\right)}{1-\omega^2 L_p C_p + \frac{j\omega L_p}{R_p}}. \quad (5.160)$$

When we examine the closed-loop function shown in Fig. 5.76, we find that the closed-loop transfer function has very little content at the fundamental frequency.

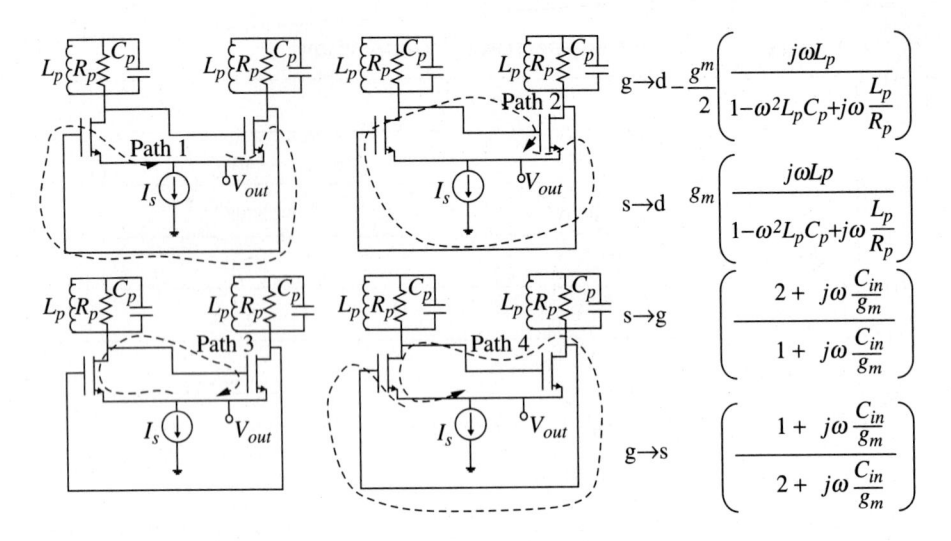

Figure 5.74 An LC tank oscillator may be used as a subharmonic oscillator.

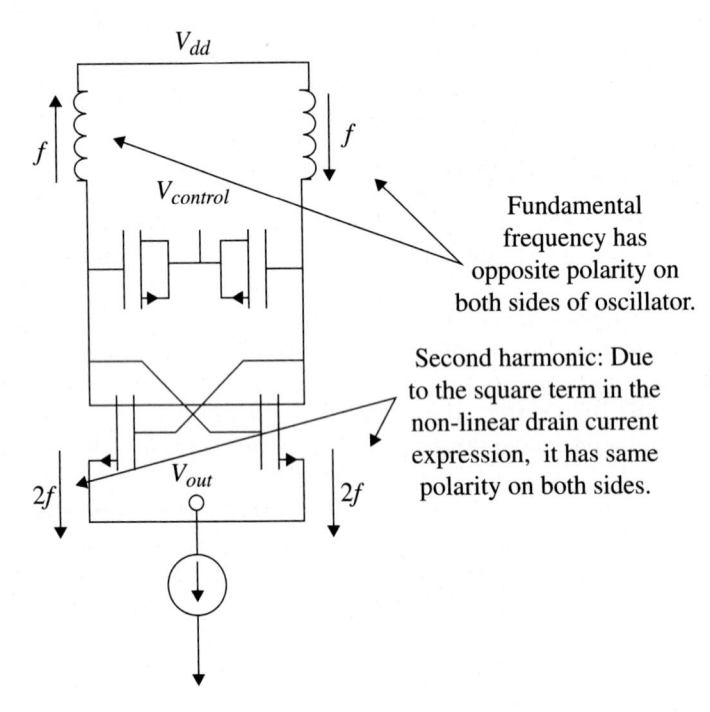

$$g \to d \quad -\frac{g_m}{2}\left(\frac{j\omega L_p}{1-\omega^2 L_p C_p + j\omega \frac{L_p}{R_p}}\right)$$

$$s \to d \quad g_m \left(\frac{j\omega L_p}{1-\omega^2 L_p C_p + j\omega \frac{L_p}{R_p}}\right)$$

$$s \to g \quad \left(\frac{2 + j\omega \frac{C_{in}}{g_m}}{1 + j\omega \frac{C_{in}}{g_m}}\right)$$

$$g \to s \quad \left(\frac{1 + j\omega \frac{C_{in}}{g_m}}{2 + j\omega \frac{C_{in}}{g_m}}\right)$$

Figure 5.75 The subharmonic oscillator takes advantage of the fact that the second harmonic can be extracted from the source node.

Fundamental frequency has opposite polarity on both sides of oscillator.

Second harmonic: Due to the square term in the non-linear drain current expression, it has same polarity on both sides.

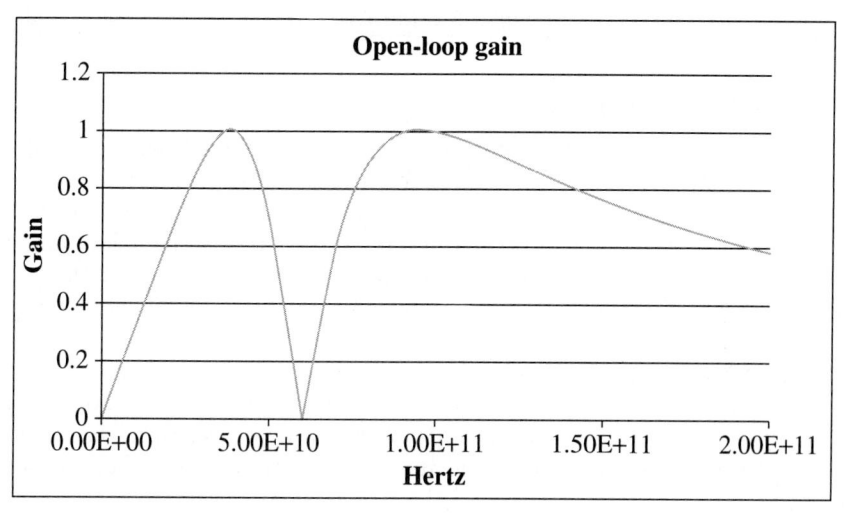

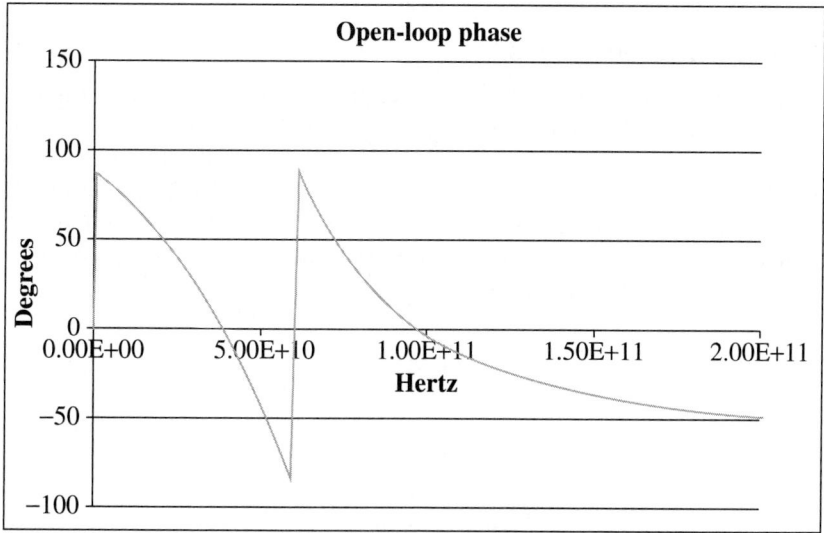

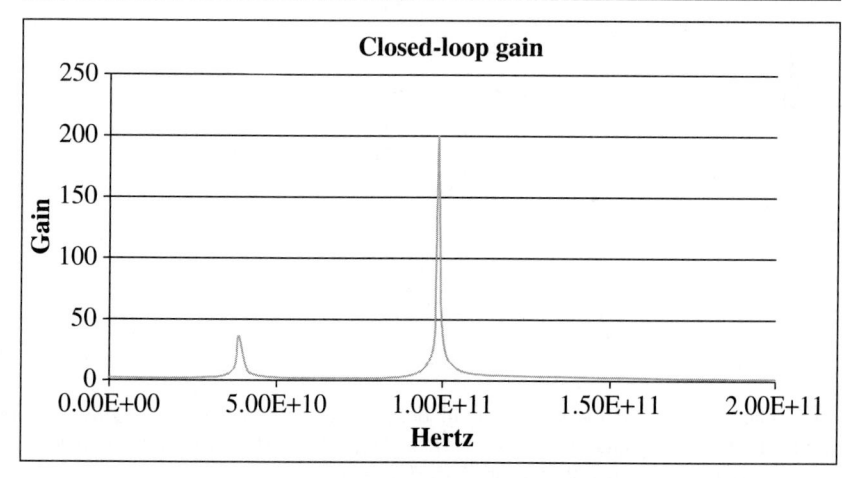

Figure 5.76 The transfer characteristics of a subharmonic oscillator.

5.11.4.4 Colpitts Oscillator

The Colpitts oscillator may be advantageous over a cross-coupled pair, especially when space is a major concern, as only one transistor is required in this format [Raz01]. The noise theory developed by Hashemi and Lee [Lee04b][HGKH05] also indicates that the narrow window of current flow per cycle in this design results in a low phase noise performance. Also, because of its design, the gate-source capacitance of the transistor used can fill the role of a varactor if a narrow tuning range is sufficient, resulting in a further space savings [HBAN07b].

Fig. 5.77 illustrates a Colpitts oscillator [Raz01, Chapter 14]. In [Raz01, Chapter 14], there is an excellent discussion of the Colpitts design. This design is not as common for mmWave processes, so we leave out a detailed analysis here. In its basic form, the Colpitts provides a greater than unity and zero-phase open-loop gain from its source to its drain.

5.11.5 Phase-Locked Loops

The frequency synthesizer or phased-lock loop (PLL) is one of the most important and challenging blocks in the design of mmWave WPAN devices [BSS+10]. PLLs are needed for production of local oscillator signals and channel selection [LTL09]. PLLs are also vital for low-jitter clock generation for very high speed (e.g., 10 Gbps) baseband signal processing [CaoADC]. The parameters that determine the merit of a PLL include output frequency range, output power, power consumption, and phase noise [BSS+10]. Barale et al. presented the following as a figure of merit [BSS+10]:

$$\text{FOM} = 10 \log \frac{\Delta f_0 \cdot P_0}{P_{DC} \cdot PN_{1\text{MHz}}} \tag{5.161}$$

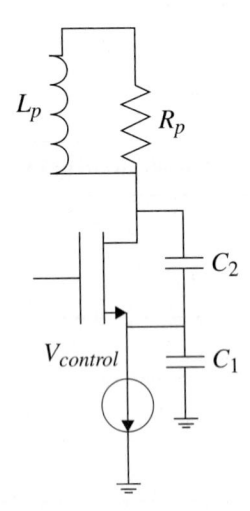

Figure 5.77 The Colpitts oscillator is a small design, but it may be challenging at mmWave frequencies due to the low gains achievable with a single transistor, which may make sufficient open-loop gains difficult to achieve [based on [Raz01, Chapter 14]].

5.11 Analog MmWave Components

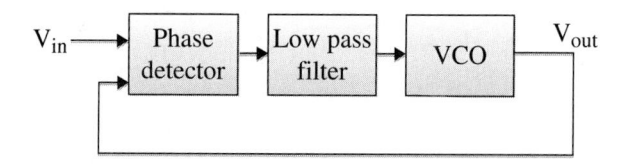

Figure 5.78 A basic PLL.

where Δf_0 is the frequency tuning range, P_0 is the output power, P_{DC} is the power consumption, and PN_{1MHz} is the phase noise at 1 MHz offset [BSS+10].

The basic operation of a PLL is illustrated in Fig. 5.78 [Raz01]. The PLL is a feedback loop that detects the phase difference of its output and input with a phase detector. The output of the phase detector is then passed through a low-pass filter to provide a DC control signal to the VCO. By changing the control signal of the VCO, the loop forces the output signal to change the rate at which it accumulates phase. If we consider a sinusoidal signal

$$V(t) = \cos(\omega t + \phi_0) \qquad (5.162)$$

the rate at which the signal accumulates phase is given by ω, and the total phase is ωt. By temporarily changing the frequency of the output signal, the loop forces the output phase to accumulate more or less quickly until the phase of the input and output are equal, at which point the frequencies of the input and output are also equal [Raz01].

$$\omega_{in} = \omega_{out} \qquad (5.163)$$

where ω_{in} and ω_{out} are the angular frequencies of the input and output. If there is any error in the loop, such as an offset voltage, then (5.163) will not hold exactly, but instead there will be some steady-state phase error [Raz01].

Fig. 5.78 does not make it apparent why a VCO is necessary, as ideally the input and output would be waveforms with the same frequency and phase. To make the loop more useful for general applications, we add a frequency divider to the loop, as shown in Fig. 5.79. This allows a low frequency sinusoid, which may be made to be a very clean sinusoid (e.g., if it is generated from a crystal oscillator), to serve as the phase reference for a high frequency sinusoid. (Note that the high frequency VCO in a PLL generally is not capable of as low a phase noise as the reference frequency generator.) This topology is often used to select the output frequency within the tuning range of the VCO — that is, to choose the output channel. This is accomplished by changing the division ratio M. If finer control is needed than can be achieved with a simple divide, then a fractional frequency divider may be used.

The phase noise of the reference signal (V_{in} in Fig. 5.78) used in a PLL will largely determine the overall PLL phase noise [BSS+10]. A major design choice is the selection of a reference signal source capable of producing tolerably low phase noise. The division ratio between output frequency and reference frequency determines the amplification of phase noise from the reference signal (Phase Noise Gain $\sim 20\log(\text{Ratio})$) to the output signal [BSS+10]. Together the reference signal frequency and reference signal frequency to output frequency ratio determine the phase-noise requirements for the reference signal.

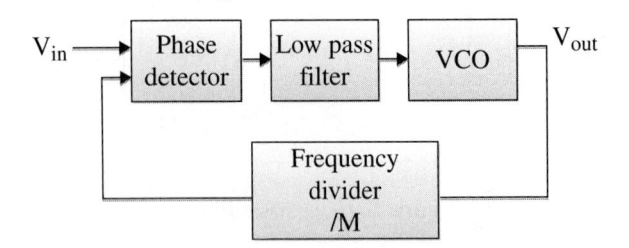

Figure 5.79 A frequency divider is inserted into the simple PLL to allow a low frequency sinusoid to serve as a phase and frequency reference for a high frequency sinusoid.

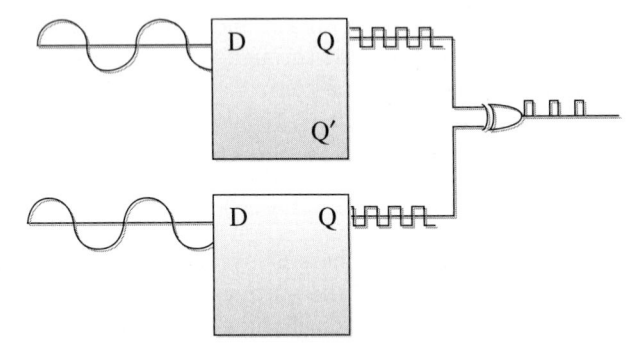

Figure 5.80 A phase detector may be constructed very simply using D-flop-flops and logic gates, such as an XOR gate at the right of the figure.

The phase detector can be constructed using simple exclusive or gates (XOR) and D-flip-flops [Raz01]. This is illustrated in Fig. 5.80, in which two D-flip-flops are used to convert the input and output sinusoids into square waves, the phases of which are compared with an XOR gate. The XOR gate provides a "1" when the input and output are different — so if the square waves are in phase the XOR gate provides zero output, and if the square waves are 180° out-of-phase the XOR output is always one.

Ideally, only one VCO would be used to minimize space and power requirements in the PLL, but in practice difficulties with the tuning range of each VCO and the frequency divider used in PLL design may require multi-VCO designs. This is because, as for 60 GHz systems, the output frequency range may exceed 15% of the carrier frequency, and a single VCO may not suffice to provide such a large tuning range. Therefore, it is likely that many PLLs will use multi-core VCOs [BSS+10].

A very simple frequency divider circuit (for a frequency division by two) is illustrated using a D-flip-flop [BAQ91] as illustrated in Fig. 5.81, which shows an asynchronous static divide-by-two architecture. There are two basic types of frequency dividers used: static frequency dividers (SFDs) and injection-locked frequency dividers (ILFDs). A static frequency divider uses basic logic gates such as the D-flip-flop shown in Fig. 5.81 to achieve frequency division [CTSMK07]. "Static" indicates that the frequency divider relies on the statically stored states of the logic gates (which change when the input signal crosses back and forth across a threshold) to achieve frequency division [CTSMK07]. An

5.11 Analog MmWave Components

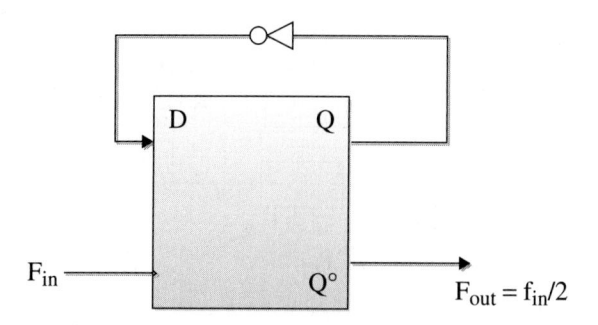

Figure 5.81 A very simple asynchronous D-flip-flop divide-by-two circuit.

injection-locked frequency divider replaces the logic-gates of a static divider with an oscillator that injection locks onto the VCO output frequency so that the output frequency is a phase-locked subharmonic of the VCO frequency [CTSMK07]. Injection locking is a phenomenon whereby an injected signal causes an oscillator to lock onto the frequency of the injected signal provided that the injected signal is strong enough and there is enough loop gain in the oscillator at the injected frequency. When designing an injection-locked frequency divider, it is necessary to consider the locking range, which may be found using [BBKH06]

$$\Delta \omega_m = \frac{\omega_0 I_{\text{inj}}}{Q I_{\text{osc}}} \quad (5.164)$$

where $\Delta \omega_m$ is the locking range, ω_0 is the fundamental frequency of the oscillator, I_{inj} is the amount of injected current, Q is the quality-factor of the LC tank used in the oscillator, and I_{osc} is the current that flows through the oscillator loop under non-locked conditions.

Static dividers are capable of higher frequency locking range and division ratios at the expense of higher power consumption and lower maximum operating frequency [SKH09][CDO07]. As illustrated in Fig. 5.82, static dividers may be most appropriate as the second divider in a frequency divider chain. If a static divider is realized as a chain of dividers (e.g., a cascade of three divide-by-two dividers for an overall divide-by-eight circuit), then it is essential that the speed/bandwidth of the input stages (i.e., the high frequency portion of the divider) meet or exceed the speed of subsequent stages [BSS+08], as the first stage will have to handle the highest input and output frequencies.

Injection-locked frequency dividers (ILFDs) are capable of higher frequency and lower power operation than static dividers [SKH09][LTL09]. The frequency division ratio of ILFDs is a key factor in determining the power consumption and locking efficiency of the ILFD, and requirements of subsequent stages in the PLL: as the division ratio increases, the locking efficiency should decrease, the locking range should decrease, and power requirements of subsequent stages should decrease (largely because they are operating at lower frequencies with a higher frequency ratio) [SKH09]. There are various oscillator topologies suitable for injection locking. Any topology with a fundamental frequency ground point with low capacitance to ground may be suitable for use in an ILFD. Fig. 5.83 illustrates a common method of performing injection locking using an LC tank VCO. The

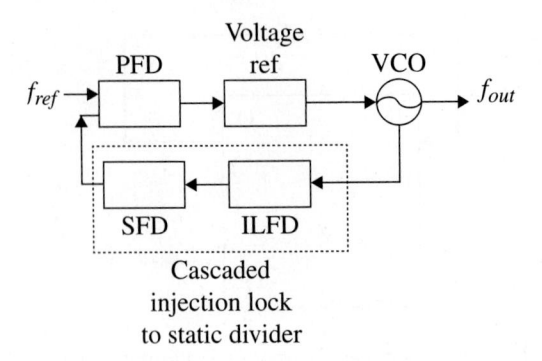

Figure 5.82 An injection-locked frequency divider (ILFD) followed by a static frequency divider (SFD) is a very popular arrangement for high speed, locking range, and programmability.

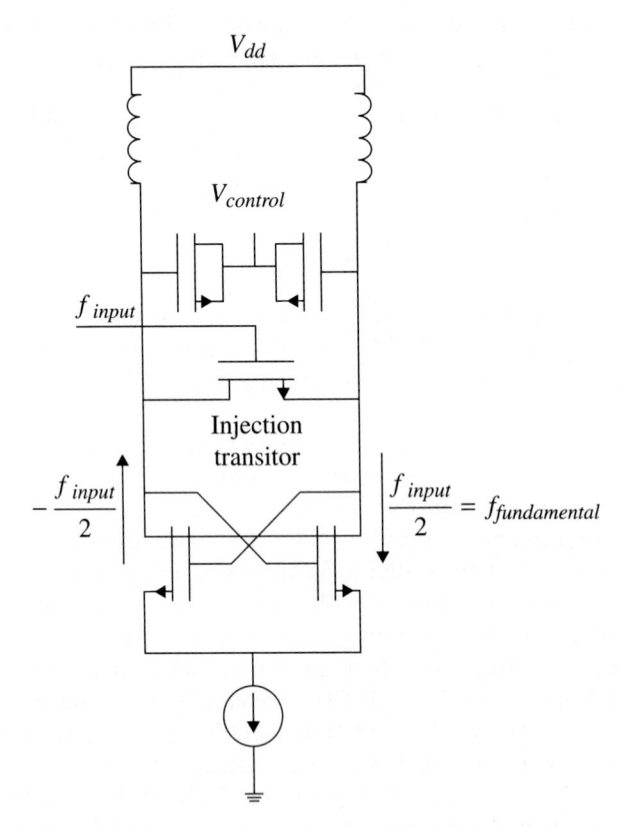

Figure 5.83 The input point is a ground point for the fundamental frequency. The fundamental frequency is injection locked to half the input frequency. The locking range will improve with increasing transconductance of the injection transistor [LTL09]. The varactors can be used to tune the tank resonant frequency to f_{input}.

5.11 Analog MmWave Components

varactors are not strictly necessary, but they can be used to tune the resonant frequency of the LC tank to the input frequency to improve locking range [HCLC09]. We will now consider several case studies of frequency dividers used for mmWave applications.

In 2008, Barale et al. presented the design of a latch programmable frequency divider with a maximum input frequency of 3.5 GHz and a minimum input frequency of 640 MHz for a division ratio of 24, a power dissipation of 4.5 mW, and division ratios of 24, 25, 26, and 27 [BSS+08]. The minimum input frequency is explained as the result of leakage current discharging the state of the flip-flops. They made use of a combination of D-flip-flop and half-transparent JK-flip-flops for the best combination of speed and programmability. JK-flip-flops with lower speed capability but improved ability to implement arbitrary division ratios were used after a high-speed D-flip-flop input stage [BSS+08]. Division ratio selection was implemented using a 4-input Mux [BSS+08].

In 2009, Sim et al. [SKH09] presented a ring-oscillator based ILFD in 0.18 μm CMOS for an output frequency up to 30.95 GHz. Their use of a ring oscillator was based on the smaller footprint and greater possible division ratio for this type of oscillator versus LC tank oscillators [SKH09]. Rather than use a standard approach of injecting the input frequency at a single point in the ring oscillator, Sim et al. made use of a balanced injection scheme in which coupling capacitors coupled the input frequency to every element of the ring oscillator [SKH09].

In 2009, Hsu et al. presented an LC tank ILFD in 0.13 μm CMOS. They used a shunt inductor connected to the source node of a cross-coupled pair LC tank oscillator to resonate with the parasitic capacitance of the tail current source and thus improve locking range [HCLC09]. Also, to improve tuning range at the expense of lower quality factor and higher phase noise, they opted to implement their inductors using microstrip lines rather than spiral inductors [HCLC09]. They compared the locking rage of their implementation to implementations that forego the use of shunt inductors and tank varactors. Designs without a shunt peaking inductor or tank-varactors had the worst locking range. The addition of a shunt peaking inductor substantially improved locking range, while use of both a shunt inductor and tank varactors resulted in the best tuning range (with an input power of 5 dBm, use of both techniques resulted in a tuning range of 8.8 GHz, compared to approximately 4 GHz without either technique [HCLC09]).

Lastly, we will consider several mmWave PLL case studies. In 2010, Barale et al. presented a phase-locked loop in 90 nm CMOS compatible with 60 GHz standards that made use of a 27 MHz reference signal [BSS+10]. A dual-core injection-locked frequency divider design with the outputs of both VCOs shorted together was used to provide adequate tuning range for the PLL. The actual output of the PLL was divided into three channels from the dual-core LC tank VCO at 49.68-51.84 GHz, 51.84-54 GHz, and 54-56.16 GHz. Buffers were used at the outputs of each VCO to improve isolation [BSS+10]. VCOs with differential outputs also improved isolation.

In 2007, C. Cao et al. [CDO07] presented a 50 GHz PLL in 0.13 μm CMOS that made use of an injection-locked frequency divider to overcome challenges with static frequency dividers. They relied on a single-core LC tank VCO for their design. The divide-by-two injection-locked frequency divider relied on the addition of a fundamental common-mode shunt transistor as shown in Fig. 5.84. The overall design consumed only

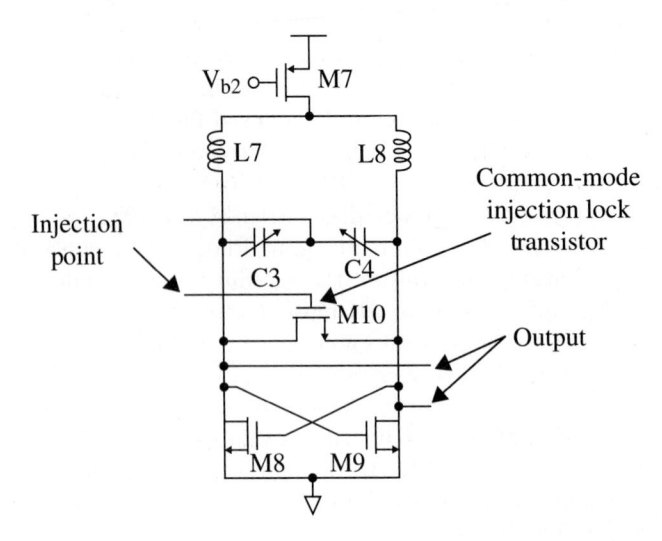

Figure 5.84 Cao et al. [CDO07] relied on a shunt-transistor to perform injection locking. They found that a wider transistor improved locking range but decreased output power [based on part of a figure from [CDO07] © IEEE].

57 mW, approximately an order of magnitude lower than SiGe PLLs demonstrated up to that date [CDO07].

In 2007, Hoshino et al. presented a phase-locked loop in 90 nm CMOS with an output frequency from 61 to 63 GHz [HTM+07]. To avoid the use of area-intensive transmission lines or inductors, Hoshino et al. used a ring-oscillator design for an injection-locked frequency divider that preceded two static D-flip-flop divide-by-16 dividers (the ring oscillator occupied an area of 80 μm × 40 μm). The oscillator had three stages and an output buffer and provided a frequency ratio of 4 [HTM+07]. An LC tank VCO with transmission line inductors was used for its ability to provide a differential output [HTM+07]. A source follower to common-source follower cascade was used at the output of the VCO to ensure adequate signal power to the injection-locked frequency divider.

5.12 Consumption Factor Theory

As mmWave communication systems and circuits proliferate, the ability to design, analyze, and compare systems and circuit implementations for maximum power efficiency (e.g., improved battery life) will become increasingly important. The *consumption factor theory* has been developed to allow analysis and design of power efficient communication systems. The consumption factor theory is very general and may be applied to any cascaded communication system, including the case of relays in propagation channels.

The *consumption factor* (CF) is defined as the maximum ratio of data rate to the total power consumed [MR14b]. The CF may be defined for an individual circuit, for a complete cascaded system, or for the cascade of a few components in any communication system. The CF theory allows an engineer to understand how the parameters of individual

5.12 Consumption Factor Theory

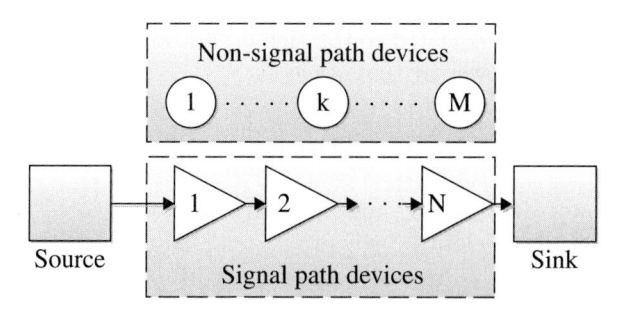

Figure 5.85 A general communication system comprising many circuit components.

circuits or components in a communication system impact overall power efficiency while delivering a desired data rate through the system. Basically, maximizing the CF allows one to maximize the power efficiency of a communication circuit, system, or device for a particular data rate.

There are a few intermediate steps required to derive CF, and these steps provide their own useful metrics. Consider a general cascaded communication system, as shown in Fig. 5.85, in which information is generated at a source and sent as a signal down a signal path to a sink. *Signal path components* such as amplifiers and mixers are responsible for transmitting the information signal (e.g., signal power) to the sink. In addition, there may be components *off the signal path*, such as voltage regulation equipment, microprocessors, and smartphone screens that do not participate directly in the signal path but do consume power and provide a needed function for the device. The *efficiency* of the ith component or device along the signal path, the ith signal path component, may be written as

$$\eta_i = \frac{P_{\text{sig}_i}}{P_{\text{sig}_i} + P_{\text{non-sig}_i}} \quad (5.165)$$

where P_{sig_i} is the total signal power delivered by the ith stage to the $(i+1)$th stage, and $P_{\text{non-sig}_i}$ is the signal power used by the ith stage component but not delivered as signal power. The total power consumed by the ith stage may be written

$$P_{\text{consumed}_i} = P_{\text{non-sig}_i} + P_{\text{added-sig}_i} \quad (5.166)$$

where $P_{\text{added-sig}_i}$ is the total signal power added by the ith component, which is the difference in the signal power *delivered by* the ith component and the signal power *delivered to* the ith component. We can sum up the signal power contributions from all of the components along the signal path (from left to right in Fig. 5.85) to find

$$\sum_{i=1}^{N} P_{\text{added-sig}_i} = P_{\text{sig}_N} - P_{\text{sig}_{\text{source}}} \quad (5.167)$$

where $P_{\text{sig}_{\text{source}}}$ is the signal power from the source, and P_{sig_N} is the signal power delivered by the Nth (and last stage) signal-path component. Now, from (5.165), the total wasted power of the kth stage not dedicated to the signal path may be related to the efficiency and total delivered signal power by that stage

$$P_{\text{non-sig}_k} = P_{\text{sig}_k}\left(\frac{1}{\eta_k} - 1\right). \tag{5.168}$$

Also, the signal power delivered by the kth stage may be related to the total power delivered to the sink, thus yielding

$$P_{\text{non-sig}_k} = \frac{P_{\text{sig}_N}}{\prod_{i=k+1}^{N} G_i}\left(\frac{1}{\eta_k} - 1\right) \tag{5.169}$$

where G_i is the gain of the ith stage. We may therefore compute the total power consumed by the communication system as the power consumed by the source, and the three additional power consumption terms that represent the amount of power (a) contributed by, and (b) wasted by, the in-path cascaded components, and (c) the power consumed by the non-signal path components

$$P_{\text{consumed}} = P_{\text{sig\_source}} + \sum_{i=1}^{N} P_{\text{added-sig}_i} + \sum_{k=1}^{N} P_{\text{non-sig}_k}$$

$$+ \sum_{k=1}^{M} P_{\text{non-path}_k} \tag{5.170}$$

$$= P_{\text{sig}_N}\left(1 + \sum_{k=1}^{N} \frac{1}{\prod_{i=k+1}^{N} G_i}\left(\frac{1}{\eta_k} - 1\right)\right) + \sum_{k=1}^{M} P_{\text{non-path}_k}. \tag{5.171}$$

To incorporate the efficiencies $\eta_{\text{non-path}_k}$ of the devices that are off the signal path, we may simply add them (e.g., consider them at the end of the cascade) and assume they have unity gain since they have nothing to do with the transmission of power through the cascade, and their efficiencies are related to how well they perform their intended function when compared with their power consumption. For example, one merely writes the total power consumption of the kth non-path block $P_{\text{non-path}_k}$ in terms of its efficiency $\eta_{\text{non-path}_k}$ (the ratio of its usefully dissipated power to its total power expenditure) in relation to the usefully dissipated power P_{u_k} (power that directly contributes to its intended functionality)

$$P_{\text{non-path}_k} = \frac{P_{u_k}}{\eta_{\text{non-path}_k}}. \tag{5.172}$$

We may then re-write (5.171) as

$$P_{\text{consumed}} = P_{\text{sig}_N}\left(1 + \frac{1}{P_{\text{sig}_N}}\sum_{k=1}^{M} \frac{P_{u_k}}{\eta_{\text{non-path}_k}}\right.$$
$$\left.+ \sum_{k=1}^{N} \frac{1}{\prod_{i=k+1}^{N} G_i}\left(\frac{1}{\eta_k} - 1\right)\right). \tag{5.173}$$

5.12 Consumption Factor Theory

Now, if we define H as the *power-efficiency factor* for all cascaded components on the signal path, we note from (5.171) that

$$P_{\text{consumed}} = \frac{P_{\text{sig}_N}}{H} + P_{\text{non-path}}. \tag{5.174}$$

From (5.171) we may define the power efficiency of the entire communication system η_{cs} as

$$\eta_{cs} = \frac{P_{\text{sig}_N}}{P_{\text{consumed}}} = \frac{1}{\frac{1}{H} + \frac{1}{P_{\text{sig}_N}} \sum_{k=1}^{M} P_{\text{non-path}_k}} \tag{5.175}$$

where, in (5.174) and (5.175), we defined the *system power-efficiency factor H* of all cascaded components (on the signal path) as

$$H = \left\{ 1 + \sum_{k=1}^{N} \frac{1}{\prod_{i=k+1}^{N} G_i} \left(\frac{1}{\eta_k} - 1 \right) \right\}^{-1} \tag{5.176}$$

where H ranges between 0 and 1 and may be considered to be a figure of merit, the efficiency of the entire signal path cascade (where $H = 1$ is perfect power efficiency). Note also that H^{-1} is also a figure of merit, ranging from one to infinity, where $H^{-1} = 1$ is perfect power efficiency with no wasted power. This is similar to the theory of Noise Figure [MR14b]. A key result from (5.176) is that the gains of the components that follow (e.g., are to the right of elements in Fig. 5.85) a particular component "degenerate" the efficiency of the cascade. An implication of this is that *the efficiencies of devices that handle the most power are most important in terms of the power-efficiency factor of the entire cascade.* In addition to having obvious utility in analyzing circuits and cascaded systems, and comparing their power efficiency using CF, another non-obvious but interesting use of CF is to determine the power-efficiency factor of a wireless channel, since a channel is part of a cascaded system. In order for the power-efficiency factor and consumption factor analysis to produce results that are consistent with expectation for power consumption, we must impose the following condition:

$$\lim_{G \to 0} \frac{G}{H} = 1. \tag{5.177}$$

This is necessary to prevent the conclusion that zero power is consumed by a cascaded system with a block that has zero gain. In particular, by choosing

$$H_{\text{channel}} = G_{\text{channel}} \tag{5.178}$$

we ensure that we can never calculate zero power consumption for a communication system communicating through a channel with zero gain.

Consider also the case of two cascaded sub-systems whose Hs have already been characterized, where sub-system #2, with power-efficiency factor $H_{\text{sub-system 2}}$ and gain $G_{\text{sub-system 2}}$ follows sub-system #1 with power-efficiency factor $H_{\text{sub-system 1}}$. We can show that the power-efficiency factor of the entire cascade, $H_{\text{cascaded system}}$, may be written much like the classic noise figure theory

$$H_{\text{cascaded system}}^{-1} = H_{\text{sub-system 2}}^{-1} + \frac{1}{G_{\text{sub-system 2}}}\left(H_{\text{sub-system 1}}^{-1} - 1\right). \qquad (5.179)$$

Consider now the case in which a wireless channel exists between a transmitter and receiver. The channel can be treated as a simple 100% efficient component between the transmitter and receiver cascade. The overall power-efficiency factor of the entire transmitter-receiver pair can be found from (5.178) and (5.179) as [MR14b]

$$\begin{aligned}H_{\text{link}}^{-1} = H_{RX}^{-1} &+ \frac{1}{G_{RX}}\left(\frac{1}{G_{\text{channel}}} - 1\right) \\ &+ \frac{1}{G_{RX}G_{\text{channel}_i}}\left(H_{TX}^{-1} - 1\right)\end{aligned} \qquad (5.180)$$

where we have implicitly indicated that the power-efficiency factor of the wireless channel is simply equal to the channel gain. In other words, (5.168) and (5.176) yield $H_{\text{channel}} = G_{\text{channel}}$. Note from (5.180) that if the receiver gain is much smaller than the expected channel path loss, the cascaded power-efficiency factor will be very small and on the order of the product of the channel gain with the receiver gain.

5.12.1 Numerical Example of Power-Efficiency Factor

To better illustrate the use of the power-efficiency factor, consider a simple scenario of a cascade of a baseband amplifier, followed by a mixer, followed by an RF amplifier. We will consider two different examples of this cascade scenario, where different components are used, in order to compare the power efficiencies due to the particular specifications of components. Assume that for both cascade examples, the RF amplifier is a commercially available MAX2265 power amplifier by Maxim Integrated with 37% efficiency [Maxim]. In both cases, the mixer is an ADEX-10L mixer by Mini-Circuits with a maximum conversion loss of 8.8 dB [MC]. In the first case, the baseband amplifier (the component farthest to the left in Fig. 5.85 if in a transmitter, and farthest to the right if in a receiver) is an ERA-1+ by Mini-Circuits, and in the second case the baseband amplifier is an ERA-4+ [MC], also by Mini-Circuits. The maximum efficiencies of these parts are estimated by taking the ratio of their maximum output signal power to their dissipated DC power. As the mixer is a passive component, its gain and efficiency are equal. Table 5.8 summarizes the efficiencies and gains of each component in the cascade. Using (5.175), the power-efficiency factor of the first scenario is

$$H_{\text{scenario 1}} = \frac{1}{\frac{1}{0.37} + \frac{1}{16.17}\left(\frac{1}{0.36} - 1\right) + \frac{1}{0.36*16.17}\left(\frac{1}{0.1165} - 1\right)} = 0.2398, \qquad (5.181)$$

whereas the power-efficiency factor of the second scenario is

$$H_{\text{scenario 2}} = \frac{1}{\frac{1}{0.37} + \frac{1}{16.17}\left(\frac{1}{0.36} - 1\right) + \frac{1}{0.36*16.17}\left(\frac{1}{0.1836} - 1\right)} = 0.2813. \qquad (5.182)$$

5.12 Consumption Factor Theory

Table 5.8 An example of the use of the power-efficiency factor to compare two cascades of a baseband amp, mixer, and RF amp [from [MR14b] © IEEE].

Component	Gain	Efficiency(H)
Example 1		
MAX2265 RF Amp	24.5 dB (voltage gain of 16.17)	37%
ADEX-10L Mixer	−8.8 dB	36%
ERA-1+ BB Amp	10.9 dB	11.65%
Example 2		
MAX2265 RF Amp	24.5 dB	37%
ADEX-10L Mixer	−8.8 dB	36%
ERA-1+ BB Amp	13.4 dB	18.36%

Therefore, we see that the second scenario offers a superior power efficiency compared to the first scenario, due to the better efficiency of the baseband amplifier, but it falls far short of the ideal power-efficiency factor of unity. Using different components and architectures, it is possible to characterize and compare, in a quantitative manner, the power-efficiency factor and consumption factor of cascaded components.

As a second example, consider the cascade of a transmitter power amplifier communicating through a free space channel with a low noise amplifier at the receiver. Let us assume that the cascade, in the first case, uses the same RF power amplifier as in the previous example (MAX2265), while the LNA is a Maxim Semiconductor MAX2643 with a gain of 16.7 dB (6.68 absolute voltage gain) [MR14b]. We will assume this LNA has 100% efficiency for the purposes of illustrating the impact of the PA's efficiency and the channel (i.e., here we ignore the LNA's efficiency, although this can easily be done as explained above). For a carrier frequency of 900 MHz, now consider the cascade for a second case where the MAX2265 RF power amplifier is replaced with a hypothetical RF amplifier device having 45% power efficiency (a slight improvement). Assume the link is a 100 m free space radio channel with gain of −71.5 dB. Since the propagation channel loss greatly exceeds the LNA gain, equation (5.180) applies, where the first two terms are smaller than the third term such that

$$H_{\text{link}} \approx G_{\text{Rx}} G_{\text{channel}} H_{\text{Tx}} \tag{5.183}$$

where H_{Tx} is the efficiency of the RF amplifier, so that in the first case using the MAX2265 amplifier (37% efficiency), the power-efficiency factor of the cascaded system is 173.5e-9, whereas in the second case (using an RF power amplifier with 45% efficiency), the power-efficiency factor is 211.02e-9. The second case has an improved power-efficiency factor commensurate with the power-efficiency improvement of the RF amplifier stage. These simple examples demonstrate how the power-efficiency factor may be used to compare and quantify the power efficiencies of different cascaded systems and demonstrate the importance of using higher efficiency RF amplifiers for improved power efficiency throughout a transmitter-receiver link.

5.12.2 Consumption Factor Definition

We now define the *consumption factor*, CF, and *operating consumption factor* for a general communication system where

$$CF = \left(\frac{R}{P_{\text{consumed}}}\right)_{\text{max}} = \frac{R_{\text{max}}}{P_{\text{consumed,min}}} \quad (5.184)$$

and

$$\text{operating } CF = \frac{R}{P_{\text{consumed}}} \quad (5.185)$$

where R is the operational or actual data rate (in bits-per-second or bps) and R_{max} is the maximum data rate supportable by the communication system. The notation *operating CF* denotes a particular operating point of an actual system that may not be realizing the maximum possible efficiency, due to the fact that it may have a particular operating signal-to-noise ratio (SNR), bit rate, or power consumption. We use the actual operating point to determine the margin from an optimal power consumption level or optimum bit rate perspective, as shown below. Our goal is to maximize the consumption factor. Further analysis based on only maximizing R or minimizing P_{consumed} is also pertinent to system optimization. For a very general communication system in an AWGN channel, R_{max} may be written using Shannon's information theory according to the SNR:

$$R_{\text{max}} = \text{Channel Capacity} = B \log_2(1 + \text{SNR}). \quad (5.186)$$

Or, for frequency selective channels

$$R_{\text{max}} = \int_0^B \log_2\left(1 + \frac{P_r(f)}{N(f)}\right) df = \int_0^B \log_2\left(1 + \frac{|H(f)|^2 P_t(f)}{N(f)}\right) df \quad (5.187)$$

where $P_r(f)$, $P_t(f)$, and $N(f)$ are the power spectral densities of the received power, the transmitted power, and the noise power at the detector, respectively. $H(f)$ is the frequency response of the channel and any blocks that precede the detector. Note that equations (5.184) and (5.185) make no assumptions about the signaling, modulation, or coding schemes used by the communication system. To support a given spectral efficiency η_{sig} (bps/Hz), there is a minimum SNR assuming an AWGN channel

$$\frac{\text{SNR}}{M_{\text{SNR}}} = \text{SNR}_{\text{min}} = 2^{\eta_{\text{sig}}} - 1. \quad (5.188)$$

The operating SNR of the system and the margin of the operating SNR (M_{SNR}) above the minimum SNR$_{\text{min}}$ may be used with equations (5.174), (5.184), and (5.185) to find the consumption factor and operating consumption factor given as

$$CF = \frac{B \log_2(1 + \text{SNR})}{P_{\text{non-path}} + \left(\frac{\text{SNR}}{M_{\text{SNR}}}\right) \times \frac{P_{\text{noise}}}{H}} \quad (5.189a)$$

$$CF = \frac{B \log_2(1 + M_{\text{SNR}}(2^{\eta_{\text{sig}}} - 1))}{P_{\text{non-path}} + (2^{\eta_{\text{sig}}} - 1) \times \frac{P_{\text{noise}}}{H}}. \quad (5.189b)$$

5.12 Consumption Factor Theory

$$\text{Operating } CF = \frac{B \log_2 \left(\frac{\text{SNR}}{M_{\text{SNR}}} + 1 \right)}{P_{\text{non-path}} + \text{SNR} \times \frac{P_{\text{noise}}}{H}} \quad (5.190a)$$

$$\text{Operating } CF = \frac{B \eta_{\text{sig}}}{P_{\text{non-path}} + M_{\text{SNR}}(2^{\eta_{\text{sig}}} - 1) \times \frac{P_{\text{noise}}}{H}}. \quad (5.190b)$$

Shannon's limit describes the minimum energy-per-bit-per-noise spectral density required to achieve arbitrarily low probability of bit error through proper coding scheme selection:

$$\frac{E_b}{N_0} = \ln(2). \quad (5.191)$$

This limit is generally found by using Shannon's capacity theorem and allowing the code used to occupy an infinite bandwidth [Cou07].

As was shown using the consumption factor theory, the maximum ratio of data rate to power for a communication system may be written as in (5.190a)

$$CF = \frac{B \log_2(1 + \text{SNR})}{P_{NP} + \text{SNR}_{\min} \times \frac{P_{\text{noise}}}{H}}. \quad (5.192)$$

Let us take the limit of (5.192) as the bandwidth approaches infinity, assuming AWGN. First, recall that the SNR may be written in terms of the energy-per-bit, the time required to transmit a single bit T_b, the noise spectral density N_0, and the bandwidth of the system B

$$\text{SNR} = \frac{\frac{E_b}{T_b}}{N_0 B}. \quad (5.193)$$

In the limit as B approaches infinity, the SNR clearly weakens to approach the minimum acceptable SNR. Therefore

$$\lim_{B \to \infty} CF = \frac{1}{E_{bc,\min}} = \lim_{B \to \infty} \left\{ \frac{(B \log_2(1 + \text{SNR}_{\min}))}{P_{NP} + \text{SNR}_{\min} \times \frac{P_{\text{noise}}}{H}} \right\}$$

$$= \lim_{B \to \infty} \left\{ \frac{\left(B \log_2 \left(1 + \frac{\frac{E_b}{T_b}}{N_0 B} \right) \right)}{P_{NP} + \frac{\frac{E_b}{T_b}}{N_0 B} \times \frac{N_0 B}{H}} \right\} \quad (5.194)$$

where $E_{bc,\min}$ is the minimum energy-per-bit that must be consumed by the communication system, and E_b is the minimum energy-per-bit that must be present in the signal carried by the communication system and delivered to the receiver's detector. $E_{bc,\min}$ and E_b are generally not equal, as $E_{bc,\min}$ must always be greater than E_b, and equality holds only when the receiver is noiseless. Note that the denominator is no longer a function of bandwidth. We may therefore apply the result from Shannon's bound from [Cou07] to find

$$\frac{1}{E_{bc,\min}} = \frac{E_b}{N_0 T_b \ln(2)} \times \frac{1}{P_{NP} + \frac{E_b}{T_b H}} \quad (5.195)$$

$$E_{bc,\min} = \frac{\ln(2) N_0}{E_b} \frac{P_{NP}}{C} + \frac{N_0 \ln(2)}{H} = \frac{P_{NP}}{C} + \frac{\ln(2) N_0}{H} \quad (5.196)$$

where we have made the substitution $C = \frac{1}{T_b}$, that is, in the limit the bit-rate approaches the channel capacity C. Equation (5.196) should be interpreted as the minimum energy that must be expended per bit over the noise spectral density in order to obtain arbitrarily low error rate. This interpretation should not be confused with the interpretation of the original Shannon limit, which relates to the bit energy per noise spectral density within the signal that flows through the communication system. Note that if the system is 100% efficient on the signal path, and no power is used off the signal path, then (5.196) will yield the same result as Shannon's limit, indicating that, in effect, the communication system and the signal it carries have become identical. In general, however, (5.196) will require more energy than Shannon's limit due to inefficiencies. Equation (5.196) indicates the tremendous importance of the efficiency of a communication system in determining the energy cost of a single bit. Note also that the total power consumption to send a single bit is given by $P_{bc,\text{bit}}$:

$$P_{bc,\text{bit}} = P_{NP} + \frac{E_b C}{H} = C E_{bc,\min} \qquad (5.197)$$

It is instructive to estimate the required power consumption per bit as a function of the cell radius for a single user at the edge of the cell, as bits delivered to the edge of the cell are expected to be the most costly from an energy perspective.

Recall first that the power-efficiency factor over a wireless link may be written

$$\begin{aligned} H_{\text{link}}^{-1} &= H_{RX}^{-1} + \frac{1}{G_{RX}} \left(\frac{1}{G_{\text{channel}}} - 1 \right) \\ &+ \frac{1}{G_{RX} G_{\text{channel}_i}} \left(H_{TX}^{-1} - 1 \right) \end{aligned} \qquad (5.198)$$

$$\begin{aligned} H_{\text{link}} &= (H_{RX} H_{TX} G_{RX} G_{\text{channel}}) / (H_{TX} G_{RX} G_{\text{channel}} \\ &+ H_{RX} H_{TX} (1 - G_{\text{channel}}) + H_{RX} (1 - H_{TX})) \end{aligned} \qquad (5.199)$$

where G_{channel} is the link channel gain, $G_{RX,\text{ANT}} G_{TX,\text{ANT}}$ is the product of the transmitter and receiver antennas (which is bound physically to be less than or equal to the channel gain), H_{RX} is the power-efficiency factor of the receiver, H_{TX} is the power-efficiency factor of the transmitter, and G_{RX} is the gain of the receiver. Using (5.199) in (5.196), we find

$$\begin{aligned} E_{cb,\min} &= \frac{P_{NP}}{C} + \ln(2) N_0 \Big\{ H_{RX}^{-1} + \frac{1}{G_{RX}} \left(\frac{1}{G_{\text{channel}}} - 1 \right) \\ &+ \frac{1}{G_{RX} G_{\text{channel}}} \left(H_{TX}^{-1} - 1 \right) \Big\} \end{aligned} \qquad (5.200)$$

If we factor out the inverse of the channel gain, we find that

$$E_{cb,\min} = \frac{P_{NP}}{C} + \frac{\ln(2) N_0}{G_{RX} G_{\text{channel}}} \left\{ G_{RX} G_{\text{channel}} H_{RX}^{-1} + H_{TX}^{-1} - G_{\text{channel}} \right\}. \qquad (5.201)$$

5.12 Consumption Factor Theory

In the limit of very small channel gains (which is reasonable given the large path loss at mmWave frequencies), this approaches

$$H_{\text{link}} \to H_{TX} G_{RX} G_{\text{channel}} \tag{5.202}$$

$$E_{b,\min} = \frac{P_{NP}}{C} + \frac{\ln(2) N_0}{G_{RX} G_{\text{channel}} H_{TX}}. \tag{5.203}$$

Here is the interpretation that should be used with regard to the approximation for the system power-efficiency ratio: For cases in which $G_{RX} G_{\text{channel}}$ is much smaller than unity, a highly attenuating stage such as a wireless channel exists between two components, and it is best if the stage immediately after the attenuating stage has high gain, so that the stage immediately before does not need to have an extremely high output power that would result in increased loss. Secondly, we see the importance of the efficiency of the power amplifier, as it must overcome the loss incurred in the channel.

Equation (5.203) confirms that the energy cost of a single bit does indeed increase as the channel gain decreases. Several examples of the use of equation (5.203) are shown in Figs. 5.86 through 5.89. Figs. 5.86 and 5.87 show an example for a 20 GHz system with path loss modeled according to a large-scale propagation model from equation (3.2) in Chapter 3

$$G_{\text{channel}} = \frac{k}{d^\alpha} \tag{5.204}$$

where k is a constant

$$k = \left(\frac{\lambda_c}{4\pi \times 5m} \right)^2 \tag{5.205}$$

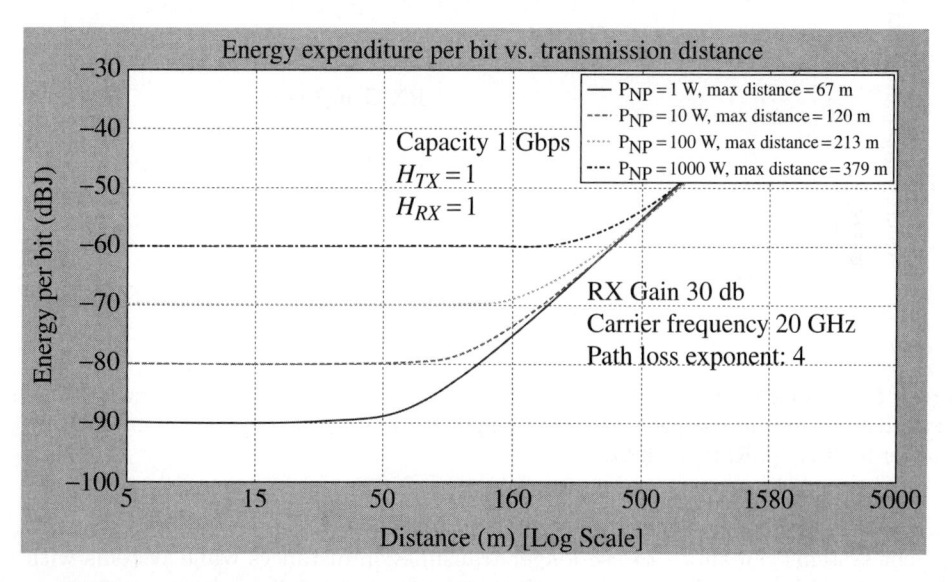

Figure 5.86 For a system with high signal path efficiency and high non-path power consumption, we see that the energy expenditure per bit is dominated by non-path power, indicating little advantage to shortening transmission distances [from [MR14b] © IEEE].

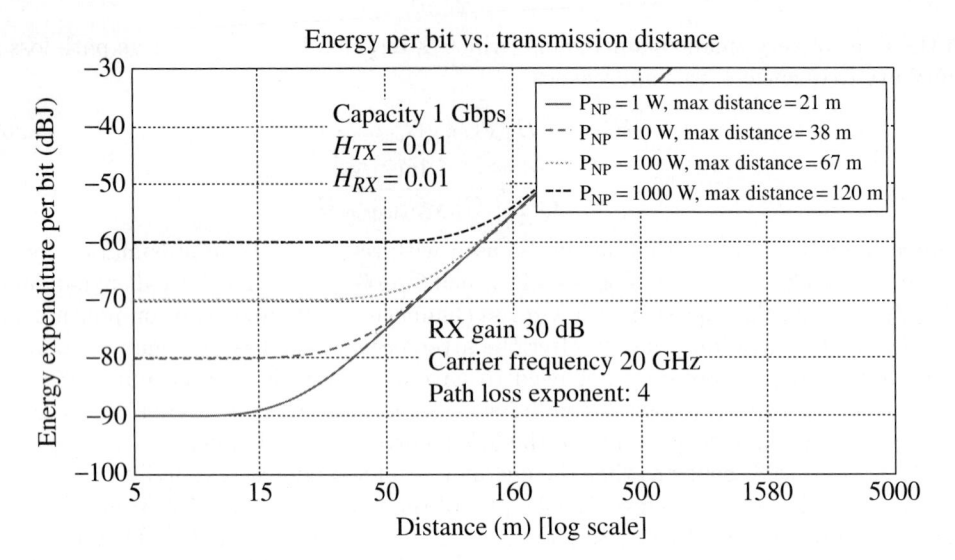

Figure 5.87 When signal path components are less efficient, as illustrated here, then shorter transmission distances start to become advantageous, as signal path power starts to represent a larger portion of the power expenditure per bit [from [MR14b] © IEEE].

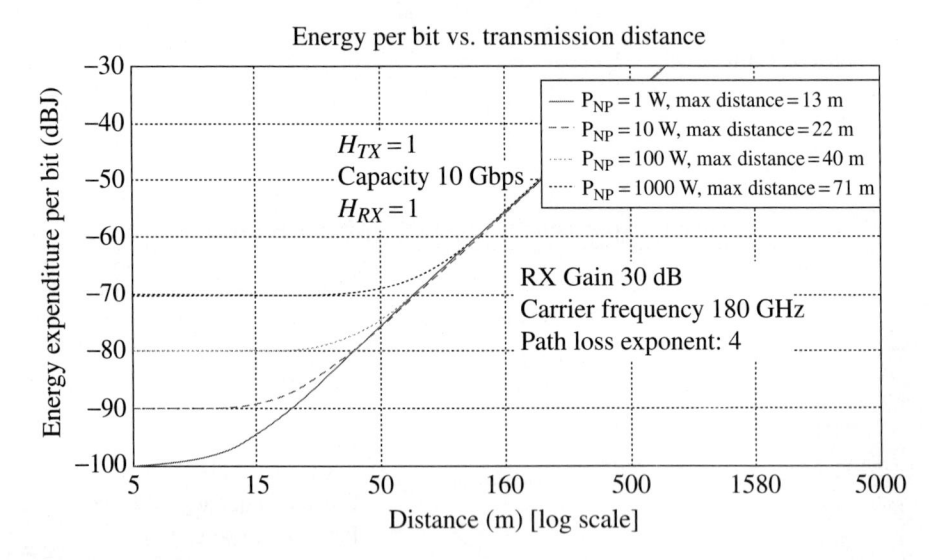

Figure 5.88 A higher frequency system that can provide a much higher bit rate capacity without substantially increasing non-path power consumption may result in a net reduction in the energy price per bit [from [MR14b] © IEEE].

and λ_c is the carrier wavelength. Contrasting Fig. 5.86 and Fig. 5.87 indicates that highly efficient systems can afford to use longer transmission distances while systems with less efficient signal path components should use shorter transmission distances (the decrease in efficiency is reflected in the change in H_{TX} and H_{RX} between the two plots). Figs. 5.88

5.12 Consumption Factor Theory

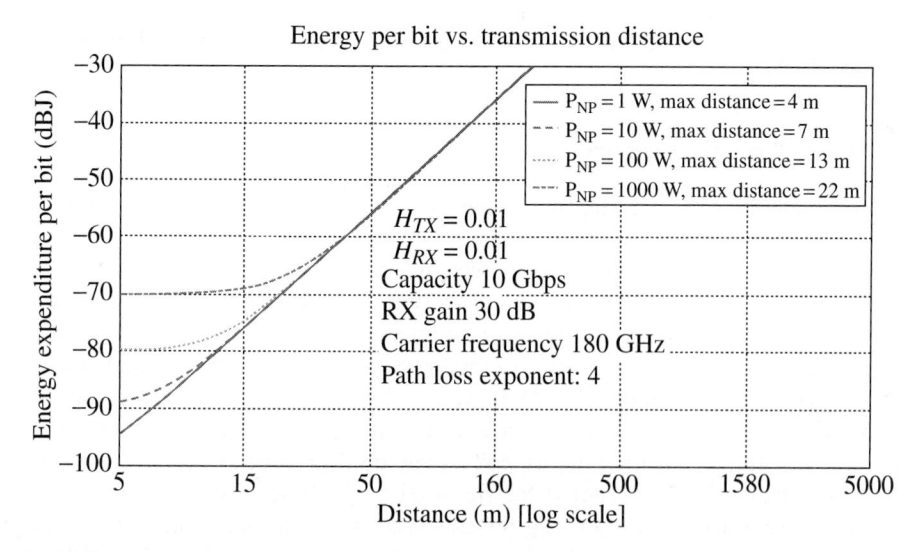

Figure 5.89 Lower efficiencies of signal path components motivate the use of shorter transmission distances [from [MR14b] © IEEE].

and 5.89, in which the carrier frequency has been increased to 180 GHz, indicate that shorter transmission distances should be used for higher carrier frequencies.

We may easily determine with equation (5.200) the maximum transmission distance for which non-path power dominates the power expenditure per bit, and hence the maximum transmission distance for which each bit becomes progressively more energy-expensive:

$$\frac{P_{NP}}{C} > \frac{\ln(2)N_0}{G_{RX}G_{\text{channel}}} \left(\left(\frac{G_{RX}}{H_{RX}} - 1 \right) G_{\text{channel}} + H_{TX}^{-1} \right) \quad (5.206)$$

$$G_{\text{channel}} > \frac{N_0 C \ln(2)}{H_{TX} \left(P_{NP} G_{RX} + N_0 C \ln(2) \left(1 - \frac{G_{RX}}{H_{RX}} \right) \right)}. \quad (5.207)$$

If we model the channel gain as

$$G_{\text{channel}} = \frac{k}{d^\alpha} \quad (5.208)$$

where k is a constant that indicates the path loss at some reference distance, d is the link distance, and α is the path loss exponent, then

$$\frac{k}{d^\alpha} > \frac{N_0 C \ln(2)}{H_{TX} \left(P_{NP} G_{RX} + N_0 C \ln(2) \left(1 - \frac{G_{RX}}{H_{RX}} \right) \right)} \quad (5.209)$$

$$d < \left\{ \frac{H_{TX} k}{\ln(2) N_0 C} \left(P_{NP} G_{RX} + N_0 C \ln(2) \left(1 - \frac{G_{RX}}{H_{RX}} \right) \right) \right\}^{\frac{1}{\alpha}}. \quad (5.210)$$

If $P_{NP} < \frac{N_0 C \ln(2)}{H_{RX}}$, then (5.210) is unlikely to have a positive solution due to the small value of $N_0 C \ln(2)$.

5.13 Chapter Summary

This chapter has provided in-depth knowledge and background on many vital aspects of analog millimeter wave transistors, their fabrication, and important circuit design approaches for the basic building blocks within a transceiver. The treatment of analog circuits has intentionally included details of CMOS and MOSFET semiconductor theory, in sufficient detail to allow communications engineers to appreciate and understand the fundamentals and challenges of mmWave analog circuits, and to understand the capabilities and approaches used to create circuits that will enable the mmWave revolution. Key mmWave parameters used to characterize active and passive analog components, such as S-parameters and Y-parameters, as well as key qualities of merit, such as third order intercept, are defined and presented for the reader, as these terms and parameters are vital for measuring and characterizing devices used in communication systems. Popular circuit and electromagnetic simulators and typical tests done by engineers to understand semiconductor manufacturing processes have been taught in this chapter, to allow exploration and design in future semiconductor technologies. The chapter also includes key design approaches for transmission lines, amplifiers (both power amplifiers and low-noise amplifiers), frequency synthesizers and voltage-controlled oscillators (VCOs), and frequency dividers. The chapter concludes with a new and powerful theory, called the consumption factor theory, that allows an engineer to quantify and compare the power efficiency of any device or cascade of devices. This is a new figure of merit that will likely be important for understanding the power versus bandwidth tradeoff in very broadband wireless systems of the future.

Chapter 6

Multi-Gbps Digital Baseband Circuits

6.1 Introduction

This chapter provides detailed specifications for and recent advances in digital baseband processing circuitry and offers insights into the challenges and approaches that will enable multi-gigabit per second (Gbps) data transmission rates in future wireless mmWave communication systems. We present the fundamental concepts and challenges associated with the implementation of baseband circuitry within integrated circuits, and consider both analog-to-digital converters (ADCs) and digital-to-analog converters (DACs), crucial components of the baseband circuitry in modern transceivers.

ADCs are the gateway between analog and digital portions of a chip, as they convert continuous analog signals into discrete digital signals. Discrete digital signals may be processed by a digital processor, passed along on digital busses and multiplexers within a chip, and stored in discrete/digital memory. ADCs are key components of receiver baseband circuits, where information coming from the RF/IF stages of the receiver are demodulated, processed, stored, and ultimately used by the human users or computers in real time or in the future. ADCs also are used in feedback paths, where analog signals within a circuit are monitored, digitized, and sent to a processor as a digital word to be processed for feedback (e.g., an ADC may be used to digitize signals for a biasing circuit within a mixer that must be balanced or must have its error or offset minimized, or a transmitter power amplifier that must be set at a particular level, etc.). The performance characteristics of the ADC in a communications device play a major role in determining the overall performance of the device. For example, the bandwidth of the ADC directly limits the bandwidth of processed information within the device. The *dynamic range* of the ADC, in terms of the number of bits created to represent each sample, determines the fidelity of reproduction of an analog signal as it is represented by the digitized signal, and thus sets the limit of dynamic range in the entire device. Modern devices are increasingly facing the challenge of enabling high data rates on the order of several gigabits per second (Gbps) while using as little power as possible. The power draw of the ADC substantially impacts the power of the entire device; the speed and dynamic range of the ADC determine the amount and fidelity of data that can be converted from the analog signal domain into the digital domain. In this chapter, basic ADC architectures and the components used in the ADC, such as comparators and track-and-hold amplifiers, are introduced. Also, this chapter provides modern design trends for ADCs as related to mmWave radio systems of the future.

DACs convert the digital representations of signals (that may exist in memory, are processed in signal processors, or are carried in digital form within signaling busses and digital multiplexers) into analog signals that are used in the analog signal chain of a communications system. For example, baseband in-phase and quadrature (I and Q) modulated signals are often generated digitally by a baseband modem or a digital signal processor (DSP). DACs are then used to convert the digital I and Q signals into analog I and Q signals that are processed by the mixers and the IF/RF stages of a transmitter. Millimeter wave (mmWave) digital-to-analog converters (DACs) for first-generation mmWave communication devices require only moderate resolutions (i.e., moderate dynamic ranges) due to the low power spectral densities used in these early devices and the relatively small amounts of dynamic signal fluctuations due to the much greater RF bandwidths (and thus much less instantaneous signal fading) in mobile communications. Moderate or low dynamic range systems are also likely to be advantageous in terms of power consumption compared with higher dynamic range systems. The sampling rates of mmWave DACs, however, will need to be high enough to be operable with very wide channel bandwidths and at baseband bandwidths up to and exceeding 2.16 GHz [ZS09b][WPS08].

A number of excellent resources exist in textbooks (and on the Internet) for understanding the performance metrics and design basics of ADCs and DACs, and many integrated circuit manufacturers offer guidelines and illustrative tutorials [Max02].

6.2 Review of Sampling and Conversion for ADCs and DACs

The sampling theorem, discussed in Section 2.2, is the most important fundamental law in baseband signal processing and dictates the behavior of ADCs and DACs. Signals in analog circuits are accurately described by continuous functions of time. These continuous functions do not occupy two different values in two distinct time slots without first occupying an intermediate value. Further, these continuous functions have a value for every point in time during an observation interval. In contrast, the signals used in a digital processor are discrete, are digital in nature, and have one of only a finite number of values or states. Digital signals are discrete because their value is not tied directly to time, but rather they are indexed according to the state of the clock period. Converting between analog and digital signals therefore involves two processes: sampling an analog signal so that it has a relatively constant value over a single clock period (e.g., taking a sample of the analog signal, often called "sample-and-hold"), and then digitizing the analog signal sample so that it may take on only one of a finite set of values (e.g., quantizing the sample and representing it digitally during the clock cycle).

The role of an ADC is to capture a sample of an analog signal over a very small time aperture (the sample time interval) and to accurately produce a digital codeword that represents the analog signal in a finite number of bits that may be stored in memory

6.2 Review of Sampling and Conversion for ADCs and DACs

or processed by a computer. The ADC continually samples the analog signal, producing many codewords over time. The ability to digitize the analog waveform and represent the voltage as a series of bits is critical to modern wireless communications. The DAC process is the inverse of the the ADC process. The DAC converts a digital word into an analog voltage during the clock cycle. Because a DAC produces many sharp voltage transitions for time-varying codewords, the DAC must use a low-pass filter to smooth the many discrete voltage transitions produced by the time-varying codewords.

Let us consider an arbitrary time signal as shown in Fig. 6.1. The signal has frequency content that is centered at baseband and that occupies a bandwidth of 100 MHz, as shown in Fig. 6.2. We may write the arbitrary signal and its frequency domain representation as:

$$w(t) \leftrightarrow W(f) \tag{6.1}$$

where $w(t)$ is the arbitrary signal and $W(f)$ is its Fourier transform

$$W(f) = \int_{-\infty}^{\infty} w(t) e^{-j\omega t} dt. \tag{6.2}$$

Later, we review key aspects of the sampling theorem as they relate to ADC and DAC processing. See Section 2.2 for further details. When a signal is sampled, its sample values

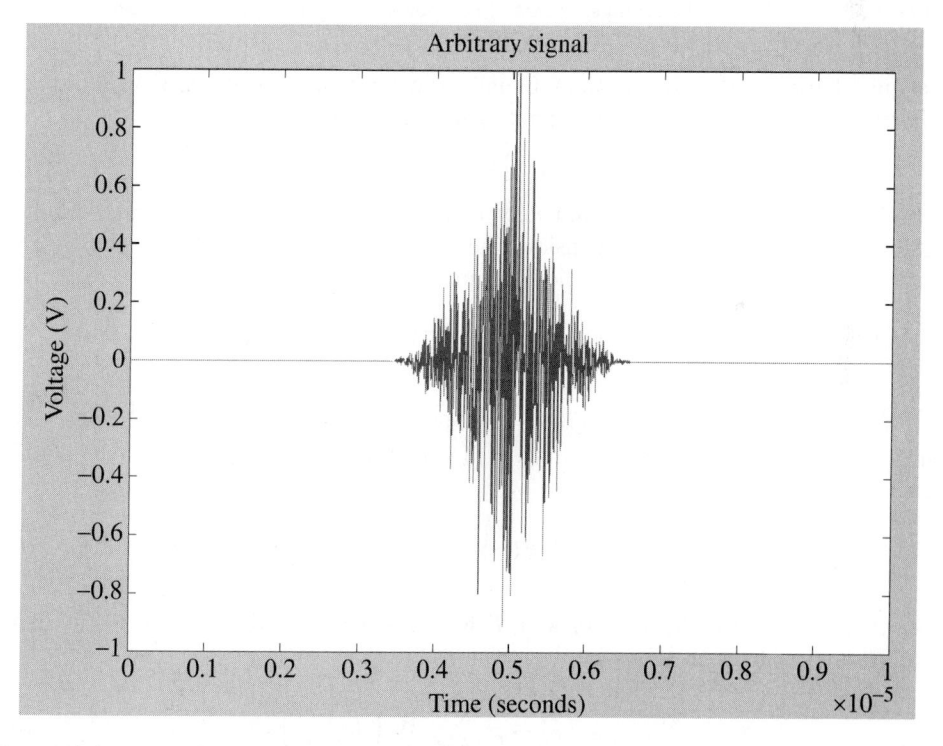

Figure 6.1 An arbitrary baseband analog signal having an approximate bandwidth of 100 MHz.

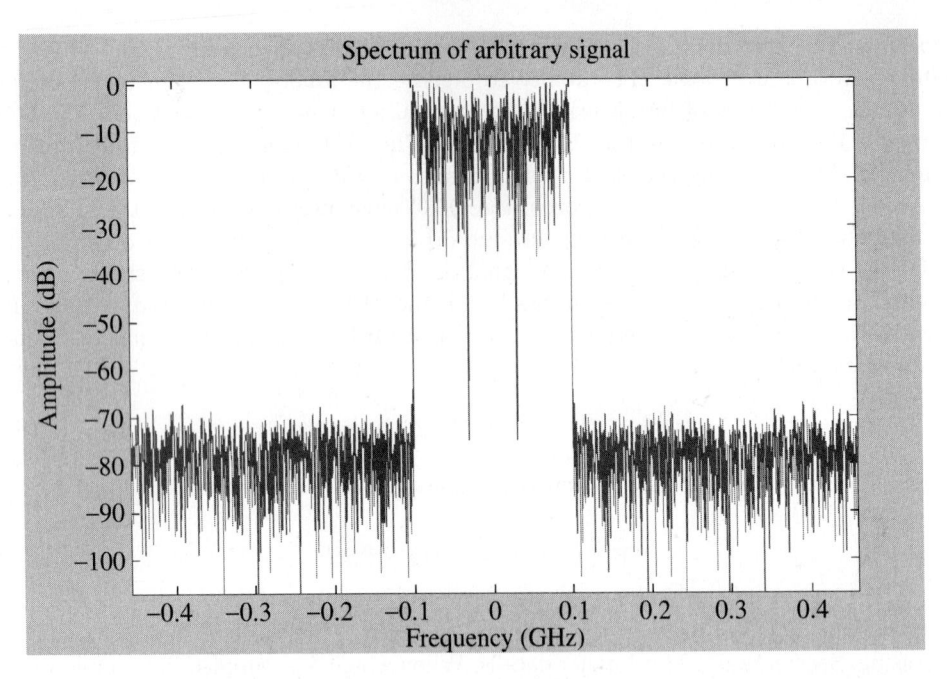

Figure 6.2 The spectrum of the arbitrary analog waveform shown in Fig. 6.1. The spectrum has been normalized such that the strongest spectral component has an amplitude of 0 dB.

are acquired at distinct points in time. Usually, the points in time are periodically spaced, so that we may write the sampled discrete time signal as

$$w[n] = w(nT) \tag{6.3}$$

where T is the sampling period and $w[n]$ is a discrete-time signal. To understand the sampling process, it is useful to consider the spectrum of an analog signal that is given by multiplication of $w(t)$ with a periodic impulse train

$$w_{\text{sampled}}(t) = w(t) \sum_{k=-\infty}^{\infty} \delta(t - kT) \tag{6.4}$$

where $\delta(t - kT)$ is an impulse centered at time $t = kT$. In the frequency domain, the result is a convolution of the spectrum $W(f)$ with the Fourier transform of the periodic impulse train:

$$\mathcal{F}\left\{ \sum_{k=-\infty}^{\infty} \delta(t - kT) \right\} = \frac{1}{T} \sum_{k=-\infty}^{\infty} \delta\left(f - \frac{k}{T}\right) \tag{6.5}$$

where $\mathcal{F}$ is the Fourier transform and T is the sampling interval. The result of the convolution of (6.5) with $W(f)$ is

$$W_{\text{sampled}}(f) = \frac{1}{T} \sum_{k=-\infty}^{\infty} W\left(f - \frac{k}{T}\right). \tag{6.6}$$

6.2 Review of Sampling and Conversion for ADCs and DACs

The analog sampled signal has a frequency domain representation that is periodic with a period given by the sampling frequency $1/T$. The discrete-time Fourier transform (DTFT) of $w[n]$ relates to the Fourier transform through the relationship

$$W_{\text{DTFT}}\left(e^{j2\pi f}\right) = \frac{1}{T} \sum_{k=-\infty}^{\infty} W\left(\frac{f}{T} - \frac{k}{T}\right). \qquad (6.7)$$

Therefore, the spectrum of $w_{\text{sampled}}(t)$ in (6.6) and the spectrum of $w[n]$ in (6.7) are related through a rescaling of the frequency axis. For convenience, most analog designers focus on characterizing the original $w_{\text{sampled}}(t)$, whereas baseband digital designers focus on $w[n]$.

Fig. 6.3 illustrates the sampled signal for a sampling rate of 400 MHz. Fig. 6.4 shows the spectrum of the sampled signal. Comparing Fig. 6.2 and Fig. 6.4, it is clear that the noise floor has been elevated in the sampled case compared with the original signal. This is the result of aliasing, in which some of the high-frequency noise energy from the original spectrum adds multiple times in the sampled spectrum. Aliasing is a critical design issue and is discussed next.

When the signal is sampled sufficiently fast to exceed the Nyquist sampling limit (i.e., when the sampling rate is much greater than twice the maximum frequency of the analog signal), then the copies of the spectrum do not overlap, as shown in Fig. 6.4. If the sampling rate is lower than twice the baseband bandwidth of the signal, the copies of the spectrum will overlap and distort the signal. This phenomenon, in which high

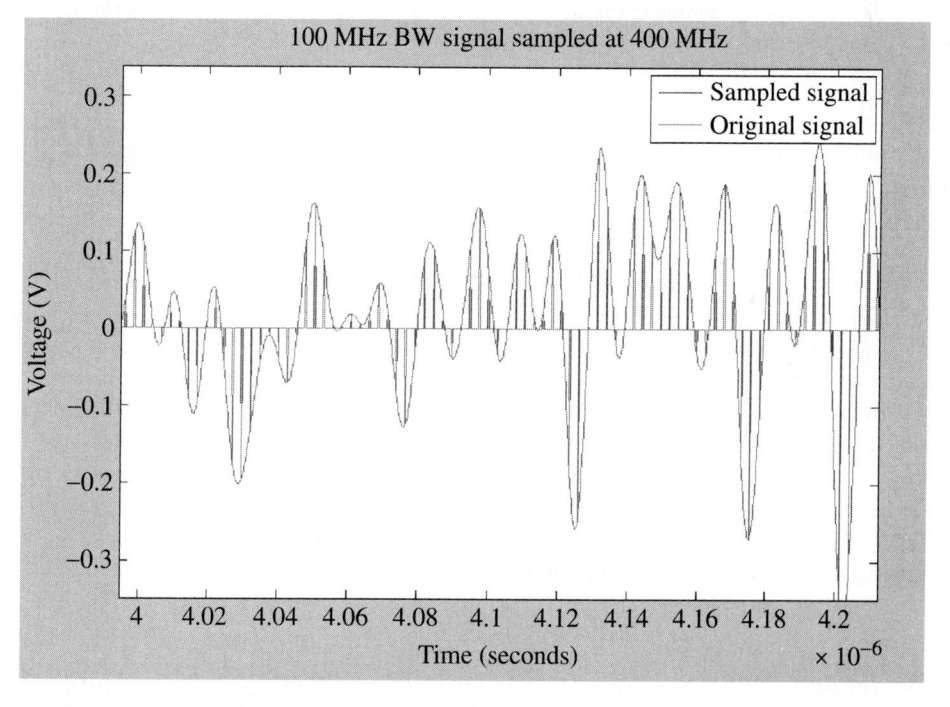

Figure 6.3 A zoomed-in version of the arbitrary signal in Fig.6.1 showing how it has been sampled. The bandwidth (BW) of the original signal is 100 MHz, and the sampling rate is 400 MHz.

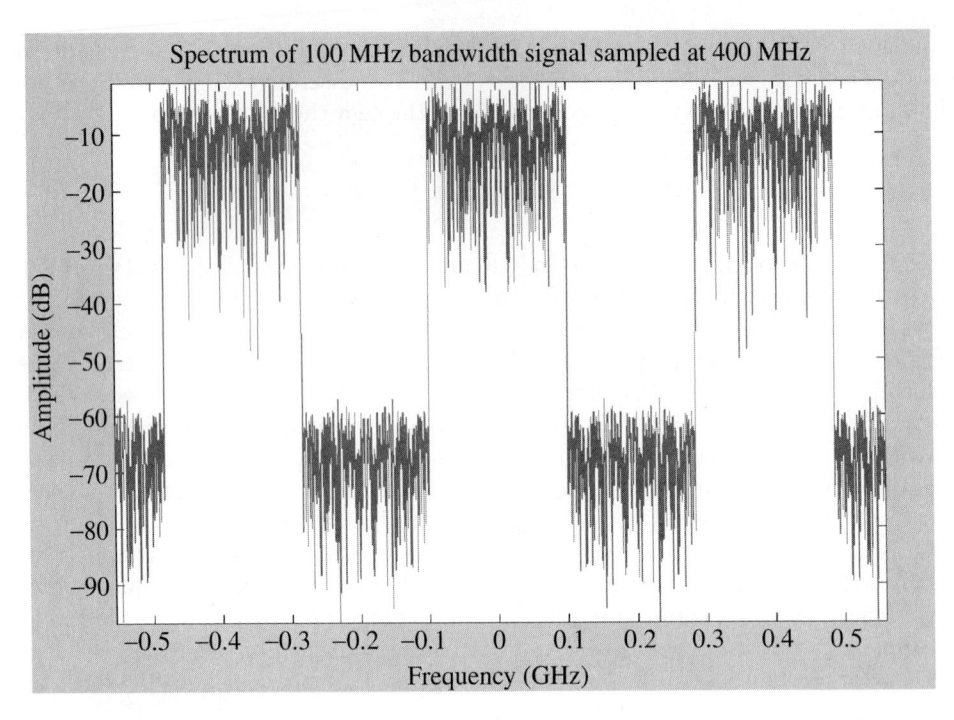

Figure 6.4 The result of sampling the signal in Fig. 6.1 in the time domain is to make the spectrum periodic in the frequency domain.

frequency components masquerade as lower frequency components, is known as *aliasing*. Fig. 6.5 illustrates the result of sampling at 100 MHz and shows clearly that aliasing has resulted. To avoid aliasing, the sampling rate must be at least greater than or equal to twice the signal's baseband bandwidth. This minimum sampling frequency is known as the *Nyquist frequency*.

In addition to sampling a signal at certain points in time (a process called *digitization*), an ADC also maps each of those signal time samples to a set of discrete values (a process known as *quantization*). The number of quantization values for each sampled point of the signal is usually described in terms of the number of bits that the ADC uses to encode each sample. The maximum possible dynamic range of the digitzed and quantized signal as produced by the ADC is dependent on the number of bits used to quantize the signal. The dynamic range is the ratio of the largest signal range that may be represented by the ADC, to the smallest quantifiable signal increase (step-size) that may be represented by the ADC. For an ADC that can represent signals from $V_{\min}$ to $V_{\max}$ using n bits, it is known that the dynamic range is given by

$$Dynamic\ Range\ [dB] = -20 \log_{10} \left(\frac{V_{\max} - V_{\min}}{\frac{V_{\max} - V_{\min}}{2^n}} \right) = 20 \log_{10}(2) n \approx 6.0 n\ [\text{dB}]. \quad (6.8)$$

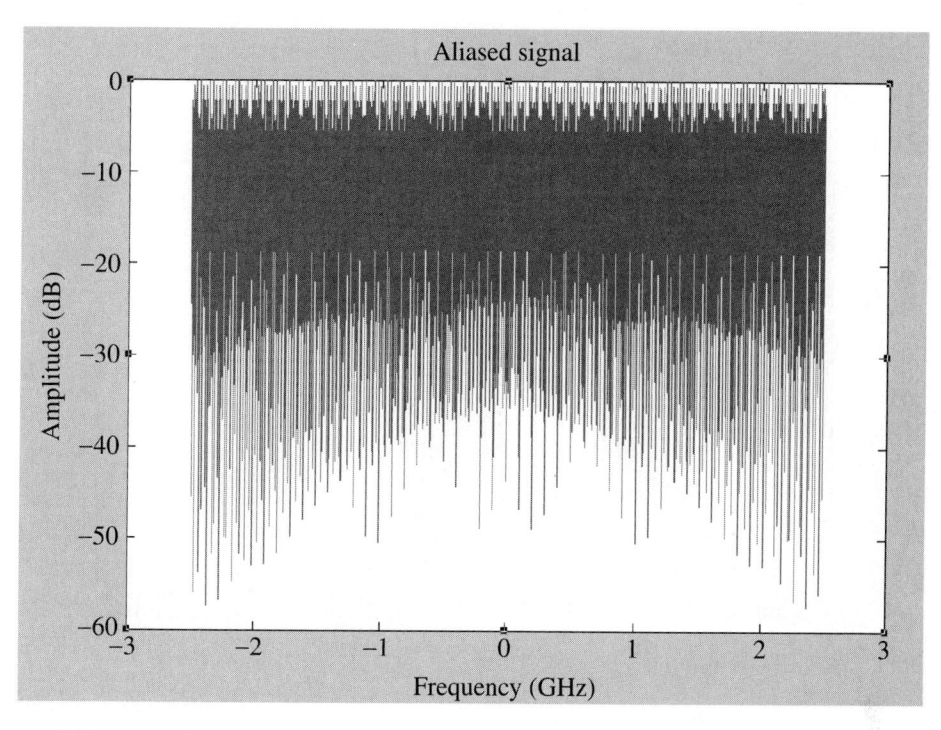

6.2 Review of Sampling and Conversion for ADCs and DACs

Figure 6.5 If the signal of Fig. 6.1 with a baseband bandwidth of 100 MHz is sampled at 100 MHz, the result is an aliased signal. In the frequency domain, overlapping copies of the original signal's spectrum completely distort the resulting sampled signal.

Therefore, we expect a 6 bit ADC to provide 36 dB of dynamic range. The dynamic range of a signal in the frequency domain may be ascertained by examining its spectrum and, in general, is found by noting the difference between the maximum magnitude (the spectrum peak) and the minimum magnitude (the noise floor). For example, Fig. 6.6 shows the resulting spectrum of the signal of Fig. 6.1 after each sample is quantized using 4 bits. We see in Fig. 6.6 that the dynamic range has been reduced to 24 dB.

When working on circuit designs, it is more common to define dynamic range in terms of root mean square (RMS) values. The RMS value of an energy waveform $v(t)$ is found by:

$$V_{\rm rms} = \sqrt{\int_{-\infty}^{\infty} v(t)^2 \, dt}. \tag{6.9}$$

The RMS value of a periodic sinusoid is found by changing the limits of integration in (6.9) to be over a single period, and then dividing by the period while taking the limit of the period to infinity, and it is given by:

$$v(t) = A\sin(\omega t) \Rightarrow V_{\rm rms} = \frac{A}{\sqrt{2}}. \tag{6.10}$$

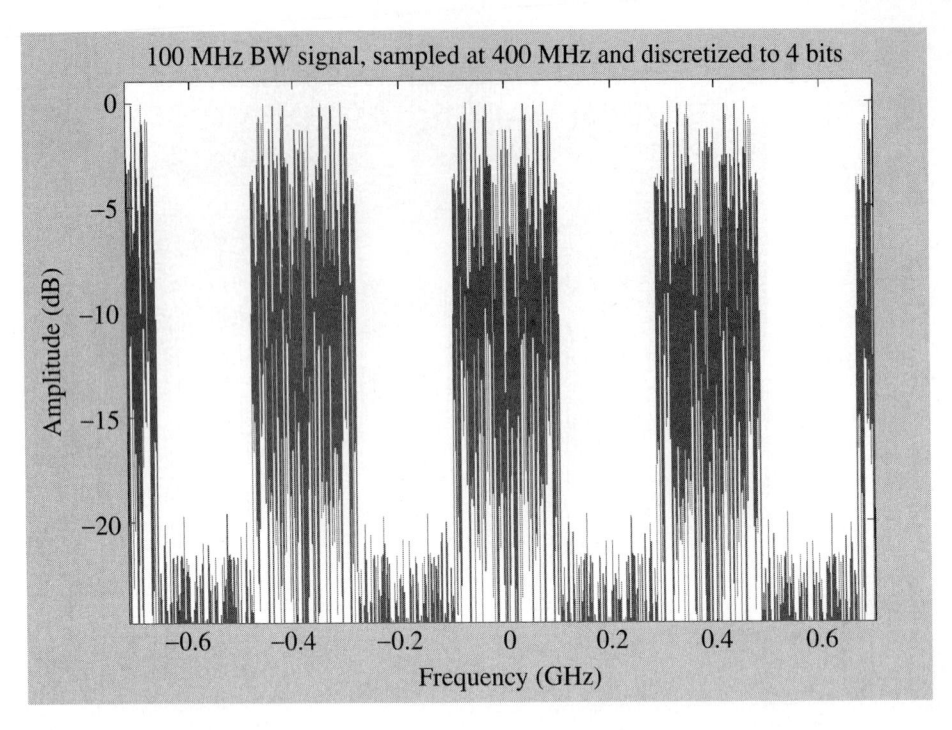

Figure 6.6 Quantizing the signal of Fig. 6.1 with 4 bits reduces the dynamic range to 24 dB.

If we quantize this signal to N bits and assume that $A = \frac{1}{2}(V_{\max} - V_{\min})$, then we find that the RMS value of this full-scale sinusoid is:

$$V_{\text{rms}} = \frac{V_{\max} - V_{\min}}{2\sqrt{2}}. \tag{6.11}$$

When a signal is quantized, the goal of the ADC is to produce a unique n-bit codeword for each small step change in applied input voltage, so that the infinite possible voltage values within a continuum of voltages in between $V_{\min}$ and $V_{\max}$ at the input of the ADC are mapped to a finite number of n-bit codewords that are reasonable digital representations of the particular applied analog voltage. The RMS value of the smallest analog signal that can be detected and differentiated from another analog signal, to yield different codewords, is determined by the quantization error. The quantization error is determined by the voltage represented by the *least significant bit* (LSB) of the ADC, where, assuming uniform quantization bins (i.e., quantiles) over all signal values, the LSB voltage is given by the ratio of the range of the ADC to the number of quantization levels

$$\text{LSB} = \frac{V_{\max} - V_{\min}}{2^n}. \tag{6.12}$$

6.2 Review of Sampling and Conversion for ADCs and DACs

The quantization error is what determines the RMS value of the smallest detectable signal. In general, we assume that the quantization error is uniformly distributed about 0 from $\frac{-\text{LSB}}{2}$ to $\frac{\pm\text{LSB}}{2}$. The probability density function of the quantization error is

$$f_{\mathbf{X}}(x) = \frac{1}{\text{LSB}} \text{ for } \left\{|x| \leq \frac{\text{LSB}}{2}\right\} \tag{6.13}$$

and $f_{\mathbf{X}}(x) = 0$ otherwise for the random variable $\mathbf{X}$, which characterizes the quantization error within a quantization bin. From (6.13) we can determine the RMS value of the expected quantization error signal

$$\text{RMS Quantization Error} = \sqrt{\int_{\frac{-\text{LSB}}{2}}^{\frac{\text{LSB}}{2}} \frac{x^2}{\text{LSB}} dx} \tag{6.14}$$

and simplifying

$$\text{RMS Quantization Error} = \frac{\text{LSB}}{\sqrt{12}} = \frac{V_{\max} - V_{\min}}{2^n \sqrt{12}}. \tag{6.15}$$

Equation (6.13) is a uniform distribution to model the assumption that the digitized signal has an equal likelihood of having any value between two successive quantization thresholds (i.e., within a quantile). This is generally true for full-scale sinusoids and most random signals. However, it is possible for some signals to have a non-uniform distribution within a quantile: For example, a constant signal would result in a constant quantization error (e.g., if the constant signal is 1/4 LSB from a bit boundary, then it would have a quantization error of 1/4 LSB). Using equations (6.11) and (6.15), we find the maximum possible dynamic range, considering only quantization error (also known as the maximum *SQNR*, or *signal-to-quantization-noise-ratio*) in dB to be

$$\text{Dynamic Range}_{\text{SQNR}} = 6.02n + 1.7 \text{ dB}. \tag{6.16}$$

In addition to the effects of quantization, there are many sampling phenomena that can result in a decrease in the signal quality and dynamic range in both an ADC and DAC. For example, up to this point we have assumed that all of the samples occur at regular intervals. But, in a real-world ADC there will be some amount of randomness in the timing when each sample is taken, and there will be some timing error when a DAC produces an analog signal. The resulting uncertainty in the sampling time is referred to as *jitter* and is a result of noise in the sampling circuit. Fig. 6.7 illustrates an arbitrary signal that has been sampled with jitter. In the figure, we see that the time between samples is not uniform. The impact of jitter depends on how quickly the original signal changes relative to the error in sampling time. Very slow signals will not be greatly impacted by jitter, whereas very fast signals will be severely impacted. Therefore, we expect the error in signal quality and dynamic range to be a function of the uncertainty in the sampling time in addition to the input signal's frequency [TI13]. The maximum

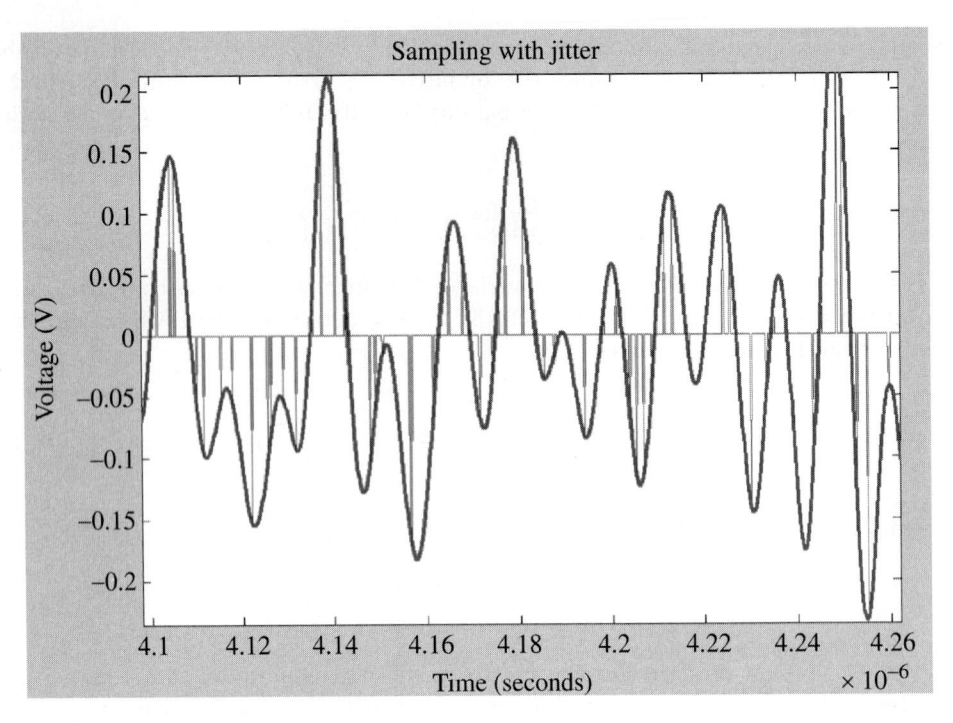

Figure 6.7 An example of sampling with jitter, where uncertainty in the time interval between samples results in decreased dynamic range.

possible dynamic range considering jitter effects (also known as the maximum *SJNR*, or *signal-to-jitter-noise ratio*) of the resulting discrete signal is given by [SAW90]

$$\text{Dynamic Range}_{\text{SJNR}} = -20\log_{10}(\tau f_{in}) - 5.17 \text{ dB}, \quad (6.17)$$

where τ is the maximum timing jitter between all samples, and f_{in} is the frequency of the signal to be sampled. Note that [TI13] uses the standard deviation of timing jitter, rather than the maximum timing jitter, in which case the constant term on the far right of (6.17) will be larger (15.96 dB) and where τ in (6.17) would be the standard deviation, and not peak, timing jitter. The constant term in (6.17) may also vary as a function of the applied waveform, and may be smaller if a deterministic (e.g., a sine wave) input signal is used. The degradation of dynamic range due to jitter may be ignored when the timing jitter τ is much less than the reciprocal of the input signal frequency or bandwidth.

Equations (6.16) and (6.17) both provide limits on achievable signal quality and dynamic range of an ADC, based on the number of bits and the amount of timing jitter, respectively. These sources of error can occur independently and the more severe of the two sources of error governs the overall limit in the achievable signal quality and dynamic range of an ADC. In today's practice, the jitter of the sampling clock is engineered to be low enough so that (6.16) governs the achievable dynamic range of the ADC. However, at millimeter wave frequencies with ultrawideband data streams, it is more difficult to design very low jitter sampling clocks, suggesting that (6.17) may become more important for multi-Gbps mmWave systems.

6.3 Device Mismatches: An Inhibitor to ADCs and DACs

Mismatch refers to the random variations in performance and physical characteristics among transistors and other fabricated circuit elements that occur during the silicon chip fabrication process. Mismatches occur even when the circuit elements of a chip have identical design, layout, and manufacturing processes. Causes of mismatch can occur in various stages of the wafer fabrication process and include ion-implantation variation, mobility variations, fixed and mobile oxide charges, line edge and width roughness changes, and other oxide variations across a wafer [PDW89][LZK+07]. Mismatch is a major limiter on the performance of analog-to-digital converters and digital-to-analog converters, especially flash analog-to-digital converters and current steering digital-to-analog converters, discussed later in this chapter (see Section 6.8) [PDW89][US02][WPS08]. The best-known and most widely used model for MOSFET mismatch was developed by Pelgrom, Duinmaijer, and Welbers [PDW89]. The theory covers mismatch between devices on a single wafer but was not originally developed to describe mismatch statistics between wafers or batches [PDW89]. The theory describes both the geographical mismatch between any two points on the wafer and process-dependent mismatch related to particular steps in the fabrication process. One of the key findings of the model is that variance in device parameters is given by [PDW89]

$$\sigma^2\left(\Delta P\right) = \frac{A_p^2}{WL} + S_p^2 D_x^2, \qquad (6.18)$$

which states that the variance in a parameter is related to the ratio of an area parameter to the minimum area of interest $\frac{A_p^2}{WL}$, assumed to be the area of a single transistor, plus a spacing parameter. At first glance, this equation would seem to indicate inevitable increases in device mismatch as technology scaling (shortening of the transistor gate length) continues. If the parameter A_p remains constant then that would be the case, but a more thorough examination of each fabrication process must be made, as the foundry may have taken steps to reduce variation in their more advanced processes. It may be most useful to think of (6.18) from the perspective of the foundry, as an indication of the maximum area over which the foundry can allow process variations to occur and still deliver a wafer with acceptable mismatch. Equation (6.18) indicates that a foundry cannot allow large variations to take place over an area substantially larger than the minimum feature size of the technology (indicated by the width W and length L of the smallest available transistor in the process). From the perspective of a circuit designer, (6.18) indicates that smaller devices will exhibit greater mismatch, but this should be considered separately in the light of each process technology. (As an analogy, the reader is encouraged to speculate on the differences in size variance between a thousand small 1 mm × 1 mm squares drawn with a child's crayon and a thousand small 1 mm × 1 mm squares drawn with a very fine-tipped pen).

General circuit design rules for mismatch include using parallel devices in cases in which mismatch should be minimized (e.g., in a differential pair or in the comparator of an ADC). Devices that are spaced farther apart on a chip are more likely to be mismatched [PDW89], so spacing should be minimized between two devices whose matching is critical. Using differential circuit designs, as opposed to single-ended designs, is another way to use common mode rejection to mitigate voltage offsets or noise induced by mismatch in some circuit designs. Averaging, such as using passives in parallel to average out mismatch, is another commonly used technique. As semiconductor fabrication techniques continue to make transistor gates much smaller, into the deep sub-micron region (smaller than 20 nm), mismatch will become a more challenging issue as bandwidths increase and voltage levels decrease to under 1 V.

In 2007, Lim et al. of Nanyang University studied the effect of multi-finger transistor layouts on mismatch in a 0.13 μm and 90 nm CMOS [LZK+07]. They found that multi-finger designs outperformed single-finger designs in terms of mismatch in both 0.13 μm and 90 nm CMOS. The effect of finger interleaving on multi-finger designs was also studied, and the effect of interleaving was found to be dependent on device type (NMOS or PMOS) and size [LZK+07]. They also found that the use of multiple metal layers (e.g., in a sophisticated BEOL process) had nearly no impact on device mismatch versus the use of only a single metal layer. In addition, they found that adding a gate protection diode slightly increased device mismatch (such diodes are only needed if very long pieces of metal are connected to the transistor gate).

The transistors near one another on a chip will exhibit correlated mismatches while the mismatches between two widely separated devices will be less or only weakly correlated [WT99]. As gate lengths have continued to shrink, it has become increasingly popular to use digital processes for analog purposes. Unfortunately, many of the process innovations developed to improve digital performance have also hurt analog performance [TWMA10]. As carrier frequencies continue to increase to meet demand for bandwidth, it is likely that analog circuits will continue to be constructed in more recent iterations of digital processes. In certain cases, the high-speed operation of a deeply scaled process may be needed in only certain portions of a chip or circuit, while slower operation can be tolerated elsewhere. Certain techniques have been developed to improve transistor mismatch in such circumstances. In general, it is not advisable from the standpoint of analog circuit design and mismatch to use minimum length transistors in the latest process generation, as new technology nodes are generally improved over time.

In cases in which longer-than-minimum gate lengths are used in portions of an analog circuit, it may be advantageous to lay out long-channel transistors as series of short-channel devices to improve mismatch [TWMA10]. This technique does, however, hurt other performance measures such as drive current and voltage headroom [TWMA10].

In 2010, Tuinhout et al. studied the effects of segmentation (i.e., the use of a series of very short-channel devices to realize long-channel devices) on MOSFET mismatch [TWMA10]. Their interest was in the mismatch performance of long-channel devices (greater than 1 μm) constructed in deeply scaled processes (e.g., 45 nm CMOS). They found that segmentation can result in a substantial improvement in mismatch but can also deteriorate other performance metrics, such as substantially reducing drive current.

6.4 Basic Analog-to-Digital Conversion Circuitry: Comparators

The most basic building block of an ADC is a *comparator*, which is a device that is used in the digitization process of an analog signal; it compares two voltages and delivers a high or low output voltage (i.e., a digital 1 or 0) based on which of its two input signals

6.4 Basic Analog-to-Digital Conversion Circuitry: Comparators

is highest. Fig. 6.8 shows the basic symbol of a comparator. In an ADC, many parallel comparators are used to simultaneously compare the incoming analog signal. For each comparator C, one of the applied input voltages is a known reference voltage produced internally by the ADC, and the other applied input voltage is the analog signal which is to be compared. By determining whether or not the analog input signal exceeds the reference threshold, N parallel comparators using N different internal reference voltage levels enable the ADC to quickly sample and quantize the analog signal in parallel while producing an N-bit word that represents the analog voltage. The selected comparator design for an ADC must be well matched to the selected reference voltages to ensure proper conversion. The *metastability window* of a comparator is the smallest voltage at its inputs that can trigger the comparator to make a decision [SVC09]. Improper selection of reference voltages relative to the comparator metastability window may result in increased non-linearity.

The basic operation of a comparator is explained by Fig. 6.9, based on [LLC06, Chapter 12]. The switches S_1-S_4 are switched periodically to realize the sampling operation. In each period, S_4 is first switched to drain the capacitor of charge. Next, switches 2 and 3 are switched on while 1 and 4 remain off. Switching on S_3 forces the input voltage and output voltage to be equal. In addition, this forces the input voltage to equal the threshold voltage of the inverter [LLC06], defined as the input voltage level

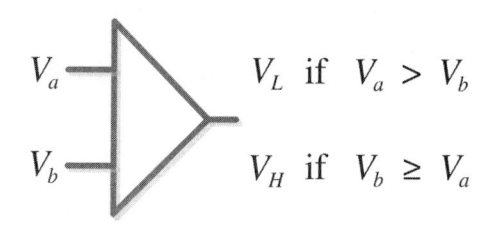

V_a ⟶ ▷ ⟶ V_L if $V_a > V_b$
V_b ⟶   ⟶ V_H if $V_b \geq V_a$

Figure 6.8 A comparator is perhaps the most basic building block of all analog-to-digital converters. Many comparators are used in parallel to perform analog-to-digital conversion.

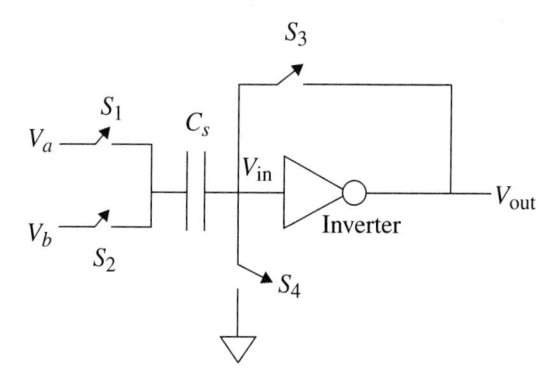

Figure 6.9 A basic implementation of a comparator, which serves as a core building block of an ADC.

that forces the output to transition from low to high. By switching on S_2, we force the charge on C_s to be given by $Q_s = C_s(V_b - V_{in}) = C_s(V_b - V_{threshold})$. Next, S_2 and S_3 are opened and S_1 is closed. But, because the capacitor has not drained, the charge on the capacitor cannot have changed. Therefore, $V_b - V_{threshold} = V_a - V_{in}$, and we find that the new V_{in} is given by $V_{in} = V_a - V_b + V_{threshold}$. The output of the inverter at this point in the cycle is therefore equal to $V_{out} = AV_{in} = A(V_a - V_b + V_{threshold})$. Next, the cycle just described repeats itself for as long as the comparator operates.

In the comparator circuit of Fig. 6.9, the inverter is one of the most important circuit blocks. The inverter amplifier used in a comparator should be high gain and should have a sufficiently high bandwidth relative to the input signals and sampling speed. A basic CMOS inverter is illustrated in Fig. 6.10. Historically, op-amps were often used to implement comparators. While the description of operation given previously for Figure 6.9 is specific to a *switched capacitor* comparator, it is a simple matter to build a comparator circuit out of operational amplifiers (op-amps) or high gain rail-to-rail amplifiers. For example, a Schmitt trigger comparator applies positive feedback to the non-inverting input of a differential amplifier and may be used in place of the switched capacitor design of Figure 6.9. However, today, switched capacitor comparators are increasingly displacing op-amps in ADCs due to the difficulty in realizing the stacked transistor designs needed for op-amps. This difficulty arises from the reduced supply voltages now being required by modern technology processes [Mat07]. Since switching comparators may operate with much lower supply voltages than op-amps — as low as 0.5 V — switched-cap comparators will be important for mmWave ADCs [Mat07]. In practice, the choice of comparator design will hinge on whether or not the ADC is integrated with the mmWave radio. If

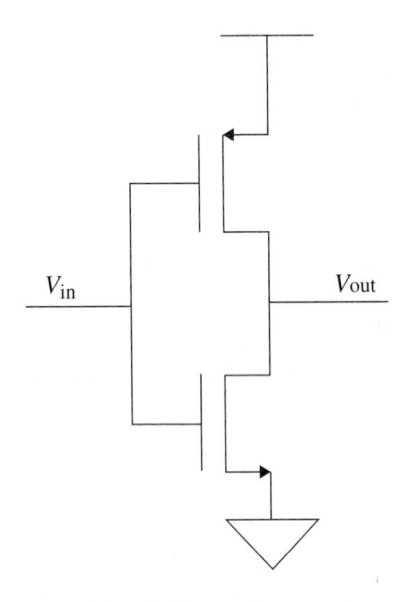

Figure 6.10 A basic inverter.

6.4 Basic Analog-to-Digital Conversion Circuitry: Comparators

it is integrated within a single chip, then a high enough supply voltage will likely be available to use op-amp comparators, since the analog portion of the mmWave radio will use an older technology process with greater supply voltage, however, if integrated on a separate chip with the strict power budget of Fig. 6.9. Positive feedback techniques may be used to increase comparator gain, thus providing a way to increase resolution and reduce errors as digitization speeds increase [Mat07].

6.4.1 Basic ADC Components: Track-and-Hold Amplifiers

Another basic building block of an ADC is a *track-and-hold (T/H)* or *sample-and-hold (S/H)* amplifier. The track-and-hold amplifier is at the very first input stage of the ADC, and serves to ensure stability for sampling by the comparator circuits. It acts to maintain a constant analog signal to a parallel bank of comparators over a very small interval of time, to prevent mismatches in the sampling times of individual comparators. The individual comparators rely upon the track-and-hold or sample-and-hold amplifier to keep the incoming analog signal fixed while comparing the incoming analog signal with known internal reference voltages of the ADC. These internal reference voltages may be produced by an internal DAC or by using simple *resistor divider* circuits to create a number of accurate reference voltages that may be compared with voltage-divided versions of the incoming signal. Because many comparators are used in parallel to compare different voltage-divided versions of the analog input signal, there can be conversion errors, called "*bubbles*," that occur when the input to a bank of comparators changes before all comparators can finish the decision and sampling process. Note that the track-and-hold amplifier must track, or allow the analog input signal to naturally vary, just before the discretization operation is performed by the comparators. Alternatively, the sample-and-hold amplifier locks the voltage of the analog signal prior to discretization by the comparators. The possibility of bubbles and the very large baseband bandwidths that are likely to be used in mmWave systems make low dynamic range, and hence low comparator count ADCs, attractive. Such ADCs will not suffer greatly from bubble errors and will consume less power than large dynamic range ADCs.

The operation of a sample-and-hold amplifier is illustrated in Fig. 6.11, whereas the operation of a track-and-hold amplifier is illustrated in Fig. 6.12. From a system perspective, a sample-and-hold amplifier is simply a cascade of two track-and-hold amplifiers [PDW89]. Fig. 6.13 illustrates the cascade of a track-and-hold amplifier, a bank of comparators, and an encoder, and it may be considered the basic layout of a flash ADC. Sample-and-hold and track-and-hold amplifiers can substantially increase the input bandwidth of an amplifier to near the Nyquist rate when used correctly [Koe00]. When used, sample-and-hold or track-and-hold circuitry sets the upper bound on analog-to-digital speed or bandwidth conversion performance [CJK+10][LKL+05]. Lower-resolution ADCs with resolutions less than 6 bits often forego the use of a sample-and-hold circuit to increase speed while reducing power consumption [SV07]. Sample-and-hold circuitry is necessary, however, when high resolutions are needed, even though their use may result in decreased speed performance [SVC09]. The input bandwidth of track-and-hold circuitry must exceed the ADC clock rate to be effective [SVC09].

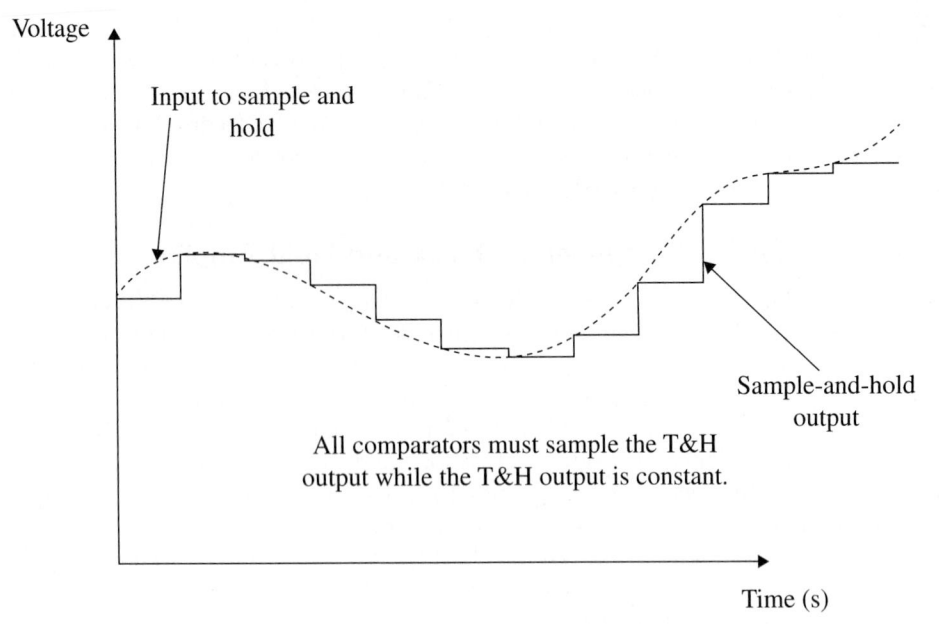

Figure 6.11 Example of an analog input signal and the resulting discrete version of the signal produced by an ADC using a sample-and-hold amplifier.

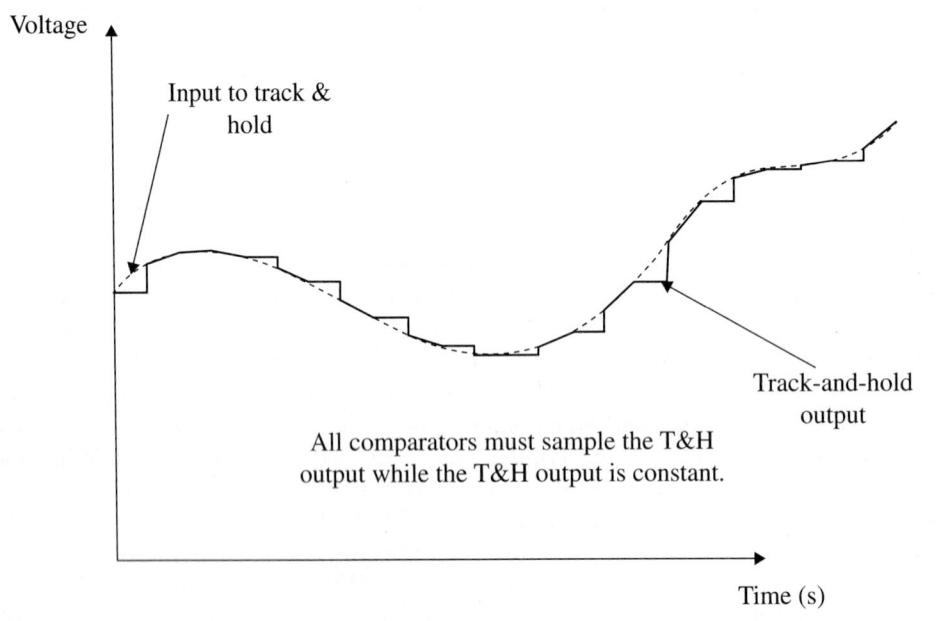

Figure 6.12 Example of an analog input signal and the resulting discrete version of the signal produced by an ADC using a track-and-hold amplifier. A track-and-hold circuit follows the input naturally when it is not in hold mode, unlike the sample-and-hold, which tracks the input at instantaneous times only.

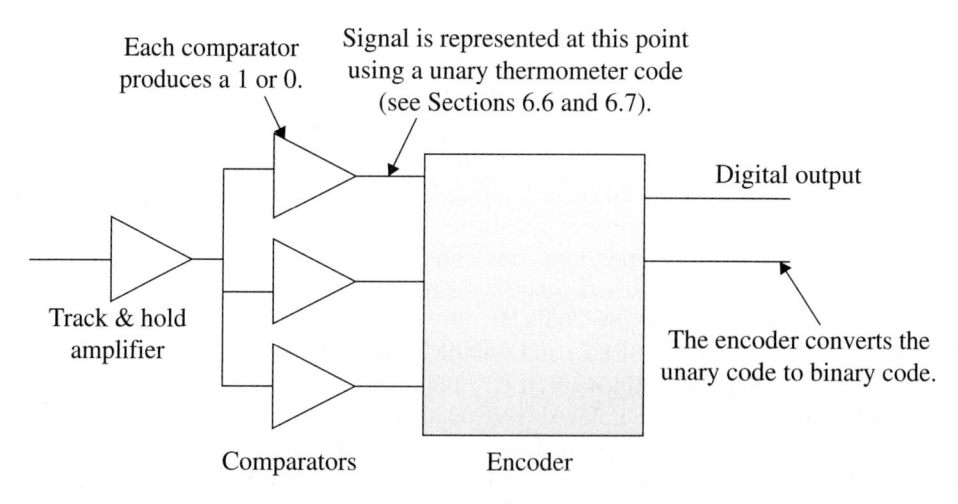

6.4 Basic Analog-to-Digital Conversion Circuitry: Comparators

Figure 6.13 The track-and-hold amplifier offers a constant signal, so that voltage dividers or a DAC (not shown) may offer voltages of varying significance as reference inputs (not shown) to the comparators. The held input signal is simultaneously compared in parallel with different reference voltages across the comparators to determine the most to least significant bit values of the digitized ADC output.

Mathematically, a sample-and-hold amplifier may be described as an ideal impulse sampler followed by a low-pass filter whose frequency response is given by a sinc function [Bak09]:

$$V_{\text{out}}(f) = T \left[\sum_{n=-\infty}^{\infty} V_{\text{in}}(f - nf_s) \right] \text{sinc}(\pi T f) \qquad (6.19)$$

where T is the sample period and is equal to the inverse of the sample frequency f_s. The input signal in the frequency domain is $V_{\text{in}}(f)$ and the output is $V_{\text{out}}(f)$, and f is the frequency. As a result of the $\text{sinc}(\pi T f)$ factor, the output signal is somewhat distorted when compared with the input signal – that is, higher-input frequencies closer to the Nyquist frequency are attenuated relative to input frequencies closer to DC [Bak09]. In fact, the Nyquist frequency $f_s/2$ is 3.9 dB down compared with frequencies very close to DC [Bak09].

A useful overview of the track-and-hold operation is provided in [Bak09]. If the tracking and hold times in a track-and-hold amplifier are equal, then the output of the track-and-hold amplifier will be given by [Bak09]

$$V_{\text{out}}(f) = \sum_{n=-\infty}^{\infty} \left\{ \frac{1}{2} \text{sinc}\left(\frac{\frac{\pi}{2} f}{f_s}\right) e^{-\frac{j\pi}{2} \frac{f}{f_s}} + \frac{1}{2} \text{sinc}\left(\frac{\pi}{2} n\right) e^{-j \frac{3\pi}{2} n} \right\} V_{\text{in}}(f - nf_s), \qquad (6.20)$$

which has the advantage of being flatter in the band from DC to $f_s/2$ than the sample-and-hold amplifier — at $f_s/2$ the frequency response is only down by 1.1 dB.

A track-and-hold amplifier should offer low total harmonic distortion at a high sampling rate, a high input bandwidth, and not consume excessive power. The harmonic distortion at the output will determine the usability of the amplifier for a given resolution. The input bandwidth should be as large as the ADC channel in which the amplifier is used.

The basic design of most open-loop track-and-hold or sample-and-hold circuits is a preamplifier followed by a switched-capacitor circuit with an optional output amplifier to improve driving capability [SCV06][CJK+10]. A very common approach is to use a switched emitter follower (SEF) [LKL+05][SCV06][LKC08] illustrated in Fig. 6.14. This architecture offers high linearity [LKL+05], with potentially high noise figure if large resistive degeneration is employed [SVC09]. Low isolation during hold mode due to feedthrough from the parasitic base-emitter capacitance in the switched emitter follower (SEF) can also degrade performance and is often addressed by means of feedthrough capacitor [LKC08]. A possible and common implementation of this design is to use a

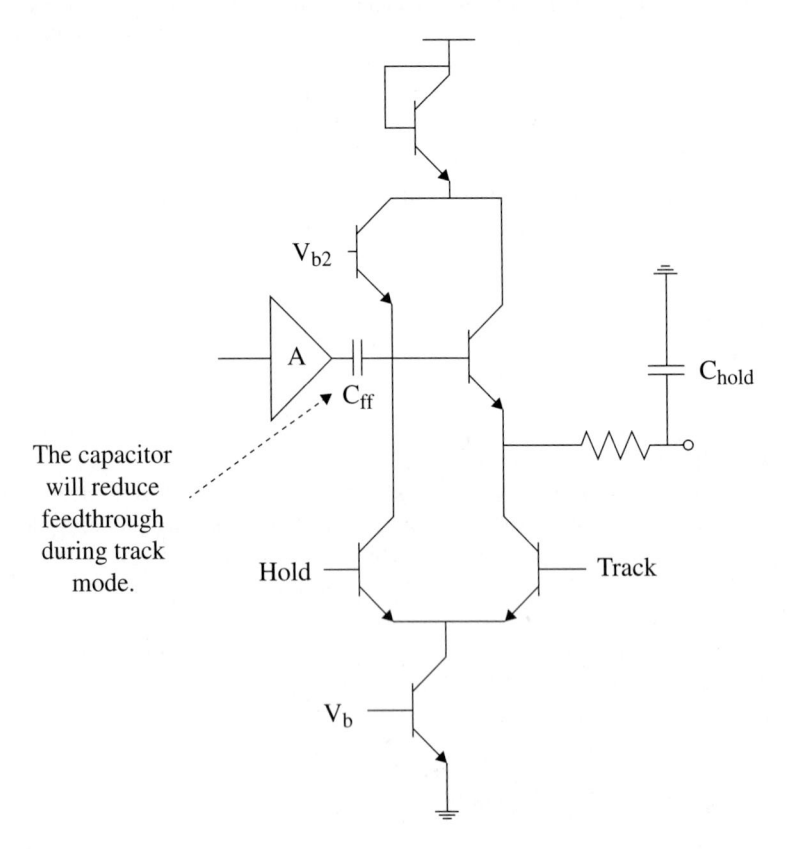

Figure 6.14 A switched emitter follower is a very common open-loop design for high-speed track-and-hold or sample-and-hold circuitry. Separate track-and-hold clocks control switching between the two modes.

6.4 Basic Analog-to-Digital Conversion Circuitry: Comparators

differential pair preamplifier with resistive degeneration to feed a switched capacitor emitter follower output stage [SCV06][CJK+10][LKC08]. Most very high speed track-and-hold amplifiers use open-loop designs to achieve maximum speed [LKL+05].

There are several points to keep in mind when designing a track-and-hold or sample-and-hold amplifier. The input capacitance of a track-and-hold amplifier should be low to allow for high-speed operation and linear to avoid signal distortion [LvTVN08]. The bandwidth of a track-and-hold circuit used in an n-bit ADC must be greater than $(n+1) \cdot \ln(2) \cdot 2 \cdot f_s/2\pi$ to avoid problems due to long settling times of sample signals [LvTVN08], and the input capacitance should be low enough to meet this requirement. Additionally, the buffer used to drive the switched capacitor in a track-and-hold or sample-and-hold amplifier has a major impact on performance [LKL+05]. The selection of a switch topology will affect the selection of a holding capacitor. In Fig. 6.14, the switched signal is applied to the ports labeled "Hold" and "Track." If a current-mode topology is used, then the hold capacitor should be reduced to allow for full discharging between samples. If a voltage-based topology is used, then this requirement is relaxed and a larger capacitance for reduced signal droop may be used [CJK+10].

The selection of a hold capacitor size in track-and-hold or sample-and-hold amplifiers is very important because it determines the droop rate of the circuit [CJK+10]. A low hold capacitance is beneficial for low distortion and bandwidth, but it may also result in excessive feedthrough of unwanted signals due to coupling through parasitic capacitances. It may also increase the speed of signal droop and decrease the ability to prevent errors from clock-skew [SCV06][CJK+10].

The duty cycle used in sample-and-hold circuitry may affect the performance of the ADC. The *duty cycle* of a track-and-hold amplifier refers to the percentage of the time that the amplifier tracks/samples the input or holds its current output value. For example, Chu et al. made use of a 25-75 sample-and-hold duty cycle for the sub-ADCs in a four-channel time-interleaved 40 gigasamples per second (Gs/s) 4 bit flash ADC [CJK+10]. This duty cycle (25% sample, 75% hold) increased the time allowed for conversion of each sub-ADC. When selecting a duty cycle, all timing requirements should be satisfied (e.g., setup and hold times) before substantially shifting from a more standard 50-50 duty cycle.

There are several caveats to consider when implementing a track-and-hold amplifier. Although sample-and-hold or track-and-hold circuitry can help to eliminate problems caused by clock skew and jitter, it may also have the effect of increasing distortion due to its highly non-linear transfer function [SVC09]. For this reason, the decision to implement a track-and-hold or sample-and-hold circuit in an ADC should be made only after verifying that the benefit from reduced clock skew outweighs the impacts of increased distortion, and after verification of adequate input frequency bandwidth of the track-and-hold circuitry [SVC09]. A necessary consideration when implementing track-and-hold or sample-and-hold circuitry is proper synchronization between the clock path used to switch the hold circuitry, and the clock path used to switch the comparator bank [SVC09]. Poor synchronization of these two paths will render the track-and-hold circuitry useless. Lastly, if the output signal of a track-and-hold circuit is not stable during hold mode (i.e., if it "droops") then it will not be effective in reducing the effects of clock jitter [CJK+10].

Holding circuits may benefit substantially from a differential design. For example, Shahramian et al. [SCV06] found that a differential implementation for a track-and-hold amplifier significantly reduced common-mode clock feedthrough.

It is useful to consider several examples from the literature to see how ADC speeds are being increased dramatically and the state-of-the-art performance specifications that are being developed. In 2006, Shahramian et al. presented a 40 gigasamples per second (Gs/s) track-and-hold amplifier in 0.18 μm SiGe BiCMOS for use in high-speed ADCs [SCV06]. The amplifier consumed 540 mW from 3.6 V and had a 3-dB bandwidth of 43 GHz. The architecture of the amplifier consisted of a low-noise amplifier followed by an emitter-follower differential pair that fed the track-and-hold circuit. The track-and-hold circuit was switched using a clock distribution tree. The output of the track-and-hold block drove a differential-pair driver cascade. The use of a low-noise preamplifier improved performance [SCV06]. All blocks used differential implementations. The input low-noise amplifier relied on a transimpedance architecture to offer superior noise performance compared with a emitter follower. In 2010, Cao et al. used a two-stage quasi-differential source-follower design for a track-and-hold amplifier in a 6-bit 10 Gs/s time-interleaved flash ADC [CZS+10]. The amplifier had two stages: The first stage used a pair of differential NMOS source followers and the second stage was a pair of differential PMOS source followers. Low gain source followers were advantageous for this design because their high bandwidth helped to mitigate settling time differences among the paths of the ADC [CZS+10].

In 2008, Louwsma et al. [LvTVN08] presented the design of a track-and-hold amplifier used in a 16-channel time-interleaved ADC with successive approximation sub-ADCs. The design was based on a boot-strapped sample switch for good linearity followed by a linear source buffer of two cascaded source followers. A cascade of source-followers, which was also used by Cao et al. [CZS+10], is useful for improving the gain and linearity of minimum-length transistors — resulting in a more linear overall design. In order to achieve acceptable bandwidth from each track-and-hold amplifier without significantly increasing power, a switch was added to the output of each buffer. In 2010, Chu et al. [CJK+10] presented a sample-and-hold circuit designed to operate up to 60 Gs/s but measured up to only 40 Gs/s. The design relied on a switched emitter follower similar to that used by Shahramian et al. [SCV06]. Differential degeneration was used on an input differential pair to improve linearity. An output emitter follower circuit was used to improve output driving capability [CJK+10]. In 2005, Lu et al. [LKL+05] presented the design of an 8-bit 12 Gs/s track-and-hold circuit in 0.25 μm SiGe BiCMOS [LKL+05]. The design was based on a pseudo-differential (suitable for a differential clock input) architecture with inductive peaking for improved bandwidth. The design made use of auxiliary paths to prevent capacitive loading and to improve isolation [LKL+05]. A cascaded output stage consisted of two buffer amplifiers. The first provided extra current injection from a current mirror to improve signal droop, and the second featured a higher supply voltage to improve linearity. In 2008, Li et al. [LKC08] presented a 40 Gs/s fully differential track-and-hold circuit in 130 nm SiGe BiCMOS with a switched emitter follower design. The authors employed a current compensation circuit to reduce the droop rate of the hold capacitor. In addition, the authors used a feedthrough compensation scheme more complex than the standard feedthrough suppression capacitor: A feedthrough attenuation network with 180° phase shift from the output of the main path was added to cancel the main path in hold mode. The switched emitter follower used a stacked emitter follower design that resulted in an additional degree of isolation between input and output during hold mode.

6.5 Goals and Challenges in ADC Design

There are several goals and challenges associated with the design of analog-to-digital converters (ADCs). From a very high level, it is desirable to have ADCs that are high speed, low power, and high resolution [VKTS09]. Common figures of merit include the energy per conversion, power required for a given sampling rate, bandwidth-accuracy products, and figures of merit (FOMs) in terms of bandwidth, accuracy, and power [US02][IE08][CLC+06].

Aperture delay (t_{AD}) in an ADC is the interval between the sampling edge of the clock signal (i.e., the rising edge of the clock signal) and the instant when the sample is taken. The sample is taken when the ADC's track-and-hold goes into the hold state. It is desirable to reduce aperture delay. Aperture jitter (t_{AJ}) is the sample-to-sample variation in the aperture delay. Typical ADC aperture jitter values are much smaller than those of aperture delay.

For practical mmWave communication system design, it is valuable to consider the tradeoffs in ADC design and the goals of the overall mmWave system in context with the ADC design. This requires knowledge of the modulation scheme used, the bandwidth of the sampled signal, the amount of channel equalization required, and the overall power budget of the device. As summarized by Uyttenhove et al., ADC designs involve tradeoffs between the desired speed and accuracy (i.e., resolution) with power consumption and device mismatch [US02].

In very high speed converters, clock jitter becomes a major challenge, and high levels of clock jitter will directly degrade the achievable SNR or ENOB (as defined below) of a converter [SVC09]. Clock jitter is undesirable because it causes timing mismatches in the sampling time of comparators, resulting in erroneous conversions. Clock jitter will increase in its impact proportionally to the signal frequency [LvTVN08]. For ADCs in which the clock must be distributed to many different points (e.g., in "high-resolution" flash ADCs), clock distribution circuitry becomes a major design challenge. Shahramian et al. used a tapered differential pair approach for very high speed data sampling (35 Gs/s) [SVC09], but such an approach will increase the footprint and power requirements of the ADC.

The *effective number of bits* (ENOB) is a reflection of the "accurate" resolution of the ADC with a given signal input frequency or resolution, and it can be calculated as [VKTS09]

$$\text{ENOB} = \frac{\text{SNDR} - 1.76}{6.02} \qquad (6.21)$$

where SNDR is the *signal-to-noise-and-distortion ratio* and is defined as the ratio of output signal power to the sum of noise power and distortion power [Has09]. At high input frequencies, we expect the number of error-free conversions (or ENOB) to decrease because the difference between the signal period and the inherent timing mismatches in the circuit (i.e., *skew and jitter*) becomes smaller, thus leading to more conversion errors. [SV07][ARS06][SVC09][CJK+10]. At higher resolutions, we may expect the difference between ENOB and the stated resolution to increase due to an aggregation of errors across more comparators [ARS06]. The degradation of ENOB with increasing frequency may be slowed or eliminated through the use of track-and-hold circuitry, though in practice the effectiveness of track-and-hold circuitry in improving ENOB is limited by the bandwidth of the track-and-hold circuitry [SVC09]. Voltage offsets between comparators or sampling time offsets, caused by non-uniform delays, for example, can result in erroneous ADC outputs (bubbles). In certain designs, for example, flash ADCs, bubble errors caused by circuit mismatch are a pressing design concern [US02][PDW89]. Certain ADC designs are more prone to bubble errors than others, and in these cases it is often necessary to use bubble correction or prevention circuits that can slow operation [SV07]. At each stage of the ADC it is possible to implement a circuit to help reduce bubble errors. For example, clock buffers and track-and-hold circuitry may be used to reduce bubble errors caused by sampling time mismatches.

It is instructive to consider the input bandwidth of an ADC, as this sets a limit on the useful baseband sampling frequency. A simple analysis, in which the ADC is considered as a simple 1-pole system followed by an ideal sampling circuit, makes clear the need for a high input bandwidth requirement. Fig. 6.15, based on a discussion of time-interleaved ADCs by Kurosawa et al. [KMK+00], illustrates this point. In some cases (such as for emerging broadband wireless systems) in which a very wide bandwidth is required, it may be necessary to use pipelined ADC designs [SV07]. The input bandwidth cutoff will be reflected in the degradation of ENOB with increasing input frequency.

The scaling of transistors and the physics that govern a scaled device are the primary drivers of ADC trends in recent years (and of most portions of the mmWave radio). The results of scaled transistors include faster operation, lower supply voltage, increased

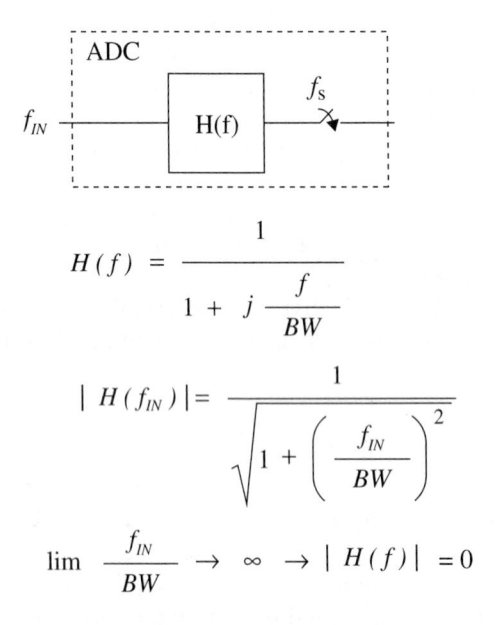

$$H(f) = \frac{1}{1 + j\dfrac{f}{BW}}$$

$$|H(f_{IN})| = \frac{1}{\sqrt{1 + \left(\dfrac{f_{IN}}{BW}\right)^2}}$$

$$\lim \frac{f_{IN}}{BW} \to \infty \to |H(f)| = 0$$

Figure 6.15 A simple circuit model for an ADC. The bandwidth of the ADC needs to exceed the bandwidth of the incoming signal to avoid loss in signal magnitude.

6.5 Goals and Challenges in ADC Design

operation in weak or moderate inversion regimes, and increased mismatch. The implications of these trends for ADCs are faster conversion rates, lower dynamic ranges/lower resolutions and SNRs [Mat07], new designs to accommodate the increased non-linearity of small-signal operation in moderate or weak inversion (i.e., the drain-source voltage is more variable as it is not fully saturated), and new architectures needed to achieve high resolutions with low supply voltages [Mat07]. Successful mmWave devices will evolve in the same way as the ultra-fast ADCs needed in mmWave systems in order to be successful. They will be very high bandwidth but will have low resolutions, allowing them to achieve high data rates at relatively low levels of power consumption. Uyttenhove et al. presented a very general equation (6.22) that can be used to study many of the trends in ADC design [US02]:

$$\text{Speed} \times \frac{\text{Accuracy}^2}{\text{Power}} \approx 1/(C_{ox} A_{vt}^2) \qquad (6.22)$$

where speed is proportional to the transit frequency of active devices in the ADC, accuracy is proportional to $\sqrt{WL}$ (W = transistor width and L = gate length), and power is proportional to V_{dd} (the supply voltage). C_{ox} is the gate oxide capacitance, which is dependent mainly on gate oxide thickness and gate oxide material, and A_{vt} is a technology-dependent parameter that describes mismatch between transistors. From the equation it is clear that as supply voltage decreases, either speed or accuracy must also decrease. For very high speed converters, reductions in supply voltage must be accommodated by reductions in accuracy. It would also seem possible to mitigate supply voltage decreases by reducing A_{vt}, which describes the statistics of mismatch across many transistors for a given technology. Unfortunately, A_{vt} actually increases with increased doping levels, indicating that with increased scaling A_{vt} should actually increase [US02]. A_{vt} is given by [US02]:

$$A_{vt} = \frac{q}{\sqrt{2}\epsilon_{ox}} t_{ox} \sqrt{D_{\text{total}}} \qquad (6.23)$$

where D_{total} is the doping concentration, t_{ox} is the oxide thickness, q is the charge of an electron, and ϵ_{ox} is the gate oxide permittivity. Doping concentrations will continue to increase with device scaling to control such effects as punch-through and to keep source and drain well isolated. t_{ox} is no longer scaling down with technology scaling due to the high gate leakage currents associated with extremely thin gates. We see that the best means of decreasing A_{vt} then for modern processes is through the use of "high-k" (i.e., high relative permittivity) gate oxides. Fortunately, (6.23) indicates that increasing ϵ_{ox} will help the speed-accuracy-power tradeoff more than a proportional increase in doping concentration will hurt the tradeoff.

The lower dynamic ranges possible for high-speed designs due to reduced supply voltages is in some ways synergistic with mmWave use cases — at least initially. The large bandwidths available at these frequency ranges, and the desire to make devices inexpensive to increase demand and adoption, mean that low spectral efficiencies are almost preferable. With consumers' current expectations, a spectral efficiency of 1 (bit/sec)/Hz would be more than adequate to feel like a wireless revolution with an RF channel size of $\sim$ 2 GHz. Low spectral resolutions and small dynamic changes in the broadband wireless channel also mean that lower-cost ADCs may be used, resulting in lower overall costs.

But, in the long run, this will change as consumers become accustomed to multi-Gbps wireless data rates and begin to demand tens of Gbps to hundreds of Gbps and even terabits per second (Tbps) data rates. It is clear that current consumer expectations and technological ability are in a "sweet spot," making rapid adoption likely. However, in the long run, the serious challenge of low supply voltage versus achievable resolution will need to be solved, as technology nodes become smaller and smaller.

6.5.1 Integral and Differential Non-Linearity

Differential non-linearity (DNL) and *integral non-linearity* (INL) are non-linear effects caused by imperfections in the sample-and-hold process when converting an analog signal voltage to a digital codeword, and they are not removable by a calibrating circuit. These non-linearities play a role in both ADCs and DACs, and they are parameters that are measured and used to compare different ADCs. For ADCs, it is critical for each individual voltage sample to be converted to a unique digital codeword without error or confusion, even in the face of noise, distortion, cross talk, and spurious signals in the analog source. To guarantee no missing codes and a monotonic transfer function as the input voltage is increased, an ADC's DNL must be less than one least significant bit (1LSB). Higher values of DNL are generally what limit the ADC's performance in terms of signal-to-noise ratio (SNR) and *spurious-free dynamic range* (SFDR). See Fig. 6.16 [Max01].

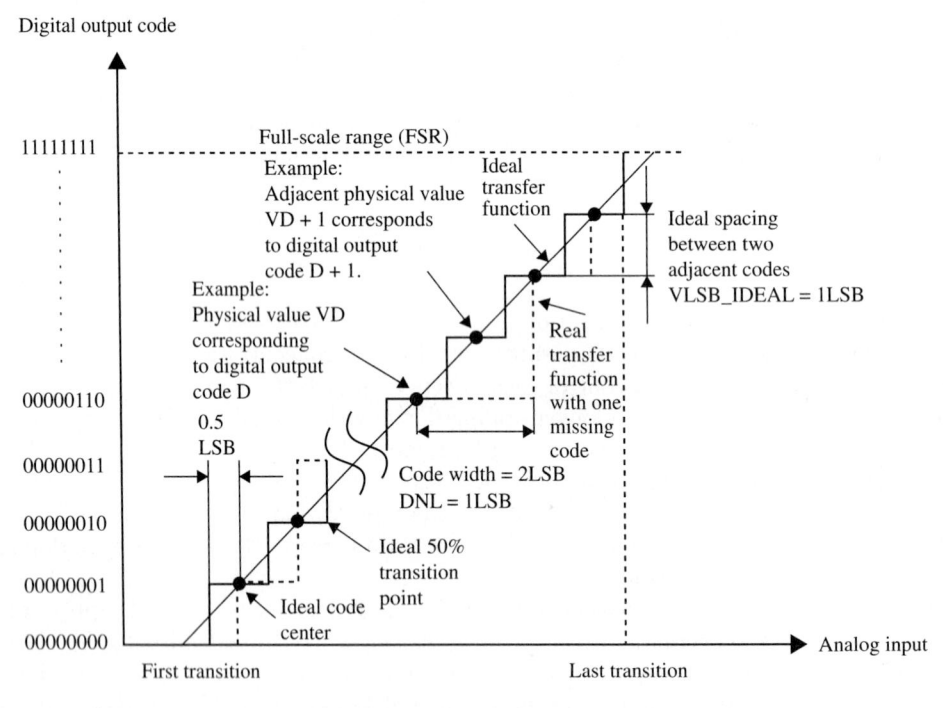

Figure 6.16 To guarantee no missing codes and a monotonic transfer function, an ADC's differential non-linearity (DNL) must be less than 1LSB [from [Max01] © Maxim Integrated].

6.6 Encoders

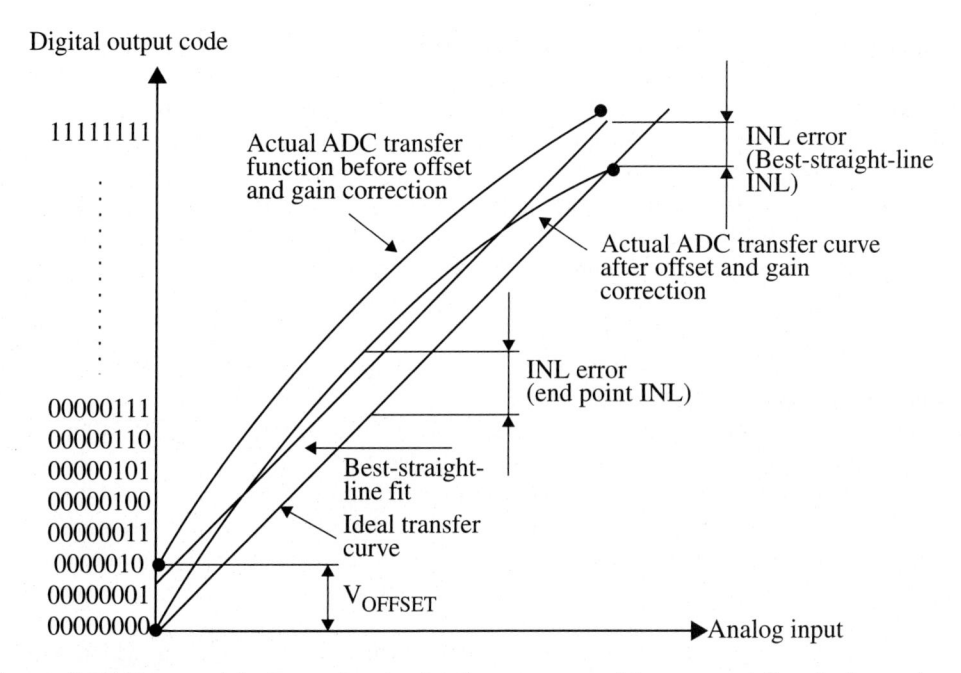

Figure 6.17 Best-straight-line and end-point fit are two possible ways to define the integral non-linearity characteristic (INL) of an ADC [from [Max01] © Maxim Integrated].

INL error is described as the deviation, in LSB or percent of full-scale range (FSR), of an actual transfer function from a straight line. The INL-error magnitude then depends directly on the position chosen for this straight line. At least two definitions are common: "best-straight-line INL" and "end-point INL" (see Fig. 6.17).

6.6 Encoders

An encoder is a circuit that converts the format of data for use in other portions of a circuit. In an ADC, an encoder may consist of one or more circuits, including multiplexers or buffers, that reformat several different digital results (say, from different comparators in parallel), so that the resulting digital information may be represented in a standard format, for use in a serial stream of data for use in a circuit serial bus, or put into parallel digital words that may be presented on a parallel bus for use by a DSP or memory. The encoder used in an ADC can greatly impact speed, accuracy, and power. In certain ADC designs for extremely high data rates, for example, flash ADCs, the digital encoder can account for as much as 70% of the ADC power usage [SV07].

As an example of encoding, refer to Fig. 6.13, in which an ADC uses a parallel bank of comparators, each using different internal reference voltages, to determine the digital representation of the incoming analog signal. Efficient encoding that can detect bubble errors and errors from individual comparators are an important requirement in ADC design, as will be apparent in many practical ADC architectures discussed in Section 6.7. As the analog input voltage exceeds the reference voltages across the bank of parallel comparators at the ADC input, the number of 1s out of the bank of comparators increases. The name "thermometer code" is derived from the fact that at each comparator, if the analog input voltage is larger than the reference voltage for that comparator, the comparator produces a logical "1" output, and hence, across the bank of parallel comparators, larger voltages will produce more 1s at the output of the comparator bank. The analogy here is that a larger input voltage signal will produce more 1s out of the bank of comparators, just like hotter temperatures cause the mercury to rise in a thermometer. The encoder must include a circuit to convert the parallel thermometer code outputs from the bank of comparators into useful binary data that represents the input signal voltage in standard binary words [MRM09][SV07]. Each comparator provides a unary output, a simple "1." to denote whether or not the input signal to the comparator exceeds the internal reference voltage applied to that particular comparator. By using this unary coding across the comparators and counting up the ordered number of 1s out of the bank of comparators, the quantized voltage value may be easily determined and represented in a code, such as a thermometer code.

Encoding techniques used in communications, such as Gray coding, in which sequential code words differ by only one bit in the word, can be used to detect and correct errors in groups of individual comparator outputs with minimal computation speed and costs, and such approaches are used to detect bubble errors. Individual samples from individual comparators may be corrected, and in the case of the MSBs, may be ignored if the variation between successive samples is too great to be believable or achievable with known input bandwidth limitations. Previous samples (repetition codes) may be used in place of errant samples if error correction is not possible in order to avoid spurious sample words.

Methods to reduce the physical size of the encoder, for example, through the use of *multiplexer* (mux)-based[1] thermometer-to-binary encoders in flash ADCs, offer promise for future mmWave systems having multi-Gbps data rates [VKTS09][MRM09]. Also, selection of the encoder should take into account the resiliency of each design to bubbles — errors caused by voltage offsets, sampling time mismatches, or circuit metastability [SV07]. Section 6.7 illustrates how encoders and decoders are used in modern ADCs.

In addition to finding an encoder design capable of low power, high speed, and error resilient operation, there are a few necessities that must be possessed by the encoder. These include a uniform delay for all outputs and output stability for the required hold time of the preceding stages in the digital circuit. One way to help ensure a uniform delay is to design the encoder such that all signals pass through the same number of gates on the way to the output [SV07]. Certain encoder designs, for example, fat-tree encoders

---

1. A multiplexer is a circuit or device that selects signals from several inputs and places the signals onto a single line.

used in flash, will suffer from synchronization issues because each output bit must pass through a different number of gates [ARS06].

6.7 Trends and Architectures for MmWave Wireless ADCs

There exist several competing trends for future ADCs: to increase the dynamic range (and thus the number of bits and achievable SNR) while increasing the sampling speeds, and to decrease power consumption. Meeting the goals of both of these trends at the same time is impossible, as this involves the classic power-bandwidth tradeoff. Additionally, two main "power" trends for ADC design include the trend to increase (yes, we said *increase*) power consumption to overcome mismatch in deeply scaled sub-micron technologies to achieve better performance (e.g., in bit conversions per second), and the trend to reduce power consumption and achieve faster sampling speeds per individual ADC for systems-on-a-chip to keep the overall power allotted to analog-to-digital conversion reasonable [US02].

The high data and clock rates needed in mmWave systems result in increased challenges caused by clock skew rate due to differences in clock and data propagation times [SVC09]. As data rates increase, an initial method for reducing skew will be to use higher power preamplifiers and perhaps tapered input buffers that will also have the effect of increasing power consumption.

The increased reliance on comparators as opposed to op-amps and the need for high sampling rates in such applications as WPANs have increased the popularity of comparator-based designs such as successive approximation ADCs, pipeline ADCs, and sub-ranging ADCs [Mat07][CLC$^+$06]. For first-generation mmWave circuits, the need for high speed and the lower spectral resolutions required make flash ADCs (see Section 6.7.4) an attractive design, as these are generally the fastest ADCs [ARS06].

Matsuzawa summarized the trend in dynamic range and achievable SNRs with the equation that gives the dynamic range of a differential sample-and-hold circuit: SNR $= \frac{CV_{pp}}{4kT}$, where V_{pp} is the peak-to-peak voltage of the sampled signal [Mat07]. As supply voltages decrease, the peak-to-peak voltage of sampled signals must also decrease so that the transistor will not enter the linear-region of operation (i.e., region before pinch-off of channel). Matsuzawa points out that the SNR can be kept high if the sampling capacitance increases [Mat07], but this clearly also has the effect of a lower sampling rate.

6.7.1 Pipeline ADC

The concept of a pipelined ADC is illustrated in Fig. 6.18. At the input of the ADC, a track-and-hold (T/H) circuit is used to perform sampling of the input signal, as described in Sections 6.4 and 6.5. Each sub-ADC has a resolution lower than the aggregate ADC. The top ADC provides the most significant bit outputs. Subsequent ADCs receive an amplified residual to allow a course resolution to be used to generate the LSBs of the original signal. This design is not tied to a particular ADC architecture — each sub-ADC may be constructed using, for example, a flash ADC or a successive approximation ADC, as was done by Louwsma et al. [LvTVN08]. The number of bits of the first ADC should

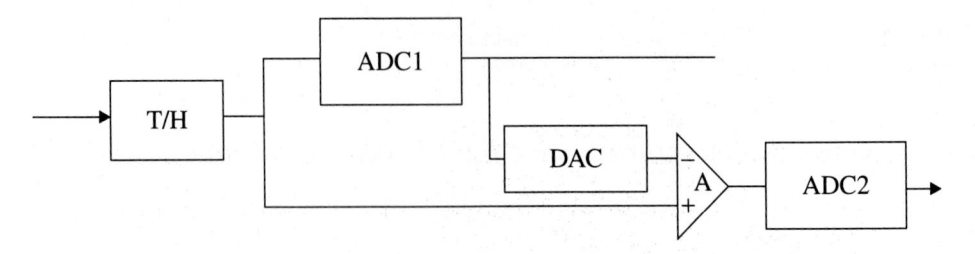

Figure 6.18 A pipelined ADC relies on multiple sub-ADCs.

be carefully selected for realizable requirements of the intermediate DAC [LvTVN08]. In 2008, Louwsma et al. used pipelined successive approximation ADCs as the sub-ADCs of a time-interleaved, 10-bit, 1.35 Gs/s ADC.

6.7.2 Successive Approximation ADCs

Successive approximation (SA) ADCs offer the advantages of high power efficiency and low number of required comparators [LvTVN08]. Individually, successive approximation ADCs historically are not among the fastest architectures, and therefore are not likely to be used in mmWave devices in the near future. But, if incorporated into a time-interleaved architecture, it may be possible to use successive approximation ADCs in mmWave systems. In 2008, Louwsma et al. [LvTVN08] presented a time-interleaved ADC that made use of successive approximation sub-ADCs. They used three techniques to make the successive approximation architecture useful for a high-speed design: successive approximation pipelining (each sub-channel used a pair of pipelined SA ADCs), a single-sided overrange technique, and look-ahead logic [LvTVN08]. Overranging is a technique that reduces the amount of time that must be allocated to DAC settling time in each clock cycle, and it can also be used to substantially reduce power per conversion [LvTVN08]. The look-ahead logic scheme is a simple means of reducing the bottleneck that is represented by the digital control logic in a successive approximation design. Because a binary search like algorithm is used, there are only two possible values for the DAC after each conversion. Lousma et al. program the control logic to calculate these in advance before a new input to the control logic is received [LvTVN08].

The basic operation of a successive approximation ADC is illustrated in Fig. 6.19. The operation is similar to searching through a sorted list using a binary search. The input controls the output level of the DAC, which is compared to the input signal. Each iteration of the loop is used to generate a bit of the output, and the value out of comparator is used by the control logic to determine the next DAC output. For an N-bit ADC, N iterations of the loop must complete within the sample period.

The digital-to-analog converter (DAC) used in the successive approximation computation is a vital component of a successive approximation ADC. Increasing the sampling rate of successive approximation ADCs improves their utility for mmWave designs. A possible means of making successive approximation ADCs faster is through the use of asynchronous operation, in which the settling-time for each sampled bit is set to a different value based on the bits' importance to increase operating speed [Mat07].

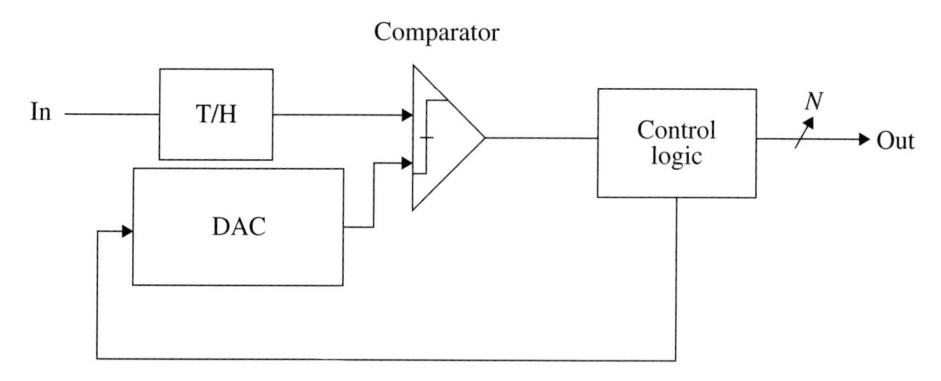

6.7 Trends and Architectures for MmWave Wireless ADCs

Figure 6.19 The loop must iterate N times for each sampling period.

6.7.3 Time-Interleaved ADC

Time-interleaved architectures offer one approach for achieving high sampling rates, resolution, and power efficiency, but they have the disadvantage of increasing area and power consumption, and possibly requiring extensive calibration [SVC09][LvTVN08] [CJK+10][KMK+00]. The basic approach of a time-interleaved ADC is to use a parallel bank of sub-ADCs, each with a differently phased sampling [SVC09], as shown in Fig. 6.20. Each sub-ADC requires a sampling rate of the f_s/M, where f_s is the sampling rate of the entire ADC and M is the degree of parallelism (i.e., the number of parallel ADCs). The amount of parallelism is limited by the maximum allowable input capacitance to the ADC, the achievable phase resolution in sampling clocks, and the amount of complexity and space allotted to the design (both increase with increasing frequency) [CZS+10][LvTVN08][CJK+10].

An advantage of a time-interleaved design is the ease and low speed at which output data may be multiplexed. This is advantageous in cases in which very high speed serial data must be processed by slower parallel technology [CJK+10]. The high degree of parallelism in time-interleaved ADCs makes them susceptible to mismatch errors. Possible mismatches between paths include gain mismatch, bandwidth mismatch, and offset and clock phase mismatch (clock skew) [CZS+10]. In practice, the amount of clock skew is a limiting factor to the amount of parallelism, and hence the sampling rate, of time-interleaved ADCs [CZS+10][CJK+10]. When the number of channels is large, the SNR of the ADC for a given standard deviation timing offset is given by $\frac{1}{(\sigma(\Delta t) \cdot 2\pi \cdot f_{\text{in}})}$ [LvTVN08], indicating that as the timing offset increases relative to the sampling rate, the signal-to-noise ratio will also decrease. This formula can be used to estimate the maximum allowable timing mismatch between sub-channels. Clock skew will produce spurious frequency peaks at $f_{\text{spurious}} = \frac{kf_s}{M} \pm f_{\text{in}}$, the same frequencies of spurs caused by gain mismatches [KMK+00]. In addition to systematic clock skew, time-interleaved ADCs will suffer from random clock jitter just as in all ADC architectures [KMK+00].

In addition to phase mismatch, gain mismatch is a major concern in time-interleaved ADCs. Gain mismatch between the sub-channels of a time-interleaved ADC creates spurious tones at multiples of the sub-channel sampling frequencies plus an offset equal to

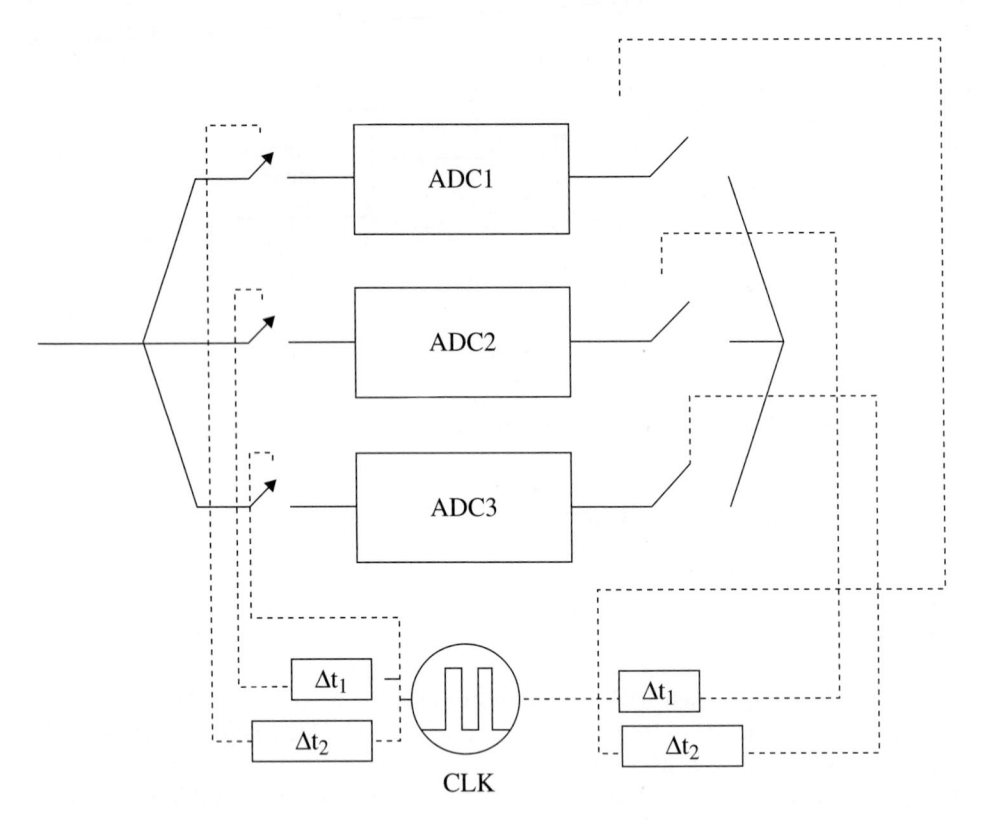

Figure 6.20 Parallel ADC architecture that allows time interleaving for higher accuracy A/D conversion at much higher sampling rates than the rate at which a single ADC operates. Δt_1 is the time delay between when the clock edge is delivered to ADC2 and ADC3, while Δt_2 is the time delay between when the clock is delivered to ADC1 and ADC3.

the input frequency: $f_{\text{spurious}} = \frac{kf_s}{M} \pm f_{\text{in}}$ [LvTVN08][CZS+10][KMK+00], where f_s is the sample frequency of the entire ADC and M is the degree of parallelism. Gain and phase mismatch of time-interleaved ADCs will interact and are not separate processes [KMK+00]. These tones are due to the modulation of gain-mismatch effects by the input signal [CZS+10].

One method of compensating for the effects of gain offsets in signal paths of a time-interleaved flash ADC is to use variable gain amplifiers (VGAs) at the inputs of each ADC [CZS+10]. The VGAs can be tuned to keep the signal strengths into each ADC nearly equal, and they can be tuned digitally, for example, by a DSP [CZS+10]. In practice, the gains of VGAs used in this manner should be kept small so as not to exacerbate DC offsets [CZS+10]. This indicates that VGAs cannot be used to compensate for a poorly laid out structure that introduces heavy gain mismatches.

DC offsets between sub-channels produce spurious tones similar to gain and phase mismatches, but the spurs are located periodically in frequency at $\frac{kf_s}{M}$ where M is the degree of parallelism and f_s is the sampling rage of the entire ADC

[CZS+10][LvTVN08][KMK+00]. Intuitively, this can be understood if the output of the ADC with a constant input is considered under conditions of differing channel offsets: each channel will produce a spurious tone at multiples of the channel's sampling rate [KMK+00]. These spurs are due to offset modulations of the ADC output [CZS+10].

The bandwidth required by each channel in an interleaved ADC must exceed the sampling rate of the sub-channel (by a factor of $N \ln(2)/\pi$, where N is the number of channels) [CZS+10][LvTVN08] to avoid settling time issues. Bandwidth mismatch between channels results in different settling times — causing both phase and gain differences — among the sub-ADCs used in a time-interleaved architecture [CZS+10][LvTVN08][KMK+00]. This can be seen by treating each sub-ADC as a simple single-pole system, in which a change in the bandwidth (i.e., the location of the pole) directly translates into different channel gains and phases for each sub-channel. Note that because the sampling time of each sub-channel is scaled by a factor equal to 1 over the parallelism factor that each channel requires, a substantially smaller bandwidth is required than for the single channel of a non-time interleaved ADC. This is the major advantage of a time-interleaved ADC.

6.7.4 Flash and Folding-Flash ADC

Flash ADCs do not require feedback and are fully parallel, and therefore they are suitable for very fast conversion rates [VKTS09][OD09]. Their resolutions are limited, though, by the fact that the number of comparators needed scales exponentially with required resolution, resulting in potentially very large input capacitances, which limit the analog input bandwidth [CLC+06][IE08]. The resolution required for many wireless applications does not exceed the capabilities of flash architectures (up to 8 bits) [PA03], so this architecture is likely to be used in many mmWave systems. Fig. 6.21 illustrates a standard flash ADC. Flash ADC performance is greatly impacted or even dominated by the level of mismatch that can be achieved by differential circuits used in comparators [US02]. Therefore, the trend in mismatch with continued device scaling is a means of predicting the evolution of flash ADCs. Other challenges for flash design are reducing area and power consumption [MRM09].

The flash architecture operates by comparing the input to a set of reference voltages with a bank of comparators. It is composed of a bank of 2^{N-1} comparators followed by a temperature-to-binary decoder [ARS06] (where N is the number of bits of resolution). A thermometer code is the result of the output of the bank of comparators. Note that as the analog input voltage exceeds the reference voltages on the resistor reference voltage ladder, the number of 1s out of the bank of comparators increases. The name *thermometer code* is derived from the fact that at each comparator, if the analog input voltage is larger than the reference voltage, the comparator produces a logical "1" output, and hence, across the bank of parallel comparators, larger voltages will produce more 1s at the output of the comparators. The analogy is that a larger input voltage signal will produce more 1s out of the bank of comparators, just like hotter temperatures cause the mercury to rise in a thermometer. In this way, a thermometer code may be considered to be a unary code. A decoder is needed to take the parallel outputs of the comparator bank (the thermometer code) and convert those results to binary data that represents

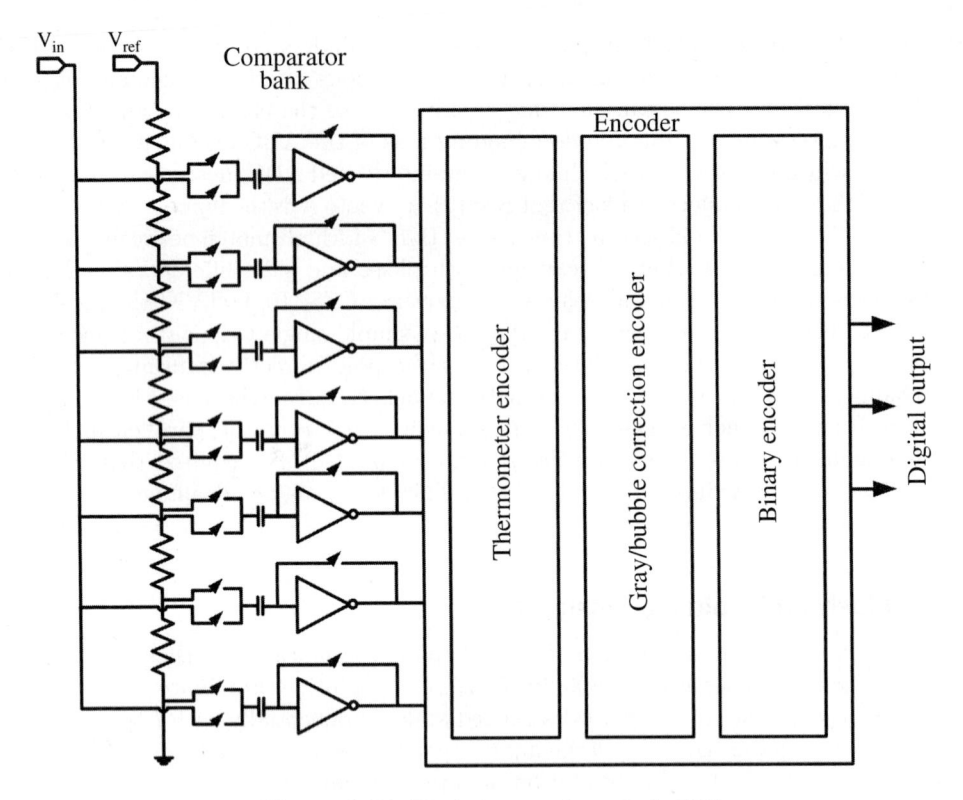

Figure 6.21 Block diagram for a flash ADC.

the input signal voltage in standard binary words. The decoder uses these outputs to produce an N-bit binary code [MRM09][SV07].

At the input of the comparator bank that produces the thermometer code, flash ADCs will often make use of a bank of preamplifiers and track-and-hold (or sample-and-hold) circuits whose role is to reduce the effects of input referred voltage offsets [IE08] and skew [SVC09], respectively. This is often necessary, as voltage offsets can dominate the performance degradation of a flash ADC [US02]. In order to increase data rates, it may be acceptable to supplement track-and-hold circuitry with a tapered buffer or amplifier tree. This is a fairly brute-force method for increasing sampling rates, as it can dramatically increase power consumption of the ADC [SVC09].

Input referred offsets are voltage offsets used to describe random voltage errors at the inputs of comparators, due to noise, transistor imperfections, or device mismatches, that transfer to output error. Input referred offsets cause conversion errors in the digital outputs of the ADC and are a major challenge for flash ADCs [US02][IE08]. Although preamplifiers can mitigate the effects of offset to a degree, their impact is limited. Often at the output of a preamplifier bank, a flash ADC will employ a resistive averaging network to further enhance the offset mitigating effects of a preamplifier array [PA03][IE08]. A resistive averaging network reduces the effects of voltage offset currents and voltages by averaging out their contributions (which is possible because the offsets are statistically

6.7 Trends and Architectures for MmWave Wireless ADCs

zero mean and uncorrelated due to process variations) [PA03]. This can be understood by recalling that the offset sources are random and uncorrelated in nature, thus when they are summed in the *resistive network*, their sum does not grow as quickly as the sum of the amplified and correlated signals that also add in the resistive network [PA03]. Averaging may also be performed using capacitors or split transistors [PA03]. As a result of the offset mitigating effects of averaging networks, smaller devices may be used while not suffering reduced accuracy (i.e., conversion errors are suppressed). This may help to shrink the size of the ADC [PA03], provided that the extra space on the chip required by the resistors and dummy preamplifiers does not overwhelm the space saved through use of smaller devices. Selection of the correct resistor values is important when designing a resistor network — usually this comes down to selecting the correct ratio of lateral to longitudinal resistors [IE08][PA03] — to obtain the best performance improvement with a resistive network.

Interpolating flash ADCs use resistive averaging of preamplifier outputs to help reduce the number of preamplifiers that feed the comparators [IE08][PA03]. The resistive network is essentially identical to the network used in an averaging network, with the difference being in how the resistive network is used. Unlike an averaging ADC, in which the resistive network is used to average out random offset currents or voltages, the network is used in an interpolating ADC to approximate the output of a preamplifier that has been removed from the preamplifier array [PA03]. This indicates that for averaging purposes low "isolation" between the array elements is beneficial, whereas for interpolating purposes the points of the array require high isolation to preserve the zero crossing points of the removed preamplifiers [PA03]. This is usually specified by an *interpolating factor*, which gives the ratio of the number of preamplifiers in the interpolating case compared with the non-interpolating case (e.g., a factor of 2 would indicate that one-half of the preamplifiers from the non-interpolating flash are used in the interpolating flash) [IE08]. Reducing the number of preamplifiers through interpolating results in a reduced input capacitance and can therefore improve speed. In addition to allowing reduced input capacitance, interpolating averaging flash ADCs will usually result in reduced effects from input-referred voltage offsets (even if input capacitance is not reduced, e.g., if half as many preamplifiers with twice the input capacitance are used compared to a full flash ADC), allowing smaller devices to be used and hence improving bandwidth [IE08].

Flash designs that use resistive banks between the preamplifier bank and comparator bank may suffer from voltage-pulling of preamplifiers near the voltage rail edges (near ground and the supply voltage) toward the rest of the preamplifier array [IE08]. This results in increased contributions to non-linearity from edge preamplifiers [PA03]. This is often solved by the addition of dummy preamplifiers that serve to absorb this pulling effect but do not actually contribute to output [IE08][PA03]. These dummy preamplifiers, though necessary, have the effect of reducing voltage headroom, so their use should be limited to the minimum necessary. This indicates that resistive networks allow an increase in either speed or accuracy (or smaller increases in both) at the expense of input voltage range.

A set of reference voltages must be generated for comparison to the input voltage. Reference voltage generation may require a substantial portion of the ADC's power budget [TK08], unless very large resistances are used, in which case the design may require a large area on chip and may have more parasitics. A *resistive ladder* is frequently used to generate these reference voltages. This is a straightforward approach in which the resistor values are chosen for correct incremental voltage drops to produce the correct reference voltages. A challenge with this design is the inherent uncertainty in resistance values for passive resistors. Rather than implementing separate resistors, it may be possible to simply periodically tap a metal line, which may result in space savings at the expense of very large current consumption, unless a sample-and-hold approach is used for the reference voltages generated from the metal line. If power consumption is a concern, therefore, metal resistors are unlikely to be a viable option for the reference voltage ladder. Voltage references may be generated without resistor ladders by using active devices. For example, Shahramian et al. use a set of cascode offset amplifiers to generate a set of reference voltages for input to their comparator array [SVC09].

The type of encoder used in flash ADCs is generally a thermometer to binary encoder (this is sometimes called a decoder). The encoder for flash ADCs can be a major bottleneck for both power and speed [ARS06][MRM09]. Different methods exist for converting (decoding) the thermometer code to binary such as the *ones-counter*, multiplexed-based (mux-based), *Wallace tree*, *fat-tree*, or logic-based methods [MRM09][SV07].

Ones-counter decoders work by counting the number of ones in the thermometer code to convert to binary [SV07]. Various ones-counter topologies may be used, but the Wallace tree counter is a popular choice [SV07][MRM09]. Bubble errors are handled and suppressed globally, resulting in lower timing bubble error susceptibility than read-only memory (ROM) or mux decoders [MRM09][SV07]. Wallace tree decoding may be pipelined for improved speed.

The ROM decoder works by using the thermometer code from the comparators to address into a ROM with the correct outputs for each input already pre-stored for memory look up [MRM09]. This technique suffers from bubble error susceptibility because a small error in the thermometer code results in the reading of the wrong row in the ROM [MRM09][SV07][ARS06]. An implication of this susceptibility is the requirement for a bubble error correction circuit whose complexity increases as the sampling rate increases [SV07]. The advantages of the ROM decoder include easy layout and design [SV07] in addition to speed.

The logic-based decoder is based purely on logic gates (e.g., AND and XOR gates) [MRM09]. While simple and fast, an asynchronous logic-based design will be susceptible to errors (such as random glitches to 0) caused by timing mismatches, especially as the number of bits increases. This design is not appropriate unless a very small number of bits is used.

Mux-based decoders are based on a recursive comparison algorithm in which the thermometer code is successively split into smaller thermometer codes [MRM09]. Mux-based methods provide some level of bubble error suppression but are not as bubble resilient as ones-counter decoders [SV07]. They have the advantage of requiring smaller area footprints on the chip (up to 40% less area than a ones counter [SV07]) and a smaller number of transistors than ROM, fat-tree, or ones-counter decoders [MRM09].

The fat-tree decoder is the fastest and one of the least power hungry methods for converting the thermometer code to a binary code, and it is based on a conversion to a 1-of-N code followed by a conversion to binary using all OR gates [MRM09][ARS06]. For example, a 5-bit unary code of 0u00111 (where the leading "0u" denotes a unary code) would first be used, with the 1s counted across the code representing the unary value.

6.7 Trends and Architectures for MmWave Wireless ADCs

Here the code yields three 1s in the unary code (i.e., the thermometer code), represented as 0u00111. Then, the thermometer code would be converted to a binary code to represent the actual voltage sample. For this example, the thermometer code would be converted by a fat-tree decoder comprised of numerous gates and converted into a binary code of 0b11 to enable it to be processed, as it is conventional to process data once it is in the digital domain in binary form (or in a closely related form, such as hexadecimal). The fat-tree decoder does not rely on a clock circuit [ARS06]. A disadvantage of this design is its layout complexity involving many OR gates and susceptibility to synchronization errors that may require a pipelining approach for very high sampling rates [ARS06]. These synchronization errors will result in an increased number of bubble errors and faster roll-off of ENOB with increasing input signal frequency [ARS06]. Although not widespread, there may be an advantage to using experimental decoder designs, such as neural networks, in certain circumstances as wideband mmWave wireless systems proliferate [ARS06].

Folding-flash ADCs are flash ADCs in which each comparator may be used to provide multiple thermometer code outputs, resulting in fewer required comparators for a given resolution compared with a standard flash design [SV07][OD09]. Folding-flash ADCs may be advantageous for lower power consumption and smaller area compared to a regular flash ADC [OD09], but they have the disadvantage of lower conversion rates [CLC+06]. Folding designs require the use of folding amplifiers between preamplifiers and comparators [OD09]. The thermometer-to-binary encoder used in a folding-flash design must take this dual-use nature into account, for example, through the use of a folding Wallace tree design [SV07]. The basic operation of a folding flash ADC is presented in Fig. 6.22. The input voltage is input to both a coarse ADC that produces the most significant bits and to a folding amplifier and fine ADC cascade to produce the least significant bits.

The folding circuit in the folding-flash ADC can be thought of in terms of a modulo mathematical operation, which also explains the advantages of the flash ADC design. For example, suppose there is a need to convert signals from 0 to 2 V with a resolution of 0.1 volts. Ordinarily, this would require $\log_2(2/.1) = 5$ bits of resolution; and $2^5 - 1 = 31$ comparators would be needed with a standard flash architecture. Now, with a folding-flash design, suppose the coarse ADC converted the same inputs from 0 to 2 V with a resolution of 0.5 volts, requiring $\log_2(2/.5) = 2$ bits of resolution and $2^2 - 1 = 3$

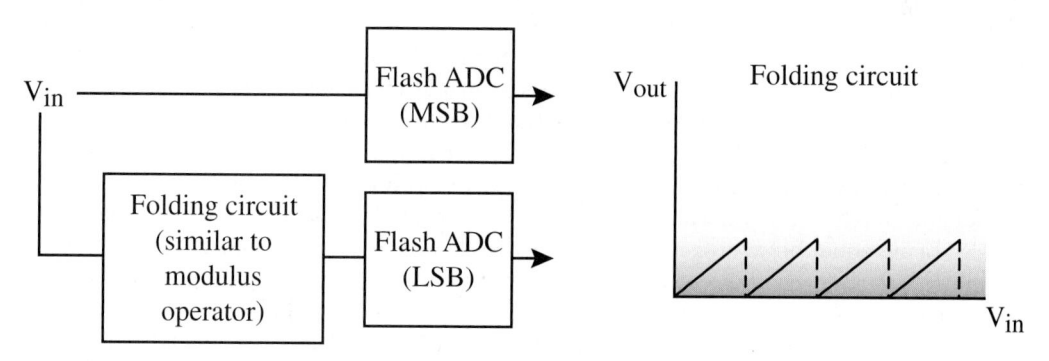

Figure 6.22 The basic design of a folding-flash ADC.

comparators. The folding amplifier and fine ADC provide the rest of the conversion as follows: The folding circuit indicates in the analog domain the value of the input, modulo the resolution of the course ADC, or modulo 0.5 volts in this example. The fine ADC converts the output of the folding amp to the digital domain with a resolution of 0.1 volts, requiring $\log_2(0.5/0.1) = 3$ bits of resolution and $2^3 - 1 = 7$ comparators. The folding-flash design therefore requires a total of only $3 + 7 = 10$ comparators versus 31 for the original design, a substantial savings of more than 60%.

The *folding factor* of the folding amplifier, which is equal to the peak voltage input divided by the resolution of the coarse ADC, indicates how many comparators are needed in the folding-flash design. The lowest number of comparators for a folding-flash ADC is upper bounded by $\frac{\sqrt{\Delta V}}{V_r}$ where ΔV is the overall desired maximum voltage range of the input signal and V_r is the smallest voltage resolution desired by the ADC. The folding factor also determines the amount of preprocessing hardware required in the form of folding amplifiers, with higher folding factors requiring more preprocessing.

The transfer function of the folding amplifier indicates that it is highly non-linear and will result in frequency upconversion of large-scale signals [OD09]. This results in a reduced maximum input frequency for a folding flash compared with a standard flash ADC [OD09]. Traditional methods of implementing the folding amplifier using differential pairs of alternating polarity may be unrealizable in deeply scaled processes with low supply voltages [OD09]. Stacked current steering designs have also been suggested and implemented in processes with gate lengths as small as 0.13 μm CMOS with supply voltages of 1.5 V [OD09]. These have demonstrated the potential for low power, low supply voltage operation.

The input capacitance of a flash ADC grows exponentially with the number of comparators [IE08], which is one reason that this design is not suitable for very high frequency input signals or very high-resolution ADC implementations. Fig. 6.23 summarizes the various errors that can be encountered by a flash ADC.

The effect of transistor offsets can dominate the performance of flash ADCs [US02][CJK+10]. Offsets result in random fluctuations in the reference voltage seen by each comparator, and large offsets therefore directly increase conversion error. The most important consideration is the offset relative to the size of the least significant bit (the offset needs to be smaller than the LSB). This indicates that the resolution of a flash ADC is determined largely by achievable offset (e.g., lower offsets result in lower allowable values for the least significant bit and hence higher resolutions [PA03] in addition to acceptable input capacitance). To obtain a high-yield flash ADC design, it is essential that the standard deviation in offset across multiple chips be much smaller than the least significant bit [US02]. This indicates that as supply voltages decrease, voltage offsets must also decrease if high yields are to occur for ADCs in fabrication. Unfortunately, the standard deviation between transistor threshold voltages, which is largely responsible for mismatch, tends to scale as $1/\sqrt{WL}$, indicating that as devices become smaller, all else being equal between generations, mismatch will increase [US02][TWMA10]. The mismatch affects each stage of the ADC from preamplifiers to comparators, so that at each stage errors can accumulate. Another challenge to design is that as devices operate closer to moderate or even weak inversion, the *Pelgrom mismatch model* predicts that they will also exhibit greater mismatch. Since supply voltages have fallen more quickly than threshold voltages, transistors are more likely to operate in moderate or

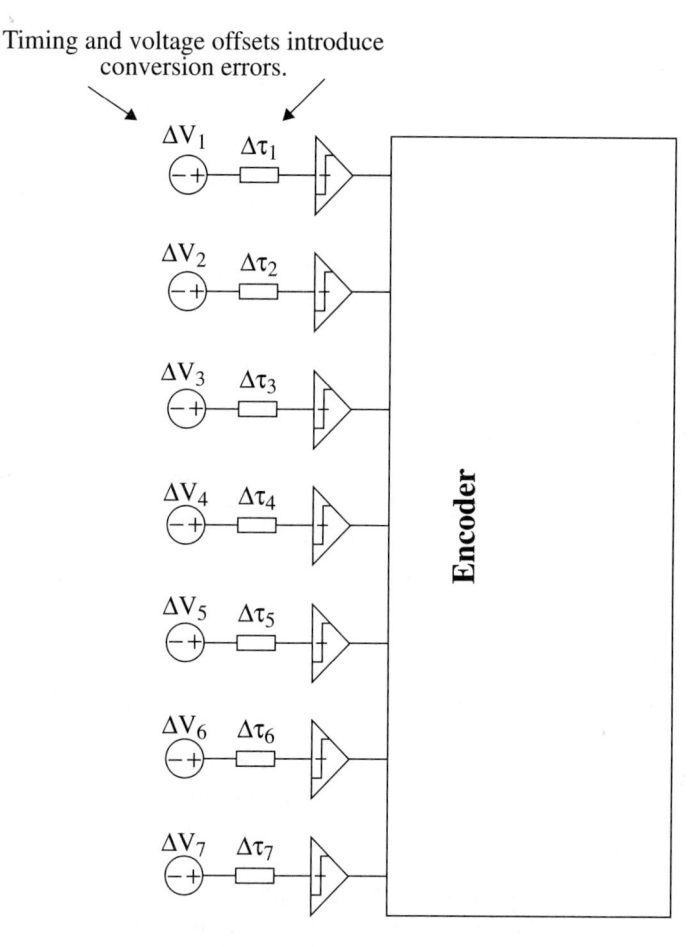

6.7 Trends and Architectures for MmWave Wireless ADCs

Figure 6.23 Timing and voltage offsets are the sources of conversion errors in a flash ADC.

weak inversion [Yos07]. Decoupling capacitors may be used to improve resilience to offset or mismatch errors [CLC+06], but these inevitably reduce speed or increase power consumption.

The equations governing voltage offsets due to device mismatch [US02][PDW89] make it clear that flash ADCs should not be constructed with minimum size transistors, but rather should use larger transistors to achieve high-accuracy conversions. An implication of this is that the input transistors to a flash ADC may offer high capacitance, reducing the bandwidth capability of the design. One technique that can be used to allow smaller input transistors, while reducing the effects of mismatch, is to use resistive averaging [IE08]. Folding and interpolating flash designs also result in a decrease in input capacitance compared with standard flash ADCs [IE08], thus increasing the bandwidth. Such techniques that help facilitate the use of smaller transistors will increase the frequency of the highest frequency signal that may be converted with a flash ADC. Auto-zeroing approaches, which are not covered here, may also be used to reduce mismatch sensitivity of flash ADC designs.

Increasing the resolution of a flash ADC requires an increase in the power or a decrease in the speed of the ADC. In either case, increased resolution is achieved by adding more comparators. Research into means of reducing power consumption for high-resolution flash ADCs is ongoing. Proposed methods include reducing the number of times a comparator must operate without affecting the output (e.g., if a comparator adds a leading zero as in 010 vs. 10) and variable resolution designs in which resolution is increased only in response to low received signal power at the ADC input [VKTS09]. Another possible means of reducing current or power consumption of flash ADCs is by using very high density resistors for the generation of voltage references. Folding and interpolating flash ADCs offer the means of achieving higher accuracy with lower power consumption than a standard ADC [OD09].

6.7.5 ADC Case Studies

To understand the recent evolution of ADCs, and promising approaches for future ultra-wideband mmWave wireless systems, it is instructive to consult recent research contributions. In 2009, Veeramachanen et al. [VKTS09] presented a 4-bit ADC with a sampling rate of 1-2 gigasamples per second (Gs/s) simulated in 65 nm CMOS. The design was based on the *Threshold Inverter Quantization (TIQ)* method, in which rather than making use of a resistor ladder, the input transistors of the comparators was fabricated with various aspect ratios (W/L) to have different threshold voltages. The comparators were based on a cascaded inverter design, in which the ratio of PMOS widths to NMOS widths set the threshold voltage of the inverter (so that it could operate as a comparator). The ADC used a variable resolution design to facilitate lower average power usage. This is achieved through the use of a peak detector to determine the peak voltage of the input signal and to enable or disable input inverter-based comparators [VKTS09].

In 2006, Cho et al. presented a 6-bit interpolating flash ADC in 0.18 μm CMOS that made use of wideband track-and-hold amplifiers to improve accuracy [CLC+06]. The design made use of built-in current and voltage references to improve noise performance. The track-and-hold amplifier was based on a PMOS source follower design. The amplifiers made use of both the sample clock and the sample clock complement to switch between sampling and holding operations [CLC+06]. The flash comparators made use of a differential difference amplifier followed by a latch.

In 2009, Shahramian et al. [SVC09] implemented a 35 Gs/s 4-bit flash ADC in 0.18 μm SiGe BiCMOS. In order to achieve such a high sampling rate, they used a tapered buffer technique to drive the comparator capacitive loads. This technique was required to drive the comparators fast enough to achieve the desired sampling rate, but resulted in very high power consumptions and large amounts of space dedicated to the amplifier tree. Also, this technique, if not used carefully, will greatly expand the thickness of metal traces needed to supply the currents needed by the largest amplifiers in the tree. The high power consumption of this ADC of 4.5 W reflects the power needed to drive a capacitive load at such a high rate and fits expectations from (6.22). The design did not include a thermometer-to-binary encoder, but rather directly reconverted the thermometer code out of the comparator bank to analog using a high-speed DAC for testing purposes.

In 2010, Cao et al. [CZS+10] presented a complete analog front-end that included a 10 Gs/s time-interleaved flash ADC. Each sub-ADC in the time-interleaved design relied on a full-flash architecture. The design included 6.25-bit current DACs at the input of each comparator for offset calibration. The DAC offset compensator for each ADC was capable of providing more than 10.8 LSB of offset compensation.

In 2008, Louwsma et al. presented a time-interleaved 10-bit, 1.35 Gs/s ADC composed of successive approximation sub-ADCs [LvTVN08]. The ADC used 16 sub-ADC channels for an acceptable tradeoff between input capacitance and increased sampling rate through parallelism. This fairly high degree of parallelism allowed the use of successive approximation sub-ADCs even with an aggregate sampling rate in the Gs/s range.

In 2010, Chu et al. [CJK+10] presented a 40 Gs/s, 4-bit time-interleaved ADC fabricated in IBM 8HP SiGe technology. Four sub-channels were used to reduce the sampling rate of each channel to 10 Gs/s. A two-level sample-and-hold design was used in which a 40 Gs/s sample-and-hold amplifier was followed by a 10 Gs/s sample-and-hold amplifier on each sub-channel. Each sub-ADC following was a 4-bit differential flash ADC for maximum speed. The number of preamplifiers used for each sub-ADC was minimized to reduce input capacitance through the use of interpolation. The encoder used a one-of-n architecture.

In 2008, Tu et al. [TK08] presented an 8-bit 800 MS/s time-interleaved ADC with flash sub-ADCs in 65 nm CMOS. Each sub-ADC had 4 bits of resolution. The dynamic latched comparators for each channel used shared preamplifiers to improve speed and prevent kickback noise. This comparator design allows for fast regeneration time. In 2010, Greshishchev et al. [GAB+10] of Nortel presented a 40 Gs/s ADC in 65 nm CMOS that relied on an interleaved architecture. The case studies listed above give insights into future architectures and design approaches that may be used to offer tens of Gbps data rates in wireless communication systems of the future.

6.8 Digital-to-Analog Converters (DACs)

Many of the same concepts taught in ADCs are also applicable to the design and performance limitations of digital-to-analog converters (DACs). DACs translate digital signals from memory, from a data bus, or from a processor into analog signals that may be passed onto the analog signal chain of a transmitter. Millimeter wave (mmWave) digital-to-analog converters (DACs) for first-generation mmWave communication devices will likely only require moderate resolutions (i.e., moderate dynamic ranges) due to the low spectral densities used in these early devices. The sampling rates of mmWave DACs, however, will need to be high enough to be operable with very wide channel bandwidths, and this presents particular challenges. Digital values may reside in memory and can be processed in computers, but they must be converted into analog signals such as analog in-phase (I) and quadrature (Q) baseband waveforms, or as polar baseband waveforms, which are then modulated onto a mmWave carrier and transmitted over the air. Thus, the role of the DAC is to take digital codewords from a digital processor, from a multiplexer, or from memory, and convert the codewords into analog signals that may be applied to the analog transmitter and sent through the antenna into the channel. Once a digital signal processor or memory provides the particular digital words in a fashion that properly represents the desired time-varying signal at the desired clock rate, the DAC then converts the digital codewords into analog signals using two distinct processes: reading the incoming digital codeword, and then mapping the incoming digital codeword to an analog voltage signal with a constant voltage over a particular clock period. Generally, a third process, low-pass filtering, is applied to the output of the DAC, often with a software-controllable low-pass filter based on the particular data rates or modulation bandwidth specifications to be supported by the analog output signal of the DAC.

The role of a DAC is to capture a digital codeword and to convert the codeword into an analog signal over a very small time aperture (the sample time interval), so as to produce, as accurately and faithfully as possible, a brief duration of an analog signal represented by the digital codeword. The DAC continually reads in digital codewords and produces many sequential analog finite-time waveforms over time. The ability of a DAC to reconstruct an analog waveform with high fidelity is critical to modern wireless communications, particularly for generating waveforms at the transmitter that comply with standard requirements and spectral mask requirements. The switching time (and the sample time) of the DAC is a function of the smallest bit or symbol rate used in a transmitter modulation scheme, thus it follows directly that future mmWave multi-gigabit per second (Gbps) transmission rates will require DACs that can create analog waveforms that have changing signal amplitudes within time spans that are much less than a nanosecond. This requirement means that slew rates must be extremely high, in the gigahertz range, and time apertures must be extremely small, in the sub-nanosecond range.

6.8.1 Basic Digital-to-Analog Converter Circuitry: The Current DAC

The most basic DAC is the *current DAC*, which relies on digital switches to switch a current through a resistor on or off, thus changing the voltage across the resistor. This is illustrated in Fig. 6.24. Millimeter wave (mmWave) digital-to-analog converters (DACs) will need sampling rates high enough to be operable with channel bandwidths and signaling frequencies in excess of several gigahertz [ZS09b][WPS08]. A resistor ladder, in which the resistor in Fig. 6.24 is expanded to a ladder configuration, is an approach for implementing a DAC [LvTVN08].

The DAC should be capable of supporting an adequate sampling rate to, at a minimum, support a Nyquist sampling criterion. Often, however, it is advantageous to use oversampling to overcome the impact of device mismatches on the dynamic output of the DAC. Thus, oversampling is an effective way to nullify the dynamic range limitation and resolution errors to due mismatch [WT99].

A high ratio of *signal power to noise and distortion power* (SNDR, also called SINAD) is a very important performance metric of high-speed DACs [SWF00]. The SNDR indicates the output power at the fundamental harmonic for a sinusoidal input relative to the power of all distortion and spurious signals within the Nyquist frequency [SWF00].

A high *spurious-free dynamic range* (SFDR) is one of the most important design goals for a high-speed DAC [RSAL06][SWF00]. The SFDR reflects the linearity of the device and indicates the difference between the RMS power of the fundamental harmonic

6.8 Digital-to-Analog Converters (DACs)

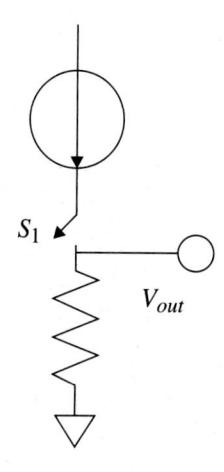

Figure 6.24 A current DAC is one of the most basic DAC designs.

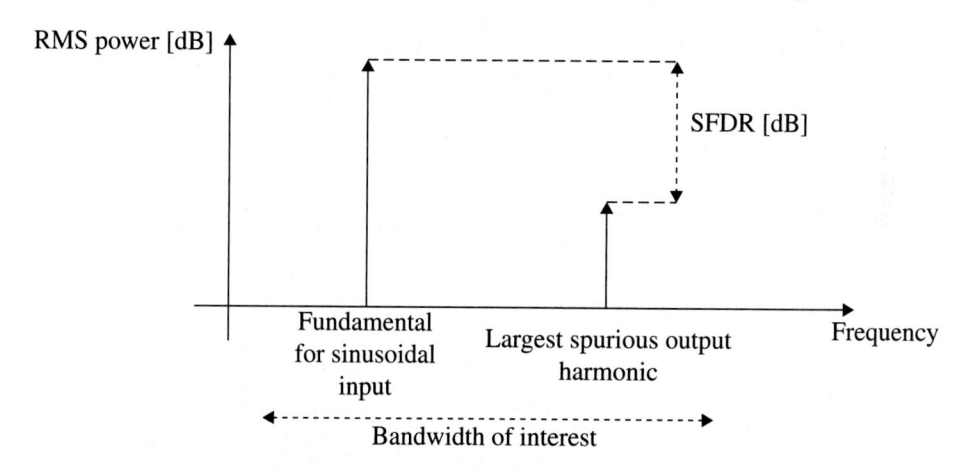

Figure 6.25 Spurious free dynamic range (SFDR) indicates the relative power of the fundamental output frequency and the largest spurious output signal [SWF00]. The bandwidth of interest is often specified as the Nyquist frequency of the DAC based on its conversion rate.

for a sinusoidal input and the largest spurious harmonic within a certain bandwidth of the fundamental harmonic [SWF00][WT99]. This is illustrated in Fig. 6.25. The major challenges to overcome for DAC designs to achieve high SFDR include finite output impedance, gain-dependent output impedance, major carry glitches, charge feedthrough, common switch voltage spikes, and supply noise [RSAL06][WT99]. Differential designs can be used to improve SFDR [WT99].

High-speed DACs require circuitry with fast settling times so they are not limited by static error performance. DACs are susceptible to both dynamic and static errors. Static performance metrics for DACs include offset error, gain error, integral non-linearity (INL), and differential non-linearity (DNL). Static errors indicate the errors in DAC outputs after the output has settled [WT99].

Dynamic DAC errors such as glitches are extremely important in high-speed DAC design [CRW07][SWF00] and outweigh the importance of static errors [WT99]. Glitch errors occur when the timing of the DAC is such that the applied codeword (i.e., the state of the DAC) is not completely settled in the DAC circuitry, resulting in a seemingly random signal output for an intended codeword. A *glitch* is a dynamic effect caused by uncertainty of relative switch outputs (i.e., different transition times for switches in the DAC) after digital control signals have been sent, and it is undesirable because it adds considerable noise to the output [Koe00][CRW07][WT99]. Dynamic errors such as glitches relate to transient errors in the DAC output following a change in the DAC state [WT99]. In DAC design, it is desirable to have as low glitch operation as possible [SB09][YS02], and the amount of glitch limits the output signal accuracy of high-speed DACs [CRW07]. A glitch is illustrated in Fig. 6.26, in which a transition from zero output to a mid-level output is preceded by a very rapid voltage peak. For current steering designs, the glitch energy (usually specified in V·sec) largely determines the noise performance of the DAC [YS02]. The glitch energy is a reflection of the time and strength of voltage variations on a transistor switch after it has been instructed to change stage. Most current steering designs will rely on high-speed deglitching circuits to provide clean outputs. Because binary designs are most susceptible to glitches (as opposed to unary[2] current steering DACs), it is also common to use a unary or segmented (i.e., partially unary) design to improve immunity to glitches [CRW07]. In order to achieve low glitch, it is advisable to minimize the amount of signal feedthrough from digital portions of the DAC to the output current sources [SB09].

A less important dynamic error than glitch error is settling error. Dynamic settling errors indicate the difference in the DAC output from the correct value before the output has completely settled [CRW07]. Fig. 6.27 illustrates dynamic settling errors. We may

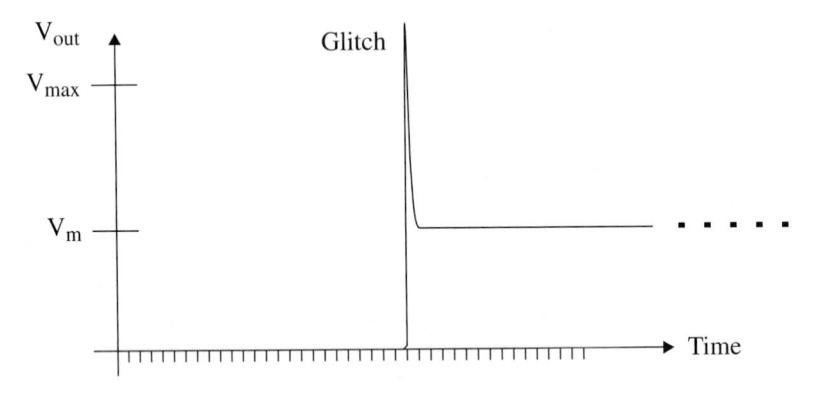

Figure 6.26 Glitches are caused by non-uniform switching times of DAC switches [CRW07].

---

2. Unary weighting is such that each bit is weighted equally. For example, in a unary DAC design, the control byte 0u11111111 would correspond to providing 8 times the current as byte 0u00000001 (where "0u" stands for unary).

6.8 Digital-to-Analog Converters (DACs)

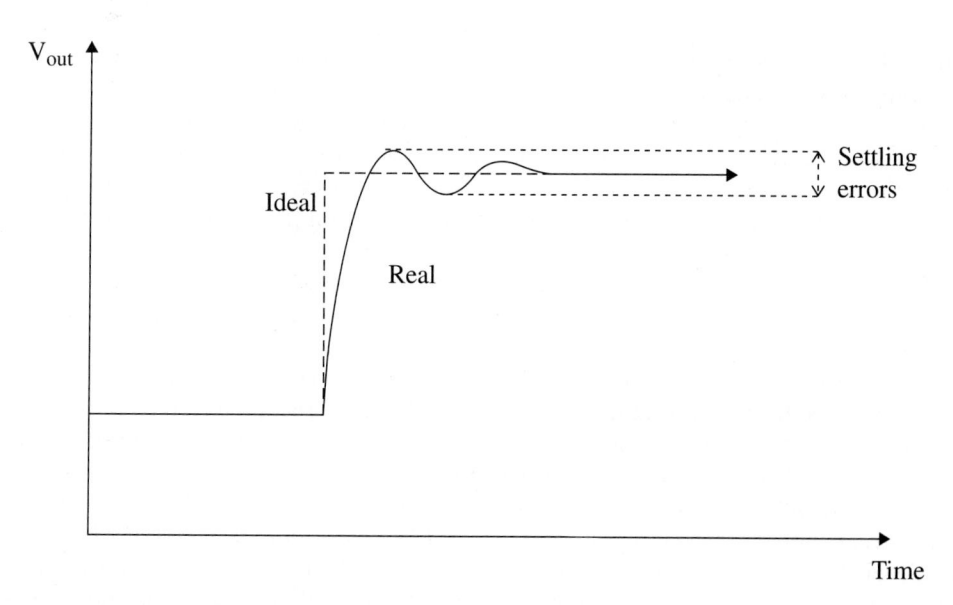

Figure 6.27 *Settling dynamic errors* refers to settling errors before the output has reached its final value.

associate an RC time constant with the time required to settle to the final value, and this time constant should be less than the sample period of the DAC to avoid large output errors [WT99]. Minimizing the glitch energy in the DAC will have the added benefit of reducing settling errors because large glitches can increase the settling time [WT99].

Fluctuations in output voltage due to switching noise from digital components in the DAC is undesirable, as this degrades frequency-domain performance [SWF00]. In certain circumstances, it is advantageous to use isolating driver circuits to reduce the impact of digital noise or loading on the analog output [SWF00]. These circuits may include, for example, extra cascaded elements to add a series capacitance between the digital input and analog output to result in reduced coupling capacitance.

As in many analog circuit designs, using a differential design, as opposed to single-ended circuity, is favorable for DACs because it facilitates common mode rejection, eliminating noise due to voltage offsets or mismatch, while reducing output impedance requirements [RSAL06]. The device fabrication of DACs have important impacts on DAC performance at extremely fast clock rates. Long-channel MOSFET devices suffer from much less channel length modulation than very short channel devices [Raz01, Chapter 2]. This results in increased output impedance for long-channel devices, making them preferable to very short-channel devices for the current sources in a DAC [RSAL06]. Long-channel devices have the disadvantage of possibly limiting the output swing [RSAL06].

The current steering DAC is a common architecture known for its high speed, high resolution, and acceptable power consumption, and it is commonly used in communication circuits [SB09][WPS08]. This stems in part from the ease with which current outputs may be scaled (i.e., by the aspect ratio, "W/L," of a transistor) and added [WT99]. The design is also advantageous for high-speed operation because it is open-loop and does not require high impedance switches [WPS08]. The basic design of a current steering DAC is to base the number of current sources feeding an output load based on the digital value entering the DAC [RSAL06]. Current steering DAC performance is limited largely by voltage fluctuations in current source output nodes from timing errors, feedthrough of control signals to output lines, and the degree of synchronization that exists between control signals that control different switching transistors to control current sources [YS02].

A high output impedance for the current sources used in DACs is extremely important because it has a direct impact on SFDR and SNDR of the DAC — both can be estimated with expressions containing only the output impedance, the load resistance, and the value of the digital input [SWF00][WT99][WPS08]. Increasing output resistance will usually improve both SFDR and SNDR [SWF00][WT99]. Techniques to increase output voltage, such as gain boosting or operating at increased drain-source saturation voltages for output transistors in current steering DACs, may be used to improve performance [SWF00]. The output impedance of an actual current source is not infinite, and the non-ideal current source can be modeled as shown in Fig. 6.28 (figure from [SWF00]) [WT99]. The SFDR for a binary current steering DAC can be estimated if the ratio of the current source output impedance to load impedance is a large decibel value according to (6.24) [WT99]. R_{ratio} is the ratio of the unit output impedance (i.e., the output impedance of the smallest current source in the DAC) to the load impedance. X_{ac} is the ac magnitude of the input signal. The equation assumes that the dc input magnitude is roughly equal

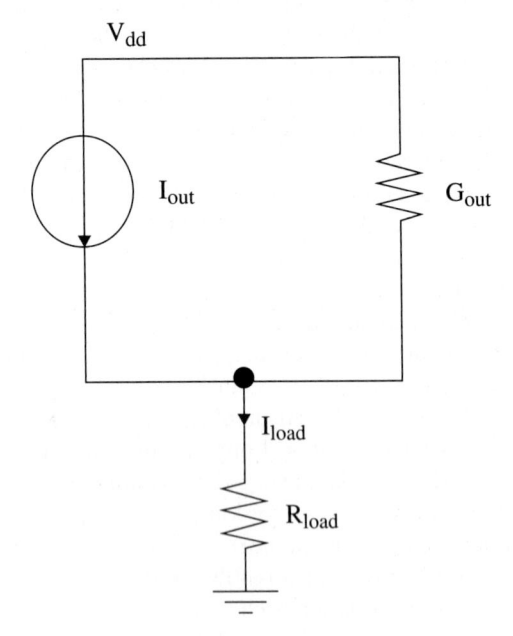

Figure 6.28 A model for a non-ideal current source with a non-infinite output impedance [from [SWF00] © IEEE].

6.8 Digital-to-Analog Converters (DACs)

to the magnitude of the ac input.

$$\text{SFDR (dB)} \approx 20 \log R_{\text{ratio}} - 20 \log \frac{X_{ac}}{2} \qquad (6.24)$$

The source output impedance of a MOSFET is approximated by $\frac{1}{j\omega C_s + g_m}$, where C_s is the source capacitance and g_m is the device transconductance. The expression indicates that a high source capacitance will result in a low output impedance for the MOSFET at high frequencies, therefore MOSFET current sources used in DACs should have minimized source capacitance [WPS08].

Mismatch between devices used in current sources determines achievable DAC resolution and large mismatch will degrade the linearity of a DAC's output — degrading INL and DNL [WT99][WPS08]. The signal-to-noise-and-distortion ratio (SNDR) and spurious-free dynamic range (SFDR) of the DAC output will also decrease with increasing mismatch (and therefore with shrinking device size [PDW89]) [WT99]. Smaller devices, depending on the quality of manufacturing, will suffer from greater mismatch [PDW89]. This is one reason why it is usually not advisable to use minimum-size devices for current sources. To select transistor size, the acceptable current mismatch and average current from current sources should be specified. Well-known equations exist that relate output current and mismatch to device size [PDW89][WPS08], and with these equations and specifications it is possible to select a starting size for design.

Long-channel device current sources will be less susceptible to voltage fluctuations than short-channel devices, and will also have higher output impedances. In certain cases it may not be possible to use long-channel devices, but a cascoded current source may be used instead for the same benefit [YS02][WPS08].

The switches used to "activate" or "deactivate" a current switch must be well designed to avoid introducing non-linear errors. A key is to determine and design for the correct control signal swing to keep all switches as linear as possible by maintaining a high output impedance of the DAC up to the DAC's Nyquist rate [WPS08]. The sizes and biasing of MOSFETs used for switching should be carefully selected with regard to the output voltage swing and load impedance to ensure that they do not introduce non-linear distortion [WPS08].

In current steering DACs, it is desirable to have low switch-on resistance and parasitic capacitance [SB09]. Therefore, an optimum device size should be found for optimum resistance and capacitance. The current steering DAC architecture relies on current sources proportional to the value of each bit position in a binary word (i.e., each successive current source is twice the previous current source) [RSAL06][WPS08]. Fig. 6.29 illustrates the basic design of a binary DAC. The binary digital data is converted by mapping each bit position to a switch controlling its corresponding current source. An advantage for binary designs is lower area and fewer control signals, which increases speed and reduces area of the DAC [WPS08]. Although this design is not area intensive, it can suffer from poor dynamic and static linearity due to mismatch errors among the current sources because mismatch errors do not scale in the same way as output current (current scales as W/L and mismatch as $1/\sqrt{WL}$) [SB09][WPS08]. Pseudo-segmented designs in which

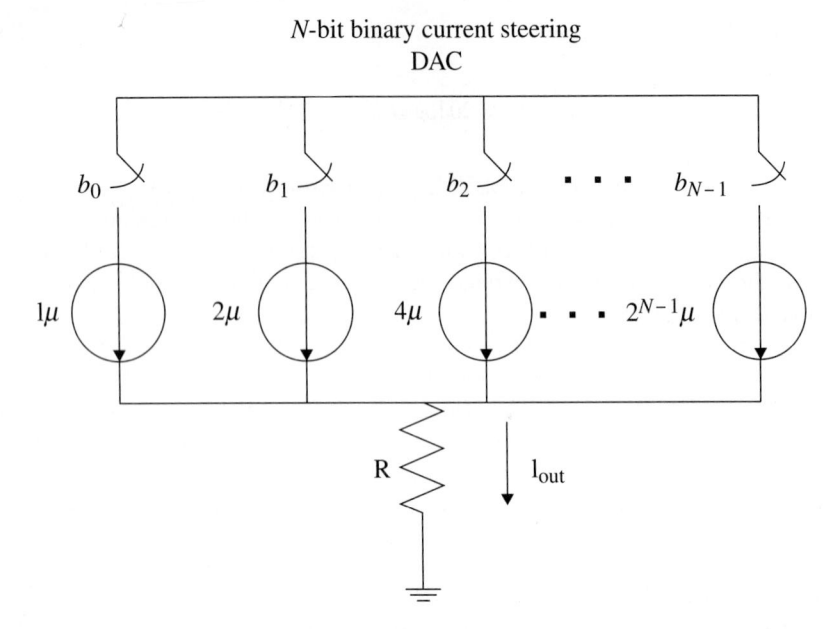

Figure 6.29 The basic design of a binary DAC. Each current source is sized proportionally according to the significance of the binary bit that controls it.

the binary-scaled current sources are composed of aggregations of uniform current sources may be used to improve the performance of this design [WPS08].

The delay from the input to all different outputs of the decoder used in unary or segmented DACs should be as short and uniform as possible [YS02]. Non-uniform delay from input to output current source control switches will severely degrade linearity [WPS08].

The typical DAC architecture is basically the inverse process of the ADC. The DAC first must convert a digital binary word to a thermometer code. Each bit of the thermometer code is used to switch one of a uniform array of current sources [RSAL06]. This uniformity results in simple analog layout compared with a binary DAC, but the need for a decoder increases the digital complexity significantly [WPS08]. Uniformity also has the advantage of improving matching and linearity and reducing the error caused by any one switching error [RSAL06][SB09][WPS08]. The current sources and their glitch performance have a large role in determining the performance of the DAC [YS02]. The disadvantages of unary DACs are increased area, reduced speed, and high power consumption compared with a binary DAC due to the need for a decoder [WPS08].

A segmented approach uses binary encoding to switch the least significant bits and unary encoding for the most significant bits [RSAL06]. This design represents an intermediate tradeoff between analog and digital complexity and improved linearity versus increased area and power [SB09][WPS08]. Use of a decoder to convert from binary to thermometer code for the most significant bits adds considerable complexity to the digital design [RSAL06]. The performance of this design is strongly impacted by the number of

6.8 Digital-to-Analog Converters (DACs)

bits converted using binary coding and the number converted using temperature unary coding [SB09]. As the percentage of bits using unary coding — the percentage segmentation — increases, the area dedicated to digital components increases exponentially, but the area required for a given differential non-linearity (DNL) decreases exponentially [SB09]. These two competing trends, together with the insensitivity of integral non-linearity (INL) to device area, can be used to determine an optimal segmentation percentage [SB09]. In order to achieve uniform delay between binary and unary portions of the DAC, delay circuits should be added to the binary path to match the delay caused by the binary-to-unary decoder used in the unary path. The decoder used in unary or segmented DACs to convert from binary to unary coding schemes can be a major bottleneck for speed improvements [SWF00].

6.8.2 Case Studies of DAC Circuit Designs

It is instructive to review recent contributions in the field of DACs to gain insights into methods for improving speeds to accommodate tens of Gbps in the future. As mentioned earlier, glitches are a troublesome dyamic source of error as speeds are increased within a DAC. In addition to the use of specialized de-glitching analog circuitry within a DAC, it is possible to use digital approaches for partial (band-limited) glitch correction [CRW07]. This technique relies on modifying the digital representation of the output to include de-glitching digital signals [CRW07]. A digital approach to glitch correction is limited by the speed of the digital clock and will usually require a delay to perform the digital de-glitch processing [CRW07]. The approach may be implemented by inserting the DAC into an error-correcting negative feedback loop in which the difference between the DAC output and a smoothed version of the DAC output (i.e., the output after passing through a low-pass filter) are minimized [CRW07].

In 2009, Shahramian et al. [SVC09] implemented a 35 Gbps 15-bit thermometer-code DAC in 0.18 μm SiGe BiCMOS technology. The design relied on a voltage summing technique based on a tree of current mode inverters. The DAC consumed 0.5 W from a 5 V supply and provided an output swing of 3 volts peak-to-peak with a 200 mV resolution.

In 2010, Chu et al. [CJK+10] presented a 4-bit current steering DAC in IBM 8HP SiGe BiCMOS [CJK+10]. The design consisted of a bank of parallel differential pairs with uniform emitter-degenerated tail current sources and "R-2R" resistive current summing loads for differential output.

In 2006, Radiom et al. presented a folded current steering segmented DAC with gain boosting [RSAL06]. To equalize the delay in the conversion of most significant bits and least significant bits, a dummy decoder was added for the least significant bits to match the delay caused by the thermometer decoder used to convert the most significant bits. The architecture was intended for high linearity and high output impedance of current sources. The folded design incorporated common mode feedback and adjustable saturation drain-source voltage of folded transistors. The latter allowed for increased voltage headroom for current sources by reducing the saturation V_{DS} of the folded transistors and resulting in improved PSRR. The common mode feedback provided for compensation of temperature effects.

In 2009, Sarkar et al. presented the design of an 8-bit segmented current steering DAC in 0.18 μm CMOS [SB09]. Sarkar et al. used a segmentation of 6-2: the 6 most significant bits were converted using a unary approach and the 2 least significant bits were converted using a binary approach for least chip area with an integral non-linearity (INL) and differential non-linearity (DNL) of 1 least significant bit (LSB) and 0.5 LSB, respectively. The unary current sources were arranged in a matrix structure with four sub-matrices with independent biasing for most efficient use of area and low INL and DNL. This also allows for reduced complexity of the binary-to-unary encoders. The current source matrix used dummy cells and shuffling to reduce edge-mismatch and gradient mismatches in the cell.

In 2002, Yoo et al. [YS02] presented the design of a 1.8 V, 10-bit 300 Ms/s 8+2 segmented DAC in 0.25 μm CMOS. The DAC featured low spurious noise deglitching current cells for low-noise performance. The binary-to-unary decoder utilized a Binary Decision Diagram (BDD) methodology for improved speed and area compared with a standard decoder.

In 2000, Seo et al. presented the design of a 14-bit 1 gigasamples per second (Gs/s) DAC that relied on an R-2R architecture and gain boosting for high output impedance of current sources [SWF00]. The design also incorporated a double-segmented architecture in which two separate binary-to-thermometer decoders were used rather than one. Seo et al.'s design is shown in Fig. 6.30 and resulted in a substantial reduction in the number of total decoder cells [SWF00]. The eight most significant bits were decoded using multiple emitter coupled logic (MEL) for minimum delay through the decoder.

In 2008, Wu et al. [WPS08] presented the design of a 3 Gs/s 6-bit DAC in 130 nm CMOS. The DAC used a binary weighting pseudo-segmentation architecture. This technique relies on a binary combination of uniform current sources rather than binary-scaled current sources to improve performance compared with a standard binary DAC [WPS08]. Wu et al. used dummy current sources to mitigate edge effects and a common-centroid switching scheme to combat gradient-induced mismatches. These steps helped to improve the resolution of the DAC.

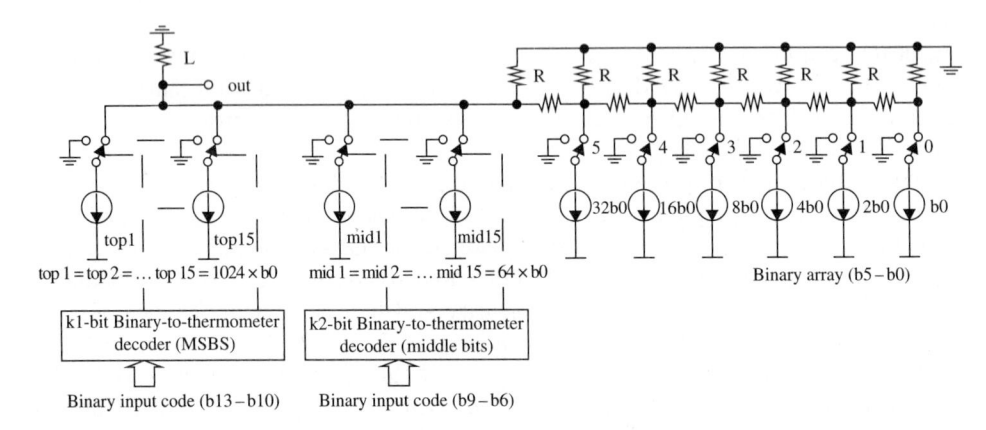

Figure 6.30 A double-segmented DAC approach results in a substantial space savings versus a single-segmentation design in which only one decoder would be present [from [SWF00] © IEEE].

6.9 Chapter Summary

Baseband circuits in future mmWave wireless communication systems will rely upon multi-Gbps data rates, and most likely up to several tens of Gbps by the year 2025. Techniques described in this chapter will surely be parallelized and improved upon over time, and new approaches will evolve, as well. Key to enabling such data rates are reliable, high-fidelity digital-to-analog (DAC) and analog-to-digital (ADC) converters. This chapter has provided a wide range of technical design issues and references to help the reader understand the fundamentals of ADCs and DACs, and the challenges with reaching such high bandwidths. The chapter has also provided many references to help the reader develop a deeper understanding regarding particular circuit design issues that are crucial for enabling the mmWave wireless communications future.

Part III

MmWave Design and Applications

Part III

MmWave Design and Applications

Chapter 7

MmWave Physical Layer Design and Algorithms

7.1 Introduction

In this section we survey physical layer (PHY) signal processing for communication at millimeter wavelengths. Specific examples from existing 60 GHz standards (WPANs and WLANs) are included to enhance this discussion, but the concepts are applicable to many new mmWave systems that will be developed in the future.

Traditionally, PHY algorithm design and selection is completed through simulation and analysis of the linear system model with impulse-response characterizations of the wireless channel [RF91][RHF93][FRT93][TSRK04]. Analog and RF engineers ensure that wireless hardware efficiently maintains an accurate approximation of the linear system model. At millimeter wavelengths, the game has changed dramatically. Although accurate approximation of the linear system model is possible, it may not be cost- or energy-effective. The high carrier frequency, the use of directional and adaptive antennas, and the massive bandwidth all create significant challenges for hardware engineers, and compromises between cost and performance must be made.

Smart selection of PHY algorithms at mmWave carrier frequencies must now take into account various transceiver impairments. The "best" modulation strategy will be found through a performance evaluation in the context of transceiver impairments of varying degrees and the cost associated with correcting these impairments. To enable such a study, Section 7.2 summarizes the salient transceiver impairments and techniques for modeling these impairments in the mmWave linear system model. Next, Section 7.3 presents the potential features of a high-throughput link with high spectral efficiency in the PHY along with the performance of these features in the context of the wireless channel impulse response and the aforementioned millimeter wave (mmWave) transceiver impairments (see Section 3.7.1.2). Afterward, Section 7.4 summarizes possible modulation architectures that enable very low-complexity transceivers, albeit at the cost of spectral efficiency, particularly in the high SNR regime. Finally, Section 7.5 previews PHY designs that could influence future designs if found to be practically achievable and cost-effective.

7.2 Practical Transceivers

Although we do not discuss every potential impairment, this section includes a thorough list of all salient impairments for mmWave wireless transceivers. This includes imperfections such as analog-to-digital conversion (ADC), power amplifier non-linearity characteristics, and frequency reference instability (which produces phase noise). Each subsection will include a discussion of potential consequences for each impairment as well as simple models to enable PHY analysis and simulation. As a preview, consider Fig. 7.1, which shows a complete PHY simulation block diagram that includes the important transceiver impairments discussed in this section.

7.2.1 Signal Clipping and Quantization

Digital communications, as surveyed in Chapter 2, enable the highly configurable, capacity-optimized baseband processing that has produced high throughput links in the modern era of wireless communications. In the receiver, as elaborated in Chapter 6, quantization circuitry in an ADC provides a digital representation of the analog baseband signal. The bit resolution of ADCs in modern wireless receivers (typically 8-12 bits [SPM09]) is sufficient to render very minimal performance degradation. Hence, modulation, equalization, and detection strategies are not commonly evaluated in the context of the number of quantization bits dedicated to each received digital sample.

Chapter 6 discussed the myriad design trade-offs and limitations for mmWave ADC circuits. Because of these concerns, ADC parameters have become an important design consideration. Walden, in his survey of ADC capabilities, was one of the first to recognize that ADC performance is increasingly becoming a design bottleneck [Wal99]. He discovered several important design trends in ADCs including:

- Multi-giga-sample-per-second ADCs with high bit resolution require parallel architectures, such as flash conversion [DSI$^+$08] and time-interleaved conversion

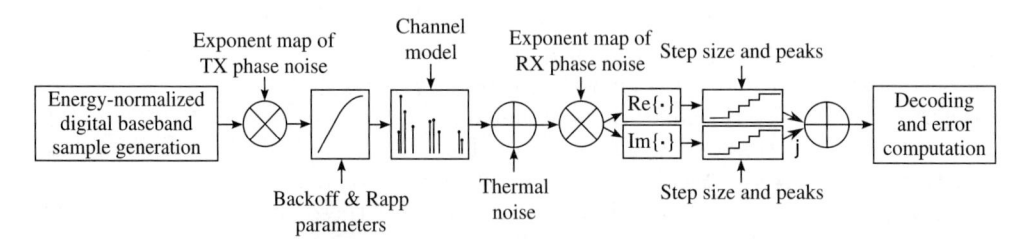

Figure 7.1 Complete PHY simulation with important mmWave transceiver impairments. To create error-rate simulations as a function of SNR, the thermal noise variance is adapted and large-scale channel effects are removed from the candidate set of impulse responses, as described in Section 3.7.1.2. To create error-rate simulations as a function of distance, the thermal noise variance is fixed according to noise figure properties of the receiver, and the channel model includes large-scale path loss as a function of distance, as described in Section 3.7.1.1. Simulations will implement PA non-linearity, phase noise, and quantization effects assuming normalized input power (scaling factors before phase noise insertion and quantization must be included for error-versus-distance simulations because the channel model does not have normalized-energy impulse responses).

[PNS+03]. Parallel architectures are inherently inefficient in terms of device space and power dissipation.

- If one considers the entire suite of available aftermarket ADCs, it can be observed that doubling the sample rate leads to approximately 1 bit less resolution. This is effectively equivalent to stating that as the sample rate increases, the "ENOB" of an ADC (as defined in Chapter 6) decreases due to reduced SNDR.

- Adding a single bit of resolution without reducing the sampling rate or performance of the ADC takes roughly 5 years of technology evolution.

These trends have led many to conclude that "ADC technology is not scaling up at a rate commensurate with that demanded by emerging multi-gigabit systems, thus making it necessary to explore system design with relaxed ADC specifications." [SPM09] As a more specific example, the MAX109, Maxim Integrated Products, Inc.'s multi-gigabit-per-second ADC provides 8-bit resolution at 2.2 giga-samples-per-second, but consumes 6.8 W [Max08]. Hence, we expect that PHY algorithms that are compatible with lower precision ADCs (< 8 bits) can yield substantial savings in power and device cost. It may even make sense to design PHYs that use fewer ADCs (as in the hybrid beamforming architecture [AELH14][EAHAS+12]) or PHYs that work with ultra-low precision ADCs [SDM09][DM10][MH14]. Clearly, it is in the interest of the PHY algorithm designer to consider the impact of bit resolution on the mmWave transceiver that exploits unparalleled spectrum bandwidth.

Two processes model the resultant effect of increasing or decreasing the number of bits in the ADC: amplitude thresholding (clipping) and quantization, as shown in Fig. 7.2. If we assume that quantization is uniform,[1] that is, the distance between each quantization level in the signal amplitude is equal, the resolution of the ADC is determined by the number of bits and the bounds on the signal for both the real and imaginary components (I/Q dimensions are sampled separately). These bounds are determined by the fixed peak-to-peak voltage swing in the ADC for both the I and Q signals *and* the automatic gain control (AGC) algorithm that controls the signal gain before the ADC through a variable gain amplifier (VGA) or variable attenuator. Note that the AGC algorithm (analog or digital) is typically tuned to the type of communication waveform expected by the receiver. For example, waveforms that do not display large signal peaks in both the in-phase and quadrature channels (assuming zero mean), such as constant envelope modulation, have signal amplitudes that are well behaved [AAS86]. The AGC algorithm for constant envelope waveforms can use larger gain settings before the ADC, allowing signals to have larger RMS voltages before quantization. Note that this is a desirable setting because it increases the effective dynamic range of the receiver, that is, this decreases the distance between useable quantization levels, relative to the RMS signal voltage, in the communication waveform and allows higher order modulation symbols to be detected without being impacted by receiver noise.

---

1. It should be noted that non-uniform quantization becomes more useful as the number of bits decreases in the ADC. Unfortunately, not enough is known yet about the performance of a complete wireless receiver (e.g., with AGC, synchronization, channel estimation, equalization, and detection) after non-uniform quantization in the ADC to warrant coverage in this chapter.

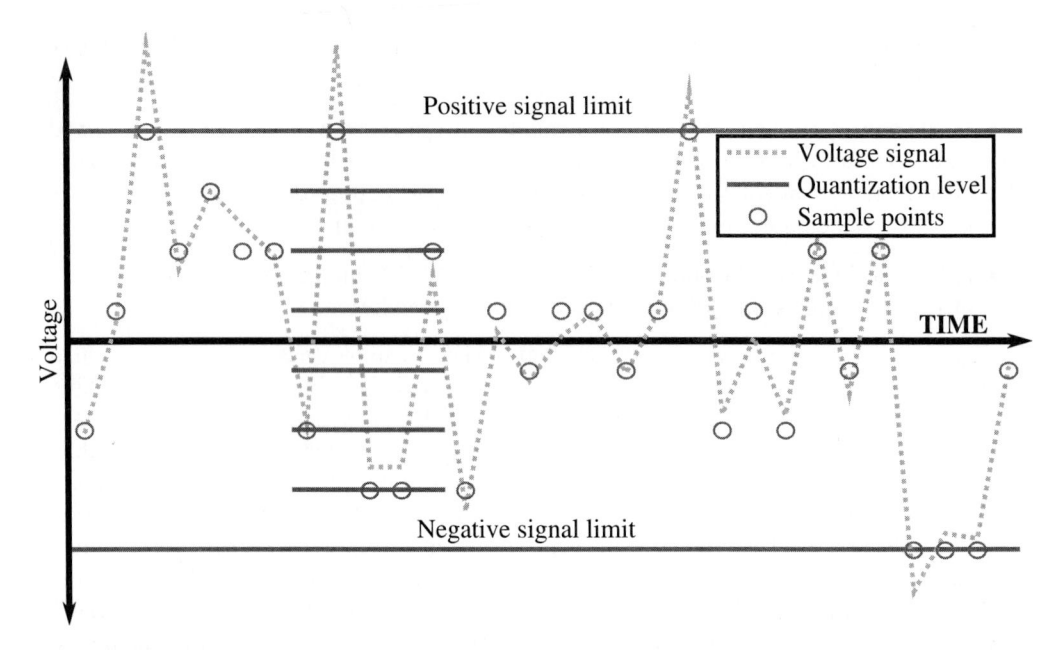

Figure 7.2 Amplitude thresholding and quantization for a 3-bit ADC with uniform quantization levels in the receiver. In a wireless communications receiver, ADCs quantize both the in-phase and quadrature channels independently. Automatic gain control (AGC) is used to normalize the energy of the received complex baseband signal so that the ADC thresholds and quantization levels can be fixed (i.e., do not depend on fading). As shown in Chapter 3, fading will be less pronounced with directional antennas.

Digital simulation of ADC effects is straightforward through ADC models from Chapter 6. Simulations on digital computers already operate on quantized data, although the bit resolution is very high (typically 32 bits) with precise floating point encoding. Transformation of received complex baseband samples to represent b-bit fixed-point wireless receivers (ADCs with ASIC or FPGA baseband implementations) is a simple process, as follows.

1. Estimate the sample energy, $\overline{\mathcal{E}}$, over some window of time relative to the first sample in each packet.[2] Ideally, the sample energy estimate is evaluated over the entire packet. In this ideal case, if the channel impulse response energy is normalized, this step can be simplified through direct calculation. For completely accurate AGC representation, however, energy can only be evaluated in a window of time before

---

2. Modeling ADC effects in wireless communication receivers makes the most sense in the context of framed or packetized data. Training data in each frame or packet allows the AGC algorithm to estimate the signal strength and fix the gain settings for the duration of the packet (before critical functionality such as synchronization commences).

the first sample of a packet due to sample latency in the ADC, computational latency in the AGC, or gain control latency in the VGA (especially with digital settings).

2. Compute positive and negative bounds as a function of $\overline{\mathcal{E}}$. Typically, ADCs expect I/Q signals with normalized RMS voltage, so we assume that the AGC will try to set the RMS voltage as a scalar multiple of the maximum signal voltage, that is, $-\alpha\sqrt{\overline{\mathcal{E}}}/\alpha\sqrt{\overline{\mathcal{E}}}$ are the lower/upper bounds used for clipping, respectively, in both the I and Q channels of the baseband data.

3. Separate the real and imaginary components of each input sample. Compare each real and imaginary sample coefficient to the lower and upper bounds. If any coefficient exceeds/is below the upper/lower bound, it is clipped at that upper/lower bound, respectively.

4. For an ADC with b quantization bits, compute $2^b - 2$ equally spaced quantization levels between the lower and upper bounds. With the previously defined amplitude bounds, this would result in quantization level intervals equal to $2\alpha\sqrt{\overline{\mathcal{E}}}/(2^b - 1)$.

5. Compute the Euclidean distance (absolute value) between each clipped real and imaginary coefficient and all the quantization levels. Each coefficient is mapped to the quantization level with minimum distance.

6. Convert I and Q (real and imaginary) coefficients back into complex samples.

Note that this ADC model is non-linear and cannot be inverted.

7.2.2 Power Amplifier Non-linearity

Chapter 5 showed how power amplifiers (PAs) synthesize the desired output power of RF waveforms. As discussed in Chapter 3, wireless propagation is challenged by path loss, making power amplifiers important to satisfy link budget requirements. Chapter 5 also showed, however, that mmWave amplifiers are plagued with limited gain and low efficiency, especially when implemented in CMOS, encouraging operation closer to the saturation region. Unfortunately, wideband PHY strategies are sensitive to non-linearities and, further, power amplifiers also tend to exhibit more severe non-linearity as the operating frequency increases [DES+04]. For all of these reasons, non-linear PA models are of paramount importance to mmWave PHY algorithm designers, such that high fidelity analysis and simulation may provide accurate tradeoffs between the degree of PA non-linearity and the performance of a candidate PHY. The reader is encouraged to reference Chapter 5 for a complete discussion of mmWave PA non-linearity.

The modified Rapp model is used to represent non-linear effects discussed in Chapter 5 [Rap91][HH98]. It was employed for performance benchmarking in the standardization of both IEEE 802.15.3c and IEEE 802.11ad 60 GHz standards. This model characterizes the two important properties of amplifier non-linearity, amplitude-modulation-input to amplitude-modulation-output distortion (AM-AM), and amplitude-modulation-input to phase-modulation-output distortion (AM-PM). Fig. 7.3 shows the AM-AM and AM-PM distortion profile for a 60 GHz CMOS PA in 65 nm along with the modified Rapp model

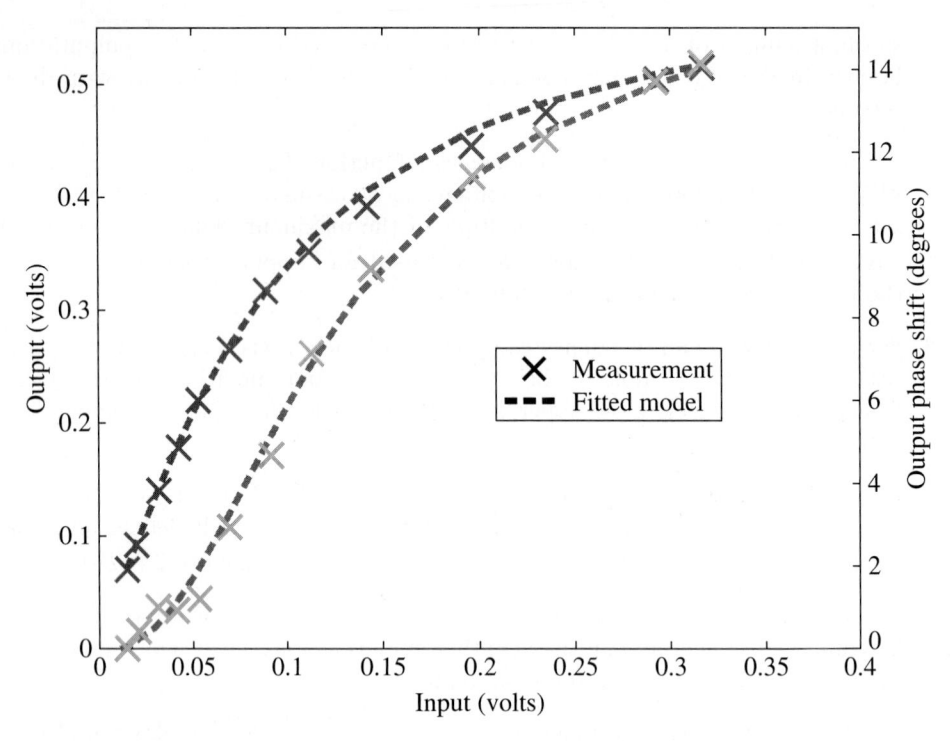

Figure 7.3 AM-PM and AM-PM measurements and modified Rapp model for a CMOS 65 nm PA. For more detail on mmWave PA statistics, please consult the tables provided in Chapter 5 [plot created with data from [EMT+09] © IEEE].

representation [Per10]. To characterize each of these effects, the modified Rapp model uses the following equations for AM-AM and AM-PM distortion, respectively, as a function of input magnitude A_{in}.

$$A_{\text{out}} = \frac{gA_{\text{in}}}{\left(1 + \left(\frac{gA_{\text{in}}}{A_{\text{sat}}}\right)^{2p}\right)^{\frac{1}{2p}}} \tag{7.1}$$

$$\Delta\theta_{\text{out}} = \frac{\alpha A_{\text{in}}^{q_1}}{1 + \left(\frac{A_{\text{in}}}{\beta}\right)^{q_2}} \tag{7.2}$$

Here, A_{out}, $\Delta\theta_{\text{out}}$, g, and A_{sat} represent the output signal magnitude, the output phase distortion, the amplifier small signal gain, and the saturation amplitude of the amplifiers, respectively, while p, α, β, q_1, and q_2 are curve fitting parameters that are matched to measurements of (7.1) and (7.2). For example, for the CMOS 60 GHz PA shown in Fig. 7.3, the modified Rapp curve fitting parameters are $p = 0.81$, $\alpha = 2560$, $\beta = 0.114$, $q_1 = 2.4$, and $q_2 = 2.3$ as determined through least square fitting from PA output RMS voltage measurements and phase measurements given $g = 4.65$ and $A_{\text{sat}} = 0.58$ [EMT+09]. These parameters may vary significantly between devices. For comparison, the NEC GaAs 60 GHz PA featured in [C+06] is characterized by $p = 0.81$, $\alpha = -48000$, $\beta = 0.123$, $q_1 = 3.8$, and $q_2 = 3.7$ given $g = 19$ and $A_{\text{sat}} = 1.4$.

7.2.3 Phase Noise

Chapter 2 and Chapter 5 showed how transmitted and received RF signals are up-/down-converted to/from millimeter wavelengths through operations on local oscillator (LO) signals. Ideally, the LO is a stable, clean sinusoid, mathematically represented by

$$V_0 \cos(2\pi f_c t) \tag{7.3}$$

where t is the time reference variable, V_0 is the local oscillator voltage amplitude, and f_c is the desired carrier (center) frequency of the RF signal. As shown in Chapter 5, through discussion of voltage-controlled oscillator (VCO) and phase-locked loop (PLL) mmWave circuits, practical LOs are distorted due to unstable sources that cannot be entirely mitigated. This is modeled mathematically in the local oscillator through

$$V_0 (1 + v(t)) \cos(2\pi f_c t + \theta(t)) \tag{7.4}$$

where $v(t)$ and $\theta(t)$ are the amplitude and phase distortion of the LO, respectively, as a function of time. The amplitude distortion effect at the outset of PLL packages is typically assumed to be much less severe than the phase distortion effect, leaving us with the standard LO model for phase-only distortion (phase noise), that is, $V_0 \cos(2\pi f_c t + \theta(t))$.

To analyze phase-noise effects at baseband, a more convenient model is needed. Note that the following discussion is also partially presented in Chapter 5. In this chapter, however, we focus more on aspects related to linear system modeling for physical layer (PHY) algorithm design. If we use the standard assumption that phase distortion is well approximated through a sum of K sinusoids in frequency-modulation (FM) form [Dru00], then

$$\theta(t) \approx \beta \sum_{k=0}^{K-1} \alpha_k \cos(2\pi f_k t + \phi_k) \tag{7.5}$$

where β is the maximum phase deviation, α_k are the amplitude coefficients of each FM sinusoid component, f_k are the sinusoid frequencies ($f_k \ll f_c$), and ϕ_k are the phase offsets of each FM sinusoid component. Next, we substitute (7.5) back into our general phase noise model and determine that

$$\cos(2\pi f_c t + \theta(t)) = \sin(2\pi f_c t) \cos(\theta(t)) + \sin(\theta(t)) \cos(2\pi f_c t) \tag{7.6}$$

$$\approx \sin(2\pi f_c t) + \theta(t) \cos(2\pi f_c t) \tag{7.7}$$

$$\approx \sin(2\pi f_c t) + \beta \sum_{k=0}^{K-1} \alpha_k \cos(2\pi f_k t + \phi_k) \cos(2\pi f_c t) \tag{7.8}$$

$$= \sin(2\pi f_c t) + \sum_{k=0}^{K-1} \frac{\beta \alpha_k}{2} \cos(2\pi (f_c + f_k) t + \phi_k)$$

$$+ \sum_{k=0}^{K-1} \frac{\beta \alpha_k}{2} \cos(2\pi (f_c - f_k) t - \phi_k) \tag{7.9}$$

where (7.6) results from the equality $\cos(x+y) = \sin(x)\cos(y) + \sin(y)\cos(x)$, (7.7) results from small phase approximations on sinusoids ($\theta(t) \ll 1$), and (7.9) results from the equality $\cos(x)\cos(y) = \frac{1}{2}(\cos(x-y) + \cos(x+y))$. Hence, it is possible to approximate phase noise effects through sinusoidal components weighted symmetrically about the carrier frequency. To determine the weights of each sinusoid and characterize phase noise, we must measure the power spectral density (PSD) of the carrier signal. While the ideal oscillator is equivalently represented in the frequency domain by two impulses symmetrically located with respect to the 0 Hz axis, the PSD of the practical oscillator is represented with sidebands as illustrated in Fig. 7.4. The PSD itself often reveals many characteristics about the observed phase noise. In the log-scale, different phase noise sources yield different slopes. Leeson was among the first to recognize this ability [Lee66]. In Leeson's model, he attributes f^{-4} effects or -40 dB/decade slopes to random walk noise caused by environmental factors such as mechanical shock, vibration, and temperature. Further, Leeson attributes f^{-3} effects to flicker frequency modulation noise caused by resonator noise and active component noise in oscillators, f^{-2} effects to white frequency modulation noise due to amplifier stages, f^{-1} effects to flicker phase noise caused by all active components, and f^0 effects to white phase noise caused by broadband thermal noise [Lee66]. An illustration of these real-world impairments is shown in Fig. 7.5.

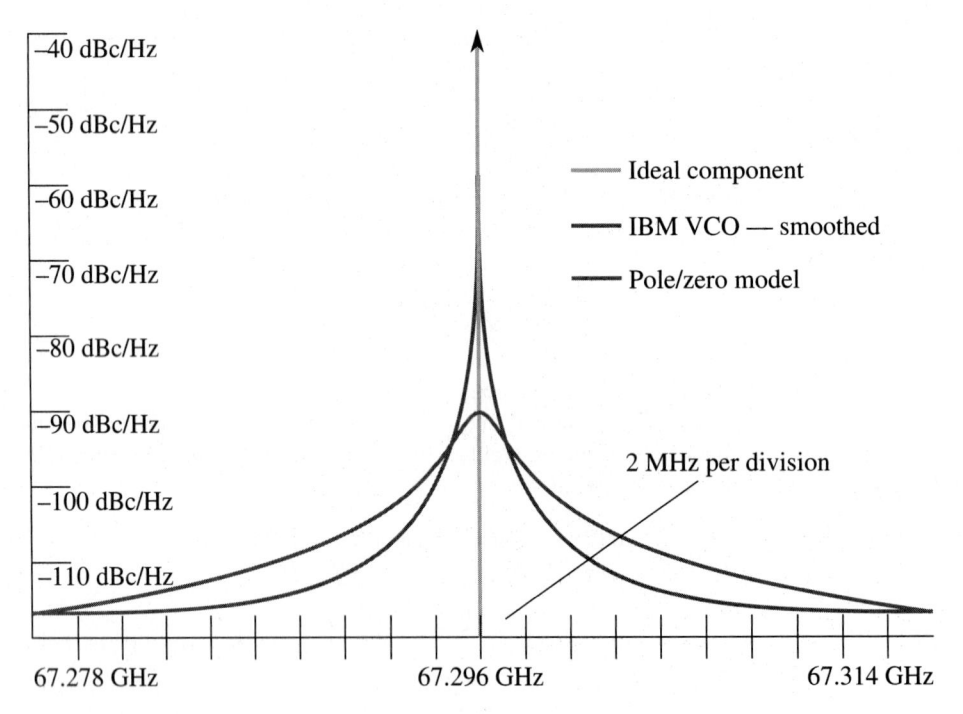

Figure 7.4 Measured power spectral density of single sideband (SSB) VCO output for desired signal at 67.3 GHz. PSD measurements normalized to desired carrier output power (represented by dBc/Hz). A similar figure is shown in Chapter 5. Here, however, we have included a comparison to a pole/zero model that facilitates physical layer performance simulations [data taken from [FRP$^+$05] and smoothed to create the plot © IEEE].

7.2 Practical Transceivers

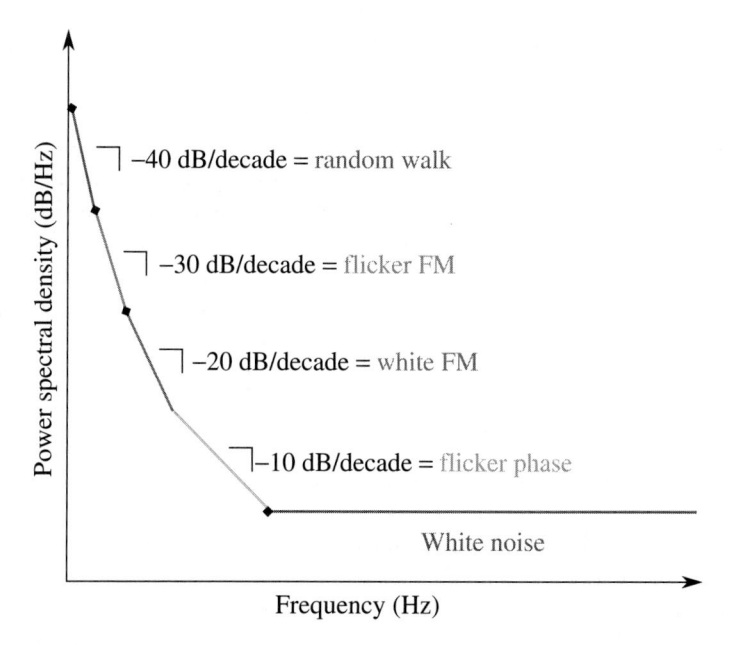

Figure 7.5 Different phase noise effects and their contribution to the power spectral density in the Leeson model [Lee66].

Many models have been created for adding phase noise to complex baseband simulations [DMR00]. The most common model, which is also supported by IEEE 802.11ad and IEEE 802.15.3c standards, is the colored additive white Gaussian noise (AWGN) noise model, whose effects are generated as follows.

- First, zero-mean, unit-variance, white, and complex-Gaussian random variables are generated for each signal sample.
- Next, the frequency response of the white noise samples are shaped according to phase noise PSD measurements to produce colored noise. This shaping can occur through convolution in the time domain with phase noise filters that match the spectrum of PSD measurements or through multiplication in the frequency domain followed by the inverse fast Fourier transform (IFFT).
- Then, the colored noise is directly added to the phase of the transmitted wireless samples (after pulse shaping, but before power amplifier non-linearity effects are added).
- Finally, the first three steps are repeated at the receiver, after channel effects, but before thermal noise is added.

Note that phase noise occurs both at the transmitter and at the receiver because local oscillators are used on both devices to transform the baseband/passband signal to passband/baseband frequencies, respectively. Sometimes the model is placed only at the transmitter because it can capture both TX and RX device effects in channels without large delay spread.

The sidebands of the oscillator lead to a multitude of problems including adjacent channel interference and signal masking [Rou08], inter-carrier interference (ICI) with OFDM modulation [PVBM95][ETWH01], and general performance degradation due to increased noise power [CW00][Rou08]. The last two effects are of significant concern in mmWave links and, consequently, standard phase noise PSD measurements have been created to allow PHY designers to test the robustness of transceiver algorithms. To model the PSD of phase noise at both the transmitter and the receiver, the IEEE 802.15.3c and IEEE 802.11ad 60 GHz standards, for example, use a simple 1-pole, 1-zero filter with response

$$L(f) = \gamma \left(\frac{1 + \left(\frac{f}{f_z}\right)^2}{1 + \left(\frac{f}{f_p}\right)^2} \right) \quad (7.10)$$

where f is frequency (in Hertz), $f_p = 1$ MHz is the pole frequency, $f_z = 100$ MHz is the zero frequency, and $\gamma = -90$ dBc/Hz is the close-in phase noise strength. Hence, the time domain equivalent impulse response of this filter should be convolved with the white noise samples added at the transmitter and receiver. If we compare this response to the smoothed phase noise PSD in Fig. 7.4 we note that the close-in phase noise strength has been truncated at -90 dBc/Hz (which occurs at roughly 1 MHz) because phase noise components < 1 MHz do not significantly degrade signal integrity in IEEE 802.15.3c and IEEE 802.11ad links.

7.3 High-Throughput PHYs

Throughout the previous chapters, we have emphasized that gain in the communication channel is important to achieve reasonable link margins. There are two general methods to provide these gains: (1) frequency or time spreading and (2) beamforming. With frequency or time spreading, the transmitted signal exhibits redundancy at the symbol level, allowing processing gain or coding gain to be captured before bits are decoded [Pro01]. With beamforming, configuration of the beams should only occur after links are established to minimize signaling overhead, to avoid wasted power, and to avoid unneeded interfering transmissions. The advantage of frequency spreading is that it does not require any special purpose hardware or configuration (such as antenna arrays with correct phase arguments). The redundancy of spreading signals over bandwidth or time, however, substantially reduces the overall transmission rate. Hence, spreading does not leverage the vast spectrum resources in an efficient way because the dominant motivation for moving to mmWave spectrum is to enable higher data rates and larger capacity. Furthermore, since mmWave systems using directional beams are more likely to be interference limited rather than noise limited, and are likely to use time division duplexing (TDD) rather than Frequency Division Duplexing (FDD), the interference resistance properties and processing gains offered by frequency spread techniques are less beneficial (although many spread spectrum techniques are readily combined with beamforming methods [LR99a]). Because of this, mmWave spread spectrum communication is likely to be preferred primarily for control communication before or during the negotiation of beamforming, and not after the beams are negotiated. With beamforming, as shown in Chapter 3, it is possible to achieve moderate to high SNR at the receiver over surprisingly large distances without spreading in both LOS and NLOS channels [SNS+14][SR14][RRE14][Rap14][Rap14a]. System designers are interested in PHY algorithms that maximize performance in all of these scenarios.

7.3.1 Modulation, Coding, and Equalization

Information theory tells us that the spectral efficiency of communication, that is, the number of bits/second/Hz (bps/Hz), is maximized through the use of infinite symbol block lengths and Gaussian symbol coding.[3] Modern digital wireless communications approximate this format with quadrature amplitude modulation (QAM), error correcting codes, and finite code block length, making implementation complexity more manageable (both at the transmitter and at the receiver) [Gol05].

Early cellular and satellite communication links were represented by frequency-flat, narrowband channels where the optimal decoding was straightforward [Rap02] [PBA03][RF91][FRT93][DR98]. In narrowband channels, the receiver determines the maximum likelihood bit sequence from the received symbols as a function of the QAM constellation, error correcting code, thermal noise (imparted from analog circuit components), memoryless channel magnitude, and the memoryless channel phase. Modern wideband communication links (including the mmWave links studied in this text), however, are no longer memoryless. Indeed, as established in Chapter 3 (see Sections 3.7.1.2 and 3.8.3), mmWave channels may have multiple taps in the channel impulse response. In wideband channels the receiver optimally selects bit estimates using the maximum likelihood criterion as a function of all channel taps. Unfortunately, this maximum likelihood approach, as implemented through Viterbi decoding, exhibits exponential scaling of the processing complexity as a function of the number of channel taps. The number of taps in a mmWave communication channel is often very large due both to the giga-sample-per-second symbol rate and the large number of reflective paths in indoor and outdoor channels (although narrow beamwidths will reduce the multipath components). Because implementation complexity is a bottleneck at such high symbol rates, the maximum likelihood approach (over many multipath taps) is not typically considered for implementation. To circumvent this complexity bottleneck, receivers commonly feature channel equalization before detection such that Viterbi decoders only need to consider a single-tap response (effectively rendering a narrowband channel in the eyes of the detector).

In general, as previewed in Chapter 2, the wireless design engineer must select from a suite of equalization algorithms including: linear zero-forcing equalizers (which reduce peak distortion) [Cio11], linear minimum mean-square-error equalizers [Mon84], decision feedback equalizers (which use soft-output received symbols through linear feed-forward equalizers and fed back through linear feedback equalizers) [RHF93][PS07], non-linear equalizers [BB83], and adaptive equalizers (which continuously improve equalizer

---

3. Assuming a linear, time-invariant channel with AWGN.

realizations as more symbols are received) [Qur82].[4] Table 7.1 lists advantages and disadvantages of these equalization options (also see [Rap02, Ch 6]).

The equalization algorithms discussed in Table 7.1 have traditionally operated in the time domain. Recently, as discussed in Chapter 2, frequency domain equalization (FDE) has become popular because frequency domain processing can lead to reduced processing complexity. With FDE, each tone (or frequency domain sub-channel) can completely mitigate wideband channel effects with a simple single-tap linear equalizer, since each sub-channel is narrowband (undergoes flat fading). Consequently, most mmWave implementations that mitigate wideband channel effects are expected to feature linear single-tap equalization algorithms implemented on individual "narrowband" frequency sub-channels. To optimize complexity, the fast Fourier transform (FFT) is exploited to enter the frequency domain. To enable frequency domain transformation through the FFT at the receiver, waveforms must be formatted through blocks of symbols prefixed with guard intervals to preserve cyclic convolution in the event of the worst-case multi-path propagation (guard intervals are often realized through a cyclic prefix or repeated training sequence; see Chapter 2 for more details).

Although the operation of equalization in the frequency domain appears to be almost universally preferred for mmWave wideband channels, this is not the case for modulation and coding. For example, the choice of modulation for 60 GHz indoor standards was divided into two camps: (1) OFDM, and (2) SC-FDE (see Chapter 2 for more information on these modulation classes). This, in part, has resulted in multiple physical layer architectures within 60 GHz standards, including ECMA-387, IEEE 802.15.3c, and

Table 7.1 Advantages and disadvantages of several popular equalization algorithms. *Processing complexity* refers to the complexity of the equalizer once its realization is known, whereas *discovery complexity* refers to the complexity of discovering the equalizer realization as a function of the channel response.

Equalizer Type	Advantages	Disadvantages
Linear (non-adaptive)	Low discovery and processing complexity	Moderate performance
Non-linear (non-adaptive)	Performance advantage over linear	Higher discovery and processing complexity
DFE (linear+non-linear)	Low discovery complexity, improved performance	Increased processing complexity, error propagation
Adaptive (linear+non-linear)	Compatible with evolving channel estimates	Increased processing complexity, convergence time

---

4. Technically, adaptive equalizers are not a separate category since linear, non-linear, and decision feedback equalizer (DFE) equalizers can all be made adaptive. Modern commercial wireless links, however, do not typically feature adaptive equalizers due to the disadvantages discussed in Table 7.1. Because of this, we place adaptive equalizers in a separate category so that we do not have to discuss this topic further.

7.3 High-Throughput PHYs

IEEE 802.11ad (see Chapter 9 for standard details). Early outdoor enhanced local access networks are also considering variations of these modulations [Gho14]. In frequency-flat channels with perfect transceivers, OFDM and SC-FDE exhibit the same performance. In frequency selective mmWave channels, however, this is not the case.

OFDM, in theory, is better suited for frequency selective channel operation. By transmitting in the orthogonal frequency domain components (sub-carriers), and assuming perfect linear transceivers, infinite FFT block sizes with Gaussian coding, and optimal per sub-carrier power allocation through water-filling, OFDM is capacity optimal [Gol05]. In practice, FFT block sizes must be small to manage complexity, since power allocation per sub-carrier incurs exorbitant overhead [Dar04], transceivers at mmWave will have many imperfections [Raz98], and high-rate codes (which are efficient in the capacity sense) may not be sufficient for severe frequency selectivity [WMG04]. Motivated by these practical concerns, for example, several mmWave hardware vendors have been backing SC-FDE with claims of better performance on practical platforms. This has also sparked a debate in the research community. Unfortunately, a clear preference for either OFDM and SC-FDE cannot be stated because there are preferable operation scenarios for each. This is the focus of the next section.

7.3.2 A Practical Comparison of OFDM and SC-FDE

The simulations in this section use the parameters and models in Tables 7.2, 7.3, and 7.4. First, we consider the performance of OFDM and SC-FDE in terms of the wireless channel impulse response, without power amplifier non-linearity, quantization, or phase noise through simulation. As a baseline, Fig. 7.6 illustrates the symbol error

Table 7.2 Parameters and models common to both SC-FDE and OFDM for the simulations in this section. The transceiver impairment models were guided by recommendations from the IEEE 802.15.3c and IEEE 802.11ad standard committees (see Section 3.8.3). A spectral mask filter, matching IEEE 802.15.3c specifications, is applied at the transmitter to ensure that out-of-band signal contributions are not enhancing performance. The constellation and coding set were chosen to enable high spectral efficiency and high reliability in both OFDM and SC-FDE.

Parameter/Model	Value
Baseband sample rate	5.28 Gbps
Spectral mask pre-filter	IEEE 802.15.3c
Channel model	IEEE 802.15.3c Gold Set
Symbol constellation	QPSK and 16-QAM
Error control coding	RS(255,239), LDPC(672,432), LDPC(672,336)
Channel estimation	Noiseless, time-domain
Synchronization	Perfect (time and frequency)
Bit-decoding	Viterbi decoding with LLR inputs
Power amplifier	Modified Rapp model (CMOS)
Analog-to-digital conversion	Limiter with uniform quantization
Jitter/phase noise	IEEE 802.15.3c 1-zero/1-pole response

Table 7.3 OFDM and SC-FDE specific simulation parameters. The symbol rate for OFDM and SC-FDE is unequal because they both meet the spectral mask requirements differently (OFDM reduces rate through guard tones). We choose the symbol rates to be simple fractions of the IEEE 802.15.3c AV PHY sample rate (5.28 Gbps) and include 172 null tones in OFDM to fairly compare data rates. An excessive cyclic prefix/pilot word of length 256 symbols is used to ensure that unprotected inter-symbol interference does not distort results.

Parameter	OFDM	SC-FDE
Symbol rate	2.64 GHz	1.76 GHz
Pulse shaping	None	Root-raised cosine ($\beta = 0.25$)
Cyclic prefix	256 symbols	256 symbols
FFT size	512 symbols	512 symbols
Null tones	171 guard, 1 DC	N/A
BPSK rate	1.17 Gbps	1.17 Gbps
FD equalizer	ZF	MMSE
Soft-outputs	Tone-weighted LLR	Equal-weighted LLR

Table 7.4 Specific channel model characteristics. For more discussion on these channel models, the reader is encouraged to reference Chapter 3, specifically Section 3.8.3.

Channel Feature	Value
Framework	Saleh-Valenzuela cluster model
TX antenna beamwidth	30 (CM1.3), 60 (CM2.3)
RX antenna beamwidth	30 (CM1.3), 60 (CM2.3)
Impulse response energy	Normalized to 1
RMS delay spread	50 ns (CM1.3), 100 ns (CM2.3)
Propagation scenario	Office (CM1.3), home theater (CM2.3)

rate of OFDM and SC-FDE in AWGN and the CM1.3 LOS channel model (given in Section 3.8.3) with narrow beamforming (office) for QPSK constellations and variable coding rates/configurations. This plot shows that, as expected, the performance of OFDM and SC-FDE is identical in AWGN. When a spectral mask is added to both OFDM and SC-FDE, the performance does not change. Fig. 7.6 also shows us that practical CM1.3 channels do represent a noticeable loss in performance due to frequency selectivity from indirect weak multipath contributions. When robust coding is available (e.g., LDPC(672,336)), the performance loss is minimized with OFDM because it is more capable than SC-FDE of mitigating frequency selectivity. Conversely, SC-FDE outperforms OFDM when only light coding is available (e.g., RS(255,239)) because OFDM depends on coding to protect against faded sub-carriers. SC-FDE balances frequency selective fading over all symbols within a block, resulting in less coding requirements.

Fig. 7.7 shows the same simulated performance of OFDM and SC-FDE as Fig. 7.6, but with 16-QAM constellations. It is interesting to note that the same performance trends are observed, albeit shifted to the right by $\approx$ 6 dB. Hence, we can conclude

7.3 High-Throughput PHYs

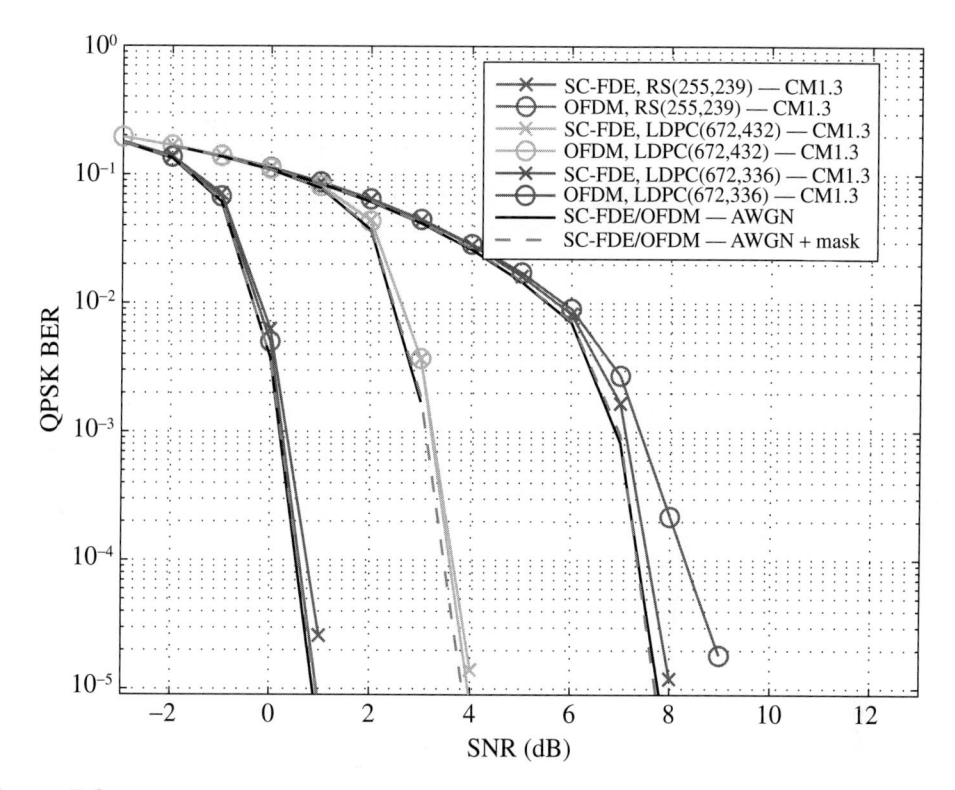

Figure 7.6 Bit error rate as a function of SNR for OFDM and SC-FDE in AWGN and LOS channels with QPSK constellations. The coding options are RS(255,239) and LDPC(672,336). The 802.15.3c spectral mask is added to the AWGN channel to demonstrate bandwidth conservation of each modulation strategy. Hardware impairments are not considered.

that the mmWave high-throughput PHY framework does not need to be tailored to each constellation, at least in the absence of hardware impairments or severe frequency selectivity.

Next, we investigate the addition of substantial multipath-induced frequency selectivity, as typically experienced in the NLOS home theater application with broad beamforming and modeled through CM2.3. Fig. 7.8 shows the performance of QPSK constellations. Note how the tradeoff between OFDM and SC-FDE has been magnified. In severe frequency selectivity, OFDM requires heavy coding for successful communication. OFDM cannot provide robust links with only RS(255,239). Alternatively, when heavy coding is available, the performance advantage of OFDM over SC-FDE grows due to its superior frequency selective fading mitigation capabilities. Intuitively, this is due to OFDM's ability to exploit the sub-carriers with good channels and protect the bad sub-carriers with coding, rather than be subject to the combined channel performance over all sub-carriers for each QAM symbol, as with SC-FDE. In SC-FDE, every QAM symbol is partially impaired by frequency selective fading, whereas in OFDM only a small subset of the QAM symbols are impaired by frequency selective fading. Further, FDE colors the noise in single-carrier transmission, reducing the performance of our Viterbi algorithm.

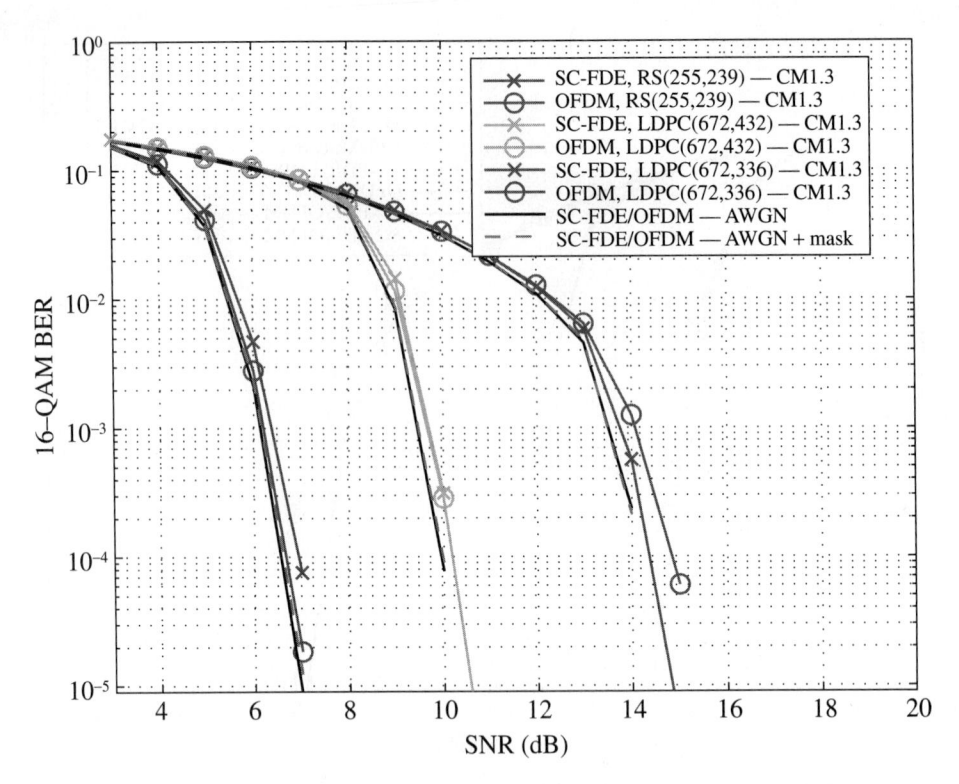

Figure 7.7 Bit error rate as a function of SNR for OFDM and SC-FDE in LOS CM1.3 channels with 16-QAM constellations. Hardware impairments are not considered.

Fig. 7.9 plots the same performance of 16-QAM constellations with frequency selectivity. Although the tradeoff between OFDM and SC-FDE is once again magnified, the same trends are observed. We can strengthen our previous conclusion that constellation selection does not influence the chosen high-throughput PHY framework in the absence of hardware impairments.

Given the current discussion in this subsection, which has not yet addressed practical transceiver impairments, we already find that different operational scenarios provide preference for OFDM and SC-FDE. SC-FDE is desired when high rate codes are used. Because high rate codes coupled with intelligent adaptive modulation and coding are typically better at minimizing the capacity gap in frequency flat channels and are often implemented with less digital complexity (due to fewer parity bits and lower required modulation order), it appears that SC-FDE is well suited for LOS channels or when implementation complexity is at a premium. OFDM, however, is more apt at addressing frequency selectivity, although this requires more intensive error control coding and high constellation orders. This discussion is not complete, however, because we must ensure that these design tradeoffs hold in real transceiver hardware with real impairments. Additionally, comparisons of power consumption for practical implementations of OFDM and SC-FDE implementations must be carefully considered, as well. As we will see in the next

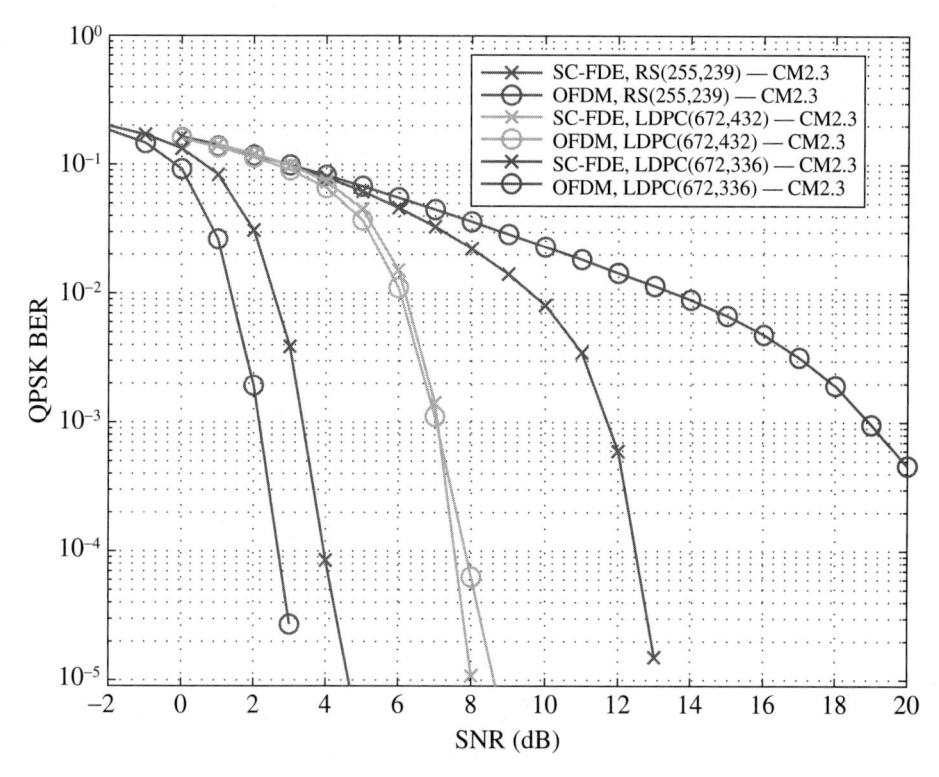

Figure 7.8 Bit error rate as a function of SNR for OFDM and SC-FDE in NLOS CM2.3 channels with QPSK constellations. No hardware impairments are considered.

simulation results, OFDM struggles more and is more heavily impacted with hardware impairments. Choosing the modulation and coding strategy is not as straightforward as it appears at first glance.

We begin our investigation of performance degradation with hardware impairments through phase-noise additions using the 802.15.3c/802.11ad pole/zero model. Our investigation did not show a remarkable difference in performance with the inclusion of phase noise, although there was a marginal performance loss in LOS channels (CM1.3). While we did not consider an extensive array of closely spaced co-channel or adjacent channel interference scenarios, we conclude that the performance loss associated with frequency selectivity outweighs the effects of phase noise. Intuitively, if thermal noise is comparable to or stronger than phase noise at the receiver, then the SNR after equalization (which includes residual intersymbol interference) is overwhelmingly lower than the signal-to-phase-noise ratio. Note that this is not a general conclusion. As demonstrated previously, the impact of phase noise will depend on the joint analysis of all operational parameters in the mmWave link, including adjacent channel interference effects. Also, if future mmWave networks use time division duplexing and a small number of users per cell or access point, the deleterious effects of adjacent channel interference and phase noise may be further reduced.

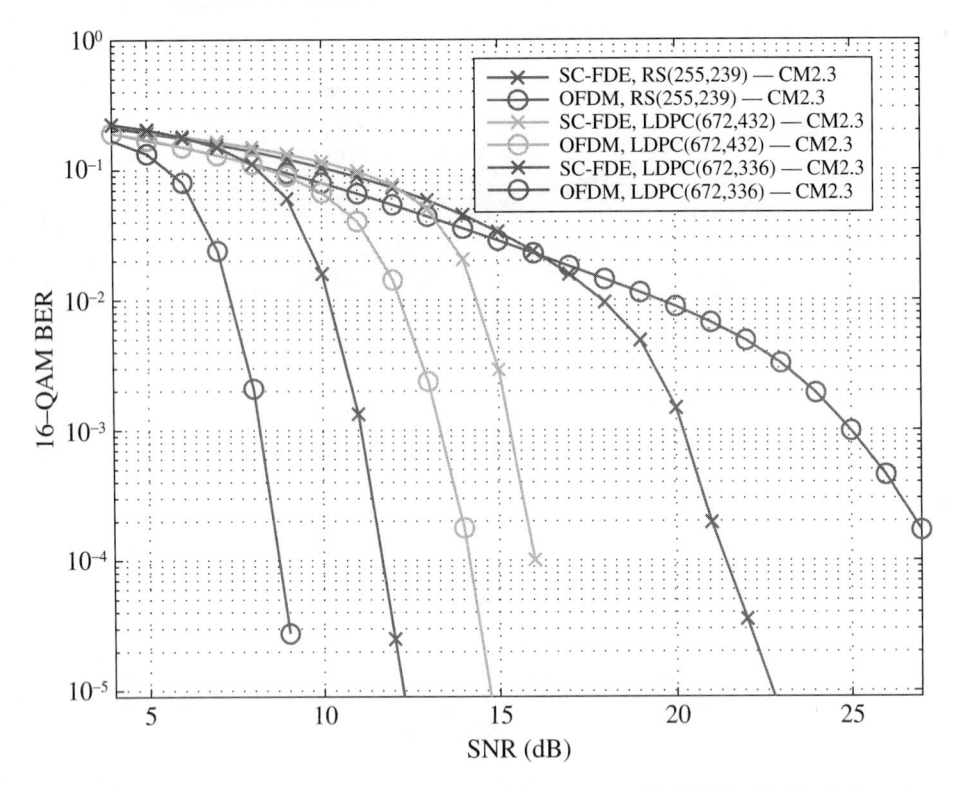

Figure 7.9 Bit error rate as a function of SNR for OFDM and SC-FDE in NLOS CM2.3 channels with 16-QAM constellations. No hardware impairments are considered.

Next, we consider an additional hardware impairment, PA non-linearity. Fig. 7.10 shows the performance with PA non-linearity in LOS CM1.2 channels for QPSK constellations. With PA non-linearity, we have disrupted the previously observed design tradeoffs between OFDM and SC-FDE. With less amplifier backoff, we see that SC-FDE is preferable in all scenarios, regardless of coding robustness. Consequently, we determine that SC-FDE is more robust to PA non-linearity and may operate with less backoff, resulting in more power-efficient communication and a larger observed SNR. This determination is strengthened by Fig. 7.11, which shows that these effects are magnified for higher order constellations, which require high SNR conditions (fundamentally limited by amplifier non-linearity). In fact, we note that OFDM requires > 5 dB backoff whereas SC-FDE only requires ≈ 3 to 4 dB backoff. In general, we will notice a minimum of 2-3 dB SNR gain due to increased transmit power with SC-FDE.

We might have expected that frequency selectivity in the channel would strengthen the non-linear PA advantage of SC-FDE even further, because frequency selective channels require higher total SNR for adequate BER performance. However, Figs. 7.12 and 7.13, which show the BER performance of QPSK and 16-QAM in CM2.3 NLOS channels, suggest that this is not the case. It turns out that the superior frequency selectivity

7.3 High-Throughput PHYs

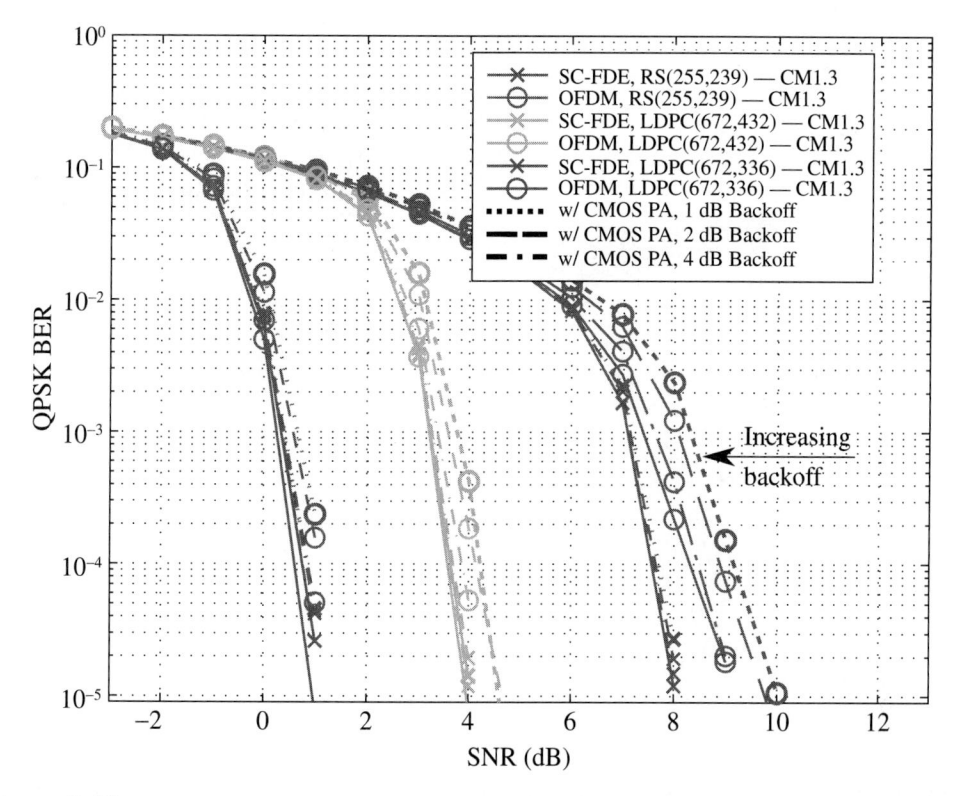

Figure 7.10 Bit error rate as a function of SNR for OFDM and SC-FDE in LOS CM1.3 channels and CMOS PA non-linearity with QPSK constellations.

mitigation capabilities of OFDM offset the PA non-linearity benefits of SC-FDE, at least when robust LDPC coding is used. Consequently, in frequency selective environments, OFDM is at least on par with SC-FDE in the face of amplifier non-linearity, especially with higher order constellations and heavy coding. Hence, there is still a case for both modulation formats in different operating regimes.

Our final consideration is ADC bit precision. Prior work suggests that SC-FDE has superior performance with less bit precision [LLS+08]. This is intuitive because SC-FDE signals feature a lower *peak-to-average-power ratio* (PAPR). Intuition is confirmed in Figs. 7.14 and 7.15, which show the performance of QPSK and 16-QAM constellations, respectively, in LOS CM1.3 channels with 5-bit ADC precision. We also see from these plots that OFDM needs to make sure that the peak magnitude of the ADC needs to be 2σ (2σ peak mag $\Rightarrow$ each I/Q ADC has a maximum amplitude of $2\sigma/\sqrt{2}$) to achieve close to ideal performance, where σ is the RMS voltage of the received packet. SC-FDE, however, only needs the peak magnitude of the ADC to be 1.5σ. Hence, the AGC can operate more aggressively with SC-FDE and needs to provide more headroom for OFDM, thereby decreasing the SNR and decreasing the effective precision of OFDM samples.

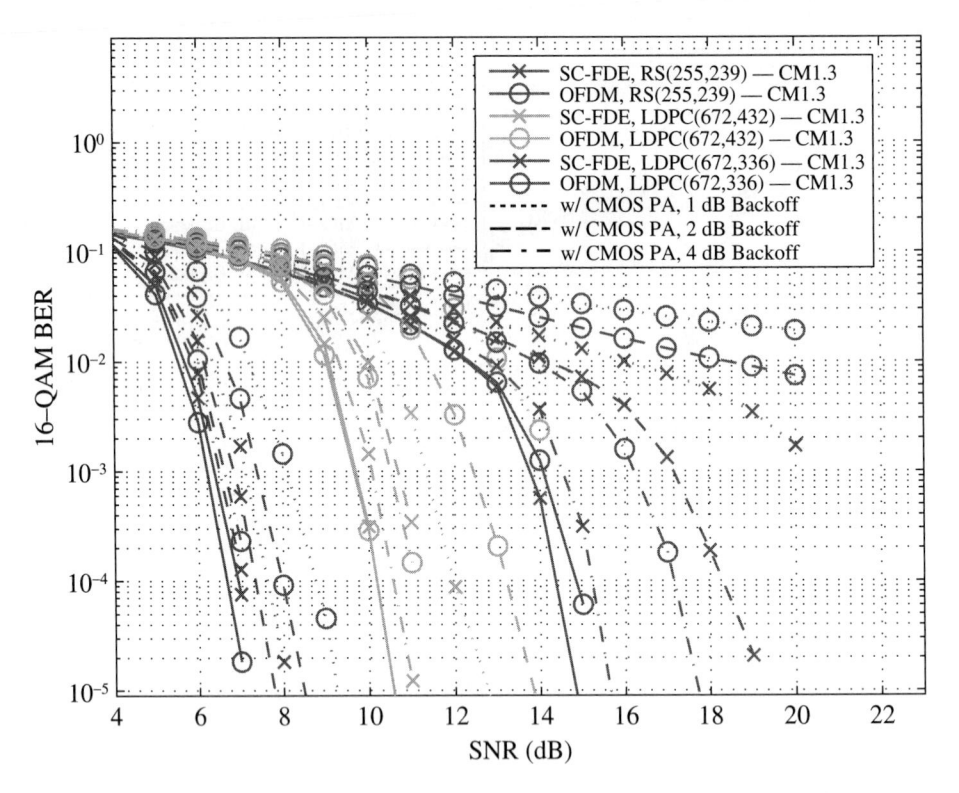

Figure 7.11 Bit error rate as a function of SNR for OFDM and SC-FDE in LOS CM1.3 channels and CMOS PA nonlinearity with 16-QAM constellations.

7.3.2.1 Null Cyclic Prefix Single Carrier (NCP-SC) Modulation

OFDM and SC-FDE have found use in fourth-generation (4G) LTE cellular systems, and as described earlier, the single tap narrowband approach offered by OFDM is attractive from a complexity and implementation standpoint. However, the multiplexing of users in frequency, as is performed in today's 4G 3GPP-LTE cellular systems in the UHF/microwave bands, is not necessarily a useful feature for the much wider bandwidths expected at mmWave carrier frequencies.

There are many reasons why frequency multiplexing of users may not be important for mmWave mobile communications. First, as shown in Chapter 3, outdoor mmWave systems will likely be deployed in small cells with very small time slot sizes (e.g., 100 μs or less), meaning very few users will need to transmit within a time slot at a particular access point or base station. Second, the high bandwidth and small cell coverage suggest the use of extremely small OFDM or SC-FDE symbol times (e.g., on the order of a fraction of a nanosecond, as symbol rates become several Gigasymbols per second) and propagation delays on the order of a few hundred nanoseconds, meaning that the active users could just as efficiently be multiplexed in time, rather than frequency [RRE14][Gho14]. Lastly, mmWave systems will need large antenna arrays for at least one end of the link to overcome path loss. Full digital beamforming may be impractical, at least initially,

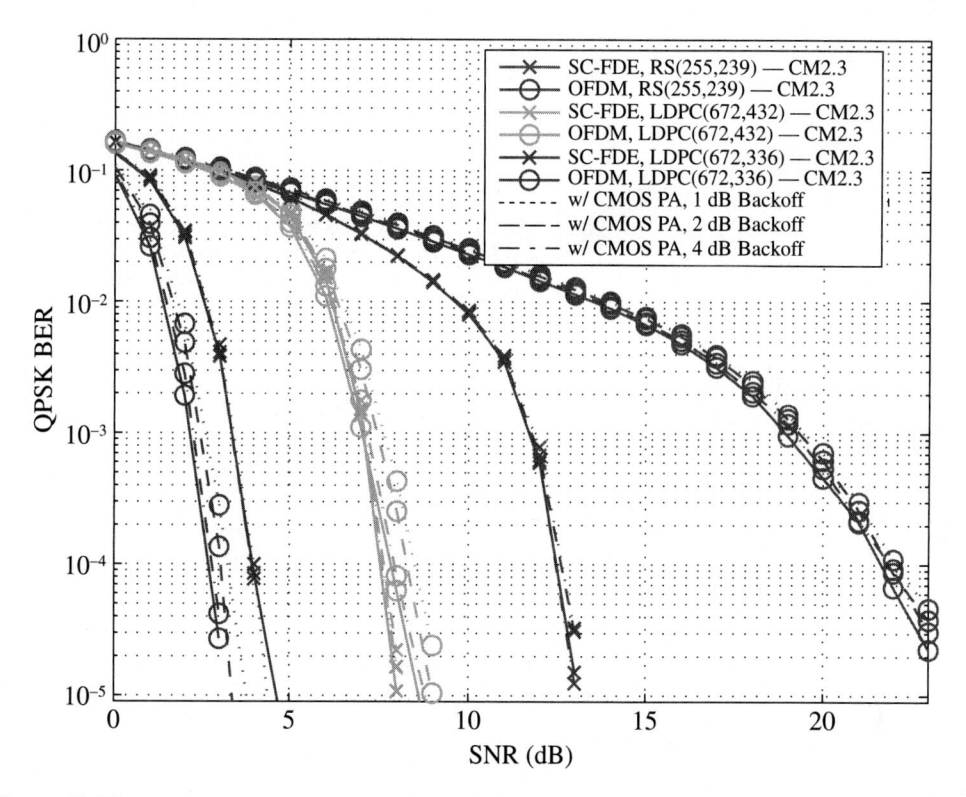

Figure 7.12 Bit error rate as a function of SNR for OFDM and SC-FDE in NLOS CM2.3 channels and CMOS PA non-linearity with QPSK constellations.

since the DACs and ADCs that are needed behind each of the antennas and operating at multi-gigahertz bandwidths will consume a large amount of power. This drives initial mmWave systems to use RF beamforming, which requires only a single ADC and DAC behind the entire array of antennas [Gho14][RRC14]. The use of RF beamforming implies that only a single beam (or at most two or three) per polarization will be created at any given time, thus suggesting that users be separated in time, rather than frequency, since each user could use a unique beam that provides the best gain for that user.

Since the ability to multiplex users in frequency will not likely be critical for mmWave access communications, it may not be necessary to accept some of the drawbacks of OFDM, such as the high PAPR that, as discussed in Chapter 5 and earlier in this chapter, results in less efficient power amplifiers (PAs) and shrinks the expected range, while negatively impacting out-of-band emissions.

Ghosh et al. [Gho14] recently proposed a form of cyclic prefix (CP) single-carrier (SC) modulation, where the regular CPs are replaced with null CPs. This concept is called *null CP SC* (NCP-SC) [CGK+13], it has near-constant-envelope properties during data transmission, and it is inherently efficient, like the reverse link in 3GPP-LTE [3GPP09], which uses DFT-spread OFDM (DFT-S-OFDM). NCP-SC also shares the advantages of

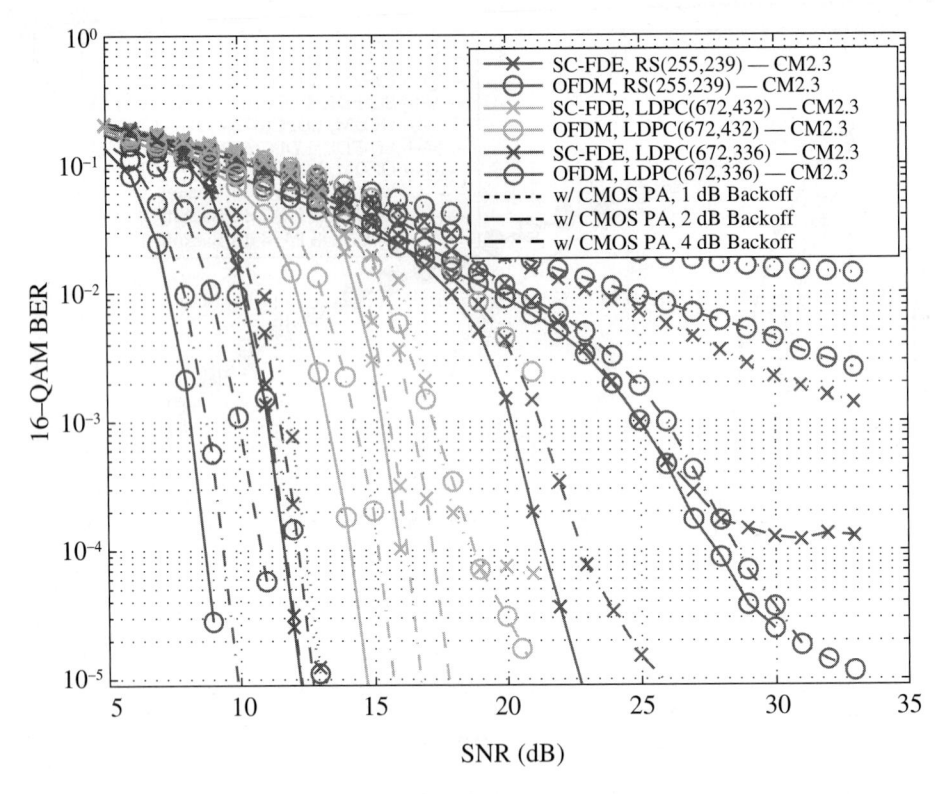

Figure 7.13 Bit error rate as a function of SNR for OFDM and SC-FDE in NLOS CM2.3 channels and CMOS PA non-linearity with 16-QAM constellations.

SC-FDE, shown in Section 7.2.1, for frequency selective channels and non-linear PAs (see Fig. 7.11).

For both OFDM and single-carrier modulation, as described above and in Chapter 2, a cyclic prefix is the repetition of the last L_{CP} symbols in a FFT block and enables efficient frequency-domain equalization by turning time-domain linear convolution with a channel into circular convolution. The idea in [Gho14] is to create a CP-SC signal but with nulls replacing the usual CPs. Work in [Gho14] proposes QAM symbol periods of 0.65 ns (1.536 Gs/s), with a null time of 0.67 μs (1024 QAM symbols). This is reasonable for urban multipath channels, as shown in Chapter 3. NCP-SC works by appending N_{CP} null symbols at the end of a block of N symbols, where the null symbols of one block are effectively the CP for next block of symbols. The number of data symbols per block is given by $N_D = N - N_{CP}$. Note that the size of the symbol block, N, is the same regardless of the CP size. As a result, the CP length can adaptively change the number of data bits per block on a per-user basis without altering the frame timing simply by puncturing the data.

The frame structure of the NCP-SC modulation can easily be designed to meet a 1 millisecond (ms) latency requirement, where a super frame (e.g., 20 ms) could be

7.3 High-Throughput PHYs

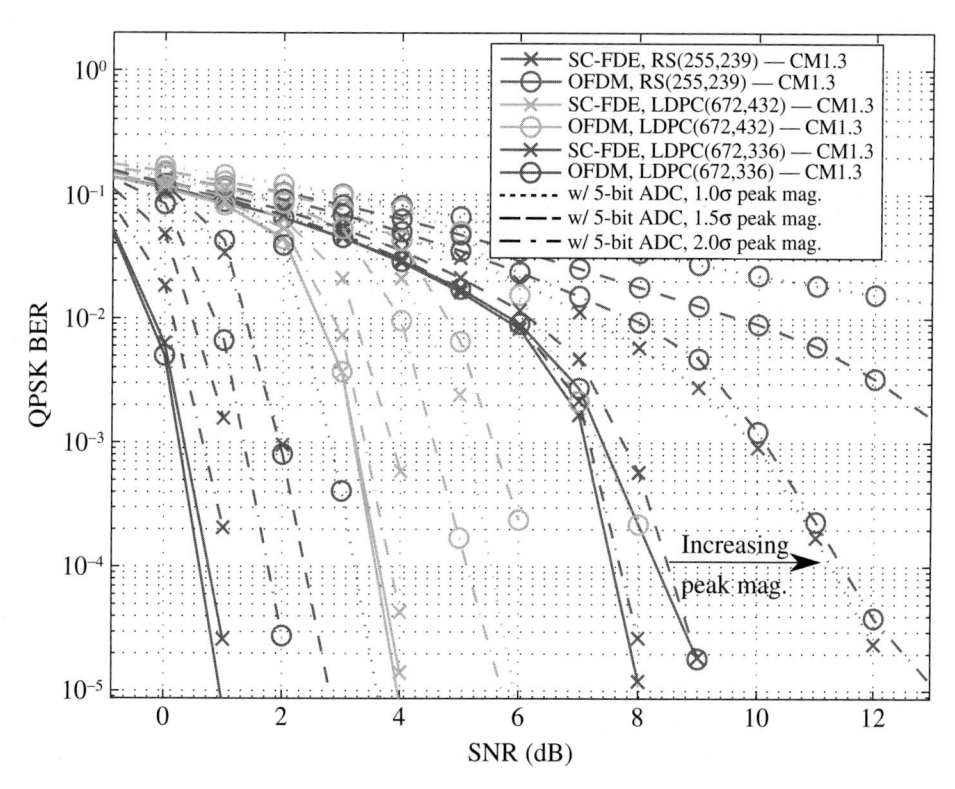

Figure 7.14 Bit error rate as a function of SNR for OFDM and SC-FDE in LOS CM1.3 channels and 5-bit ADC samples with QPSK constellations.

broken up into subframes (e.g., 500 μs each), and a subframe would contain some number of slots (e.g., 5 slots of length 100 μs). TDD operation would allow each subframe to be either a mobile uplink or downlink, or used as backhaul, such that each base station could dynamically use TDD however needed during a subframe [Gho14]. The decoding of the data will be similar between OFDM, SC-FDE, and NCP-SC, and it is possible that the complexity of the coding could dominate the computational complexity given the high data rates of these mmWave systems.

The following three advantages of NCP-SC are interesting to note when compared to current modulations being used for current-day LTE cellular and IEEE 802.11ad standards: 1) the null cyclic prefixes provide a dead time for ramping down and ramping up antenna beams so that the RF beams can be changed in between NCP-SC symbols without destroying the cyclic prefix property nor requiring additional guard time. This allows efficient switching between users and antenna beams within a time slot; 2) the null CP provides an easy way to estimate post-equalizer noise plus interference (i.e., in the null portion of the symbol, no desired signal energy would be present); and 3) NCP-SC has lower PAPR and better out-of-band emissions than OFDM, and is comparable to SC-FDE [Gho14]. The frequency-domain receiver can then operate by taking an

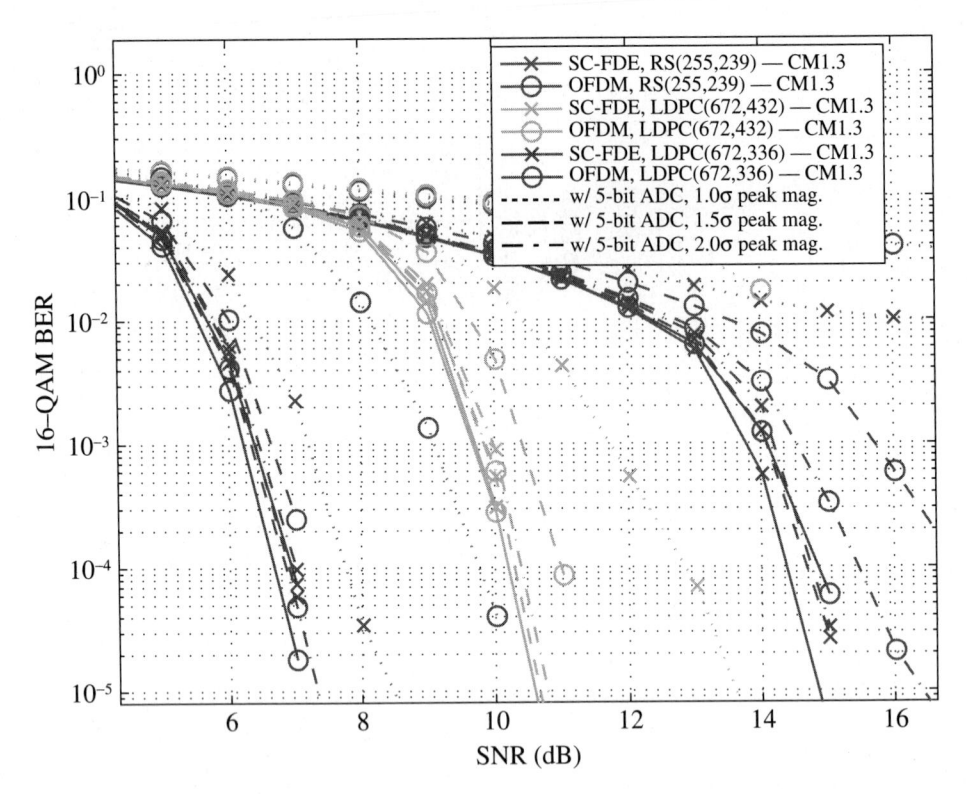

Figure 7.15 Bit error rate as a function of SNR for OFDM and SC-FDE in LOS CM1.3 channels and 5-bit ADC samples with 16-QAM constellations.

over-sampled FFT (e.g., a $2N$ point FFT) of the appropriate received block of N symbols. Note that [Gho14] uses a null CP system where the transmitter is turned off very briefly, instead of the pilot word insertion method used to generate a training sequence for the SC-FDE mode, or a cyclic prefix copied from the data in the OFDM mode of IEEE 802.11ad [802.11ad-11], which instead of null symbols uses the same known training symbols as CPs on each block.

It should be noted that the training prefix in conventional OFDM, such as used in the IEEE 802.11ad standard, is useful for compensating and tracking residual phase errors of low cost mmWave oscillators in the long packets, thus making the 802.11ad MAC more efficient with very low-cost hardware. The absence of this training prefix in NCP-SC may be compensated for by measuring data symbols or pilots, depending on the quality of hardware and the length of packets in an NCP-SC link. Training for phase and frequency offset correction may be performed by monitoring data streams, as is done in today's 60 GHz standards (as discussed in Section 7.3.3.3 and in [RCMR01]).

While much still remains to be studied, the preceding NCP-SC concept is an example of how future mmWave PHY layer designs may be able to achieve multi-Gbps mobile data rates by modifying present-day methods to exploit fast beamsteering between users, as well as new types of interference channels, time division duplexing, and multiple access schemes [Gho14].

7.3.3 Synchronization and Channel Estimation

One often-overlooked aspect of PHY algorithm design is synchronization. Before packets are able to be decoded, relevant channel parameters must be estimated. Some parameters, such as frequency offset and symbol timing, may be evaluated at lower frequencies in multi-band links (see Chapter 8 for a discussion on mmWave multi-band communication). Other parameters, including SNR and channel impulse response, must be evaluated in the frequency band actually used for communication. Once these parameters are estimated, the frequency offset is removed and the packet is decoded. As described in Chapter 2, channel parameter estimation exploits known training sequences (at the beginning of a packet in the *preamble* or periodically through *pilots*). Next, we highlight new synchronization features in mmWave links that do not take advantage of multi-band synchronization capabilities. The primary motivation for these features is complexity reduction.

7.3.3.1 Channel Estimation

Operations on Golay complementary sequences, which have been made available in most 60 GHz preambles (including IEEE 802.15.3c and IEEE 802.11ad), represent the current innovation of mmWave channel estimation. A complementary pair of binary, length-N Golay sequences, for example $\mathbf{a}_N$ and $\mathbf{b}_N$, exhibit the desirable property that the sum of their separate cyclic auto-correlations is zero except when auto-correlation is perfectly aligned. That is, if $\mathbf{a}_N = \{a_n\}_{n=0}^{N-1}$ and $\mathbf{b}_N = \{b_n\}_{n=0}^{N-1}$ for $a_n, b_n \in \{-1, +1\}$ $\forall n$, then

$$\sum_{n=0}^{N-1} a_n a^*_{\mathrm{mod}[n+k,N]} + b_n b^*_{\mathrm{mod}[n+k,N]} = \delta_K\left[\mathrm{mod}\left[k, N\right]\right] \qquad (7.11)$$

where $\delta_K[\cdot]$ is the Kronecker delta function, which equals 1 at argument 0 and is 0 for every other argument, and $\mathrm{mod}[\cdot, N]$ is the modulo-N operation [Gol61]. This zero-sidelobe property is desirable because the sidelobes of cyclic auto-correlation with all known *single* binary sequences *are never zero* [LSHM03]. Hence, cyclic auto-correlation with Golay sequences can be used to uncover each tap in the time-domain channel impulse response without degradation due to sidelobe effects. For example, consider the design of a Golay-based training sequence for channel estimation, as illustrated in Fig. 7.16 [RJ01][KFN+08].

This Golay-based auto-correlation for channel estimation procedure exhibits significantly lower processing delay and complexity than alternative methods. For example, most popular channel impulse response estimation methods, such as least-squares channel estimation, must store all channel-distorted training samples before estimation begins ($2 \times N \times N_s + 3N_{\mathrm{CP}}$ sample delay in Fig. 7.16), whereas the Golay correlation procedure for channel estimation can begin halfway through the training sequence ($N \times N_s + 3N_{\mathrm{CP}}$ sample delay in Fig. 7.16) and produces small total delay such that minimal buffering is needed to begin symbol decoding. Further, the unique properties of Golay sequences have been shown to substantially reduce the complexity of correlators such that only $\log_2(N)$ multiplications and $2\log_2(N)$ additions are required for each sequence correlation, whereas generic sequence correlations require N multiplications and $N-1$ additions, respectively, without compromising memory requirements [Pop99]. Because delay,

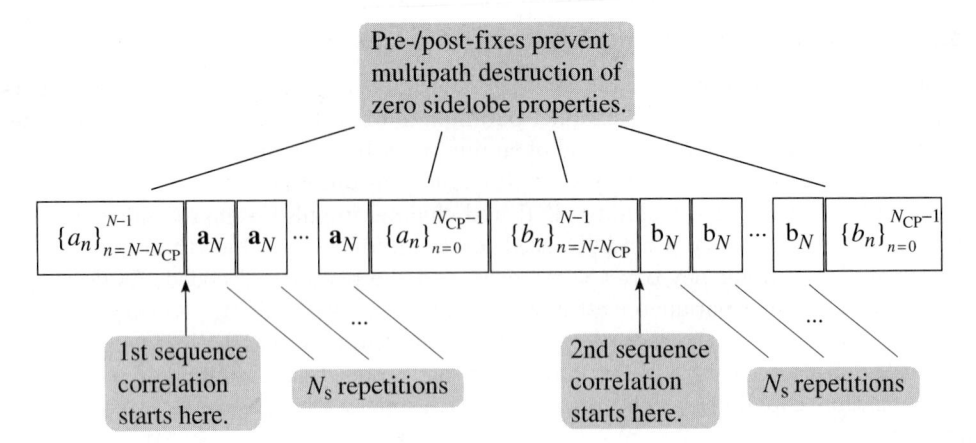

Figure 7.16 Illustration of how length-N complementary Golay sequence pair, $\mathbf{a}_N$ and $\mathbf{b}_N$, may be used to construct a training sequence that enables channel impulse response estimation through complementary correlation. Pre- and post-fixes are added before and after N_s repetitions of each complementary sequence to prevent excess multipath from disrupting zero-sidelobe properties. At the receiver we correlate with the channel distorted versions of each N-length complementary sequence and add them together to yield an estimate of a single tap. Delayed correlations are computed for each tap, and each of the N_s repetitions is used to improve estimate robustness in the presence of noise.

storage, and processing complexity are at a premium in mmWave devices, the advantage of Golay sequences for synchronization is significant.

While Golay sequences have a large impact on complexity reduction in 60 GHz WLAN and WPAN, it is unclear whether Golay sequences can be fully utilized in mmWave cellular networks for channel estimation. If beam steering or precoding is applied simultaneously to multiple users over separate time/frequency allocations, channel estimation sequences must be independently transmitted for each user. To minimize overhead associated with channel estimation, channel estimation reference sequences would be delivered on a per-user basis only on the resources allocated to each user. This scenario, however, requires more time and frequency domain flexibility than is currently provided by Golay sequence channel estimation processes.

7.3.3.2 Packet Detection

Packet detection can also take advantage of Golay sequences to reduce complexity. The complexity of standard correlation-based detectors for mmWave communication is excessive due to massive sample rates and increased pass loss when searching for the proper beams with low gain antennas. In addition to 1-bit processing with Golay sequences, correlation windowing and forgetting factor (γ) inclusion limit complexity. For example, consider the 1-bit (and, hence, energy normalized) channel-distorted receive data sequence $\{r_k\}_k$ and the packet detection reference sequence $\{s_n\}_{n=0}^{N-1}$. The standard correlation-based detector evaluates the positive metric $z_k = |\sum_{n=0}^{N-1} r_{k+n} s_n^*|$ with respect to a packet detection threshold for each index k. Even though Golay sequences enable reduced-complexity correlation, the packet detection sequences for mmWave communication may be very long (e.g., thousands of symbols for sufficient processing gain), leading to a large number of multiplications and additions. Further, the accumulation of each product, even in pipelined stages, is nontrivial due to large N (since the sum must be represented by at least $\lceil \log_2(N+1) \rceil + 1$ signed bits). It has been shown that buffer size may be reduced to $M < N$ through the approximation $z_k \approx (1-\gamma) |\sum_{n=M}^{N-1} r_{k+n} s_n^*| + \gamma z_{k-1}$, for $0 < \gamma < 1$ where it is assumed that the full correlation was computed at some point in the past [BWF+09].

7.3.3.3 Adaptive Training

In outdoor cellular networks with centralized (base station) control, adaptive training (for synchronization functionality) through user-specific pilots is already available. WLAN and WPAN links, however, normally provide a single preamble training structure for all users in all operating conditions. Unfortunately, millimeter-wave preambles require substantially more sequence repetitions compared with their microwave counterparts. As discussed in Chapter 2 and earlier in this section, sequence repetitions are necessary for packet detection, synchronization, and channel estimation processing gain to overcome excessive path loss before compensation methods, for example, antenna beamforming, are negotiated. The optimal number of sequence repetitions, however, depends on the link application. Consequently, preamble length adaptation may prove important for efficiency in mmWave links. For example, 60 GHz standards provide an adaptive preamble that can reduce overhead and/or increase robustness in special operational scenarios [BWF+09]. During multimedia streaming, for example, packets are sent continuously over a large duration. In this case, assuming no link disruptions through shadowing, antenna beamforming can be tuned periodically during the stream rather than being completely reconfigured from scratch periodically, reducing the necessary spreading factor of the preamble and data in the middle of the stream. Further, synchronization parameters (such as frequency offset) may be evaluated in a control loop throughout the stream, leading to less training required for its estimation during the stream. Alternatively, some traffic scenarios cannot make overhead reduction assumptions, such as in broadcast scenarios, when critical control data is being exchanged, or when data has not been exchanged between the transmitter and targeted receiver for a large period of time. In these scenarios, more robust preambles with higher spreading factors are employed.

7.4 PHYs for Low Complexity, High Efficiency

OFDM and SC-FDE (and their multi-user multiple access counterparts OFDMA and SC-FDMA) are expected to remain as the top two modulation and equalization candidates for mmWave communication in the near future. Alternative modulation strategies, however, have been considered, some of which are available as optional features in 60 GHz standards. In this section we summarize these alternative strategies, giving special focus to the low-complexity or high-efficiency features that make them attractive. For example, some modulation strategies are highly spectral-efficient at low SNR and/or are able to tolerate non-linearities in the power amplifier whereas other strategies are able to be implemented with a small device footprint and operate with extremely small power consumption. All alternative strategies, however, suffer greatly in terms of spectral efficiency, especially at high SNR where higher constellation orders are used with OFDM and SC-FDE for massive data rate capabilities. Further, mitigation of frequency selectivity is not necessarily straightforward or efficient in low-complexity designs. Despite these drawbacks, the modulation strategies presented in this section remain important because one advantage of mmWave communications is that, because of the massive spectrum that is available, spectral efficiency is not as critically important as it is in microwave bands. These low-complexity PHY strategies will likely be very important for many early mmWave wireless implementations.

7.4.1 Frequency Shift Keying (FSK)

FSK has the advantage that its implementation is extremely simple, especially with binary frequency shift keying (BFSK). Consider that modulation can be performed directly through a VCO, where the voltages are appropriately tuned to map input data to discrete frequencies. There are several simple candidate demodulation techniques that result in much simpler implementation than baseband I/Q sampling (resulting from superheterodyne or direct conversion receivers, as well as ultra-simple non-coherent detection). Fig. 7.17 shows an example FSK demodulator used in [NNP+02] at 60 GHz to accomplish 1 Gbps data rates.

Despite the simplistic implementation, there are many FSK drawbacks. While FSK has been shown to tolerate multipath more than un equalized Amplitude Shift Keying (ASK is discussed next), FSK still suffers in highly frequency selective channels [AB83][Rap02]. Further, FSK exhibits significantly less spectral efficiency than QAM/PSK constellations [DA03]. The most critical drawback of FSK, however, is that the simple demodulator architecture removes many of the advanced features that are enabled by baseband I/Q sampling. For example, without the complex baseband representation at the receiver, beamforming strategies are not directly available. The absence of the complex baseband model also means that training sequences must be evaluated after constellation demapping and, consequently, packet synchronization may need to be partially handled by analog circuits, making packet-based communication very cumbersome. As a result, FSK has only appeared in special purpose streaming scenarios on dumb, cheap, low-power devices.

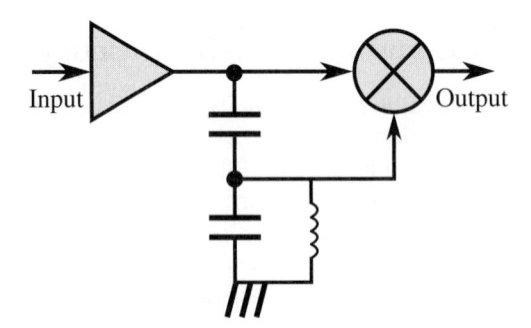

Figure 7.17 Example of an FSK demodulator drawn from a description in [NNP+02]. This stage is followed by a low pass filter to produce total demodulator output.

7.4.2 On-Off, Amplitude Shift Keying (OOK, ASK)

The motivation and advantage of ASK is very similar to FSK. With ASK, we can directly generate wireless communication waveforms by adjusting the amplitude of the carrier according to the input data sequence. Although any number of levels can be considered, the largest reduction in complexity is observed through On-Off Keying (OOK), that is, ASK with two amplitude levels: (1) full power, and (2) zero power.[5] Hence, the receiver determines if there is a signal or if there is not a signal, and maps this to a 1 or 0, accordingly. OOK transceivers can be implemented without baseband sampling and may use ultra-simple non-coherent detection methods. Clearly, this will suffer from the same drawbacks as FSK. Baseband sampling can be used to provide more implementation flexibility, but this will sacrifice many of the complexity savings that make ASK/OOK attractive in the first place [BFE+04].

7.4.3 Continuous Phase Modulation

Continuous phase modulation (CPM) waveforms are modeled, generally, in continuous-time at baseband by

$$x\left(t, \{\alpha[n]\}_{n=0}^{N-1}\right) = \sqrt{\frac{2E_s}{T_s}} e^{j\phi\left(t, \{\alpha[n]\}_{n=0}^{N-1}\right)} \tag{7.12}$$

where E_s is the energy per symbol, T_s is the symbol time, and $\alpha[n] \in \{\pm 1, \pm 3, \ldots, \pm(M-1)\}$ is the nth M-ary data symbol. As the name implies, CPM waveforms exhibit continuous phase functions

$$\phi\left(t, \{\alpha[n]\}_{n=0}^{N-1}\right) = 2\pi h \sum_{n=0}^{N-1} \alpha[n] q\left(t - nT_s\right) \tag{7.13}$$

given that

$$q(t) = \int_{-\infty}^{t} g(\tau) d\tau \tag{7.14}$$

where $h \in \mathbb{R}^+$ is a fixed parameter called the *modulation index* and

$$g(t) = \begin{cases} 0 & t \notin [0, LT_s] \\ \text{continuous} & \text{otherwise} \end{cases} \tag{7.15}$$

is the pulse function with constraint

$$\int_{-\infty}^{\infty} g(t) dt = \frac{1}{2}. \tag{7.16}$$

---

5. OOK transmitters may also utilize alternate mark inversion (AMI), in which the transmitted symbols are successively phase shifted. This phase structure can be leveraged to improve receiver sensitivity without I/Q baseband sampling [AF04].

The advantages of CPM for mmWave communication are as follows.

1. CPM signals feature constant amplitudes, making them unaffected by non-linearity in power amplifiers.

2. The smoothness of CPM signal phase makes for waveforms that exhibit low spectral spillover into adjacent channels.

3. Linear basis decompositions (originally provided by Laurent) make modern equalization practices feasible (including frequency domain equalization) with I/Q baseband digital sampling [MM95][Dan06][PV06].

4. CPM is more tolerant than single-carrier QAM of hardware impairments, including low-precision ADCs and phase noise [NVTL$^+$07].

Unfortunately, although the performance of binary CPM communication is very competitive at low SNR, as evidenced by its presence in the GSM cellular communication links (GMSK) and in deep space communication links, CPM communication for $M > 2$ suffers from severely degraded BER when compared to M-QAM for a fixed SNR [AAS86]. Consequently, CPM is only expected to be applied to mmWave links operating in the low SNR regime. Future outdoor cellular mmWave links that hope to maximize range in power-limited environments — or when synchronizing or searching for proper beam directions — may find CPM attractive for this purpose.

7.5 Future PHY Considerations

In this section, we preview two non-conventional PHY design strategies that may support lower device cost and power consumption (Section 7.5.1) and may multiply rates beyond what is currently achievable in mmWave links (Section 7.5.2).

7.5.1 Ultra-Low ADC Resolution

Section 7.2.1 discussed the advantage of low-precision ADCs, with respect to power consumption and device complexity, as the sampling rates of wireless communication signals reach many billions of samples-per-second. Since wireless networks are expected to continue to stretch into larger bandwidth allocations at ever-increasing operating frequencies, this concern will not quickly be alleviated, as was shown in Chapter 6. In Section 7.3.2, we compared the performance of OFDM and SC-FDE modulation and equalization for high throughput communications as a function of ADC resolution and concluded that SC-FDE offers better performance with less ADC complexity. This is an important discussion for the already-available 60 GHz standards, but a more proactive approach is needed for the future of mmWave communication. Although some work has designed ADCs tailored to specific modulation formats [NLSS12], a more effective approach is to redesign the PHY for low ADC resolution. Specifically, we need to better understand the fundamental limits of communication with low ADC precision and, if these fundamental limits are reasonable, design practical algorithms suitable for low ADC precision that approach these limits.

7.5 Future PHY Considerations

Basic capacity studies have already provided significant promise for this approach [SDM09]. Fig. 7.18 demonstrates that capacity losses for 3-bit ADCs are small, assuming reasonable SNR values at the receiver and perfect link synchronization. Despite this promise, there are also many hurdles to overcome. Estimation of real/complex-valued parameters during synchronization (e.g., frequency offset estimation, channel estimation) is very sensitive to bit precision. Fundamental studies of general parameter estimation as a function of bit precision have been conducted with less promising results [PWO01][DM10]. Acceptable performance is only available, at least in the standard parameter estimation scenario, with closed-loop ADCs that feature dithering and post-dithering gain control.[6] It is not known, however, whether such closed-loop functionality will be compatible with the operation of an entire wireless communication receiver or if parameter estimation can be isolated when training data enters the ADC. The application of low precision ADCs for mmWave becomes more promising when there are additional receive antennas, which effectively expand the received constellation space [MH14], allowing high spectral efficiency.

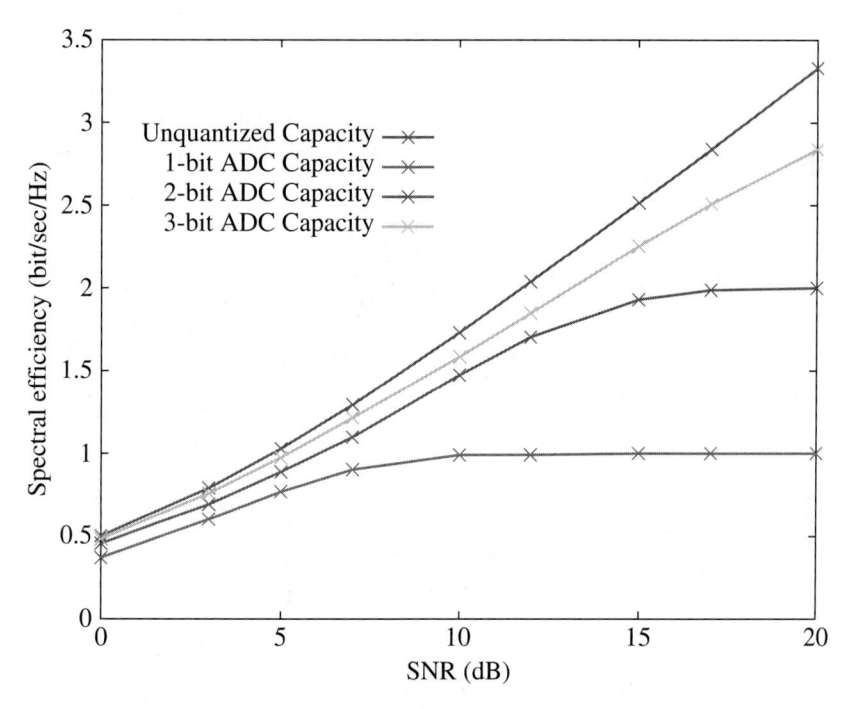

Figure 7.18 Capacity comparison with 1-, 2-, 3-, and, ∞-bit ADC precision for a discrete memoryless channel with perfect synchronization [SDM09].

---

6. Dithering is a process by which random noise is added to ADC inputs in order to prevent differential non-linearity effects. Post-dithering gain control is not to be confused with AGC, which normalizes the variance of the received signal. In the context of parameter estimation, the amplitude of the dither noise and the post-dither gain weight are controlled through a feedback loop.

Our earlier comparisons in Section 7.3.2 suggest that equalization of frequency selective channels will also be very challenging at very low bit resolution. One approach to overcome multipath-induced frequency selective fading is to design an altogether new modulation and equalization digital processing architecture that is less sensitive to bit precision. An alternative approach, which has seen interest recently, is to perform equalization procedures in analog, or at least partially in analog, through mixed signal equalization [SB08][FHM10][HRA10]. Example analog and mixed equalization architectures are shown in Fig. 7.19.

7.5.2 Spatial Multiplexing

MIMO spatial multiplexing recently provided a major capacity enhancement for microwave links, perhaps best demonstrated by WLAN standard IEEE 802.11n, (at 2.4 and 5 GHz carrier frequencies) [802.11-12], next-generation WLAN standard IEEE 802.11ac [802.11-12], and mobile broadband standard 3GPP-LTE (at carrier frequencies ranging from 700 MHz to 3.5 GHz) [3GPP09]. MIMO adds multiple antennas, complete with analog and digital processing chains, to each transmit and receive unit. As discussed in Chapter 2, if the impulse response between each transmit and receive antenna pair is sufficiently different (often characterized through statistical correlation), independent data can be transmitted over $N_s = \min\{N_r, N_t\}$ independent spatial streams through spatial multiplexing where N_r and N_t are the number of receive and transmit antennas, respectively [AAS+14]. At microwave frequencies, favorable channel correlation

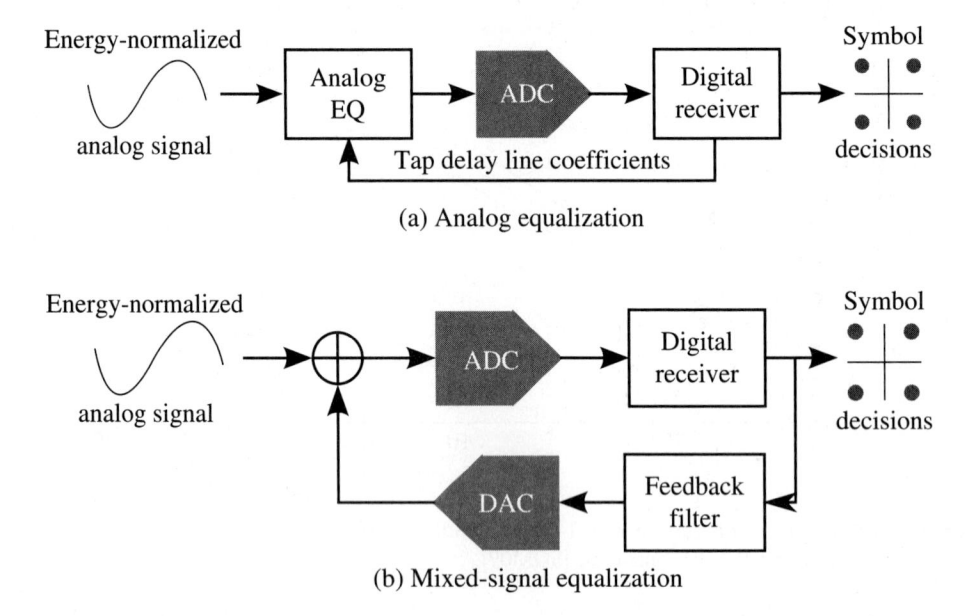

Figure 7.19 Analog and mixed signal equalization architectures can reduce the ADC bit resolution of the overall receiver (assuming that synchronization and other receiver functionality can be maintained). Mixed signal equalization can be considered a DFE with an analog feedback filter and a digital feedforward filter.

7.5 Future PHY Considerations

properties (which allow MIMO to produce sizable capacity gains) occur in NLOS channels where a rich number of scatterers provide many different path lengths at varying angles of arrival at the receiver. In LOS channels, work in [SWA+13] showed that all antennas transmit and receive from roughly the same angles with roughly the same path lengths. Hence, the channels are not sufficiently different and MIMO spatial multiplexing does not work well in LOS channels at microwave or UHF frequencies.

Thus far, there has not been a significant push to consider MIMO spatial multiplexing at mmWave frequencies for two main reasons. First, the challenge of multi-gigabit-per-second mmWave communication is daunting enough without the addition of spatial multiplexing, which adds significant complexity (for example, MIMO increases digital baseband complexity and ADC bit precision requirements). Second, a common perception has been that the high path loss and the lack of scattering or reflections make mmWave channels less favorable for MIMO. Early work showed that there are only a few scatterers available in indoor channels [Smu02][XKR02], and antenna resources must be used to compensate for weak paths through beamforming [PWI03]. However, as shown in Chapter 3, recent works in urban and suburban outdoor channels [SWA+13][AWW+13][RSM+13][RDBD+13][SR14][AAS+14][RQT+12] show that non-LOS (NLOS) as well as LOS mmWave channels may be able to exploit MIMO due to the surprisingly large number of reflections and scattering sources in outdoor environments.

It should be clear that there is a case to be made for mmWave MIMO in the LOS regime (where it is largely unfavorable for microwave frequencies), as well as for NLOS outdoor channels. To better understand why, consider a LOS link using uniform linear MIMO antenna arrays (ULAs) at the transmitter and receiver, as illustrated in Fig. 7.20. If we further assume frequency flat channels with normalized and equal path loss between each antenna pair (reasonable in LOS channels if $r \gg L_t$ and $r \gg L_r$), then the channel impulse response of each transmit-receive antenna pair is captured through matrix $\mathbf{H} \in \mathbb{C}^{N_r \times N_t}$ where $[\mathbf{H}]_{k,\ell} = \exp(j2\pi r_{k,\ell}/\lambda)$ represents the only non-zero impulse response coefficient between the ℓth transmit antenna and the kth receive

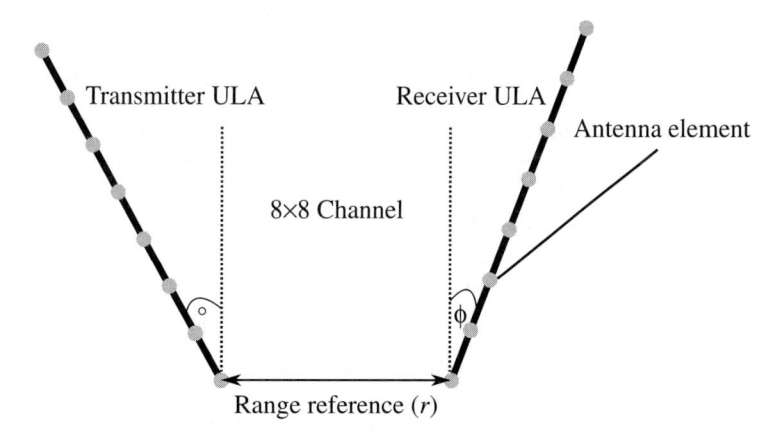

Figure 7.20 LOS MIMO channel with arbitrary uniform linear array (ULA) alignment and $N_r = N_t = 8$ elements on each ULA. The range reference of the link is denoted by r, and the total antenna array lengths at the receiver and transmitter are L_r and L_t, respectively.

antenna, λ is the operating wavelength, and $r_{k,\ell}$ represents the shortest distance between transmit antenna index ℓ and receive antenna index k. This formulation reveals an important observation: *MIMO LOS channels become less correlated as the operating frequency increases*, assuming fixed ULA and range dimensions.

Prior work on this ULA model has shown, in fact, that perfect MIMO channels with orthogonal rows and columns can be created by satisfying the condition

$$\frac{L_t L_r}{(N_t - 1)(N_r - 1)} = \frac{\lambda r}{\min\{N_r, N_t\} \cos(\theta) \cos(\phi)} \quad (7.17)$$

through the use of closed form approximations of $r_{k,\ell}$ [BOO07]. If we consider the simple, yet practical case where $L_t = L_r := L$, $N_t = N_r := N$, and $\theta = \phi = 0$ then (7.17) reduces to

$$L = (N-1)\sqrt{\lambda r / N} \quad (7.18)$$

which suggests that the primary tradeoff for LOS MIMO operation is between range (r), wavelength/frequency, and array length. Table 7.5 shows some example configurations that meet this condition. In other words, LOS MIMO was not feasible at microwave frequencies because the antenna dimensions required to produce *good* MIMO channels were so large that they were not practical (see Table 7.5). We can also reformulate (7.18) in terms of range to find

$$r = \frac{L^2 N}{\lambda (N-1)^2} \quad (7.19)$$

$$= \frac{r_R N}{\pi} \quad (7.20)$$

where r_R is the well-known *Rayleigh range* parameter, which is used in optics and antenna theory "to characterize the distance a beam propagates without spreading appreciably"

Table 7.5 Example antenna array dimensions, range, and frequency configurations that enable perfect orthogonal MIMO channels with (5, 10, and 50) antenna elements. Cases in which r and $L_t = L_r$ are of the same order of magnitude (e.g., 2.45 GHz at 10 m) are not considered practical because the model assumptions used to derive the orthogonal MIMO channel condition are violated.

Frequency	$L_t = L_r$	r
2.45 GHz	(1.98 m, 3.15 m, 7.67 m)	10 m
2.45 GHz	(6.26 m, 9.96 m, 24.25 m)	100 m
2.45 GHz	(19.8 m, 31.5 m, 76.7 m)	1000 m
60.00 GHz	(0.40 m, 0.64 m, 1.55 m)	10 m
60.00 GHz	(1.26 m, 2.01 m, 4.90 m)	100 m
60.00 GHz	(4.00 m, 6.36 m, 15.5 m)	1000 m
240.00 GHz	(0.20 m, 0.32 m, 0.77 m)	10 m
240.00 GHz	(0.63 m, 1.00 m, 2.45 m)	100 m
240.00 GHz	(2.00 m, 3.18 m, 7.75 m)	1000 m

[GW01]. Note that, for a fixed array dimension and number of antenna elements, an order of magnitude increase in range requires an order of magnitude increase in operating frequency. For a fixed operating frequency and number of antenna elements, an order of magnitude increase in array dimension leads to an increase of two orders of magnitude in range. This makes LOS MIMO attractive for applications in mmWave cellular like fronthaul or backhaul.

MmWave communication could exploit practical LOS MIMO strategies, leading to wireless links with staggering rates (hundreds of gigabits per second or more). One major limitation of LOS MIMO, however, is the reduced set of operational scenarios that satisfy the LOS model and the LOS MIMO condition, although potential scenarios could include outdoor cellular, backhaul, and fixed home installations. Further limitations include the need for antenna arrays that adapt their mechanical or electrical configuration according to operational range [MSN10] and the large complexity required to implement standard MIMO receiver protocols, especially on high-dimensional platforms with giga-sample-per-second ADCs [PNG03]. One advantage of the LOS MIMO configuration, however, is the orthogonality of the MIMO channel with equal magnitude elements. This could lead to low-complexity analog spatial equalization to separate the streams before digital conversion [Mad08][SST+09]. These tradeoffs will be further explored in the near future. In 2013, DARPA launched the 100G program to "build and test an airborne-based communications link with fiber-optic-equivalent capacity and long reach that can propagate through clouds and provide high availability" through LOS MIMO mmWave links [DAR].

7.6 Chapter Summary

PHY mmWave algorithm design features drastically different design constraints when compared with microwave links. Because analog circuit technology is being pushed to its limit and because link budgets are more constrained than ever before, impairments outside of the commonly assumed linear system with additive Gaussian noise model are likely. Consequently, PHY algorithm designers must learn to explicitly take into account ADC bit precision, PA non-linearity, and phase noise. In this context, equalization will likely continue through frequency domain representation, although whether modulation strategies remain in the frequency domain (OFDM) or switch to the time domain (SC-FDE) remains to be seen. As an extensive analysis in this chapter showed, both modulation formats have merit in disparate operating regimes, and new modulation schemes based on the existing formats are likely to develop as new mmWave products and services are explored. More understated features of communication such as synchronization have also changed; the complexity bottleneck and higher path loss of mmWave links, combined with the added benefit of steerable antennas and beamforming, will require innovation for channel estimation and packet detection. As we look toward the future, many more innovations are likely for mmWave PHYs, including ultra-low receiver bit precision to further reduce complexity requirements, and MIMO arrays that work in both NLOS and LOS channels to multiply data rates even higher.

Chapter 8

Higher Layer Design Considerations for MmWave

8.1 Introduction

As mmWave communication is primarily a physical layer (PHY) communication technology, much of the focus in this book has been on the physical layer and below, including circuits, antennas, and baseband considerations. Specific features of mmWave communication, however, have influenced the design of higher layers, especially the data link layer and the medium access control (MAC) protocol sub-layer (see Section 2.10 from Chapter 2 for background on the layer terminology). For example, the wide use of highly directional and adaptive antennas (see Chapters 3 and 4) requires special MAC protocols. This chapter surveys higher layer issues for future communications at mmWave frequencies.

We begin the chapter with a review of challenges for networking in mmWave devices in Section 8.2. We highlight specific issues that pertain to the mmWave PHY layer and how they impact the higher layers. Then, throughout the chapter, we discuss specific topics in more detail. Since beam steering is important in mmWave, we explain how beam steering has been incorporated into some mmWave MAC protocols in Section 8.3. Multi-room coverage is a consideration for certain indoor mmWave systems due to high penetration loss and little diffraction. This can be achieved by multi-hop operation, specifically relays, as described in Section 8.4. Multimedia communication is one of the most important applications for mmWave, especially at 60 GHz. Consequently, 60 GHz systems support functionality for multimedia communication exposed by the physical layer, especially unequal error protection as described in Section 8.5. Coverage is a challenge in mmWave systems given the use of directional antennas. In Section 8.6, we describe multi-band strategies for mmWave systems where coverage is improved by taking advantage of lower frequencies. We conclude the chapter with some highlights from recent coverage and capacity analysis of mmWave cellular systems in Section 8.7. Because many features of MAC protocols are specific to standards, we use some examples from standards but leave more detailed discussion to Chapter 9. Although many examples are drawn from 60 GHz, throughout the chapter we speculate on how these concepts may be applied to 5G mmWave cellular and backhaul.

8.2 Challenges when Networking MmWave Devices

Building a network with mmWave devices offers new challenges that arise from characteristics of the mmWave communication channel. In this section we review key challenges, focusing on the short range links that will be used in PAN and WLAN applications. Most of the difficulty faced when networking at mmWave results from using beamforming at the transmitter and the receiver. This makes certain functions performed by the MAC protocol more difficult, for example, collision detection, but also offers the potential for high performance, for example, through spatial reuse. This section reviews the concept of directional transmission and reception, and then explains its impact on device discovery, collision detection/avoidance, channel reliability, and spatial reuse.

8.2.1 Directional Antennas at the PHY

MmWave systems will make great use of directional transmission to provide higher link quality. To explain the motivation for directional transmission, we use the example in [SMM11][SZM+09]. Consider a simple communication link in free space. Referring to the Friis free space equation (3.2), the received power between isotropic antennas in free space scales as λ^2. In terms of decibels, the power loss incurred is $20 \log_{10} \lambda$. For example, in 60 GHz systems, $\lambda = 5$ mm. Compared with the ISM band of 2.4 GHz, where $\lambda = 12.5$ cm, and ignoring excess path loss due to oxygen absorption or environmental factors, 60 GHz systems using isotropic (omnidirectional) antennas have an additional free space path loss of $20 \log_{10} 0.125 - 20 \log_{10} 0.005 = 28$ dB. This means that 60 GHz systems require 28 dB of additional gain to make up the free space path loss difference. This was also shown in Section 3.7.1, (see Figs. 3.25 and 3.30) where outdoor mmWave channels were measured, and resulting large-scale propagation models for omnidirectional antennas were compared at 1.9 GHz, 28 GHz, and 73 GHz. Assuming similar allowable transmit power limitations at both mmWave and microwave frequencies, higher power cannot be used to close this gap. Higher transmit powers would also result in prohibitively low battery life for mobile applications.

Antennas provide gain that can overcome loss predicted by the free space equation, as discussed in Chapter 3. The gain of an antenna is the amount of additional power that is captured (when used either to transmit or receive) compared with an ideal isotropic antenna. As shown in Chapter 3, the antenna gain is related to a quantity known as the *aperture*, which is the effective area (or the electrical area) of the antenna, often directly proportional but not necessarily identical to the physical area of the antenna. The gain (in linear scale) is proportional to the antenna aperture, and it scales inversely with λ^2. Fixing its physical size, the gain of an antenna at 60 GHz is exactly 28 dB higher than the gain of an antenna at 2.4 GHz. Consequently, as shown in Chapter 3, if the antennas can consume the same amount of area, the free space path loss gap can be closed, since the additional free space path loss at higher frequencies is precisely made up by the additional gain provided by an antenna with an equivalent size. Assuming the same size is used at both the transmitter and receiver, the 60 GHz link will actually have

8.2 Challenges when Networking MmWave Devices

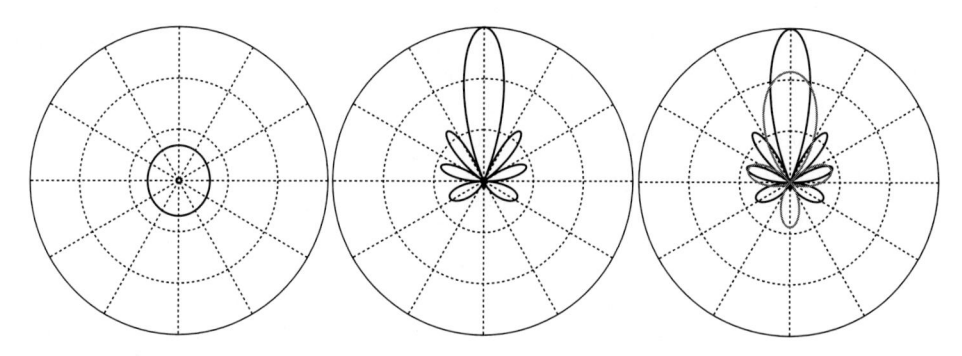

Figure 8.1 The concept of beamforming. (left) An omnidirectional beam pattern. (middle) A directional beam pattern where a main lobe has high gain presumably in the direction of transmission or reception. (right) Variation of the beam pattern. Wider beams tend to have lower gain than narrower beams but cover a wider area.

28 dB + 28 dB − 28 dB = 28 dB more gain in free space compared with the 2.4 GHz link! As was shown in Chapters 3 and 4 and [RRE14], we can more than overcome mmWave path loss by increasing the antenna size at either the transmitter or receiver. In practice, these directional antenna gains may be limited when there are effective isotropic radiated power constraints due to governmental regulations, thus limiting the maximum antenna gain that may be allowed in practice.

A directional antenna is one way to achieve high gain. As illustrated in Fig. 8.1, a directional antenna has a large gain in a particular direction and a lower gain in other directions. An (ideal) omnidirectional/isotropic antenna has a similar gain in all directions. Several different directional antenna designs were discussed in Chapter 4.

In a line-of-sight (LOS) channel, the most effective and predictable communication with a directional antenna occurs when the transmit directional antenna is pointed toward the receiver and the receive directional antenna is pointed at the transmitter – this is called *boresight alignment*[1]. If the antenna has a fixed beam pattern, the transmitter and receiver would have to be physically pointed to one another. This is reminiscent of how a dish antenna for satellite TV is pointed very accurately toward the satellite in space. In non-line-of-sight (NLOS) channels, the pointing of a directional antenna is more complicated. The best pointing may be toward one or more dominant reflections within a single beam, as shown in Chapter 3 [SR14a]. In NLOS channels, it is desirable to have more complicated beam patterns that can place energy on multiple propagation paths, which requires some adaptivity on the part of the antenna.

Antenna arrays are an approach for implementing a flexible adaptive antenna. As discussed in Chapters 2 and 4, antenna arrays consist of multiple antenna elements; the set of elements defines the aperture of the antenna. On the receive side, for example, phasing and summing the outputs of each antenna and summing the signals gives an effective directional beam. By changing the phase in the summation, it is possible to steer the main lobe of the antenna in different directions. This is called a *phased*

---

[1]. Early evidence shows that even in mmWave LOS channels, there may be some value in combining beams that are on boresight with other beams that involve reflections, as Chapter 3 shows that radar cross sections of large surface scatterers could yield paths that are even stronger than free space.

array. Phased arrays, and more generally, adaptive arrays in which phase and amplitude weights are employed, have been the subject of much research for many years [Muh96] [Ron96][LR99b][Tso01][MHM10][R+11]. They are widely used in radar systems [Fen08]. Because of the small wavelength of mmWave, it is easy to imagine antenna arrays with hundreds of elements or more [RRC14].

Directional transmission and reception impacts several aspects of networking mmWave devices. The main reason, as pointed out in [SMM11] and [SCP09], is that there is (most likely) no *omni-directional transmission mode in mmWave systems* since antenna gain at both the transmitter and receiver is required to achieve good performance on the link. This is in contrast to lower frequency work on networks with directional transmission and reception [TMRB02][CYRV06][KSV00][GMM06], in which either a separate omnidirectional transmission is incorporated [KR05] or out-of-band busy tones (a signal on one carrier that indicates a channel on another carrier is occupied) were employed [HSSJ02][YH92]. Consequently, the entire network needs to operate assuming that some degree of directional transmission and reception is always required. Alternatively, multi-band operation is required where coordination occurs over a lower frequency (e.g., 2.4 GHz in WLAN or 1.9 GHz in cellular). Assuming the network operates entirely at mmWave, this raises several important questions:

- What happens if a new node enters a network? How do other nodes know to point their beams to a new node? This is the *device discovery*, or *neighbor discovery*, problem.

- What happens if a third node wants to talk with two nodes that are already communicating? Since the nodes are communicating with directional antennas, the third node sends packets but does not get a reply, assuming that there is a collision, backing off, and retransmitting its packets. This is known as the *deafness* problem [CV04].

- What happens if an interfering node is not prevented from transmitting? This is the *hidden node* problem.

- What happens if a node is prevented from communicating even when it will not interfere with other nodes in the network? This is the *exposed node* problem.

- What happens if the direct transmission path is temporarily blocked? This could happen with the deployment of transmitters and receivers in clutter. This is known as the *path blockage* problem.

- If pairs of nodes cause little interference to each other, shouldn't they be able to communicate at the same time? This is known as the *underutilization* problem.

The remainder of this section provides insight into how these questions, applicable to many wireless systems, are being answered for mmWave systems. The focus is on specific issues pertaining to WPAN and WLAN 60 GHz systems, but the implications on 5G mmWave cellular are also mentioned. Note that as backhaul networks may be multi-hop, many of the insights from WPAN (which operates in an ad hoc network) may also apply to backhaul networks.

8.2.2 Device Discovery

Device discovery is a general term for identifying neighbors in a network. An example device discovery scenario in a personal area network (PAN) is illustrated in Fig. 8.2. In some cases device discovery is done manually, as in the case of a download of a file from a kiosk. In general, however, device discovery must be performed automatically by the network without manual user intervention. Device discovery is performed frequently to allow for new additions to the network and changes in the network topology.

Device discovery is challenging at mmWaves due to the use of directional antennas. In a conventional network without directional antennas, a new device, upon powering up, might simply listen on several frequencies for beacons sent from other devices to find neighbors. In a mmWave network, directional antennas at both ends of the link require that in order to detect a neighbor, the new device must point its antenna in the direction of the neighbor while at the same time the neighbor sends a beacon in the direction of the new device. The inclusion of highly directional antennas, therefore, suggests that it may take a long time for devices to "sync up" and discover one another.

There are several different approaches for device discovery and beamforming in directional networks. We summarize the approach taken in [SYON09] for IEEE 802.15.3c. This is discussed in more detail in Section 9.3. There are three phases: neighbor discovery, beam discovery, and beam tracking. Periodically, the piconet coordinator (such as an access point) sends out beacons on a predetermined number of spatial beams in what is known as *quasi-omni transmission mode*. This is the least directional (i.e., widest beam) mode of transmission. The mobile device listens for these beacons during each quasi-omni transmission mode, and successively uses each of its available directional receive modes and directions. In this manner, the device is able to determine the best combination of beams based on the transmissions from the piconet coordinator and the link quality measured using its own receive beams. The device then sends this information back to the access point, or PNC (see Fig. 8.2), in the appropriate contention access period corresponding to the best quasi-omni transmission mode. If access is granted by the PNC, an additional beam discovery phase may proceed in which several beam training cycles are used by the piconet coordinator and the device to refine their selection of an optimum beam. This allows the device to use a fine directional beamforming mode with

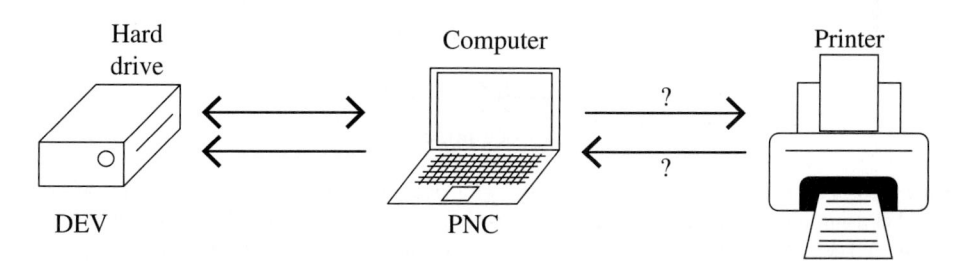

Figure 8.2 Device discovery in a piconet. The piconet coordinator (PNC), in this case a laptop, is communicating with a device (DEV), in this case a hard drive. A new printer is added to the network. Device discovery is the process by which the printer learns about other devices in the network and/or other piconet coordinators and the process by which the piconet coordinator and other devices learn of the existence of the printer.

higher gain. After the beam discovery phase is complete, beam tracking is initialized at the device and/or at the PNC, where the optimality of the fine directional beamforming modes is checked over time to account for slow variations in the channel. This multi-phase approach provides an effective solution for both device discovery and beamforming.

There are other techniques for device discovery based on different system concepts. For example, the approach in [PKJP12] leverages the availability of a low frequency 2.4 GHz connection with wide range and omnidirectional transmission to assist in device discovery. This omnidirectional communication mode avoids the challenge of needing to simultaneously train the beamformers.

The example above shows that device discovery will be an issue in mmWave cellular systems. The impact of device discovery depends critically on the availability of other frequencies to establish the communication link. For example, cellular systems already support multiple radio access technologies, potentially on different frequencies, with provisions for handoff between frequencies. Interesting cellular concepts like the *phantom cell* have been proposed [KBNI13][MI13][Rap14a], in which the control planes and data planes are split: the low data rate control information is sent by high-power nodes at existing narrowband UHF/microwave frequencies, and high-speed user data is sent by lower-power nodes at mmWave. Having a microwave system (such as a legacy cellular or WiFi network) available significantly reduces the complexity of device discovery, as the device can register first with the lower frequency base stations. Then, a specific discovery procedure could be implemented by the mmWave base station to determine and maintain coverage for the mobile device.

8.2.3 Collision Detection and Collision Avoidance

Directional transmission and reception complicates the operations that must be performed in conventional networks as devices attempt to detect and avoid collisions. An illustration of key concepts is provided in Fig. 8.3 using terminals A through F.

- The problem of deafness occurs when terminal C is trying to communicate with another terminal A that is transmitting in a different direction. Because of deafness,

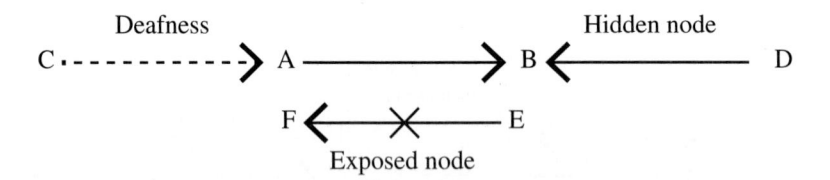

```
        Deafness                              Hidden node
   C ·----------> A ──────────> B <────────── D
                  F <─────✗───── E
                     Exposed node
```

Figure 8.3 The problems of hidden terminal and deafness. Terminal A is communicating with terminal B using directional transmission and directional reception. Terminal C cannot hear that A is busy, yet it keeps transmitting to A, possibly backing off, then transmitting again. This is called *deafness*. Terminal D wants to send a message to terminal B. Because directional antenna patterns are not perfect, there will be some leakage of the interference signal from D into B, reducing the performance or completely corrupting the communication link from A to B. This is the *hidden terminal* problem. Terminal E wants to send a message to Terminal F. Because of its close directional proximity to B, with quasi-omni directional reception, E may think the medium is busy and not transmit, even though with more directionality transmission terminals E and F would be able to communicate with minimal interference on the A to B link. This is known as the *exposed node* problem.

terminal C continues attempting to transmit (with possible backoffs) until it is able to establish a communication link with terminal A. During this time, other nodes may sense the medium is occupied and be prevented from transmitting.

- The hidden terminal problem occurs when terminal D is trying to communicate with terminal B, which is listening to a communication in a different direction. If terminal D is close to terminal B, then some of D's signal will still be received by terminal B due to nonidealities in the design of antenna patterns. This may create an error on the link from A to B causing retransmission. The hidden terminal problem corrupts ongoing transmissions and may also prevent other new nodes from transmitting, as well.

- The exposed node problem occurs when terminal E believes the medium to be busy yet it could support simultaneous highly directional transmissions. The exposed node problem reduces transmission opportunities and, thus, fails to fully exploit spatial reuse.

There are different ways to manage directional antenna networks. One solution is to modify conventional protocols, for example, carrier sense multiple access with collision avoidance (CSMA/CA), as used in IEEE 802.11. It is common for CSMA/CA protocols to include both a *request to send* (RTS) message and a *clear to send* (CTS) message to avoid problems due to hidden terminals. The basic CSMA/CA protocol assumes that nodes send and receive using omnidirectional antennas. Early extensions to directional antennas used some modifications to create a similar setting in which either a separate omnidirectional transmission was incorporated [KR05] or out-of-band busy tones were employed [HSSJ02][YH92]. A better solution for mmWave systems is to design the protocol assuming only directional antennas and in-band communication is available [KJT08][SCP09]. For example, from [KJT08], a directional RTS message may be used and successively repeated in several different quasi-omni directions. Similarly, a CTS message may also be sent in several quasi-omni directions. Terminals listening for access to the channel would need to sense around them rapidly enough to sense one of the RTS or CTS messages. Problems of spatial efficiency could be improved by having every node maintain a table with directional information [KSV00] that indicates the directions that are currently in use and the directions that are available for transmission. When combined with directional information obtained from neighbor discovery, terminals can then exploit spatial reuse by transmitting to nodes in available directions.

Another strategy for designing a MAC protocol for directional transmission is to take a more centralized approach as used, for example, in IEEE 802.15. In this case, the network is divided into piconets with each piconet having one piconet controller. The piconet controller would then receive requests for transmission during a directional contention access period. Information could be transmitted from a device, to the piconet terminal, to another device, or the piconet terminal could establish a direct device-to-device connection. Since requests are made to the piconet controller after the process of neighbor discovery, all terminals need to transmit to the controller using their directional transmission mode, and they need to listen using the directional receive mode appropriate for the controller. This approach is suitable for networking a number of devices that are relatively close to one another, such as in a personal area network (PAN), but is not suitable for a mesh of devices that covers a large area where propagation distances are several tens or hundreds of nanoseconds (i.e., hundreds of feet apart).

It is also important to note that for the case of the unlicensed mmWave spectrum, as Chapter 3 shows, is likely to be in the 180, 330, and 380 GHz bands, in addition to today's available 60 GHz band, multiple radio technologies may be present, and they may not be compatible. This was the case for the 2.4 GHz ISM band in the early days of Bluetooth and WiFi and similar to what we now find with WiGig and WirelessHD in the 60 GHz band. Therefore, collision detection and avoidance should also be designed with coexistence in mind. Other system functions could also be modified to improve performance. For example, while codebook-based beamforming may not be designed today to take advantage of nulls in the pattern, it could be used to maximize network throughput (i.e., beamforming configuration is not entirely based on providing the highest received signal strength). This would enable better coexistence.

Random access is used in cellular systems only for special functions, for example, the uplink random access channel. In a typical protocol, multiple users signal their requests to the base station over the random access channel. Collisions are resolved through timing back-off. No attempt is made to perform carrier sensing due to the structure in cellular systems (only certain time and frequency resources are available at any given point for random access). It is possible that future mmWave cellular systems will be designed differently than microwave systems. For example, base stations might communicate with other base stations in the same frequency using what is called *inband backhaul* (a type of relaying as discussed in Section 8.4 and contemplated in Section 7.3.2.1). In this case, the topology of the network becomes more ad hoc and users seeking to access the channel may compete with transmission between base stations. Further research is required to develop protocols that support seamless interaction between base stations and subscribers in such a network.

8.2.4 Channel Reliability Due to Human Blockage

Directional transmission and reception offers high link quality since multipath fading is reduced by virtue of beamforming along a single dominant path. Unfortunately, highly directional beam steering may be sensitive to blockages of the beam. For example, Fig. 8.4 shows how a human moving in a room might block the main path between a multimedia source and a high definition display. Because of the high data rates of mmWave communication links, blockage of the main path could lead to an outage of a second or more, resulting in the loss of gigabits of data.

Blockage due to humans has been a significant area of research in mmWave for consumer applications. In [CZZ04], measurement results in an office environment were reported to show 20 dB of attenuation due to a human blocking the direct path and (depending on the amount of human activity) median blockage durations on the order of several hundred milliseconds and one to eight blockages per minute. In [BDRQL11], outdoor measurements were performed with human blockage where it was found that the human body does not simply absorb the signal (it is reflective). This means that reflected paths from the human body may still be usable by a smart receiver. We should note, however, that the presence of reflective human bodies also increases the effective

8.2 Challenges when Networking MmWave Devices

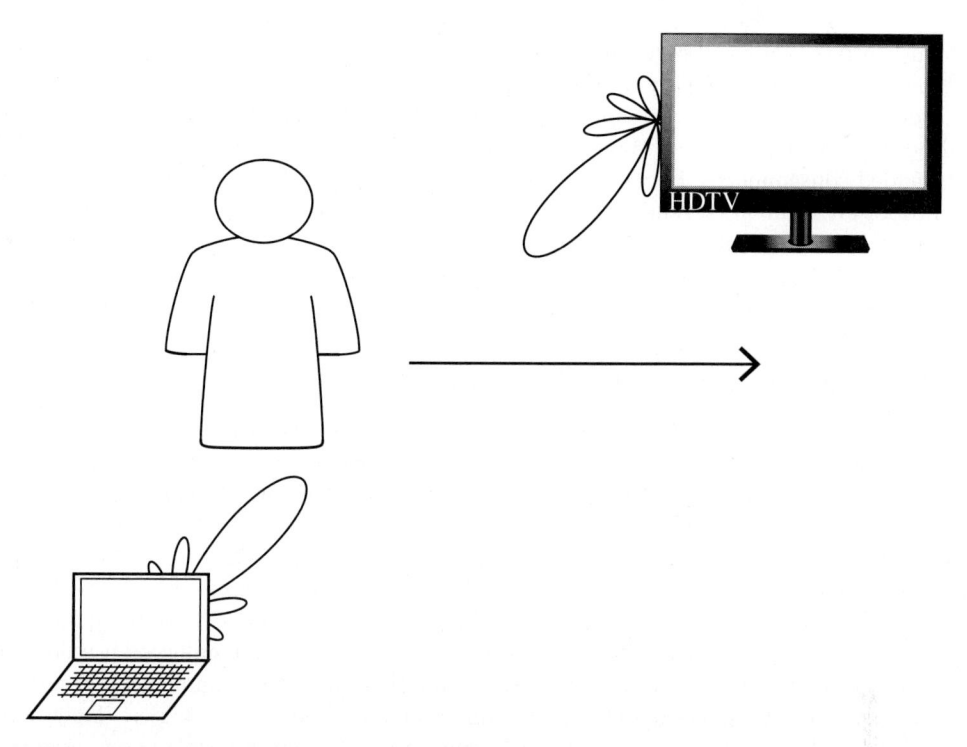

Figure 8.4 With highly directional transmission and reception, as is required to satisfy link budgets for indoor and outdoor communication, blockage of the main propagation path will reduce the link quality and likely create an outage.

delay spread, which becomes a further source of link errors since the equalizer would need to be updated [JPM$^+$11]. The inclusion of human effects in indoor channel models was discussed in Section 3.8.3.

From a MAC perspective, there are several approaches to reduce the impact of human blockage. The direct approach would be to reinitialize the link by performing beam discovery. For the PAN considered in [WLwP$^+$09], the beamforming setup time is around 30 ms, which may be significant depending on the duration of the blockage. An alternative includes multi-beam transmission [SNS$^+$07], in which multiple beams are maintained between the transmitter and receiver that are separated enough in angle such that if one of the beams is in blockage, then the other beams are not. This requires a protocol that can perform beam discovery and tracking in multiple angles. Another approach is multi-hop communication, in which other nodes are used to route signals around blockages when the direct path is blocked [LSW$^+$09][SZM$^+$09]. This requires the MAC to maintain information about potential relay paths and to switch to a relay when the direct link is blocked. Instead of simply being responsive to errors, the efficiency of the aforementioned adaptation approaches can be improved if active probing is employed [TP11]. The idea is to periodically monitor the dips in accumulated signal power to identify rotated, displaced, and blocked states, so that the appropriate action can be taken.

Human blockages will certainly be an issue in mmWave cellular systems. The extent of blockages depends strongly on the assumed infrastructure environment. For example, for indoor networks, ceiling-mounted picocells will shower information down upon users and perhaps be less sensitive than wall-mounted picocells. Humans will also cause other unintended consequences. Because humans are reflective [BDRQL11] and may be moving in outdoor environments, reflected paths from moving humans will become an additional source of multipath and Doppler spread in the received signal. This means that more complicated adaptive equalizers may be required in crowded urban areas (e.g., stadiums with an excited crowd at a rock concert).

8.2.5 Channel Utilization and Spatial Reuse

One of the benefits of highly directional transmission and reception is the reduction in co-channel interference. This means that in a distributed network, more transmit-receive pairs can communicate simultaneously, creating what is known as spatial reuse. The concept of *spatial reuse* is illustrated in Fig. 8.5. An efficient MAC protocol will allow many simultaneous transmit and receive opportunities when possible.

Spatial reuse is interesting in cases where there are multiple sources that have information to transmit to multiple destinations. An example where spatial reuse is useful is the typical office environment considered in [PG09], where multiple employees use 60 GHz to establish a link between their computers and their monitors. Simulation results in [PG09] compare the performance of four different coordination strategies. They show average video losses of 12% to 25% for a baseline time division multiple access strategy, randomized scheduling, and measure and reschedule protocols. In contrast, the loss with scheduling by a super controller that performs joint scheduling based on complete channel state information, including radio node locations for purposes of spatial reuse, has almost negligible loss. This shows that the typical office environment, even with directional antenna interference, is still rich enough that high levels of coordination in the MAC are beneficial.

Spatial reuse is a critical issue in mmWave cellular systems. The area spectral efficiency of a cellular system (an efficiency metric) requires that frequencies be reused aggressively [Rap02]. This can be achieved by having all base stations reuse the same carrier frequency. This seems very plausible in mmWave networks as the use of beamforming tends to reduce the impact of co-channel interference [LR99b] [CR01a] [CR01b] [BH13b][BH13c][RRE14]. Even higher spectral reuse can be achieved by allowing the base

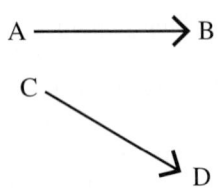

A ⟶ B
C ⟶ D

Figure 8.5 With highly directional transmission and reception, it is possible for more links to communicate simultaneously. This is known as *spatial reuse*.

station to beamform to serve multiple users at the same time [AEAH12]. In [AEAH12] it was shown that three users could be supported with advanced spatial-reuse beamforming at more than twice the sum rate as a single user with simple directional beamforming; higher gains with more advanced kinds of beamforming may be possible.

8.3 Beam Adaptation Protocols

As discussed in Section 8.2.1 and in Chapter 4, directional antennas are an important component of mmWave communication systems. Systems with fixed transmit and receive locations, for example, point-to-point communication or wireless backhaul connections, may use a special antenna design to achieve high gain and a narrow beamwidth. In systems where devices are not fixed, antenna gain will be achieved through adaptive antenna arrays.

There are different ways to perform beam steering, as discussed in Chapter 4 and [LR99b]. In an adaptive array, the beamforming weights may be adaptively adjusted using, for example, the least mean squares algorithm [WMGG67] to maximize received power. A more sophisticated approach may be used where the direction of arrival (or departure) is estimated using an algorithm like MUSIC (as discussed in Chapter 4) [Sch86][PRK86][Muh96]; then, the weights are tuned based on the directions. More sophisticated applications of antenna arrays may also involve canceling interference [LR99b][Ron96][CEGS10][LPC11] or simultaneously receiving from (or transmitting to) multiple users using a smart antenna. Another approach for beamforming is a switched beam system. In this case, the system picks the best beamforming weights from a predetermined set of beamforming vectors, each corresponding to a particular beam. Switched beams are popular for 60 GHz systems, as described in [802.11ad-10], [802.15.3-03], and [ECMA10]. Switched beam systems do not necessarily require estimating the direction of arrival information (or information about the array geometry) and may therefore have shorter startup times [WLwP+09].

This section reviews protocols to support beam steering for transmission and reception. The emphasis is on how the transmitter and receiver determine how to configure their arrays for effective communication. This is a complex task that requires cooperation between the physical layer and higher layers to achieve the best performance. A particular challenge is that both the transmitter and receiver must adapt their beams simultaneously and must periodically re-adapt. The extent of adaptation is a function of the type of device and the activity. For example, collecting data on a smartphone with 100 ms intervals, [TP11] reports 6° to 36° of angular variation when reading or Web browsing and 72° to 80° of angular variation for other activities that involve more frequent device repositioning. The extent of adaptation required is expected to be lower for laptops or home electronics since they are less frequently repositioned. The frequency of adaptation also depends on the array geometry employed. For example, the results in [PP10] show that linear arrays require more frequent adaptation than rectangular or square arrays, for the same number of elements. In general, we would expect that for an array with a given number of elements, the arrangement that minimizes the array perimeter will require the least-frequent re-adaptations. Adaptation is an important task for the MAC protocol.

8.3.1 Beam Adaptation in IEEE 802.15.3c

The approach for beam adaptation in IEEE 802.15.3c makes a good representative example of how beam adaptation works in a mmWave system. This section reviews the key principles and discusses recent advancements in the area. A more detailed treatment of the protocol is found in Chapter 9.

IEEE 802.15.3c uses a codebook-based beamforming protocol, proposed and described in more detail in [WLwP[+]09]. The protocol is built around a specially designed set of beamforming codebooks. A beam codebook is a set of antenna weights. Each codeword is a vector that consists of the amplitude and phase weight for an antenna. The codewords correspond to different beam patterns. The protocol in IEEE 802.15.3c uses three groups of beam patterns categorized from widest beamwidth to narrowest: quasi-omni, sector, and beam. The patterns are illustrated in Fig. 8.6. The beamforming protocol is implemented after other network functions have been performed, for example, neighbor discovery and identification of a piconet controller (using, for example, quasi-omni beam patterns, as shown in Figure 8.2).

The patterns are used as part of a three-stage protocol. The protocol begins when the piconet controller determines that the *first stage* is the device-to-device linkage. The objective of this stage is to identify the best quasi-omni patterns for both transmission and reception by each device. It is important to note that reciprocity is not assumed, so the best receive pattern may be different from the best transmit pattern if, for example, different antennas are used for transmit and receive. Essentially, one device transmits a training sequence successively on all of its quasi-omni patterns while the other device receives successively on all of its quasi-omni patterns, in such a way that every possible transmit and receive pair is considered. The best transmit patterns are fed back from the receiver to the transmitter. The *second stage* is the sector-level search that proceeds along the best quasi-omni pattern as identified in the first stage. This stage proceeds like the device-to-device stage, with the objective of determining the best transmit and receive sectors for each device. Control information is fed back over the quasi-omni patterns. The *final stage* is the beam-level search. This stage proceeds as with the sector-level search, using the results from the previous search that indicated the best sector to find

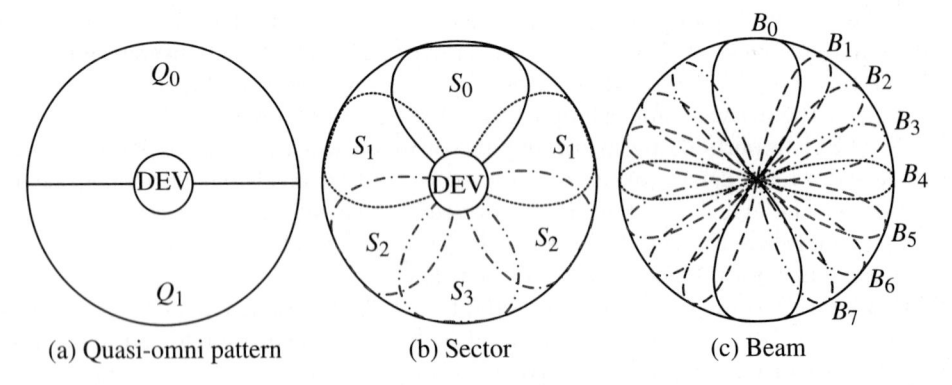

(a) Quasi-omni pattern   (b) Sector   (c) Beam

Figure 8.6 By adjusting the beamforming weights, different patterns can be achieved that trade off array gain for wider beamwidth [modified from [WLwP[+]09, Figure 5] © IEEE].

the sharpest beam for transmission and reception, with control information sent over the sector-level beams.

Beam tracking is an optional feature. If enabled and supported by the devices, the transmit and receive beams will periodically be switched so that the performance can be measured on other beams, and possibly switched. This information can also be stored and used for fast beam switching if there is a blockage.

A proactive beamforming variant of [WLwP+09] was proposed in [WLP+10]. This variant still has a three-stage protocol, but the first stage is performed as part of the beacon process, so that several devices may train their receive beamformers prior to association. This allows devices to associate using a directional pattern, thereby improving coverage. The group training proposed in [HK11] allows multiple devices to train their beamformers at the same time using information available at the piconet controller. Being proactive and leveraging side information in the network allows the network to reduce the overhead associated with beam training, thus improving overall efficiency.

8.3.2 Beam Adaptation in IEEE 802.11ad

Beam adaptation in IEEE 802.11ad is similar to that in IEEE 802.15.3c. The major steps are sector sweep, refinement, and tracking. The sector sweep can contain up to four components: transmit sector sweep (potentially in both link directions, i.e., transmitting from both stations (STAs)), receive sector sweep (potentially in both link directions, i.e., receiving from both STAs), feedback, and acknowledgment. Once sector-level training is completed, the beams may be further refined in the refinement phase, over several iterations. Finally, tracking may be performed over a small set of antenna configurations to keep the best beams up-to-date. IEEE 802.11ad allows the beam pattern to change during transmission of a single packet. This means that multiple beam patterns can be explored at once instead of requiring sequential transmissions on separate patterns. This allows the MAC to explore more beam patterns with lower overall overhead. It is important to note that while beam steering is not mandatory, MAC protocols for beam adaptation are required even in designs that do not incorporate beam steering, and they may communicate with stations that do have beam steering.

One improvement to the beam training in IEEE 802.11ad was suggested in [TPA11]. The idea — *beam coding* — is to send different coded training information messages simultaneously from several beam patterns. This allows the receiver to estimate the channels from multiple directions simultaneously with lower overhead than if successive beam training is employed over time. Because of analog limitations, the process of training multiple beam patterns is achieved in [TPA11] by modulating the analog phase shifters. Beam coding is an interesting approach to improve efficiency and allow more sophisticated beamforming algorithms to be employed.

A codebook-based beamforming approach, as used in IEEE 802.15.3c, was proposed in [ZO12]. In this protocol, discrete Fourier transform (DFT)-based codebooks are used that support $\pi/2$ degree phase resolution. A specific structure is proposed that supports angular rotations in the refinement phase. Simulation results in [ZO12] show that the proposed protocol improves adaptation time. In general, codebook-based approaches are useful in improving beam training efficiency.

8.3.3 Beam Adaptation for Backhaul

Backhaul is another important application of mmWave communication. Backhaul is used in cellular networks to connect base stations together and to connect base stations back to the wired network infrastructure. Obtaining high data rate backhaul connections at conventional cellular frequencies is challenging in urban areas and may be cost prohibitive. Consequently, there has been interest in using mmWave for providing backhaul solutions, especially in the 28, 38-40, and 73 GHz bands, and more recently in the USA unlicensed 60 GHz band, where the EIRP limits were greatly increased in 2013 (see Chapter 2). Although it is possible to use highly directional fixed antennas, for example, a parabolic dish antenna, [HKL+11]'s authors argue that the solution is sensitive to wind-induced misalignment. Therefore, they suggest that adaptive arrays should also be used for the mmWave backhaul application. Consequently, beam adaptation is still required.

From a MAC perspective, protocols for backhaul links are typically less complicated than for IEEE 802.15.3c or IEEE 802.11ad. The reason is that only two links are communicating, not a network, thus functions like neighbor discovery do not need to be performed. Further, the adaptation times may be longer, depending on the exact deployment. Due to wind, most likely only small refinement will be required. If the main communicating path is blocked, then the link will need to be retrained. Fast adaptation is still required for backhaul applications.

Backhaul solutions tend to be proprietary; the detailed implementations vary. We summarize the proposed backhaul adaptation solution in [HKL+11], which uses a codebook-based beamforming approach. Multilevel beamforming codebooks are constructed in [HKL+11] that permit a tree search in the angular domain. An example of the beam patterns created by such codebooks are illustrated in Fig. 8.7. The beamforming codebooks are used at both the transmitter and receiver. The system searches for the best transmit and receiver beam in an iterative fashion. The transmitter sounds different transmit beam patterns at a particular level of refinement. The receiver (which may adjust its receive beam patterns during this time as well) responds back with the best transmit beam. The search then proceeds using a finer resolution of the codebook. Tracking can be used to make refinements in the beam patterns once the link is established. The results in [HKL+11] show that beamforming gain achieved approaches that were achieved by an exhaustive search at high SNR. Therefore, hierarchical beam adaptation shows promise in backhaul settings.

8.3.4 Beam Adaptation through Channel Estimation

Beam adaptation is a transceiver technique in which the transmit and receive arrays are steered to achieve high communication. The mechanisms for beam adaptation are coordinated by the MAC protocol. Beam adaptation, however, is not the only way to train the transmit and receive beamforming vectors. An alternative is to estimate the channel directly, then to devise the beamformers from that estimated channel (see, e.g., the discussion in Section 2.8.3).

Beam adaptation based on the channel requires channel estimation. This may be done in a single shot or in an iterative fashion. First, the propagation channel is estimated between the transmitter and receiver, as seen by the receiver. Then, the preferred beams

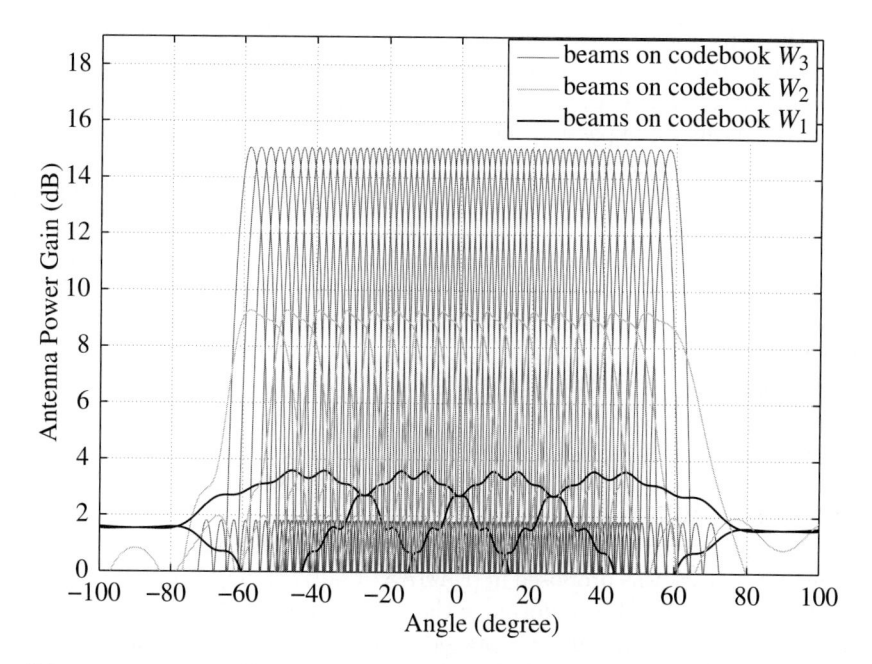

Figure 8.7 A multilevel codebook proposed in [HKL+11] for wireless backhaul. Higher levels of the codebook have narrower beams, thus enhanced resolution [from [HKL+11, Figure 2] © IEEE].

are determined based on the propagation channel state information, for example, the directions of arrivals or directions of departures of strong paths in the channel.

Channel estimation for mmWave is a challenging task because of the large number of antennas. It is further complicated by the use of analog beamforming at the transmitter and receiver. This makes it hard to estimate the channel directly, as typically performed in a MIMO communication system, by sending independent training on each antenna. Consequently, research has pursued alternative techniques for channel estimation. These approaches may be especially valuable in mmWave cellular systems where multiple mobile stations can share the training signals sent from the base station.

One approach for estimation and beam adaptation was proposed in [RVM12]. The proposed protocol uses two phases. The first phase uses what is called *compressive estimation*, which draws on concepts from compressive sensing [Bar07][CW08][BAN14]. The idea is for the receiver to first make several measurements with random beamforming directions. These measurements are then combined and refined using Newton's method to obtain estimates of the angles of arrival of the multipath components. The approach exploits the observation that the number of paths in a mmWave channel are typically much less than the number of antennas in the array. For example, perhaps only four dominant multipaths exist (a reasonable number for mmWave channels, as shown in Chapter 3), but there may be 64 antennas. After estimating the angles of arrival, the second phase of the protocol uses quantized beam steering to choose the weights and phases for the antenna array to maximize a ratio of the received signal power and the power from the undesired multipath components. Knowledge of the array manifold (the structure of the array) is exploited in this algorithm. The phases of the antenna weights are quantized due to hardware constraints, as in other beam-steering algorithms. The approach in [RVM12] thus exploits sparsity to measure the channel and then estimate the beamforming weights, bypassing the need for iterative beam adaptation.

Another approach for estimation and beam adaptation is to use a hybrid approach that combines both digital and analog beamforming, as illustrated in Fig. 8.8 and discussed in [Gho14] and [RRC14]. With a hybrid approach, each antenna is connected to multiple beamformers. The beamformers may each be connected to their own separate RF and baseband hardware, allowing for a further step of digital beamforming. The number of baseband chains is kept small, at two or four, to reduce excess hardware requirements. The use of hybrid beamforming allows multi-stream MIMO communication techniques and the use of multiuser MIMO processing. It was shown in [EAHAS+12] that such an approach provides near optimum performance in sparse channels. Additional approaches that offer promise for forming beams and exploiting MIMO in the face of multipath for various antenna structures and operating scenarios include works in [CGR+03], [ZXLS07], [TPA11], [BAN13], [PWI13], and [SR14].

A sparse beamforming approach for beam estimation and adaptation in the hybrid beamforming framework was proposed in [EAHAS+12a][EAHAS+12]. With hybrid beamforming, there are two sets of beamformers: the ones in the analog domain and the ones in the digital baseband domain. Further, precoding to support multi-stream MIMO transmission is also possible. It is difficult to apply traditional MIMO precoding algorithms because conventional approaches for sending training data from all transmit antennas are unlikely to be applied for the same reasons as in the case of analog beamforming

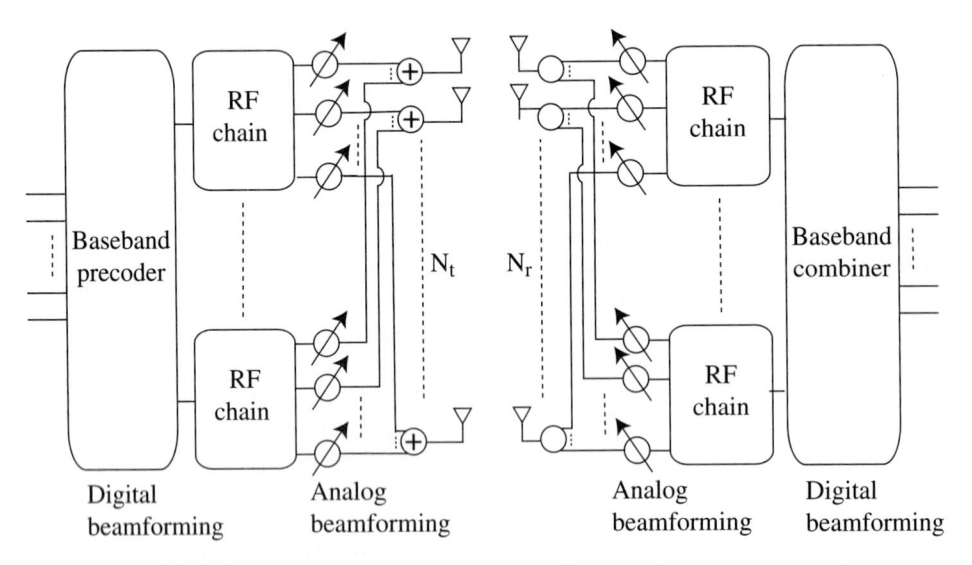

Figure 8.8 The hybrid precoding concept using the system model from [ERAS+14]. The outputs of a digital baseband precoder are each applied to an analog beamformer and summed at the RF. At the receiver, the outputs from the receive antennas are split, then beamformed, in analog. The beamformed outputs are converted to baseband and then further processed with digital combining.

dicussed earlier (i.e., the goal in practical mmWave implementations will be to limit complexity by using a hybrid approach such that several antenna elements are combined for one RF beamformer, thus limiting the number of ADCs required [Gho14][RRC14]). This makes channel estimation followed by beam selection the preferred architecture with the hybrid beamforming approach. An algorithm for estimating the beamformers is provided in [ERAS+14], assuming that the channel has already been estimated. The problem is formulated as a sparse reconstruction problem. Essentially, the unconstrained solution is found, and then a set of close hybrid precoders is found that approximate its performance. The work in [ERAS+14] sets the stage for further investigation of hybrid precoding techniques.

MAC protocols for beam adaptation would require modification to make them suitable for beam adaptation through channel estimation. The current beam adaptation approaches emphasize techniques that allow the transmit and receiver pair to zoom in on a desired set of beamforming weights, from a codebook of possible weights. With channel estimation, the protocol would need to sound the channel by sending training data so that the receiver can estimate the parameters and compute the best beamforming weights, but this must be done in a way that incurs low overhead.

An approach to combine both precoding and channel estimation is proposed in [AELH13]. The proposed algorithm in [AELH13] uses partial information about the direction of arrivals and direction of departures to estimate the path gains in the channel and then compute the hybrid precoding solution. A generalization is found in [AELH14] that combines some aspects of adaptive compressed sensing and array processing in estimating mmWave channels. A sequence of beam patterns is adaptively designed using hybrid analog/digital hardware to jointly estimate the directions of arrival/departure and the path gains of multipath mmWave channels. The approaches in [AELH13] and [AELH14] provide a foundation for the development of more sophisticated techniques for hybrid precoding in the future.

8.4 Relaying for Coverage Extension

MmWave communication works best over LOS communication links. Unfortunately, LOS communication is not always possible. For example, an access point may be located in one room while a transmitter is in another room, the direct communication path blocked by a wall. Even if located in the same room, the LOS path may suffer from human blockages that cause multi-second outages as described in Section 8.2.4. A natural approach to communication is to "burn" through the obstacle (to borrow the explanation from [SZM+09]). The penalty of going through the obstacle is a reduction of the received signal power of 10 to 30 dB or more, depending on the material and propagation scenario as discussed in Chapter 3. The net result is a lower data rate. An alternative to going through the blockage is to go around the blockage by creating alternative propagation paths using a relay. In this way, direct communication between a transmitter and receiver is replaced with multi-hop communication.

A *relay* (also known as an *intelligent repeater* that has selective detection and forwarding capabilities) receives an input signal, performs some signal processing, and then transmits a new signal. A relay may be another communication device in the network that is either generating or receiving its own information at the time, or it may be a dedicated piece of hardware. A repeater, used in WiFi networks to extend coverage in large houses, is one example of a relay. Relaying methods are well suited for future wideband wireless communication networks that will use smaller coverage ranges [Rap11a][Hea10] [PH09][PPTH09].

Different types of relay operations are illustrated in Fig. 8.9. Thus far in mmWave systems, most work has focused on multi-hop relays. The relays operate in half duplex mode, meaning that they either receive or transmit but not both at the same time. Normally, relays alternately receive a block of data then transmit a block of data. Though relaying is a type of multi-hop communication, the term *relaying* usually refers to the use of multi-hop to help transmission between a single source and destination and is typically performed at layers 1 and 2 of the network stack. In ad hoc or mesh networks, multi-hop communication is performed by all nodes in the network as a layer 3 function and may involve additional operations like routing and addressing. Relays and multi-hop communication are important mechanisms to deal with blockages in mmWave systems.

Relays can be classified based on the signal processing applied to generate the transmit signal, see for example, Fig. 8.10. The two most popular types of intelligent repeater operations are decode-and-forward (DF) and amplify-and-forward (AF) [LTW04]. In DF operation, the relay decodes the received signal completely including making decisions on the bits that were transmitted, then re-encodes it before retransmitting. In AF operation, the relay may perform some signal processing on the received signal but does not make decisions on the transmitted bits. An analog repeater is a primative case of AF operation in which the received signal is amplified and retransmitted [Dru88]. Comparisons between DF and AF have been made for a variety of configurations in prior work. For example, results from [KGG05], for a simple path loss model, show that DF gives higher performance when the relay is closer to the transmitter, whereas AF gives higher performance when the relay is closer to the receiver. A strategy called *compress-and-forward* has been shown to provide higher performance but has not been as widely studied in practice [KGG05]. A strategy called *demodulate-and-forward* makes decisions on the constellation symbols sent but does not perform

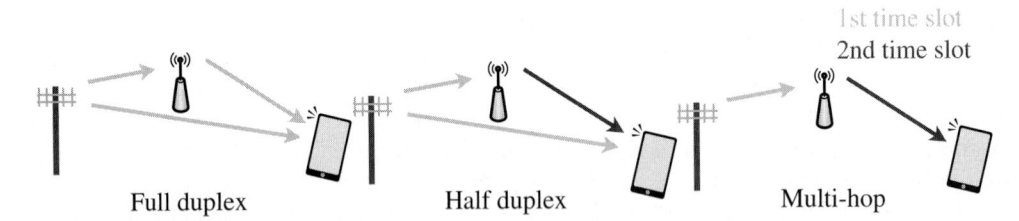

Figure 8.9 Different relay configurations with source, relay, and destination. In theory, a communication link with a relay may exploit both the direct link from the source to the destination and the indirect link through the relay. With a full duplex relay, the relay listens and retransmits at the same time. A practical example of a full duplex relay is a repeater. With a half duplex relay, the relay either transmits or receives and communication may be broken into two phases: transmission from the source, and transmission from the relay. In a multi-hop channel, the relay is half duplex and the source to destination link is not exploited (from [Hea10]).

8.4 Relaying for Coverage Extension

$$r \longrightarrow \bigotimes \longrightarrow \tilde{r} = Gr$$

Amplify-and-forward relay

$$r \longrightarrow \boxed{\text{DeMOD}} \longrightarrow \boxed{\text{Detect}} \longrightarrow \boxed{\text{MOD}} \longrightarrow \tilde{r}$$

Decode-and-forward relay

$$r \longrightarrow \boxed{\text{Compress}} \longrightarrow \tilde{r} = Q(r)$$

Compress-and-forward relay

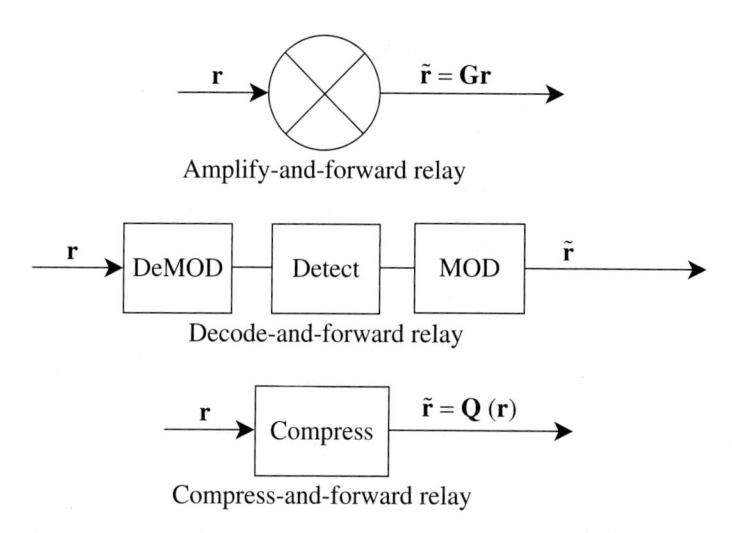

Figure 8.10 Conceptual illustration of different kinds of relay operation (**r** represents the input signal). An amplify-and-forward relay rescales and retransmits the received signal r to produce a scaled signal $\tilde{r} = Gr$ where G is a gain. A decode-and-forward relay demodulates, detects, and remodulates the signal r to create a new modulated signal $\tilde{r}$ prior to transmission. A compress-and-forward relay retransmits a compressed version of the received signal [from [Hea10]].

the error control decoding, reducing the latency of the relay at the expense of reliability [CL06]. Both DF and AF functionality is supported in IEEE 802.11ad. Relaying is not explicit in IEEE 802.15.3c because multi-hop communication is built into the MAC protocol. ECMA 387 supports AF relaying from Type A devices (the higher functionality devices). Relaying and multi-hop communication — to avoid blockages and improve coverage interference, or security — is an ongoing topic of research in mmWave systems. For outdoor cellular scenarios, repeaters/relays will likely become more pervasive due to the need for dense infrastructure deployment [BH13b] [BH14a] [Rap10] [BH13c] [RSM+13] [RRC14] [SNS+14].

An in-room architecture that uses multi-hop communication to provide resilience to blockages is proposed in [SZM+09]. The proposed architecture consists of an access point and multiple wireless terminals; the access point is assumed to have a nominal unobstructed LOS connection to the terminals. Directional beamforming is used on all communication links. In an initial discovery phase, the access point discovers all of its wireless terminals and mutual beam training is performed. Then, the access point asks each wireless terminal to perform neighbor discovery and beam training to its neighbors, sending information about its neighbors back to the access point. During normal operation, the access point frequently polls its wireless terminals. If it detects that a communication link is broken, the access point initiates a lost node discovery phase to establish a relay path through an alternative terminal that also has a connection to the lost terminal. Through that process, a two-hop link is established between the access point, a wireless terminal acting as a relay, and the obstructed wireless terminal. Even with the relay connection, the access point continues to poll the lost terminal to detect when the direct connection reappears. Simulations show that the strategy proposed in [SZM+09] achieves throughput rates that are about 80% of the values for the unobstructed case and dramatically improves the network connectivity.

In networks with multiple simultaneous connections, the use of relaying can reduce spatial reuse: the relayed connection may improve the link throughput of one link at the expense of a reduction of the system throughput. A relaying approach called *deflection routing* where timeslots for the direct and relay paths are shared is advocated in [LSW+09]. The idea is illustrated in Fig. 8.11 — spatial reuse is employed to allow the second hop to coexist with another link in the network. The proposed strategy in [LSW+09] works with an 802.15.3c style MAC protocol and uses decode-and-forward (DF) relay capability. There are two phases in the protocol: building the channel gain table and normal operation. To build the channel gain table, all of the devices take turns probing the channel while the other devices listen and record their channel gains. This

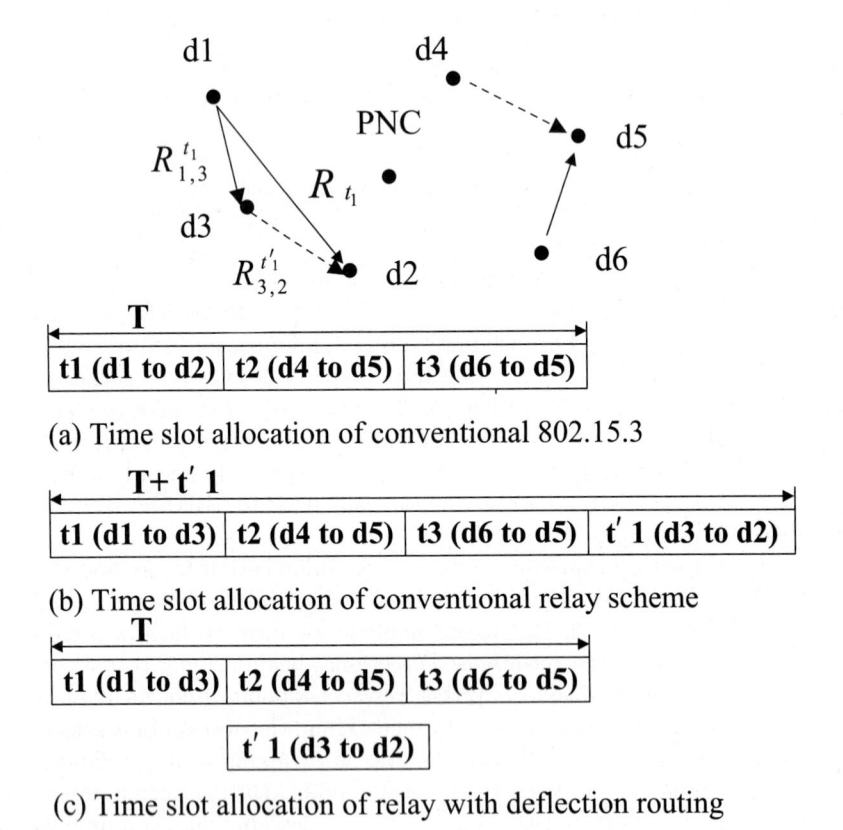

(a) Time slot allocation of conventional 802.15.3

(b) Time slot allocation of conventional relay scheme

(c) Time slot allocation of relay with deflection routing

Figure 8.11 The principle of deflection routing, in which spatial reuse is used to improve the overall efficiency of relaying. The d_k notation refers to device k, while the t_k refers to time slot k. The relay transmission occurs during $t'1$. The figure shows that using a deflection routing technique allows the relay transmission to happen at the same time as other transmissions in the network with minimal interference, thus reducing the demand for network resources [from [LSW+09, Figure 1] © IEEE].

information is forwarded to the piconet node controller (PNC) or access point by each device so that the PNC can build a table with the average channel gains between all pairs of devices. The PNC then runs a routing algorithm to determine which node it should use as a relay, in the event that its path is blocked; the algorithm is based on attempting to maximize some notion of effective sum link throughput to find the appropriate relay and time slot that attempts to maximize the system throughput. During normal operation, if a transmitter communicating with a receiver detects a problem with the link (based on feedback from the receiver, such as lost ACK messages), the device switches immediately to use its relay (this is called *deflection*). The rapid switching is designed to allow for delay-sensitive traffic like video to be supported. The simulations in [LSW+09] show system throughput gains in the range of 20% to 35% greater than throughputs achieved for a conventional relay approach.

A protocol for more aggressive spatial reuse through relaying is proposed in [QCSM11]. The idea is to break long hop links into multi-hop communication links that are scheduled concurrently. The premise is that long links consume more spatial resources, which can be freed up and used for multiple simultaneous connections that overall take less time. As shown in the last section of Chapter 5, several shorter links (i.e., relays) may also be advantageous from a power savings perspective, especially in channels with high path loss exponents [MR14b]. Different than [SZM+09] and [LSW+09], in [QCSM11] relaying is proposed as a way to further increase system throughput, and not just as a reaction to link blockage. The protocol in [QCSM11] is built around an 802.15.3c style MAC with a PNC. The PNC periodically collects network topology and traffic information from all the nodes and uses it to implement a hop selection algorithm. When devices wish to transmit, they communicate their intentions to the PNC, which then schedules transmissions in a way to maximize concurrency, possibly breaking transmissions into multiple hops. Simulations show improvements in network throughput of 20% to 100% gain compared to scheduling single hops concurrently but not breaking up into multiple hops.

A decode-and-forward (DF) strategy with incremental redundancy, which adapts to LOS or NLOS conditions, is proposed in [LNKH10]. When there is a good path between the source and destination, half duplex relaying is used instead of just two-hop communication (see Fig. 8.9). No directional beamforming is assumed at the source, relay, or destination. The proposed relaying strategy is based on the concept of incremental redundancy using a serial concatenated code that consists of a Reed-Solomon code and a convolutional code. It assumes that the relay knows the conditions of the direct link so it can adapt its strategy accordingly. The source broadcasts a coded packet to both the destination and the relay. Assuming that the packet is decoded correctly, if a LOS link exists, it forwards additional parity bits to the destination. If a NLOS link exists, it re-encodes the entire packet and forwards it to the destination. The benefit of the proposed approach is that when the LOS link is available, only a few additional parity bits are sent, making the second phase of the relay operation very short and thus improving overall efficiency. In mixed LOS and NLOS environments, simulations show that the proposed relay transmission strategy increases average throughput around 50%. The baseline for comparison used in [LNKH10] is a relay strategy in which a rate compatible punctured convolutional code is used but the relay does not adapt its behavior based on the direct link.

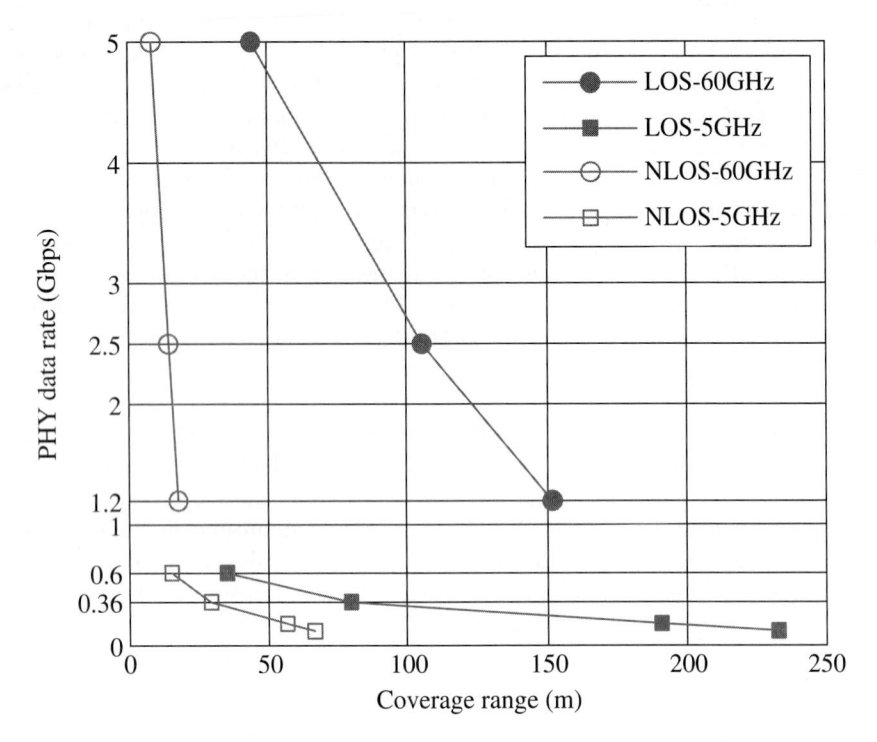

Figure 8.12 Coverage range and data rate at the physical layer for a 5 GHz link and 60 GHz link under two different channel conditions: LOS and NLOS. It can be seen that the microwave link provides higher coverage at the expense of smaller data rates. A multi-band protocol could obtain the rate benefits of 60 GHz and the coverage benefits of lower frequencies [from [YP08, Figure 1] © IEEE].

The performance of DF and AF multi-hop strategies with directional antennas in the presence of interference is evaluated in [LPCF12]. End-to-end signal-to-interference-plus-noise ratios (SINRs) are developed that incorporate a Nakagami fading channel, sidelobes of the directional antenna, and potential differences in the noise variance. The distribution of the resulting SINR expressions are calculated using results on sums of Gamma distributions.[2] The resulting outage probability simulations show that DF outperforms AF by about 2 dB in the considered simulation scenario, and using directional beamforming provides 13 dB gains over the use of omnidirectional antennas. This confirms that directional antennas are useful for multi-hop relays.

Multi-band diversity is only one benefit of a multi-band MAC protocol. Another benefit is range extension and spatial reuse through multi-hop communication or mesh network concepts [YP08]. As illustrated in Fig. 8.12, a multi-hop 60 GHz communication link can outperform a single-hop 5 GHz link to provide higher range, at the expense of latency and end-to-end reliability. The handover model can be generalized in the

---

2. Another accurate yet simpler approach is to compute SINRs based on the sum of log-normal power levels from interferers [CR01a] [CR01b].

mesh configuration. For example, as described in [YP08], a multi-band WLAN access point might be connected to a number of 60 GHz access points that do not have a wired Ethernet connection. The multi-band access point could serve multi-band users by routing traffic over multiple hops through the 60 GHz access points. Thanks to the highly directional antennas and spatial reuse, it may be possible to have multiple links active simultaneously, further improving the overall system performance.

Relays have been studied for application in cellular networks operating at microwave frequencies, but they have not been widely deployed (except for the primative repeater). Relays were considered in IEEE 802.16 in subgroup j [802.16-09][PH09] but did not end up in WiMax. The performance of relays in cellular systems has been evaluated numerically using system-level simulators [HYFP04][VM05][SW07][BRHR09][YKG09], idealized terrain [DWV08], ray tracing software applied to particular urban areas [ID08][SZW08], and experiments with an LTE-Advanced testbed [WVH$^+$09]. A general conclusion from prior work is that relays in cellular systems are sensitive to interference. Comparing different relay approaches in a cellular environment [PPTH09], it was found that a shared multiple antenna relay could overcome some of these issues. The conclusions in [PH09], [HYFP04], [VM05], [SW07], [BRHR09], [YKG09], [DWV08], [ID08], [SZW08], [WVH$^+$09], and [PPTH09] were drawn based on propagation models and general configurations of microwave cellular systems. Relays may prove to be important in mmWave cellular systems because coverage is a more acute problem given large difference between NLOS and LOS propagation as discussed in Chapter 3. Also, many of the conclusions drawn in prior work did not consider highly directional antenna arrays at the relays, as would be found in mmWave cellular systems. This will reduce the impact of interference and the sensitivity of relays to out-of-cell interference. Consequently, relays remain an active area of interest for mmWave cellular systems and may be particularly useful for coordinating service coverage for mobile users within buildings, vehicles, or as they move from outdoors to indoors.

8.5 Support for Multimedia Transmission

High-quality audio, video, and display are important usage models for mmWave links because microwave spectrum generally cannot support the high data rates required by uncompressed video sources. As a result, supporting multimedia applications has been a priority in the development of WPAN and WLAN standards and has influenced design decisions at other layers.

Supporting uncompressed video is a requirement if mmWave are to be used for cable replacement for computer workstations or multimedia centers. Examples of cable connections that can be replaced include high-definition multimedia interface (HDMI) and digital video interface (DVI). For the cable replacement application, mmWave communication systems should be agnostic about the content, providing high quality for movies, gaming, and general computer displays.

To provide some guidelines on performance, a review of different features and characteristics of HDMI video is provided. There are several different versions of HDMI that support different screen sizes, frame rates, and resolutions per pixel, not to mention multiple channels for audio content. HDMI 1.3, for example, supports 60 frames per second at a resolution of 1920 × 1200 (known as 1080p) with up to 48 bits per pixel. The number of bits per pixel is divided depending on the color space. For example, with RGB and 48 bits total, 16 bits would be allocated to each color: red, green, and blue. Typical values for the bits per pixel are 24, 30, 36, and 48, with 36 or greater denoting deep color. The video bandwidths for 1920 × 1200 with 60 frames per second vary from about 3.3 to 6.6 Gbps depending on the number of bits per pixel. The rates can be reduced by using fewer frames per second; 24 frames per second is common in movies, whereas at least 60 frames per second is typical for a computer display. HDMI 1.3 and 1.4 support a total maximum throughput of 8.16 Gbps.

Many video sources are actually compressed to reduce storage requirements and bandwidth requirements for video streaming over the Internet. One of the most popular standards for compression is H.264/MPEG-4 Part 10, which is commonly used on Blu-ray Discs. Various techniques are used to exploit spatial and temporal redundancy to enable high compression. In H.264, the compressed video sequence is organized into a group of pictures (GOP). The GOP contains different kinds of frames, like an I-frame (intra-coded picture that can be decoded on its own), a P-frame (forward predicted picture that is the quantized prediction from usually an I-frame), and a B-frame (bidirectionally predicted picture that may be predicted or interpolated from earlier and later frames). There are various other techniques used as well, including variable size motion compensation, spatial prediction, and entropy coding. Data rates for high definition video from a Blu-ray Disc, for example, may be in the range from 25 to 50 Mbps, which is much lower than the uncompressed HDMI data rates. Despite this fact, current 60 GHz systems are designed to operate on the uncompressed video. The main reason is that the uncompressed video is only available at the application layer, and customizing the PHY and MAC for compressed video would involve a cross-layer design problem and a substantial amount of interaction between hardware and software. Dealing with uncompressed video also avoids the need for implementing an additional codec, does not require transcoding, and maintains a constant bit rate of traffic.

There are several different ways to support uncompressed video in WLAN and WPAN. Although video can be treated as just another data source, uncompressed video has redundancy (or correlation) in space and time. Because the end user for most envisioned video applications is a human, this redundancy can be exploited to create error concealment algorithms that minimize perceptual distortion. This reduces the impact of errors and makes the viewing experience more enjoyable.

The most common way to implement video aware processing in 60 GHz systems is to classify bits at the pixel level based on their impact on video quality and then treat each class differently during transmission. Although there may be two or more classes, two classes are common in current implementations, for example, ECMA-387 and IEEE 802.15.3c. The classification is normally made based on the most significant bits (MSBs) and least significant bits (LSBs) for each color of each pixel. The classification between MSB and LSB may be flexible and communicated over a control channel or it may be fixed. For example, if a color depth of 24 bits is supported, then red, green, and blue would each be represented with 8 bits. A fixed classification would allocate 4 bits for the MSB and 4 bits for the LSB for each pixel.

Unequal error protection (UEP) is most commonly used to create data paths with different reliability for the MSB and LSB classes. The idea is to use some combination of different modulation and coding to send the MSBs over a channel with a lower bit

8.5 Support for Multimedia Transmission

error rate than seen by the LSBs. UEP should be contrasted with equal error protection (EEP) where different classes are treated equally.

There are different forms of UEP that have been studied for multimedia transmission in WPAN and WLAN. UEP-by-coding occurs when each data path receives a different forward error correction (FEC) code rate but are sent using the same modulation format. UEP-by-MCS occurs when each data path receives a different modulation and coding scheme (MCS), which consists of a potentially different combination of modulation order and FEC code rate. UEP-by-modulation occurs when a skewed constellation is used to transmit each data path with the more reliable bits mapping to constellation points with larger minimum distance.

Error concealment is an important component of video aware communication. The idea is to exploit correlation in the source to reduce the perceptual impact of errors. To implement error concealment, some knowledge of when an error occurs is required. Most commonly with UEP, this is achieved by attaching different cyclic redundancy check (CRC) codes to MSB and LSB portions of packets. In this way, it becomes possible to detect if a packet was not decoded correctly and either request that it be resent or implement some form of error concealment. If the LSB is lost then only the MSB portions of the pixel data will be displayed. To enable concealment, pixel partitioning may be used to exploit the spatial redundancy between pixels, that is, the high likelihood that adjacent pixels have a similar color. Then, if a subpacket corresponding to one of the partitions is lost, one of the other partitions can be copied in its place, resulting in less perceptual distortion than if the pixels were simply not displayed.

Supporting video for mmWave is an ongoing topic of research. A good review of uncompressed video transmission in 60 GHz is provided in [SOK+08], which reviews many of the concepts discussed in the section. Several concepts from [SOK+08] are illustrated in Fig. 8.13. One other idea that is highlighted is uncompressed video automatic repeat request (ARQ). The idea is that feedback about both the decoding status of the MSB and LSB might be used to determine which packets require retransmission. If the MSB packet is not decoded correctly then it may be retransmitted with a lower error modulation and coding rate, whereas if the MSB is received correctly, irrespective of the LSB status, then no retransmission is requested. Another approach for error concealment is also discussed in [SOK+08]. The idea is to use some properties of Reed-Solomon codes at the physical layer to provide feedback from the PHY to the MAC layers. Because of certain properties of Reed-Solomon codes, a typical Reed-Solomon decoder will output either a decoding failure (more likely) or the wrong codeword (less likely). Decoding failures can be passed to the MAC layer and used like a CRC failure but with more granularity. Pixel partitioning can be used to replace failed packets with adjacent good packets. If a codeword from another partition is not available to replace the lost codeword, then another Reed-Solomon codeword that is close is used instead. This procedure is called *code swapping* and is a part of ECMA-387.

Another approach that leverages the Reed-Solomon decoder to improve transmission of uncompressed video is described in [MMGM09]. If a pixel is classified as a decoding failure, the receiver only attempts to decode the MSBs for that pixel. Assuming the Reed-Solomon code is systematic (the resulting code consists of the data or systematic portion and the parity bits), only the systematic portion is provided during a decoding failure. An "all in consensus" rule flips the MSBs of incorrectly decoded codewords when consensus

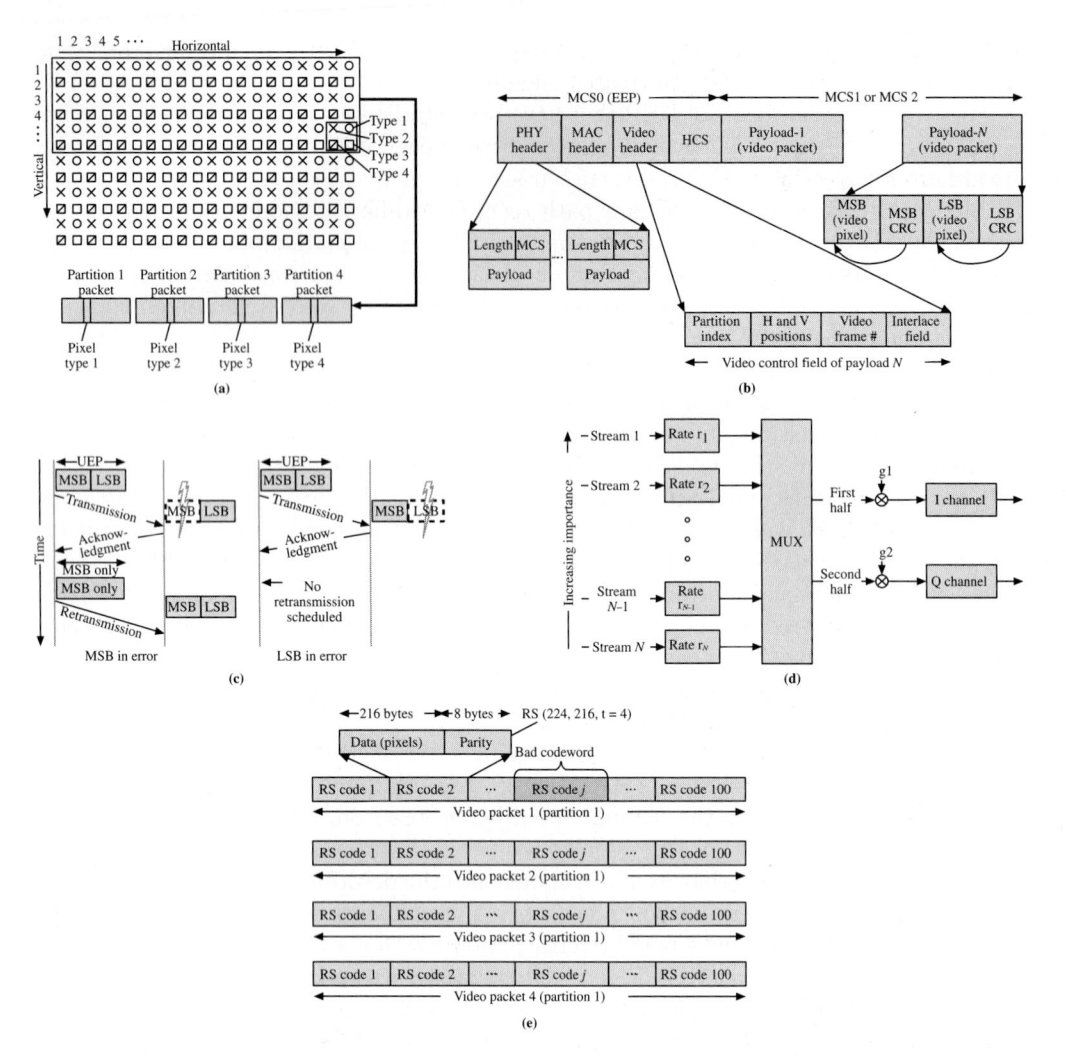

Figure 8.13 Different strategies for supporting video in mmWave systems. (a) Pixel partitioning. (b) Frame format. (c) Uncompressed video with automatic repeat request. (d) Unequal error protection. (e) Error concealment using Reed-Solomon codes. [From [SOK+08, Figure 13] © IEEE]

is reached with the neighboring pixel positions. The resulting algorithm provides gains of 7 dB in terms of peak signal-to-noise ratio (PSNR).

A more flexible partitioning of MSBs and LSBs is proposed in [HL11]. The idea is to recognize that, for certain color spaces like YCrCb, one component has more perceptual importance. In [HL11], multiple levels of priority are suggested with the Y component being high priority, the MSBs of Cr and Cb becoming medium priority, and the LSBs of Cr and Cb becoming low priority. The resulting approach provides higher performance than simpler UEP approaches applied to the YCrCb format.

Support of UEP for compressed video is considered in [LWS+08]. The idea is to assign priority to the different parts of the compressed video differently, based on their importance in decoding. For example, following the H.264 terminology, control information and I-frames would be classified as high priority whereas P-frames and B-frames would be classified as low priority. Using UEP-by-modulation, the results in [LWS+08] show that compressed video can also enjoy the benefits from UEP.

There have been some other approaches for facilitating multimedia transmission that use multiple beams to send video information with different priorities. A multi-beam solution to avoid the impact of human blockages is suggested in [SNS+07]. It is argued that blockages are important especially for video because of the high data rates, real-time requirement, limitations of in-chip buffers, and long video format switching times. The proposed solution uses pixel partitioning, multi-beam selection, and fast format selection. The idea of multi-beam selection is to use multiple beams pointing in different directions to provide multiple channels of communication between the transmitter and receiver. There may be a primary beam and a secondary beam. The secondary beam may be used as a backup or for retransmissions, or it may carry one pixel partition for error concealment. Fast video format adaptation involves maintaining a constant resolution output, for example, 1080p, even though the reduced resolution may be used on the PHY layer to accommodate low data rate channels. Fast video format adaptation avoids potentially long delays associated with format changes in HDTVs.

A framework for video aware relaying for IEEE 802.11ad was proposed in [KTMM11]. Three different modes of operation are considered: non-relaying, a single amplify-and-forward relay, and a single decode-and-forward relay, with adaptation possible between modes based on channel conditions. Adaptive video coding is considered to match the compressed video source rate to the changing channel conditions. The main result of [KTMM11] is an algorithm that jointly chooses the relay configuration and the amount of video compression based on channel conditions. The use of adaptive relaying provides throughput gains of $2\times$ or more compared with non-relay operation.

Quality-of-service considerations for supporting video in wireless networks are reviewed in [PCPY09]. Using the observation that video streaming will be a significant user of 60 GHz resources, but not the only application that will be running, [PCPY09] suggests that a MAC protocol that combines time division multiple access (TDMA) with CSMA or poling may provide better channel utilization efficiency in a multi-user network with different applications.

8.6 Multiband Considerations

Microwave frequencies are widely deployed in commercial wireless links. For example, WLAN systems like IEEE 802.11n use unlicensed bands at UHF/microwave carrier frequencies of 2.4 and 5 GHz. IEEE 802.11n provides throughputs of hundreds of megabits per second, in 20 or 40 MHz of bandwidth, over a typical range of up to 100 m, much larger than the likely range of IEEE 802.11ad. Similarly, cellular systems like 3GPP LTE use licensed microwave carrier frequencies such as 850 MHz, 1.9 GHz, or 2.1 GHz to provide coverage for 10 km or more. A major challenge for mmWave systems is to provide comparable coverage and quality to microwave systems while at the same time providing higher data rates.

The ability of mmWave to provide high-quality links and high coverage is impacted by the propagation environment. People are a major source of link outage while walls are a major source of additional attenuation. In [PCPY08], a comparison is provided between transmission ranges at 5 and 60 GHz under certain assumptions. It was found that in ideal conditions 60 GHz can provide higher range due to the array gain from the transmit and receive arrays, also discussed in Chapter 3 and Section 8.2.1. With penetration through multiple walls or blockage by multiple people, the range of 60 GHz reduced below 5 GHz. The results in [PCPY08] are based on a link budget calculation; further degradations in performance will be observed for moving humans [CZZ04][JPM+11].

A way to ensure that mmWave systems maintain the coverage and link quality of microwave systems is through multi-band operation, also known as *band diversity*. The idea is to support some level of hybrid transmissions at microwave and at mmWave carrier frequencies. To understand the concept, consider the data rate versus coverage range plot in Fig. 8.12, provided for a microwave and a 60 GHz link operating in each case of LOS and NLOS under certain assumptions (see [YP08] for details). In this plot, it can be seen that the 60 GHz link provides high data rates for short distances but cannot serve long distances. Alternately, the microwave link is able to support moderate data rates over a relatively larger range. A multi-band protocol would exploit this property by leveraging 60 GHz for high data rates when available and otherwise falling back on lower data rate microwave links. The frequency of this adaptation depends on the extent of integration; it could happen fast on the order of milliseconds or slowly on the order of hundreds of milliseconds or seconds.

Different levels of multi-band integration have been considered in the research community [PCPY08][KOT+11]. Two of the extreme levels of integration are provided from [PCPY08].

- *Handover model* — In this case, communication is supported at both microwave and mmWave carrier frequencies, potentially simultaneously. The MAC protocol supports transfer from the microwave link to the mmWave link and vice versa. For example, if a link blockage is detected, the communication link could fall back on a lower-speed microwave link. It is argued in [SHV+11] that falling back to the low frequency is relatively easy and that the challenge lies in transitioning efficiently to the shorter range high frequency link (performing device discovery and beam discovery efficiently, for example).

- *Full MAC* — In this case, the microwave and mmWave carriers are completely integrated and one MAC coordinates transmissions on both carriers. There might be a high degree of coordination in the physical layer operations including synchronization and joint operations of some functions like link adaptation.

Another perspective on multi-band integration is provided in [KOT+11], which suggests partitioning data and control on the different bands. For example, mmWave may be supported only in one direction, for example, the downlink direction in the kiosk download application. In this case, a conventional microwave transmission would be supplemented by a unidirectional mmWave transmission from the kiosk to the device. After setting up the link, the mmWave transmission could be used just for the high data rate transmissions to the device, with MAC information like packet acknowledgments sent over the microwave link.

Multi-band operation has been considered in several 60 GHz efforts. For example, in the IST Broadway project [IST05], the WLAN HIPERLAN/2 standard was augmented with

8.6 Multiband Considerations

60 GHz operation and a high level of RF integration. IST Broadway supported functions such as switching between bands of operation, especially to deal with 60 GHz link blockages. In ECMA-387 [ECMA10, Section 19], out-of-band control channels are supported using IEEE 802.11g to provide "WPAN management and control with omni-directional transmission." In this case, the 60 GHz link is supported in only one direction with the microwave link used for bidirectional communication; multi-band operation is used to avoid some of the challenges of directional MAC protocols and to facilitate spatial reuse. IEEE 802.11ad [802.11ad-10] supports multi-band operation where it is possible to switch sessions from one carrier frequency to another. Tighter integration was proposed prior to the start of IEEE 802.11ac/ad but was not pursued by the standards [DH07]. This will be of particular interest if the ad hoc mode, called the *personal base station mode* (PBSS), is employed, which will facilitate peer-to-peer 60 GHz connections.

It seems likely that mmWave cellular systems will coexist with microwave systems. Although mmWave and microwave can be treated as separate channels, like 850 MHz and 1.9 GHz in today's cellular systems, it seems promising to consider a co-design that accounts for the inherit differences in communication at these different frequencies. An important difference is the requirement for directional transmission and reception in mmWave links. For example, random access incurs overhead in a directional communication system (where the receiver has to listen in many directions for a transmission) but is relatively easier in a system that supports omnidirectional signals. Of course, directional mmWave links offer much higher data rates than available at present with existing microwave spectrum allocations. This motivates the design and analysis of co-designed mmWave and microwave communication systems to determine good architectures and to analyze the implications of different design tradeoffs.

Umbrella architectures that use *phantom cells* [KBNI13][MI13][SSB$^+$14][Rap14a], are illustrated in Fig. 8.14 and are a promising approach for future hybrid cellular systems. The phantom cell concept allows mobile devices to be served by two types of wireless infrastructure simultaneously, without knowing the precise nature of the serving cell. This

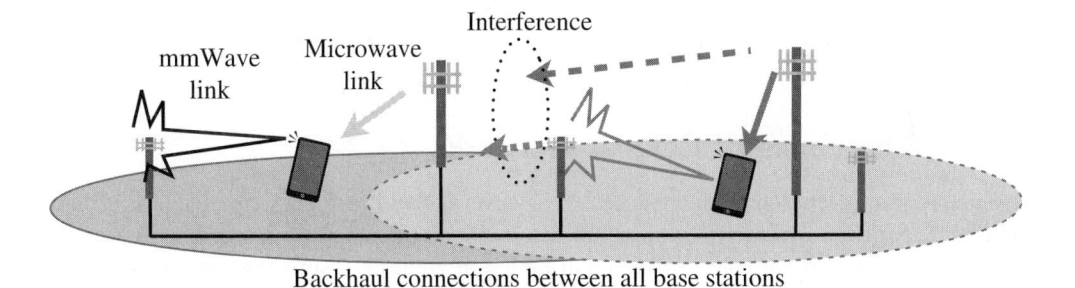

Figure 8.14 A model for coexistence between mmWave and microwave cellular where the microwave cellular network forms an umbrella network to facilitate the management of many mmWave communication links and to simplify functions like handoff. On the left side, a mobile device may connect either to a microwave or mmWave base station or to both simultaneously using the phantom cell concept. Interference on the microwave frequencies comes from other microwave base stations and on the millimeter wave frequencies from other millimeter wave small cells. As shown in Chapter 3 and elsewhere, the directionality of the beam patterns reduces the impact of mmWave interferers [BAH14][SBM92][RRE14][RRC14][ALS$^+$14].

is viewed as an attractive way to gracefully and seamlessly migrate from current 4G-LTE cellular networks that use UHF/microwave frequencies, up to 5G mmWave networks. The approach taught in [SSBK014] is to split the control plane and the user plane (C/U split) that serves the mobile users such that control plane data would be communicated using existing macrocells on UHF/microwave frequencies, while multi-Gbps user data would be carried by newer, closer-spaced cells using mmWave frequencies. As more mmWave cells are deployed, greater capacity would be seamlessly provided. While co-location of mmWave and microwave base stations seems like a reasonable starting point, the base stations may or may not be co-located and may have different densities since mmWave base stations may be smaller and easier to deploy with wireless backhaul (fronthaul) to nearby UHF/microwave base stations [BH14a].

In a sense, the mmWave base stations look like a tier of small cells in a heterogeneous network [ACD+12][Gho14][RRC14], but this is not exactly the case. An important difference is that the mmWave base stations will share the same frequency but will not experience cross-tier interference from the macrocell. This is also different from the WiFi offloading case [LLY+12][ACD+12][LBEY11] because WiFi access points are uncontrolled by the cellular operator in an unlicensed band and are shared with non-cellular users. There are many interesting related research challenges. For example, the microwave frequency control channels could be used to facilitate backhaul of limited feedback information [LHL+08] to allow close mmWave transmitters to schedule in a way that minimizes out-of-cell interference. Similar infrastructure-aided interference management appears in the device-to-device literature [DRW+09][FDM+12][LZLS12]. A primary consideration in [DRW+09], [FDM+12], and [LZLS12] is that the mmWave base stations are fixed with potentially higher power and longer range, assuming narrow beams are used to communicate.

It should be noted that multi-band operation is not likely to just be limited to selecting between microwave and mmWave frequencies, as mmWave networks will likely support multiple mmWave bands, for example 28, 38-40, and 72-86 GHz. Future mobile and infrastructure devices will therefore require new types of antenna arrays that support multiband operation [Rap13]. No matter whether microwave or mmWave frequencies are used for control or delivery of traffic, many new research challenges lie ahead in supporting multi-band operation in mmWave cellular systems.

8.7 Performance of Cellular Networks

In this section, we present some results on the coverage and rate performance of mmWave cellular networks. Considering the short range and high density of mmWave cells, the deployment of mmWave base stations will be less regular as macro base stations. Hence, we apply a stochastic geometry model to simulate the locations of mmWave base stations as a Poisson point process (PPP), which is reasonable in small cell scenarios [ABKG11].

The mmWave channel varies in quality depending on whether the path is LOS or NLOS [RGBD+13]. It is useful to incorporate this observation into the analysis of mmWave cellular systems. This section summarizes some results on the application of an analytical blockage model proposed in [BVH14] to mmWave cellular systems. The blockage model in [BVH14] provides a concise way to model random distributed objects

8.7 Performance of Cellular Networks

in space leveraging a mathematical concept called *random shape theory*. The Boolean scheme is the simplest process of objects in random shape theory [Cow89]. In a Boolean scheme, the centers of objects form a PPP, and each object is allowed to have independent shape, size, and orientations according to certain distributions. In [BVH14], the random located buildings are modeled as a Boolean scheme of rectangles, and the LOS probability of a link, that is, the probability that a link is not blocked by buildings, is shown to be a negative exponential function of the link length. Besides the exponential function in [BVH14], other forms of LOS probability function can be also found in [ALS[+]14]. Noting the fact that the longer the path, the more likely it will be intersected by blockages, a general LOS probability function needs to be a non-increasing function of the link length. Correlations of shadowing between links are also ignored in the analysis, which has been shown to cause negligible errors in the aggregate SINR evaluation [BVH14]. The blockage model has been shown to be a good match with real scenarios by fitting the parameters appropriately [BVH14]. Given one realization of blockage locations, we will determine whether a base station is LOS or NLOS to a user, and we will apply different channel models correspondingly to take account the difference in path loss between LOS and NLOS links.

We summarize the key items in the proposed mmWave cellular network model from [BH13b] as follows.

1. *Blockage process* — We use random shape theory to model the random blockages as a stationary process of objects, for example, the Boolean scheme model used in [BVH14], on the plane. Based on the statistics of the blockage process, the LOS probability function of a link $p(r)$ can be derived as a function of the link length r. The probability $p(r)$ is assumed to be independent for all links by neglecting correlations of shadowing. The LOS probability function $p(r)$ can be further simplified as a step function, in which only the base stations located inside a fixed ball centered at the user are considered as LOS [BH14].

2. *Base station PPP* — The base stations form a homogeneous PPP Φ on the same plane. Due to the presence of blockages, a base station can be located either indoor or outdoor. From an outdoor user perspective of view, the outdoor base stations can be further divided into two sub-processes: the LOS base stations and the NLOS base stations. As the LOS probability for each base station is assumed to be independent and dependent on the length of the link, the LOS and NLOS base stations form two non-homogeneous PPPs on the plane, to which different path loss laws are applied.

3. *Outdoor users* — A typical user is assumed to be outdoor and located at the origin of the plane. By the stationarity of the blockage and base station processes, the downlink performance received at the typical user is representative of the aggregated downlink performance in the network. The typical user is associated with the base station that provides the smallest path loss.

4. *Directional beamforming* — Directional beamforming is applied at both base stations and mobile stations. The typical user and its associated base station are

assumed to estimate the channels and adjust their steering orientation to exploit the maximum possible directionality gain. The steering angles of the interfering base stations are randomly posed. The actual array patterns are approximated by a sectoring model in the analysis, where constant directivity gains are assumed for the main lobe and the sidelobes. Using the sectoring model, the antenna pattern is fully characterized by the main lobe gain, the main lobe beamwidth, and the front-back ratio of the array.

In the proposed system model, the downlink single-to-interference-plus-noise-ratio (SINR) in a mmWave network can be expressed as

$$\text{SINR} = \frac{h_0 G_0 PL(|X_0|)}{\sigma^2 + \sum_{\ell > 0: X_\ell \in \Phi} h_\ell G_\ell PL(|X_\ell|)} \quad (8.1)$$

where h_ℓ is the fading to the user, G_ℓ is the combined gain of the transmitter and receiver beamforming, $PL(\cdot)$ is the path loss to the user, σ^2 is the noise power, X_ℓ denotes the location of the base stations, and X_0 denotes the base station with the smallest path loss to the user. Note that different path loss formulae, for example, with different LOS and NLOS path loss exponents, are used to compute the path loss $PL(\cdot)$, whether the base station is LOS to the user or not. Leveraging concepts of stochastic geometry, efficient expressions to evaluate the SINR distribution have been derived in [BH14].

Even suffering from blockage effects and high path loss, mmWave systems can achieve acceptable SINR coverage when base stations are sufficiently dense. In Fig. 8.15, we compare the SINR coverage probability with different base station densities, where we fix the LOS range (a quantity that depends on the building perimeter and density) at 200 m and change the base station density. Parameters of the simulations are listed in Table 8.1. Simulations show that the coverage of mmWave networks is very dependent on the density of base stations. Specifically, the results in Fig. 8.15 indicate that mmWave

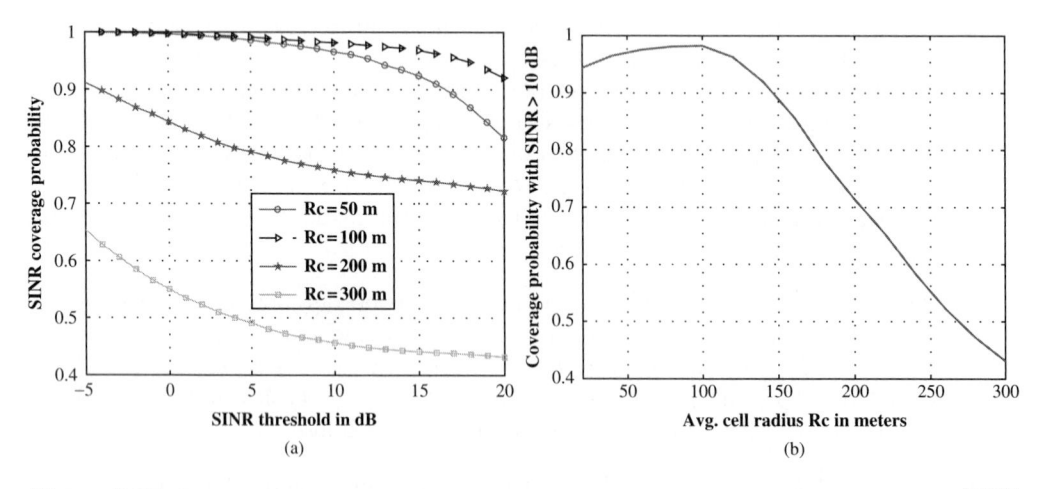

Figure 8.15 SINR coverage probability with different base station densities, where $R_c = \sqrt{1/\pi\lambda}$ and λ is the density of base stations.

8.7 Performance of Cellular Networks

Table 8.1 Parameters for the mmWave cellular system simulations.

Simulation Parameter	Value/Variable
Carrier frequency	28 GHz
Bandwidth	100 MHz
Base station TX power	30 dBm
Noise figure	7 dB
Beamforming pattern	2D sectoring antenna pattern
LOS probability function	Negative exponential function
Average LOS range	225 m
Average cell radius	R_c
LOS path loss exponent	1.96
NLOS path loss exponent	3.86
Small-scale fading	Nakagami fading of parameter 3 for NLOS
Average user per cell	10
Scheduler	Total fairness

cellular networks need a dense deployment of base stations to achieve acceptable coverage probability. Increasing base station density in a sufficiently dense network, however, may not improve SINR coverage, as shown in Fig. 8.15b. We present an intuitive explanation of the curves as follows. When increasing the base station density, the typical user at the origin will observe as if all base stations were squeezed toward it on the plane. Considering the presence of blockages, the user can only observe a finite number of base stations inside its LOS region. Increasing base station density will equivalently squeeze more base stations into the LOS region and render them LOS. As shown in Fig. 8.15, when the base station density is low, increasing the base station density may help improve the aggregate SINR by avoiding the case that few base stations are LOS. Instead, when the base stations are already sufficiently dense, squeezing more base stations into the LOS region will increase the number of strong LOS interferers and thus increase the total interference power, which might cause a drop of the SINR coverage. The decrease of SINR coverage in the dense network regime also indicates that dense mmWave networks may work in a interference-limited regime, and its performance can be further improved by allowing cooperation between base stations and switching base stations on/off smartly.

MmWave cellular can further apply multi-user beamforming to increase cell throughput. We compare the cell throughput in a multi-user mmWave network with the rates in microwave networks in Fig. 8.16. To make a fair comparison, we also use the stochastic model network model to simulate the microwave networks, and we assume that the overhead takes up 20% of the total bandwidth. In the microwave networks, the downlink bandwidth is 20 MHz, and average cell radius is 200 m. In the single user (SU)-MIMO microwave network, we assume that base stations and mobile stations each have four antennas to perform spatial multiplexing with zero forcing (ZF)-precoder. In the massive MIMO simulation, we apply the asymptotic rate derived in [BH13a], which upper bounds

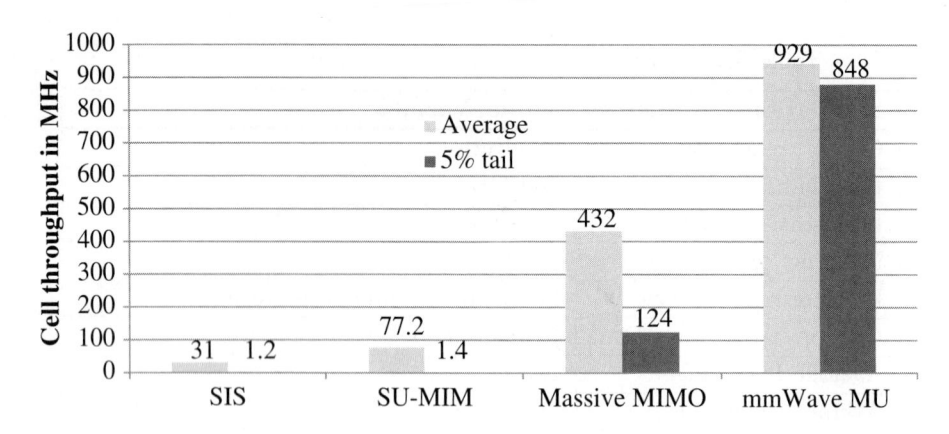

Figure 8.16 Comparison of cell throughput of mmWave networks and microwave networks.

the achievable rate, and assume the base stations serve 10 users simultaneously. In the mmWave simulation, the average cell radius is 100 m. We also assume the base stations randomly choose two users, and perform hybrid beamforming [ERAS+14] to reduce intra-cell interference. Simulation results show that with the large bandwidth at mmWave frequencies, mmWave systems outperform the conventional microwave network and are promising to reliably support 1 Gbps transmission. Other recent work similarly shows that multi-Gbps data rates, and remarkably good performance at the edge of the cell boundries, will be feasible in mmWave cellular networks [RRE14][Gho14][SSB+14].

The simulations in this section and others in [BH13b] and [BH13c] show that dense mmWave networks can provide comparable SINR coverage and much higher achievable rates compared with conventional microwave networks. These conclusions have been confirmed in other more comprehensive simulation frameworks [ALS+14][RRE14][Gho14]. There are still many research challenges that need to be addressed. For example, as dense mmWave networks will be interference-limited, the performance can be potentially improved by better management of interference, such as cooperating between base stations and switching base stations on/off in an intelligent way. More technical details of mmWave performance, such as applying a theoretical model to derive the SINR and rate distribution, are available in [BH13b][BH13c].

8.8 Chapter Summary

This chapter summarized many of the challenges related to designing higher layer protocols to support key features of mmWave communication technology. Much of the chapter focused specifically on how the MAC protocol copes with the mmWave PHY. One of the challenges of effective communication is to allow the transmitter and receiver to adapt their beams to achieve a high-quality communication link. This is especially difficult when new devices are associated in existing networks. This chapter reviewed several techniques for beam adaptation including iterative techniques and approaches that leverage channel estimation. Achieving high coverage is important but challenging

8.8 Chapter Summary

in mmWave systems due to attenuation from blockages. A solution is the use of multi-hop communication, also known as *relaying*. This chapter reviewed several relay concepts and explained how they fit in to a mmWave network framework to improve coverage. Multimedia is one of the main applications of the high data rates offered by mmWave networks. This chapter described some of the cross-layer techniques used to support video with an emphasis on unequal error protection (UEP). The idea is to map more resilient features of a multimedia signal to more robust features of the transmission technique, in this way protecting the more important parts of the multimedia signal. This chapter continued with a discussion of multi-band techniques, explaining how mmWave systems might coexist and evolve from lower frequency systems. Using multiple bands makes some of the control functions simpler in mmWave networks, for example, simplifying neighbor discovery and resource allocation.

The chapter concluded with some preliminary results on the coverage and capacity potential of mmWave 5G cellular networks. The fact that mmWave has a great potential as a technology in cellular systems was established. MmWave may be useful for both backhaul (as briefly discussed in this chapter and in Chapter 3) as well as for the access link, and it may provide the impetus to cause cellular and WiFi to blend into a more unified mobile and portable access architecture. MmWave offers a backhaul solution that provides a high data rate connection to the base station without the cost associated with wired spectrum. Using mmWave for cellular is very exciting and introduces many new challenges. Cellular systems will likely use licensed spectrum, not unlicensed spectrum as in WPAN and WLAN, so the unification of cellular and WiFi may not happen naturally or immediately, but given the fact that over 50% of today's cellular network data is offloaded onto unlicensed WiFi spectrum in large cities, the idea is not so far-fetched. Users will expect to receive high-quality service in cellular systems and are likely to be less forgiving of outages. Cellular systems need to support mobility and provide coverage over large geographic areas, supporting functions such as handoff. Overall, the future is bright for mmWave at the higher layers, with great opportunities to make breakthroughs using spatial processing while exploiting the massive bandwidths available.

Chapter 9

MmWave Standardization

9.1 Introduction

The mmWave wireless field is nascent and will rely upon an international standard setting process to ensure that a global mass market, with worldwide interoperability, will exist for consumers throughout the world. Standards activities for mmWave 5G cellular systems have not yet begun, but they are anticipated to begin in 2015–2016. Since the early 2000s, however, there have been standardization efforts for unlicensed 60 GHz products for personal area and local area networks. These products will soon become pervasive throughout the world [AH91][CF94]. Spectrum at 60 GHz became internationally available in the 1990s, well ahead of the commercial viability of mass market mmWave products.

It took about ten years for low-cost commercial products to evolve from 1998, when the USA became the first country in the world to authorize low-power unlicensed 60 GHz operations [Rap02]. Early work in a wide range of technical areas helped pave the way for eventual commercialization of the 60 GHz band [TM88][SL90][AH91][BMB91][SR92][PPH93] [CP93][CF94][LLH94][ARY95][CR96][MMI96][SMI$^+$97][ATB$^+$98][KSM$^+$98][GLR99]. The pace of 60 GHz research accelerated at the start of the new millennium, as engineers throughout the world gained knowledge and perspective that would be incorporated into various 60 GHz standards, prototypes, and products [XKR00][MID$^+$00][XKR02][AR$^+$02] [OMI$^+$02] [NNP$^+$02] [Smu02] [Sno02] [CGR$^+$03] [AR04] [BFE$^+$04] [EDNB04] [EDNB05] [DENB05] [VKR$^+$05] [ACV06] [BBKH06] [BNVT$^+$06] [RFP$^+$06] [KKKL06] [YGY$^+$06] [SHW$^+$06] [AKD$^+$07] [HBAN07a] [LCF$^+$07] [LHCC07] [NFH07] [PR07] [DR07] [PG07a] [AKBP08][LSE08][LS08][CRR08][BDS$^+$08][DSS$^+$08][NH08][MTH$^+$08][VKKH08][CRN09] [PR09][SB09a][SUR09][BSS$^+$10][RMGJ11]. Early standardization efforts did occur at the beginning of the millennium, but these activities were limited to country-specific standards [ARI00][ARI01], and very early efforts to create international standards did not adequately address and exploit the unique characteristics of 60 GHz channels [802.16-01].

In Chapter 1, we discussed how short-range wireless networks provide the most relevant applications to take advantage of unlicensed 60 GHz spectrum, as well as other frequencies in the mmWave band. In Chapters 3, 4, 5, and 6, we showed how advances have been made in understanding the channel [ECS+98] and creating the antennas, transmitters, and receivers needed to make low-cost mmWave multi-Gbps products for both indoor and outdoor WiFi and cellular networks. In Chapters 7 and 8 we discussed design decisions for the physical layer and above in the context of mmWave propagation, antennas, and circuit characteristics. In this final chapter, we focus on the recent short range wireless communication standards at 60 GHz, whose features are summarized in Table 9.1. The characteristics of 60 GHz standards will likely be shared by other developing or future mmWave standards due to the fundamental nature of the mmWave channel. At the time of this writing, the only advanced standardization efforts for mmWave wireless products have occurred for personal and local area network products that exploit the globally available unlicensed 60 GHz band. Our intent for this chapter is to elucidate the standardization details, with the belief that insights will be gleaned by the reader that can be applied to many other applications, bands, and use cases for future mmWave communication systems, for both short-range and long-range use that are sure to emerge in the coming years.

As shown in Table 9.1, there are currently five international standards that deal with mmWave WLAN and PAN applications in the global unlicensed 60 GHz band. WirelessHD focuses on the major market driver for 60 GHz wireless: streaming of uncompressed video. In contrast, ECMA-387 and IEEE 802.15.3c provide the architecture for all potential WPAN network topologies, of which wireless streaming of uncompressed video is a subset. Further, WiGig and IEEE 802.11ad focus on exploiting 60 GHz spectrum for WLAN network topologies While not all of these standards are likely to be a commercial success, there is much to be learned about the practice of mmWave wireless communication from the standards themselves.

The organization of this chapter is as follows. We first summarize the international spectrum regulations at 60 GHz in Section 9.2. The remaining sections (with the exception of the chapter summary in Section 9.8) are each dedicated to a single standard from Table 9.1. We note that IEEE 802.15.3c, as discussed in Section 9.3, and IEEE 802.11ad, as discussed in Section 9.6, are covered with greater depth than the other standards in this chapter due to the accessibility of IEEE standard contributions combined with the influence of IEEE standards (as evidenced by the global success of WiFi and Bluetooth) on the commercial wireless market.

Table 9.1 Important 60 GHz standards and their features.

Standard	Bandwidth	Rates	Topology	Approval Date
WirelessHD	2.16 GHz	3.807 Gbps	WVAN	01/2008
ECMA-387	2.16 GHz	$\leq$ 6.35 Gbps	WPAN	12/2008
IEEE 802.15.3c	2.16 GHz	$\leq$ 5.28 Gbps	WPAN	09/2009
WiGig	2.16 GHz	6.76 Gbps	WLAN	12/2009
IEEE 802.11ad	2.16 GHz	6.76 Gbps	WLAN	12/2012

9.2 60 GHz Spectrum Regulation

In this section, we summarize the availability of unlicensed 60 GHz spectrum worldwide. It is important to consider that spectrum allocation is controlled by governing bodies. As such, most of the smaller countries adopt the policies of the "big players." From a historical perspective, Europe (in particular, the United Kingdom), the United States, and Japan were the first to show interest in license-free 60 GHz use [UK688][Eur10][npr94][Ham03].

9.2.1 International Recommendations

The International Telecommunication Union (ITU) is an international organization whose main purpose is to aid governing bodies in the standardization of radio and telecommunications. Recommendations are published by the ITU regarding issues that it believes these governing bodies should follow. The ITU is an agency of the United Nations and, consequently, the Recommendations issued carry significant weight.

The ITU has suggested (in the ITU Radio Regulations) that the 55.78-66 GHz bands be available for all types of communication (although the 64-66 GHz band excludes aeronautical mobile applications). Two Recommendations from the ITU are relevant to millimeter-wave (mmWave) communication [ITU]:

- ITU-Radio (ITU-R) Recommendation 5.547
- ITU-R R-5.556

These Recommendations apply to fixed wireless services. ITU-R R-5.547 of the ITU Radio Regulations describes the 55.78-59 GHz and 64-66 GHz as suitable for high-density applications for fixed wireless services.

9.2.2 Regulations in North America

In the United States of America, the Federal Communications Commission (FCC) regulates radio transmissions. In 1995, the FCC allocated the 57-64 GHz frequency band for unlicensed communication under Part 15 in Title 47 of the Code of Federal Regulations [Fed06]. The largest power output under any circumstance limits the peak output power to less than 500 mW (27 dBm). Other conditional regulations include the following:

- The average power density (during the transmit interval) shall not exceed 9 $\mu W/cm^2$ (about 23 dBm with isotropic antennas) 3 m from the radiating structure.

- The peak power density (during the transmit interval) shall not exceed 18 $\mu W/cm^2$ (about 25 dBm with isotropic antennas) 3 m from the radiating structure.

- If the total emission bandwidth (BW) in Hertz is less than 1×10^5, the peak power is limited to $\left(0.5 \times BW \times 10^{-5}\right)$ W.

Public use of this band, however, specifically excludes the use of this spectrum in aircraft, for mobile field disturbance sensors (which probe to sense changes in the spectrum response), and for satellite applications. Satellite crosslink applications are an interesting application, but currently the FCC does not permit this for the public since the military uses 60 GHz for satellite crosslinks (see http://space.skyrocket.de/doc_sdat/milstar-1.htm). Fixed field disturbance sensors, such as those used to automatically open doors, have additional regulations (see FCC Part 15 guidelines for details).

In August 2013, the FCC made an important rule change to increase the EIRP for the 60 GHz unlicensed band for outdoor backhaul applications, through the FCC Report and Order 13-112 [Fed13]. The new FCC rule dramatically increased the total EIRP of unlicensed 60 GHz systems in the USA from 40 dBm average EIRP, and 43 dBm peak EIRP, up to 82 dBm and 85 dBm, respectively, for systems with antenna gains equal to or greater than 51 dBi. This increased EIRP limit, combined with higher receiver antenna gain, enables 60 GHz mmWave systems to more than offset the increased path loss associated with 60 GHz. The EIRP limit is reduced by 2 dB for every decibel that the antenna gain is below 51 dBi. Thus, FCC 13-112 allowed outdoor 60 GHz systems to achieve much greater coverage distances through the use of higher gain antennas, with as much as 42 dB more gain and EIRP than before. This rule enables outdoor 60 GHz unlicensed systems to be used for backhaul over distances as great as 2 km, even when considering the oxygen absorption described in Chapter 3.

In Canada, the Spectrum Management and Telecommunications Sector (S/T) of Industry Canada (IC) provides guidelines for spectrum usage [Can06]. Canada allows "low-power, license-exempt" communications in the 57-64 GHz band. This regulation, as detailed in Radio Standards Specification 210, adopts the FCC regulation parameters under Part 15.

9.2.3 Regulations in Europe

European Conference of Postal and Telecommunication Administrations (CEPT), which includes 48 European nations, organizes communication regulations in most European countries. CEPT has published European Telecommunications Standards Institute (ETSI) standard 302 217-3 for fixed point-to-point (backhaul) communication in the 57-66 GHz band and 302 567 for broadband radio access networks in the 57-66 GHz band intended for license-exempt short range devices [ETSI12a][ETSI12b]. European Commission decisions for harmonizing the 60 GHz spectrum within Europe have also been published [ECd13]. Compatibility requirements include 40 dBm maximum EIRP for broadband short range device operation.

9.2.4 Regulations in Japan

Japan is regulated by the Ministry of Internal Affairs and Communications, which allows unlicensed radiation at 59-66 GHz [Min06] and licensed radiation at 54.25-59 GHz. The specific terms of the unlicensed spectrum use between 59-66 GHz include the following:

- Maximum antenna gain less than 47 dBi
- Maximum average output power less than 10 mW
- Maximum occupied bandwidth less than 2.5 GHz

9.2.5 Regulations in Korea

The Millimeter Wave band Frequency Study Group (mmW FSG) was organized under the Korea Radio Promotion Association (RAPA) in 2005 for the purpose of organizing 60 GHz communication in Korea [KKKL06]. The following regulatory requirements are found in the 57-64 GHz band:

- All applications permitted
- Maximum transmission power of 0.01 mW (-20 dBm) for outdoor (point-to-point) applications between 57-58 GHz
- Maximum transmission power of 10 mW (10 dBm) for all other applications
- Antenna gain for outdoor applications limited to 47 dBi
- Antenna gain for indoor applications limited to 17 dBi
- Out of band emission measurement for all applications equal or less than -26 dBm at 1 MHz resolution

9.2.6 Regulations in Australia

In 2005, the Australian Communications and Media Authority (ACMA) provided free "class licenses" for low power interference devices (LIPDs) at 59.4-62.9 GHz. The terms of this regulation are as follows [dAD04]:

- Maximum EIRP of 150 W (approximately 51.7 dBm)
- Peak transmitted power of 10 mW (10 dBm)
- Terrestrial applications only

9.2.7 Regulations in China

The Chinese government has released spectrum between 59-64 GHz for short range wireless communications [Xia11]. Specific radiation restrictions include the following:

- Peak transmit power of 10 mW (10 dBm)
- Maximum antenna gain of 37 dBi

9.2.8 Comments

The preceding regulations imposed by different national entities indicate significant common unlicensed spectrum worldwide at 60 GHz as summarized in Fig. 9.1. It should be noted that other nations will regulate 60 GHz spectrum under some subset of the provided regulations when market demand for 60 GHz devices is considered sufficient (if not already regulated).

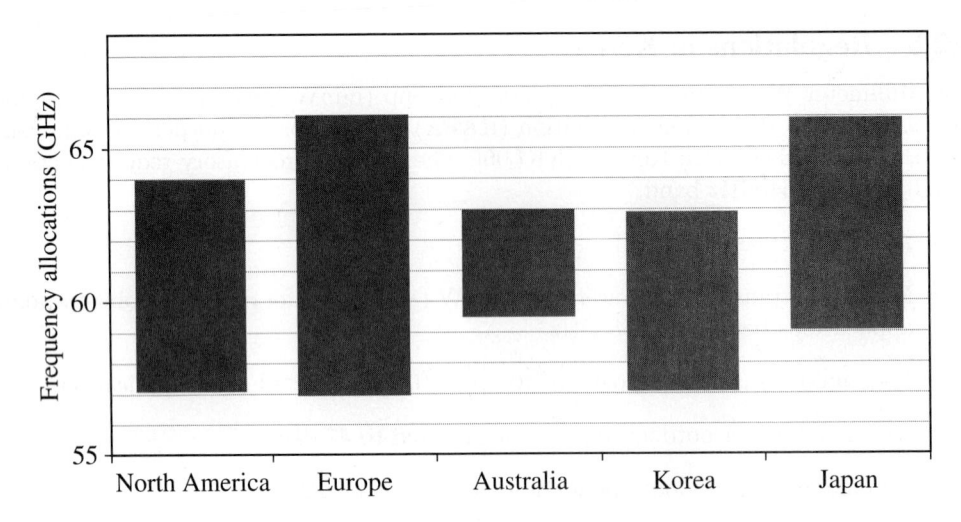

Figure 9.1 International frequency allocation for 60 GHz wireless communication systems.

9.3 IEEE 802.15.3c

IEEE 802.15.3c represents a comprehensive 60 GHz standard for unlicensed short range, high bandwidth applications through WPANs. Since IEEE 802.15.3c is *primarily* a 60 GHz physical layer extension of the IEEE 802.15.3 WPAN, we first discuss the IEEE 802.15.3 MAC.

9.3.1 IEEE 802.15.3 MAC

The IEEE 802.15.3 standard defines the WPAN ad hoc network through *piconets* [802.15.3-03][802.15.3-06]:

> A piconet is distinguished from other types of data networks in that communications are normally confined to a small area around the person or object that typically covers at least 10 m in all directions and envelops the person or thing whether stationary or in motion.
>
> This is in contrast to local area networks (LANs), metropolitan area networks (MANs), and wide area networks (WANs), each of which covers a successively larger geographic area, such as a single building or a campus that would interconnect facilities in different parts of a country or of the world.

Each piconet contains two classes of participants: the single coordinator (PNC) described in Chapter 8 and the remaining devices (DEV). As illustrated in Figs. 9.2 and 9.3, the PNC provides network synchronization through beacons, and data transmission transpires between any two DEVs in the piconet. The role of the PNC is to control access to the piconet, to provide time-slotted communication through network synchronization, to manage the quality of service (QoS), to enable coexistence with other networks, and to collect the information that enables these features.

9.3 IEEE 802.15.3c

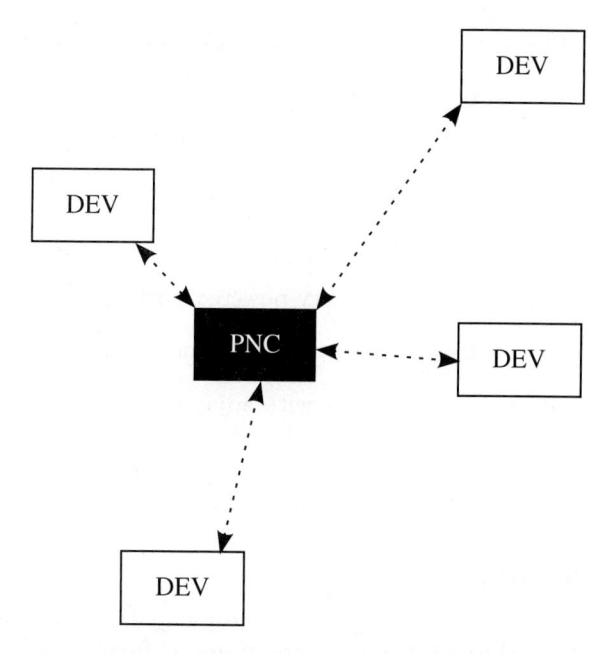

Figure 9.2 Piconet structure for WPAN ad hoc networking. Each piconet is assigned a coordinator (PNC). Control information is communicated to the remaining devices (DEV) in the piconet through beacons. Dotted lines represent beacon communications.

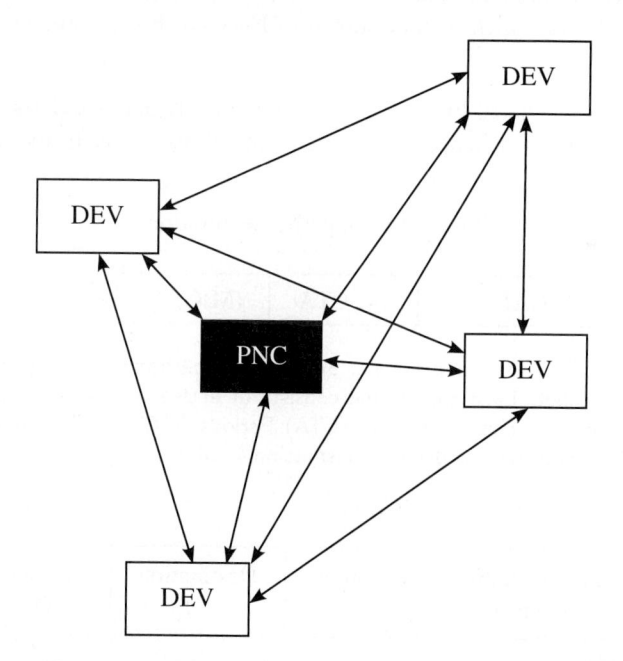

Figure 9.3 Piconet structure for WPAN ad hoc networking. Solid lines represent potential data communication, excluding control information, which is transmitted as shown in Fig. 9.2.

The remainder of this section summarizes the salient features of the IEEE 802.15.3 MAC in further detail. The reader is encouraged to consider that this standard was designed to deliver a number of important features not present, at least not concurrently, in competing IEEE wireless standards. These features include the following:

- Fast device connections (less than 1 second [Ree05])
- Ad hoc, on-demand, networking
- Efficient power management for battery-powered portable devices
- Easy entrance and exit from network (dynamic membership)
- Interference handling and coexistence with adjacent networks
- Information security
- Efficient transfer of large bandwidth data

9.3.1.1 Frame Structure

At its highest level, IEEE 802.15.3 MAC data are segmented into superframes, which are allocated continuously back-to-back. Each superframe is further broken down into the Beacon Period, the Contention Access Period (CAP), and the Channel Time Allocation Periods (CTAPs) as illustrated in Fig. 9.4. In each of these periods within the superframe, MAC frames are communicated between DEVs using a fixed frame format. Every frame includes the header with format shown in Fig. 9.5. Each segment of the header is summarized as follows.

- *Frame Control* — The 16-bit Frame Control field is transmitted first and determines the frame type (i.e., a beacon frame, an immediate ACK frame, a delayed ACK

Single MAC superframe duration					
Beacon	CAP	(M)CTA	(M)CTA	...	(M)CTA

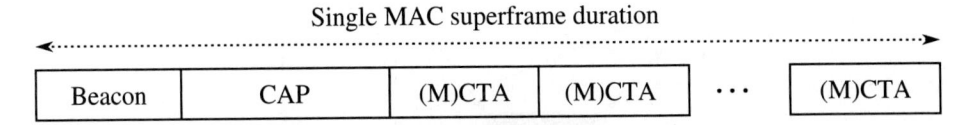

Figure 9.4 IEEE 802.15.3 MAC superframe. Successive superframes are transmitted throughout the duration of the piconet. Each superframe consists of a Beacon Period, a Contention Access Period (CAP), and Channel Time Allocation (CTA) Periods (CTAP). CTAs may be reserved for management operation (MCTA). Order of transmission is from left to right (Beacon transmitted first).

Stream Index	Fragmentation Control	Source ID	Destination ID	Piconet ID	Frame Control

Figure 9.5 IEEE 802.15.3 MAC frame header. Order of transmission is from right to left (Frame Control transmitted first).

frame, a command frame, or a data frame), whether security is enabled in the frame, whether this frame is a re-transmission, the ACK policy (i.e., no ACK, immediate ACK, delayed ACK, or delayed ACK request), and whether more data are to be transmitted by the same DEV during the Channel Time Allocation (CTA).

- *Piconet ID* — The 16-bit Piconet ID field provides a unique identifier for the piconet of operation.

- *Source/Destination ID* — The 8-bit Source/Destination fields uniquely identify the source/destination device for the IEEE 802.15.3 data frame. The PNC assigns these IDs and there are reserved IDs for specific circumstances including IDs for the PNCs, IDs for neighbor piconet DEVs, destination IDs for multicast frames, IDs for unassociated DEVs attempting to join the piconet, and for broadcast frames.

- *Fragmentation Control* — The 24-bit Fragmentation Control field aids in the fragmentation and reassembly of MAC frames.

- *Stream Index* — The 8-bit Stream Index field is transmitted last in the header and determines the stream index of isochronous data. Asynchronous data and MCTA traffic have reserved identifiers as well.

The format of the body of every MAC frame is illustrated in Fig. 9.6. Each segment of the body is summarized as follows. The secure session ID, the secure frame counter, and the integrity code are only necessary fields when the MAC frame is transmitted with security features.

- *Secure Session ID* — The 16-bit Secure Session ID field is transmitted first and identifies the key used to protect the secure MAC payload.

- *Secure Frame Counter* — The 16-bit Secure Frame Counter field ensures the uniqueness of the cryptographic nonce in secure MAC frames. A DEV initializes this counter to zero for the first frame sent and increments for each successive frame. If a DEV receives a new key, the DEV sets the counter to zero.

- *Payload* — The variable-length Payload field contains MAC data information with or without security protection.

- *Integrity Code* — This 64-bit field contains an Integrity Code that cryptographically protects the MAC header and Payload.

- *Frame Check Sequence* — The 32-bit frame check sequence allows the MAC at a receiving DEV to verify the integrity of the MAC payload through a cyclic redundancy check (CRC) according to ANSI X3.66-1979.

Frame Check Sequence	Integrity Code	Payload	Secure Frame Counter	Secure Session ID

Figure 9.6 IEEE 802.15.3 MAC frame body. Order of transmission is from right to left (Secure Session ID transmitted first).

9.3.1.2 Information Exchange

Data exchanges between DEVs and PNCs in a piconet occur in the Beacon Period, the CAP, or the CTAP in the superframe. The Beacon Period is used by the PNC to disperse management information about the piconet or to issue commands from the PNC to DEVs in the piconet. The CAP allows for on-demand access to the network through carrier sense multiple access with collision avoidance (CSMA/CA) in a similar manner to IEEE 802.11. This access method allows for efficient, low latency asynchronous data communication between DEVs or command communication between the DEV and PNC. Due to the overhead associated with CSMA/CA, however, the CAP is not desired for the transfer of a large quantity of data. New DEVs that wish to join the piconet must also use the CAP to communicate with the PNC.

The CTAP uses scheduled time slots, allocated by the PNC, for all communication. The PNC specifies all of the CTAs that are allocated to a single CTAP within a superframe. If the source or destination is also the PNC, this is designated as a management CTA (MCTA). Typically, the difference between a standard CTA and a MCTA is strictly a matter of nomenclature and not of functionality. As a special case, however, slotted aloha communication is provided whenever the source/destination DEV ID is left unspecified by the PNC. This special case is provided to allow for future compatibility with PHYs that may not be able to perform clear channel assessment with the same proficiency as IEEE 802.15.3 PHYs.

DEVs can request access within the CTAP in two ways: (1) *isochronously* or a CTA that is allocated to a DEV on a periodic basis (may be more or less frequent than once per superframe) without specification of the total time allocated. Isochronous streams are given a distinct index for each source/destination pair and must be ended by the source when no data are left for transmission. (2) *asynchronously* or a CTA that is allocated for a fixed time amount as early as possible.

9.3.1.3 Starting a Piconet

A piconet begins when a PNC-capable DEV begins transmitting beacon messages (not all DEVs are capable of PNC operation). The transmission of beacon messages notifies other DEVs, within range, of the piconet creation. Before the self-promotion of a DEV to PNC and before transmission of beacon messages, however, the DEV must scan all available frequency channels. If the DEV determines that a channel is available and free from interference, which is determined by the period of time in which no interference is measured, it may consider itself the PNC and begin transmitting beacon messages.

As 60 GHz networks become more popular, it is anticipated that channel resources will become increasingly scarce. If all channels are in use, the PNC-capable DEV is not allowed to create a new piconet. Instead, the PNC-capable DEV may join one of the pre-existing piconets without PNC classification. Since the reason that a PNC-capable DEV desires to create a piconet is to associate with any DEVs that may have compatible and potentially secure data sources that need coordination, simple piconet membership may not be sufficient if the PNC of the joined piconet cannot provide controlled access to the data sources. Fortunately, even if the PNC-capable DEV is not able to create its own piconet, the PNC-capable DEV can still act as a PNC of a child piconet that operates under the guidelines of the PNC of the parent, newly joined piconet.

9.3.1.4 Nested Piconets

Each piconet can also contain nested piconets through the formation of a child piconet. In the first level of the nested piconet, a child PNC is assigned to a DEV in the original piconet.[1] Access in the child piconet is subject to the time-slotted allocations provided by the PNC in the parent piconet. Theoretically, there is no limit to the depth of the nested piconet tree or the number of children PNCs per level; however, physical limitations on the available bandwidth will keep child and parent PNCs from indefinitely creating child piconets.

Security is maintained within each of the nested piconets. Furthermore, the allocation of bandwidth resources within the child piconet is the responsibility of the child PNC as long as the child piconet maintains these resources within the entire allocation given the child PNC within the parent piconet. In other words, peer-to-peer communication is only available for DEVs that belong to the same piconet on the same level of the nested piconet tree. An additional classification is available to a child piconet. A *neighbor* piconet is a child piconet where the child PNC is not available for peer-to-peer communication with the other DEVs in the parent piconet. Essentially, the neighbor piconet is purely a coexistence mechanism to allow sharing of spectrum resources. It is distinguished from other child piconets in that the child was only created because no free channels were available for the neighbor PNC to create its own piconet. As an illustration of a neighbor piconet, consider the following example.

Fig. 9.7 shows the formation of a child/neighbor piconet where the neighbor PNC is a personal computer with associated portable entertainment DEVs. The neighbor piconet is created under an existing parent PNC in a home entertainment set-top box which itself is associated with a high-definition television (HDTV) and high-fidelity (Hi-Fi) stereo.

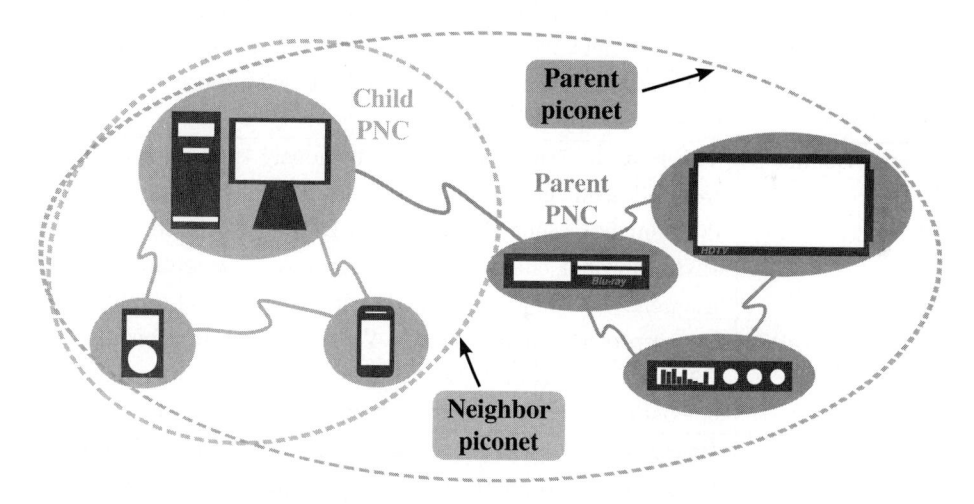

Figure 9.7 Neighbor piconet. The parent piconet (controlled by the set-top box) manages coexistence with the neighbor piconet (controlled by the personal computer).

---

1. Nested piconets are piconet structures with multiple levels of child piconets.

The neighbor (PC) piconet is formed because all available frequency channels were occupied by other piconets. The personal computer PNC is a child PNC under the set-top box, but because it does not desire communication between the set-top box, the HDTV, and the Hi-Fi, it requests classification as a neighbor PNC. The set-top box parent PNC determines the time/bandwidth allocated to the HDTV, the Hi-Fi, and the PC.

9.3.1.5 Interference and Coexistence

Along with the formation of neighbor piconets, other mechanisms are available in the IEEE 802.15.3 MAC to reduce interference and improve coexistence with other IEEE 802.15.3 devices. IEEE 802.15.3 allows for dynamic channel selection to minimize the interference contribution to/from adjacent non-standard networks and transmit power selection to minimize the interference footprint of the piconet. Table 9.2 summarizes the interference reduction and coexistence strategies available to IEEE 802.15.3 networks.

Table 9.2 IEEE 802.15.3 MAC interference avoidance and coexistence strategies.

Strategy	Description
Neighbor Piconet	If no free channels are available, coexistence is maintained by creating a neighbor piconet where bandwidth/resources are shared under the parent piconet. The ability of piconets to provide children depends on data demands of DEVs in each piconet.
PNC Handover	If the current PNC is at the edge of the piconet, it may be contributing and receiving excessive interference from adjacent networks. The PNC may request a handover of its responsibilities to another PNC with more desirable coexistence properties.
Dynamic Channels	The PNC has the ability to continuously track interference from adjacent networks. PNCs may continuously scan the different frequency channels, and DEVs associated with the PNC can offer channel status information by scanning all frequency channels or by evaluating the channel quality between other piconet DEVs. With this information, the PNC may determine that another channel provides better coexistence and may switch the piconet to that channel.
Power Control	Power control operates on two levels: (1) the PNC limits the transmit power to reduce the footprint of the piconet and, hence, the maximum distance for connectivity; (2) each DEV can increase or decrease the transmit power by single increments of 2 dB such that only enough power is used to maintain sufficient link performance.

9.3.1.6 Energy Efficiency

Minimizing the transmit power not only reduces interference and improves coexistence, but it also increases the energy efficiency of IEEE 802.15.3 devices since less source (e.g., battery) drain occurs when the transmit power is reduced. The best method for increasing energy efficiency in wireless devices, however, is to place devices in sleep mode when they are not in use. In much the same way as in IEEE 802.11 networks, beacons allow DEVs to determine sleep and wake intervals. IEEE 802.15.3 provides four power management modes: (1) device-synchronized power save, (2) piconet-synchronized power save, (3) asynchronous power save, and (4) active or no power management modes. Table 9.3 summarizes each of the first three power management modes.

9.3.1.7 Association and Disassociation

DEV association in a piconet requires authentication from the PNC. Once the authentication is processed successfully, a new DEV may join the piconet. A number of reasons may be given, however, by a PNC for denial of association including:

- Maximum number of DEVs are already served by PNC.
- Not enough bandwidth is available for all DEVs if new DEV is added.
- Channel quality is not good enough for DEV in the piconet.
- The PNC is turning off and there are no other PNCs to serve the piconet.

Table 9.3 IEEE 802.15.3 power management modes.

Power Management Mode	Description
Piconet-Synchronized (PS)	A DEV requests PS mode and the PNC determines the sleep interval by informing the DEV of wake beacons. All DEVs in PS power management must be synchronized to and listen to all wake beacons.
Device-Synchronized (DS)	In DS mode, many DEVs set their sleep periods together. A DS power save set means that all of the DEVs have the same time interval between wake periods. DS mode allows all DEVs to wake-up together and exchange data. This enables DEVs to be informed when other DEVs are awake.
Asynchronous (AS)	In AS mode, DEVs enters sleep mode for periods of time. The only responsibility of the DEV in AS mode is to communicate with the PNC before the association timeout period (ATP) expires. Unlike the other modes, all devices act independently.

- The DEV is requesting association as a neighbor and the PNC does not allow neighbor piconets.

- The frequency channel is being switched by the PNC.

- The PNC is being changed in the piconet, so association cannot be processed at this time.

- Any other unspecified reason for the PNC to not associate with the DEV exists.

Similarly, a DEV may be disassociated by the PNC. If the DEV confirms the request for disassociation, it will be completed. The PNC, however, may also proceed unilaterally with a disassociation of a DEV for the following reasons:

- The association timeout period (ATP) expired without contact from the DEV to the PNC.

- The channel quality of the associated DEV is not good enough to exist in the piconet.

- The PNC is turning off and there are no other PNCs to serve the piconet.

- The PNC is not compatible with the DEV or vice versa.

- Any other unspecified reason for the PNC to disassociate with the DEV exists.

9.3.1.8 PNC Handover and Ending a Piconet

If another PNC-capable DEV exists in the piconet, the PNC may request handover of control to the PNC-capable DEV. All existing resource allocations are maintained during the handover process. Generally this occurs because either (1) a more capable DEV is discovered in the network, or (2) the PNC is ceasing operation in the piconet.

A piconet ends when the PNC ceases operation and there are no other PNC-capable DEVs to take over the role of PNC. If a PNC abruptly ceases operation without notification, the piconet officially ends when the ATP expires. A PNC-capable DEV may then start a new piconet in the frequency channel of the piconet where PNC notifies the end of operation or the ATP expires. If there are child piconets present under the parent PNC whose operation is ceasing, only one piconet may continue operating. The remaining piconet is specified by the PNC of the top-level piconet. All resources in the frequency channel will now be dedicated to this piconet. When a child piconet ends, this does not affect the operation of a parent piconet.

9.3.2 IEEE 802.15.3c MmWave PHY

The IEEE 802.15.3 standard provides a specification for a physical layer (PHY) with quadrature amplitude modulation (QAM) constellations and trellis-coded modulation at 2.4 GHz [802.15.3-03]. The data rates provided by the 2.4 GHz PHY are comparable to IEEE 802.11a/g with a maximum PHY data rate of 55 Mbps. The mmWave PHY, which is defined in the IEEE 802.15.3c amendment, provides significantly higher rates by exploiting 60 GHz unlicensed spectrum. Initially, IEEE 802.15.3c was proposed as a PHY-only amendment, but due to the unique properties of 60 GHz networks, such as beam steering, MAC functionality has also been added in the approved document [802.15.3-09].

The mmWave PHY offers a common mode and three different operating modes: single carrier mode (SC-PHY), high-speed interface mode (HSI PHY), and audio/visual mode (AV PHY). The common mode supports low-directionality (or omnidirectional) antennas, high multipath environments, and provides the ability to "find" weaker stations using scanning beams at the cost of relatively low data throughputs. Three different operating modes are provided because a consensus could not be reached on a single PHY implementation that addressed all desired operation properties including low implementation complexity, high efficiency operation in multipath channels, and flexibility with streaming media. The SC-PHY, which is the most flexible mode with low-complexity implementation and high efficiency in low-multipath channels, supports data rates up to 5 Gbps through single-carrier modulation with low-complexity Reed-Solomon (RS) coding or high-complexity low-density parity check (LDPC) coding. In contrast, the HSI PHY involves the highest complexity and is implemented with OFDM modulation with LDPC-coding over sub-carriers. The HSI PHY is best suited for high-multipath environments that result, for example, from omnidirectional antennas in factories and open-plan buildings [Rap89][SR91]. The final operating mode, the AV PHY, is best suited for high-definition audio/video streaming through its OFDM PHY and concatenated convolutional/Reed-Solomon (inner/outer) codes.

Only one of the operating modes is required for a standard-compliant IEEE 802.15.3c device. Each piconet typically uses a single PHY for all member DEVs. It is possible, however, that a dependent piconet uses a different operating mode than the parent piconet, assuming that both the parent PNC and the child PNC are compatible with each distinct operating mode. Moreover, it is possible that multi-mode DEVs within a piconet may use any operating mode that they share within the CTA. To enable coordination of devices that implement different PHYs with distinct operating modes, it is desirable for each IEEE 802.15.3c device to offer common mode signaling such that beacon messages may be exchanged. Note, however, that the common mode signaling (discussed in Section 9.3.2.2) is not required for AV PHY-compliant devices.

9.3.2.1 Channelization and Spectral Mask

Fig. 9.8 shows the channelization of the mmWave PHY in IEEE 802.15.3c. Four 2.16 GHz channels are provided at the center frequencies 58.32, 60.48, 62.64, and 64.8 GHz. The carrier frequencies chosen are based on even integer-multiples of commonly available frequency references. For coexistence and adherence to the international spectral regulations that can vary from country to country, Fig. 9.9 shows the mmWave PHY spectral mask. The transmission power is normalized and all devices are allowed to transmit at full power according to the relevant local regulatory standards for the operating device.

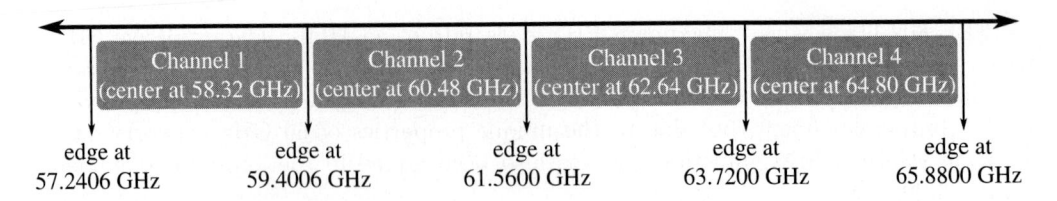

Figure 9.8 Channelization in IEEE 802.15.3c provides four different channels for mmWave PHY.

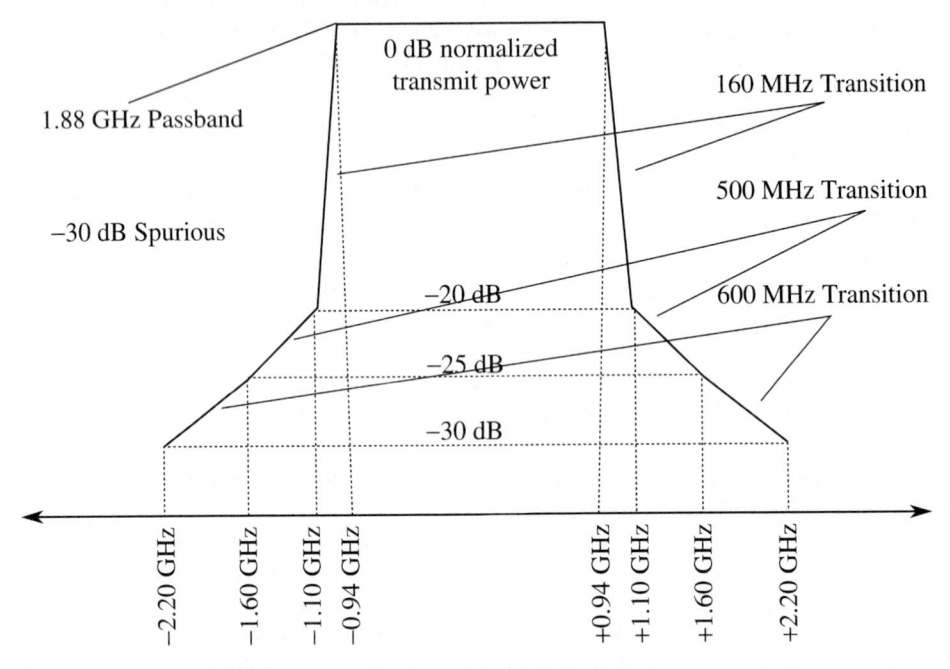

Figure 9.9 Normalized spectral mask for mmWave PHY transmissions in IEEE 802.15.3c. Transmissions must be attenuated 20 dB ± 0.94 GHz from center frequency, by 25 dB ± 1.10 GHz from center frequency, and by 30 dB ± 2.20 GHz from center frequency. To measure the spectral mask, a resolution bandwidth of 3 MHz and a video bandwidth of 300 kHz is specified in the standard.

9.3.2.2 Common Mode Operation

As discussed throughout this text, particularly in Chapters 3, 7, and 8, phased arrays with beam steering can be used to combat path loss, multipath delay spread, and Doppler induced fading in mmWave wireless channels. Phased array processing, however, is not desirable to achieve link gain in the common mode in IEEE 802.15.3c since it needs to be compatible with broadcast and link negotiation messages. Consequently, the common mode in IEEE 802.15.3c is a single carrier transmission mode that compensates for the significant channel and device losses of 60 GHz channels through data spreading, as it assumes that omnidirectional antennas, or near-omnidirectional antennas, are used to form all links. The loss in spectral efficiency that results from data spreading is not a large concern since the common mode is not intended to handle large data transfers, but rather is intended to provide initial device discovery and initialization of communications before beamforming and high data rate transfers are provided to devices.

Common mode signaling (CMS) frames are formatted as illustrated in Fig. 9.10. The CMS preamble provides for Automatic Gain Control (AGC), frame/symbol synchronization, and channel estimation at the receiver. The preamble is composed of the fields shown in Fig. 9.11. The breakdown of each preamble field is summarized in Table 9.4. The CMS frame header (which includes both the PHY header and the MAC header) and PHY payload are each protected by Reed-Solomon (RS) FEC with mother

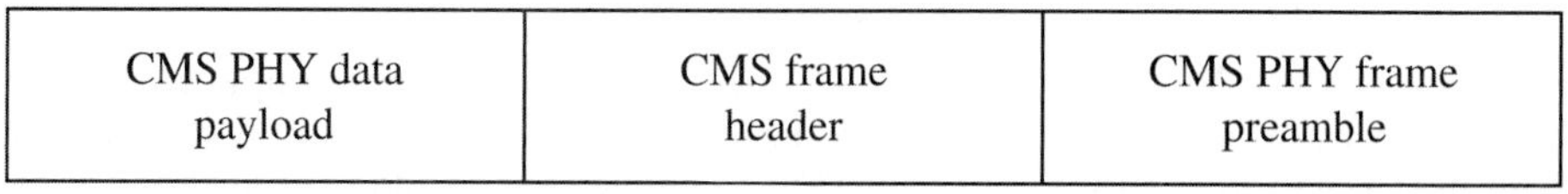

CMS PHY data payload	CMS frame header	CMS PHY frame preamble

Figure 9.10 Common mode signaling (CMS) PHY frame format. Order of transmission is from right to left (preamble transmitted first).

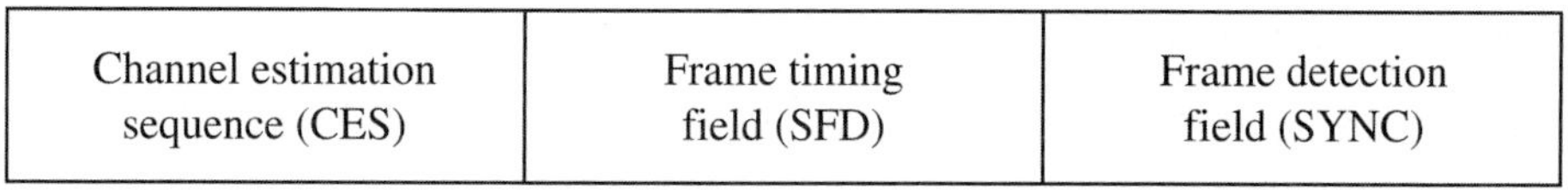

Channel estimation sequence (CES)	Frame timing field (SFD)	Frame detection field (SYNC)

Figure 9.11 Fields in the CMS preamble. Order of transmission is from right to left (SYNC transmitted first).

Table 9.4 Sequences used in the CMS PHY frame. Golay sequences are defined in hexadecimal format. All sequences use LSB-first transmission rules.

Item	Description
SYNC	48 repetitions of $\mathbf{b}_{128}$ used for frame detection
SFD	$[+1 -1 +1 +1 -1 -1 -1] \otimes \mathbf{b}_{128}$; used for frame and symbol synchronization (i.e., start frame delimiter)
CES	$[\mathbf{b}_{128}\mathbf{b}_{256}\mathbf{a}_{256}\mathbf{b}_{256}\mathbf{a}_{256}\mathbf{a}_{128}]$; used for estimation of channel coefficients
$\mathbf{a}_{64}$	Complementary 64-bit Golay sequence 1: 0x63AF05C963500536
$\mathbf{b}_{64}$	Complementary 64-bit Golay sequence 2: 0x6CA00AC66C5F0A39
$\mathbf{a}_{128}$	Complementary 128-bit Golay sequence 1: 0x0536635005C963AFFAC99CAF05C963AF
$\mathbf{b}_{128}$	Complementary 128-bit Golay sequence 2: 0x0A396C5F0AC66CA0F5C693A00AC66CA0
$\mathbf{a}_{256}$	$[\mathbf{a}_{128}\mathbf{b}_{128}]$
$\mathbf{b}_{256}$	$[\mathbf{a}_{128}\overline{\mathbf{b}}_{128}]$
$\overline{\mathbf{b}}_{128}$	Binary complement of $\overline{\mathbf{b}}_{128}$

code RS($n+16,n$) for n octet frame header and RS(255,239) for the PHY payload.[2] The header and payload are additionally spread with 64-bit Golay sequences $\mathbf{a}_{64}$ and $\mathbf{b}_{64}$. Each bit is spread with $\mathbf{a}_{64}$ or $\mathbf{b}_{64}$ depending on the output of a linear feedback shift register, which generates a pseudo-random bit sequence at the same rate as the original coded bits and not the 64× spreading rate. A complete description of the code spreading operation is provided in Section 12.1.12.2 of the IEEE 802.15.3c standard [802.15.3-09].

9.3.2.3 SC-PHY

Although the SC-PHY in IEEE 802.15.3c does not provide many application-specific functions, it does provide the most flexible PHY in terms of the breadth of applications that can be served, especially applications that demand low implementation complexity. The serviced applications are separated into three classes:

1. Low-power and low-cost for the least demanding applications

2. Moderate complexity

3. High complexity for the most demanding applications

Within each class there are a multitude of modulation and coding schemes (MCSs) that determine the symbol constellation, modulation format, spreading gain, and forward error correction (FEC). Table 9.5 summarizes the configuration of each MCS for the SC-PHY. All SC-PHYs must implement both MCS 0 and MCS 3, the mandatory PHY rate (MPR). Hence, it is only mandatory for a SC-PHY to have Class 1 compatibility. Class 1 uses the most robust modulation format through $\pi/2$ BPSK with a variable spreading factor. RS (mandatory) and LDPC codes (optional) provide FEC for Class 1. Class 2 provides more aggressive data rates by doubling the constellation order. All Class 2 MCS, except MCS 11, require LDPC FEC. Class 3 provides the most aggressive data rates of up to 5.28 Gbps with yet another increase in the constellation order. Note that RS FEC is not available for Class 3 operation.

Several observations about the flexibility of the SC-PHY can be made from Table 9.5. First, because it uses single-carrier constellations, it offers high power amplifier efficiency for low-complexity devices, as shown in Chapter 7. Further, low-complexity RS FEC is allowed to provide rates up to 3.3 Gbps. Low-complexity RS FEC does not work well in the other popular modulation format at 60 GHz (OFDM) since OFDM is highly sensitive to frequency selectivity caused by multipath in the channel. To compensate for path loss, variable spreading gain is allowed, which offers much less device cost than adaptive antenna arrays. Finally, very high rate video streaming can also be handled by added LDPC FEC compatibility with higher order constellations. As discussed later in this section, the SC-PHY data symbols are fragmented in blocks to allow low-complexity frequency-domain equalization for highly frequency selective channels.

Similar to the CMS frame format in Fig. 9.10 (CMS is also an MCS in the SC-PHY), the SC-PHY frame format can be broken down into the PHY preamble (with CES,

---

2. The header and payload are also scrambled before FEC operation as prescribed in Section 12.1.12.3 of the IEEE 802.15.3c standard. The RS poly configuration is described in Section 12.1.12.1 of the IEEE 802.15.3c standard [802.15.3-09].

Table 9.5 Modulation and coding schemes in IEEE 802.15.3c SC-PHY. L_{SF} denotes the length of the spreading sequence. Note: The data rate is calculated for zero pilot word length.

	MCS	Data Rate	Modulation	FEC
Class 1	0 (CMS)	25.8 Mbps	$\pi/2$ BPSK ($L_{SF}=64$)	RS(255,239)
	1	412 Mbps	$\pi/2$ BPSK ($L_{SF}=4$)	RS(255,239)
	2	825 Mbps	$\pi/2$ BPSK ($L_{SF}=2$)	RS(255,239)
	3 (MPR)	1,650 Mbps	$\pi/2$ BPSK	RS(255,239)
	4	1,320 Mbps	$\pi/2$ BPSK	LDPC(672,504)
	5	440 Mbps	$\pi/2$ BPSK ($L_{SF}=2$)	LDPC(672,336)
	6	880 Mbps	$\pi/2$ BPSK	LDPC(672,336)
Class 2	7	1,760 Mbps	$\pi/2$ QPSK	LDPC(672,336)
	8	2,640 Mbps	$\pi/2$ QPSK	LDPC(672,504)
	9	3,080 Mbps	$\pi/2$ QPSK	LDPC(672,588)
	10	3,290 Mbps	$\pi/2$ QPSK	LDPC(1440,1344)
	11	3,300 Mbps	$\pi/2$ QPSK	RS(255,239)
Class 3	12	3,960 Mbps	$\pi/2$ 8-PSK	LDPC(672,504)
	13	5,280 Mbps	$\pi/2$ 16-QAM	LDPC(672,504)

Table 9.6 Breakdown of each field in the SC-PHY preamble. Note that the common mode (MCS 0) follows Table 9.4. Sequences are transmitted LSB first.

Item	Description
SYNC	14 repetitions of $\mathbf{a}_{128}$
SFD (MR)	$[+1 -1 +1 -1] \otimes \mathbf{a}_{128}$;
SFD (HR)	$[+1 +1 -1 -1] \otimes \mathbf{a}_{128}$;
CES	$[\mathbf{b}_{128} \mathbf{b}_{256} \mathbf{a}_{256} \mathbf{b}_{256} \mathbf{a}_{256}]$; used for estimation of channel coefficients
$\mathbf{a}_{128}$	Complementary 128-bit Golay sequence 1: 0x0536635005C963AFFAC99CAF05C963AF
$\mathbf{b}_{128}$	Complementary 128-bit Golay sequence 2: 0x0A396C5F0AC66CA0F5C693A00AC66CA0
$\mathbf{a}_{256}$	$[\mathbf{a}_{128} \mathbf{b}_{128}]$
$\mathbf{b}_{256}$	$[\mathbf{a}_{128} \overline{\mathbf{b}}_{128}]$
$\overline{\mathbf{b}}_{128}$	Binary complement of $\overline{\mathbf{b}}_{128}$

SFD, and SYNC fields), the frame header, and the PHY payload. The breakdown of each preamble field is summarized in Table 9.6. Note that the header is not transmitted using the same configuration as the CMS. Three header configurations are available: CMS rate, medium rate (MR), and high rate (HR). These rates are differentiated by their spreading factors (64, 6, and 2, respectively) and their data block formatting (to be discussed later). The header is always modulated with $\pi/2$-BPSK and encoded with RS(n+16,n) where n is the number of octets in the total header. A MAC sub-header is also optionally available for the advanced MAC features of IEEE 802.15.3c (also to be discussed later). Sub-headers are only available with MR and HR configurations. If the base header uses HR, the sub-header will also use HR, otherwise the sub-header will use MR (non-SC-PHY CMS does not have sub-header capability).

As summarized in Fig. 9.12, to create the SC-PHY payload, each binary MAC data frame is scrambled,[3] encoded (RS or LDPC), padded with stuff bits (for evenly built SC-PHY blocks), spread (optionally), mapped to the appropriate symbol constellation ($\pi/2$-BPSK, $\pi/2$-QPSK, $\pi/2$-8-PSK, or $\pi/2$-16-QAM), and finally formatted into SC-PHY blocks. The encoding process in Fig. 9.12 occurs through RS/LDPC coding as shown in Table 9.5.[4] The spreading process is included in this figure, but only Class 1 MCS 0, 1, 2, and 5 have non-trivial spreading operations (spreading factor $L_{\text{SF}} > 1$).

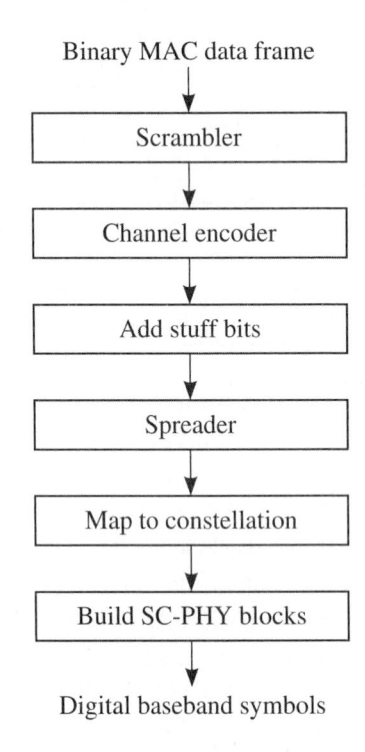

Figure 9.12 Block diagram of payload building process in SC-PHY.

---

[3]. The scrambler uses the same linear feedback shift register (LFSR) as the spreading operation with generator polynomial $x^{15} + x^{14} + 1$. The seed of the LFSR, however, is random and communicated between the transmitter and receiver (note that the seed of the scrambling operation in the header is always known). See 12.2.2.10 of the standard for more details.

[4]. MCS 0-3 and 11 use the RS(255,249) (coding rate = 0.93725) block code in GF(2^8). The remaining MCS use irregular LDPC block codes with rates 3/4, 1/2, 1/2, 3/4, 7/8, 14/15, 3/4, and 3/4 for MCS 4, 5, 6, 7, 8, 9, 10, 12, and 13, respectively. For specific encoder architectures, see 12.2.2.6 of the standard [802.15.3-09].

9.3 IEEE 802.15.3c

As mentioned before, the CMS mode (MCS 0) uses complementary Golay sequences for spreading. The remaining MCS, however, use pseudo-random binary sequences resulting from a linear feedback shift register with generator polynomial $x^{15} + x^{14} + 1$ and seed [010100000011111] at chip rate of $R_c = 1{,}760$ MHz resulting from the input encoded and stuffed bit rate R_c/L_{SF}.

The constellations used for the SC-PHY are non-standard, especially the $\pi/2$-BPSK constellation. As mentioned in Chapter 2, $\pi/2$-BPSK constellations in [802.15.3-09] are formed by mapping successive binary data elements to BPSK constellations where the constellation map is successively rotated by $\pi/2$ radians. This minimizes the phase transition between adjacent symbols to provide bandwidth efficiency and more accurate waveforms with practical hardware. For simplicity and consistency, all of the remaining constellations borrow this successive $\pi/2$ rotation operation after constellation mapping, although the operation is not necessarily beneficial as in $\pi/2$-BPSK. All constellations are Gray-coded to minimize average bit error rate.

The block transmission of SC-PHY data is illustrated in Fig. 9.13. After constellation mapping, the complex data symbols are formatted into $N_b + 1$ SC-PHY blocks. Within each block, except for the last block, are 64 sub-blocks. Each sub-block contains 512 data symbols, forming a single data chunk. Before each data chunk (see Fig. 9.14), a pilot word composed of known training symbols is inserted. The pilot word serves two primary functions. First, through its periodic insertion, the pilot word enables simple tracking of clock phase and frequency. Second, the pilot word allows for the cyclic convolution property to hold, which in turn provides for efficient frequency domain equalization (with

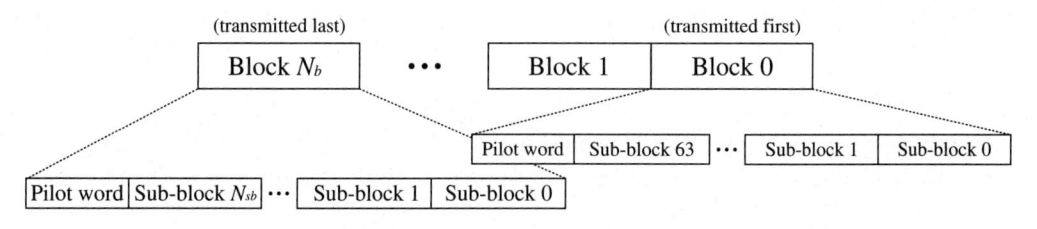

Figure 9.13 Block transmission for SC-PHY data.

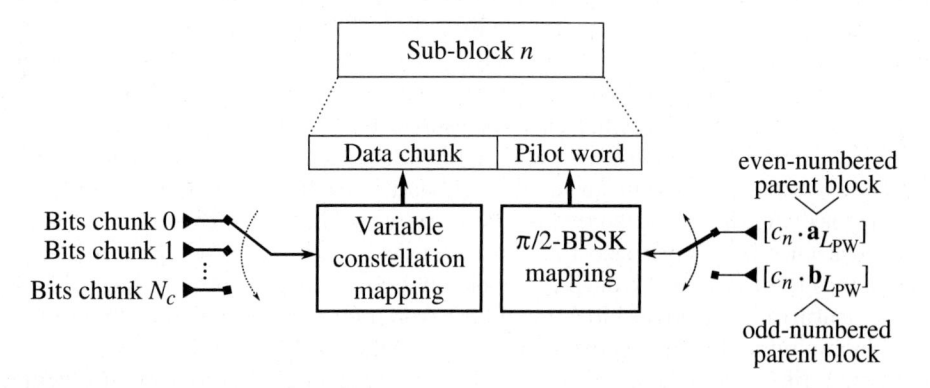

Figure 9.14 Breakdown of sub-block components in SC-PHY waveforms. Each block is composed of sub-blocks, where each sub-block is composed of $Nc + 1$ chunks (symbols). Each of these chunks are mapped from bit chunks.

Table 9.7 Pilot words in SC-PHY. $L_{\text{PW}} = 8$ is optionally supported.

Pilot Word	Description
$\mathbf{a}_8$	0xEB
$\mathbf{b}_8$	0xD8
$\mathbf{a}_{64}$	(see Table 9.4)
$\mathbf{b}_{64}$	(see Table 9.4)

I/FFT size of 512). Note that a pilot word suffix must also be inserted at the end of each data block to preserve cyclic convolution in the last sub-block. Pilot words are allowed to have variable length, $L_{\text{PW}} = 0, 8$, or 64 (see Table 9.7), and are always modulated with $\pi/2$-BPSK. A polarity, $c_n = \pm 1$, is assigned to each sub-block (indexed by n) and alternates for each adjacent sub-block throughout the block ($c_{n+1} = -c_n$). The pilot word is one of two complementary Golay sequences, $\mathbf{a}_{L_{\text{PW}}}$ for even-numbered blocks and $\mathbf{b}_{L_{\text{PW}}}$ for odd-numbered blocks.

The formatting procedure of each sub-block is shown in Fig. 9.14. The spread binary data are evenly divided in $N_c + 1$ chunks (stuff bits ensure that the output of the constellation mapper is a multiple of ($512 \times N_{\text{BPS}}$) where N_{BPS} is the number of bits represented in each constellation mapped symbol) before it is mapped to constellations and assigned to sub-blocks in round-robin fashion. Optionally, pilot channel estimation sequences (PCESs) may be necessary to track changes in the wireless channel or to train low-cost oscillators used in consumer products. In this event, the PCES are inserted before the first sub-block of each block, wherein the PCES is precisely the same as the CES field in the preamble.

9.3.2.4 HSI PHY

The high-speed interface (HSI) PHY in IEEE 802.15.3c is geared toward high performance to accommodate a rapid dump of data, over a short distance, in a short period of time, for example, the kiosk transfer or docking station scenario. Because of this focus, implementation complexity and multipath delay spread tolerance have not been as acutely considered as in the SC-PHY. Consequently, the HSI PHY uses QAM constellations, OFDM modulation, and LDPC coding. Table 9.8 summarizes the configuration of each MCS for the HSI PHY. All HSI PHY compatible IEEE 802.15.3c devices must support MCS 1 and either MCS 0 or the CMS mode in Section 9.3.2.2. The modulation formatting parameters in Table 9.8 only correspond to the PHY payload.

FEC in the HSI PHY uses two parallel LDPC encoders. This operation is further explained in Section 9.3.2.6, but for now we will assume that all binary source data, after scrambling, is demultiplexing into two separate streams. These two streams are separately LDPC block encoded (block size 672 with identical rates) followed by a multiplexer that combines the FEC binary output by selecting bits in round-robin fashion. The HSI PHY also features an optional block bit interleaver (block size equal to 2,688 bits) with adjustable minimum distance between interleaved bits (1, 2, 4, and 6).

Following FEC and bit interleaving, bits are transformed into complex symbols through M-QAM constellations for $M = 4, 16$, and 64. In standard operation, all constellations are typical gray-coded QAM constellations. After constellation mapping, the

Table 9.8 Modulation and coding schemes in HSI PHY. L_{SF} denotes the length of the spreading sequence. Note: MCS with equal error protection only listed (UEP MCS will be described in Section 9.3.2.6).

MCS	Data Rate	Modulation	FEC
0	32.1 Mbps	QPSK ($L_{SF} = 48$)	LDPC(672,336)
1	1,540 Mbps	QPSK	LDPC(672,336)
2	2,310 Mbps	QPSK	LDPC(672,504)
3	2,695 Mbps	QPSK	LDPC(672,588)
4	3,080 Mbps	16-QAM	LDPC(672,336)
5	4,620 Mbps	16-QAM	LDPC(672,504)
6	5,390 Mbps	16-QAM	LDPC(672,588)
7	5,775 Mbps	64-QAM	LDPC(672,420)

symbols are spread and interleaved before mapping onto sub-carriers. For MCS not equal to 0, the spreading operation is trivial. For MCS 0, however, the processing gain equals 48. This is accomplished by first grouping the complex baseband QAM symbols into groups of 7. Then, each group of 7 symbols is Kronecker-multiplied by the length 24 complex spreading vector $[+1, +j, -1, +j, +j, +1, -1, +j, -j, +j, -1, -j, -1, +1, +1, +1, +j, -j, -1, -1, -1, +j, -j, +j]$, resulting in groups of length 168. To achieve the last factor of 2, the order is reversed and the complex conjugate of each group is concatenated, creating complex data blocks of size 336 (the number of data sub-carriers per OFDM symbol). For a thorough treatment of the spreading operation, the reader is encouraged to consult 12.3.2.7.2 of the IEEE 802.15.3c standard.

All HSI PHY MCSs use OFDM modulation. The total number of OFDM sub-carriers in the HSI PHY is 512, which is the same as the block size of the SC-PHY single-carrier data chunks. An IEEE 802.15.3c device may share the FFT in the receiver for frequency domain equalization between the SC-PHY and the HSI PHY. The sub-carrier formatting, complete with guard and pilot sub-carriers, is summarized in Fig. 9.15. Sixteen pilot sub-carriers are equally spaced every 22 sub-carriers (first sub-carriers are distance 12 from the DC tone, edge sub-carriers are distance 166 from the DC tone). To prevent spectral periodicity and potential spectrum inefficiency, all pilots are pseudo-random QPSK symbols. Pilots provide characterization of phase and frequency offsets as well as evolutionary channel estimation. Before mapping, all sub-carriers, including guard, reserve, and DC tones, are block-interleaved according to 12.3.2.8 of the IEEE 802.15.3c standard where the block size corresponds to 512, the number of total sub-carriers in the OFDM symbol (equivalently the FFT order).

Each OFDM symbol must be preceded by a length-64 cyclic prefix to preserve cyclic convolution in channels with memory. Given the sample time, this suggests that the maximum delay multipath that can be correctly equalized is approximately 24 ns. Because the tolerance to delay spread is low, the HSI PHY cannot tolerate NLOS scenarios without extremely directional antennas. As with the SC-PHY, the HSI PHY may optionally insert the 128-length PCES, consisting of the sequence $\mathbf{c}_{128}$, to provide full re-estimation of the channel in the middle of a packet. When activated, the PCES will be inserted every 96 OFDM symbols, suggesting that PCES is only necessary for very long packets.

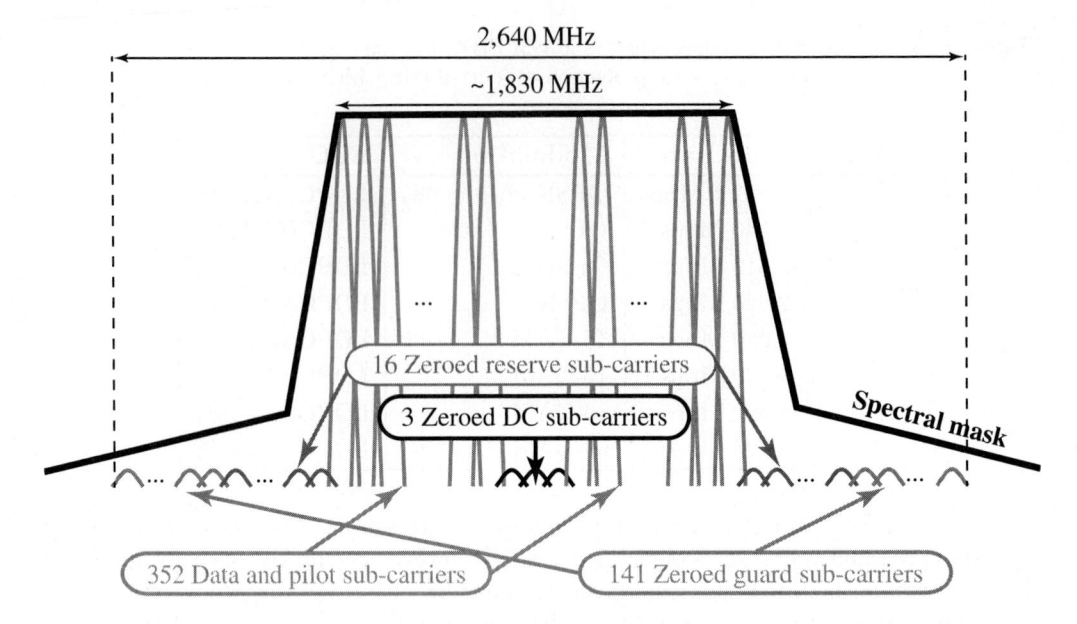

Figure 9.15 OFDM symbol formatting in the HSI PHY in IEEE 802.15.3c. The sub-carrier frequency spacing is 5.15625 MHz for all 512 sub-carriers. Three null DC tones prevent carrier feed through as well as ADC/DAC offset problems. The guard tones are usually nulled to meet spectral mask requirements, although customized guard tone values may optimize front end effects.

Table 9.9 Modulation and coding schemes in HSI PHY header. Optional rates are the result of decreased overhead.

Payload MCS	Data Rate	Sub-carrier Constellation	FEC
0	16.8 Mbps (29.6 Mbps optional)	QPSK ($L_{SF} = 48$)	LDPC(672,336)
1+	587 Mbps (1,363 Mbps optional)	QPSK	LDPC(672,336)

The frame format of HSI PHY systems is similar to the SC-PHY: preamble, header, and payload. Similarly, the header frame format is the same: the main header (which includes the PHY header and MAC header) and the optional header (which includes the MAC sub-header). The preamble of the HSI PHY is formatted in two ways: short (optional, to reduce overhead) or long. The short preamble data are identical to the SC-PHY preamble data. The long preamble data are the same as the CMS preamble data, albeit with HSI PHY MCS 0.

The header is not modulated in the same way as the payload (Table 9.8). There are two formatting procedures corresponding to when the payload is formatted with MCS 0 and otherwise, as illustrated in Table 9.9. The OFDM formatting procedure is very similar to the payload formatting procedure described above except that the cyclic prefix length is increased to 128 in the header.

9.3.2.5 AV PHY

The IEEE 802.15.3c AV PHY, like the HSI PHY, does not focus on reduced algorithm complexity in the physical layer. It does, however, reduce implementation complexity by exploiting the asymmetric nature of the targeted application: multimedia streaming. Multimedia devices typically only act as a source or sink, hence, limited transmit/receive functionality can be issued to a sink/source, respectively. Consider, for example, the display in a home theater. It only requires high complexity in receive mode, not in transmit mode, as the data flow is highly asymmetric.

Functionally, the IEEE 802.15.3c AV PHY is separated into two operating modes: high-rate (HRP) and low-rate (LRP). Table 9.10 summarizes the configuration of each MCS for the AV PHY: An AV PHY-compatible device must have, at a minimum, one of the four configurations as follows:

1. *HR0* — The lightest AV PHY device with only LRP transmit/receive functionality
2. *HRRX* — The high-rate sink with HRP receive functionality as well as LRP transmit/receive functionality
3. *HRTX* — The high-rate source with HRP transmit functionality as well as LRP transmit/receive functionality
4. *HRTR* — The flexible AV PHY device with HRP and LRP transmit/receive functionality

HRP functionality requires compatibility with modes HRP 0 and HRP 1 (see Table 9.10). LRP functionality requires compatibility with modes LRP 0, 1, and 2. The LRP modes will be used in a similar manner to CMS for broadcast and multicast.

Both the HRP and LRP use OFDM modulation. The HRP OFDM parameters are set according to the spectrum mask of Fig. 9.16. The LRP OFDM formatting parameters are subject to an additional spectral mask illustrated in Fig. 9.17. To meet these additional spectrum requirements, the LRP OFDM parameters use a smaller frequency spacing and 128 subcarriers, illustrated in Fig. 9.18. The channelization of the LRP is such that three

Table 9.10 Modulation and coding schemes in AV PHY. L_{SF} denotes the length of the spreading sequence. Note: MCS with equal error protection only listed (UEP MCS is described in Section 9.3.2.6).

	Index	Data Rate	Modulation	FEC
LRP	0	2.5 Mbps	BPSK ($L_{SF} = 8$)	BCC (rate 1/3)
	1	3.8 Mbps	BPSK ($L_{SF} = 8$)	BCC (rate 1/2)
	2	5.1 Mbps	BPSK ($L_{SF} = 8$)	BCC (rate 2/3)
	3	10.2 Mbps	BPSK ($L_{SF} = 4$)	BCC (rate 2/3)
HRP	0	0.952 Gbps	QPSK	RS+BCC (rate 1/3)
	1	1.904 Gbps	QPSK	RS+BCC (rate 2/3)
	2	3.807 Gbps	16-QAM	RS+BCC (rate 2/3)

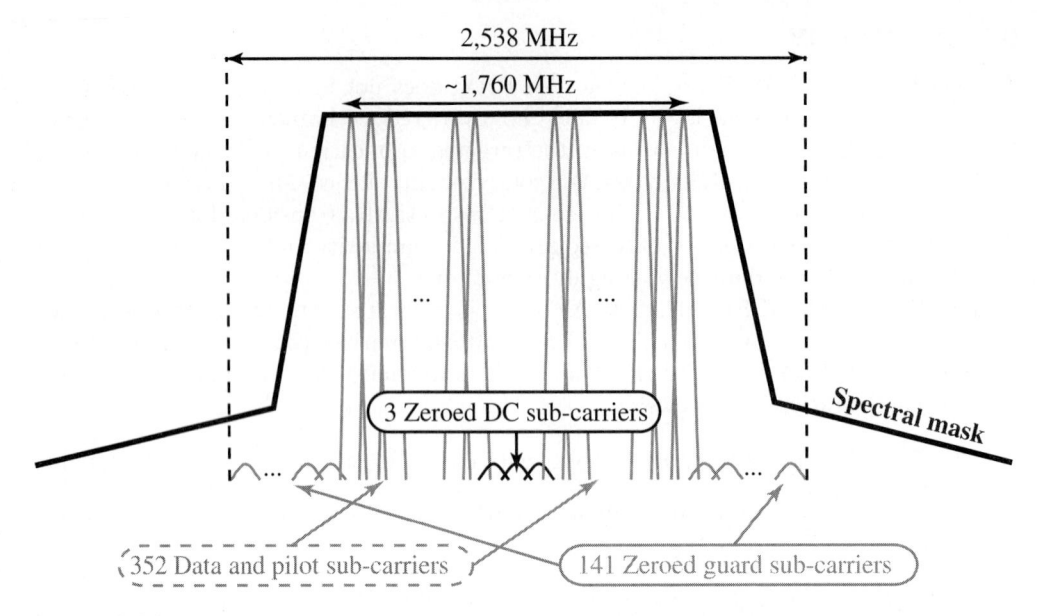

Figure 9.16 OFDM symbol formatting for the HRP. The sub-carrier frequency spacing is ≈ 4.96 MHz for all 512 sub-carriers. Three DC tones and all guard tones are nulled.

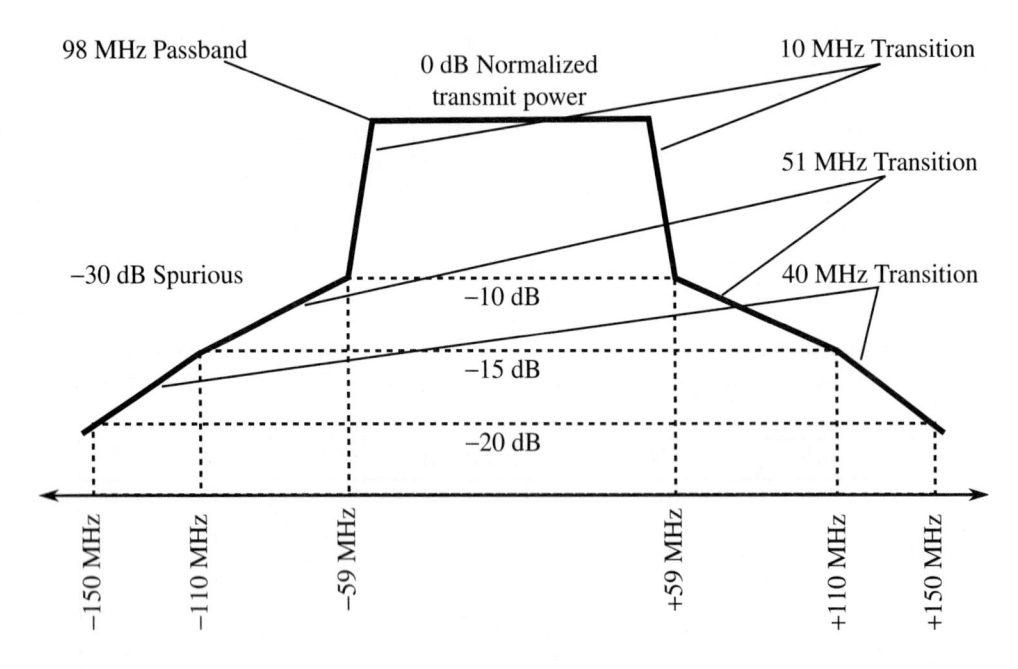

Figure 9.17 Normalized spectral mask for LRP in IEEE 802.15.3c AV PHY. The spectral mask in Fig. 9.9 overlays this spectral mask to allow multiple LRP channels to occupy a single IEEE 802.15.3c channel allocation.

9.3 IEEE 802.15.3c

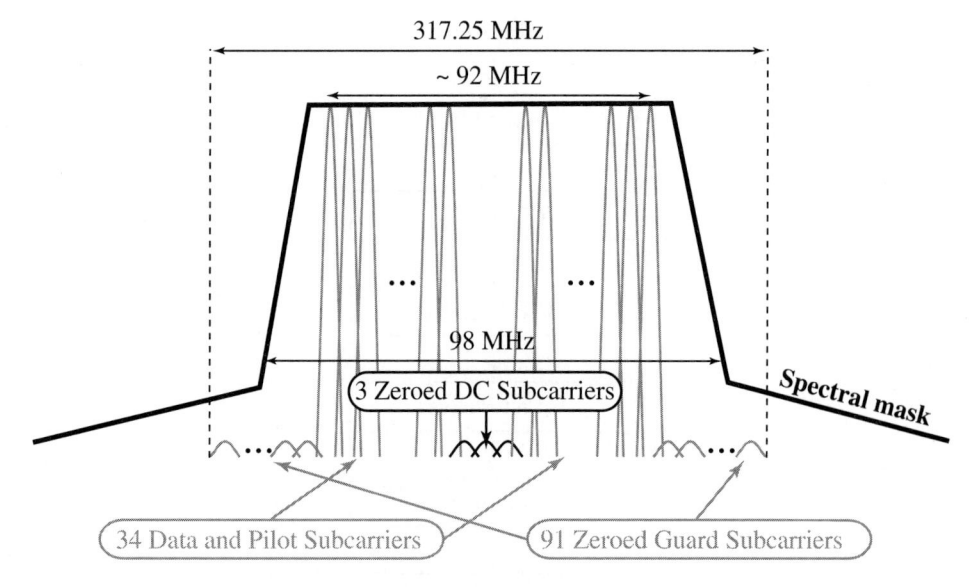

Figure 9.18 OFDM symbol formatting for the LRP. The sub-carrier frequency spacing is 2.48 MHz for all 128 sub-carriers. There are 37 data and null subcarriers, each with a subcarrier width of 2.48 MHz, resulting in an occupied bandwidth of 91.76 MHz ($\sim$ 92 MHz). The specified spectral mask passband bandwidth (at 10 dB down) is 98 MHz, allowing for roll-off in the LRP mode. Three DC tones and all guard tones are nulled.

LRP channels can fit within each 802.15.3c channel. For example, if the center frequency of the SC-PHY, HSI PHY, or HRP is f_c, the LRP may be centered at:

1. $f_c - 158.625$ MHz
2. f_c
3. $f_c + 158.625$ MHz

representing LRP channels 1, 2, and 3, respectively.

Fig. 9.19 shows the block diagram for creating AV PHY HRP frames. After scrambling (as described in 12.4.2.4 of [802.15.3-09]), the data are first (224,216) Reed-Solomon outer encoded, which provides further protection against bursty errors produced by Viterbi decoding of the punctured binary convolutional inner codes (BCCs) with rate 1/3 mother code generated by polynomials $(133)_8$, $(171)_8$, and $(165)_8$ (octal representation resulting in constraint length = 7). For rates 1/2, 4/7, 2/3, and 4/5 the puncturing matrices are

$$\begin{bmatrix} 1 \\ 1 \\ 0 \end{bmatrix}, \begin{bmatrix} 1 & 1 & 1 & 1 \\ 1 & 0 & 1 & 1 \\ 0 & 0 & 0 & 0 \end{bmatrix}, \begin{bmatrix} 1 & 1 \\ 1 & 0 \\ 0 & 0 \end{bmatrix}, \text{ and } \begin{bmatrix} 1 & 1 & 1 & 1 \\ 1 & 0 & 0 & 0 \\ 0 & 0 & 0 & 0 \end{bmatrix},$$

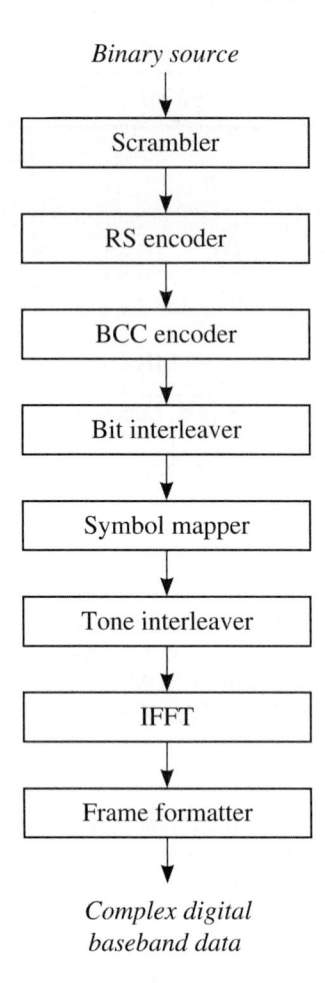

```
                Binary source
                      │
                      ▼
              ┌───────────────┐
              │   Scrambler   │
              └───────────────┘
                      │
                      ▼
              ┌───────────────┐
              │   RS encoder  │
              └───────────────┘
                      │
                      ▼
              ┌───────────────┐
              │  BCC encoder  │
              └───────────────┘
                      │
                      ▼
              ┌───────────────┐
              │ Bit interleaver│
              └───────────────┘
                      │
                      ▼
              ┌───────────────┐
              │ Symbol mapper │
              └───────────────┘
                      │
                      ▼
              ┌───────────────┐
              │Tone interleaver│
              └───────────────┘
                      │
                      ▼
              ┌───────────────┐
              │     IFFT      │
              └───────────────┘
                      │
                      ▼
              ┌───────────────┐
              │Frame formatter│
              └───────────────┘
                      │
                      ▼
              Complex digital
               baseband data
```

Figure 9.19 Block diagram showing the HRP PHY frame formatting procedure. The HRP features concatenated RS block and convolutional encoding and tone interleaving.

respectively. After mapping the interleaved bits to QPSK/16-QAM constellations (pilot data use BPSK constellations) and onto the sub-carriers, a tone interleaver prevents adjacent sub-carriers from corresponding to adjacent bits in the BCC output.[5] In the HRP, the sub-carrier location of pilots changes for each successive OFDM symbol, although there are always exactly 22 sub-carriers between each pilot. The BPSK data for

---

5. This is due to outer and inner bit interleavers that are not highlighted in this section nor in Fig. 9.19. In actuality, the RS encoder is preceded by a multiplexer, which allows for parallel RS and BCC encoding operations to provide UEP. Since this will be explained in more detail in Section 9.3.2.6, we have not explained these operations here for simplicity of presentation.

9.3 IEEE 802.15.3c

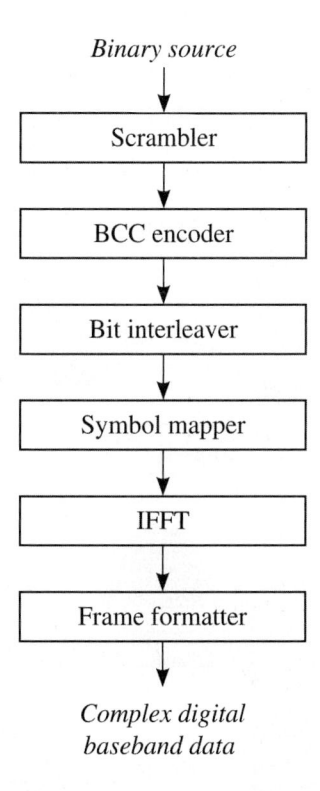

```
Binary source
     ↓
 Scrambler
     ↓
 BCC encoder
     ↓
 Bit interleaver
     ↓
 Symbol mapper
     ↓
    IFFT
     ↓
 Frame formatter
     ↓
Complex digital
 baseband data
```

Figure 9.20 Block diagram showing the LRP PHY frame formatting procedure.

the pilot tones are borrowed from the preamble channel estimation data discussed later in this section.

As illustrated in the block diagram in Fig. 9.20, the main formatting procedures that distinguish the HRP PHY from the LRP PHY are the outer Reed-Solomon block codes, BPSK constellations (no support for QPSK or 16-QAM), tone interleaving, and static pilot locations (4 pilots per OFDM symbol) in the LRP. The LRP frame formatter also performs additional OFDM symbol repetition, which can serve two purposes: processing gain and spatial diversity. Spatial diversity can be especially valuable when trained beam steering is not available (for more on beam steering, see Section 9.3.2.7). For example, in the common mode (see Section 9.3.2.2), where omni-directional antennas are assumed for the operating mode as illustrated in Fig. 9.21, each repeated OFDM symbol, including the cyclic prefix, is sent in a different beam steering direction in order to ensure that all potential directional paths are covered. Symbol repetition can also be used in directional mode, where each OFDM symbol repetition uses the same antenna direction. Note that there are no constraints on the actual directions for each repetition used in omni and directional modes. For example, in directional mode, the actual antenna radiation pattern may be isotropic.

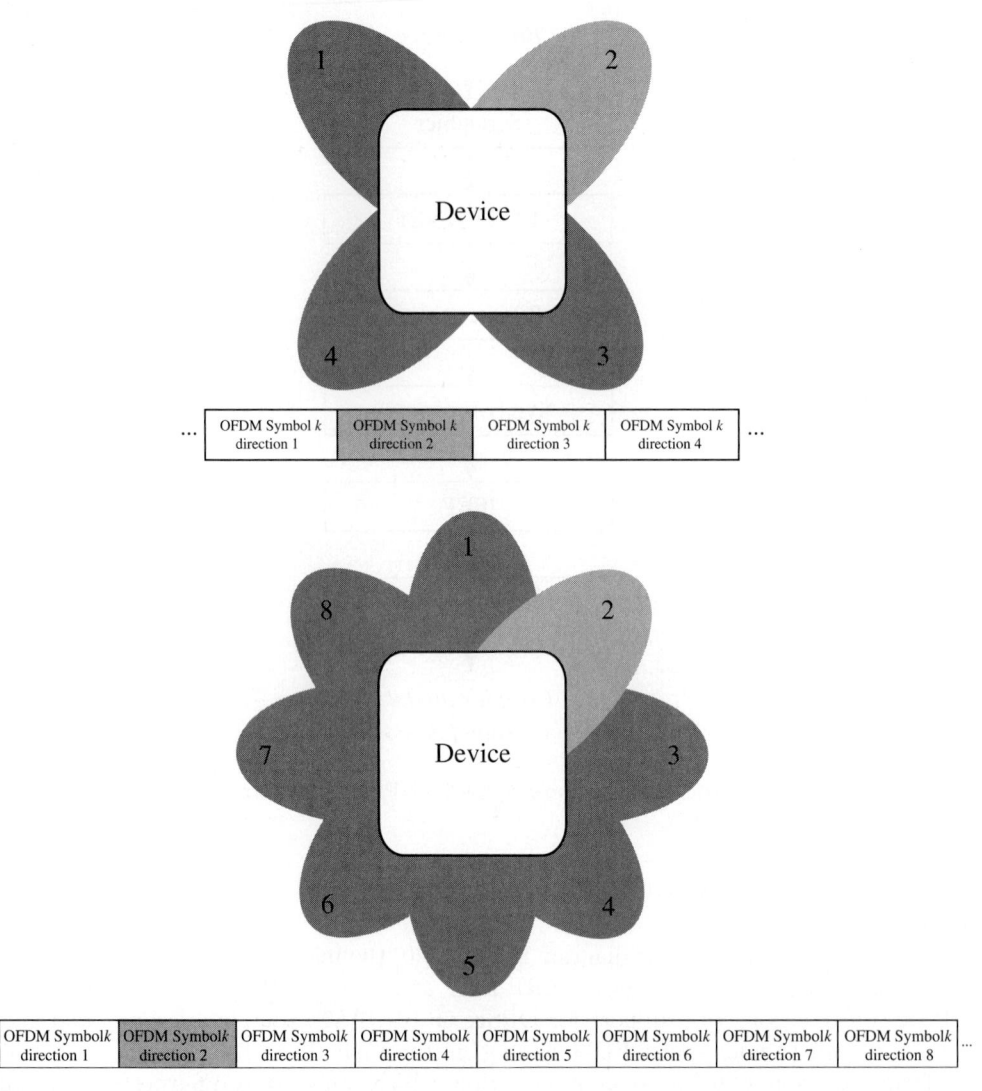

Figure 9.21 LRP OFDM symbol repetition allows for different antenna radiation patterns for each repetition. AV PHY LRP MCS support both 4 and 8 order symbol repetition where $L_{SF} = 4$ or 8 (see Table 9.10).

The frame format of the AV PHY is structurally similar to the SC and HSI PHYs with a preamble followed by a PHY header, a MAC header, a header check sequence (HCS), and the MAC frame (payload). For the HRP, the breakdown of the preamble is shown in Table 9.11.

9.3 IEEE 802.15.3c

Table 9.11 Breakdown of each field in HRP preamble. Note that the first four preamble symbols have samples equal to $1+j$ or $-1-j$, resulting in the same signals on I and Q channels as well as 3 dB higher output power.

Item	Description
Preamble (symbols 1-4)	$\{\mathbf{av}_{1.5}, \mathbf{av}_{1.5}, \mathbf{av}_{1.5}, \mathbf{av}_{1.5}, \mathbf{av}_{1.5}, \overline{\mathbf{av}}_1, \text{(sufficient zeros)}\}$ together length is $(512+64) \times 4 = 2304$
Preamble (symbols 5-6)	Concatenate two IFFTs of $2 \times \text{bin}(\{\mathbf{bv}_1, \mathbf{0}_{38}, \mathbf{bv}_2\}) - \mathbf{1}_{512}$ and add 128-length cyclic prefix (sub-carrier symbols = +1/-1)
Preamble (symbols 7-8)	Concatenate two IFFTs of $-2 \times \text{bin}(\{\mathbf{bv}_1, \mathbf{0}_{38}, \mathbf{bv}_2\}) - \mathbf{1}_{512}$ and add 128-length cyclic prefix (sub-carrier symbols = -1/+1)
$\mathbf{av}_r$	Rate r upsampling of $\sqrt{2}e^{j\frac{\pi}{4}}(2 \times \boldsymbol{\alpha}\mathbf{v} - \mathbf{1}_{255})$
$\boldsymbol{\alpha}\mathbf{v}$	255-length binary output from shift register with polynomial $x^8 + x^7 + x^2 + x + 1$ and seed $\{11111111\}$
$\mathbf{bv}_1$ (in hex)	{08E55930668EFB3227E5C4429BDABF04FB5ACAED75CA8}
$\mathbf{bv}_2$ (in hex)	{0DA1858D2794837B8FA3EFA25F3A5C30C7572DFAE7910}
$\mathbf{0}_k$	k-length sequence of zeros
$\mathbf{1}_k$	k-length sequence of ones
bin($\mathbf{x}$)	Sequence conversion from hex to binary

Channel estimation field	Second AGC field	Diversity training field	Fine CFO/Timing estimation field	Coarse CFO/Timing estimation field	Frame detection & AGC field

Figure 9.22 Fields of long preamble.

The frame format for the LRP is much different due to the availability of spatial diversity. In general, there are two types of LRP frame formats: omni and directional. Omni formatting is required for broadcast and multicast frames (e.g., beacons, CTA messages). Omni LRP frames offer both short and long preambles. Short preambles are used for the first frame in a CTA as well as with ACKs. Long preambles are used with beacons.

First, we illustrate the long omni preamble format in Fig. 9.22, which is broken down in Table 9.12. Since the LRP omni frame provides for spatial diversity by allowing different transmit directions, the preamble must also allow for different transmit directions. Hence, in each field, after N_{switch} symbols, the transmitter may change the antenna configuration. N_{switch} in the long omni preamble equals $78, 234, 80, 640, 64,$ and 156 symbols for each of the fields in order of transmission, respectively.

Table 9.12 Breakdown of each LRP omni preamble in HRP preamble. The first AGC, the coarse CFO/timing estimation, and fine CFO/timing estimation fields must be additionally mapped onto $\pi/4$-QPSK (or OQPSK) constellations. This is processed by mapping the chip/binary sequence onto BPSK constellations for both the I and Q channels. In OQPSK the Q branch is delayed by 1/2 symbol. Note the presence of the 13-chip Barker sequence in the first AGC and coarse CFO/timing estimation field [RG71]. The first four fields may be transmitted with up to 3 dB larger or smaller power than the second AGC field, the channel estimation field, and the payload.

Item	Description
Frame detection and AGC field	$\mathbf{1}_{26} \otimes [-1-1+1] \otimes$ $[-1-1-1-1-1+1+1-1-1+1-1+1-1]$
Coarse CFO/timing estimation field	$\mathbf{1}_{9} \otimes [-1+1-1+1+1+1-1-1-1+1] \otimes$ $[-1-1-1-1-1+1+1-1-1+1-1+1-1]$
Fine CFO/timing estimation field	1,440-length binary shift register output with polynomial $x^{12}+x^{11}+x^{8}+x^{6}+1$ and seed $\{101101010000\}$
Diversity training field	2,560-length binary shift register output with polynomial $x^{6}+x^{5}+1$ and seed $\{010111\}$
Second AGC field	$\mathbf{1}_{20} \otimes \mathrm{IDFT}(\{[0-1-1+1-1], \mathbf{1}_{23}, [+1+1+1-1]\})$
Channel estimation field	$\mathbf{1}_{32} \otimes \{[\mathbf{cv}]_{101:128}, \mathbf{cv}\}$
cv	$\mathrm{IDFT}(\{0,0,+1,-1,+1,-1,-1,+1,-1,+1,+1,+1$ $+1,+1,+1,-1,-1,+1,+1,\mathbf{0}_{91},-1,+1,+1,+1,+1,$ $+1,-1,-1,+1,-1,+1,-1,-1,-1,+1,-1,+1,0\})$

9.3.2.6 Unequal Error Protection

Until very recently, wireless standards have operated under the assumption that every bit holds the same intrinsic value for the application the wireless network serves. As shown in Chapter 8, several applications, most notably video streaming, are better serviced if certain bits are guaranteed more protection from impairments in the wireless channel. For IEEE 802.15.3c, this unequal error protection (UEP) of bits is implemented in various ways. IEEE 802.15.3c supports three different UEP types, which all accept two classes of data: (1) most significant bits (MSBs), which are more critical to the data in the serviced application, and (2) least significant bits (LSBs), which are less critical to the serviced application. The manner in which the three UEP types handle MSBs and LSBs is described as follows:

- *UEP Type 1* — Type 1 protection operates at the MAC level, assuming the MAC has been given frames that are segmented into sub-frames containing only MSBs or LSBs. The MAC designates MCSs for each sub-frame that consist of the same modulation parameters (except for the coding rate). In this way, lower coding rates may be assigned to the MSB sub-frames to provide higher protection. UEP Type 1 is optionally supported by SC-PHY DEVs.

- *UEP Type 2* — Similar to Type 1, Type 2 UEP is accomplished by sending MSB sub-frames with higher protection. In Type 2, however, higher protection may be offered by changing the constellation for different sub-frames in addition to the FEC rate. UEP Type 2 is optionally supported by SC-PHY DEVs.

- *UEP Type 3* — Type 3 is handled in the PHY by either providing more robust coding rates to the MSBs or by skewing the constellation dimensions such that the minimum distance between constellation points is greater on the axis mapped to the MSBs (e.g., the skewed constellations mentioned in Chapter 2). UEP Type 3 is optionally supported by all DEVs.

Fig. 9.23 demonstrates the operation of UEP Type 3 through split coding for the SC-PHYs complete with a combined bit multiplexer/interleaver that provides a single bitstream such that the remainder of SC-PHY waveform formatting is the same as discussed in Section 9.3.2.3. Table 9.13 summarizes the different configurations for each branch of the encoder, the modulation formats, and the resultant data rate. For MCS 7, the joint multiplexer/interleaver in the SC-PHY assembles 10 bits at a time, 6 from the MSB

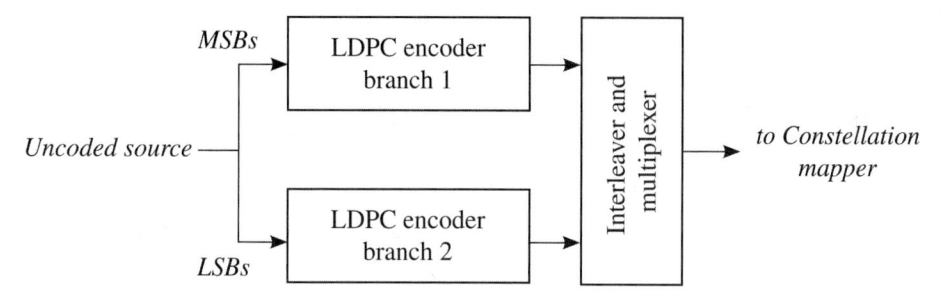

Figure 9.23 UEP Type 3 through split MSB/LSB encoder branches in the SC-PHY (LDPC encoding). The SC-PHY joint multiplexing/interleaving configuration depends on the UEP MCS implemented.

Table 9.13 UEP MCS for the SC-PHY. All three UEP types are supported. Note that UEP does not offer spreading gain. Data rates shown are approximated to three significant figures with zero pilot word length. Note that skewed constellations are limited to QPSK in the SC-PHY.

MCS	Data Rate	Modulation	FEC	Compatibility
UEP 1	1,420 Mbps	$\pi/2$ BPSK	RS(255,239)	UEP Types 1 and 2
UEP 2	756 Mbps	$\pi/2$ BPSK	LDPC(672,336)	UEP Types 1 and 2
UEP 3	1,130 Mbps	$\pi/2$ BPSK	LDPC(672,504)	UEP Types 1 and 2
UEP 4	1,510 Mbps	$\pi/2$ QPSK	LDPC(672,336)	UEP Types 1 and 2
UEP 5	2,270 Mbps	$\pi/2$ QPSK	LDPC(672,504)	UEP Types 1 and 2
UEP 6	2,650 Mbps	$\pi/2$ QPSK	LDPC(672,588)	UEP Types 1 and 2
UEP 7	2,040 Mbps	$\pi/2$ QPSK	LDPC(672,336) (MSB) LDPC(672,504) (LSB)	UEP Type 3
UEP 8	2,650 Mbps	$\pi/2$ QPSK	LDPC(672,504) (MSB) LDPC(672,588) (LSB)	UEP Type 3

branch ($\{a_1, a_2, a_3, a_4, a_5, a_6\}$) and 4 from the LSB branch ($\{b_1, b_2, b_3, b_4\}$) such that the joined bit-stream becomes $\{a_1, b_1, a_2, b_2, a_3, a_4, b_3, a_5, b_4, a_6\}$. Similarly, for MCS 8, the SC-PHY assembles 13 bits (7 from MSBs, 6 from LSBs) to form the joined bit-stream $\{a_1, b_1, a_2, b_2, a_3, b_3, a_4, b_4, a_5, b_5, a_6, b_6, a_7\}$. With skewed constellations, the bits are selected from each branch in a round-robin fashion. Note that because the SC-PHY operates on a single carrier, heavy interleaving is not needed within encoded blocks (channels are assumed to fade slowly within a frame).

The SC-PHY also provides for skewed constellations by increasing the energy in a single symbol dimension by a factor of 1.25. This is demonstrated in Fig. 9.24 with 16-QAM. This provides a different source of UEP by mapping more important bits to the skewed dimensions. In the SC-PHY, variable rate branches and skewed constellations are not used concurrently, thus skewed constellations use a single LDPC rate, although encoding is still done on both branches in parallel.

The HSI PHY also supports UEP Type 3 through skewed constellations or more robust coding of MSBs (non-concurrent), allocated as MCS 8-11 (see Table 9.14). The HSI PHY processes UEP with variable rates over two branches of LDPC coding in much the same manner as in the SC-PHY (Fig. 9.23). However, because the HSI PHY uses OFDM, it requires heavy protection of frequency selective fades. Note that the HSI PHY multiplexes the bits before interleaving (not jointly operated). Also, unlike the

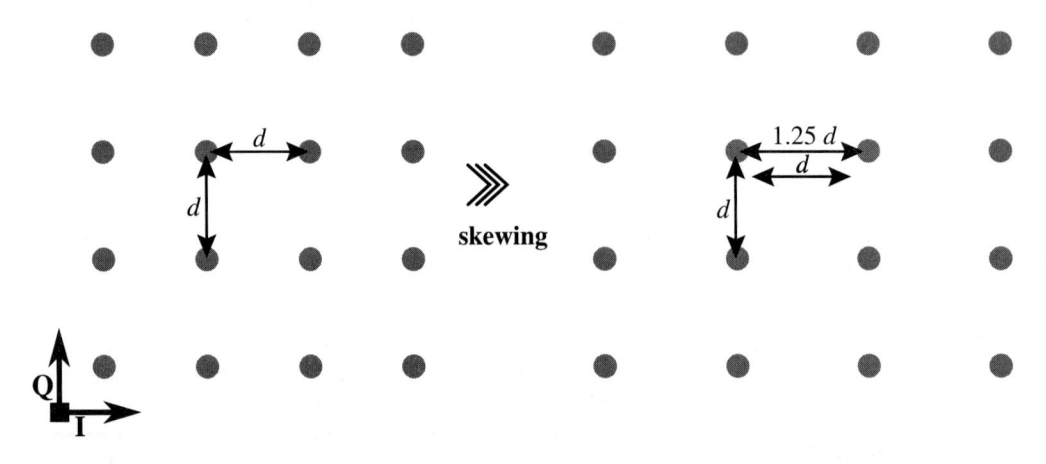

Figure 9.24 UEP Type 3 through skewing of 16-QAM. Here, the minimum distance between in-phase constellation points (where MSB bits are mapped) is increased by a factor 1.25.

Table 9.14 UEP MCS for the HSI PHY. Only UEP Type 3 is supported. Through constellation skewing, UEP is not restricted to just MCS 8-11 here, but may also be used with MCS 0-7 in Table 9.8. Note that HSI MCS 8-11 do not offer spreading gain.

MCS	Data Rate	Modulation	FEC (MSB)	FEC (LSB)
8	1,925 Mbps	QPSK	LDPC(672,336)	LDPC(672,504)
9	2,503 Mbps	QPSK	LDPC(672,504)	LDPC(672,588)
10	3,850 Mbps	16-QAM	LDPC(672,336)	LDPC(672,504)
11	5,005 Mbps	16-QAM	LDPC(672,504)	LDPC(672,588)

SC-PHY, all non-UEP MCSs use two LDPC encoder branches, hence constellation skewing of the in-phase dimension through the 1.25 factor is available on MCS 0-7. The operation of the encoder was previously abstracted in Section 9.2.3.4 so as not to deal with UEP concerns until appropriate. Fig. 9.25 shows the operation of the encoder for MCS 0-11 in the HSI PHY. The UEP data multiplexer does not operate in a round-robin fashion (as with non-UEP MCS 0-7), but instead assembles 5 bits total for MCS 8 and 10 yielding $\{a_1, b_1, a_2, b_2, a_3\}$ and 13 bits total for MCS 9 and 11 yielding $\{a_1, b_1, a_2, b_2, a_3, b_3, a_4, b_4, a_5, b_5, a_6, b_6, a_7, b_7, a_8\}$ following the notation described for joint multiplexing/interleaving of the SC-PHY.

UEP in the AV PHY occurs in the HRP mode, with added MCS modes (HRP 3-6 shown in Table 9.15). As with the HSI PHY, we neglected to discuss the encoder configuration of the HRP in Section 9.3.2.5. In Fig. 9.26, we illustrate the two-branch bit processing, which includes scrambling, encoding (each branch with one RS encoder and four BCC encoders), an outer interleaver, a demultiplexer, and a multiplexer. The source bits are parsed into upper and lower branches differently for different MCS, depending on whether UEP is available. For HRP 3-6, the upper and lower branches correspond to the MSBs and LSBs, respectively. Otherwise, for HRP 0-2, the pattern illustrated in Fig. 9.27 is used to parse bits into upper and lower branches. The outer interleavers segment the RS-encoded data into four parallel BCCs, labeled A-D for the upper branch and E-H for the lower branch. For payload data, four octets are assigned to each BCC in a round-robin fashion. Each BCC uses generators (octal format) $(133)_8$, $(171)_8$, and $(165)_8$ to generate the rate 1/3 mother code that is punctured for higher rates, where

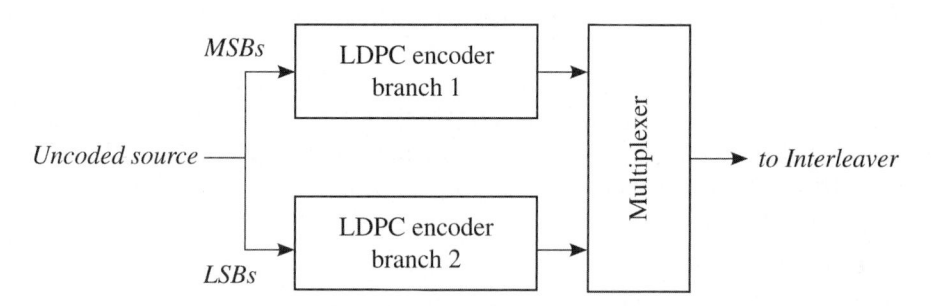

Figure 9.25 HSI encoder, which has two branches for all MCS 0-11. Note the addition of a branch demultiplexer (MSB/LSB parsing) as a fundamental component of the encoder for all MCS (including non-UEP).

Table 9.15 UEP MCS for the HRP of the AV PHY. All coding use Reed-Solomon outer codes with variable rate inner binary convolutional codes. Note that HRP 5 and HRP 6 do not encode LSBs and send only the vital MSB information to the receiver.

MCS	Data Rate	Modulation	FEC (MSB)	FEC (LSB)
HRP 3	1,940 Mbps	QPSK	Rate 4/7	Rate 4/5
HRP 4	3,807 Mbps	16-QAM	Rate 4/7	Rate 4/5
HRP 5	952 Mbps	QPSK	Rate 1/3	(No LSB)
HRP 6	1,904 Mbps	QPSK	Rate 2/3	(No LSB)

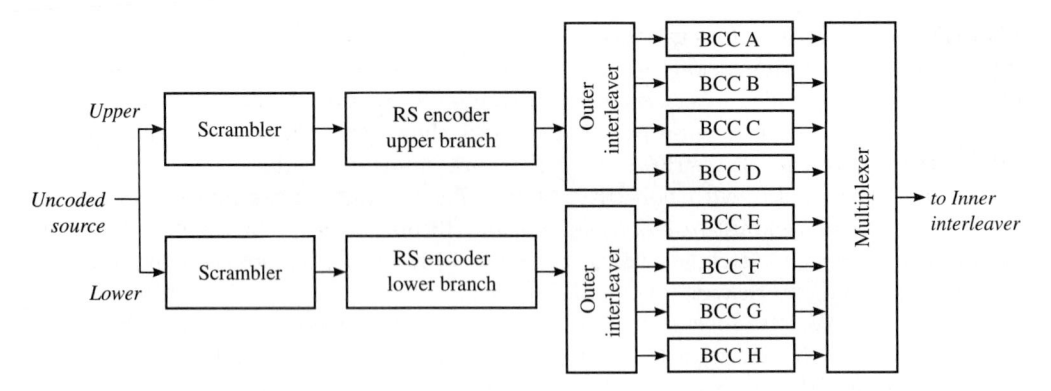

Figure 9.26 Encoder for the HRP mode of the AV PHY. A single source bit-stream is segmented into upper and lower branches, each of which are outer encoded with RS(224,216) and inner encoded with 8 punctured rate 1/3 binary convolutional encoders (labeled A-H).

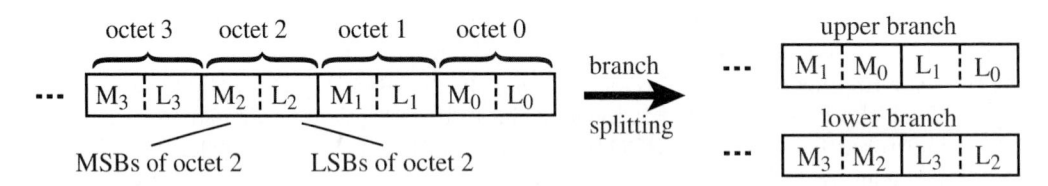

Figure 9.27 Parsing pattern into upper and lower branches for non-UEP HRP modes in the AV PHY. Every two octets the upper and lower branch swap from LSBs to MSBs and MSBs to LSBs, respectively.

different branches may use different puncturing matrices with UEP without constellation skewing.[6] The BCC output for all 8 encoders in both branches is combined into a single bit-stream for interleaving using the data multiplexer, which has three modes of operation as follows:

1. *Non-UEP Operation* — All BCCs use the same puncturing matrix and produce the same number of bits for a fixed encoder input. The multiplexer processes 6 bits at a time from each BCC in round-robin fashion. Hence, there is a 42-bit gap between the end of a 6-bit sequence and the beginning of the next 6-bit sequence from a single BCC. Each 48-bit period (representing 6 bits from all BCCs) is also processed by the inner interleaver using a block interleaver where, if $k \in \{0, 1, \ldots, 47\}$ is the index of the bit within the block before interleaving, $(6\lfloor k/6 \rfloor - 5(k)_6)_{48}$ is the index of the bit within the block after interleaving.

2. *UEP through Variable Branch Coding Rates* — Each of the branches produces different numbers of bits depending on the coding rates of each branch. For the

---

6. Note that the header data, in comparison to the payload, has slightly different HRP encoding parameters, although conceptually the process is the same [802.15.3-09].

UEP MCS in HRP, this yields rate 4/7 BCC on the upper branch and rate 4/5 on the lower branch. Hence, the multiplexer cycle processes a different number of bits depending on the branch. For the upper branch, 7 bits are selected from each BCC whereas 5 bits are selected from each BCC on the lower branch. The order of the round-robin processing from each BCC alternates for each 48-bit block. In the first block, round-robin processing operates as before by selecting bit outputs from BCCs A→H, in order. In the second block, the multiplexer does not start with BCC A in the upper branch nor BCC E in the lower branch during the round-robin cycle. Instead, the BCCs are processed B→D A F→H E. The multiplexer alternates between these two operations for the remainder of the packet. After multiplexing, the same interleaver as in non-UEP operation processes the single bit-stream.

3. *UEP through Constellation Skewing* — The AV PHY does not support concurrent operation of constellation skewing and variable coding rates per branch. Hence, the multiplexer and interleaver should operate in the same fashion as with non-UEP operation. However, it is important to remember that the lower branch consists of MSBs and that all the MSBs must be mapped to the in-phase dimension of the QPSK/16-QAM constellation. The modulator therefore preserves the branch origin of bits and processes the I/Q constellation mappers independently, each of which only selects bits (in order after interleaving) from lower/upper branch of the encoder, respectively.

9.3.2.7 Antenna Beamforming and Tracking

The framework for optional antenna beamforming in the IEEE 802.15.3c standard is unique. Standards at microwave frequencies, such as IEEE 802.16e/m, 3GPP-LTE, and IEEE 802.11n, select beamforming antenna configurations through algorithms that exploit exhaustive estimation of channel state information. The high gains of 60 GHz antennas that can be provided in small physical dimensions allow commercial device platforms to offer large antenna arrays. Even with recent limited feedback advances that compress the overhead needed to perform exhaustive training of large antenna arrays, overhead still grows with the square of the antenna dimensions per device. Consequently, there is significant motivation to reduce the number of antennas that need to be trained in large 60 GHz antenna arrays. This is achieved by creating antenna array codebooks that exploit the gain geometry of different antenna configurations. First, we consider the general beamforming system model.

1. *Beamforming Model* — Fig. 9.28 shows the reference model for beamforming in the forward link. Actual implementation may be at baseband, IF, or RF at the vendor's discretion. Note that the antenna configuration and dimensionality is not necessarily equal to the antenna configuration and dimensionality of the reverse link. Hence, the beamforming protocol must be completed for both the forward and reverse link unless the system is operating with a *symmetric antenna set* (SAS). In the SAS scenario the same antennas are used to transmit and receive signals on each device such that $N_t^{(1)} = N_r^{(1)}$, $N_t^{(2)} = N_r^{(2)}$, $\mathbf{v}^{(1)} = \mathbf{w}^{(1)*}$, and $\mathbf{v}^{(2)} = \mathbf{w}^{(2)*}$.

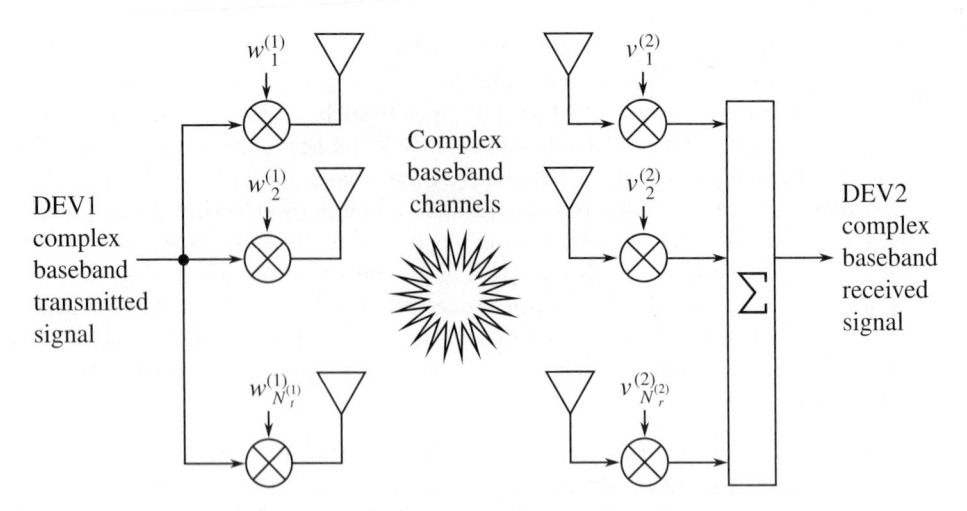

Figure 9.28 Beamforming complex baseband model. The identifier $^{(i)}, i \in \{1,2\}$ is used to index the device. Vectors $\mathbf{w}^{(1)}$, $\mathbf{w}^{(2)}$ define the complex transmit beamforming coefficients, whereas vectors $\mathbf{v}^{(1)}$, $\mathbf{v}^{(2)}$ define the complex receive beamforming coefficients, for devices 1 and 2, respectively.

2. *Antenna Geometry Configurations* — Fundamentally, the larger the antenna gain, the narrower the radiation/reception pattern of the antenna. To maximize the link margin in directional 60 GHz channels, a narrow beam focused on the strongest channel path is desired. We can efficiently determine a high-gain antenna configuration by progressively narrowing the antenna beam. Four antenna pattern resolutions are defined in IEEE 802.15.3c codebooks, as illustrated in Fig. 9.29. The codebook, antenna array design and implementation details, and pattern definitions are not specified in the IEEE 802.15.3c standard and are left up to the vendor. Predefined codebooks are available, however, for codebooks with uniform linear/planar arrays and half-wavelength spacing, as discussed later in the *pattern estimation and tracking* (PET) protocol.

3. *Beamforming Protocol* — The beamforming protocol (previously discussed in Section 8.3.1) determines the best antenna configuration for communication between two devices. Although four levels of resolution are defined for beam patterns, the beamforming protocol only searches over two levels: sectors and beams (please see Section 8.3.1 for a more general description of multi-level beamforming). Initially, link negotiations provide the best quasi-omni direction. Then, the beamforming protocol finds the best beam pattern. The best high-resolution (hi-res) beam pattern will be found through further tracking after the beamforming protocol is completed. For the general antenna configuration between DEV 1 and DEV 2, the following beamforming protocol is implemented sequentially in time.

    (a) *Sector Training* — A training cycle is sent from each sector within the chosen quasi-omni antenna direction at the transmitting DEV. Within each cycle, a training sequence is repeated for each possible sector within the chosen

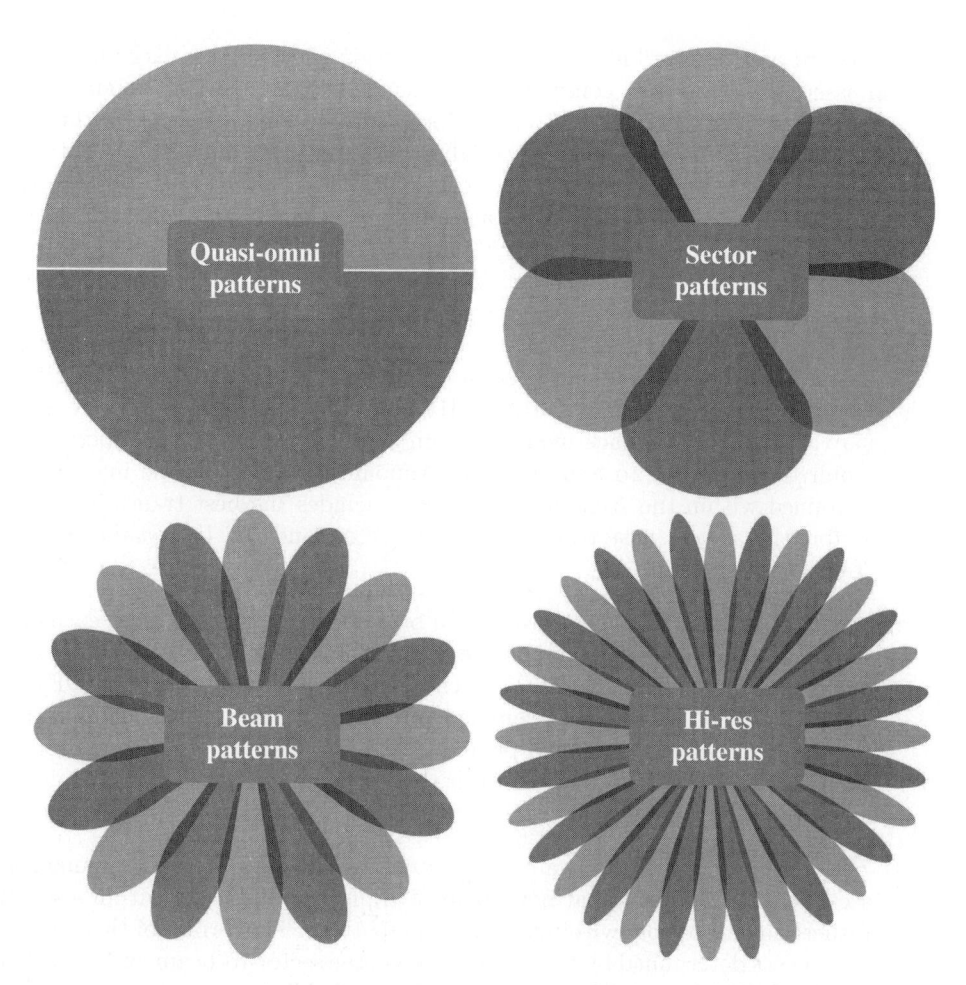

Figure 9.29 Four levels of patterns in antenna beamforming codebook for eight-element uniform linear array (patterns visualized on the azimuthal plane (top view) for a vertical array orientation) [802.15.3-09].

quasi-omni direction at the receiver. In the SC-PHY and HSI PHY, the sector training sequence is the CMS preamble and the AV PHY uses the HRP preamble. For each sequence, the receiver uses a distinct antenna configuration corresponding to a different sector within the quasi-omni direction. After sector training is completed in one direction, for example, from DEV 1 to DEV 2, the process is completed in the reverse direction by swapping roles of transmitter and receiver, for example, from DEV 2 to DEV 1.

(b) *Sector Feedback* — We continue the protocol description, now assuming that communication from DEV 1 to DEV 2 is the forward link and communication from DEV 2 to DEV 1 is the reverse link. At the completion of sector training, DEV 2 and DEV 1 determine the best antenna configuration for the forward and reverse link, respectively. To begin the feedback exchange, DEV 1 issues an Announce command to exchange the reverse link antenna configuration. This command is repeated sequentially for each sector within the specified quasi-omni pattern since DEV 1 does not yet know the best transmit antenna configuration for the forward link. Simultaneously, DEV 2 sets its receive antenna configuration according to calculations completed during sector training and waits for DEV 1 to send the Announce command from the desired transmit antenna configuration. Finally, DEV 2 responds with its own Announce command using the best antenna configuration delivered by DEV 1 during the first Announce command and DEV 1 simultaneously listens on the best receive antenna configuration. For the SC-PHY and HSI PHY, feedback uses CMS while the AV PHY uses the LRP mode. The AV PHY may also set the feedback mode to omni, in which case only one repetition is required for DEV 1 to issue the first Announce command. The information contained within the Announce command includes the best transmit/receive sector, the second best transmit/receive sector, and the *link quality indices* (LQIs) of each.[7]

(c) *Sector-to-Beam Exchange* — Next, the beamforming protocol must exchange the beam capabilities of each device. First, DEV 1 sends an Announce command to DEV 2 and then DEV 2 sends an Announce command to DEV 1. The information within these commands includes the number of beams within the selected sector, the type of preamble to be used, and PET information (if available, to be discussed later). After this exchange, the devices are ready for beam training.

(d) *Beam Training* — Beam training proceeds in virtually the same manner as sector training, except that now we must train over all beams within a sector rather than all sectors within a quasi-omni pattern. The format of the training sequence is determined by the previous stage, the sector-to-beam exchange. For the AV PHY, this will always be the HRP preamble.

(e) *Beam Feedback* — Beam feedback proceeds in virtually the same manner as sector feedback, except that now feedback defines the best beam within a sector rather than the best sector within a quasi-omni pattern. Beam feedback follows the same formatting as sector feedback. The information contained within this Announce command includes the best transmit/receive beam, the second best transmit/receive beam, the LQIs, and pattern estimation and tracking (PET) phase/amplitude information.

(f) *Beam-to-Hi-Res-Beam Exchange* — This final stage is optional, depending on whether tracking is supported by both devices. If it is, this exchange follows the

---

7. Three LQIs are predefined in Section 12.1.8.3 of the IEEE 802.15.3c standard [802.15.3c-09]. They include signal-to-noise ratio (SNR), signal-to-interference-plus-noise ratio (SINR), and received signal strength indicator reference (RSSIr).

same procedure as the sector-to-beam exchange, but with mappings between beams and hi-res beams instead of sectors and beams. The information within the Announce commands includes the number of hi-res beams within the selected beam, the type of synchronization sequence to use for beam tracking, the beam clusters, and *pattern estimation and tracking* (PET) information (described later in this section).

In the symmetric antenna (SAS) case, the control overhead in the beamforming protocol is drastically reduced. Sector and beam training only needs to be completed in one direction, followed by a single feedback exchange from the receiver.

Beam tracking is supported between devices in IEEE 802.15.3c. Because devices are mobile, the standard does not support training of a single hi-res beam through the beamforming protocol. Instead, clusters of hi-res beams are incorporated into a single beam. When tracking is enabled between two communicating devices, the best and second best cluster, as determined initially by the beamforming protocol described on the previous page, are evaluated quasi-periodically, where the frequency of each evaluation and update is determined during the beam-to-hi-res-beam exchange. The frequency of best cluster tracking will always be higher than the frequency of second best cluster tracking.

Beam tracking is accomplished by setting the Beam Tracking flag in a PHY header, followed by high-resolution beam training sequences for each hi-res beam in the cluster. Because a cluster is essentially composed of a single hi-res beam and its neighboring beams, as illustrated in Fig. 9.30, finding the best cluster is the same as finding the best hi-res beam. In this way, even though the initial beamforming protocol does not find the best hi-res beam, tracking will eventually identify it. Additionally, if tracking occurs with sufficient frequency, beamforming can adapt to changing channel conditions.

Pattern estimation and tracking (PET) is supported in IEEE 802.15.3c for more structured uniform antenna arrays, that is, half-wavelength antenna element spacing in

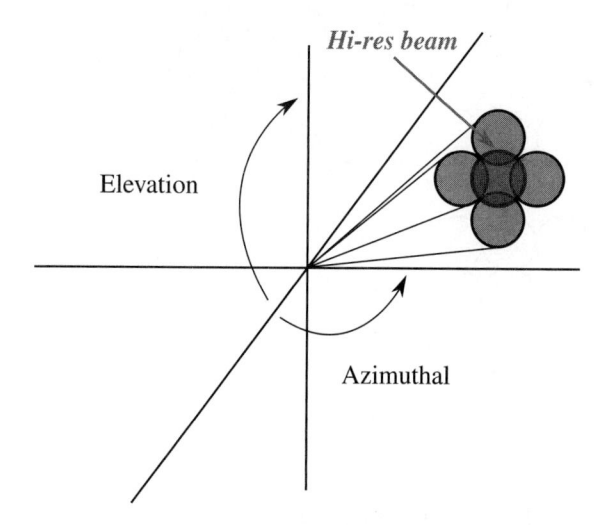

Figure 9.30 High-resolution beam cluster in 3-dimensional space.

one or two dimensions (linear or planar). Because of this assumed antenna structure, codebooks and clusters are more rigidly defined, reducing overhead. Explicitly, for arbitrary DEV i, the kth element of transmit beamforming vector is given by

$$\left[\mathbf{w}^{(i)}\right]_k = j^{\lfloor \frac{4N_t^{(i)}}{K} \times \text{mod}(k+K/2,K) \rfloor} \tag{9.1}$$

for 1-dimensional uniform arrays with $K \geq N_t^{(i)}$ beam patterns. For a fixed K, the amount of beam pattern overlap is reduced by increasing the antenna dimensions. Note that the receive beamforming vector follows the same form. The 2-dimensional codebook uses the 1-dimensional codebook for both the elevation and azimuth angles. Because of the uniformity in the array, clusters can be defined explicitly by the number of hi-res beam neighbors per dimension, as shown in Fig. 9.31, allowing a single cluster to be characterized by just one byte.

Consider the byte represented by $\{b_7 b_6 b_5 b_4 b_3 b_2 b_1 b_0\}$. Assuming a fixed center hi-res beam, bits b_2, b_1, and b_0 determine the number of neighbor beams in the elevation angle whereas bits b_5, b_4, and b_3 determine the number of neighbor beams in the azimuth angle. Finally, bits b_7 and b_6 determine the geometry of the cluster (whether neighbors extend past elevation or azimuth angles).

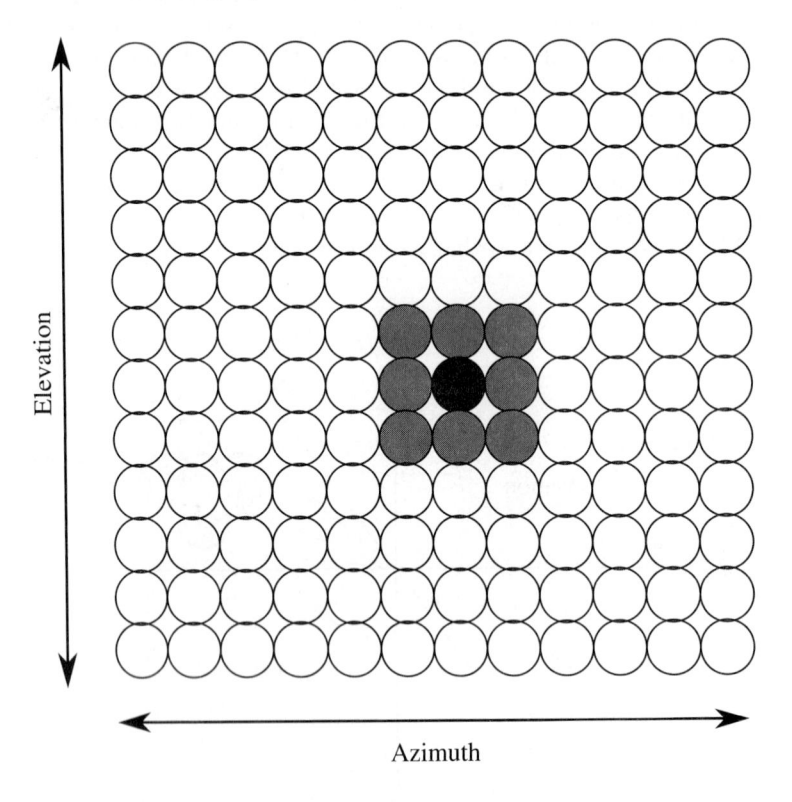

Figure 9.31 Clusters in PET. Each circle represents a hi-res beam. An example cluster is colored in gray with the center hi-res beam colored in black.

9.3.2.8 On-Off Keying

As discussed in Chapter 7, OOK, although limited in its spectral efficiency and maximum throughput, may significantly reduce the complexity of transceiver implementation [DC07][SUR09][MAR10b][KLP+10]. IEEE 802.15.3c optionally supports OOK through the SC-PHY. Because OOK's primary application is on extremely low complexity devices that may use non-coherent demodulation, it can only be used for communication within child piconets. If an OOK-capable DEV wishes to act as a PNC, it must be able to communicate through CMS. Non-PNC OOK-capable devices are not required to maintain compatibility with CMS. Similar to OOK, *dual alternate mark inversion* (DAMI) signaling is used for very low complexity communication, with the same aforementioned OOK piconet constraints. Fig. 9.32 shows the comparison between OOK and DAMI constellations. DAMI signaling uses polarity, not just amplitude. Further, DAMI is differential, which allows it to maintain a 50% duty cycle. Both OOK and DAMI signaling are coupled with RS(255,239) FEC. OOK allows spreading up to $L_{SF} = 2$ with spreading sequence [+1 + 1]. DAMI also incorporates single sideband modulation, represented at passband through the Hilbert transform.

OOK uses PHY preambles that include three fields based on repeated Golay sequences as follows (in the order listed):[8]

1. *SYNC* — For frame detection and AGC, SYNC is represented by 16 repetitions of $\mathbf{a}_{128}$ as defined in Table 9.6.

2. *SFD* — For frequency and frame synchronization, SFD is represented by 4 repetitions of $\mathbf{a}_{128}$. The sign of each repetition indicates information about the frame to be transmitted. The polarity of the first repetition indicates whether or not a

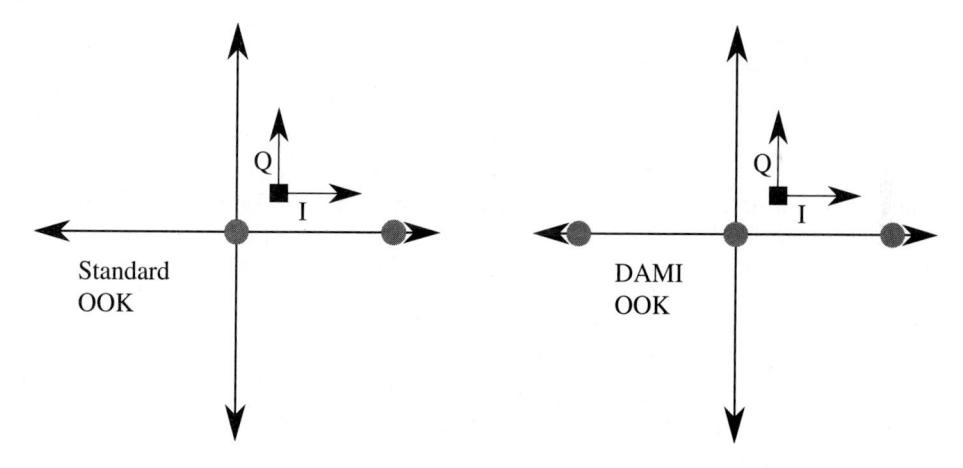

Figure 9.32 OOK and DAMI constellations for optional use within the SC-PHY [802.15.3-09].

---

8. Instead of a preamble, DAMI includes two pilot tones, one at the center frequency and one in the center of the nulled sideband.

channel estimation sequence (CES) is included in the preamble. The polarity of the second and third repetition together signal the spreading factor ($L_{\text{SF}} = 1$ or 2) to be used in the frame.

3. *CES* — Used for channel estimation and equalization, CES is constructed from the four predefined Golay sequences: $\mathbf{a}_{128}, \mathbf{b}_{128}, \overline{\mathbf{a}}_{128},$ and $\overline{\mathbf{b}}_{128}$. Before modulating each sequence, 64-bit cyclic prefix and postfix are added to each Golay sequence, doubling the CES length.

In the payload, each frame is formatted into blocks of length $508 \times L_{\text{SF}}$ followed by $4 \times L_{\text{SF}}$ pilot symbols ([1010] for $L_{\text{SF}} = 1$ and [11001100] for $L_{\text{SF}} = 2$). The header is transmitted in the same way as the payload, but with the inclusion of an additional $16\times$ spreading factor resulting from the same spreading sequence in SC-PHY header.

9.4 WirelessHD

Whereas IEEE 802.15.3c represents an *open* 60 GHz WPAN standard, WirelessHD represents a *private* or *closed* 60 GHz WPAN standard. WirelessHD is an alliance of technology corporations that was the first to focus on developing the 60 GHz device marketplace. The WirelessHD specification has focused on developing standards specifically designed for short range high-definition multimedia streaming. Backed by major 60 GHz wireless market players, including Broadcom, Intel, LG Electronics, Panasonic, NEC, Samsung, SiBEAM, Sony, and Toshiba, among others, WirelessHD 1.0 was first released in January 2008 (for private distribution) followed by the release of its compliance specification in January 2009. Since then, the retail market has seen multiple products integrated with WirelessHD.

9.4.1 Application Focus

While IEEE 802.15.3c provides a suite of implementation options to offer optimized performance for all mmWave WPANs, WirelessHD specifically targets high-definition multimedia streaming applications. The audio/video options of explicit interest, along with their required data rate and desired latency, are listed in Table 9.16.[9] Further, to ensure common interoperability for all devices with this application focus, all compliant WirelessHD systems must provide the following requirements [Wir10]:

- A WirelessHD source and sink with 59.94/60 Hz 480p video
- A WirelessHD source and sink with 50 Hz 576p video
- A WirelessHD sink for HDTV compatibility with 59.94/60 Hz 720p or 1080i video
- A WirelessHD sink for HDTV compatibility with 50 Hz 720p or 1080i video

---

9. For the video options, WirelessHD targets a pixel error ratio of less that 1 part per billion, which requires the application layer to produce bit error rates below 4×10^{-11}.

Table 9.16 WirelessHD 1.0 targeted applications and their network requirements [Wir10].

Target Application	≈ Data Rate	Desired Latency
Compressed 5.1 surround sound	1.5 Mbps	2 msec
Uncompressed 5.1 surround sound	20 Mbps	2 msec
Compressed 1080p HDTV	20-40 Mbps	2 msec
Uncompressed 7.1 surround sound	40 Mbps	2 msec
Uncompressed 480p HDTV	500 Mbps	2 msec
Uncompressed 720p HDTV	1,400 Mbps	2 msec
Uncompressed 1080i HDTV	1,500 Mbps	2 msec
Uncompressed 1080p HDTV	3,000 Mbps	2 msec

- A WirelessHD source and sink that supports linear PCM audio formats with 16 bits per sample at 32 kHz, 44.1 kHz, and 48 kHz sampling frequencies

- All WirelessHD devices support RGB 4:4:4, YCbCR 4:2:2, and YCbCr 4:4:4 color spaces with 24 bits per pixel color depth in all supported resolutions

By servicing these above-listed target applications, WirelessHD supports device implementation on a multitude of platforms. Platforms that act as multimedia sources include the following:

- Recorded media players, including Blu-ray Disc and HD DVD players

- Multimedia receivers, including broadcast HDTV and HD radio receivers

- Personal devices, including notebook PCs, digital video cameras, and digital audio players

Platforms that act as multimedia sinks include the following:

- Video displays, including HDTVs and LCD monitors

- Multimedia recorders, including Blu-ray Disc and HD DVD recorders

- Personal devices, including digital video players, digital audio players, and digital cameras

- Audio devices, including amplifiers and speakers

For a detailed breakdown of the use case for each source/sink combination and the network requirements, please refer to the technical specification [Wir10].

9.4.2 WirelessHD Technical Specification

A primary motivation for the WirelessHD private standard activity, conducted outside of the IEEE 802.15.3c standard body, was to circumvent the slow political process of IEEE standards. Although both standards were being developed simultaneously, WirelessHD was completed and released almost two years before IEEE 802.15.3c. WirelessHD products have been in full-scale production for several years, making it the first widespread commercial mmWave wireless standard.

As the world's first global wireless standard for the nascent 60 GHz band, the member corporations of WirelessHD realized the importance of offering compliance with major open standard organizations. This led to the inclusion of major portions of WirelessHD as the AV PHY option in the IEEE 802.15.3c standard in 2013. Since the details of the AV PHY within IEEE 802.15.3c are presented in Section 9.3.2.5, we refrain from repeating this information here. Instead, this section presents the unique technical features of the WirelessHD standard that are not considered or incorporated in IEEE 802.15.3c.

9.4.2.1 Source Packetizer and Layering Model

The layering model for WirelessHD devices is illustrated in Fig. 9.33, where the adaptation sub-layer, the MAC sub-layer, and the PHY layer are explicitly highlighted. The audio/visual (A/V) controller in the adaptation sub-layer is responsible for device control, connection control, as well as handling device capabilities. A/V data are formatted

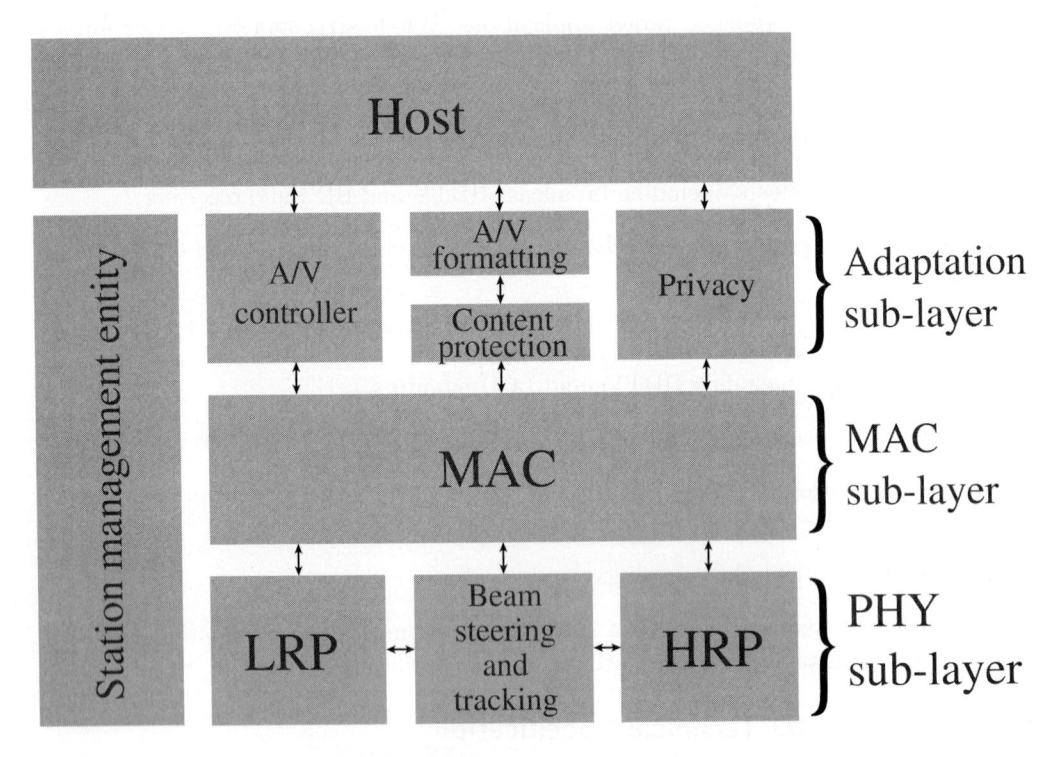

Figure 9.33 Layering of WirelessHD device.

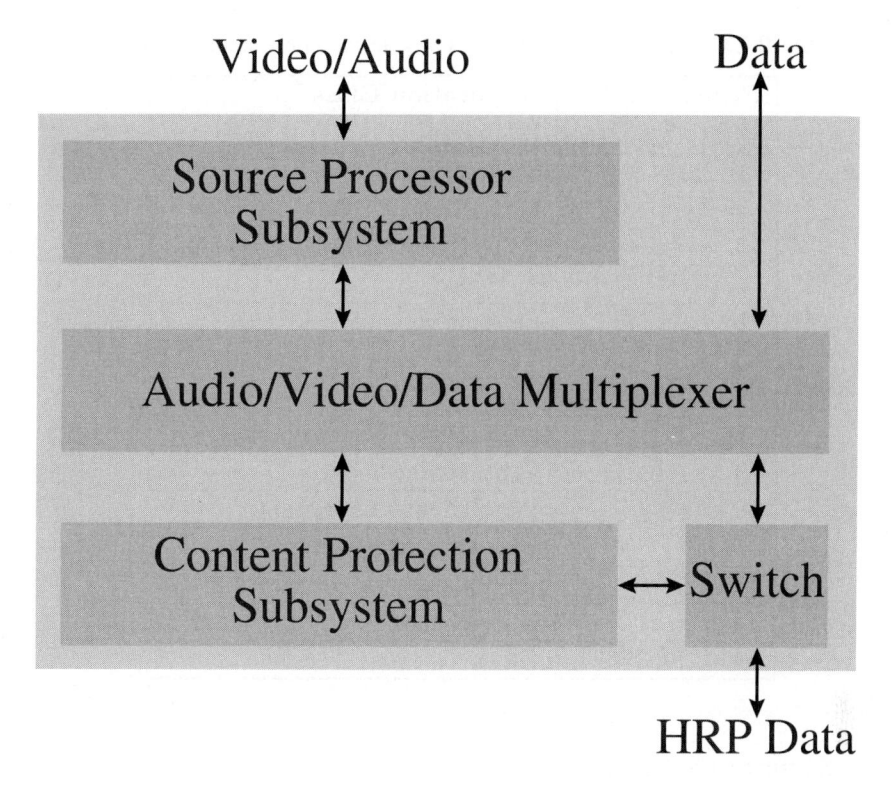

Figure 9.34 WirelessHD packetizer diagram [Wir10].

and packetized also in this adaptation sub-layer. The A/V packetizer is shown in more detail through Fig. 9.34. Before A/V data are forwarded to the MAC, the source processor subsystem formats the A/V data to allow compatibility with enhanced features such as UEP. Content protection is also enabled to ensure that digital media rights are preserved during wireless transit.

9.4.2.2 Handover Rules

In Section 9.3.1.8, we discussed the procedures for handing over piconet coordinator roles to devices with enhanced capabilities. In WirelessHD, the rules for coordinator election and handover are precisely defined according to the device application class. Upon start-up, the first device that turns on becomes coordinator by default. Hand-off of coordinator responsibilities to new devices in the piconet depends on the priority index (order ID) of the device class. If the piconet contains a device with smaller priority index, the coordinator must hand over coordinator responsibilities. Table 9.17 shows the device classes and their associated priority indices.

Table 9.17 Priority indices (order ID) of application classes in WirelessHD.

Priority Index	Application Class
0	Digital television
1	Set-top box
2	DVD/Blu-ray/HD-DVD player
3	DVD/Blu-ray/HD-DVD recorder
4	A/V receiver
5	Personal computer
6	Video projector
7	Gaming console
8	Digital video camera
9	Digital still camera
10	Personal digital assistant
11	Personal multimedia player
12	MP3 player
13	Cell phone
14	Other

9.4.2.3 Security Features

WirelessHD provides two different types of data security: personal and media. Personal security occurs at the adaptation sub-layer and ensures the privacy of data by preventing unwarranted access to transmissions. In general, personal security is accomplished through three main functions, as follows:

- Four-pass public key exchanges
- Encryption
- Cryptographic integrity evaluation

Media security ensures that the original, raw multimedia content is extractable from WirelessHD devices. WirelessHD authorizes digital transmission content protection specified by the Digital Transmission License Administrator, LLC.

9.4.3 The Next Generation of WirelessHD

In May 2010, WirelessHD announced the release of standard version 1.1. While maintaining backwards compatibility to version 1.0, the next generation standard stretches data rates from the original 3.807 Gbps peak data transmission rates up to 28 Gbps. This increased data rate will enable wireless multimedia streaming with increased video resolution (e.g., 4K resolution [OKM+04]), increased color depth in frames (e.g., Deep Color™ [CV05]), higher frame rates [MSM04], and 3-dimensional video formatting [YIMT02].

9.5 ECMA-387

ECMA International is an organization dedicated to the standardization of information and communication systems. ECMA-387 has not received the same level of attention as IEEE 802.15.3c and WirelessHD. Nevertheless, ECMA-387 represents an important 60 GHz standard. It was released almost a full year before IEEE 802.15.3c, so it is not surprising that many of the design concepts in IEEE 802.15.3c are also present in ECMA-387, especially in the PHY. In fact, the HRP of IEEE 802.15.3c and WirelessHD closely follows the OFDM PHY included in Type A devices defined below. ECMA-387 also supports channel bonding, which multiplies the throughput up to a factor of four, enabling data rates that exceed 25 Gbps.

9.5.1 Device Classes in ECMA-387

Similar to the three different mmWave PHYs in IEEE 802.15.3c, ECMA-387 provides specifications for three distinct device classes, identified A-C. In contrast to IEEE 802.15.3c, however, the ECMA-387 device classes display less operational overlap, and the usage scenarios for each device class are much better defined. The device classes for ECMA-387 and their operational scenarios are listed as follows:

- *Type A Devices* — Type A is the high performance device class for multimedia streaming and general ultra broadband WPAN. Transmission range extends up to 10 m with multipath in NLOS links. PHY features include beam steering with antenna arrays, single-carrier and optional OFDM modulation (both compatible with frequency domain equalization through a cyclic prefix), UEP, concatenated codes, 2× symbol spreading (base rate), and channel bonding. Rates without channel bonding extend up to 6.35 Gbps. Type A devices also support discovery mode operation, which is very similar to the common mode in IEEE 802.15.3c, where omnidirectional antennas and poor channel conditions are assumed, such that symbol repetition is used to compensate for insufficient array gain before antennas are trained.

- *Type B Devices* — Type B is the medium performance device class with significantly lower complexity and power consumption than Type A devices. Transmission range extends up to 3 m in LOS links. PHY features include simple single-carrier modulation, low-complexity Reed-Solomon codes, UEP, 2× symbol spreading (base rate), and channel bonding. Rates without channel bonding extend up to 3.175 Gbps.

- *Type C Devices* — Type C is the low performance device class with the minimum implementation complexity and power consumption. Transmission range is limited to 1 m in LOS links. PHY features include amplitude only modulation (ASK and OOK), low-complexity Reed-Solomon codes, and 2× symbol spreading (base rate). Rates extend up to 3.2 Gbps.

Notice that Type A and B devices in ECMA-387 support channel bonding by adding up to three adjacent additional channels to the original channel for data rate multiplication. All device classes coexist and interoperate under the standard, although each

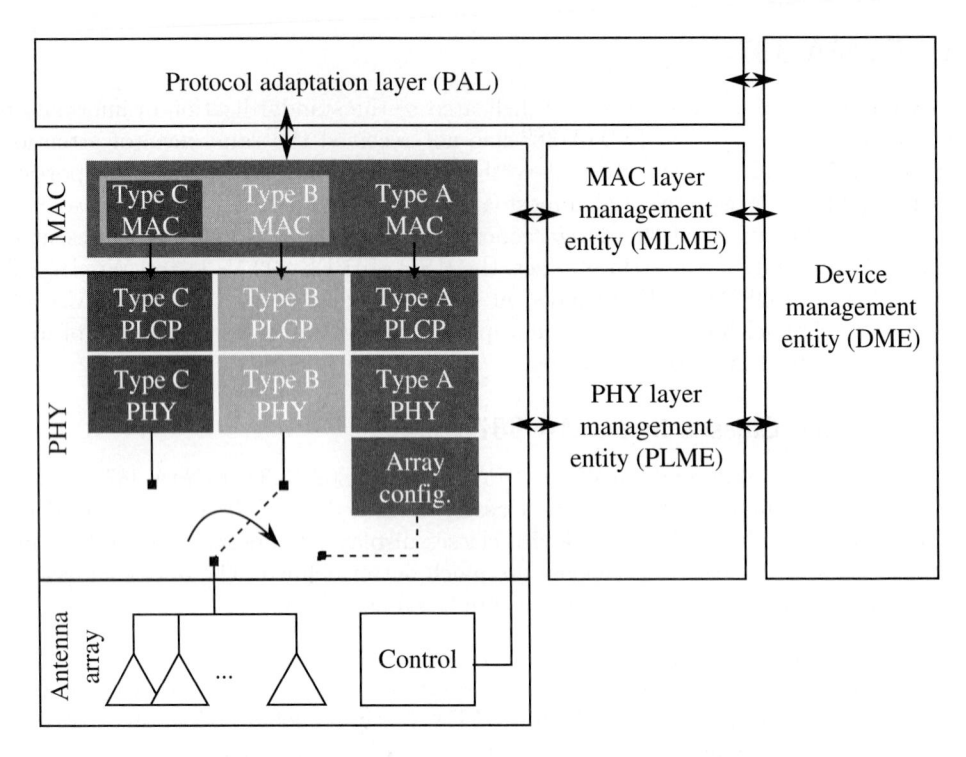

Figure 9.35 Protocol structure of ECMA-387 [ECMA08].

device class does not necessarily require the presence of another device class for standard operation. This is highlighted in Fig. 9.35, which is taken from the standard specification. Each multiplexer decides which PHY to operate under. Note that the MAC for Type B devices is a subset of the MAC used for Type A devices. Similarly, the MAC for Type C devices is a subset of the MAC for Type C devices. The HDMI protocol adaptation layer allows the ECMA-387 link to stand between the HDMI source and sink.

9.5.2 Channelization in ECMA-387

ECMA-387 and IEEE 802.15.3c both support channels 1-4, as illustrated in Fig. 9.8. With channel bonding, ECMA-387 also supports channels 5-10 which represent bonded (i.e., concatenated) radio channels that are various combinations of the original four 2.16 GHz channels shown in Figure 9.8. The configuration of these channels is listed in Table 9.18. The channel selection process is not completed until a device finds another for communication within the discovery channel, which is Channel Index 3. Once two devices determine that they wish to communicate, all potential channels will be scanned. Once a channel is selected, a channel switch procedure may also occur following another channel scan.

9.5 ECMA-387

Table 9.18 Bonded channels available to ECMA-387.

Channel Index	Bonded Channels	Lower Frequency	Center Frequency	Upper Frequency
5	1,2	57.24 GHz	59.40 GHz	61.56 GHz
6	2,3	59.40 GHz	61.56 GHz	63.72 GHz
7	3,4	61.56 GHz	63.72 GHz	65.88 GHz
8	1,2,3	57.24 GHz	60.48 GHz	63.72 GHz
9	2,3,4	59.40 GHz	62.64 GHz	65.88 GHz
10	1,2,3,4	57.24 GHz	61.56 GHz	65.88 GHz

9.5.3 MAC and PHY Overview for ECMA-387

Although many of the PHY features in IEEE 802.15.3c and WirelessHD mirror ECMA-387, the piconet MAC framework is not included in the ECMA-387 standard. Instead, ECMA-387 provides a distributed contention-based access protocol without coordinators. This contention-based access occurs in two forms, depending on the class of the MAC frame and operational intent. Only beacons, control frames, antenna training, and command frames are allowed to use the *distributed contention access* (DCA) protocol. DCA communicates with neighbors through the standard carrier sense and random back-off mechanisms. DCA operates on the discovery channel, whose PHY implementation is more robust in Type A devices. The *distributed reservation protocol* (DRP) is required for application data communication, where synchronized neighbor devices request and grant reserved time slots for a single link. DRP must be set up using beacons with the DCA and provides higher rate communication between devices once it is initiated. DRP transmission is formatted through superframes, where each superframe consists of a variable length Beacon Period followed by 256 medium access slots reserved for device communication.

The ECMA-387 MAC frame format includes many of the standard features in state-of-the-art wireless standards such as frame fragmenting, frame aggregation, block acknowledgments, and link quality feedback (frame error statistics, transmit power, and link quality metric measurements). ECMA-387 also includes frame formatting procedures for specific 60 GHz applications, such as control frames tailored to video streaming and unequal error protection. The ECMA-387 MAC also features power control through transmit power adjustments and hibernation procedures, coexistence and interoperability of different PHY classes, secure communication through encryption and data authentication, and special compatibility with wired HDMI interfaces through a protocol adaptation layer. ECMA-387 preserves the content protection scheme built into HDMI.

Optional features of the ECMA-387 MAC include out-of-band (OOB) control channels at 2.4 GHz enabled by IEEE 802.11g devices, whose architecture is illustrated in Fig. 9.36, and amplify-and-forward relaying. The OOB control channel can be very effective for avoiding interference of neighboring devices undiscovered at 60 GHz and for fast recovery of sudden 60 GHz link failure. Amplify-and-forward relaying allows for extended range of 60 GHz links and is compatible with antenna training at both the transmitter and receiver of the relay.

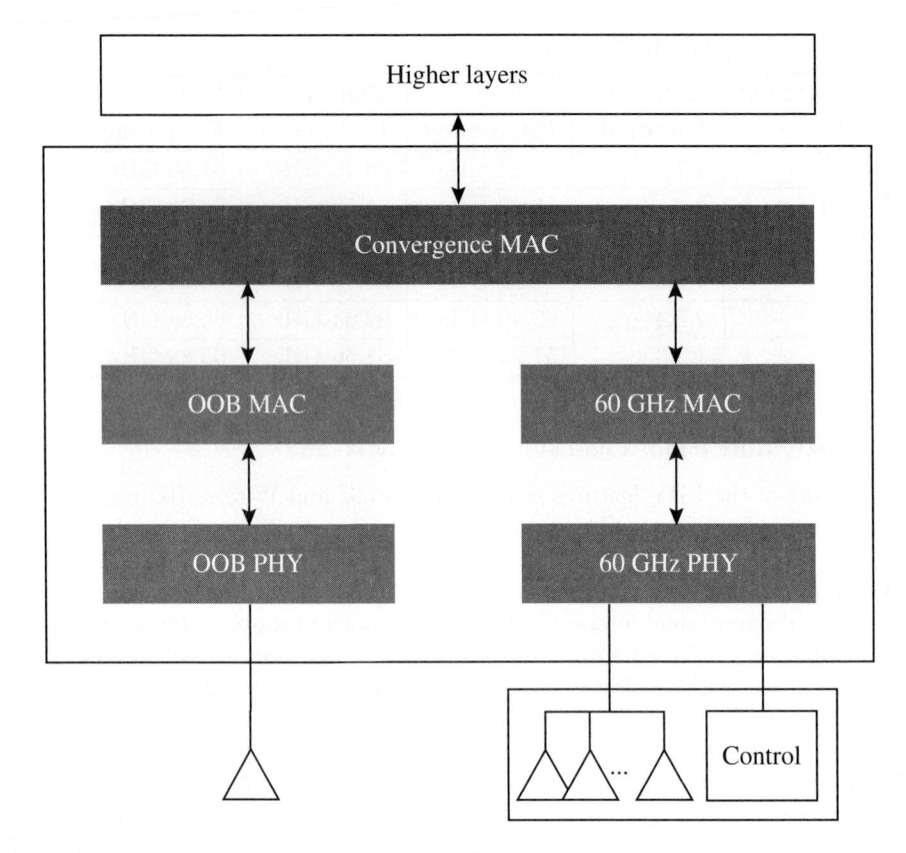

Figure 9.36 Out-out-band (OOB) control channel layered architecture in ECMA-387 [ECMA08].

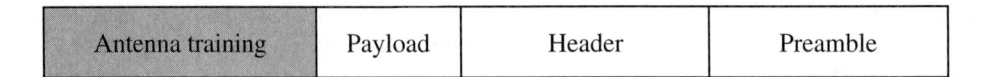

| Antenna training | Payload | Header | Preamble |

Figure 9.37 General frame format for ECMA-387 PHY frames. Order is right to left.

All PHY frames in ECMA-387 follow the format shown in Fig. 9.37. First, a preamble is transmitted to provide receive gain control, synchronization, and channel estimation. Next, a header is transmitted to inform the receiver of the included PHY parameters. The header is followed by the payload, which includes all data for the serviced application, passed down through the OSI stack. The final, and optional, portion of the PHY frame is the *antenna training sequence* (ATS). Each PHY formats the preamble, header, and payload for each PHY slightly differently; however, the ECMA-387 standard includes a discovery mode preamble that is used for device discovery. The discovery mode preamble is made more robust through repetition of order up to 128.

ATS is only available in Type A and Type B frames and is used to perform antenna beamforming in devices with antenna arrays. The ATS consists of the 256-length

Frank-Zadoff (FZ) sequence defined in Section 10.2.2.3 of the ECMA-387 standard [ECMA08]. In the ATS field, the FZ sequence is repeated for each transmit antenna configuration and then each receive antenna configuration that desires to be trained (64 different transmit and receive antennas configurations are possible). Hence, the FZ sequence is used to estimate channel information for each configuration.

9.5.4 Type A PHY in ECMA-387

Type A PHYs may include up to 22 modulation and coding schemes, as illustrated in Table 9.19. The symbol rate of all SC MCS is 1.728 GHz times the number of bonded

Table 9.19 Modulation and coding schemes for Type A PHYs in ECMA-387. All data rates assume that no bonding occurs and nominal cyclic prefix lengths. *Note that MCS A0 uses spreading with $L_{SF} = 2$. All ECMA-387 Type A devices must support MCS A0 without channel bonding (other modes optional). Additionally, all Type A devices must support the base rates of Type B and Type C devices (MCS B0 and C0 from Tables 9.20 and 9.21, respectively). Discovery mode is compatible with MCS A0-A7, where each symbol is repeated up to 128 times, depending on what processing gain is needed to compensate for insufficient antenna array gain.

MCS	Modulation	FEC	Data Rate
A0	SC, BPSK	RS + rate 1/2 BCC	397 Mbps*
A1	SC, BPSK	RS + rate 1/2 BCC	794 Mbps
A2	SC, BPSK	RS	1.588 Gbps
A3	SC, QPSK	RS + rate 1/2 BCC	1.588 Gbps
A4	SC, QPSK	RS + rate 6/7 BCC	2.722 Gbps
A5	SC, QPSK	RS	3.175 Gbps
A6	SC, NS8-QAM	RS + rate 5/6 TCM	4.234 Gbps
A7	SC, NS8-QAM	RS	4.763 Gbps
A8	SC, TCM16-QAM	RS + rate 2/3 TCM	4.763 Gbps
A9	SC, 16-QAM	RS	6.350 Gbps
A10	SC, QPSK	RS + rate 1/2 BCC (MSB only)	1.588 Gbps
A11	SC, 16-QAM	RS + rate 4/7,4/5 BCC (UEP)	4.234 Gbps
A12	SC, QPSK (UEP)	RS + rate 2/3 BCC	2.117 Gbps
A13	SC, 16-QAM (UEP)	RS + rate 2/3 BCC	4.234 Gbps
A14	OFDM, QPSK	RS + rate 1/3 BCC	1.008 Gbps
A15	OFDM, QPSK	RS + rate 2/3 BCC	2.016 Gbps
A16	OFDM, 16-QAM	RS + rate 2/3 BCC	4.032 Gbps
A17	OFDM, QPSK	RS + rate 4/7,4/5 BCC (UEP)	2.016 Gbps
A18	OFDM, 16-QAM	RS + rate 4/7,4/5 BCC (UEP)	4.032 Gbps
A19	OFDM, QPSK (UEP)	RS + rate 2/3 BCC	2.016 Gbps
A20	OFDM, 16-QAM (UEP)	RS + rate 2/3 BCC	4.032 Gbps
A21	OFDM, QPSK	RS + rate 2/3 BCC (MSB only)	2.016 Gbps

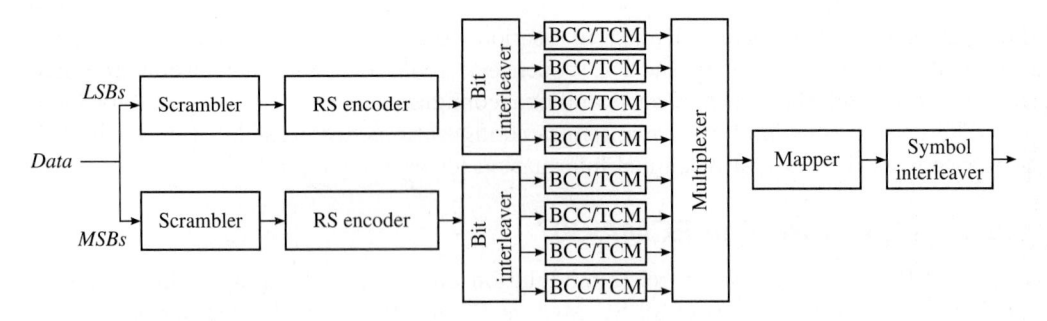

Figure 9.38 Formatting procedure for SC MCS in Type A devices.

channels, whereas the symbol rate for all OFDM MCS is 2.592 GHz (channel bonding is not available with OFDM). The Type A PHY features block transmission of symbols in the time and frequency domains (SC and OFDM, respectively). Note that the OFDM MCS are formatted using the same procedure as the AV PHY HRP mode of IEEE 802.15.3c, although specific parameters may vary slightly. In ECMA-387, OFDM modes use 512 total sub-carriers with 360 data sub-carriers, 3 DC sub-carriers, 16 pilot sub-carriers, 133 guard sub-carriers, and a cyclic prefix of length 64.

The general formatting procedure for SC MCS in Type A devices is shown in Fig. 9.38. All SC MCS use RS(255,239) codes followed by length-48 bit interleaving, while select modes concatenate with punctured binary convolutional codes (generator polynomials $(23)_8$ and $(35)_8$). In UEP modes (with constellation skewing or variable coding rates), two branches are needed with four inner codes each, while non-UEP modes only use a single branch. MCS A6 and A8 also feature trellis coded modulation (TCM).[10] In MCS A0, all data symbols are repeated before block formatting. Each block of 252 data symbols appends 4 pilot symbols before the cyclic prefix is added. Available cyclic prefix lengths include 0, 32, 64, and 96.

9.5.5 Type B PHY in ECMA-387

Type B PHYs may include the modulation and coding schemes shown in Table 9.20. The symbol rate and the block size for Type B SC MCS are the same as Type A SC MCS, at 1.728 GHz and the 252 samples, respectively. Each block is also followed by four pilots, although no cyclic prefix is offered in Type B devices, limiting their utility in multipath channels. The general formatting procedure for SC MCS in Type B devices is shown in Fig. 9.39. The functional block performs virtually the same operations as in Type A SC devices, although the BPSK/QPSK constellations are rotated differently. Differential encoding is processed after symbol interleaving.

MCS B4 offers DAMI in the same manner as in the SC-PHY of IEEE 802.15.3c, as illustrated in Fig. 9.32. The sample rate of MCS B4 is 3.456 GHz, and MCS B4 also includes RS encoding before DAMI mapping.

---

10. TCM is implemented through bit demultiplexing into two groups before convolutional coding. See Section 10.2.2.5.1.6 of [ECMA08] for more details.

9.5 ECMA-387

Table 9.20 Modulation and coding schemes for Type B PHYs in ECMA-387. All data rates assume that no bonding occurs. *Note that MCS B0 uses spreading with $L_{SF} = 2$. All ECMA-387 Type B devices must support MCS B0 without channel bonding (other modes optional). Additionally, all Type B devices must support the base rates of Type C devices (MCS C0 from Table 9.21).

MCS	Modulation	FEC	Data Rate
B0	SC, DBPSK	RS	794 Mbps*
B1	SC, DBPSK	RS	1.588 Gbps
B2	SC, DQPSK	RS	3.175 Gbps
B3	SC, QPSK (UEP)	RS	3.175 Gbps
B4	DAMI	RS	3.175 Gbps

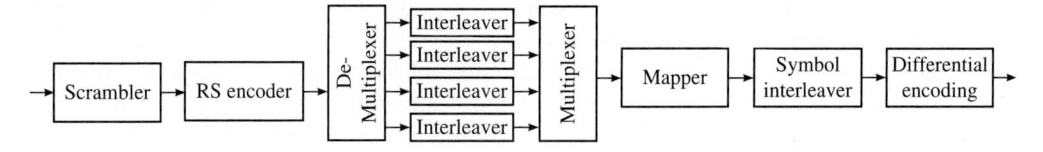

Figure 9.39 Formatting procedure for SC MCS in Type B devices.

Table 9.21 Modulation and coding schemes for Type C PHYs in ECMA-387. *Note that MCS C0 uses spreading with $L_{SF} = 2$. All ECMA-387 Type C devices must support MCS C0 without channel bonding (other modes optional).

MCS	Modulation	FEC	Data Rate
C0	SC, OOK	RS	800 Mbps*
C1	SC, OOK	RS	1.6 Gbps
C2	SC, 4ASK	RS	3.2 Gbps

9.5.6 Type C PHY in ECMA-387

Type C PHYs may include the modulation and coding schemes shown in Table 9.21. The symbol rate for Type C MCS is again 1.728 GHz, and each block includes 508 data symbols followed by 4 pilot symbols for MCS C1 and C2. MCS C0 processes blocks of 1,016 data symbols followed by 8 pilots.

9.5.7 The Second Edition of ECMA-387

A revision of the ECMA-387 standard was published in December 2011. The focus of the standard has been sharpened by eliminating less useful modes for increased harmony of standard-compliant devices. Specifically, Type C devices do not appear in version 2 of the standard.

9.6 IEEE 802.11ad

IEEE 802.11ad and IEEE 802.11ac are the future of wireless local area networking (WLAN). IEEE 802.11ac focuses on carrier frequencies between 5 and 6 GHz, where less channel bandwidth is available, whereas IEEE 802.11ad focuses on the 60 GHz spectrum, where many gigahertz of global spectrum are allocated (see Fig. 9.1). Proponents of IEEE 802.11ac believe that many gigabit WLAN applications are better served by maximizing the spectral efficiency at lower carrier frequencies through large-dimension MIMO processing with up to 8 antennas, high-order constellations (e.g., 256-QAM), and efficient forward error correction. Alternatively, proponents of IEEE 802.11ad believe that the best solution is to move to the much higher mmWave frequency band of 60 GHz where more bandwidth is available, focusing less on the spectral efficiency. On the one hand, IEEE 802.11ac is less risky and is seen as a more direct extension of IEEE 802.11n by using a microwave band with less path loss and with a solid base of highly integrated circuits. IEEE 802.11ad must exploit cutting-edge CMOS millimeter-wave device technology, consume massive bandwidths, introduce mmWave adaptive beam steering, and implement directional MACs, all of which are not sufficiently proven in the market. On the other hand, IEEE 802.11ad offers a more direct transition to future networks, which will inevitably scale carrier frequencies to match scaling bandwidth demands. One limitation of both standards is that the range of gigabit transfers will be substantially reduced compared with megabit transfers of legacy IEEE 802.11a/b/g/n devices due to the large number of streams and the large constellation order in IEEE 802.11ac, or the increased mmWave path loss in IEEE 802.11ad if large antenna gains are not used.

This section summarizes the salient features of IEEE 802.11ad, giving special attention to changes from IEEE 802.11n. In general, IEEE 802.11ad is a multi-band (2.4/5/60 GHz) solution that includes many of the features from WPAN 60 GHz standards (e.g., IEEE 802.15.3c) including beam steering, beam tracking, relaying, and directional MAC operation.

9.6.1 IEEE 802.11 Background

Table 9.22 shows an overview of past improvements to the IEEE 802.11 standard. The initial IEEE 802.11 legacy standard, approved in 1997, along with its 2 Mbps spread spectrum PHY, provides the original and current *carrier sense multiple access with collision avoidance* (CSMA/CA) MAC — the now-classical *request-to-send* (RTS), *clear-to-send* (CTS), *data* (DATA), and *acknowledgment* (ACK) exchange depicted in Fig. 9.40. If no traffic is detected for a specified period of time, the source node sends the RTS control frame to request access to the channel. If the destination node hears the RTS message and does not measure any other traffic, it replies with a CTS control frame. Next, the source node transmits the data payload. If the destination successfully receives the data, it replies with a positive acknowledgment. The RTS and CTS messages are optional (transmission may begin with the data payload if no traffic is detected by the transmitter). The RTS and CTS messages are encouraged in environments where the hidden node problem is likely to occur.

The 802.11 standard did not capture a significant market share until the a and b amendments in 1999 [Rap02]. IEEE 802.11b provides 11 Mbps of PHY throughput through a modulation technology known as complementary code keying. IEEE 802.11a

9.6 IEEE 802.11ad

Table 9.22 Overview of past 802.11 standards/amendments.

	Pub. Date	New Bands	Data Rates	MAC Additions
-	07/97	2.4 GHz	1-2 Mbps	CSMA/CA
		3.3 PHz	1-2 Mbps	
a	09/99	5 GHz	6-54 Mbps	-
b	09/99	2.4 GHz	1-11 Mbps	-
d	07/02	-	-	Regulatory domains
g	06/03	2.4 GHz	-	-
h	10/03	5 GHz	-	Europe, coexistence
i	08/04	-	-	Security (WPA/WPA2)
j	11/04	5 GHz	-	Japan
e	11/05	-	-	Quality of service (QoS)
k	06/08	-	-	Measurements and management
r	07/08	-	-	Fast/secure connectivity
y	11/08	3.7 GHz	-	USA, contention
w	09/09	-	-	Management frame security
n	10/09	2.4/5 GHz	6.5-600 Mbps	Aggregation
p	07/10	5.9 GHz	1.5-54 Mbps	Vehicular access
z	10/10	-	-	Direct link setup
u	02/11	-	-	External networks
v	02/11	-	-	Client configuration
s	09/11	-	-	Mesh, low-cost data

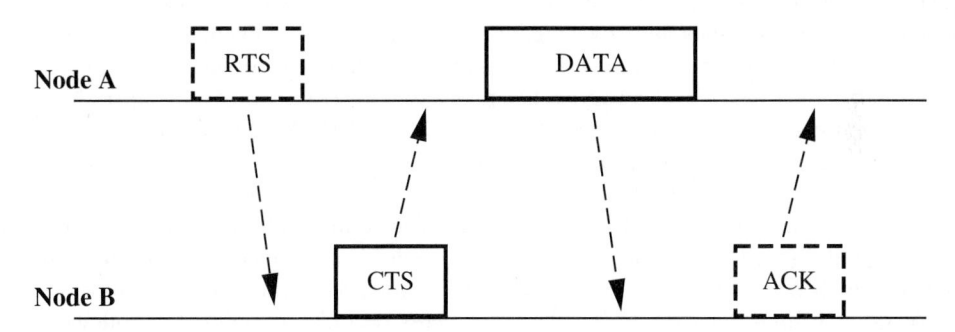

Figure 9.40 The traditional RTS/CTS/DATA/ACK exchange for IEEE 802.11 medium access control (MAC).

extends PHY throughput to 54 Mbps with OFDM modulation (which has since become a popular modulation choice). Whereas 802.11b operates on the lower 2.45 GHz industrial, scientific, and medical bands, 802.11a uses the UNI bands at 5 GHz. IEEE 802.11g was created to transition the 802.11a PHY to 2.45 GHz carrier frequencies. IEEE 802.11n includes several MAC amendments (e.g., IEEE 802.11e) and several new MAC enhancements, through frame aggregation, block acknowledgments, data piggybacking on ACKs, and link adaptation. IEEE 802.11n also offers major PHY enhancements, primarily through multiple antenna processing (up to four antennas) and optional 40 MHz channels for rates up to 600 Mbps.

9.6.2 Important IEEE 802.11ad MAC Features

There are several notable new components of the MAC in the IEEE 802.11ad amendment. They are each described in this section.

9.6.2.1 Directional Multi-Gigabit Access in IEEE 802.11ad

The core feature of IEEE 802.11ad is the new directional multi-gigabit (DMG) PHY (described in the next section) with gigabit-per-second data transfer capabilities through multiple-antenna beamforming in 60 GHz spectrum. The standard IEEE 802.11 *distributed coordination function* (DCF) with CSMA/CA spectrum access method cannot adequately service the DMG PHY due to higher path loss (if not compensated with antenna gains) and directional links. Instead, in a procedure similar to medium access in IEEE 802.15.3c, DMG access is granted to users by a coordinator or, as defined in IEEE 802.11ad, either a *personal basic service set control point* (PCP) or *access point* (AP). PCPs are necessary for IEEE 802.11ad devices that operate in an ad hoc mode where no centralized control was available in the past. In this section, we will follow IEEE notation in which each device in IEEE 802.11ad will also be referred to as a *station* (STA) regardless of whether it is operating as a PCP/AP.

The MAC architecture for DMG STAs is substantially different from the distributed coordination function discussed in Chapter 8, as shown in Fig. 9.41. The chief difference for DMG STAs is the replacement of DCF with DMG Channel Access. Note that DCF is still available for DMG STAs, but the rules for its operation are modified to suit the constraints of 60 GHz PHY operation underneath DMG channel access. For example, because of overhead issues and the exaggerated hidden node problem in directional links (as discussed in Chapter 8), DCF transmissions are advised to point in the direction of the intended receiver (direction must be known before transmission occurs). Also, the transmission back-off rules are different for DMG STAs. Back-off constraints are relaxed on DMG STAs to allow increased spatial reuse.

DMG channel access in IEEE 802.11ad is contained within beacons intervals, an example of which is illustrated in Fig. 9.42. Each beacon interval is formatted into distinct access periods by a PCP or AP, forming a schedule. Communication is distinct within each period. If a STA is not a PCP or AP, it cannot access the DMG medium without poll or grant frames from the PCP/AP. The access periods are defined as follows:

- *Beacon Transmission Interval (BTI)* — This access period is used for a PCP or AP to transmit beacons. In IEEE 802.11ad, beacons are used to establish the beacon interval and access schedule, to synchronize the network, and to exchange access and capability information, as well as for beamforming training. It is not necessary that all beacon intervals contain a BTI. A non-PCP/non-AP STA is not allowed to transmit beacons during the BTI.

9.6 IEEE 802.11ad

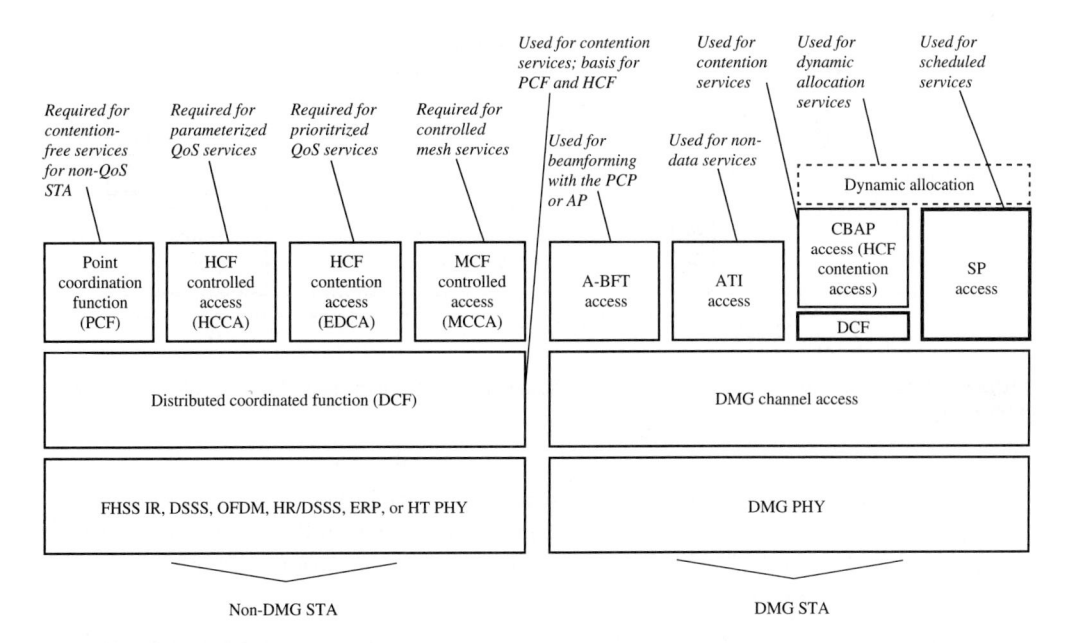

Figure 9.41 MAC architecture block diagram to illustrate differences between DMG STAs (at 60 GHz) and non-DMG STAs. Note that the fundamental dependence on DCF is not present with DMG STAs. © IEEE [802.11-12]

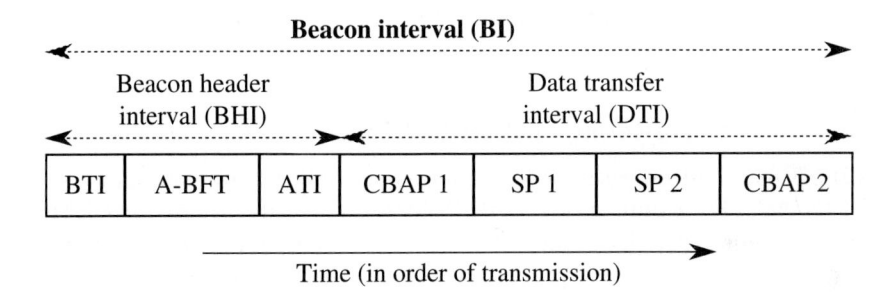

Figure 9.42 Example beacon interval format.

- *Association Beamforming Training (A-BFT)* — A-BFT is a beamforming training period reserved to train the PCPs or APs that transmitted a beacon during the BTI. The A-BFT is also not necessarily present in beacon intervals.

- *Announcement Transmission Interval (ATI)* — During the ATI, request and response frames are exchanged between the PCP/AP and the other STAs. The PCP/AP will initiate all frame exchanges, as demonstrated in Fig. 9.43. The ATI is also optionally present in the beacon interval.

- *Data Transfer Interval (DTI)* — The DTI is typically the most important period in the beacon interval for most STAs. It is used for frame exchanges between all

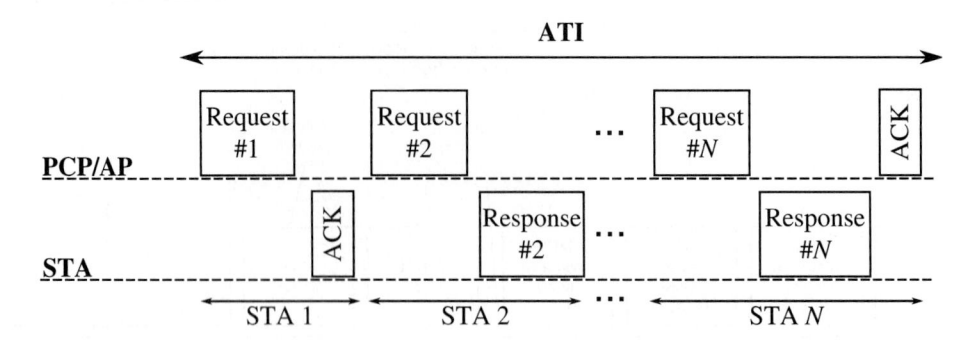

Figure 9.43 Communication within the Announcement Transmission Interval (ATI). © IEEE [802.11-12]

STAs. The DTI is further subdivided into (potentially) multiple iterations of two periods: the contention-based access period (CBAP), and the scheduled service period (SP). The PCP/AP determines the presence, duration, and number of iterations of CBAPs and SPs.

- *Scheduled Service Period (SP)* — Each SP is assigned to a STA-to-STA link by the PCP or AP. For high data rate transmission, the SP is preferred since STAs are allowed to use all available link features (e.g., beamforming).
- *Contention-Based Access Period (CBAP)* — The CBAP is the DMG region for CSMA/CA random access, that is, DCF. STAs may use the CBAP to request SP allocations.

9.6.2.2 PCP/AP Clustering in IEEE 802.11ad

60 GHz IEEE 802.11ad networks reduce interference and improve spatial reuse by clustering PCPs/APs. For example, if two PCPs/APs coordinate two distinct DMG access schedules simultaneously, they must exist in non-overlapping time periods to prevent interference. To create non-overlapping schedules, a new level is created in the access hierarchy through synchronization PCPs (S-PCPs) and synchronization APs (S-APs). S-APs/S-PCPs are responsible for creating clusters of APs/PCPs with non-overlapping access schedules in two ways, as follows:

- *Decentralized Clustering* — All DMG PCPs/APs within a cluster are coordinated by a single S-PCP/S-AP. Promotion to a S-PCP/S-AP occurs through beacons. Generally, the first PCP/AP to announce the role of S-PCP/S-AP will be observed as such by other PCPs/APs with which it communicates. The S-PCP/S-AP of a decentralized cluster creates beacon service periods (beacon SPs) to allow the cluster PCPs/APs to exchange beacons with other STAs in the cluster (the PCPs/APs and their associated STAs). A PCP or AP may join the decentralized cluster if it has clustering capabilities and detects an empty beacon SP not occupied by another PCP/AP. Note that the member PCPs/APs align their beacon interval with the beacon interval of the S-PCP/S-AP, as illustrated in Fig. 9.44.

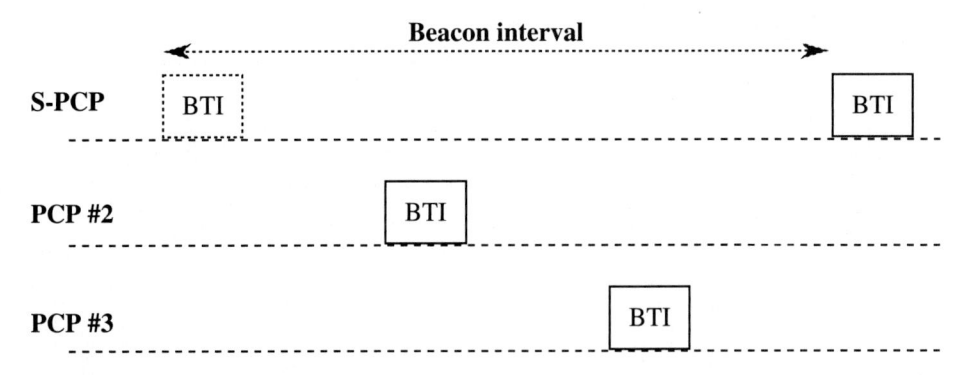

Figure 9.44 Transmission of BTIs for 3 PCPs within a decentralized cluster. Note that the overall beacon interval is set by the S-PCP.

- *Centralized Clustering* — A centralized cluster contains (potentially) multiple, coordinated S-APs within a single cluster. Centralized clusters are formed by a STA (S-AP capable) through two steps: (1) configuration and (2) verification. The configuration step provides operational information (such as scheduling information and frequency of transmission) to the STA. In the verification step, the channel is monitored for beacons and existing clusters. If no clusters exist, the cluster may be successfully formed and the STA is promoted to an S-AP. Beacon SPs are assigned in a similar manner to decentralized clustering to allow member PCPs/APs to interact with STAs in the network. The chief difference between decentralized and centralized clustering is that the centralized cluster may contain multiple S-AP devices, which increases cluster data transfer capabilities, but also adds significant complexity in the cluster design.

Note that provisions in the standard allow for S-PCP/S-AP hand-off and cluster maintenance for both decentralized and centralized clustering.

9.6.2.3 Beamforming in IEEE 802.11ad

The placement of training data and the configuration of the antenna array is provided by the PHY; however, the process of antenna training and configuration is managed by the MAC (see Chapter 8 for more details). Beamforming in IEEE 802.11ad shares many of the same features as beamforming in IEEE 802.15.3c, although many details are different. Training antennas for beamforming is broadly separated into two processes: (1) sector sweeping, and (2) beam refinement, as demonstrated in Fig. 9.45.

- *Sector Sweep* — The sector sweep is the first step toward training the antennas at the transmitter and receiver. Consider the concept of a sector in Fig. 9.29. Note that a single antenna on a DMG PHY may have at most 64 sectors. The total number of sectors across all antennas cannot exceed 128.[11] A sector sweep contains four

---

11. What constitutes an antenna is at the discretion of the vendor. For example, a multi-element antenna array that forms sectors through different phase combinations of each antenna element could,

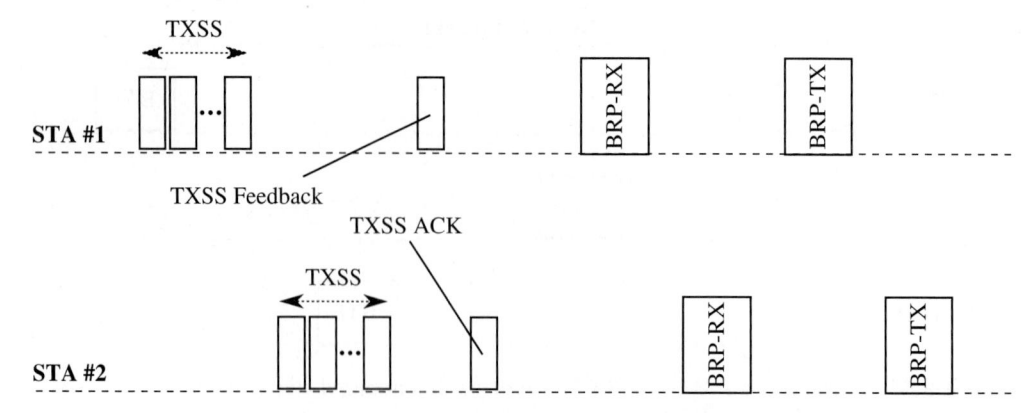

Figure 9.45 The first two processes of beamforming training in 802.11ad (sector sweep and refinement). The transmit or receive sector sweep (TXSS or RXSS) allows for estimation of coarse directionality at the transmitter or receiver. This may be refined through the beam refinement protocol (BRP). During typical operation, TXSS is executed to provide low-rate PHY operation. BRP is then executed on the low-rate PHY packets to improve the signal quality and boost PHY rates.

components: initiator transmit or receive sector sweep (TXSS or RXSS), responder TXSS or RXSS, feedback, and acknowledgment (ACK). In TXSS, CPHY formatted frames (discussed in Section 9.6.3) are transmitted for each combination of the transmitting STA sector and each *antenna* (with one fixed sector) on the receiving STA. This allows the receiver to estimate the best sector configuration at the transmitter with respect to all possible receive antennas. For RXSS, the roles are reversed. RXSS frames are sent with a fixed sector for each combination of transmit antenna and each *sector* at the receiving STA. Sector sweeps are concluded with feedback and ACK messages, which serve two purposes: (1) to exchange best-sector information, and (2) to confirm a successful sector sweep operation.

- *Beam Refinement* — Once sector-level training is completed, the beams may be further refined, if desired, through the beam refinement protocol (BRP). Beam refinement is enabled by two types of BRP packets: (1) BRP-RX, and (2) BRP-TX. BRP packets are PHY packets with antenna training appended (discussed in the PHY section sequel and illustrated in Fig. 9.51). BRP-RX packets use TRN-R training sequences, which enable receive antenna evaluation, whereas BRP-TX packets use TRN-T training sequences, which enable transmit antenna evaluation. BRP-TX packets may be transmitted with multiple antenna configurations since the transmit antenna configuration is allowed to change for each TRN-T sequence within the BRP-TX packet. The number of TRN-R/TRN-T sequences in each BRP

---

for the purposes of the standard, be considered a single antenna with multiple sectors (according to various phase combinations).

packet is determined by the number of antenna array settings desired to be trained at the receiver/transmitter.

Note that more time and effort spent in beam refinement means less effort spent in sector sweeping, and vice versa. The best balance depends on the vendor antenna configuration and the supported PHY options.

The IEEE 802.11ad standard also provides beam tracking to ensure that antenna configurations do not become stale as the propagation environment changes. Beam tracking and beam refinement are virtually the same procedures, with beam tracking referring to periodic refinement over a smaller set of antenna configurations. Beam tracking adds the same antenna training in the same format as in the BRP.

While the sector sweep and beam refinement processes are required for beamforming training, their placement within the 802.11ad MAC process depends on the DMG channel access method. In the BTI, the PCP/AP first initiates the TXSS through DMG beacons with different sector configurations. The A-BFT, which follows the BTI, allows a non-PCP/non-AP STA to respond to beamforming training initiated in the BTI. Beamforming training occurs between non-PCP/non-AP STAs in the DTI by issuing a beamforming training request to the PCP/AP.

9.6.2.4 Relaying in IEEE 802.11ad

As discussed in Chapter 8 and in Section 9.5, relays can be very helpful at 60 GHz to extend transmission range, provide network monitoring and security, and protect against link disruption. We discussed in Chapter 3 that link disruption can happen from many factors, including human movement. For 802.11ad, a link with a relay-capable STA and relay-compatible STAs may establish relay operation. Because of the schedule in DMG channel access provided by the PCP/AP, a STA knows when to become a relay and/or when to operate in each relay mode (if available). Relay operation takes on two forms: link switching and link cooperating.

In the link switching relay operation, the relay will only be used to provide an alternative electromagnetic path. Link switching can use either full-duplex amplify-and-forward or half-duplex decode-and-forward relays. Full-duplex amplify-and-forward link switching relays have two frame transmission modes: normal and alternation. Both modes fundamentally depend on the link change interval time period.

In normal mode, as illustrated in Fig. 9.46, the STA-to-STA link will continue until the link is disrupted. At the transmitter, link disruption is defined as the absence of the ACK response message for data transmitted at the beginning of the link change interval. At the receiver, *link disruption* is defined as the absence of receive data within a fixed time period after the link change interval begins. Once the link is disrupted, each STA will adapt its configuration (including beamforming characteristics) to communicate through the relay, that is, the STA-to-relay-to-STA link. If the link is disrupted again, the STAs will switch to the STA-to-STA link. This toggling process continues while in normal mode. In alternation mode, the link toggles every link change interval, regardless of link disruption, as illustrated in Fig. 9.47. Because the alternation period is fast (i.e., the link

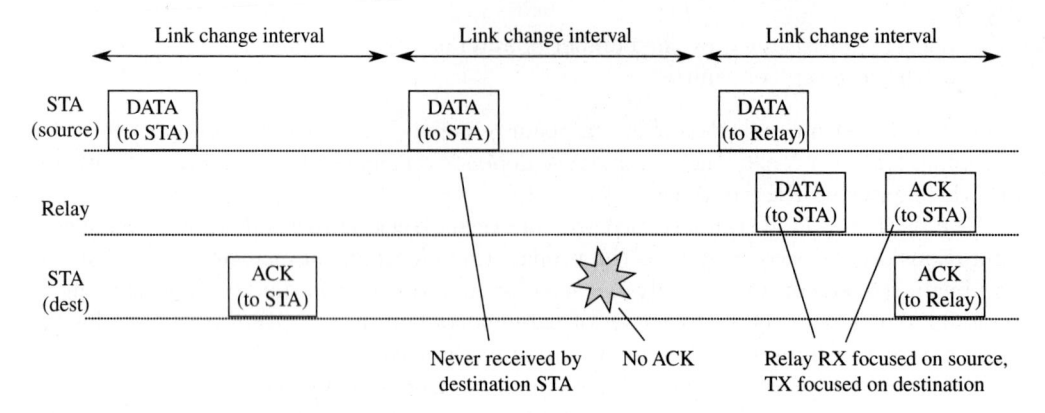

Figure 9.46 Transmissions by source STA, amplify-and-forward relay, and destination STA in a link switching example in normal mode.

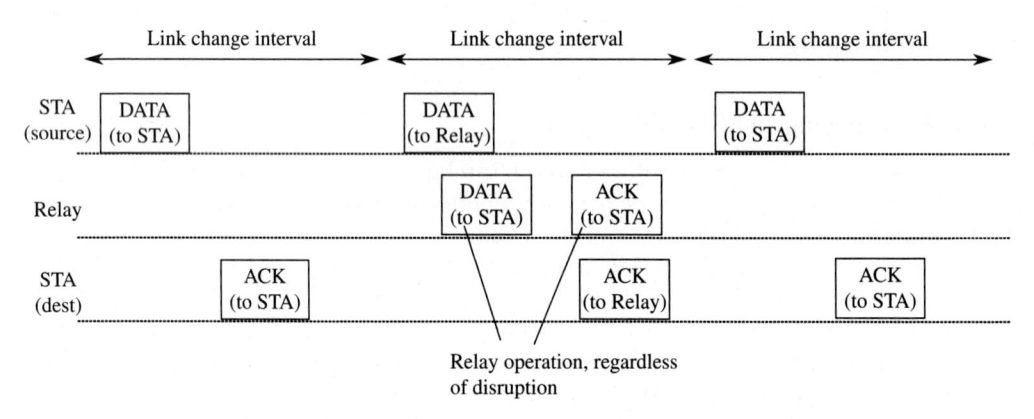

Figure 9.47 Transmissions by source STA, amplify-and-forward relay, and destination STA in a link switching example in alternation mode.

change interval is short), this also provides robustness to link disruption. Decode-and-forward relays operating with link switching always use the normal mode, although the transmission must be broken up into two periods.

In the link cooperating relay operation, the relay link is always activated as a means to consistently improve the communication. Each transmission has two stages. In the first stage, data are transmitted to the relay where, if beamforming is present, the transmitting STA is steered toward the relay. In the second stage, the data are sent from the transmitting STA again, only this time the antennas are steered toward the receiving STA (if beamforming present). Simultaneously, the relay sends the same data to the receiving STA, which is also steered toward the receiving STA (beamforming present). Fig. 9.48 illustrates this process.

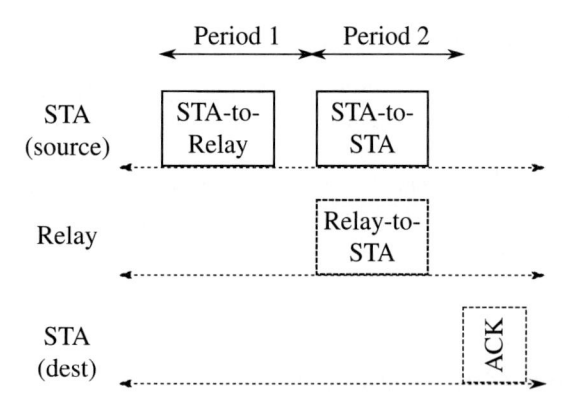

Figure 9.48 Transmissions by source STA, decode-and-forward relay, and destination STA in a link cooperation relay example.

9.6.2.5 Multi-band Operation in IEEE 802.11ad

An IEEE 802.11ad STA may be capable of supporting more than one band, that is, the DMG band and any combination of the bands supported by other IEEE 802.11 standards. Broadly, these bands are classified into three groups: (1) the low band (LB), 2.4 to 2.4835 GHz; (2) the high band (HB), 4.9 to 5.825 GHz; and (3) the ultra band (UB), 57 to 66 GHz. IEEE 802.11ad allows any of these bands to be used either sequentially or concurrently, that is, data transfer can occur in more than one band simultaneously. Further, for a single session, the operation can arbitrarily be transferred from one band to another. Each band can operate with the same MAC address (transparent) or with different MAC addresses (non-transparent) in each band. Since MAC management information is shared between each band, a high level of flexibility in multi-band operation is enabled, although PHYs cannot directly share information between different bands.

9.6.2.6 Link Adaptation in IEEE 802.11ad

The process of link adaptation through channel feedback was first introduced in IEEE 802.11n. Any STA in the DMG network may request link margin information from any other STA with which it intends to communicate. In response, feedback is offered on the recommended MCS, the expected SNR, and the expected SNR margin in the required SNR for the recommended MCS. The link margin information is useful for transmit power control as well as fast session transfer. Note that STAs may feed back unsolicited link margin information at its own discretion. The link margin information is important since DMG PHYs operate over massive bandwidths with potentially large frequency selectivity. Since a simple link quality metric is not available in these scenarios, the standard lets the receiver determine the recommended MCS for the frequency selective channel as well as the offset in average SNR from the required channel quality.

9.6.2.7 Security in IEEE 802.11ad

IEEE 802.11ad continues to use the advanced encryption standard (AES) encryption algorithm. Previously, IEEE 802.11 utilized the counter mode *cipher block chaining message authentication code protocol* (CCMP) to accomplish AES processing. Scaling to gigabit rates and to very large frames, however, has proven challenging for CCMP. Consequently, IEEE 802.11ad supports the Galois counter mode with *Galois message authentication code protocol* (GCMP). GCMP both reduces the computation complexity and allows for greater parallelism in its implementation without compromising security.

9.6.3 Directional Multi-Gigabit PHY Overview for IEEE 802.11ad

The directional multi-gigabit (DMG) PHY specification formats 60 GHz waveforms in IEEE 802.11ad WLANs. It consists of four formatting procedures:

1. *Control PHY (CPHY)* — For compatible operation of all DMG IEEE 802.11ad devices, regardless of the vendor implementation

2. *OFDM PHY* — For multiple carrier operation and maximal spectral efficiency

3. *Single-Carrier (SC) PHY* — For the best tradeoff between spectral efficiency and implementation complexity

4. *Low-power (LP) SC-PHY* — For minimal implementation complexity and power consumption

IEEE 802.15.3c similarly defined four PHY formats with CMS, SC-PHY, HSI PHY, and AV PHY. IEEE 802.11ad, however, better separates the usage scenarios for each PHY (as was the case in ECMA-387). Further, the DMG PHY in IEEE 802.11ad provides better coexistence between different PHY modes since the CPHY (MCS 0) is mandatory (along with MCS 1-4), whereas CMS need not be implemented for AV PHY devices in IEEE 802.15.3c.

9.6.3.1 Common PHY Aspects of IEEE 802.11ad

Along with a more precise operational distinction between different PHY formats (in comparison to IEEE 802.15.3c), IEEE 802.11ad also provides more uniformity in the formatting procedures for each PHY. In this section, we discuss the common aspects between all DMG PHYs. The DMG PHY uses channels 1-4 in the 60 GHz band, as defined in Fig. 9.8, which is the same for all the previously described standards. The spectral mask for the DMG PHY in IEEE 802.11ad varies slightly from IEEE 802.15.3c, as shown in Fig. 9.49.

Rather than defining separate MAC/PHY interfaces for each DMG PHY, IEEE 802.11ad defines a single MAC interface, meaning that each PHY format is only specified through the modulation and coding scheme. Table 9.23 shows the 32 MCSs that encompass all four PHYs.

9.6 IEEE 802.11ad

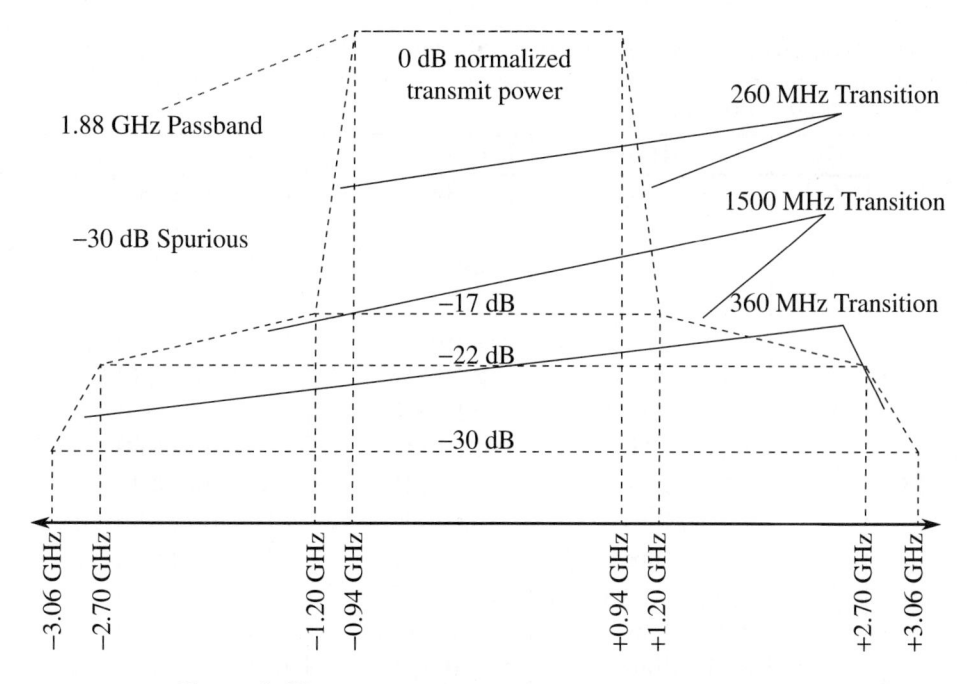

Figure 9.49 Spectral mask of DMG PHY in IEEE 802.11ad.

| Sync (STF) | CE training | Header | Data payload | Antenna training |

Figure 9.50 General DMG PHY frame format.

The general PHY frame structure in Fig. 9.50 is used for each of the MCS in Table 9.23. The short training field (STF) is used for the AGC, time synchronization, and frequency synchronization. The channel estimation (CE) training sequence is used to remove fading effects on received waveforms. To eliminate residual frequency and channel estimate offsets in the payload, receivers are expected to train channel gain and phase through pilots (selected sub-carriers in the OFDM PHY and the cyclic prefix in the SC-PHY). Note that the STF and CE fields are based on Golay sequences. See Table 9.24 for a list of Golay sequences used in 802.11ad.

The STF is a repeated Golay sequence whose polarity is inverted only in the last repetition. A continuously running Golay correlator at the receiver will provide a negative spike at the end of the STF to signal the start of an 802.11ad PHY packet. After a packet is detected, each PHY may be identified before header decoding through the CE field (which is different for SC and OFDM PHYs). After PHY identification, the receiver has two opportunities to estimate the channel through the CE field. Further, the amount of delay spread in the channel estimate is configurable since Golay sequences can be

Table 9.23 Modulation and coding schemes in the IEEE 802.11ad DMG PHY. L_{SF} denotes the length of the spreading sequence. Note: The data rate is only calculated for the data payload and does not include other PHY frame fields.

	MCS	Data Rate	Modulation	FEC
C	0	27.5 Mbps	DBPSK ($L_{SF}=32$)	LDPC(672,504) (shortened)
SC	1	385 Mbps	$\pi/2$ BPSK ($L_{SF}=2$)	LDPC(672,336)
	2	770 Mbps	$\pi/2$ BPSK	LDPC(672,336)
	3	962.5 Mbps	$\pi/2$ BPSK	LDPC(672,420)
	4	1,155 Mbps	$\pi/2$ BPSK	LDPC(672,504)
	5	1,251.25 Mbps	$\pi/2$ BPSK	LDPC(672,546)
	6	1,540 Mbps	$\pi/2$ QPSK	LDPC(672,336)
	7	1,925 Mbps	$\pi/2$ QPSK	LDPC(672,420)
	8	2,310 Mbps	$\pi/2$ QPSK	LDPC(672,504)
	9	2,502.5 Mbps	$\pi/2$ QPSK	LDPC(672,546)
	10	3,080 Mbps	$\pi/2$ 16-QAM	LDPC(672,336)
	11	3,850 Mbps	$\pi/2$ 16-QAM	LDPC(672,420)
	12	4,620 Mbps	$\pi/2$ 16-QAM	LDPC(672,504)
OFDM	13	693 Mbps	SQPSK	LDPC(672,336)
	14	866.25 Mbps	SQPSK	LDPC(672,420)
	15	1,386 Mbps	QPSK	LDPC(672,336)
	16	1,732.5 Mbps	QPSK	LDPC(672,420)
	17	2,079 Mbps	QPSK	LDPC(672,504)
	18	2,772 Mbps	16-QAM	LDPC(672,336)
	19	3,465 Mbps	16-QAM	LDPC(672,420)
	20	4,158 Mbps	16-QAM	LDPC(672,504)
	21	4,504.5 Mbps	16-QAM	LDPC(672,546)
	22	5,197.5 Mbps	64-QAM	LDPC(672,420)
	23	6,237 Mbps	64-QAM	LDPC(672,504)
	24	6,756.75 Mbps	64-QAM	LDPC(672,546)
LP	25	626 Mbps	$\pi/2$ BPSK	RS(224,208)+OBC(16,8)
	26	834 Mbps	$\pi/2$ BPSK	RS(224,208)+OBC(12,8)
	27	1,112 Mbps	$\pi/2$ BPSK	RS(224,208)+OBC(9,8)
	28	1,251 Mbps	$\pi/2$ QPSK	RS(224,208)+OBC(16,8)
	29	1,668 Mbps	$\pi/2$ QPSK	RS(224,208)+OBC(12,8)
	30	2,224 Mbps	$\pi/2$ QPSK	RS(224,208)+OBC(9,8)
	31	2,503 Mbps	$\pi/2$ QPSK	RS(224,208)+OBC(8,8)

9.6 IEEE 802.11ad

Table 9.24 Golay sequence definitions used in IEEE 802.11ad. The descriptions represent the complex baseband representation of single-carrier wireless signals through hexadecimal notation where the MSB is transmitted first, bit 0 is mapped to -1, and bit 1 is mapped to $+1$. Note that, while Golay sequences can be efficiently computed, 802.11ad transceiver implementations are likely to use look-up tables for each of these components.

Item	Description
Ga_{32}	0xfa39c90a
Gb_{32}	0x05c6c90a
Ga_{64}	0x28d8282728d8d7d8
Gb_{64}	0xd727d7d828d8d7d8
Ga_{128}	0xc059cf563fa6cf56c059cf56c05930a9
Gb_{128}	0x3fa630a9c05930a9c059cf56c05930a9

concatenated to double order. Estimating the channel through larger sequences (but same total block size) provides more delay spread resolution, but also increases noise in the estimate.

The header of each 802.11ad PHY frame informs the receiver about DMG PHY formatting procedures for the data payload. PHY header information includes the following:

- *Scrambler Initialization* — for seeding the scrambler; found in all DMG PHYs

- *MCS* — as listed in Table 9.23; not needed in CPHY

- *Frame Length* — the number of source octets (bytes) encapsulated in the attached PHY packet; found in all DMG PHYs

- *Appended Packet* — indicates next packet will be delivered without standard inter-frame spacing; not found in CPHY

- *Packet Type* — indicates whether the packet includes training fields and what type; found in all DMG PHYs

- *Training Length* — if included, defines the amount training included (number of ANT fields); found in all DMG PHYs

- *Aggregation* — indicates aggregation of multiple MAC frames into a single PHY packet; not found in CPHY

- *Beam Tracking Request* — indicates a request for beam tracking in future packets; not found in CPHY

- *Tone Pairing Type* — indicates static tone pairing (STP) or dynamic tone pairing (DTP); only for OFDM PHY

- *DTP Indicator* — indicates that the DTP mapping is should be updated; only for OFDM PHY

- *Last RSSI* — defines the value of the channel quality metric of last transmission; not found in CPHY

- *Header Check Sequence* — used to check validity of header information at receiver; found in all DMG PHYs

Finally, the *antenna training field* (ANT) enables beamforming training for 60 GHz antenna arrays through the BRP and beam tracking protocols. ANT field composition for all PHYs includes a similar structure as illustrated in Fig. 9.51. Note that each ANT field uses a separate transmit or receive antenna configuration (depending on TRN-T or TRN-R configuration, for example) in order to find the best antenna weights. The CE sequence, which is injected into every fifth symbol after the AGC is completed, allows the receiver to coarsely determine the taps with largest magnitude.

The final common PHY element is the general procedure for payload formation. Fig. 9.52 shows the common PHY formatting procedure through scrambling, encoding, modulation, and spreading. Note that the scrambler shift register realization is also identical for all PHYs, as represented by the polynomial $x^7 + x^4 + 1$ with transmitter-defined seed.

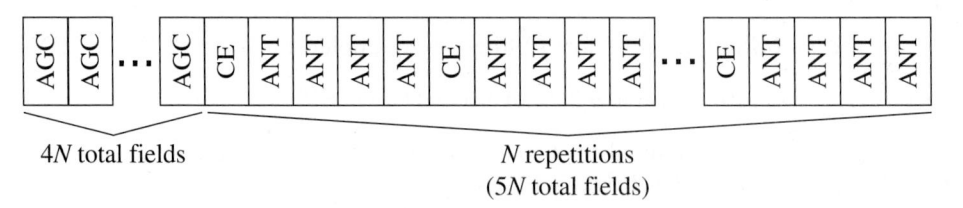

Figure 9.51 General composition of antenna training field that is appended to PHY packets when antenna beam refinement is requested. The N repetitions consist of the N different antenna configurations that will be trained at either the transmitter *or* the receiver. The AGC field is used to set the AGC for each antenna configuration and is equal to $[1_5 \otimes \mathbf{Ga}_{64}]$ for the SC-PHY and OFDM and $[1_5 \otimes \mathbf{Gb}_{64}]$ for the CPHY, both modulated with $\pi/2$-BPSK. The CE sequence is the same as the CE sequence in the preamble (which is also PHY dependent) with $\pi/2$-BPSK modulation. Finally, the antenna configuration training sequence (ANT) is equal to the sequence $[\mathbf{Ga}_{128}, -\mathbf{Gb}_{128}, \mathbf{Ga}_{128}, \mathbf{Gb}_{128}, \mathbf{Ga}_{128}]$ with $\pi/2$-BPSK modulation for all PHYs.

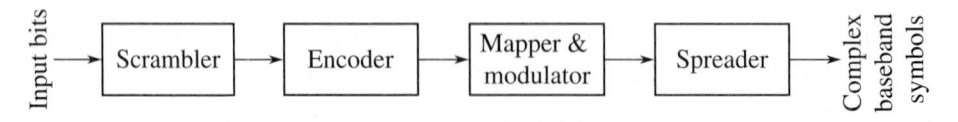

Figure 9.52 Frame formatting procedure common to all PHYs.

9.6.3.2 Control PHY (CPHY) in IEEE 802.11ad

The preamble elements in the CPHY are shown in Table 9.25. Both the header and payload data are each formatted in the same manner within the control PHY. After scrambling, the binary data are encoded using the shortened LDPC(672,504), where shortening reduces the effective code rate to 1/2. The encoded data use single-carrier differential binary shift keying (DBPSK) where, assuming encoded bit sequence $\{c_0, c_1, c_2 \ldots\}$, the resulting DBSPK complex baseband sequence becomes $\{d_0, d_1, d_2, \ldots\} = \{2c_0 - 1, (2c_1 - 1) \times (2c_0 - 1), (2c_2 - 1) \times (2c_1 - 1) \times (2c_0 - 1), \ldots\}$. Following DBPSK modulation, the data are spread with the phase-rotated 32-length Golay sequence $\mathbf{a}_{32}$ resulting in $\{d_0, d_1, d_2, \ldots\} \otimes (\mathbf{a}_{32} \odot (\mathbf{1}_8 \otimes [+1, +j, -1, -j]))$. The symbol rate of control PHY data are 1,760 MHz.

9.6.3.3 SC-PHY in IEEE 802.11ad

The preamble elements in the SC-PHY are shown in Table 9.26. All SC-PHY MCS, that is, MCS 1-12, follow the same general formatting procedure for the header and payload data. First, the 64-bit header data are zero-padded to create a 504-bit block to be encoded through LDPC(672,504). One hundred sixty of the 168 parity bits from this encoding are concatenated with the scrambled header bits in two different ways to form two 224-bit sequences $\mathbf{h}_1$ and $\mathbf{h}_2$. The concatenation of $\{\mathbf{h}_1, \mathbf{h}_2\}$ (448 bits) is $\pi/2$-BPSK mapped and guard symbols are pre-pended in the same format as the payload. A second, identical, but polarity-inverted SC-block (data symbols are inverted, not guard symbols) is attached to achieve a spreading factor of 2.

Table 9.25 Breakdown of each field in CPHY preamble in IEEE 802.11ad. The descriptions represent the complex baseband representation of single-carrier wireless signals.

Item	Description
STF	$[([\mathbf{1}_{48}, -1] \otimes \mathbf{Gb}_{128}), -\mathbf{Ga}_{128}] \odot (\mathbf{1}_{1600} \otimes [+1, +j, -1, -j])$
CE	$[\mathbf{Gu}_{512}, \mathbf{Gv}_{512}, -\mathbf{Gb}_{128}] \odot (\mathbf{1}_{288} \otimes [+1, +j, -1, -j])$
$\mathbf{Gu}_{512}$	$[-\mathbf{Gb}_{128}, -\mathbf{Ga}_{128}, \mathbf{Gb}_{128}, -\mathbf{Ga}_{128}]$
$\mathbf{Gv}_{512}$	$[-\mathbf{Gb}_{128}, \mathbf{Ga}_{128}, -\mathbf{Gb}_{128}, -\mathbf{Ga}_{128}]$

Table 9.26 Breakdown of each field in SC-PHY preamble in IEEE 802.11ad. The descriptions represent the complex baseband representation of single-carrier wireless signals. Note that the CE field is the same as the CE field of the CPHY.

Item	Description
STF	$([\mathbf{1}_{16}, -1] \otimes \mathbf{Ga}_{128}) \odot (\mathbf{1}_{544} \otimes [+1, +j, -1, -j])$
CE	$[\mathbf{Gu}_{512}, \mathbf{Gv}_{512}, -\mathbf{Gb}_{128}] \odot (\mathbf{1}_{288} \otimes [+1, +j, -1, -j])$

The payload data formatting procedure is straightforward. After scrambling, the payload data are LDPC encoded to produce coded-bit block size of 672. For MCS 1, where $L_{SF} = 2$, the information bits are repeated (not the parity bits) to reduce the effective code rate and provide 2× processing gain at the receiver. The encoded bits are then modulated with the $\pi/2$-rotated constellations in Table 9.23, which are the same as described in the SC-PHY of IEEE 802.15.3c. The resultant complex baseband sequence is segmented into blocks of size 448 and the $\mathbf{Ga}_{64}$ Golay sequence is pre-pended. These Golay guard sequences serve as pilots and also enable frequency domain equalization. Note that the SC-PHY in IEEE 802.15.3c also features Golay sequences for pilots. The symbol rate of the SC-PHY data is 1,760 MHz.

9.6.3.4 OFDM PHY in IEEE 802.11ad

The preamble elements in the OFDM PHY are shown in Table 9.27. The header data, which consist of 64 bits, are formatted through a straightforward procedure. First, the header data are zero-padded to create a 504-bit block to be encoded through LDPC(676,504). One hundred sixty of the 168 parity bits from this encoding are concatenated with the scrambled header bits in three different ways to form three 224-bit sequences $\mathbf{h}_1$, $\mathbf{h}_2$, and $\mathbf{h}_3$. The concatenation of $\{\mathbf{h}_1, \mathbf{h}_2, \mathbf{h}_3\}$ is QPSK mapped and OFDM modulated with pilots through the same procedure as the data payload.

Like all IEEE 802.11ad MCS (except the LP SC-PHY), the payload data are first scrambled and then LDPC encoded. Next, the data are mapped to the MCS-specific constellation from Table 9.23. OFDM formatting in IEEE 802.11ad assumes the parameters in Table 9.28. If we assume that the index 0 refers to the DC sub-carrier, the pilot sub-carrier will be inserted on sub-carriers indices $\{-150, -130, -110, -90, -70, -50, -30, -10, 10, 30, 50, 70, 90, 110, 130, 150\}$ such that all pilots are spaced 20 tones apart. The pilot sequence for these sub-carriers is $[-1, 1, -1, 1, 1, -1, -1, -1, -1, -1, 1, 1, 1, -1, 1, 1]$. This pilot sequence is also pseudo-randomly polarized for each OFDM symbol where the polarity is generated by the scrambler shift registers with the all-one seed.

Table 9.27 Breakdown of OFDM PHY preamble components in IEEE 802.11ad. The descriptions represent the complex-baseband representation of single-carrier wireless signals. Note that the STF field is the same as the STF field of the SC-PHY. The filter coefficients are used to effectively up-sample the STF and CE to the higher sample rate of OFDM, which is 1.5× higher than SC. First, the STF and CE are up-sampled by factor 3. Then the up-sampled preamble is convolved with the filter. Finally, the filtered signal is down-sampled by factor 2 to produce the OFDM preamble.

Item	Description
STF	$([\mathbf{1}_{16}, -1] \otimes \mathbf{Ga}_{128}) \odot (\mathbf{1}_{544} \otimes [+1, +j, -1, -j])$
CE	$[\mathbf{Gv}_{512}, \mathbf{Gu}_{512}, -\mathbf{Gb}_{512}] \odot (\mathbf{1}_{288} \otimes [+1, +j, -1, -j])$
Filter	[1,0,1,1,-2,-3,0,5,5,-3,-9,-4,10,14,-1,-20,-16,14,33,9,-35,-42,11,64,40,-50, -96,-15,120,126,-62,-256,-148,360,985,1267,985,360,-148,-256,-62,126, 120,-15,96,-50,40,64,11,-42,-35,9,33,14,-16,-20,-1,14,10,-4,-9,-3,5,5,0,-3, -2,1,1,0,-1] $\times \sqrt{12}/4047$

9.6 IEEE 802.11ad

Table 9.28 OFDM formatting parameters in IEEE 802.11ad.

OFDM Parameter	Value
Symbol Rate	2,640 MHz
Total Sub-carriers	512
DC Sub-carriers	3
Pilot Sub-carriers	16
Data Sub-carriers	336
Guard (Edge) Sub-carriers	157
Cyclic Prefix Size	128

The header formatting in OFDM and all data formatting in non-OFDM MCS in IEEE 802.11ad use standard formatting procedures; however, the payload formatting in the OFDM PHY of IEEE 802.11ad has several novel features in the constellation mapping. Spread QPSK (SQPSK) in MCS 13-14 is a constellation mapping procedure in which spreading is accomplished by repeating conjugate versions of symbols. Consider the encoded bit sequence to be mapped onto a single OFDM symbol, $\{c_0, c_1, c_2, c_3, \ldots\}$. With SQPSK, each pair of bits is mapped to a single constellation point, resulting in the sequence $\{d_0, d_1, \ldots\} = \{((2c_0 - 1) + j(2c_1 - 1))/\sqrt{2}, ((2c_2 - 1) + j(2c_3 - 1))/\sqrt{2}, \ldots\}$ as with standard QPSK. However, this only yields half of the data sub-carriers. The remaining sub-carriers are generated by conjugating to produce the remaining SQPSK symbols, resulting in the final SQPSK sequence to be mapped onto sub-carrier pairs $\{\tilde{d}_0, \tilde{d}_1, \tilde{d}_2, \tilde{d}_3, \ldots\} = \{d_0, d_0^*, d_1, d_1^*, \ldots\}$.

For MCS 15-17, spreading of QPSK symbols is not used, although the QPSK mapping is still non-standard. First, the encoded bit sequence $\{c_0, c_1, c_2, c_3, \ldots\}$, for a single OFDM symbol, is mapped to QPSK constellation points to produce the complex-baseband sequence $\{d_0, d_1, \ldots\}$ where $\{d_0, d_1\} = \{((2c_0 - 1) + j(2c_2 - 1))/\sqrt{2}, ((2c_1 - 1) + j(2c_3 - 1))/\sqrt{2}\}$. Second, every QPSK symbol pair undergoes a matrix transformation with matrix

$$\mathbf{T} = \begin{bmatrix} 1/\sqrt{5} & 2/\sqrt{5} \\ -2/\sqrt{5} & 1/\sqrt{5} \end{bmatrix} \quad (9.2)$$

such that the resultant QPSK sequence (for sub-carrier pair mapping) becomes equal to $\{[\mathbf{T}[d_0, d_1]^T]^T, [\mathbf{T}[d_2, d_3]^T]^T, \ldots\}$. The constellation of the QPSK sequence after this transformation appears the same as the 16-QAM constellation. Essentially, rather than have two separate QPSK constellation points on two distinct sub-carriers, we have two 16-QAM constellation points that are shared over two sub-carriers. This sharing is sometimes called dual carrier modulation (DCM). It allows the receiver to capture frequency diversity across QPSK symbols without traditional spreading.

SQPSK and QPSK constellations produce correlation in each (S)QPSK complex symbol pair. To maximize performance in frequency selective channels, the pairs are separated and mapped to distant sub-carriers through *static tone pairing* (STP) and *dynamic tone pairing* (DTP). Since each OFDM symbol consists of 336 data sub-carriers, STP maximizes the minimum pair distance by separating each pair by $336/2 = 168$ symbols. One undesired effect of STP is that the predefined static mapping may not be suited to the current sub-carrier channel quality profile (e.g., if both elements in a pair are mapped to frequency nulls). Given more information about the sub-carrier channel profile (through, e.g., the channel impulse response), the transmitter may be able to make better decisions about where to map sub-carrier pairs in an attempt to balance the channel quality among pairs. DTP, which is optionally implemented, adds this capability by dividing the 336 (S)QPSK symbols per OFDM symbol into 42 groups, each of which consists of 4 (S)QPSK symbol pairs. It is then at the vendor's discretion to determine the sub-carrier offset of each pair within each group. STP and DTP operations are demonstrated in Figs. 9.53 and 9.54, respectively. These operations directly follow the (S)QPSK mapping in a single

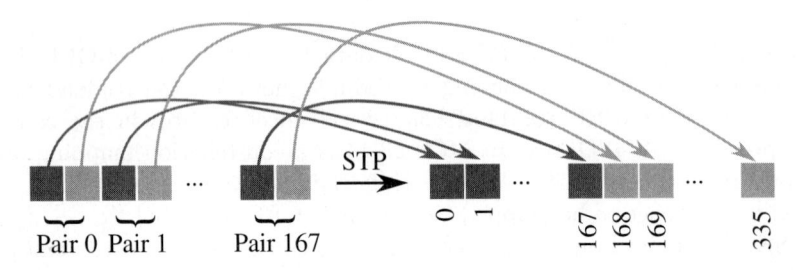

Figure 9.53 Static tone pairing (STP) in MCS 13-17. Even and odd sub-carriers are paired (out of 336 sub-carriers total) and mapped to maximize the minimum sub-carrier distance between even and odd sub-carriers (168 sub-carrier distance).

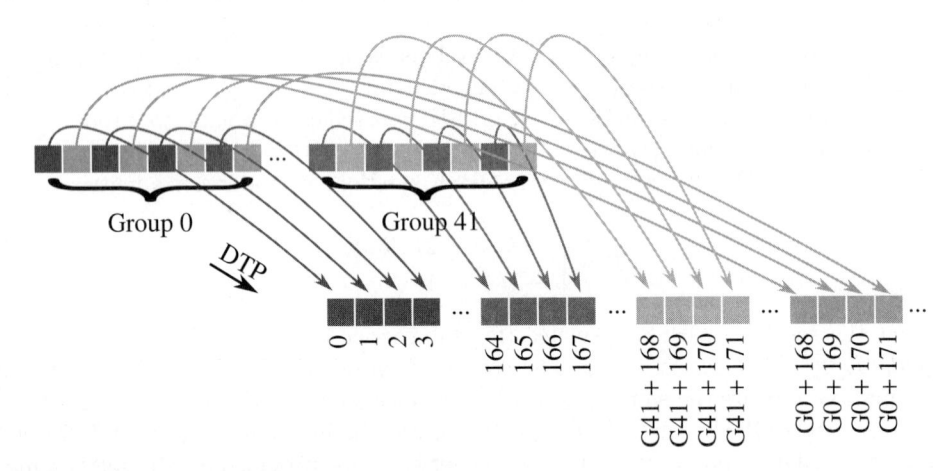

Figure 9.54 Dynamic tone pairing (DTP) in MCS 13-17. For DTP, the *group pair index* (GPI) is given to the PHY and is defined by the transmitter. The mapping GPI is hence a permutation where GPI : $\{0, 1, \ldots, 41\} \to \{0, 1, \ldots, 41\}$ and GPI : $k \mapsto G_k$. Note that although the even elements of each group have a fixed mapping, the odd elements may be mapped more generally. In other words, G_k may vary for a fixed k, depending on the link configuration. The ends of the DTP transformed sub-carriers are not shown to maintain generality, although one of the 42 groups must be mapped to the last DTP group in practice.

OFDM symbol (operations act on $\{d_0, d_0^*, \ldots\}$ for SQPSK and $\{[\mathbf{T}[d_0, d_1]^T]^T, \ldots\}$ for QPSK). STP must be implemented by OFDM PHY vendors, whereas DTP is optional. Vendors are likely to design a finite set of fixed mappings for DTP for a finite set of channel realizations.

The 16-QAM and 64-QAM constellations of MCS 18-21 and MCS 22-24, respectively, also follow non-standard mapping operations, although STP and DTP are not implemented because correlation does not exist across sub-carriers after constellation mapping. Notice that, until now, interleaving has not been integrated into symbol formatting. With SQPSK and QPSK, the maximum number of coded bits per OFDM symbol is 336 and 672, respectively, which is always an integer factor of the block length of the LDPC code. Hence, because block codes are not sensitive to burst errors, interleaving within a single OFDM symbol does not affect performance. With 16-QAM and 64-QAM, however, the numbers of coded bits per symbol are 1,344 and 2,016, respectively. Therefore interleaving in MCS 18-24, which is performed during the constellation mapping, will help to randomize the effective channel quality for each bit through all code blocks.

16-QAM incorporates interleaving by alternating between adjacent code blocks for a single OFDM symbol. For example, if we consider the encoded bit sequence $\{c_0, c_1, c_2, \ldots\}$, then the first 16-QAM symbol in the first OFDM symbol will map from bits $\{c_0, c_1, c_2, c_3\}$, the second 16-QAM symbol in the first OFDM symbol will map from bits $\{c_{672}, c_{673}, c_{674}, c_{675}\}$, the third 16-QAM symbol in the first OFDM symbol will map from bits $\{c_4, c_5, c_6, c_7\}$, the fourth 16-QAM symbol in the first OFDM symbol will map from bits $\{c_{676}, c_{677}, c_{678}, c_{679}\}$, and so on.

Whereas 16-QAM needs to interleave over 2 LDPC code blocks, 64-QAM must interleave over 3 LDPC code blocks. Hence, 64-QAM creates symbols by alternating over the three code blocks in a round-robin fashion. For example, the first 64-QAM symbol in the first OFDM symbol will map from bits $\{c_0, c_1, c_2, c_3, c_4, c_5\}$, the second 64-QAM symbol in the first OFDM symbol will map from bits $\{c_{672}, c_{673}, c_{674}, c_{675}, c_{676}, c_{677}\}$, the third 64-QAM symbol in the first OFDM symbol will map from bits $\{c_{1344}, c_{1345}, c_{1346}, c_{1347}, c_{1348}, c_{1349}\}$, the fourth 64-QAM symbol in the first OFDM symbol will map from bits $\{c_6, c_7, c_8, c_9, c_{10}, c_{11}\}$, and so on.[12]

9.6.3.5 Low-Power (LP) SC-PHY in IEEE 802.11ad

The preamble elements in the LP SC-PHY are shown in Table 9.29. As with the SC-PHY, the LP SC-PHY MCS, that is, MCS 25-31, follow the same general formatting procedure for the header and payload data. After scrambling, header data are outer encoded with a shortened RS(24,8) code and inner encoded with an octet block code (OBC(16,8)) to

---

12. Note that interleaving is not considered outside of a single OFDM symbol. This makes IEEE 802.11ad OFDM waveforms sensitive to impulsive noise, impulsive interference, or time-selective fading (within a packet). Given the sampling rate of 2,640 MHz, however, these phenomena are not expected to be significantly observed. Further, the complexity increase of total packet interleaving is substantial especially in terms of memory.

Table 9.29 Breakdown of each field in low-power SC-PHY preamble in IEEE 802.11ad. The descriptions represent the complex baseband representation of single-carrier wireless signals. Note that the STF and CE field are exactly the same as the STF and CE field in the SC-PHY.

Item	Description
STF	$(\mathbf{1}_{15}, -1] \otimes \mathbf{Ga}_{128}) \odot (\mathbf{1}_{512} \otimes [+1, +j, -1, -j])$
CE	$[-\mathbf{Gb}_{128}, -\mathbf{Ga}_{128}, \mathbf{Gb}_{128}, -\mathbf{Ga}_{128}, -\mathbf{Gb}_{128}, \mathbf{Ga}_{128}, -\mathbf{Gb}_{128},$ $-\mathbf{Ga}_{128}, -\mathbf{Gb}_{128}] \odot (\mathbf{1}_{288} \otimes [+1, +j, -1, -j])$

produce an effective coding rate of $1/6$.[13] After the OBC, the header data are block-interleaved (7 rows written, 8 columns read) and modulated with $\pi/2$-BPSK. Finally, the header is segmented into blocks of size 512 with guard intervals in the same manner as the payload and then $\pi/2$-BPSK modulated.

The payload data are formatted similarly, although the coding rates and modulation order differ according to the MCS configurations in Table 9.23 and all MCSs use the RS(224,208) inner code. The encoded and modulated data are formatted into blocks of length 512, where each block contains 8 sub-blocks of length 64. The first sub-block is $\mathbf{Ga}_{64}$. The remaining sub-blocks consist each of data blocks of length 56 with a cyclic postfix of length 8 (which is the last 8 symbols of $\mathbf{Ga}_{64}$). Hence, each 512-length block consists of 392 data symbols. The last block of data in a packet will be followed by $\mathbf{Ga}_{64}$, also.[14] The symbol rate of low-power SC-PHY data is 1,760 MHz.

9.7 WiGig

Like WirelessHD, WiGig was created through a private industry consortium that included Advanced Micro Devices, Inc., Atheros Communications Inc., Broadcom Corporation, Cisco Systems, Inc., Dell Inc., Intel Corporation, Marvell Technology Group Ltd., Mediatek Inc., Microsoft Corporation, NEC Corporation, Nokia Corporation, Nvidia Corporation, Panasonic Corporation, Samsung Electronics Co. Ltd., Toshiba Corporation, and Wilocity Ltd. WiGig and WiFi have consolidated technology and certification development, such that WiGig is now completely part of the activities of the WiFi Alliance. Despite being essentially the same as IEEE 802.11ad, WiGig includes additional support for specific protocol adaptation layers including HDMI, DisplayPort, USB, and PCIe.

---

13. Octet block codes sequentially process each octet block of binary data. An OBC($N, 8$) produces $N \geq 8$ bits for every octet input. For example, an OBC($9, 8$) is essentially a block code that appends a parity bit onto each octet input. In the standard, this example is referred to as single parity check (SPC) coding.

14. Note that the received data may be equalized with a block of length 512 or length 64, depending on the available complexity at the receiver (performance will be significantly better in more multipath with the larger block size).

9.8 Chapter Summary

In this chapter, we first previewed the international spectrum regulations for 60 GHz unlicensed communication. Then, we discussed the emerging international standards that exploit 60 GHz unlicensed spectrum with WPAN topologies (ECMA-387, IEEE 802.15.3c), video streaming network topologies (WirelessHD), and WLAN topologies (IEEE 802.11ad and WiGig). These state-of-the-art wireless standards feature a host of cutting-edge technologies to enhance efficiency, for example, through single-carrier block transmissions, on-off-keying, and/or unequal error protection, and they also combat excessive path loss at 60 GHz, for example, through relays, high-dimensional phased arrays, and multi-band coordination.

Due to the inherent nature of mmWave frequencies and the technical aspects treated throughout this text, many emerging or future mmWave wireless products and standards (such as 5G mmWave cellular, intervehicular communications, and backhaul/fronthaul communications standards) are likely to share characteristics with the 60 GHz WPAN/WLAN standards discussed in this chapter. Our hope is that this chapter — and the entire text — will be a useful guide, providing the basic principles, concepts, and tools that will empower generations of engineers to bring new mmWave products and services to the wireless communications field.

Bibliography

[3GPP03] V6.1.0, 3GPP TR 25.996, "Spatial Channel Model for Multipath Input Multiple Output (MIMO) Simulations," Sep. 2003.

[3GPP09] 3GPP TS 36.201, "Evolved Universal Terrestrial Radio Access (E-UTRA); Physical Channels and Modulation (Release 8)," 2009 (R9-2010, R10-2010, R11-2012).

[802.11ad-10] IEEE, "802.11ad Draft 0.1." IEEE, Jun. 2010, http://www.ieee802 .org/11/Reports/tgad update.htm.

[802.11ad-11] IEEE, P802.11ad/D5, "IEEE Draft Standard for Local and Metropolitan Area Networks - Specific Requirements - Part 11: Wireless LAN Medium Access Control (MAC) and Physical Layer (PHY) Specifications - Amendment 3: Enhancements for Very High Throughput in the 60 GHz Band," IEEE, Dec. 2011.

[802.11-12] IEEE 802.11, "Wireless LAN Medium Access Control (MAC) and Physical Layer (PHY) Specifications," IEEE, Apr. 2012. doi:10.1109/IEEESTD.2012. 6178212.

[802.11-12a] "802.11-2012 - IEEE Standard for Information Technology - Telecommunications and Information Exchange between Systems Local and Metropolitan Area Networks - Specific Requirements Part 11: Wireless LAN Medium Access Control (MAC) and Physical Layer (PHY) Specifications," IEEE, Dec. 2012.

[802.11-12-VHT] "IEEE Standard for Information Technology - Telecommunications and Information Exchange between Systems - Local and Metropolitan Area Networks - Specific Requirements. Part 11: Wireless LAN Medium Access Control (MAC) and Physical Layer (PHY) Specifications. Amendment 3: Enhancements for Very High Throughput in the 60 GHz Band," IEEE, Dec. 2012.

[802.15.3-03] "802.15.3-2003 - IEEE Standard for Information Technology - Telecommunications and Information Exchange between Systems - Local and Metropolitan Area Networks - Specific Requirements Part 15.3: Wireless Medium Access Control (MAC) and Physical Layer (PHY) Specifications for High Rate Wireless Personal Area Networks (WPANs)," IEEE, 2003.

[802.15.3-06] "IEEE Standard for Information Technology - Telecommunications and Information Exchange between Systems - Local and Metropolitan Area Networks - Specific Requirements Part 15.3: Wireless Medium Access Control (MAC) and Physical Layer (PHY) Specifications for High Rate Wireless Personal Area Networks (WPANs). Amendment 1: MAC Sublayer," IEEE, 2006.

[802.15.3-09] "IEEE Standard for Information Technology - Telecommunications and Information Exchange between Systems - Local and Metropolitan Area Networks - Specific Requirements. Part 15.3: Wireless Medium Access Control (MAC) and Physical Layer (PHY) Specifications for High Rate Wireless Personal Area Networks (WPANs). Amendment 2: Millimeter-wave-based Alternative Physical Layer Extension," IEEE, Dec. 2009.

[802.16-01] "IEEE Standard for Local and Metropolitan Area Networks – Part 16: Air Interface for Fixed Broadband Wireless Access Systems," IEEE, 2001.

[802.16-09] IEEE 802.16 Task group j, "IEEE 802.16j-2009: Standard for Local and Metropolitan Area Networks Part 16: Air Interface for Broadband Wireless Access Systems Amendment 1: Multiple Relay Specification," IEEE, Jun. 2009.

[AAS86] J. B. Anderson, T. Aulin, and C.-E. Sundberg, *Digital Phase Modulation*. New York, Springer, 1986.

[AAS$^+$14] Ansuman Adhikary, E. Al Safadi, M. K. Samimi, R. Wang, G. Caire, T. S. Rappaport, A. F. Molisch, "Joint Spatial Division and Multiplexing for mm-Wave Channels," *IEEE Journal on Selected Areas in Communications*, vol. 32, no. 6, pp. 1239–1255, Jun. 2014.

[AB83] H. Arnold and W. Bodtmann, "The Performance of FSK in Frequency-Selective Rayleigh Fading," *IEEE Transactions on Communications*, vol. 31, no. 4, pp. 568–572, 1983.

[ABKG11] J. G. Andrews, F. Baccelli, and R. Krishna Ganti, "A Tractable Approach to Coverage and Rate in Cellular Networks," *IEEE Transactions on Communications*, vol. 59, no. 11, pp. 3122–3134, Nov. 2011.

[ACD$^+$12] J. Andrews, H. Claussen, M. Dohler, S. Rangan, and M. Reed, "Femtocells: Past, Present, and Future," *IEEE Journal on Selected Areas in Communications*, vol. 30, no. 3, pp. 497–508, Apr. 2012.

[ACV06] D. Alldred, B. Cousins, and S. Voinigescu, "A 1.2V, 60 GHz Radio Receiver with On-Chip Transformers and Inductors in 90-nm CMOS," *2006 IEEE Compound Semiconductor Integrated Circuit Symposium*, IEEE, pp. 51–54, Nov. 2006.

[AEAH12] S. Akoum, O. El Ayach, and R. W. Heath Jr., "Coverage and Capacity in mmWave Cellular Systems," *2012 Asilomar Conference on Signals, Systems and Computers*, pp. 688–692, Nov. 2012.

[AELH13] A. Alkhateeb, O. El Ayach, G. Leus, and R. W. Heath Jr., "Hybrid Precoding for Millimeter Wave Cellular Systems with Partial Channel Knowledge," *2013 IEEE Information Theory and Application Workshop (ITA)*, IEEE, Feb. 2013.

[AELH14] A. Alkhateeb, O. El Ayach, G. Leus, and R. W. Heath Jr., "Channel Estimation and Hybrid Precoding for Millimeter Wave Cellular Systems," 2014, to appear in the *IEEE Journal of Selected Topics in Signal Processing*, 2014.

[AF04] N. Alic and Y. Fainman, "Data-Dependent Phase Coding for Suppression of Ghost Pulses in Optical Fibers," *IEEE Photonics Technology Letters*, vol. 16, no. 4, pp. 1212–1214, 2004.

[AH91] G. Allen and A. Hammoudeh, "Outdoor Narrow Band Characterisation of Millimetre Wave Mobile Radio Signals," *1991 IEEE Colloquium on Radiocommunications in the Range 30-60 GHz*, pp. 4/1–4/7, Jan. 1991.

[AJA+14] A. Agah, J. A. Jayamon, P. M. Asbeck, L. E. Larson, and J. F. Buckwalter, "Multi-Drive Stacked-FET Power Amplifiers at 90 GHz in 45 nm SOI CMOS," *IEEE Journal of Solid-State Circuits*, vol. 49, no. 5, pp. 1148–1157, 2014, digital object identifier: 10.1109/JSSC.2014.2308292.

[Aka74] H. Akaike, "A New Look at the Statistical Model Identification," *IEEE Transactions on Automatic Control*, vol. AC-19, pp. 783–795, Dec. 1974.

[AKBP08] S. Aloui, E. Kerherve, D. Belot, and R. Plana, "A 60GHz, 13dBm Fully Integrated 65nm RF-CMOS Power Amplifier," *2008 IEEE International Northeast Workshop on Circuits and Systems and TAISA Conference (NEWCAS-TAISA 2008)*, pp. 237–240, Jun. 2008.

[AKD+07] J. Alvarado, K. Kornegay, D. Dawn, S. Pinel, and J. Laskar, "60 GHz LNA Using a Hybrid Transmission Line and Conductive Path to Ground Technique in Silicon," *2007 IEEE Radio Frequency Integrated Circuits Symposium (RFIC 2007)*, IEEE, pp. 685–688, Jun. 2007.

[AKR83] N. G. Alexopolulos, P. B. Katehi, and D. B. Rutledge, "Substrate Optimization for Integrated Circuit Antennas," *IEEE Transactions on Microwave Theory and Techniques*, vol. 31, no. 7, pp. 550–557, Jul. 1983.

[AKS11] A. Ahmed, R. Koetter, and N. Shanbhag, "VLSI Architectures for Soft-Decision Decoding of Reed-Solomon Codes," *IEEE Transactions on Information Theory*, vol. 57, no. 2, pp. 648–667, Feb. 2011.

[AL05] A. T. Attar and T. H. Lee, "Monolithic Integrated Millimeter-Wave IMPATT Transmitter in Standard CMOS Technology," *IEEE Transactions on Microwave Theory and Techniques*, vol. 53, no. 11, pp. 3557–3561, 2005.

[Ala98] S. M. Alamouti, "A Simple Transmit Diversity Technique for Wireless Communications," *IEEE Journal on Selected Areas in Communications*, vol. 16, no. 8, pp. 1451–1458, Oct. 1998.

[ALK+09] P. Asbeck, L. Larson, D. Kimball, S. Pornpromlikit, J.-H. Jeong, C. Presti, T. Hung, F. Wang, and Y. Zhao, "Design Options for High Efficiency Linear Handset Power Amplifiers," *2009 IEEE Meeting on Silicon Monolithic Integrated Circuits in RF Systems (SiRF 2009)*, pp. 1–4, Jan. 2009.

[ALRE13] M. Akdeniz, Y. Liu, S. Rangan, and E. Erkip, "Millimeter Wave Picocellular System Evaluation for Urban Deployments," 2013. http://arxiv.org/abs/1304.3963.

[ALS+14] M. R. Akdeniz, Y. Liu, M. K. Samimi, S. Sun, S. Rangan, T. S. Rappaport, and E. Erkip, "Millimeter Wave Channel Modeling and Cellular Capacity Evaluation," *IEEE Journal on Selected Areas in Communications*, vol. 32, no. 6, pp. 1164–1179, Jun. 2014.

[AMM13] A. Artemenko, A. Maltsev, A. Mozharovskiy, R. Maslennikov, A. Sevastyanov, and V. Ssorin, "Millimeter-wave Electronically Steerable Integrated Lens Antennas for WLAN/WPAN Applications," *IEEE Transactions on Antennas and Propagation*, vol. 61, no. 4, Apr. 2013.

[ANM00] G. E. Athanasiadou, A. R. Nix, and J. P. McGeehan, "A Microcellular Ray-tracing Propagation Model and Evaluation of its Narrow-band and Wide-band Predictions," *IEEE Journal on Selected Areas in Communications*, vol. 18.3, pp. 322–335, 2000.

[AR+02] C. R. Anderson and T. S. Rappaport, "In-Building Wideband Multipath Characteristics at 2.5 and 60 GHz," *2002 IEEE Vehicular Technology Conference (VTC 2002-Fall)*, vol. 1, pp. 97–101, Sep. 2002.

[AR04] C. R. Anderson and T. S. Rappaport, "In-Building Wideband Partition Loss Measurements at 2.5 and 60 GHz," *IEEE Transactions on Wireless Communications*, vol. 3, no. 3, pp. 922–928, 2004.

[AR08] R. A. Alhalabi and G. M. Rebeiz, "High-Efficiency Angled-Dipole Antennas for Millimeter-Wave Phased Array Applications," *IEEE Transactions on Antennas and Propagation*, vol. 56, no. 10, pp. 3136–3142, Oct. 2008.

[AR10] R. A. Alhalabi and G. M. Rebeiz, "Differentially-Fed Millimeter-Wave Yagi-Uda Antenna with Folded Dipole Feed," *IEEE Transactions on Antennas and Propagation*, vol. 58, no. 3, pp. 966–969, Mar. 2010.

[ARI00] "Millimeter-Wave Video Transmission Equipment for Specified Low Power Radio Station," Association of Radio Industries and Businesses (ARIB), Dec. 2000. http://www.arib.or.jp/english/html/overview/doc/1-STD-T69v3_0.pdf.

[ARI01] "Millimeter-Wave Data Transmission Equipment for Specified Low Power Radio Station (Ultra High Speed Wireless LAN System)," Association of Radio Industries and Businesses (ARIB), May 2001. http://www.arib.or.jp/english/html/overview/doc/1-STD-T74v1_1.pdf.

[ARS06] S. M. Ali, R. Raut, and M. Sawan, "Digital Encoders for High Speed Flash-ADCs: Modeling and Comparison," *2006 IEEE North-East Workshop on Circuits and Systems*, pp. 69–72, Jun. 2006.

[ARY95] J. B. Andersen, T. S. Rappaport, S. Yoshida, "Propagation Measurements and Models for Wireless Communications Channels," *IEEE Communications Magazine*, vol. 33, no. 1, pp. 42–49, 1995.

[ATB+98] V. H. W. Allen, L. E. Taylor, L. W. Barclay, B. Honary, and M. J. Lazarus, "Practical Propagation Measurements for Indoor LANs Operating at 60 GHz," *1998 IEEE Global Telecommunications Conference (GLOBECOM 1998)*, vol. 2, pp. 898–903, 1998.

[AWW+13] Y. Azar, G. N. Wong, K. Wang, R. Mayzus, J. K. Schulz, H. Zhao, F. Gutierrez, D. Hwang, and T. S. Rappaport, "28 GHz Propagation Measurements for Outdoor Cellular Communications Using Steerable Beam Antennas in New York City," *2013 IEEE International Conference on Communications (ICC)*, pp. 5143–5147, Jun. 2013.

[Bab08] A. Babakhani, "Direct Antenna Modulation (DAM) for On-Chip Millimeterwave Transceivers," Ph.D. dissertation, California Institute of Technology, 2008.

[BAFS08] P. V. Bijumon, Y. Antar, A. P. Freundorfer, and M. Sayer, "Dielectric Resonator Antenna on Silicon Substrate for System On-Chip Applications," *IEEE Transactions on Antennas and Propagation*, vol. 56, no. 11, pp. 3404–3410, Nov. 2008.

[BAH14] T. Bai, A. Alkhateeb, and R. W. Heath Jr., "Coverage and Capacity of Millimeter Wave Cellular Networks," to appear in *IEEE Communications Magazine*, 2014.

[BAHT11] J. Baliga, R. Ayre, K. Hinton, and R. Tucker, "Green Cloud Computing: Balancing Energy in Processing, Storage, and Transport," *Proceedings of the IEEE*, vol. 99, no. 1, pp. 149–167, Jan. 2011.

[Bak09] R. J. Baker, *CMOS: Mixed-Signal Circuit Design*, 2nd ed. New York, Wiley-IEEE Press, 2009.

[Bal89] C. A. Balanis, *Advanced Engineering Electromagnetics*, 1st ed. New York, Wiley Press, 1989.

[Bal05] C. A. Balanis, *Antenna Theory: Analysis and Design*, 3rd ed. New York, Wiley-Interscience, 2005.

[BAN13] D. Berraki, S. Armour, and A. R. Nix, "Exploiting Unique Polarisation Characteristics at 60 GHz for Fast Beamforming High Throughput MIMO WLAN Systems," *2013 IEEE International Symposium on Personal, Indoor and Mobile Radio Communications (PIMRC 2013)*, pp. 233–237, Sep. 2013.

[BAN14] D. Berraki, S. Armour, and A. R. Nix, "Application of Compressive Sensing in Sparse Spatial Channel Recovery for Beamforming in MmWave Outdoor Systems," *2014 IEEE Wireless Communications and Networking Conference (WCNC)*, Istanbul, Turkey, Apr. 2014.

[BAQ91] G. Bistue, I. Adin, and C. Quemada, *Transmission Line Design Handbook*, 1st ed. Norwood, MA, Artech House, 1991.

[Bar07] R. Baraniuk, "Compressive Sensing [Lecture Notes]," *IEEE Signal Processing Magazine*, vol. 24, no. 4, pp. 118–121, 2007.

[BAS+10] T. Baykas, X. An, C.-S. Sum, M. Rahman, J. Wang, Z. Lan, R. Funada, H. Harada, and S. Kato, "Investigation of Synchronization Frame Transmission in Multi-Gbps 60 GHz WPANs," *2010 IEEE Wireless Communications and Networking Conference (WCNC)*, pp. 1–6, Apr. 2010.

[Bay+] T. Baykas et al., "IEEE 802.15.3c: The First IEEE Wireless Standard for Data Rates over 1 Gb/s," *IEEE Communications Magazine*, vol. 49, no. 7, pp. 114–121, Jul. 2011.

[BB83] S. Benedetto and E. Biglieri, "Nonlinear Equalization of Digital Satellite Channels," *IEEE Journal on Selected Areas in Communications*, vol. 1, no. 1, pp. 57–62, 1983.

[BBKH06] J. F. Buckwalter, A. Babakhani, A. Komijani, and A. Hajimiri, "An Integrated Subharmonic Couples-Oscillator Scheme for a 60 GHz Phased-Array Transmitter," *IEEE Transactions on Microwave Theory and Techniques*, vol. 54, no. 12, pp. 4271–4280, Dec. 2006.

[BC98] G. Bottomley and S. Chennakeshu, "Unification of MLSE Receivers and Extension to Time-Varying Channels," *IEEE Transactions on Communications*, vol. 46, no. 4, pp. 464–472, Apr. 1998.

[BC99] E. Bahar and P. Crittenden, "Stationary Solutions for the Rough Surface Radar Backscatter Cross Sections Based on a Two Scale Full Wave Approach," *1999 IEEE Antennas and Propagation Society International Symposium*, vol. 1, pp. 510–513, Aug. 1999.

[BCJR74] L. Bahl, J. Cocke, F. Jelinek, and J. Raviv, "Optimal Decoding of Linear Codes for Minimizing Symbol Error Rate (Corresp.)," *IEEE Transactions on Information Theory*, vol. 20, no. 2, pp. 284–287, Mar. 1974.

[BDH14] Tianyang Bai, V. Desai, and R. W. Heath Jr., "Millimeter Wave Cellular Channel Models for System Evaluation," *International Conference on Computing, Networking and Communications (ICNC)*, Honolulu, HI, Feb. 3–6, 2014.

[BDRQL11] E. Ben-Dor, T. S. Rappaport, Y. Qiao, and S. J. Lauffenburger, "Millimeter-wave 60 GHz Outdoor and Vehicle AOA Propagation Measurement Using a Broadband Channel Sounder," *2011 IEEE Global Communications Conference (GLOBECOM 2011)*, Houston, TX, Dec. 2011.

[BDS+08] J. Borremans, M. Dehan, K. Scheir, M. Kuijk, and P. Wambacq, "VCO Design for 60 GHz Applications Using Differential Shielded Inductors in 0.13 μm CMOS," *2008 IEEE Radio Frequency Integrated Circuits Symposium (RFIC 2008) IEEE*, pp. 135–138, Apr. 2008.

[Bea78] C. Beare, "The Choice of the Desired Impulse Response in Combined Linear-Viterbi Algorithm Equalizers," *IEEE Transactions on Communications*, vol. 26, no. 8, pp. 1301–1307, 1978.

[Beh09] N. Behdad, "Single- and Dual-Polarized Miniaturized Slot Antennas and Their Applications in On-Chip Integrated Radios," *IEEE iWAT 2009 International Workshop on Antenna Technology*, pp. 1–4, Mar. 2009.

[Ber65] E. Berlekamp, "On Decoding Binary Bose-Chadhuri-Hocquenghem Codes," *IEEE Transactions on Information Theory*, vol. 11, no. 4, pp. 577–579, Oct. 1965.

[BFE+04] B. Bosco, S. Franson, R. Emrick, S. Rockwell, and J. Holmes, "A 60 GHz Transceiver with Multi-Gigabit Data Rate Capability," *2004 IEEE Radio and Wireless Conference*, pp. 135–138, 2004.

[BFR+92] K. Blackard, M. Feuerstein, T. Rappaport, S. Seidel, and H. Xia, "Path Loss and Delay Spread Models as Functions of Antenna Height for Microcellular System Design," *1992 IEEE Vehicular Technology Conference (VTC 1992)*, vol. 1, pp. 333–337, May 1992.

[BG92] D. P. Bertsekas and R. Gallager, *Data Networks*, 2nd ed. Englewood Cliffs, NJ, Prentice Hall, 1992.

[BGK+06a] A. Babakhani, X. Guan, A. Komijani, A. Natarajan, and A. Hajimiri, "A 77 GHz 4-Element Phased Array Receiver with On-Chip Dipole Antennas in Silicon," *2006 IEEE International Solid-State Circuits Conference (ISSCC)*, pp. 629–638, Feb. 2006.

[BGK+06b] A. Babakhani, X. Guan, A. Komijani, A. Natarajan, and A. Hajimiri, "A 77 GHz Phased-Array Transceiver with On-Chip Antennas in Silicon: Receiver and Antennas," *IEEE Journal of Solid-State Circuits*, vol. 41, no. 12, pp. 2795–2806, Dec. 2006.

[BH02] A. Blanksby and C. Howland, "A 690-mW 1-Gb/s 1024-b, Rate-1/2 Low-Density Parity-Check Code Decoder," *IEEE Journal of Solid-State Circuits*, vol. 37, no. 3, pp. 404–412, Mar. 2002.

[BH13a] T. Bai and R. W. Heath Jr., "Asymptotic Coverage Probability and Rate in Massive MIMO Networks," http://arxiv.org/abs/1305.2233.

[BH13b] T. Bai and R. W. Heath Jr., "Coverage Analysis for Millimeter Wave Cellular Networks with Blockage Effects," *2013 IEEE Global Signal and Information Processing Conference*, Pacific Grove, CA, Nov. 2013.

[BH13c] T. Bai and R. W. Heath Jr., "Coverage Analysis in Dense Millimeter Wave Cellular Networks," *2013 Asilomar Conference on Signals, Systems and Computers*, Pacific Grove, CA, Nov. 2013.

[BH14] T. Bai and R. W. Heath Jr., "Coverage and Rate Analysis for Millimeter Wave Cellular Networks," Submitted to *IEEE Transactions on Wireless Communications*, Mar. 2014. Available at ArXiv, arXiv:1402.6430 [cs.IT].

[BH14a] T. Bai and R. W. Heath Jr., "Analysis of Millimeter Wave Cellular Networks with Overlaid Microwave Base Stations" *2014 Asilomar Conference on Signals, Systems, and Computers*, Nov. 2014.

[BHSN10] W. Bajwa, J. Haupt, A. Sayeed, and R. Nowak, "Compressed Channel Sensing: A New Approach to Estimating Sparse Multipath Channels," *Proceedings of the IEEE*, vol. 98, no. 6, pp. 1058–1076, Jun. 2010.

[BHVF08] R. Bhagavatula, R. W. Heath Jr., S. Vishwanath, and A. Forenza, "Sizing up MIMO Arrays," *IEEE Vehicular Technology Magazine*, vol. 3, no. 4, pp. 31–38, Dec. 2008.

[BKKL08] J. Brinkhoff, K. Koh, K. Kang, and F. Lin, "Scalable Transmission Line and Inductor Models for CMOS Millimeter-Wave Design," *IEEE Transactions on Microwave Theory and Techniques*, vol. 56, no. 12, pp. 2954–2962, Dec. 2008.

[BKPL09] J. Brinkhoff, K. Kang, D.-D. Pham, and F. Lin, "A 60 GHz Transformer-Based Variable-Gain Power Amplifier in 90nm CMOS," *2009 IEEE International Symposium on Radio-Frequency Integration Technology (RFIT 2009)* pp. 60–63, Jan. 2009.

[Bla03] R. E. Blahut, *Algebraic Codes for Data Transmission*. New York, Cambridge University Press, 2003.

[BM96] N. Benvenuto and R. Marchesani, "The Viterbi Algorithm for Sparse Channels," *IEEE Transactions on Communications*, vol. 44, no. 3, pp. 287–289, 1996.

[BMB91] M. Bensebti, J. P. McGeehan, and M. A. Beach, "Indoor Multipath Radio Propagation Measurements and Characterisation at 60 GHz," *1991 European Microwave Conference*, vol. 2, pp. 1217–1222, Sept. 9-12, 1991.

[BNVT+06] A. Bourdoux, J. Nsenga, W. Van Thillo, F. Horlin, and L. Van Der Perre, "Air Interface and Physical Layer Techniques for 60 GHz WPANs," *2006 Symposium on Communications and Vehicular Technology*, pp. 1–6, 2006.

[BOO07] F. Bohagen, P. Orten, and G. Oien, "Design of Optimal High-Rank Line-of-Sight MIMO Channels," *IEEE Transactions on Wireless Communications*, vol. 6, no. 4, pp. 1420–1425, 2007.

[BRC60] R. C. Bose and D. K. Ray-Chaudhuri, "On a Class of Error-Correcting Binary Group Codes," *Information and Control*, vol. 3, pp. 68–79, 1960.

[BRHR09] T. Beniero, S. Redana, J. Hamalainen, and B. Raaf, "Effect of Relaying on Coverage in 3GPP LTE-Advanced," *2009 IEEE Vehicular Technology Conference (VTC 2009-Spring)*, Barcelona, Spain, pp. 1–5, Apr. 2009.

[BS66] J. R. Biard and W. N. Shaunfield, "A High Frequency Silicon Avalanche Photodiode," *1966 International Electron Devices Meeting*, vol. 12, p. 30, 1966.

[BSL+11] T. Baykas, C.-S. Sum, Z. Lan, J. Wang, M. Rahman, H. Harada, and S. Kato, "IEEE 802.15.3c: The First IEEE Wireless Standard for Data Rates over 1 Gb/s," *IEEE Communications Magazine*, vol. 49, no. 7, pp. 114–121, Jul. 2011.

[BSS+08] F. Barale, P. Sen, S. Sarkar, S. Pinel, and J. Laskar, "Programmable Frequency-Divider for Millimeter-Wave PLL Frequency Synthesizers," *2008 European Microwave Conference*, pp. 460–463, Oct. 2008.

[BSS+10] F. Barale, P. Sen, S. Sarkar, S. Pinel, and J. Laskar, "A 60 GHz-Standard Compatible Programmable 50 GHz Phase-Locked Loop in 90 nm CMOS," *IEEE Microwave and Wireless Components Letters*, vol. 20, no. 7, pp. 411–413, Jul. 2010.

[BVH14] T. Bai, R. Vaze, and R. W. Heath Jr., "Analysis of Blockage Effects on Urban Cellular Networks," to appear in the *IEEE Transactions on Wireless Communications*, http://arxiv.org/abs/1309.4141.

[BWF+09] T. Baykas, J. Wang, R. Funada, A. Rahman, C. Sum, R. Kimura, H. Harada, and S. Kato, "Preamble Design for Millimeter-Wave Single Carrier WPANs," *2009 IEEE Vehicular Technology Conference (VTC 2009-Spring)*, IEEE, pp. 1–5, 2009.

[C+06] C.-S. Choi, Y. Shoji, H. Harada, et al., "RF Impairment Models for 60 GHz-Band SYS/PHY Simulation," IEEE P802.15 Wireless PANs, Technical Report, Nov. 2006, doc: IEEE 802.15-06-0477-01-003c.

[CAG08] V. Chandrasekhar, J. Andrews, and A. Gatherer, "Femtocell Networks: a Survey," *IEEE Communications Magazine*, vol. 46, no. 9, pp. 59–67, Sep. 2008.

[Can06] "Spectrum Management and Telecommunications," Government of Canada, 2006. http://strategis.ic.gc.ca/epic/internet/insmt-gst.nsf/en/Home.

[Cav86] R. Caverly, "Characteristic Impedance of Integrated Circuit Bond Wires (Short Paper)," *IEEE Transactions on Microwave Theory and Techniques*, vol. 34, no. 9, pp. 982–984, Sep. 1986.

[CCC+08] H.-K. Chen, H.-J. Chen, D.-C. Chang, Y.-Z. Juang, and S.-S. Lu, "A 0.6 V, 4.32 mW, 68 GHz Low Phase-Noise VCO With Intrinsic-Tuned Technique in 0.13 μm CMOS," *IEEE Microwave and Wireless Components Letters*, vol. 18, no. 7, pp. 467–469, Jul. 2008.

[CCC09] I.-S. Che, H.-K. Chiou, and N.-W. Chen, "V-Band On-Chip Dipole-Based Antenna," *IEEE Transactions on Antennas and Propagation*, vol. 57, no. 10, pp. 2853–2861, Oct. 2009.

[CDO07] C. Cao, Y. Ding, and K. K. O, "A 50 GHz Phase-Locked Loop in 0.13-μm CMOS," *IEEE Journal of Solid-State Circuits*, vol. 42, no. 8, pp. 1649–1656, Aug. 2007.

[CDY+08] C. Cao, Y. Ding, X. Yang, J. J. Lin, H. T. Wu, A. K. Verma, J. Lin, F. Martin, and K. K. O, "A 24 GHz Transmitter with On-Chip Dipole Antenna in 0.13 μm CMOS," *IEEE Journal of Solid-State Circuits*, vol. 43, no. 6, pp. 1394–1402, Jun. 2008.

[CEGS10] C.-S. Choi, M. Elkhouly, E. Grass, and C. Scheytt, "60 GHz Adaptive Beamforming Receiver Arrays for Interference Mitigation," *2010 IEEE International Symposium on Personal, Indoor and Mobile Radio Communications (PIMRC 2010)*, pp. 762–767, 2010.

[CF94] L. M. Correia and P. O. Frances, "Estimation of Materials Characteristics from Power Measurements at 60 GHz," *1994 IEEE International Symposium on Personal, Indoor and Mobile Radio Communications (PIMRC 1994)*, pp. 510–513, 1994.

[CFRU01] S.-Y. Chung, G. D. Forney Jr., T. J. Richardson, and R. Urbanke, "On the Design of Low-Density Parity-Check Codes within 0.0045 dB of the Shannon Limit," *IEEE Communications Letters*, vol. 5, no. 2, pp. 58–60, Feb. 2001.

[CG62] C. Campopiano and B. Glazer, "A Coherent Digital Amplitude and Phase Modulation Scheme," *IRE Transactions on Communications Systems*, vol. 10, no. 1, pp. 90–95, Mar. 1962.

[CGH+03] Y. Cao, R. Groves, X. Huang, N. Zamdmer, J.-O. Plouchart, R. Wachnik, T.-J. King, and C. Hu, "Frequency-Independent Equivalent-Circuit Model for On-Chip Spiral Inductors," *IEEE Journal of Solid-State Circuits*, vol. 38, no. 3, pp. 419–426, Mar. 2003.

[CGK+13] M. Cudak, A. Ghosh, T. Kovarik, R. Ratasuk, T. A. Thomas, F. W. Vook, and P. Moorut, "Moving Towards MmWave-Based Beyond-4G (B-4G) Technology," *2013 IEEE Vehicular Technology Conference (VTC 2013-Spring)*, Jun. 2-5, 2013.

[CGLS09] H. Chu, Y. X. Guo, F. Lin, and X. Q. Shi, "Wideband 60 GHz On-Chip Antenna with an Artificial Magnetic Conductor," *2009 IEEE International Symposium on Radio-Frequency Integration Technology (RFIT 2009)*, pp. 307–310, Jan. 2009.

[CGR+03] M. S. Choi, G. Grosskopf, D. Rohde, B. Kuhlow, G. Pryzrembel, and H. Ehlers, "Experiments on DOA-Estimation and Beamforming for 60 GHz Smart Antennas," *2003 IEEE Vehicular Technology Conference (VTC 2003-Spring)*, p. 1041, Apr. 2003.

[CGY10] Z. Chen, G. K. Gokeda, and Y. Yu, *Introduction to Direction-of-Arrival Estimation*, 1st ed. Norwood, MA, Artech House Books, 2010.

[CH02] Y. Cheng and C. Hu, *MOSFET Modeling and BSIM3 User's Guide*. New York, Kluwer Academic Publishers, 2002.

[Cha66] R. W. Chang, "Synthesis of Band-Limited Orthogonal Signals for Multichannel Data Transmission," *Bell System Technical Journal*, vol. 45, Dec. 1966.

[Chi64] R. Chien, "Cyclic Decoding Procedures for Bose-Chaudhuri-Hocquenghem Codes," *IEEE Transactions on Information Theory*, vol. 10, no. 4, pp. 357–363, Oct. 1964.

[Chu72] D. C. Chu, "Polyphase Codes with Good Periodic Correlation Properties, [corresp.]" *IEEE Transactions on Information Theory*, vol. 18, pp. 531–532, 1972.

[Cio11] J. M. Cioffi, "Digital Communications," Stanford, CA, Stanford University, 2011.

[CIS13] "Cisco Visual Networking Index: Global Mobile Data Traffic Forecast Update, 2012–2017," Cisco, Feb. 2013.

[CJK+10] M. Chu, P. Jacob, J.-W. Kim, M. LeRoy, R. Kraft, and J. McDonald, "A 40 Gs/s Time Interleaved ADC Using SiGe BiCMOS Technology," *IEEE Journal of Solid-State Circuits*, vol. 45, no. 2, pp. 380–390, Feb. 2010.

[CL06] D. Chen and J. N. Laneman, "Modulation and Demodulation for Cooperative Diversity in Wireless Systems," *IEEE Transactions on Wireless Communications*, vol. 5, no. 7, pp. 1785–1794, Jul. 2006.

[CLC+06] Y.-J. Cho, K.-H. Lee, H.-C. Choi, Y.-J. Kim, K.-J. Moon, S.-H. Lee, S.-B. Hyun, and S.-S. Park, "A Dual-Channel 6b 1GS/s 0.18μm CMOS ADC for Ultra Wide-Band Communication Systems," *2006 IEEE Asia Pacific Conference on Circuits and Systems (APCCAS 2006)*, pp. 339–342, Dec. 2006.

[CLH+09] C.-C. Chen, C.-C. Li, B.-J. Huang, K.-Y. Lin, H.-W. Tsao, and H. Wang, "Ring-Based Triple-Push VCOs with Wide Continuous Tuning Ranges," *IEEE Transactions on Microwave Theory and Techniques*, vol. 57, no. 9, pp. 2173–2183, Sep. 2009.

[CLL+06] F. Chen, B. Li, T. Lee, C. Christiansen, J. Gill, M, Angyal, et al., "Technology Reliability Qualification of a 65 nm CMOS Cu/Low-k BEOL Interconnect," *2006 International Symposium on the Physical and Failure Analysis of Integrated Circuits*, pp. 97–105, Jul. 2006.

[CM94] F. Classen and H. Meyr, "Frequency Synchronization Algorithms for OFDM Systems Suitable for Communication over Frequency Selective Fading Channels," *1994 IEEE Vehicular Technology Conference (VTC 1994)*, pp. 1655–1659, Jun. 1994.

[CO06] C. Cao and K. K. O, "Millimeter-Wave Voltage-Controlled Oscillators in 0.13 μm CMOS Technology," *IEEE Journal of Solid-State Circuits*, vol. 41, pp. 1297–1304, Jun. 2006.

[Cou07] L. W. Couch, *Digital and Analog Communication Systems*, 7th ed. Upper Saddle River, NJ, Prentice Hall, 2007.

[Cow89] R. Cowan, "Objects Arranged Randomly in Space: An Accessible Theory," *Advances in Applied Probability*, vol. 21, no. 3, pp. 543–569, 1989. http://www.jstor.org/stable/1427635.

[CP93] M. Chelouche and A. Plattner, "Mobile Broadband System (MBS): Trends and Impact on 60 GHz Band MMIC Development," *Electronics & Communication Engineering Journal*, vol. 5, no. 3, pp. 187–197, 1993.

[CR95] S. Czaja and J. Robertson, "Variable Data Rate Viterbi Decoder with Modified LOVA Algorithm," *1995 IEEE Region 10 International Conference on Microelectronics and VLSI (TENCON 1995)*, pp. 472–475, Nov. 1995.

[CR96] L. Correia and J. Reis, "Wideband Characterisation of the Propagation Channel for Outdoors at 60 GHz," *1996 IEEE International Symposium on Personal, Indoor and Mobile Radio Communications (PIMRC 1996)*, vol. 2, pp. 752–755, Oct. 1996.

[CR01a] P. Cardieri and T. S. Rappaport, "Application of Narrow-Beam Antennas and Fractional Loading Factor in Cellular Communication Systems," *IEEE Transactions on Vehicular Technology*, vol. 50, no. 2, pp. 430–440, Mar. 2001.

[CR01b] P. Cardieri and T. S. Rappaport, "Statistical Analysis of Co-Channel Interference in Wireless Communications Systems," *Wireless Communications and Mobile Computing*, vol. 1, no. 1, pp. 111–121, Jan./Mar. 2001.

[CR01c] M. W. Chapman and S. Raman, "A 60 GHz Uniplanar MMIC 4X Subharmonic Mixer," *2001 IEEE MTT-S International Microwave Symposium (IMS)*, pp. 95–98, May 2001.

[CR03] A. Chindapol and J. Ritcey, "Performance Analysis of Coded Modulation with Generalized Selection Combining in Rayleigh Fading," *IEEE Transactions on Communications*, vol. 51, no. 8, pp. 1348–1357, Aug. 2003.

[Cra80] R. K. Crane, "Prediction of Attenuation by Rain," *IEEE Transactions on Communications*, vol. 28, no. 9, pp. 1717–1733, Sep. 1980.

[CRdV06] J. K. Chen, T. S. Rappaport, G. de Veciana, "Iterative Water-Filling for Load-Balancing in Wireless LAN or Microcellular Networks," *2006 IEEE Vehicular Technology Conference (VTC 2006-Spring)*, pp.117–121, May 2006.

[CRdV07] J. K. Chen, T. S. Rappaport, and G. de Veciana, "Site Specific Knowledge for Improving Frequency Allocations in Wireless LAN and Cellular Networks," *2007 IEEE Vehicular Technology Conference (VTC 2007-Fall)*, pp.1431–1435, Oct. 2007.

[CRN09] D. Chowdhury, P. Reynaert, and A. Niknejad, "Design Considerations for 60 GHz Transformer-Coupled CMOS Power Amplifiers," *IEEE Journal of Solid-State Circuits*, vol. 44, no. 10, pp. 2733–2744, Oct. 2009.

[CRR08] E. Cohen, S. Ravid, and D. Ritter, "An Ultra Low Power LNA with 15 dB Gain and 4.4 dB NF in 90 nm CMOS Process for 60 GHz Phase Array Radio," *2008 IEEE Radio Frequency Integrated Circuits Syposium (RFIC 2008)*, pp. 65–68, Apr. 2008.

[CRW07] B. Catteau, P. Rombouts, and L. Weyten, "A Digital Calibration Technique for the Correction of Glitches in High-Speed DAC's," *2007 IEEE International Symposium on Circuits and Systems (ISCAS 2007)*, pp. 1477–1480, May 2007.

[CTB98] G. Caire, G. Taricco, and E. Biglieri, "Bit-Interleaved Coded Modulation," *IEEE Transactions on Information Theory*, vol. 44, no. 3, pp. 927–946, May 1998.

[CTC$^+$05] Y. Cho, M. Tsai, H. Chang, C. Chang, and H. Wang, "A Low Phase Noise 52 GHz Push-Push VCO in 0.18 μm Bulk CMOS Technologies," *2005 IEEE Radio Frequency Integrated Circuits Symposium (RFIC 2005)*, pp. 131–134, Jun. 2005.

[CTSMK07] S. Cheng, H. Tong, J. Silva-Martinez, and A. Karsilayan, "A Fully Differential Low-Power Divide-by-8 Injection-Locked Frequency Divider Up to 18 GHz," *IEEE Journal of Solid-State Circuits*, vol. 42, pp. 583–591, Mar. 2007.

[CTYYLJ07] H.-K. Chiou, Y.-C. Hsu, T.-Y. Yang, S.-G. Lin, and Y. Z. Juang, "15–60 GHz Asymmetric Broadside Coupled Balun in 0.18 μm CMOS Technology," *Electronics Letters*, vol. 43, no. 19, pp. 1028–1030, Sep. 2007.

[CV04] R. Choudhury and N. Vaidya, "Deafness: A MAC Problem in Ad Hoc Networks when using Directional Antennas," *2004 IEEE International Conference on Network Protocols*, pp. 283–292, Oct. 2004.

[CV05] S. Chang and A. Vetro, "Video Adaptation: Concepts, Technologies, and Open Issues," *Proceedings of the IEEE*, vol. 93, no. 1, pp. 148–158, 2005.

[CW00] K. Chang and J. Wiley, *RF and Microwave Wireless Systems*. Wiley Online Library, 2000.

[CW08] E. Candes and M. Wakin, "An Introduction To Compressive Sampling," *IEEE Signal Processing Magazine*, vol. 25, no. 2, pp. 21–30, 2008.

[CYRV06] R. Choudhury, X. Yang, R. Ramanathan, and N. Vaidya, "On Designing MAC Protocols for Wireless Networks Using Directional Antennas," *IEEE Transactions on Mobile Computing*, vol. 5, no. 5, pp. 477–491, May 2006.

[CZS+10] J. Cao, B. Zhang, U. Singh, D. Cui, A. Vasani, A. Garg, W. Zhang, N. Kocaman, D. Pi, B. Raghavan, H. Pan, I. Fujimori, and A. Momtaz, "A 500 mW ADC-Based CMOS AFE With Digital Calibration for 10 Gb/s Serial Links Over KR-Backplane and Multimode Fiber," *IEEE Journal of Solid-State Circuits*, vol. 45, no. 6, pp. 1172–1185, Jun. 2010.

[CZZ03a] S. Collonge, G. Zaharia, and G. E. Zein, "Experimental Investigation of the Spatial and Temporal Characteristics of the 60 GHz Radio Propagation within Residential Environments," *2003 IEEE International Symposium on Signals, Circuits, and Systems (SCS 2003)*, vol. 2, pp. 417–420, 2003.

[CZZ03b] S. Collonge, G. Zaharia, and G. E. Zein, "Influence of Furniture on 60 GHz Radio Propagation in a Residential Environment," *Microwave and Optical Technology Letters*, vol. 39, no. 3, pp. 230–233, 2003.

[CZZ04] S. Collonge, G. Zaharia, and G. E. Zein, "Influence of the Human Activity on Wide-Band Characteristics of the 60 GHz Indoor Radio Channel," *IEEE Journal of Wireless Communications*, vol. 3, no. 6, pp. 2396–2406, 2004.

[DA03] F. F. Digham and M.-S. Alouini, "Variable-Rate Variable-Power Hybrid M-FSK M-QAM for Fading Channels," *2003 IEEE Vehicular Technology Conference (VTC 2003-Fall)*, vol. 3, pp. 1512–1516, 2003.

[dAD04] G. de Alwis and M. Delahoy, "60 GHz Band Millimetre-wave Technology," Australian Communications Authority, Technical Report, Dec. 2004. http://www.acma.gov.au/webwr/radcomm/frequency_planning/radiofrequency_planning_topics/docs/sp3_04_60%20ghz%20mwt%20-%20discussion%20paper-final.pdf.

[Dan06] R. C. Daniels, "An M-ary Continuous Phase Modulated System with Coherent Detection and Frequency Domain Equalization," Master's Thesis, The University of Texas at Austin, May 2006.

[DAR] DARPA Strategic Technology Office, "100 Gb/s RF Backbone (100G)," http://www.darpa.mil/Our_Work/STO/Programs/100_Gbs_RF_Backbone_(100G).aspx.

[Dar04] D. Dardari, "Ordered Subcarrier Selection Algorithm for OFDM-Based High-Speed WLANs," *IEEE Transactions on Wireless Communications*, vol. 3, no. 5, pp. 1452–1458, 2004.

[Dav10] D. B. Davidson, *Computational Electromagnetics for RF and Microwave Engineering*, 1st ed. Cambridge University Press, 2008. A second edition was published in 2010.

[DC07] D. Daly and A. Chandrakasan, "An Energy-Efficient OOK Transceiver for Wireless Sensor Networks," *IEEE Journal of Solid-State Circuits*, vol. 42, no. 5, pp. 1003–1011, May 2007.

[DCCK08] A. Darabiha, A. Chan Carusone, and F. Kschischang, "Power Reduction Techniques for LDPC Decoders," *IEEE Journal of Solid-State Circuits*, vol. 43, no. 8, pp. 1835–1845, Aug. 2008.

[DCF94] N. Daniele, D. Chagnot, and C. Fort, "Outdoor Millimetre-Wave Propagation Measurements with Line of Sight Obstructed by Natural Elements," *Electronics Letters*, vol. 30, no. 18, pp. 1533–1534, Sep. 1994.

[DCH10] R. Daniels, C. Caramanis, and R. W. Heath Jr., "Adaptation in Convolutionally Coded MIMO-OFDM Wireless Systems Through Supervised Learning and SNR Ordering," *IEEE Transactions on Vehicular Technology*, vol. 59, no. 1, pp. 114–126, Jan. 2010.

[DEFF+97] V. Degli-Esposti, G. Falciasecca, M. Frullone, G. Riva, and G. E. Corazza, "Performance Evaluation of Space and Frequency Diversity for 60 GHz Wireless LANs Using a Ray Model," *1997 IEEE Vehicular Technology Conference (VTC 1997)*, vol. 2, pp. 984–988, 1997.

[DENB05] C. Doan, S. Emami, A. Niknejad, and R. Brodersen, "Millimeter-Wave CMOS Design," *IEEE Journal of Solid-State Circuits*, vol. 40, no. 1, pp. 144–155, Jan. 2005.

[DES+04] C. Doan, S. Emami, D. Sobel, A. Niknejad, and R. Brodersen, "Design Considerations for 60 GHz CMOS Radios," *IEEE Communications Magazine*, vol. 42, no. 12, pp. 132–140, 2004.

[DGE01] L. Deneire, B. Gyselinckx, and M. Engels, "Training Sequence Versus Cyclic Prefix—A New Look on Single Carrier Communication," *IEEE Communications Letters*, vol. 5, no. 7, pp. 292–294, Jul. 2001.

[DH07] R. Daniels and R. W. Heath Jr., "60 GHz Wireless Communications: Emerging Requirements and Design Recommendations," *IEEE Vehicular Technology Magazine,* vol. 2, no. 3, pp. 41–50, Sep. 2007.

[DHG+13] H. Dabag, B. Hanafi, F. Golcuk, A. Agah, J. F. Buckwalter, and P. M. Asbeck, "Analysis and Design of Stacked-FET Millimeter-Wave Power Amplifiers," *IEEE Transactions on Microwave Theory and Techniques,* vol. 61, no. 4, pp. 1543–1556, 2013, digital object identifier: 0.1109/TMTT.2013.2247698.

[DHH89] A. Duel-Hallen and C. Heegard, "Delayed Decision-Feedback Sequence Estimation," *IEEE Transactions on Communications,* vol. 37, no. 5, pp. 428–436, 1989.

[DH07] R. C. Daniels and R. W. Heath Jr., "Multi-Band Modulation, Coding, and Medium Access Control," 2007, contribution to VHT IEEE 802.11 Study Group, Atlanta, GA, Nov. 12, 2007. https://mentor.ieee.org/802.11/dcn/07/11-07-2780-01-0vht-multi-band-modulation-coding-and-medium-access-control.ppt.

[DLB+05] T. O. Dickson, M.-A. Lacroix, S. Boret, D. Gloria, R. Beerkens, and S. P. Voinigescu, "30-100 GHz Inductors and Transformers for Millimeter-Wave (Bi)CMOS Integrated Circuits," *IEEE Transactions on Microwave Theory and Techniques,* vol. 53, no. 1, pp. 123–133, Jan. 2005.

[DM10] O. Dabeer and U. Madhow, "Channel Estimation with Low-Precision Analog-to-Digital Conversion," *2010 IEEE International Conference on Communications (ICC),* pp. 1–6, 2010.

[dMKL+09] M. de Matos, E. Kerherve, H. Lapuyade, J. B. Begueret, and Y. Deval, "Millimeter-Wave and Power Characterization for Integrated Circuits," *2009 EAEEIE Annual Conference,* pp. 1–4, Jun. 2009.

[DMR00] A. Demir, A. Mehrotra, and J. Roychowdhury, "Phase Noise in Oscillators: A Unifying Theory and Numerical Methods for Characterization," *IEEE Transactions on Circuits and Systems I: Fundamental Theory and Applications,* vol. 47, no. 5, pp. 655–674, 2000.

[DMRH10] R. C. Daniels, J. N. Murdock, T. S. Rappaport, and R. W. Heath Jr., "60 GHz Wireless: Up Close and Personal," *IEEE Microwave Magazine,* pp. 1–6, Dec. 2010.

[DMTA96] D. Dardari, L. Minelli, V. Tralli, and O. Addrisano, "Wideband Indoor Communications at 60 GHz," *1996 IEEE International Symposium on Personal, Indoor and Mobile Radio Communications (PIMRC 1996),* vol. 3, pp. 791–794, 1996.

[DMW+11] A. Damnjanovic, J. Montojo, Y. Wei, T. Ji, T. Luo, M. Vajapeyam, T. Yoo, O. Song, and D. Malladi, "A Survey on 3GPP Heterogeneous Networks," *IEEE Wireless Communications Magazine,* vol. 18, no. 3, pp. 10–21, Jun. 2011.

[DPR97a] G. Durgin, N. Patwari, and T. S. Rappaport, "Improved 3D Ray Launching Method for Wireless Propagation Prediction," *IET (Formerly IEE) Electronics Letters,* vol. 33, no. 16, pp. 1412–1414, 1997.

[DPR97b] G. Durgin, N. Patwari, and T. S. Rappaport, "An Advanced 3D Ray Launching Method for Wireless Propagation Prediction," *1997 IEEE Vehicular Technology Conference (VTC 1997)*, pp. 785–789, 1997.

[DR98] G. Durgin and T. S. Rappaport, "Basic Relationship between Multipath Angular Spread and Narrowband Fading in Wireless Channels," *IEE (now IET) Electronics Letters*, vol. 34, no. 25, pp. 2431–2432, 1998.

[DR99a] G. D. Durgin and T. S. Rappaport, "Level-Crossing Rates and Average Fade Duration for Wireless Channels with Spatially Complicated Multipath," *1999 IEEE Global Communications Conference (GLOBECOM 1999)*, pp. 437–441, Dec. 1999.

[DR99b] G. D. Durgin and T. S. Rappaport, "Three Parameters for Relating Small-Scale Temporal Fading to Multipath Angles-of-Arrival," *1999 IEEE International Symposium on Personal, Indoor and Mobile Radio Communications (PIMRC 1999)*, pp. 1077–1081, Sep. 1999.

[DR99c] G. Durgin and T. S. Rappaport, "Effects of Multipath Angular Spread on the Spatial Cross Correlation of Received Envelope Voltages," *1999 IEEE Vehicular Technology Conference (VTC 1999)*, vol. 2, pp. 996–1000, 1999.

[DR00] G. Durgin and T. S. Rappaport, "Theory of Multipath Shape Factors for Small-Scale Fading Wireless Channels," *IEEE Transactions on Antennas and Propagation*, vol. 48, no. 5, pp. 682–693, May 2000.

[DRD99] G. D. Durgin, T. S. Rappaport, and D. A. deWolf, "More Complete Probability Density Functions for Fading in Wireless Communications" *1999 IEEE Vehicular Technology Conference (VTC 1999)*, vol. 2, pp. 985–989, May 1999.

[DRD02] G. D. Durgin, T. S. Rappaport, and D. A deWolf, "New Analytical Models and Probability Density Functions for Fading in Wireless Communications," *IEEE Transactions on Communications*, vol. 50, no. 6, pp. 1005–1015, Jun. 2002.

[Dru88] E. H. Drucker, "Development and Application of a Cellular Repeater," *1988 IEEE Vehicular Technology Conference (VTC 1988)*, pp. 321–325, Jun. 1988.

[DRU96] R. De Roo and F. Ulaby, "A Modified Physical Optics Model of the Rough Surface Reflection Coefficient," *1996 Antennas and Propagation Society International Symposium*, vol. 3, pp. 21–26, Jul. 1996.

[Dru00] E. Drucker, "Model PLL Dynamics and Phase-Noise Performance—Part 2," *Microwaves & RF*, pp. 73–82, 117, Feb. 2000.

[DRW+09] K. Doppler, M. Rinne, C. Wijting, C. Ribeiro, and K. Hugl, "Device-to-Device Communication as an Underlay to LTE-advanced Networks," *IEEE Communications Magazine*, vol. 47, no. 12, pp. 42–49, Dec. 2009.

[DRX98a] G. Durgin, T. S. Rappaport, and H. Xu, "Measurements and Models for Radio Path Loss and Penetration Loss In and Around Homes and Trees at 5.85 GHz," *IEEE Transactions on Communications*, vol. 46, no. 11, pp. 1484–1496, 1998.

[DRX98b] G. D. Durgin, T. S. Rappaport, and H. Xu, "Partition-Based Path Loss Analysis for in-Home and Residential Areas at 5.85 GHz," *1998 IEEE Global Telecommunications Conference (GLOBECOM 1998)*, vol. 2, pp. 904–909, Dec. 1998.

[DSI+08] K. Deguchi, N. Suwa, M. Ito, T. Kumamoto, and T. Miki, "A 6-bit 3.5-GS/s 0.9-V 98-mW Flash ADC in 90-nm CMOS," *IEEE Journal of Solid State Circuits*, vol. 43, no. 10, pp. 2303–2310, 2008.

[DSS+08] D. Dawn, S. Sarkar, P. Sen, B. Perumana, D. Yeh, S. Pinel, and J. Laskar, "17-dB-Gain CMOS Power Amplifier at 60 GHz," *2008 IEEE MTT-S International Microwave Symposium (IMS)*, pp. 859–862, Jun. 2008.

[DSS+09] D. Dawn, P. Sen, S. Sarkar, B. Perumana, S. Pinel, and J. Laskar, "60 GHz Integrated Transmitter Development in 90-nm CMOS," *IEEE Transactions on Microwave Theory and Techniques*, vol. 57, no. 10, pp. 2354–2367, Oct. 2009.

[DT99] D. Dardari and V. Tralli, "High-Speed Indoor Wireless Communications at 60 GHz with Coded OFDM," *IEEE Transactions on Communications*, vol. 47, no. 11, pp. 1709–1721, Nov. 1999.

[Dur03] G. D. Durgin, *Space-Time Wireless Channels,* Upper Saddle River, NJ, Prentice Hall, 2003.

[DWV08] K. Doppler, C. Wijting, and K. Valkealahti, "On the Benefits of Relays in a Metropolitan Area Network," *2008 IEEE Vehicular Technology Conference (VTC 2008-Spring)*, pp. 2301–2305, May 2008.

[EAHAS+12] O. El Ayach, R. W. Heath Jr., S. Abu-Surra, S. Rajagopal, and Z. Pi, "The Capacity Optimality of Beam Steering in Large Millimeter Wave MIMO Systems," *2012 IEEE International Workshop on Signal Processing Advances in Wireless Communications (SPAWC)*, pp. 100–104, 2012.

[EAHAS+12a] O. El Ayach, R. W. Heath Jr., S. Abu-Surra, S. Rajagopal, and Z. Pi, "Low Complexity Precoding for Large Millimeter Wave MIMO Systems," *2012 IEEE International Conference on Communications (ICC)*, pp. 3724–3729, Jun. 2012.

[ECd13] "2013/752/EU: Commission Implementing Decision of 11 December 2013 Amending Decision 2006/771/EC on Harmonisation of the Radio Spectrum for Use by Short-Range Devices and Repealing Decision 2005/928/EC," *Official Journal of the European Union,* Dec. 2013. http://eur-lex.europa.eu/legal-content/EN/TXT/?uri=uriserv:OJ.L_.2013.334.01.0017.01.ENG.

[ECS+98] R. Ertel, P. Cardieri, K. Sowerby, T. S. Rappaport, and J.Reed, "Overview of Spatial Channel Models for Antenna Array Communication Systems," *IEEE Personal Communications*, vol. 5, no. 1, pp. 10–22, Feb. 1998.

[ECMA08] Ecma International, *Standard ECMA-387: High Rate 60 GHz PHY, MAC and HDMI PAL,"* 1st ed. Geneva, Ecma International, Dec. 2008. http://www.ecma-international.org/publications/files/ECMA-ST-ARCH/ECMA-387%201st%20edition%20December%202008.pdf.

[ECMA10] Ecma International, *Standard ECMA-387: High Rate 60 GHz PHY, MAC and PALs*, 2nd ed. Geneva, Ecma International, Dec. 2010.

[EDNB04] S. Emami, C. Doan, A. Niknejad, and R. Brodersen, "Large-Signal Millimeter-Wave CMOS Modeling with BSIM3," *2004 IEEE Radio Frequency Integrated Circuits Symposium (RFIC 2004)*, pp. 163–166, Jun. 2004.

[EDNB05] S. Emami, C. Doan, A. Niknejad, and R. Brodersen, "A 60 GHz Down-Converting CMOS Single-Gate Mixer," *2005 IEEE Radio Frequency Integrated Circuits Symposium (RFIC 2005)*, pp. 163–166, Jun. 2005.

[Emr07] R. Emrick, "On-Chip Antenna Element and Array Design for Short Range Millimeter-Wave Communications," Ph.D. Dissertation, The Ohio State University, 2007.

[EMT+09] V. Erceg, M. Messe, A. Tarighat, M. Boers, J. Trachewsky, and C. Choi, "60 GHz Impairments Modeling," *IEEE P802.11 Wireless LANs*, Technical Report, 2009, doc: IEEE 802.11-09/1213r1.

[Enz02] C. Enz, "An MOS Transistor Model for RF IC Design Valid in All Regions of Operation," *IEEE Transactions on Microwave Theory and Techniques*, vol. 50, no. 1, pp. 342–359, Jan. 2002.

[Enz08] C. C. Enz, "A Short Story of the EKV MOS Transistor Model," *IEEE Solid-State Circuits Society Newsletter*, vol. 13, no. 3, pp. 24–30, summer 2008.

[EQ89] M. Eyuboglu and S. Qureshi, "Reduced-State Sequence Estimation for Coded Modulation of Intersymbol Interference Channels," *IEEE Journal on Selected Areas in Communications*, vol. 7, no. 6, pp. 989–995, 1989.

[ERAS+14] O. El Ayach, S. Rajagopal, S. Abu-Surra, Z. Pi, and R. W. Heath Jr., "Spatially Sparse Precoding in Millimeter Wave MIMO Systems," *IEEE Transactions on Wireless Communications*, vol. 13, no. 3, pp. 1499–1513, Mar. 2014.

[ETSI12a] European Telecommunications Standards Institute (ETSI), "ETSI EN 302 217-3: Fixed Radio Systems; Characteristics and Requirements for Point-to-Point Equipment and Antennas," Sep. 2012.

[ETSI12b] European Telecommunications Standards Institute (ETSI), "ETSI EN 302 567: Broadband Radio Access Networks (BRAN); 60 GHz Multiple-Gigabit WAS/RLAN Systems," Jan. 2012.

[ETWH01] M. El-Tanany, Y. Wu, and L. Házy, "Analytical Modeling and Simulation of Phase Noise Interference in OFDM-Based Digital Television Terrestrial Broadcasting Systems," *IEEE Transactions on Broadcasting*, vol. 47, no. 1, pp. 20–31, 2001.

[Eur10] The European Conference of Postal and Telecommunications Administrations, Recommendation T/R 22-03, "Provisional Recommended Use of the Frequency Range 54.25-66 GHz by Terrestrial Fixed and Mobile Systems, 1990." http://www.erodocdb.dk/docs/doc98/official/pdf/TR2203E.pdf.

[EV06] C. C. Enz and E. A. Vittoz, *Charge-based MOS Transistor Modeling*, 1st ed. New York, Wiley, 2006.

[EWA+11] S. Emami, R. Wise, E. Ali, M. G. Forbes, M. Q. Gordon, X. Guan, S. Lo, P. T. McElwee, J. Parker, J. R. Tani, J. M. Gilbert, and C. H. Doan, "A 60 GHz CMOS Phased-Array Transceiver Pair for Multi-Gb/s Wireless Communications," *2011 IEEE International Solid-State Circuits Conference (ISSCC)*, pp. 164–166, Feb. 2011.

[FABSE02] D. Falconer, S. Ariyavisitakul, A. Benyamin-Seeyar, and B. Eidson, "Frequency Domain Equalization for Single-Carrier Broadband Wireless Systems," *IEEE Communications Magazine*, vol. 40, no. 4, pp. 58–66, Apr. 2002.

[FBRSX94] M. J. Feuerstein, K. L. Blackard, T. S. Rappaport, S. Y. Seidel, and H. Xia, "Path Loss, Delay Spread, and Outage Models as Functions of Antenna Height for Microcellular System Design," *IEEE Transactions on Vehicular Technology*, vol. 43, no. 3, pp. 487–498, Aug. 1994.

[FCC88] "The Use of the Radio Frequency Spectrum Above 30 GHz: A Consultative Document," Radiocommunications Div., U.K. Dept. Trade and Industry, 1988.

[FDM+12] G. Fodor, E. Dahlman, G. Mildh, S. Parkvall, N. Reider, G. Miklos, and Z. Turanyi, "Design Aspects of Network Assisted Device-to-Device Communications," *IEEE Communications Magazine*, vol. 50, no. 3, pp. 170–177, Mar. 2012.

[Fed06] Federal Communications Commission, "Electronic Code of Federal Regulations: Part 15 - Radio Frequency Device Regulations," Washington, D.C., U.S. Government Printing Office, 2006. http://www.fcc.gov/oet/info/rules/.

[Fed13] Federal Communications Commission, FCC Report and Order 13-112, Aug. 9, 2013. http://www.fcc.gov/document/fcc-modifies-part-15-rules-unlicensed-operation-57-64-ghz-band.

[Fen08] A. J. Fenn, *Adaptive Antennas and Phased Arrays for Radar and Communications*. Norwood, MA, Artech House, 2008.

[FGR93] D. F. Filipovic, S. S. Gearhart, and G. M. Rebeiz, "Double-Slot Antennas on Extended Hemispherical and Elliptical Silicon Dielectric Lenses," *IEEE Transactions on Microwave Theory and Techniques*, vol. 41, no. 10, pp. 1738–1749, Oct. 1993.

[FGRR97] D. F. Filipovic, G. P. Gauthier, S. Raman, and G. M. Rebeiz, "Off-Axis Properties of Silicon and Quartz Dielectric Lens Antennas," *IEEE Transactions on Antennas and Propagation*, vol. 45, no. 5, pp. 760–766, May 1997.

[FHM10] X. Feng, G. He, and J. Ma, "A New Approach to Reduce the Resolution Requirement of the ADC for High Data Rate Wireless Receivers," *2010 IEEE International Conference on Signal Processing*, pp. 1565–1568, 2010.

[FHO02] B. Floyd, C.-M. Hung, and K. K. O, "Intra-Chip Wireless Interconnect for Clock Distribution Implemented with Integrated Antennas, Receivers, and Transmitters," *IEEE Journal of Solid-State Circuits,* vol. 37, no. 5, pp. 543–552, May 2002.

[FJ72] G. D. Forney Jr., "Maximum-Likelihood Sequence Estimation of Digital Sequences in the Presence of Intersymbol Interference," *IEEE Transactions on Information Theory,* vol. 18, no. 3, pp. 363–378, May 1972.

[FJ73] G. D. Forney Jr., "The Viterbi Algorithm," *Proceedings of the IEEE,* vol. 61, no. 3, pp. 268–278, Mar. 1973.

[FLC$^+$02] M. Fryziel, C. Loyez, L. Clavier, Rolland, and P. A. Rolland, "Path-Loss Model of the 60 GHz Indoor Radio Channel," *Microwave and Optical Technology Letters,* vol. 34, no. 3, pp. 158–162, 2002.

[Flo04] B. Floyd, "V-Band and W-Band SiGe Bipolar Low-Noise Amplifiers and Voltage-Controlled Oscillators," *2004 IEEE Radio Frequency Integrated Circuits Symposium (RFIC 2004),* pp. 295–298, Jun. 2004.

[FMSNJ08] M. Fakharzadeh, P. Mousavi, S. Safavi-Naeini, and S. H. Jamali, "The Effects of Imbalance Phase Shifters Loss on Phased Array Gain," *IEEE Letters on Antennas and Wireless Propagation,* vol. 7, no. 3, pp. 192–197, Mar. 2008.

[FNABS10] M. Fakharzadeh, M.-R. N.-Ahmadi, B. Biglarbegian, and J. A. Shokouh, "CMOS Phased Array Transceiver Technology for 60 GHz Wireless Applications," *IEEE Transactions on Antennas and Propagation,* vol. 58, no. 4, pp. 1093–1104, Apr. 2010.

[FNP$^+$08] M. Ferndahl, H. Nemati, B. Parvais, H. Zirath, and S. Decoutere, "Deep Submicron CMOS for Millimeter Wave Power Applications," *IEEE Microwave and Wireless Components Letters,* vol. 18, no. 5, pp. 329–331, May 2008.

[For65] J. Forney, G., "On Decoding BCH Codes," *IEEE Transactions on Information Theory,* vol. 11, no. 4, pp. 549–557, Oct. 1965.

[Fos77] G. J. Foschini, "A Reduced State Variant of Maximum Likelihood Sequence Detection Attaining Optimum Performance for High Signal-to-Noise Ratios," *IEEE Transactions on Information Theory,* vol. 23, no. 5, pp. 605–609, 1977.

[Fos96] G. J. Foschini, "Layered Space-Time Architecture for Wireless Communication in a Fading Environment When Using Multiple Antennas," *Bell Labs Technical Journal,* vol. 1, no. 2, pp. 41–59, 1996.

[FOV11] H. P. Forstner, M. Ortner, and L. Verweyen, "A Fully Integrated Homodyne Upconverter MMIC in SiGe:C for 60 GHz Wireless Applications," *2011 IEEE Meeting on Silicon Monolithic Integrated Circuits in RF Systems (SiRF 2011),* pp. 129–132, Jan. 2011.

[FPdC+04] C. Fager, J. Pedro, N. de Carvalho, H. Zirath, F. Fortes, and M. Rosario, "A Comprehensive Analysis of IMD Behavior in RF CMOS Power Amplifiers," *IEEE Journal of Solid-State Circuits*, vol. 39, no. 1, pp. 24–34, Jan. 2004.

[Fr11] "Frequency Band Review for Fixed Wireless Services," Final Report Prepared for OfCom, Document 2315/FLBR/FRP/3, Nov. 2011. http://stakeholders.ofcom.org.uk/binaries/consultations/spectrum-review/annexes/report.pdf.

[Fri46] H. T. Friis, "A Note on a Simple Transmission Formula," *Proceedings of the IRE*, vol. 34, no. 5, pp. 254–256, 1946.

[FRP+05] B. Floyd, S. Reynolds, U. Pfeiffer, T. Zwick, T. Beukema, and B. Gaucher, "SiGe Bipolar Transceiver Circuits Operating at 60 GHz," *IEEE Journal of Solid-State Circuits*, vol. 40, no. 1, pp. 156–167, Jan. 2005.

[FRT93] V. Fung, T. S. Rappaport, and B. Thoma, "Bit Error Simulation for Pi/4 DQPSK Mobile Radio Communications Using Two-Ray and Measurement-Based Impulse Response Models," *IEEE Journal on Selected Areas in Communications*, vol. 11, no. 3, pp. 393–405, Apr. 1993.

[FZ62] R. L. Frank, S. A. Zadoff, and R. Heimiller, "Phase Shift Pulse Codes with Good Periodic Correlation Properties," *IRE Transactions on Information Theory*, vol. IT-8, pp. 381–382, 1962.

[GAB+10] Y. Greshishchev, J. Aguirre, M. Besson, R. Gibbins, C. Falt, P. Flemke, N. Ben-Hamida, D. Pollex, P. Schvan, and S.-C. Wang, "A 40GS/s 6b ADC in 65nm CMOS," *2010 IEEE International Solid-State Circuits Conference (ISSCC)*, pp. 390–391, Feb. 2010.

[Gal62] R. Gallager, "Low-Density Parity-Check Codes," *IRE Transactions on Information Theory*, vol. 8, no. 1, pp. 21–28, Jan. 1962.

[GAPR09] F. Gutierrez, S. Agarwal, K. Parrish, and T. S. Rappaport, "On-Chip Integrated Antenna Structures in CMOS for 60 GHz WPAN Systems," *IEEE Journal on Selected Areas in Communications*, vol. 27, no. 8, pp. 1367–1378, Oct. 2009.

[GBS71] D. George, R. Bowen, and J. Storey, "An Adaptive Decision Feedback Equalizer," *IEEE Transactions on Communication Technology*, vol. 19, no. 3, pp. 281–293, Jun. 1971.

[GFK12] B. Grave, A. Frappe, and A. Kaiser, "A Reconfigurable 60 GHz Subsampling Receiver Architecture with Embedded Channel Filtering," *2012 IEEE International Symposium on Circuits and Systems (ISCAS)*, pp. 1295–1298, May 2012.

[GGSK11] I. Guvenc, S. Gezici, Z. Sahinoglu, and U. C. Kozat, *Reliable Communications for Short-Range Wireless Systems*. Cambridge University Press, Apr. 2011, ch. 3.

[GHH04] X. Guan, H. Hashemi, and A. Haijimiri, "A Fully Integrated 24 GHz Eight-Element Phased Array Receiver in Silicon," *IEEE Journal of Solid-State Circuits*, vol. 39, no. 12, pp. 2311–2320, Dec. 2004.

[Gho14] A. Ghosh, et al., "Millimeter Wave Enhanced Local Area Systems: A High Data Rate Approach for Future Wireless Networks," *IEEE Journal on Selected Areas in Communications,* vol. 32, no. 6, pp. 1152–1163, Jun. 2014.

[Gib07] W. C. Gibson, *The Method of Moments in Electromagnetics*, 1st ed. Boca Raton, FL, Chapman & Hall, CRC, 2007.

[GJ13] F. Gutierrez Jr., "Mm-wave and Sub-THz On Chip Antenna Array Propagation and Radiation Pattern Measurement," Ph.D. Dissertation, The University of Texas, Dec. 2013.

[GJBAS01] I. Gresham, N. Jain, T. Budka, A. Alexanian, N. Kinayman, B. Ziegner, S. Brown, and P. Staecker, "A Compact Manufacturable 76-77 GHz Radar Module for Commercial ACC Applications," *IEEE Transactions on Microwave Theory and Techniques,* vol. 49, no. 1, pp. 44, 58, Jan. 2001.

[GJRM10] F. Gutierrez Jr., T. S. Rappaport, and J. N. Murdock, "Millimeter-Wave CMOS Antennas and RFIC Parameter Extraction for Vehicular Applications," *2010 IEEE Vehicular Technology Conference (VTC 2010-Fall)*, pp. 1–6, Sep. 2010.

[GKH+07] D. Gesbert, M. Kountouris, R. W. Heath Jr., C.-B. Chae, and T. Salzer, "Shifting the MIMO Paradigm," *IEEE Signal Processing Magazine,* vol. 24, no. 5, pp. 36–46, Sep. 2007.

[GKT+09] A. Garcia, W. Kotterman, R. Thoma, U. Trautwein, D. Bruckner, W. Wirnitzer, and J. Kunisch, "60 GHz in-Cabin Real-Time Channel Sounding," *2009 IEEE International Conference on Communications and Networking in China (ChinaCOM 2009)*, pp. 1–5, Aug. 2009.

[GKZ+05] S. Gunnarsson, C. Karnfelt, H. Zirath, R. Kozhuharov, D. Kuylenstierna, A. Alping, and C. Fager, "Highly Integrated 60 GHz Transmitter and Receiver MMICs in a GaAs pHEMT Technology," *IEEE Journal of Solid-State Circuits,* vol. 40, no. 11, pp. 2174–2186, Nov. 2005.

[GLR99] F. Giannetti, M. Luise, and R. Reggiannini, "Mobile and Personal Communications in the 60 GHz Band: A Survey," *Wireless Personal Communications,* vol. 10, pp. 207–243, 1999.

[GMM06] M. X. Gong, S. Midkiff, and S. Mao, "MAC Protocols for Wireless Mesh Networks," Y. Zhang, J. Luo, and H. Hu, Eds. *Wireless Mesh Networking: Architectures, Protocols and Standards,* Boca Raton, FL, Auerbach Publications, 2006.

[GMR+12] A. Ghosh, N. Mangalvedhe, R. Ratasuk, B. Mondal, M. Cudak, E. Visotsky, T. Thomas, J. Andrews, P. Xia, H.-S. Jo, H. Dhillon, and T. Novlan, "Heterogeneous Cellular Networks: From Theory to Practice," *IEEE Communications Magazine,* vol. 50, no. 6, pp. 54–64, Jun. 2012.

[Gol61] M. Golay, "Complementary Series," *IRE Transactions on Information Theory,* vol. 7, no. 2, pp. 82–87, Apr. 1961.

[Gol05] A. Goldsmith, *Wireless Communications*. New York, Cambridge University Press, 2005.

[GRM+10] A. Ghosh, R. Ratasuk, B. Mondal, N. Mangalvedhe, and T. Thomas, "LTE-Advanced: Next-Generation Wireless Broadband Technology," *IEEE Wireless Communications*, vol. 17, no. 22, pp. 10–22, Jun. 2010.

[Gro13] S. Grobart, "Samsung Announces New '5G' Wireless Technology," May 2013. http://www.businessweek.com/articles/2013-05-13/samsung-announces-new-5g-wireless-technology.

[GS99] V. Guruswami and M. Sudan, "Improved Decoding of Reed-Solomon and Algebraic-Geometry Codes," *IEEE Transactions on Information Theory*, vol. 45, no. 6, pp. 1757–1767, Sep. 1999.

[GT06] P. Green and D. Taylor, "Dynamic Channel-Order Estimation Algorithm," *IEEE Transactions on Signal Processing*, vol. 54, no. 5, pp. 1922–1925, May 2006.

[GV05] T. Guess and M. Varanasi, "An Information-Theoretic Framework for Deriving Canonical Decision-Feedback Receivers in Gaussian Channels," *IEEE Transactions on Information Theory*, vol. 51, no. 1, pp. 173–187, Jan. 2005.

[GVS11] F. Gholam, J. Vía, and I. Santamaría, "Beamforming Design for Simplified Analog Antenna Combining Architectures," *IEEE Transactions on Vehicular Technology*, vol. 60, no. 5, pp. 2373–2378, 2011.

[GW01] G. Gbur and E. Wolf, "The Rayleigh Range of Partially Coherent Beams," *Optics Communications*, vol. 199, no. 5, pp. 295–304, 2001.

[H+94] C. M. P. Ho, et al., "Antenna Effects on Indoor Obstructed Wireless Channels and a Deterministic Image-Based Wide-Band Propagation Model for In-Building Personal Communication Systems," *International Journal of Wireless Information Networks*, vol. 1, no. 1, pp. 61–76, Jan. 1994.

[HA95] A. M. Hammoudeh and G. Allen, "Millimetric Wavelengths Radiowave Propagation for Line-of-Sight Indoor Microcellular Mobile Communications," *IEEE Journal of Vehicular Technology*, vol. 44, no. 3, pp. 449–460, 1995.

[Hag88] J. Hagenauer, "Rate-Compatible Punctured Convolutional Codes (RCPC Codes) and Their Applications," *IEEE Transactions on Communications*, vol. 36, no. 4, pp. 389–400, Apr. 1988.

[Ham03] K. Hamagushi, "Japanese Regulation for 60 GHz Band, IEEE 802.15.3/0351r1," *IEEE 802.15.3 Standard Contribution*, Sep. 2003.

[Has09] A. Hassibi, "Quantitative Metrics," *EE382V: Data Converters*, University of Texas at Austin, Spring 2009.

[Hay96] M. Hayes, *Statistical Digital Signal Processing and Modeling*. New York, Wiley, 1996.

[HBAN07a] B. Heydari, M. Bohsali, E. Adabi, and A. Niknejad, "A 60 GHz Power Amplifier in 90nm CMOS Technology," *2007 IEEE Custom Integrated Circuits Conference (CICC 2007)*, pp. 769–772, Sep. 2007.

[HBAN07b] B. Heydari, M. Bohsali, E. Adabi, and A. Niknejad, "Millimeter-Wave Devices and Circuit Blocks up to 104 GHz in 90 nm CMOS," *IEEE Journal of Solid-State Circuits*, vol. 42, no. 12, pp. 2893–2903, Dec. 2007.

[HBLK14] W. Hong, K. Baek, Y. Lee, and Y. Kim, "Design and Analysis of a Low-Profile 28 GHz Beam Steering Antenna Solution for Future 5G Cellular Applications," *2014 IEEE MTT-S International Microwave Symposium (IMS)*, Tampa, FL, Jun. 2014.

[HCLC09] W.-L. Hsu, C.-Z. Chen, Y.-S. Lin, and C.-C. Chen, "A 2 mW, 55.8 GHz CMOS Injection-Locked Frequency Divider with 7.1 GHz Locking Range," *2009 IEEE Radio and Wireless Symposium (RWS 2009)*, pp. 582–585, Jan. 2009.

[He+04] J. He et al., "Globally Optimal Transmitter Placement for Indoor Wireless Communication Systems," *IEEE Transactions on Wireless Communications*, vol. 3, no. 6, pp. 1906–1911, Nov. 2004.

[Hea10] R. W. Heath Jr., "Where are the Relay Gains in Cellular Systems?" Presentation made at the IEEE Communication Theory Workshop, 2010.

[Hei93] W. Heinrich, "Quasi-TEM Description of MMIC Coplanar Lines Including Conductor-Loss Effects," *IEEE Transactions on Microwave Theory Tech.*, vol. 41, no. 1, pp. 45–52, Jan. 1993.

[HFS08] "Left-Handed Metamaterial Design Guide," Ansoft, Feb. 2008. http://www.rfglobalnet.com/doc/left-handed-metamaterial-design-guide-0002.

[HGKH05] H. Hashemi, X. Guan, A. Komijani, and A. Hajimiri, "A 24 GHz SiGe Phased-Array Receiver-LO Phase-Shifting Approach," *IEEE Transactions on Microwave Theory and Techniques*, vol. 53, no. 2, pp. 614–626, Feb. 2005.

[HH89] J. Hagenauer and P. Hoeher, "A Viterbi Algorithm with Soft-Decision Outputs and Its Applications," *1989 IEEE Global Telecommunications Conference (GLOBECOM 1989)*, pp. 1680–1686, Nov. 1989.

[HH97] M. Hamid and R. Hamid, "Equivalent Circuit of Dipole Antenna of Arbitrary Length," *IEEE Transactions on Antennas and Propagation*, vol. 45, no. 11, pp. 1695–1696, Nov. 1997.

[HH98] M. Honkanen and S. Haggman, "New Aspects on Nonlinear Power Amplifier Modeling in Radio Communication System Simulations," *1998 IEEE International Symposium on Personal, Indoor and Mobile Radio Communications (PIMRC 1998)*, vol. 3, IEEE, 1998, pp. 844–848.

[HH09] Y. Hou and T. Hase, "Improvement on the Channel Estimation of Pilot Cyclic Prefixed Single Carrier (PCP-SC) System," *IEEE Signal Processing Letters*, vol. 16, no. 8, pp. 719–722, Aug. 2009.

[HK11] J. Haapola and S. Kato, "Efficient mm-Wave Beamforming Protocol for Group Environments," *2011 IEEE International Symposium on Personal, Indoor and Mobile Radio Communications (PIMRC 2011)*, pp. 1056–1060, 2011.

[HKL+11] S. Hur, T. Kim, D. Love, J. Krogmeier, T. Thomas, and A. Ghosh, "Multilevel Millimeter Wave Beamforming for Wireless Backhaul," *2011 IEEE GLOBECOM Workshops*, pp. 253–257, 2011.

[HKN07a] J. Hasani, M. Kamarei, and F. Ndagijimana, "New Input Matching Technique for Cascode LNA in 90 nm CMOS for Millimeter Wave Applications," *2007 IEEE International Workshop on Radio-Frequency Integration Technology (RFIT 2007)*, pp. 282–285, Dec. 2007.

[HKN07b] J. Hasani, M. Kamarei, and F. Ndagijimana, "Transmission Line Inductor Modeling and Design for Millimeter Wave Circuits in Digital CMOS Process," *2007 IEEE International Workshop on Radio-Frequency Integration Technology, (RFIT 2007)*, pp. 290–293, Dec. 2007.

[HL98] A. Hajimiri and T. Lee, "A General Theory of Phase Noise in Electrical Oscillators," *IEEE Journal of Solid-State Circuits*, vol. 33, no. 2, pp. 179–194, Feb. 1998.

[HL11] S.-E. Hong and W. Y. Lee, "Flexible Unequal Error Protection Scheme for Uncompressed Video Transmission over 60 GHz Multi-Gigabit Wireless System," *2011 IEEE International Conference on Computer Communications and Networks (ICCCN 2011)*, pp. 1–6, Aug. 2011.

[HLJ+06] F. Huang, J. Lu, N. Jiang, X. Zhang, W. Wu, and Y. Wang, "Frequency-Independent Asymmetric Double-Equivalent Circuit for On-Chip Spiral Inductors: Physics-Based Modeling and Parameter Extraction," *IEEE Journal of Solid-State Circuits*, vol. 41, no. 10, pp. 2272–2283, Oct. 2006.

[Hoc59] A. Hocquenghem, "Codes Correcteurs d'Erreurs," *Chiffres 2*, pp. 147–156, 1959.

[Hor08] A. van der Horst, "Copper Cabling Can Resolve the Cost/Power Equation," *The Data Center Journal*, Jul. 2008. http://datacenterjournal.com.

[HPWZ13] R. W. Heath Jr., S. Peters, Y. Wang, and J. Zhang, "A Current Perspective on Distributed Antenna Systems for the Downlink of Cellular Systems," *IEEE Communications Magazine*, vol. 51, no. 4, pp. 161–167, Apr. 2013.

[HR92] C. M. P. Ho and T. S. Rappaport, "Effects of Antenna Polarization and Beam Pattern on Multipath Delay Spread and Path Loss in Indoor Obstructed Wireless Channels," *1992 IEEE International Conference on Universal Personal Communications (IUCPC 1992)*, Oct. 1992.

[HR93] C. M. P. Ho and T. S. Rappaport, "Wireless Channel Prediction in a Modern Office Building Using an Image-Based Ray Tracing Method," *1993 IEEE Global Telecommunications Conference (GLOBECOM 1993)*, pp. 1247–1251, Nov./Dec. 1993.

[HRA10] K. Hassan, T. S. Rappaport, and J. G. Andrews, "Analog Equalization for Low Power 60 GHz Receivers in Realistic Multipath Channels," *2010 IEEE Global Telecommunications Conference (GLOBECOM 2010)*, pp. 1–5, Dec. 2010.

[HRBS00] X. Hao, T. S. Rappaport, R. Boyle, and J. Schaffner, "38 GHz Wide-Band Point-to-Multipoint Measurements under Different Weather Conditions," *IEEE Communications Letters*, vol. 4, no. 1, pp. 7–8, Jan. 2000.

[HRL10] E. Herth, N. Rolland, and T. Lasri, "Circularly Polarized Millimeter-Wave Antenna Using 0-Level Packaging," *IEEE Antennas and Wireless Propagation Letters*, vol. 9, pp. 934–937, Sep. 2010.

[HS99] J. Hagenauer and T. Stockhammer, "Channel Coding and Transmission Aspects for Wireless Multimedia," *Proceedings of the IEEE*, vol. 87, no. 10, pp. 1764–1777, Oct. 1999.

[HS03] J. Huang and R. Spencer, "The Design of Analog Front Ends for 1000BASE-T Receivers," *IEEE Transactions on Circuits and Systems II: Analog and Digital Signal Processing*, vol. 50, no. 10, pp. 675–684, Oct. 2003.

[HSSJ02] Z. Huang, C.-C. Shen, C. Srisathapornphat, and C. Jaikaeo, "A Busy-Tone Based Directional MAC Protocol for Ad Hoc Networks," *2002 IEEE MILCOM*, vol. 2, pp. 1233–1238, Oct. 2002.

[HTM+07] H. Hoshino, R. Tachibana, T. Mitomo, N. Ono, Y. Yoshihara, and R. Fujimoto, "A 60 GHz Phase-Locked Loop with Inductor-Less Prescaler in 90-nm CMOS," *2007 European Solid State Circuits Conference (ESSCIRC 2007)*, pp. 472–475, Sep. 2007.

[HW10] K.-K. Huang and D. D. Wentzloff, "60 GHz On-Chip Patch Antenna Integrated in a 0.13-μm CMOS Technology," *2010 IEEE International Conference on Ultra-Wideband (ICUWB 2010)*, pp. 1–4, Sep. 2010.

[HW11] K.-C. Huang and Z. Wang, *Millimeter Wave Communication Systems*. New York, Wiley-IEEE Press, 2011.

[HWHRC08] S.-S. Hsu, K.-C. Wei, C.-Y. Hsu, and H. Ru-Chuang, "A 60 GHz Millimeter Wave CPW-Fed Yagi Antenna Fabricated by Using 0.18 μm CMOS Technology," *IEEE Electron Device Letters*, vol. 29, no. 6, pp. 625–627, Jun. 2008.

[HYFP04] H. Hu, H. Yanikomeroglu, D. D. Falconer, and S. Periyalwar, "Range Extension without Capacity Penalty in Cellular Networks with Digital Fixed Relays," *2004 IEEE Global Telecommunications Conference*, Dallas, TX, pp. 3053–3057, Nov.-Dec. 2004.

[ID08] R. Irmer and F. Diehm, "On Coverage and Capacity of Relaying in LTE-advanced in Example Deployments," *2008 IEEE International Symposium on Personal, Indoor and Mobile Radio Communications (PIMRC 2008)*, pp. 1–5, Sep. 2008.

[IE08] A. Ismail and M. Elmasry, "Analysis of the Flash ADC Bandwidth-Accuracy Tradeoff in Deep-Submicron CMOS Technologies," *IEEE Transactions on Circuits Syst. II*, vol. 55, no. 10, pp. 1001–1005, Oct. 2008.

[ISO] "International Organization for Standardization (ISO)." http://www.iso.org.

[ITU] International Telecommunication Union, "Radio Recommendations." http://www.itu.int/publications/sector.aspx?lang=en&sector=1.

[IST05] "IST BroadWay: A 5/60 GHz Hybrid System Concept," *4th Concertation Meeting of IST-FP6 Communication and Network Technologies Projects and/or Associated Clusters*, Mar. 2005. ftp://ftp.cordis.europa.eu/pub/ist/docs/directorate_d/cnt/neweve/broadway_en.pdf.

[Ito00] Y. Itoh, "Microwave and Millimeter-Wave Amplifier Design Via Load-Pull Techniques," *2000 Gallium Arsenide Integrated Circuit Symposium (GaAs IC 2000)*, pp. 43–46, Nov. 2000.

[ITR09] T. Osada, M. Godwin, et al., *International Technology Roadmap for Semiconductors*, 2009 Edition, 2009. http://www.itrs.net/Links/2009ITRS/Home2009.htm.

[JA85] D. Jackson and N. Alexopoulos, "Gain Enhancement Methods for Printed Circuit Antennas," *IEEE Transactions on Antennas and Propagation*, vol. 33, no. 9, pp. 976–987, Sep. 1985.

[Jak94] W. C. Jakes, Ed., *Microwave Mobile Communications*, 2nd ed. New York, Wiley-IEEE Press, 1994.

[JEV89] D. L. Jones, R. H. Espeland, and E. J. Violette, "Vegitation Loss Measurements at 9.6, 28.8, 57.6, and 96.1 GHz Through a Conifer Orchard in Washington State," *NTIA Report*, vol. 89, no. 251, U.S. Department of Commerce, National Telecommunications and Information Administration, Oct. 1989.

[JGA+05] B. Jagannathan, D. Greenberg, R. Anna, X. Wang, J. Pekarik, M. Breitwisch, M. Erturk, L. Wagner, C. Schnabel, D. Sanderson, and S. Csutak, "RF FET Layout and Modeling for Design Success in RFCMOS Technologies," *2005 IEEE Radio Frequency Integrated Circuits Symposium (RFIC 2005)*, pp. 57–60, Jun. 2005.

[JLGH09] Y. Jiang, K. Li, J. Gao, and H. Harada, "Antenna Space Diversity and Polarization Mismatch in Wideband 60 GHz Millimeter-Wave Wireless System," *2009 IEEE International Symposium on Personal, Indoor and Mobile Radio Communications (PIMRC 2009)*, pp. 1781–1785, Sep. 2009.

[JMW72] F. Jenks, P. Morgan, and C. Warren, "Use of Four-Level Phase Modulation for Digital Mobile Radio," *IEEE Transactions on Electromagnetic Compatibility*, vol. EMC-14, no. 4, pp. 113–128, Nov. 1972.

[JPM+11] M. Jacob, S. Priebe, A. Maltsev, A. Lomayev, V. Erceg, and T. Kurner, "A Ray Tracing Based Stochastic Human Blockage Model for the IEEE 802.11ad

60 GHz Channel Model," *2011 European Conference on Antennas and Propagation (EuCAP 2011)*, pp. 3084–3088, Apr. 2011.

[JT01] H. Jafarkhani and V. Tarokh, "Multiple Transmit Antenna Differential Detection from Generalized Orthogonal Designs," *IEEE Transactions on Information Theory*, vol. 47, no. 6, pp. 2626–2631, Sep. 2001.

[JW08] T. Jiang and Y. Wu, "An Overview: Peak-to-Average Power Ratio Reduction Techniques for OFDM Signals," *IEEE Transactions on Broadcasting*, vol. 54, no. 2, pp. 257–268, Jun. 2008.

[Kah54] L. R. Kahn, "Ratio Squarer," *Proceedings of the IRE*, vol. 42, p. 1704, 1954.

[Kaj00] A. Kajiwara, "LMDS Radio Channel Obstructed by Foliage," *2000 IEEE International Conference on Communications (ICC)*, vol. 3, pp. 1583–1587, 2000.

[Kat09] R. Katz, "Tech Titan Building Boom," *IEEE Spectrum*, vol. 46, no. 2, pp. 40–54, Feb. 2009.

[Kay08] M. Kayal, *Structured Analog CMOS Design*, 1st ed. Boston, Kluwer Academic Publishing, 2008.

[Kaz08] M. K. Kazimierczuk, *RF Power Amplifiers*, 1st ed. Hoboken, NJ, John Wiley and Sons, 2008.

[KBNI13] Y. Kishiyama, A. Benjebbour, T. Nakamura, and H. Ishii, "Future Steps of LTE-A: Evolution Toward Integration of Local Area and Wide Area Systems," *IEEE Wireless Communications*, vol. 20, no. 1, pp. 12–18, Feb. 2013.

[KFL01] F. Kschischang, B. Frey, and H.-A. Loeliger, "Factor Graphs and the Sum-Product Algorithm," *IEEE Transactions on Information Theory*, vol. 47, no. 2, pp. 498–519, Feb. 2001.

[KFN+08] R. Kimura, R. Funada, Y. Nishiguchi, M. Lei, T. Baykas, C. Sum, J. Wang, A. Rahman, Y. Shoji, H. Harada, et al., "Golay Sequence Aided Channel Estimation for Millimeter-Wave WPAN Systems," *2008 IEEE International Symposium on Personal, Indoor and Mobile Radio Communications (PIMRC 2008)*, pp. 1–5, 2008.

[KGG05] G. Kramer, M. Gastpar, and P. Gupta, "Cooperative Strategies and Capacity Theorems for Relay Networks," *IEEE Transactions on Information Theory*, vol. 51, no. 9, pp. 3037–3063, Sep. 2005.

[KH96] T. Keller and L. Hanzo, "Orthogonal Frequency Division Multiplex Synchronisation Techniques for Wireless Local Area Networks," *1996 IEEE International Symposium on Personal, Indoor and Mobile Radio Communications (PIMRC 1996)*, vol. 3, pp. 963–967, Oct. 1996.

[KHR+13] A. A. Khalek, R. W. Heath Jr., S. Rajagopal, S. Abu-Surra, and J. Zhang, "Cross-Polarization RF Precoding to Mitigate Mobile Misorientation and Polarization Leakage," *2013 IEEE Consumer Communications and Networking Conference (CCNC)*, Las Vegas, NV, Jan. 10–13, 2013.

[KJT08] T. Korakis, G. Jakllari, and L. Tassiulas, "CDR-MAC: A Protocol for Full Exploitation of Directional Antennas in Ad Hoc Wireless Networks," *IEEE Transactions on Mobile Computing*, vol. 7, no. 2, pp. 145–155, Feb. 2008.

[KKKL06] K. Kim, J. Kim, Y. Kim, and W. Lee, "Description of Korean 60 GHz Unlicensed Band Allocation," *IEEE 802.15.3c Meeting Contributions*, Jul. 2006. http://www.ieee802.org/15/pub/TG3c_contributions.html.

[KL10] S. Kim and L. E. Larson, "A 44 GHz SiGe BiCMOS Phase-Shifting Sub-Harmonic Up-Converter for Phased-Array Transmitters," *IEEE Transactions on Microwave Theory and Techniques*, vol. 58, no. 5, pp. 1089–1100, May 2010.

[KLN+10] D. G. Kam, D. Liu, A. Natarajan, S. Reynolds, and B. A. Floyd, "Low-Cost Antenna-in-Package Solutions for 60 GHz Phased Array Systems," *2010 IEEE Conference on Electrical Performance of Electronic Packaging and Systems (EPEPS 2010)*, pp. 93–96, Oct. 2010.

[KLN+11] D. G. Kam, D. Liu, A. Natarajan, S. Reynolds, H.-C. Chen, and B. A. Floyd, "LTCC Packages with Embedded Phased-Array Antennas for 60 GHz Communications," *IEEE Microwave and Wireless Components Letters*, vol. 21, no. 3, p. 142, 144, Mar. 2011.

[KLP+10] K. Kang, F. Lin, D.-D. Pham, J. Brinkhoff, C.-H. Heng, Y. X. Guo, and X. Yuan, "A 60 GHz OOK Receiver With an On-Chip Antenna in 90 nm CMOS," *IEEE Journal of Solid-State Circuits*, vol. 45, no. 9, pp. 1720–1731, Sep. 2010.

[KMK+00] N. Kurosawa, K. Maruyama, H. Kobayashi, H. Sugawara, and K. Kobayashi, "Explicit Formula for Channel Mismatch Effects in Time-Interleaved ADC Systems," *2000 IEEE Instrumentation and Measurement Technology Conference. (IMTC 2000)*, vol. 2, pp. 763–768, 2000.

[KO98] K. Kim and K. K. O, "Characteristics of Integrated Dipole Antennas on Bulk, SOI, and SOS Substrates for Wireless Communications," *1998 IEEE International Interconnect Technology Conference*, pp. 21–23, Jun. 1998.

[Koe00] M. Koen, "High Speed Data Conversion," *Texas Instruments Application Note*, Texas Instruments, 2000.

[KOH+09] S. Kishimoto, N. Orihashi, Y. Hamada, M. Ito, and K. Maruhashi, "A 60 GHz Band CMOS Phased Array Transmitter Utilizing Compact Baseband Phase Shifters," *2009 IEEE Radio Frequency Integrated Circuits Symposium (RFIC 2009)*, pp. 215–219, Jun. 2009.

[KOT+11] Y. Kohda, N. Ohba, K. Takano, D. Nakano, T. Yamane, and Y. Katayama, "Instant Multimedia Contents Downloading System Using a 60 GHz-2.4 GHz Hybrid Wireless Link," *2011 IEEE International Conference on Multimedia and Expo (ICME 2011)*, pp. 1–6, Jul. 2011.

[KP11a] F. Khan and Z. Pi, "Millimeter–Wave Mobile Broadband: Unleashing 3–300 GHz Spectrum," *2011 IEEE Wireless Communications and Networking Conference (WCNC)*, May 2011.

[KP11b] F. Khan and Z. Pi, U.S. Patent WO/2011/126,266: "Apparatus and Method for Spatial Division Duplex (SDD) for Millimeter Wave Communication System," Oct. 2011.

[KPLY05] K. Kim, S. Pinel, S. Laskar, and J.-G. Yook, "Circularly & Linearly Polarized Fan Beam Patch Antenna Arrays on Liquid Crystal Polymer Substrate for V-Band Applications," *2005 Asia-Pacific Microwave Conference (APMC)*, pp. 4–7, Dec. 2005.

[KR86] D. Kasilingam and D. Rutledge, "Focusing Properties of Small Lenses," *International Journal of Infrared and Millimeter Waves*, vol. 7, no. 10, pp. 1631–1647, 1986.

[KR05] S. Kulkarni and C. Rosenberg, "DBSMA: A MAC Protocol for Multi-hop Ad-hoc Networks with Directional Antennas," *2005 IEEE International Symposium on Personal, Indoor and Mobile Radio Communications (PIMRC 2005)*, vol. 2, pp. 1371–1377, Sep. 2005.

[KR07] K.-J. Koh and G. M. Rebeiz, "An X- and Ku-Band 8-Element Linear Phased Array Receiver," *2007 IEEE Custom Integrated Circuits Conference (CICC 2007)*, pp. 761–765, Jan. 2007.

[KSH00] T. Kailath, A. H. Sayed, and B. Hassibi, *Linear Estimation*. Upper Saddle River, NJ, Prentice Hall, 2000.

[KSK+09] K. Kimoto, N. Sasaki, S. Kubota, W. Moriyama, and T. Kikkawa, "High-Gain On-Chip Antennas for LSI Intra-/Intra-Chip Wireless Interconnection," *2009 European Conference on Antennas and Propagation (EuCAP 2009)*, pp. 278–282, Mar. 2009.

[KSM77] A. R. Kerr, P. H. Siegel, and R. J. Matauch, "A Simple Quasi-Optical Mixer for 100–120 GHz," *1977 IEEE MTT-S International Microwave Symposium (IMS)*, pp. 96–98, 1977.

[KSM+98] F. Kuroki, M. Sugioka, S. Matsukawa, K. Ikeda, and T. Yoneyama, "High-Speed ASK Transceiver Based on the NRD-Guide Technology at 60 GHz Band," *IEEE Transactions on Microwave Theory and Techniques*, vol. 46, no. 6, pp. 806–810, Jun. 1998.

[KSV00] Y.-B. Ko, V. Shankarkumar, and N. F. Vaidya, "Medium Access Control Protocols Using Directional Antennas in Ad Hoc Networks," *2000 Joint Conference* of the *IEEE Computer and Communications Societies (INFOCOM 2000)*, vol. 1, pp. 13–21, 2000.

[KT73] H. Kobayashi and D. T. Tang, "A Decision-feedback Receiver for Channels with Strong Intersymbol Interference," *IBM Journal of Research and Development*, vol. 17, no. 5, pp. 413–419, Sep. 1973.

[KTMM11] J. Kim, Y. Tian, A. Molisch, and S. Mangold, "Joint Optimization of HD Video Coding Rates and Unicast Flow Control for IEEE 802.11ad Relaying," *2011 IEEE International Symposium on Personal, Indoor and Mobile Radio Communications (PIMRC 2011)*, pp. 1109–1113, Sep. 2011.

[KV03] R. Koetter and A. Vardy, "Algebraic Soft-Decision Decoding of Reed-Solomon Codes," *IEEE Transactions on Information Theory*, vol. 49, no. 11, pp. 2809–2825, Nov. 2003.

[LBEY11] F. Liu, E. Bala, E. Erkip, and R. Yang, "A Framework for Femtocells to Access Both Licensed and Unlicensed Bands," *2011 International Symposium on Modeling and Optimization in Mobile, Ad Hoc and Wireless Networks (WiOpt)*, pp. 407–411, May 2011.

[LBVM06] C. P. Lim, R. J. Burkholder, J. L. Volakis, and R. J. Marhefka, "Propagation of Indoor Wireless Communications at 60 GHz," *2006 IEEE Antennas and Propagation Society International Symposium*, pp. 2149–2152, 2006.

[LC04] S. Lin and D. Costello, *Error Control Coding*. Upper Saddle River, NJ, Prentice Hall, 2004.

[LCF+07] M. Lei, C.-S. Choi, R. Funada, H. Harada, and S. Kato, "Throughput Comparison of Multi-Gbps WPAN (IEEE 802.15.3c) PHY Layer Designs under Non-Linear 60 GHz Power Amplifier," *2007 IEEE International Symposium on Personal, Indoor and Mobile Radio Communications (PIMRC 2007)*. pp. 1–5, Sep. 2007.

[LDS+05] J.-H. Lee, G. Degean, S. Sarkar, S. Pinel, D. Lim, J. Papapolymerou, J. Laskar, and M. M. Tentzeris, "Highly Integrated Millimeter-Wave Passive Components Using 3-D LTCC System-On-Package (SOP) Technology," *IEEE Transactions on Microwave Theory and Techniques*, vol. 53, no. 6, pp. 2220–2229, Jun. 2005.

[Leb95] P. N. Lebedew, "Ueber die Dopplbrechung der Strahlen electrischer Kraft," *Annalen der Physik und Chemie*, vol. 56, no. 9, pp. 1–17, 1895.

[Lee66] D. B. Leeson, "A Simple Model of Feedback Oscillator Noise Spectrum," *Proceedings of the IEEE*, vol. 54, pp. 329–330, 1966.

[Lee04a] T. H. Lee, *The Design of CMOS Radio-Frequency Integrated Circuits*, 2nd ed. Cambridge University Press, 2004.

[Lee04b] T. H. Lee, *Planar Microwave Engineering: A Practical Guide to Theory, Measurements and Circuits*, 1st ed. Cambridge University Press, 2004.

[Lee10] D. M. W. Leenaerts, "RF CMOS Power Amplifiers and Linearization Techniques," *2010 IEEE International Solid State Circuits Conference (ISSCC)*, p. Tutorial T4, Feb. 2010.

[LFR93] O. Landron, M. Feuerstein, and T. S. Rappaport, "In situ Microwave Reflection Coefficient Measurements for Smooth and Rough Exterior Wall Surfaces," *1993 IEEE Vehicular Technology Conference (VTC 1993)*, pp. 77–80, 1993.

[LGL+04] J.-J. Lin, X. Guo, R. Li, J. Branch, J. E. Brewer, and K. K. O, "10x Improvement of Power Transmission over Free Space Using Integrated Antennas on Silicon Substrates," *IEEE 2004 Custom Integrated Circuits Conference*, pp. 697–700, Oct. 2004.

[LGL+10] J.-C. Liu, Q. Gu, T. LaRocca, N.-Y. Wang, Y.-C. Wu, and M.-C. Chang, "A 60 GHz High Gain Transformer-Coupled Differential Power Amplifier in 65nm CMOS," *2010 Asia-Pacific Microwave Conference (APMC)*, pp. 932–935, 2010.

[LGPG09] D. Liu, et al., *Advanced Millimeter-wave Technologies: Antennas, Packaging, and Circuits*. Hoboken, NJ, John Wiley and Sons, 2009.

[LH77] W. Lee and F. Hill, "A Maximum-Likelihood Sequence Estimator with Decision-Feedback Equalization," *IEEE Transactions on Communications*, vol. 25, no. 9, pp. 971–979, 1977.

[LH03] D. J. Love and R. W. Heath Jr., "Equal Gain Transmission in Multiple-Input Multiple-Output Wireless Systems," *IEEE Transactions on Communications*, vol. 51, no. 7, pp. 1102–1110, Jul. 2003.

[LH09] M. Lei and Y. Huang, "Time-Domain Channel Estimation of High Accuracy for LDPC Coded SC-FDE System Using Fixed Point Decoding in 60 GHz WPAN," *2009 IEEE Consumer Communications and Networking Conference (CCNC 2009)*, pp. 1–5, Jan. 2009.

[LHCC07] J.-X. Liu, C.-Y. Hsu, H.-R. Chuang, and C.-Y. Chen, "A 60 GHz Millimeter-wave CMOS Marchand Balun," *2007 IEEE Radio Frequency Integrated Circuits Symposium (RFIC 2007)*, pp. 445–448, 2007.

[LHL+08] D. J. Love, R. W. Heath Jr., V. K. N. Lau, D. Gesbert, B. Rao, and M. Andrews, "An Overview of Limited Feedback in Wireless Communication Systems," *IEEE Journal of Selected Areas in Communications*, vol. 26, no. 8, pp. 1341–1365, Oct. 2008.

[LHS03] D. Love, R. W. Heath Jr., and T. Strohmer, "Grassmannian Beamforming for Multiple-Input Multiple-Output Wireless Systems," *IEEE Transactions on Information Theory*, vol. 49, no. 10, pp. 2735–2747, 2003.

[LHSH04] D. Love, R. W. Heath Jr., W. Santipach, and M. Honig, "What is the Value of Limited Feedback for MIMO Channels?" *IEEE Communications Magazine*, vol. 42, no. 10, pp. 54–59, 2004.

[Lie89] H. J. Liebe, "MPM — An Atmospheric Millimeter-Wave Propagation Model," *International Journal of Infrared and Millimeter Waves*, vol. 10, no. 6, pp. 631–650, 1989.

[Lit01] A. Litwin, "Overlooked Interfacial Silicide-Polysilicon Gate Resistance in MOS Transistors," *IEEE Transactions on Electron Devices*, vol. 48, no. 9, pp. 2179–2181, 2001.

[Liu84] K. Y. Liu, "Architecture for VLSI Design of Reed-Solomon Decoders," *IEEE Transactions on Computers*, vol. C-33, no. 2, pp. 178–189, Feb. 1984.

[LKBB09] B. Leite, E. Kerherve, J.-B. Begueret, and D. Belot, "Design and Characterization of CMOS Millimeter-Wave Transformers," *2009 SBMO/IEEE MTT-S International Microwave and Optoelectronics Conference (IMOC 2009)*, pp. 402–406, Nov. 2009.

[LKC08] X. Li, W.-M. L. Kuo, and J. Cressler, "A 40 GS/s SiGe Track-and-Hold Amplifier," *2008 IEEE Bipolar/BiCMOS Circuits and Technology Meeting (BCTM 2008)*, pp. 1–4, Oct. 2008.

[LKCY10] W. Lee, J. Kim, C. S. Cho, and Y. J. Yoon, "Beamforming Lens Antennas on a High Resistivity Silicon Wafer for 60 GHz WPAN," *IEEE Transactions on Antennas and Propagation*, vol. 58, no. 3, pp. 706–713, Mar. 2010.

[LKL+05] Y. Lu, W.-M. L. Kuo, X. Li, R. Krithivasan, J. Cressler, Y. Borokhovych, H. Gustat, B. Tillack, and B. Heinemann, "An 8-bit, 12 GSample/sec SiGe Track-and-Hold Amplifier," *2005 Bipolar/BiCMOS Circuits and Technology Meeting*, pp. 148–151, Oct. 2005.

[LKN+09] E. Laskin, M. Khanpour, S. Nicolson, A. Tomkins, P. Garcia, A. Cathelin, D. Belot, and S. Voinigescu, "Nanoscale CMOS Transceiver Design in the 90-170 GHz Range," *IEEE Transactions on Microwave Theory and Techniques*, vol. 57, no. 12, pp. 3477–3490, Dec. 2009.

[LKS10] T. B. Lavate, V. K. Kokate, and A. M. Sapkal, "Performance Analysis of MUSIC and ESPRIT," *2010 International Conference on Computer Network Technology (ICCNT)*, pp. 308–311, Apr. 2010.

[LLC06] Y.-Z. Lin, Y.-T. Liu, and S.-J. Chang, "A 6-Bit 2-GS/s Flash Aanlog-to-Digital Converter in 0.18μm CMOS Process," *2006 IEEE Asian Solid-State Circuits Conference (ASSCC 2006)*, pp. 351–354, 2006.

[LLC09] T. LaRocca, J.-C. Liu, and M.-C. Chang, "60 GHz CMOS Amplifiers Using Transformer-Coupling and Artificial Dielectric Differential Transmission Lines for Compact Design," *IEEE Journal of Solid-State Circuits*, vol. 44, no. 5, pp. 1425–1435, May 2009.

[LLH94] B. Langen, G. Loger, and W. Herzig, "Reflection and Transmission Behaviour of Building Materials at 60 GHz," *1994 IEEE International Symposium on Personal, Indoor, and Mobile Radio Communications (PIMRC 1994)*, pp. 505–509, 1994.

[LLL+10] Q. Li, G. Li, W. Lee, M. Lee, D. Mazzarese, B. Clerckx, and Z. Li, "MIMO Techniques in WiMAX and LTE: A Feature Overview," *IEEE Communications Magazine*, vol. 48, no. 5, pp. 86–92, May 2010.

[LLS+08] M. Lei, I. Lakkis, C.-S. Sum, T. Baykas, J.-Y. Wang, M. Rahman, R. Kimura, R. Funada, Y. Shoji, and H. Harada, "Hardware Impairments on LDPC Coded SC-FDE and OFDM in Multi-Gbps WPAN (IEEE 802.15.3c)," *2008 IEEE Wireless Communications and Networking Conference (WCNC)*, pp. 442–446, 2008.

[LLW06] C.-M. Lo, C.-S. Lin, and H. Wang, "A Miniature V-band 3-Stage Cascode LNA in 0.13μm CMOS," *2006 IEEE International Solid-State Circuits Conference (ISSCC)*, pp. 1254–1263, Feb. 2006.

[LLY+12] K. Lee, J. Lee, Y. Yi, I. Rhee, and S. Chong, "Mobile Data Offloading: How Much Can WiFi Deliver?" *IEEE/ACM Transactions on Networking*, vol. PP, no. 99, p. 1, 2012.

[LMR08] Y. Li, H. Minn, and R. Rajatheva, "Synchronization, Channel Estimation, and Equalization in MB-OFDM Systems," *IEEE Transactions on Wireless Communications*, vol. 7, no. 11, pp. 4341–4352, Nov. 2008.

[LNKH10] W. Lee, K. Noh, S. Kim, and J. Heo, "Efficient Cooperative Transmission for Wireless 3D HD Video Transmission in 60 GHz Channel," *IEEE Transactions on Consumer Electronics*, vol. 56, no. 4, pp. 2481–2488, Nov. 2010.

[LPC11] Z. Lin, X. Peng, and F. Chin, "Enhanced Beamforming for 60 GHz OFDM System with Co-Channel Interference Mitigation," *2011 IEEE International Conference on Ultra-Wideband (ICUWB 2011)*, pp. 29–33, 2011.

[LPCF12] Z. Lin, X. Peng, F. Chin, and W. Feng, "Outage Performance of Relaying with Directional Antennas in the Presence of Co-Channel Interferences at Relays," *IEEE Wireless Communications Letters*, vol. 1, no. 4, pp. 288–291, Aug. 2012.

[LR96] J. C. Liberti and T. S. Rappaport, "Analysis of CDMA Cellular Radio Systems Employing Adaptive Antenans in Multipath Environments," *1996 IEEE Vehicular Technology Conference (VTC 1996)*, vol. 2, pp. 1076–1080, 1996.

[LR99a] X. Li and J. Ritcey, "Trellis-Coded Modulation with Bit Interleaving and Iterative Decoding," *IEEE Journal on Selected Areas in Communications*, vol. 17, no. 4, pp. 715–724, Apr. 1999.

[LR99b] J. C. Liberti and T. S. Rappaport, *Smart Antennas for Wireless Communications: IS-95 and Third Generation CDMA Applications*. Upper Saddle River, NJ, Prentice Hall, 1999.

[LRS09] L. Lianming, P. Reynaert, and M. Steyaert, "Design and Analysis of a 90 nm mm-Wave Oscillator Using Inductive-Division LC Tank," *IEEE Journal of Solid-State Circuits*, vol. 44, no. 21, pp. 1950–1958, July 2009.

[LS08] D. Liu and R. Sirdeshmukh, "A Patch Array Antenna for 60 GHz Package Applications," *2008 IEEE Antennas and Propagation Society International Symposium*, pp. 1–4, Jul. 2008.

[LSE08] Z. Liu, E. Skafidas, and R. Evans, "A 60 GHz VCO with 6 GHz Tuning Range in 130 nm Bulk CMOS," *2008 International Conference on Microwave and Millimeter Wave Technology (ICMMT 2008)*, pp. 209–211, Apr. 2008.

[LSHM03] H. Luke, H. Schotten, and H. Hadinejad-Mahram, "Binary and Quadriphase Sequences with Optimal Autocorrelation Properties: A Survey," *IEEE Transactions on Information Theory*, vol. 49, no. 12, pp. 3271–3282, 2003.

[LSV08] A. E. I. Lamminen, J. Saily, A. R. Vimpari, "60 GHz Patch Antennas and Arrays on LTCC With Embedded-Cavity Substrates," *IEEE Transactions on Antennas and Propagation*, vol. 56, no. 9, pp. 2865, 2874, Sep. 2008.

[LSW+09] Z. Lan, C.-S. Sum, J. Wang, T. Baykas, F. Kojima, H. Nakase, and H. Harada, "Relay with Deflection Routing for Effective Throughput Improvement in Gbps Millimeter-Wave WPAN Systems," *IEEE Journal on Selected Areas in Communications*, vol. 27, no. 8, pp. 1453–1465, Oct. 2009.

[LTL09] I.-T. Lee, K.-H. Tsai, and S.-I. Liu, "A 104 to 112.8 GHz CMOS Injection-Locked Frequency Divider," *IEEE Transactions on Circuits Syst. II*, vol. 56, no. 7, pp. 555–559, Jul. 2009.

[LTW04] J. N. Laneman, D. N. C. Tse, and G. W. Wornell, "Cooperative Diversity in Wireless Networks: Efficient Protocols and Outage Behavior," *IEEE Transactions on Information Theory*, vol. 50, no. 12, pp. 3062–3080, Dec. 2004.

[LvTVN08] S. Louwsma, A. van Tuijl, M. Vertregt, and B. Nauta, "A 1.35 GS/s, 10 b, 175 mW Time-Interleaved AD Converter in 0.13 μm CMOS," *IEEE Journal of Solid-State Circuits*, vol. 43, no. 4, pp. 778–786, Apr. 2008.

[LWL+09] C.-H. Lien, C.-H. Wang, C.-S. Lin, P.-S. Wu, K.-Y. Ling, and H. Wang, "Analysis and Design of Reduced-Size Marchand Rat-Race Hybrid for Millimeter-Wave Compact Balanced Mixers in 130-nm CMOS Process," *IEEE Transactions on Microwave Theory and Techniques*, vol. 57, no. 8, pp. 1966–1977, Aug. 2009.

[LWS+08] Z. Lan, J. Wang, C.-S. Sum, T. Baykas, C. Pyo, F. Kojima, H. Harada, and S. Kato, "Unequal Error Protection for Compressed Video Streaming on 60 GHz WPAN System," *2008 International Wireless Communications and Mobile Computing Conference (IWCMC 2008)*, pp. 689–693, Aug. 2008.

[LWY04] F. Liu, J. Wang, and G. Yu, "An OTST-ESPRIT Algorithm for Joint DOA-Delay Estimation," *2004 International Symposium on Communication and Information Technologies (ISCIT)*, pp. 734–739, Oct. 2004.

[LZK+07] G. Lim, X. Zhou, K. Khu, Y. K. Yoo, F. Poh, G. See, Z. Zhu, C. Wei, S. Lin, and G. Zhu, "Impact of BEOL, Multi-Fingered Layout Design, and Gate Protection

Diode on Intrinsic MOSFET Threshold Voltage Mismatch," *2007 IEEE Conference on Electron Devices and Solid-State Circuits (EDSSC 2007)*, pp. 1059–1062, 2007.

[LZLS12] L. Lei, Z. Zhong, C. Lin, and X. Shen, "Operator Controlled Device-to-Device Communications in LTE-advanced Networks," *IEEE Wireless Communications*, vol. 19, no. 3, pp. 96–104, Jun. 2012.

[M+09] A. Maltsev, et al., "Experimental Investigations of 60 GHz WLAN Systems in Office Environment," *IEEE Journal of Selected Areas on Communications*, vol. 27, no. 8, Oct. 2009.

[Mad08] U. Madhow, "MultiGigabit Millimeter Wave Communication: System Concepts and Challenges," *Information Theory and Applications Workshop*, pp. 193–196, 2008.

[Mar10a] M. Marcus, "Civil Millimeter Wave Technology and Policy," Jan. 2010. http://www.marcus-spectrum.com/page5/index.html.

[MAR10b] J. W. My, R. A. Alhalabi, and G. M. Rebeiz, "A 3 G-Bit/s W-band SiGe ASK Receiver with a High-Efficiency On-Chip Electromagnetically-Coupled Antenna," *2010 IEEE Radio Frequency Integrated Circuits Symposium (RFIC 2010)*, pp. 87–90, May 2010.

[Mas65] J. Massey, "Step-by-Step Decoding of the Bose-Chaudhuri-Hocquenghem Codes," *IEEE Transactions on Information Theory*, vol. 11, no. 4, pp. 580–585, Oct. 1965.

[Mat07] A. Matsuzawa, "Trends in High Speed ADC Design," *2007 International Conference on ASIC (ASICON 2007)*, pp. 245–248, Oct. 2007.

[Max01] "INL/DNL Measurements for High-Speed Analog-to-Digital Converters (ADCs)," Maxim Integrated, Nov. 2001. http://www.maximintegrated.com/app-notes/index.mvp/id/283.

[Max02] "ADC and DAC Glossary," Tutorial 641, Maxim Integrated, Jul. 22, 2002, http://www.maximintegrated.com/app.notes/index.mvp/id/641

[Max08] "MAX109 Datasheet: 8-Bit, 2.2Gsps ADC with Track/Hold Amplifier and 1:4 Demultiplexed LVDS Outputs," Maxim Integrated Products, Inc. rev. 1, Mar. 2008. http://datasheets.maxim-ic.com/en/ds/MAX109.pdf.

[Maxim] Maxim Integrated, "MAX2265 Data Sheet." http://www.maximintegrated.com/datasheet/index.mvp/id/2084.

[MBDGR11] J. N. Murdock, E. Ben-Dor, F. Gutierrez, and T. S. Rappaport, "Challenges and Approaches to On-Chip Millimeter Wave Antenna Pattern Measurements," *2011 IEEE MTT-S International Microwave Symposium (IMS)*, Jun. 2011.

[MBDQ+12] J. N. Murdock, E. Ben-Dor, Y. Qiao, J. I. Tamir, and T. S. Rappaport, "A 38 GHz Cellular Outage Study for an Urban Outdoor Campus Environment," *2012*

IEEE Wireless Communications and Networking Conference (WCNC), pp. 3085–3090, Apr. 2012.

[MC] Mini-Circuits, "ADEX-10L Data Sheet." http://www.minicircuits.com/pdfs/ADEX-10L.pdf.

[MC04] N. Moraitis and P. Constantinou, "Indoor Channel Measurements and Characterization at 60 GHz for Wireless Local Area Network Applications," *IEEE Journal on Antennas and Propagation*, vol. 52, no. 12, pp. 3180–3189, 2004.

[Mcl80] P. J. Mclane, "A Residual Intersymbol Interference Error Bound for Truncated-State Viterbi Detectors," *IEEE Transactions on Information Theory*, vol. 26, no. 5, pp. 548–553, 1980.

[MEP+08] A. Maltsev, V. Erceg, E. Perahia, C. Hansen, R. Maslennikov, A. Lomayev, A. Sevastyanov, A. Khoryaev, G. Morozov, M. Jacob, S. Priebe, T. Kurner, S. Kato, H. Sawada, K. Sato, and H. Harada, "Channel Models for 60 GHz WLAN Systems," doc.: IEEE 802.11-09/0334r7, May 2008.

[MEP+10] A. Maltsev, V. Erceg, E. Perahia, C. Hansen, R. Maslennikov, A. Lomayev, A. Sevastyanov, and A. Khoryaev, "Channel Models for 60 GHz WLAN Systems," IEEE Document 802.11-09/0334r8, May 2010. https://mentor.ieee.org/802.11/dcn/09/11-09-0334-08-00ad-channel-models-for-60-ghz-wlan-systems.doc.

[MF05] N. Miladinovic and M. Fossorier, "Improved Bit-Flipping Decoding of Low-Density Parity-Check Codes," *IEEE Transactions on Information Theory*, vol. 51, no. 4, pp. 1594–1606, Apr. 2005.

[MFO+07] T. Mitomo, R. Fujimoto, N. Ono, R. Tachibana, H. Hoshino, Y. Yoshihara, Y. Tsutsumi, and I. Seto, "A 60 GHz CMOS Receiver with Frequency Synthesizer," *2007 IEEE Symposium on VLSI Circuits*, pp. 172–173, Jun. 2007.

[MGFZ06] B. Motlagh, S. E. Gunnarsson, M. Ferndahl, and H. Zirath, "Fully Integrated 60 GHz Single-Ended Resistive Mixer in 90-nm CMOS Technology," *IEEE Microwave and Wireless Components Letters*, vol. 16, no. 1, pp. 25–27, Jan. 2006.

[MH14] Jianhua Mo and R. W. Heath Jr., "High SNR Capacity of Millimeter Wave MIMO Systems with One-Bit Quantization," *2014 Information Theory and Applications Workshop (ITA)*, San Diego, CA, Feb. 9–14, 2014.

[MHM10] R. A. Monzingo, R. L. Haupt, and T. W. Miller, *Introduction to Adaptive Arrays*. Raleigh, NC, SciTech Publishing, 2010.

[MHP+09] F. Mustafa, A. M. Hashim, N. Parimon, S. F. A. Rhaman, A. R. A. Rahmn, and M. N. Osman, "RF Characterization of Planar Dipole Antenna for On-Chip Integrated with GaAs-Based Schottky Diode," *2009 Asia Pacific Microwave Conference (APMC)*, pp. 571–574, Dec. 2009.

[MI13] S. Mukherjee and H. Ishii, "Energy Efficiency in the Phantom Cell Enhanced Local Area Architecture," *2013 IEEE Wireless Communications and Networking Conference (WCNC)*, pp. 1267–1272, Apr. 2013.

[MID+00] K. Maruhashi, M. Ito, L. Desclos, K. Ikuina, N. Senba, N. Takahashi, and K. Ohata, "Low-Cost 60 GHz Band Antenna-Integrated Transmitter/Receiver Modules Utilizing Multi-Layer Low-Temperature Co-Fired Ceramic Technology," *2000 IEEE International Solid-State Circuits Conference (ISSCC)*, pp. 324–325, Feb. 2000.

[Mil] "Milestones: First Millimeter-Wave Communication Experiments by J.C. Bose," http://www.ieeeghn.org/wiki/index.php/Milestones:First_Millimeter-wave_Communication_Experiments_by_J.C._Bose.

[Min06] Ministry of Internal Affairs and Communications (Japan), "Information and Communications Policy Site," 2006. http://www.soumu.go.jp/joho_tsusin/eng/laws_dt02.html.

[ML87] R. J. Meier and Y. P. Loh, "A 60 GHz Beam Waveguide Antenna System for Satellite Crosslinks," *1987 IEEE Military Communications Conference - Crisis Communications: The Promise and Reality (MILCOM 1987)*, vol. 1, pp. 0260–0264, Oct. 1987.

[MM95] U. Mengali and M. Morelli, "Decomposition of M-ary CPM Signals into PAM Waveforms," *IEEE Transactions on Information Theory*, vol. 41, no. 5, pp. 1265–1275, 1995.

[MM99] M. Morelli and U. Mengali, "An Improved Frequency Offset Estimator for OFDM Applications," *IEEE Communications Letters*, vol. 3, no. 3, pp. 75–77, Mar. 1999.

[MMGM09] M. Manohara, R. Mudumbai, J. Gibson, and U. Madhow, "Error Correction Scheme for Uncompressed HD Video over Wireless," *2009 IEEE International Conference on Multimedia and Expo (ICME 2009)*, pp. 802–805, Jul. 2009.

[MMI96] T. Manabe, Y. Miura, and T. Ihara, "Effects of Antenna Directivity on Indoor Multipath Propagation Characteristics at 60 GHz," *IEEE Journal on Selected Areas in Communications*, vol. 14, no. 3, pp. 441–448, 1996.

[MMS+10] A. Maltsev, R. Maslermikov, A. Sevastyanov, A. Lomayev, A. Khoryaev, A. Davydov, and V. Ssorin, "Characteristics of Indoor Millimeter-Wave Channel at 60 GHz in Application to Perspective WLAN System," *2010 European Conference on Antennas and Propagation (EuCAP 2010)*, pp. 1–5, 2010.

[MN96] D. MacKay and R. Neal, "Near Shannon Limit Performance of Low Density Parity Check Codes," *Electronics Letters*, vol. 32, no. 18, p. 1645, Aug. 1996.

[MN08] C. Marcu and A. Niknejad, "A 60 GHz high-Q Tapered Transmission Line Resonator in 90nm CMOS," *2008 IEEE MTT-S International Microwave Symposium (IMS)*, pp. 775–778, Jun. 2008.

[Mon71] P. Monsen, "Feedback Equalization for Fading Dispersive Channels," *IEEE Transactions on Information Theory*, vol. 17, no. 1, pp. 56–64, Jan. 1971.

[Mon84] P. Monsen, "MMSE Equalization of Interference on Fading Diversity Channels," *IEEE Transactions on Communications*, vol. 32, no. 1, pp. 5–12, 1984.

[Moo94] P. Moose, "A Technique for Orthogonal Frequency Division Multiplexing Frequency Offset Correction," *IEEE Transactions on Communications*, vol. 42, no. 10, pp. 2908–2914, Oct. 1994.

[MPRZ99] G. Masera, G. Piccinini, M. Roch, and M. Zamboni, "VLSI Architectures for Turbo Codes," *IEEE Transactions on Very Large Scale Integration (VLSI) Systems*, vol. 7, no. 3, pp. 369–379, Sep. 1999.

[MR14a] G. R. MacCartney and T. S. Rappaport, "73 GHz Millimeter Wave Propagation Measurements for Outdoor Urban Mobile and Backhaul Communications in New York City," *2014 IEEE International Conference on Communications (ICC)*, June 2014.

[MR14b] J. Murdock and T. S. Rappaport, "Consumption Factor and Power-Efficiency Factor: A Theory for Evaluating the Energy Efficiency of Cascaded Communication Systems," *IEEE Journal on Selected Areas in Communications*, vol. 32, no. 12, pp. 1–16, Dec. 2014.

[MRM09] G. Madhumati, K. Rao, and M. Madhavilatha, "Comparison of 5-bit Thermometer-to-Binary Decoders in 1.8V, 0.18 um CMOS Technology for Flash ADCs," *2009 International Conference on Signal Processing Systems*, pp. 516–520, May 2009.

[MS03] M. Mansour and N. Shanbhag, "High-Throughput LDPC Decoders," *IEEE Transactions on Very Large Scale Integration (VLSI) Systems*, vol. 11, no. 6, pp. 976–996, Dec. 2003.

[MSEA03] K. Mukkavilli, A. Sabharwal, E. Erkip, and B. Aazhang, "On Beamforming with Finite Rate Feedback in Multiple-Antenna Systems," *IEEE Transactions on Information Theory*, vol. 49, no. 10, pp. 2562–2579, Oct. 2003.

[MSM04] J. McCarthy, M. Sasse, and D. Miras, "Sharp or Smooth?: Comparing the Effects of Quantization Vs. Frame Rate for Streamed Video," *SIGCHI Conference on Human Factors in Computing Systems*, ACM, pp. 535–542, 2004.

[MSN10] M. Matthaiou, A. Sayeed, and J. Nossek, "Maximizing LoS MIMO Capacity Using Reconfigurable Antenna Arrays," *International ITG Workshop on Smart Antennas*, pp. 14–19, 2010.

[MSR14] G. MacCartney, M. Samimi, and T. S. Rappaport, "Omnidirectional Channel Models for mmWave Communications in New York City," *2014 IEEE Personal, Indoor and Mobile Communications Conference*, Washington, DC, Sep. 2014.

[MTH+08] K. Maruhashi, M. Tanomura, Y. Hamada, M. Ito, N. Orihashi, and S. Kishimoto, "60 GHz Band CMOS MMIC Technology for High-Speed Wireless Personal Area Networks," *2008 IEEE Compound Semiconductor Integrated Circuits Symposium (CSIC 2008)*, pp. 1–4, Oct. 2008.

[Muh96] R. Muhamed, "Direction of Arrival Estimation using Antenna Arrays," Master's Thesis, Virginia Tech, Jan. 1996.

[MVLP10] A. Mahanfar, R. G. Vaughan, S.-W. Lee, and A. M. Parameswaran, "Self-Assembled Monopole Antenna with Arbitrary Title Angles for System-on-Chip and System-in-Package Applications," *IEEE Transactions on Antennas and Propagation*, vol. 58, pp. 3020–3028, Sep. 2010.

[MWG+02] B. Muquet, Z. Wang, G. Giannakis, M. de Courville, and P. Duhamel, "Cyclic Prefixing or Zero Padding for Wireless Multicarrier Transmissions?" *IEEE Transactions on Communications*, vol. 50, no. 12, pp. 2136–2148, Dec. 2002.

[MZB00] H. Minn, M. Zeng, and V. Bhargava, "On Timing Offset Estimation for OFDM Systems," *IEEE Communications Letters*, vol. 4, no. 7, pp. 242–244, Jul. 2000.

[MZNR13] G. R. MacCartney, J. Zhang, S. Nie, and T. S. Rappaport, "Path Loss Models for 5G Millimeter Wave Propagation Channels in Urban Microcells," *2013 IEEE Global Communications Conference (GLOBECOM 2013)*, Dec. 2013.

[NBH10] J. Nsenga, A. Bourdoux, and F. Horlin, "Mixed Analog/Digital Beamforming for 60 GHz MIMO Frequency Selective Channels," *2010 IEEE International Conference on Communications (ICC)*, pp. 1–6, 2010.

[NFH07] A. Natarajan, B. Floyd, and A. Hajimiri, "A Bidirectional RF-Combining 60 GHz Phased-Array Front-End," *2007 IEEE International Solid-State Circuits Conference (ISSCC)*, p. 202, Feb. 2007.

[NH08] A. M. Niknejad and H. Hashemi, *Mm-Wave Silicon Technology: 60 GHz and Beyond*, 1st ed. New York, Springer, 2008.

[Nik10] A. M. Niknejad, "Siliconization of 60 GHz," *IEEE Microwave Magazine*, vol. 11, no. 1, pp. 78–85, Feb. 2010.

[NLSS12] R. Narasimha, M. Lu, N. R. Shanbhag, and A. C. Singer, "BER-Optimal Analog-to-Digital Converters for Communication Links," *IEEE Transactions on Signal Processing*, vol. 60, no. 7, pp. 3683–3691, 2012.

[NMSR13] S. Nie, G. R. MacCartney, S. Sun, and T. S. Rappaport, "72 GHz Millimeter Wave Indoor Measurements for Wireless and Backhaul Communications," *2013 IEEE International Symposium on Personal, Indoor and Mobile Radio Communications (PIMRC 2013)*, pp. 2429–2433, 2013.

[NNP+02] T. Nakagawa, K. Nishikawa, B. Piernas, T. Seki, and K. Araki, "60 GHz Antenna and 5 GHz Demodulator MMICs for More Than 1-Gbps FSK Transceivers," *2002 European Microwave Conference*, pp. 1–4, Sep. 2002.

[npr94] Federal Communications Commission, "ET Docket No. 94-124 & RM-8308: Amendments of Parts 2 and 15 of the Commission's Rules to Permit Use of Radio Frequencies Above 40 GHz for New Radio Applications," Washington, D.C., FCC, Oct. 1994. http://transition.fcc.gov/oet/dockets/et94-124/.

[NRS96] W. G. Newhall, T. S. Rappaport, and D. G. Sweeney, "A Spread Spectrum Sliding Correlator System for Propagation Measurements," *RF Design*, pp. 40–54, Apr. 1996.

[NVTL+07] J. Nsenga, W. Van Thillo, R. Lauwereins, et al., "Comparison of OQPSK and CPM for Communications at 60 GHz with a Nonideal Front End," *EURASIP Journal on Wireless Communications and Networking*, vol. 2007, 2007:doi:10.1155/2007/86206.

[NYU12] "Wireless Center for NYU Poly," *Wall Street Journal*, Anjali Athavaley, Aug. 7, 2012.

[O98] K. O, "Estimation Methods for Quality Factors of Inductors Fabricated in Silicon Integrated Circuit Process Technologies," *IEEE Journal of Solid-State Circuits*, vol. 33, no. 8, pp. 1249–1252, Aug. 1998.

[OD09] S. Oza and N. Devashrayee, "Low Voltage, Low Power Folding Amplifier for Folding & Interpolating ADC," *2009 International Conference on Advances in Recent Technologies in Communication and Computing (ARTCom 2009)*, pp. 178–182, Oct. 2009.

[OET97] Federal Communications Commission Office of Engineering Technology, "Millimeter Wave Propagation: Spectrum Management Applications," FCC Office of Engineering Technology Bulletin 70, Jul. 1997.

[OFVO11] M. Ortner, H. P. Forstner, L. Verweyen, and T. Ostermann, "A Fully Integrated Homodyne Downconverter MMIC in SiGe:C for 60 GHz Wireless Applications," *2011 IEEE Meeting on Silicon Monolithic Integrated Circuits in RF Systems (SiRF 2011)*, pp. 145–148, Jan. 2011.

[OKF+05] K. O, et al., "On-Chip Antennas in Silicon ICs and Their Application," *IEEE Transactions on Electron Devices*, vol. 52, no. 7, pp. 1312–1323, Jul. 2005.

[OKLR09] Y. Oh, S. Kim, S. Lee, and J.-S. Rieh, "The Island-Gate Varactor—A High-Q MOS Varactor for Millimeter-Wave Applications," *IEEE Microwave and Wireless Components Letters*, vol. 19, no. 4, pp. 215–217, Apr. 2009.

[OKM+04] F. Okano, M. Kanazawa, K. Mitani, K. Hamasaki, M. Sugawara, M. Seino, A. Mochimaru, and K. Doi, "Ultrahigh-Definition Television System with 4000 Scanning Lines," *2004 NAB Broadcast Engineering Conference*, pp. 437–440, 2004.

[OMI+02] K. Ohata, K. Maruhashi, M. Ito, S. Kishimoto, K. Ikuina, T. Hashiguchi, N. Takahashi, and S. Iwanaga, "Wireless 1.25 Gb/s Transceiver Module at 60 GHz Band," *2002 IEEE International Solid-State Circuits Conference (ISSCC)*, pp. 298–468, Feb. 2002.

[OS09] A. V. Oppenheim and R. W. Schafer, *Discrete-Time Signal Processing*, 3rd ed. Upper Saddle River, NJ, Prentice Hall, 2009.

[PA03] H. Pan and A. Abidi, "Spatial Filtering in Flash A/D Converters," *IEEE Transactions on Circuits and Systems II*, vol. 50, no. 8, pp. 424–436, Aug. 2003.

[PA09] K. Payandehjoo and R. Abhari, "Characterization of On-Chip Antennas for Millimeter-Wave Applications," *2009 IEEE Antennas and Propagation Society International Symposium*, pp. 1–4, Jun. 2009.

[PB02] S. Perras and L. Bouchard, "Fading Characteristics of RF Signals due to Foliage in Frequency Bands from 2 to 60 GHz," *2002 International Symposium on Wireless Personal Multimedia Communications*, vol. 1, pp. 267–271, 2002.

[PBA03] T. Pratt, C. W. Bostian, and J. E. Allnut, *Satellite Communication Systems*. Hoboken, NJ, John Wiley and Sons Inc., 2003.

[PBC99] D. S. Polydorou, P. G. Babalis, and C. N. Capsalis, "Statistical Characterization of Fading in LOS Wireless Channels with a Finite Number of Dominant Paths: Applications in Millimeter Frequencies," *International Journal of Infrared and Millimeter Waves*, vol. 20, pp. 461–472, 1999.

[PCPY08] M. Park, C. Cordeiro, E. Perahia, and L. Yang, "Millimeter-wave Multi-Gigabit WLAN: Challenges and Feasibility," *2008 IEEE International Symposium on Personal, Indoor and Mobile Radio Communications (PIMRC 2008)*, pp. 1–5, Sep. 2008.

[PCPY09] M. Park, C. Cordeiro, E. Perahia, and L. Yang, "QoS Considerations for 60 GHz Wireless Networks," *2009 IEEE GLOBECOM Workshops*, pp. 1–6, Dec. 2009.

[PCYY10] P. Park, L. Che, H.-K. Yu, and C. P. Yue, "A Fully Integrated Transmitter with Embedding Antenna for On-Wafer Wireless Testing," *IEEE Transactions on Microwave Theory and Technique*, vol. 58, no. 5, pp. 1456–1463, May 2010.

[PDK+07] P. Pepeljugoski, F. Doany, D. Kuchta, L. Schares, C. Schow, M. Ritter, and J. Kash, "Data Center and High Performance Computing Interconnects for 100 Gb/s and Beyond," *2007 Conference on Optical Fiber Communication and the National Fiber Optic Engineers Conference (OFC/NFOEC 2007)*, pp. 1–3, Mar. 2007.

[PDW89] M. Pelgrom, A. Duinmaijer, and A. Welbers, "Matching Properties of MOS Transistors," *IEEE Journal of Solid-State Circuits*, vol. 24, no. 5, pp. 1433–1439, Oct. 1989.

[Per10] E. Perahia, "TGad Evaluation Methodology," IEEE P802.11 Wireless LANs, Technical Report, 2010, doc: IEEE 802.11-09/0296r16.

[PG07a] U. Pfeiffer and D. Goren, "A 20 dBm Fully-Integrated 60 GHz SiGe Power Amplifier With Automatic Level Control," *IEEE Journal of Solid-State Circuits*, vol. 42, no. 7, pp. 1455–1463, Jul. 2007.

[PG07b] M. Piz and E. Grass, "A Synchronization Scheme for OFDM-based 60 GHz WPANs," *2007 IEEE International Symposium on Personal, Indoor and Mobile Radio Communications (PIMRC 2007)*, pp. 1–5, Sep. 2007.

[PG09] M. Park and P. Gopalakrishnan, "Analysis on Spatial Reuse, Interference, and MAC Layer Interference Mitigation Schemes in 60 GHz Wireless Networks," *2009 IEEE International Conference on Ultra-Wideband (ICUWB 2009)*, pp. 1–5, Sep. 2009.

[PH02] R. Parot and F. Harris, "Resolving and Correcting Gain and Phase Mismatch in Transmitters and Receivers for Wideband OFDM Systems," *2002 Asilomar Conference on Signals, Systems and Computers*, vol. 2, pp. 1005–1009, Nov. 2002.

[PH09] S. W. Peters and R. W. Heath Jr., "The Future of WiMAX: Multihop Relaying with IEEE 802.16j," *IEEE Communications Magazine*, vol. 47, no. 1, pp. 104–111, Jan. 2009.

[PHAH97] P. Papazian, G. Hufford, R. Achatz, and R. Hoffman, "Study of the Local Multipoint Distribution Service Radio Channel," *IEEE Transactions on Broadcasting*, vol. 43, no. 2, pp. 175–184, Jun. 1997.

[PHR09] C. H. Park, R. W. Heath Jr., and T. S. Rappaport, "Frequency-Domain Channel Estimation and Equalization for Continuous-Phase Modulations With Superimposed Pilot Sequences," *IEEE Transactions on Vehicular Technology*, vol. 58, no. 9, pp. 4903–4908, Nov. 2009.

[Pi12] Z. Pi, "Optimal Transmitter Beamforming with Per-Antenna Power Constraints," *2012 IEEE International Conference on Communications (ICC)*, pp. 3779–3784, 2012.

[PK94] A. Paulraj and T. Kailath, U.S. Patent 5345599: "Increasing Capacity in Wireless Broadcast Systems Using Distributed Transmission/Directional Reception (DTDR)," Sep. 1994.

[PK11] Z. Pi and F. Khan, "An Introduction to Millimeter-Wave Mobile Broadband Systems," *IEEE Communications Magazine*, vol. 49, no. 6, pp. 101–107, Jun. 2011.

[PKJP12] H. Park, Y. Kim, I. Jang, and S. Pack, "Cooperative Neighbor Discovery for Consumer Devices in mmWave ad-hoc Networks," *2012 IEEE International Conference on Consumer Electronics (ICCE)*, pp. 100–101, 2012.

[PKZ10] Z. Pi, F. Khan, and J. Zhang, U.S. Patent App. 12/916,019: "Techniques for Millimeter Wave Mobile Communication," Oct. 2010.

[PLK12] Z. Pi, Y. Li, and F. Khan, U.S. Patent 20,120,307,726: "Methods and Apparatus to Transmit and Receive Synchronization Signal and System Information in a Wireless Communication System," Dec. 2012.

[PLT86] R. Papa, J. Lennon, and R. Taylor, "The Variation of Bistatic Rough Surface Scattering Cross Section for a Physical Optics Model," *IEEE Transactions on Antennas and Propagation*, vol. 34, no. 10, pp. 1229–1237, Oct. 1986.

[PMR+08] U. Pfeiffer, C. Mishra, R. Rassel, S. Pinkett, and S. Reynolds, "Schottky Barrier Diode Circuits in Silicon for Future Millimeter-Wave and Terahertz Applications," *IEEE Transactions on Microwave Theory and Techniques*, vol. 56, no. 2, pp. 364–371, Feb. 2008.

[PNG+98] G. Passiopoulos, S. Nam, A. Georgiou, A. E. Ashtiani, I. D. Roberston, and E. A. Grindrod, "V-Band Single Chip, Direct Carrier BPSK Modulation Transmitter with Integrated Patch Antenna," *1998 IEEE MTT-S International Microwave Symposium (IMS)*, pp. 305–308, Jun. 1998.

[PNG03] A. Paulraj, R. Nabar, and D. Gore, *Introduction to Space-Time Wireless Communications.* New York, Cambridge University Press, 2003.

[PNS+03] K. Poulton, R. Neff, B. Setterberg, B. Wuppermann, T. Kopley, R. Jewett, J. Pernillo, C. Tan, and A. Montijo, "A 20 GS/s 8 b ADC with a 1 MB Memory in 0.18 μm CMOS," *2003 IEEE International Solid-State Circuits Conference (ISSCC)*, IEEE, pp. 318–496, 2003.

[Pop99] B. Popovic, "Efficient Golay Correlator," *Electronics Letters*, vol. 35, no. 17, pp. 1427–1428, 1999.

[Poz05] D. M. Pozar, *Microwave Engineering*, 3rd ed. Hoboken, NJ, John Wiley and Sons, Inc., 2005.

[PP10] M. Park and H. K. Pan, "Effect of Device Mobility and Phased Array Antennas on 60 GHZ Wireless Networks," *2010 ACM Workshop on MmWave Communications: From Circuits to Networks*, New York, ACM, pp. 51–56, 2010.

[PPH93] A. Plattner, N. Prediger, and W. Herzig, "Indoor and Outdoor Propagation Measurements at 5 and 60 GHz for Radio LAN Application," *1993 IEEE MTT-S International Microwave Symposium (IMS)*, vol. 2, pp. 853–856, 1993.

[PPTH09] S. Peters, A. Panah, K. Truong, and R. W. Heath Jr., "Relay Architectures for 3GPP LTE-advanced," *EURASIP Journal on Wireless Communications and Networking*, vol. 2009, no. 1, p. 618787, Jul. 2009.

[PR80] A. Peled and A. Ruiz, "Frequency Domain Data Transmission Using Reduced Computational Complexity Algorithms," *1980 IEEE International Conference on Acoustics, Speech, and Signal Processing (ICASSP 1980)*, vol. 5, pp. 964–967, Apr. 1980.

[PR07] C. Park and T. S. Rappaport, "Short-Range Wireless Communications for Next-Generation Networks: UWB, 60 GHz Millimeter-Wave WPAN, and ZigBee," *IEEE Wireless Communications*, vol. 14, no. 4, pp. 70–78, Aug. 2007.

[PR08] A. Parsa and B. Razavi, "A 60 GHz CMOS Receiver Using a 30 GHz LO," *2008 IEEE International Solid-State Circuits Conference (ISSCC)*, pp. 190–606, Feb. 2008.

[PR09] A. Parsa and B. Razavi, "A New Transceiver Architecture for the 60 GHz Band," *IEEE Journal of Solid-State Circuits*, vol. 44, no. 3, pp. 751–762, Mar. 2009.

[PRK86] A. Paulraj, R. Roy, and T. Kailath, "A Subspace Rotation Approach to Signal Parameter Estimation," *Proceedings of the IEEE*, vol. 74, no. 7, pp. 1044–1046, Jul. 1986.

[Pro01] J. Proakis, *Spread Spectrum Signals for Digital Communications*. Wiley Online Library, 2001.

[PS07] J. Proakis and M. Salehi, *Digital Communications*, 5th ed. McGraw-Hill, 2007.

[PSS+08] S. Pinel, S. Sarkar, P. Sen, B. Perumana, D. Yeh, D. Dawn, and J. Laskar, "A 90nm CMOS 60 GHz Radio," *2008 IEEE International Solid-State Circuits Conference (ISSCC)*, pp. 130–601, Feb. 2008.

[Pul10] D. Pulfrey, *Understanding Modern Transistors and Diodes*, 1st ed. Cambridge, UK, Cambridge University Press, 2010.

[PV06] F. Pancaldi and G. M. Vitetta, "Equalization Algorithms in the Frequency Domain for Continuous Phase Modulations," *IEEE Transactions on Communications*, vol. 54, no. 4, pp. 648–658, 2006.

[PVBM95] T. Pollet, M. Van Bladel, and M. Moeneclaey, "BER Sensitivity of OFDM Systems to Carrier Frequency Offset and Wiener Phase Noise," *IEEE Transactions on Communications*, vol. 43, no. 234, pp. 191–193, Feb/Mar/Apr 1995.

[PW08] P. H. Park and S. S. Wong, "An On-Chip Dipole Antenna for Millimeter-Wave Transmitters," *2008 IEEE Radio Frequency Integrated Circuits Symposium (RFIC 2008)*, pp. 629–632, Apr. 2008.

[PWI03] J. Park, Y. Wang, and T. Itoh, "A 60 GHz Integrated Antenna Array for High-Speed Digital Beamforming Applications," *2003 IEEE MTT-S International Microwave Symposium (IMS)*, vol. 3, pp. 1677–1680, 2003.

[PWO01] H. Papadopoulos, G. Wornell, and A. Oppenheim, "Sequential Signal Encoding from Noisy Measurements Using Quantizers with Dynamic Bias Control," *IEEE Transactions on Information Theory*, vol. 47, no. 3, pp. 978–1002, Mar. 2001.

[PZ02] D. Parker and D. C. Zimmermann, "Phased-Arrays Part II: Implementations, Applications, and Future Trends," *IEEE Transactions on Microwave Theory and Techniques*, vol. 50, no. 3, pp. 688–698, Mar. 2002.

[QCSM11] J. Qiao, L. Cai, X. Shen, and J. Mark, "Enabling Multi-Hop Concurrent Transmissions in 60 GHz Wireless Personal Area Networks," *IEEE Transactions on Wireless Communications*, vol. 10, no. 11, pp. 3824–3833, Nov. 2011.

[Qur82] S. Qureshi, "Adaptive Equalization," *IEEE Communications Magazine*, vol. 20, no. 2, pp. 9–16, 1982.

[Qur85] S. Qureshi, "Adaptive Equalization," *Proceedings of the IEEE*, vol. 73, no. 9, pp. 1349–1387, Sep. 1985.

[R+11] S. Rajagopal, et al., "Antenna Array Design for multi-Gbps mmWave Mobile Broadband Communication," *2011 IEEE Global Communications Conference (GLOBECOM 2011)*, Dec. 2011.

[Rap89] T. S. Rappaport, "Characterization of UHF Multipath Radio Channels in Factory Buildings," *IEEE Transactions on Antennas and Propagation*, vol. 37, no. 8, pp. 1058–1069, Aug. 1989.

[Rap91] C. Rapp, "Effects of HPA-Nonlinearity on a 4-DPSK/OFDM-Signal for a Digital Sound Broadcasting System," *The European Conference on Satellite Communications*, pp. 179–184, 1991.

[Rap98] T. S. Rappaport, *Smart Antennas: Adaptive Arrays, Algorithms, & Wireless Position Location*. New York, IEEE Press, 1998.

[Rap02] T. S. Rappaport, *Wireless Communications, Principles and Practice*, 2nd ed. Upper Saddle River, NJ, Prentice Hall, 2002.

[Rap09] T. S. Rappaport, "The Emerging World of Massively Broadband Devices: 60 GHz and Above," Keynote presentation at *2009 Virginia Tech Wireless Symposium*, Blacksburg, VA, Jun. 2009.

[Rap10] T. S. Rappaport, "Broadband Repeater with Security for Ultrawideband Technologies," U.S. Patent 7,676,194, Mar. 2010.

[Rap11] T. S. Rappaport, "Sub-Terahertz Wireless Communications: The Future Edge of the Internet," Plenary talk at *2011 IEEE International Symposium on Personal, Indoor and Mobile Radio Communications (PIMRC)*, Toronto, Canada, Sep. 2011.

[Rap11a] T. S. Rappaport, "Broadband Repeater with Security for Ultrawideband Technologies," U.S. Patent 7,983,613, Jul. 2011.

[Rap12a] T. S. Rappaport, "The Renaissance of Wireless Communications in the Massively Broadband® Era," Plenary talk at *2012 IEEE Vehicular Technology Conference (VTC 2012-Fall)*, Quebec City, Canada, Sep. 2012.

[Rap12b] T. S. Rappaport, "The Renaissance of Wireless Communications in the Massively Broadband® Era," Keynote presentation at *2012 IEEE International Conference on Communications in China (ICCC 2012)*, Beijing, China, Aug. 2012.

[Rap12c] T. S. Rappaport, "Wireless Communications in the Massively Broadband® Era," *Microwave Journal*, pp. 46–48, Dec. 2012.

[Rap13] T. S. Rappaport, "Active Antennas for Multiple Bands in Wireless Portable Devices," U.S. Patent 8,593,358, Nov. 2013.

[Rap14] T. S. Rappaport, "Millimeter Wave Cellular Communications: Channel Models, Capacity Limits, Challenges and Opportunities," Invited presentation at *2014*

IEEE Communications Theory Workshop (2014 CTW), Curacao, May 26, 2014. http://www.ieee-ctw.org/2014/slides/session1/Ted_Rappaport_CTW2014.pdf

[Rap14a] T. S. Rappaport, "Defining the Wireless Future — Millimeter Wave Wireless Communications: The Renaissance of Computing and Communications," Keynote presentation at *2014 IEEE International Conference on Communications (ICC)*, Sydney, Australia, Jun. 13, 2014. http://icc2014.ieee-icc.org/speakers_28_4101586138.pdf

[Raz98] B. Razavi, *RF Microelectronics*. Upper Saddle River, NJ, Prentice Hall, 1998.

[Raz01] B. Razavi, *Design of Analog CMOS Integrated Circuits*, 1st ed. New York, McGraw Hill, 2001.

[Raz06] B. Razavi, "A 60 GHz CMOS Receiver Front-End," *IEEE Journal of Solid-State Circuits*, vol. 41, no. 1, pp. 17–22, Jan. 2006.

[Raz08] B. Razavi, "A Millimeter-Wave Circuit Technique," *IEEE Journal of Solid-State Circuits*, vol. 43, pp. 2090–2098, Sep. 2008.

[RBDMQ12] T. S. Rappaport, E. Ben-Dor, J. Murdock, and Y. Qiao, "38 GHz and 60 GHz Angle-Dependent Propagation for Cellular & Peer-to-Peer Wireless Communications," *2012 IEEE International Conference on Communications (ICC)*, pp. 4568–4573, Jun. 2012.

[RBR93] T. Russell, C. Bostian, and T. S. Rappaport, "A Deterministic Approach to Predicting Microwave Diffraction by Buildings for Microcellular Systems," *IEEE Transactions on Antennas and Propagation*, vol. 41, no. 12, pp. 1640–1649, Dec. 1993.

[RCMR01] F. Rice, B. Cowley, B. Moran, and M. Rice, "Cramer-Rao Lower Bounds for QAM Phase and Frequency Estimation," *IEEE Transactions on Communications*, vol. 49, no. 9, pp. 1582–1591, Sep. 2001.

[RDA11] T. S. Rappaport, S. DiPierro, and R. Akturan, "Analysis and Simulation of Interference to Vehicle-Equipped Digital Receivers From Cellular Mobile Terminals Operating in Adjacent Frequencies," *IEEE Transactions on Vehicular Technology*, vol. 60, no. 4, pp. 1664–1676, May 2011.

[Reb92] G. Rebeiz, "Millimeter-Wave and Terahertz Integrated Circuit Antennas," *Proceedings of the IEEE*, vol. 80, no. 11, pp. 1748–1770, Nov. 1992.

[Ree05] J. Reed, *An Introduction to Ultra Wideband Communication Systems*. Upper Saddle River, NJ, Prentice Hall, 2005.

[Rey04] S. Reynolds, "A 60 GHz Superheterodyne Downconversion Mixer in Silicon-Germanium Bipolar Technology," *IEEE Journal of Solid-State Circuits*, vol. 39, no. 11, pp. 2065–2068, Nov. 2004.

[RF91] T. S. Rappaport and V. Fung, "Simulation of Bit Error Performance of FSK, BPSK, and $\pi/4$ DQPSK in Flat Fading Indoor Radio Channels Using a

Measurement-Based Channel Model," *IEEE Transactions on Vehicular Technology*, vol. 40, no. 4, pp. 731–740, Nov. 1991.

[RFP+06] S. Reynolds, B. Floyd, U. Pfeiffer, T. Beukema, J. Grzyb, C. Haymes, B. Gaucher, and M. Soyuer, "A Silicon 60 GHz Receiver and Transmitter Chipset for Broadband Communications," *IEEE Journal of Solid-State Circuits*, vol. 41, no. 12, pp. 2820–2831, Dec. 2006.

[RG71] A. W. Rihaczek and R. M. Golden, "Range Sidelobe Suppression for Barker Codes," *IEEE Transactions on Aerospace and Electronic Systems*, vol. AES-7, no. 6, pp. 1087–1092, 1971.

[RGAA09] T. Rappaport, F. Gutierrez, and T. Al-Attar, "Millimeter-Wave and Terahertz Wireless RFIC and On-Chip Antenna Design: Tools and Layout Techniques," *IEEE GLOBECOM Workshops*, pp. 1–7, Nov. 2009.

[RGBD+13] T. S. Rappaport, F. Gutierrez, E. Ben-Dor, J. Murdock, Y. Qiao, and J. I. Tamir, "Broadband Millimeter-Wave Propagation Measurements and Models Using Adaptive-Beam Antennas for Outdoor Urban Cellular Communications," *IEEE Transactions on Antennas and Propagation*, vol. 61, no. 4, pp. 1850–1859, Apr. 2013.

[RH91] T. S. Rappaport and D. A. Hawbaker, "Effects of Circular and Linear Polarized Antennas on Wideband Propagation Parameters in Indoor Radio Channels," *1991 IEEE Global Communications Conference (GLOBECOM 1991)*, pp. 1287–1291, Dec. 1991.

[RH92] T. S. Rappaport and D. A. Hawbaker, "Wide-band Microwave Propagation Parameters using Circular and Linear Polarized Antennas for Indoor Wireless Channels," *IEEE Transactions on Communications*, vol. 40, no. 2, pp. 240–245, Feb. 1992.

[RHF93] T. S. Rappaport, W. Huang, M. J. Feuerstein, "Performance of Decision Feedback Equalizers in Simulated Urban and Indoor Radio Channels," Invited Paper, *IEICE Transaction on Communications*, vol. E76-B, no. 2, pp. 78–89, Feb. 1993.

[Rho74] S. Rhodes, "Effect of Noisy Phase Reference on Coherent Detection of Offset-QPSK Signals," *IEEE Transactions on Communications*, vol. 22, no. 8, pp. 1046–1055, Aug. 1974.

[RHRC07] L. Ragan, A. Hassibi, T. S. Rappaport, and C. L. Christianson, "Novel On-Chip Antenna Structures and Frequency Selective Surface (FSS) Approaches for Millimeter Wave Devices," *2007 IEEE Vehicular Technology Conference (VTC 2007-Fall)*, pp. 2051–2055, Oct. 2007.

[Ris78] J. Rissanen, "Modeling by Shortest Data Description," *Automatica*, vol. 14, pp. 465–471, 1978.

[RJ01] M. Rudolf and B. Jechoux, "Design of Concatenated Extended Complementary Sequences for Inter-Base Station Synchronization in WCDMA TDD Mode," *2001 IEEE Global Telecommunications Conference*, vol. 1. IEEE, pp. 674–679, 2001.

[RJW08] U. Rizvi, G. Janssen, and J. Weber, "Impact of RF Circuit Imperfections on Multi-Carrier and Single-Carrier Based Transmissions at 60 GHz," *2008 IEEE Radio and Wireless Symposium*, pp. 691–694, Jan. 2008.

[RK95] P. Robertson and S. Kaiser, "Analysis of the Effects of Phase-Noise in Orthogonal Frequency Division Multiplex (OFDM) Systems," *1995 IEEE International Conference on Communications (ICC)*, vol. 3, pp. 1652–1657, Jun. 1995.

[RK09] J.-S. Rieh and D.-H. Kim, "An Overview of Semiconductor Technologies and Circuits for Terahertz Communication Applications," *2009 IEEE GLOBECOM Workshops*, pp. 1–6, 2009.

[RKNH06] J. Roderick, H. Kirshnaswamy, K. Newton, and H. Hashemi, "Silicon-Based Ultra-Wideband Beam-Forming," *IEEE Journal of Solid-State Circuits*, vol. 41, no. 8, pp. 1726–1740, Aug. 2006.

[RMGJ11] T. S. Rappaport, J. N. Murdock, and F. Gutierrez Jr., "State of the Art in 60 GHz Integrated Circuits and Systems for Wireless Communications," *Proceedings of the IEEE*, vol. 99, no. 8, pp. 1390–1436, Aug. 2011.

[RN88] R. L. Rogers and D. P. Neikirk, "Use of Broadside Twin Element Antennas to Increase Efficiency of Electrically Thick Dielectric Substrates," *International Journal of Infrared and Millimeter Waves*, vol. 9, pp. 949–969, 1988.

[Rog85] D. Rogers, "Propagation Considerations for Satellite Broadcasting at Frequencies Above 10 GHz," *IEEE Journal of Selected Areas in Communications JSAC*, vol. 3, no. 1, pp. 100–110, 1985.

[Ron96] Z. Rong, "Simulation of Adaptive Array Algorithms for CDMA Systems," Master's Thesis, Virginia Tech, 1996.

[Rou08] T. J. Rouphael, *RF and Digital Signal Processing for Software-Defined Radio: A Multi-Standard Multi-Mode Approach*. Burlington, MA, Newnes, 2008.

[RQT+12] T. S. Rappaport, Y. Qiao, J. I. Tamir, E. Ben-Dor, and J. N. Murdock, "Cellular Broadband Millimeter Wave Propagation and Angle of Arrival for Adaptive Beam Steering Systems," *2012 IEEE Radio and Wireless Symposium (RWS)*, pp. 151–154, Jan. 2012, invited paper and presentation.

[RR97] S. Rama and G. M. Rebiez, "Single- and Dual-Polarized Slot-Ring Subharmonic Receivers," *1997 IEEE MTT-S International Microwave Symposium (IMS)*, pp. 565–568, Jun. 1997.

[RRC14] T. S. Rappaport, W. Roh, and K. W. Cheun, "Mobile's Millimeter-Wave Makeover," *IEEE Spectrum*, vol. 51, no. 9, Sep. 2014.

[RRE14] S. Rangan, T. S. Rappaport, and E. Erkip, "Millimeter-Wave Cellular Wireless Networks: Potentials and Challenges," *Proceedings of the IEEE*, vol. 102, no. 3, pp. 366–385, Mar. 2014.

[RS60] I. S. Reed and G. Solomon, "Polynomial Codes over Certain Finite Fields," *Journal of the Society for Industrial and Applied Mathematics*, vol. 8, pp. 300–304, 1960.

[RSAL06] S. Radiom, B. Sheikholeslami, H. Aminzadeh, and R. Lotfi, "Folded-Current-Steering DAC: An Approach to Low-Voltage High-Speed High-Resolution D/A Converters," *2006 IEEE International Symposium on Circuits and Systems (ISCAS 2006)*, pp. 4783–4786, May 2006.

[RSM+13] T. S. Rappaport, S. Sun, R. Mayzus, H. Zhao, Y. Azar, K. Wang, G. N. Wong, J. K. Schulz, M. Samimi, and F. Gutierrez, "Millimeter Wave Mobile Communications for 5G Cellular: It Will Work!" *IEEE Access Journal*, vol. 1, pp. 335–349, May 2013.

[RST91] T. S. Rappaport, S. Y. Seidel, and K. Takamizawa, "Statistical Channel Impulse Response Models for Factory and Open Plan Building Radio Communicate System Design," *IEEE Transactions on Communications*, vol. 39, no. 5, pp. 794–807, May 1991.

[RU01a] T. Richardson and R. Urbanke, "Efficient Encoding of Low-Density Parity-Check Codes," *IEEE Transactions on Information Theory*, vol. 47, no. 2, pp. 638–656, Feb. 2001.

[RU01b] T. Richardson and R. Urbanke, "The Capacity of Low-Density Parity-Check Codes under Message-Passing Decoding," *IEEE Transactions on Information Theory*, vol. 47, no. 2, pp. 599–618, Feb. 2001.

[RU03] T. Richardson and R. Urbanke, "The Renaissance of Gallager's Low-Density Parity-Check Codes," *IEEE Communications Magazine*, vol. 41, no. 8, pp. 126–131, Aug. 2003.

[RU08] T. Richardson and R. Urbanke, *Modern Coding Theory*. Cambridge, UK, Cambridge University Press, 2008.

[RVM12] D. Ramasamy, S. Venkateswaran, and U. Madhow, "Compressive Adaptation of Large Steerable Arrays," *2012 Information Theory and Applications Workshop (ITA)*, pp. 234–239, 2012.

[RWD94] S. Ramo, J. Whinnery, and T. V. Duzer, *Fields and Waves in Communications*, 3rd ed. New York, John Wiley and Sons Inc., 1994.

[S+91] S. Seidel, et al., "Path Loss, Scattering, and Multipath Delay Statistics in Four European Cities for Digital Cellular and Microcellular Radiotelephone," *IEEE Transactions on Vehicular Technology*, vol. 40, no. 4, pp. 721–730, Nov. 1991.

[S+92] S. Seidel, et al., "The Impact of Surrounding Buildings on Propagation for Wireless In-Building Personal Communications System Design," *1992 IEEE Vehicular Technology Conference (VTC 1992)*, pp. 814–818, 1992.

[SA95] S. Y. Seidel and H. W. Arnold, "Propagation Measurements at 28 GHz to Investigate the Performance of Local Multipoint Distribution Service (LMDS)," *1995 IEEE GLOBECOM*, vol. 1, pp. 754–757, Nov. 1995.

[SAW90] M. Shinagawa, Y. Akazawa, and T. Wakimoto, "Jitter Analysis of High-Speed Sampling Systems," *IEEE Journal of Solid-State Circuits*, vol. 25, no. 1, pp. 220–224, 1990.

[Say08] A. H. Sayed, *Adaptive Filters*. New York, Wiley-IEEE Press, 2008.

[SB05] B. Streetman and S. Banerjee, *Solid State Electronic Devices*, 6th ed. Upper Saddle River, NJ, Prentice Hall, 2005.

[SB07] A. Seyedi and D. Birru, "On the Design of a Multi-Gigabit Short-Range Communication System in the 60 GHz Band," *2007 IEEE Consumer Communications and Networking Conference (CCNC 2007)*, pp. 1–6, Jan. 2007.

[SB08] D. A. Sobel and R. Brodersen, "A 1 Gbps Mixed-Signal Analog Front-End for a 60 GHz Wireless Receiver," *2008 Symposium on VLSI Circuits*, pp. 56–57, Apr. 2008.

[SB09] S. Sarkar and S. Banerjee, "An 8-bit 1.8 V 500 MSPS CMOS Segmented Current Steering DAC," *2009 IEEE Computer Society Annual Symposium on VLSI (ISVLSI 2009)*, pp. 268–273, May 2009.

[SB09a] D. A. Sobel and R. Brodersen, "A 1 Gbps Mixed-Signal Analog Front-End for a 60 GHz Wireless Receiver," *IEEE Journal of Solid-State Circuits*, vol. 44, no. 4, pp. 1281–1289, Apr. 2009.

[SB13] S. A. Saberali and N. C. Beaulieu, "New Expressions for TWDP Fading Statistics," *IEEE Wireless Communications Letters*, vol. 2, no. 6, pp. 643–646, Dec. 2013.

[SBB+08] K. Scheir, S. Bronckers, J. Borremans, P. Wambacq, and Y. Rolain, "A 52 GHz Phased-Array Receiver Front-End in 90 nm Digital CMOS," *IEEE Journal of Solid-State Circuits*, vol. 43, no. 12, pp. 2651–2660, Dec. 2008.

[SBL+08] C.-S. Sum, T. Baykas, M. Lei, Y. Nishiguchi, R. Kimura, R. Funada, Y. Shoji, H. Harada, and S. Kato, "Performance of Trellis-Coded-Modulation for a Multi-Gigabit Millimeter-Wave WPAN System in the Presence of Hardware Impairments," *2008 International Wireless Communications and Mobile Computing Conference (IWCMC 2008)*, pp. 678–683, Aug. 2008.

[SBM92] S. C. Swales, M. A. Beach, and J. P. McGeehan, "A Spectrum Efficient Cellular Base-Station Antenna Architecture," *1992 IEEE Antennas and Propagation Society International Symposium*, vol. 2, pp. 1069–1072, Jul. 1992.

[SC97a] T. Schmidl and D. Cox, "Robust Frequency and Timing Synchronization for OFDM," *IEEE Transactions on Communications*, vol. 45, no. 12, pp. 1613–1621, Dec. 1997.

[SC97b] P. Smulders and L. Correia, "Characterisation of Propagation in 60 GHz Radio Channels," *Electronics & Communication Engineering Journal*, vol. 9, no. 2, pp. 73–80, Apr. 1997.

[Sch86] R. Schmidt, "Multiple Emitter Location and Signal Parameter Estimation," *IEEE Transactions on Antennas and Propagation*, vol. 34, no. 3, pp. 276–280, Mar. 1986.

[SCP09] E. Shihab, L. Cai, and J. Pan, "A Distributed Asynchronous Directional-to-Directional MAC Protocol for Wireless Ad Hoc Networks," *IEEE Transactions on Vehicular Technology*, vol. 58, no. 9, pp. 5124–5134, Nov. 2009.

[SCS+08] E. Seok, C. Cao, D. Shim, D. J. Areanas, D. B. Tanner, C.-M. Huang, and K. K. O, "A 410 GHz CMOS Push-Push Oscillator with an On-Chip Patch Antenna," *2008 IEEE International Solid-State Circuits Conference (ISSCC)*, p. 472, Feb. 2008.

[SCV06] S. Shahramian, A. Carusone, and S. Voinigescu, "Design Methodology for a 40-GSamples/s Track and Hold Amplifier in 0.18-μm SiGe BiCMOS Technology," *IEEE Journal of Solid-State Circuits*, vol. 41, no. 10, pp. 2233–2240, Oct. 2006.

[SDM09] J. Singh, O. Dabeer, and U. Madhow, "On the Limits of Communication with Low-Precision Analog-to-Digital Conversion at the Receiver," *IEEE Transactions on Communications*, vol. 57, no. 12, pp. 3629–3639, Dec. 2009.

[SDR92] K. R. Schaubach, N. J. Davis, and T. S. Rappaport, "A Ray Tracing Method for Predicting Path Loss and Delay Sspread in Microcellular Environments," *1992 IEEE Vehicular Technology Conference (VTC 1992)*, vol. 2, May 1992, pp. 932–935.

[SGL+09] J. Sun, D. Gupta, Z. Lai, W. Gong, and P. Kelly, "Indoor Transmission of Multi-Gigabit-per-Second Data Rates Using Millimeter Waves," *2009 IEEE International Conference on Ultra-Wideband (ICUWB 2009)*, pp. 783–787, Sep. 2009.

[Shi96] Q. Shi, "OFDM in Bandpass Nonlinearity," *IEEE Transactions on Consumer Electronics*, vol. 42, no. 3, pp. 253–258, Aug. 1996.

[SHNT05] T. Seki, N. Honma, K. Nishikawa, and K. Tsunekawa, "Millimeter-wave High Efficiency Multi-layer Parasitic Microstrip Antenna Array on TEFLON Substrate," *IEEE Transactions on Microwave Theory and Techniques*, vol. 53, no. 6, pp. 2101–2106, Jun. 2005.

[SHV+11] H. Singh, J. Hsu, L. Verma, S. Lee, and C. Ngo, "Green Operation of Multi-Band Wireless LAN in 60 GHz and 2.4/5 GHz," *2011 IEEE Consumer Communications and Networking Conference (CCNC 2011)*, pp. 787–792, Jan. 2011.

[SHW+06] Y. Sun, F. Herzel, L. Wang, J. Borngraber, W. Winkler, and R. Kraemer, "An Integrated 60 GHz Receiver Front-End in SiGe:C BiCMOS," *2006 Meeting on*

Silicon Monolithic Integrated Circuits in RF Systems (SiRF 2006), pp. 269–272, Jan. 2006.

[Sim01] R. N. Simons, *Coplanar Waveguide Circuits, Components, and Systems*, 1st ed. New York, John Wiley and Sons, 2001.

[SKH09] S. Sim, D.-W. Kim, and S. Hong, "A CMOS Direct Injection-Locked Frequency Divider With High Division Ratios," *IEEE Microwave and Wireless Components Letters*, vol. 19, no. 5, pp. 314–316, May 2009.

[SKJ95] H. Sari, G. Karam, and I. Jeanclaude, "Transmission Techniques for Digital Terrestrial TV Broadcasting," *IEEE Communications Magazine*, vol. 33, no. 2, pp. 100–109, Feb. 1995.

[SKLG98] M. Sajadieh, F. Kschischang, and A. Leon-Garcia, "Modulation-Assisted Unequal Error Protection over the Fading Channel," *IEEE Transactions on Vehicular Technology*, vol. 47, no. 3, pp. 900–908, Aug. 1998.

[SKM+10] M. Sawahashi, Y. Kishiyama, A. Morimoto, D. Nishikawa, and M. Tanno, "Coordinated Multipoint Transmission/Reception Techniques for LTE-advanced," *IEEE Wireless Communications*, vol. 17, no. 3, pp. 26–34, Jun. 2010.

[SKX+10] J. Shi, K. Kang, Y. Z. Xiong, J. Brinkhoff, F. Lin, and X.-J. Yuan, "Millimeter-Wave Passives in 45-nm Digital CMOS," *IEEE Electron Device Letters*, vol. 31, no. 10, pp. 1080–1082, Oct. 2010.

[SL90] W. Schafer and E. Lutz, "Propagation Characteristics of Short-Range Radio Links at 60 GHz for Mobile Intervehicle Communication," *1990 SBT/IEEE International Telecommunications Symposium (ITS 1990)*, pp. 212–216, Sep. 1990.

[Sle76] D. Slepian, "On Bandwidth," *Proceedings of the IEEE*, Vol. 64, no. 3, pp. 292–300, Mar. 1976.

[SLNB90] Y. Shayan, T. Le-Ngoc, and V. Bhargava, "A Versatile Time-Domain Reed-Solomon Decoder," *IEEE Journal on Selected Areas in Communications*, vol. 8, no. 8, pp. 1535–1542, Oct. 1990.

[SLW+09] J. Sha, J. Lin, Z. Wang, L. Li, and M. Gao, "LDPC Decoder Design for High Rate Wireless Personal Area Networks," *IEEE Transactions on Consumer Electronics*, vol. 55, no. 2, pp. 455–460, May 2009.

[SM05] P. Stoica and R. L. Moses, *Spectral Analysis of Signals*. Upper Saddlebrook, NJ, Prentice Hall, 2005.

[Smi75] J. Smith, "Odd-Bit Quadrature Amplitude-Shift Keying," *IEEE Transactions on Communications*, vol. 23, no. 3, pp. 385–389, Mar. 1975.

[SMI+97] K. Sato, T. Manabe, T. Ihara, H. Saito, S. Ito, T. Tanaka, K. Sugai, et al., "Measurements of Reflection and Transmission Characteristics of Interior Structures of Office Buildings in the 60 GHz Band," *IEEE Journal of Antennas and Propagation*, vol. 45, no. 12, pp. 1783–1792, 1997.

[SMM11] S. Singh, R. Mudumbai, and U. Madhow, "Interference Analysis for Highly Directional 60 GHz Mesh Networks: The Case for Rethinking Medium Access Control," *IEEE/ACM Transactions on Networking*, vol. 19, no. 5, pp. 1513–1527, Oct. 2011.

[SMMZ06] P. Sudarshan, N. Mehta, A. Molisch, and J. Zhang, "Channel Statistics-based RF Pre-processing with Antenna Selection," *IEEE Transactions on Wireless Communications*, vol. 5, no. 12, pp. 3501–3511, Dec. 2006.

[SMS+09] S. Sankaran, C. Mao, E. Seok, D. Shim, C. Cao, R. Han, D. J. Arenas, D. B. Tanner, S. Hill, C.-M. Hung, and K. K. O, "Towards Terahertz Operation of CMOS," *2009 IEEE International Solid-State Circuits Conference (ISSCC)*, pp. 202–203a, Feb. 2009.

[SMS+14] S. Sun, G. MacCartney, M. Samimi, S. Nie, and T. S. Rappaport, "Millimeter Wave Multi-beam Antenna Combining for 5G Cellular Link Improvement in New York City," *2014 IEEE International Conference on Communications (ICC)*, Sydney Australia, Jun. 2014.

[Smu02] P. Smulders, "Exploiting the 60 GHz Band for Local Wireless Multimedia Access: Prospects and Future Directions," *IEEE Communications Magazine*, vol. 40, no. 1, pp. 140–147, Jan. 2002.

[SN07] S. M. Sze and K. K. Ng, *Physics of Semiconductor Devices*, 3rd ed. Hoboken, NJ, Wiley, 2007.

[Sno02] D. K. Snodgrass, "60 GHz Radio System Design Tradeoffs," *Microwave Journal*, vol. 45, no. 7, 2002.

[SNO08] T. Seki, K. Nishikawa, and K. Okada, "60 GHz Multi-Layer Parasitic Microstrip Array Antenna with Stacked Rings Using Multi-Layer LTCC Substrate," *2008 IEEE Radio and Wireless Symposium*, pp. 679–682, Jan. 2008.

[SNS+07] H.-R. Shao, C. Ngo, H. Singh, S. Qin, C. Kweon, G. Fan, and S. Kim, "Adaptive Multi-beam Transmission of Uncompressed Video over 60 GHz Wireless Systems," *Future Generation Communication and Networking (FGCN 2007)*, vol. 1, pp. 430–435, Dec. 2007.

[SNS+14] A. I. Sulyman, A. T. Nassar, M. K. Samimi, G. R. MacCartney, T. S. Rappaport, and A. Alsanie, "Radio Propagation Path Loss Models for 5G Cellular Networks in the 28 and 38 GHz Millimeter-Wave Bands," *IEEE Communications Magazine*, Sep. 2014.

[SOK+08] H. Singh, J. Oh, C. Kweon, X. Qin, H.-R. Shao, and C. Ngo, "A 60 GHz Wireless Network for Enabling Uncompressed Video Communication," *IEEE Communications Magazine*, vol. 46, no. 12, pp. 71–78, Dec. 2008.

[SPM09] J. Singh, S. Ponnuru, and U. Madhow, "Multi-Gigabit Communication: The ADC Bottleneck," *2009 IEEE International Conference on Ultra Wideband (ICUWB 2009)*, pp. 22–27, Sep. 2009.

[SQI01] M. Sironen, Y. Qian, and T. Itoh, "A Subharmonic Self-Oscillating Mixer with Integrated Antenna for 60 GHz Wireless Applications," *IEEE Transactions on Microwave Theory and Techniques*, vol. 49, no. 3, pp. 442–450, Mar. 2001.

[SQrS+08] H. Singh, X. Qin, H.-R. Shao, C. Ngo, C. Kwon, and S. S. Kim, "Support of Uncompressed Video Streaming Over 60 GHz Wireless Networks," *2008 IEEE Consumer Communications and Networking Conference (CCNC 2008)*, pp. 243–248, Jan. 2008.

[SR92] S. Y. Seidel, T. S. Rappaport, "914 MHz Path Loss Prediction Models for Indoor Wireless Communications in Multifloored Buildings," *IEEE Transactions on Antennas and Propagation*, vol. 40, no. 2, pp. 207–217, Feb. 1992.

[SR94] S. Y. Siedel and T. S. Rappaport, "Site Specific Propagation Predictions for Wireless In-Building Personal Communication System Design," *IEEE Transactions on Vehicular Technology*, vol. 43, no. 4, Dec. 1994.

[SR13] S. Sun and T. S. Rappaport, "Multi-Beam Antenna Combining for 28 GHz Cellular Link Improvement in Urban Environments," *2013 IEEE Global Communications Conference (GLOBECOM 2013)*, Dec. 2013.

[SR14] S. Sun and T. S. Rappaport, "Wideband MmWave Channels: Implications for Design and Implementation of Adaptive Beam Antennas," *2014 IEEE MTT-S International Microwave Symposium (IMS)*, Tampa, FL, Jun. 2014.

[SR14a] M. K. Samimi and T. S. Rappaport, "Ultra-Wideband Statistical Channel Model for Non Line of Sight Millimeter-Wave Urban Channels," *2014 IEEE Global Communications Conference (GLOBECOM 2014)*, Austin, TX, Dec. 2014.

[SRA96] R. Skidmore, T. S. Rappaport, and A. Abbott, "Interactive Coverage Region and System Design Simulation for Wireless Communication Systems in Multifloored Indoor Environments: SMT Plus," *1996 IEEE International Conference on Universal Personal Communications*, vol. 2, pp. 646–650, Sep. 1996.

[SRFT08] A. Shamim, L. Roy, N. Fong, and N. G. Tarr, "24 GHz On-Chip Antennas and Balun on Bulk Si for Air Transmission," *IEEE Transactions on Antennas and Propagation*, vol. 56, no. 2, pp. 303–311, Feb. 2008.

[SRK03] S. Shakkottai, T. S. Rappaport, and P. C. Karlsson, "Cross-layer Design for Wireless Networks," *IEEE Communications Magazine*, vol. 41, no. 10, pp. 74–80, Oct. 2003.

[SS94] N. Seshadri and C.-E. Sundberg, "List Viterbi Decoding Algorithms with Applications," *IEEE Transactions on Communications*, vol. 42, no. 234, pp. 313–323, Feb./Mar./Apr. 1994.

[SS01] C. Saint and J. Saint, *IC Layout Basics: A Practical Guide*, 1st ed. New York, McGraw-Hill Professional Publishing, 2001.

[SS04] K. Shi and E. Serpedin, "Coarse Frame and Carrier Synchronization of OFDM Systems: A New Metric and Comparison," *IEEE Transactions on Wireless Communications*, vol. 3, no. 4, pp. 1271–1284, Jul. 2004.

[SSB+14] S. Suyama, J. Shen, A. Benjebbour, Y. Kishiyama, and Y. Okumura, "Super High Bit Rate Radio Access Technologies for Small Cells Using Higher Frequency Bands," *2014 IEEE MTT-S International Microwave Symposium (IMS)*, Tampa, FL, Jun. 2014.

[SSDV+08] A. Scholten, G. Smit, B. De Vries, L. Tiemeijer, J. Croon, D. Klaassen, R. van Langevelde, X. Li, W. Wu, and G. Gildenblat, "(Invited) The New CMC Standard Compact MOS Model PSP: Advantages for RF Applications," *2008 IEEE Radio Frequency Integrated Circuits Symposium (RFIC 2008)*, pp. 247–250, Jun. 2008.

[SSM+10] E. Seok, D. Shim, C. Mao, R. Han, S. Sankaran, C. Cao, W. Knap, and K. Kenneth, "Progress and Challenges Towards Terahertz CMOS Integrated Circuits," *IEEE Journal of Solid-State Circuits*, vol. 45, no. 8, pp. 1554–1564, 2010.

[SST+09] C. Sheldon, M. Seo, E. Torkildson, M. Rodwell, and U. Madhow, "Four-Channel Spatial Multiplexing over a Millimeter-Wave Line-of-Sight Link," *2009 IEEE MTT-S International Microwave Symposium (IMS)*, pp. 389–392, Jun. 2009.

[SSTR93] S. Siedel, K. Shaubach, T. Tran, and T. S. Rappaport, "Research in Site-Specific Propagation Modeling for PCS System Design," *1993 IEEE Vehicular Technology Conference (VTC 1993)*, pp. 261–264, May 1993.

[STB09] S. Sesia, I. Toufik, and M. Baker, *LTE–The UMTS Long Term Evolution*. Wiley Online Library, 2009, vol. 66.

[STD+09] R. Severino, T. Taris, Y. Deval, D. Belot, and J. Begueret, "A SiGe: C BiCMOS LNA for 60 GHz Band Applications," *2009 IEEE Bipolar/BiCMOS Circuits and Technology Meeting (BCTM 2009)*, pp. 51–54, Oct. 2009.

[STLL05] J. R. Sohn, H.-S. Tae, J.-G. Lee, and J.-H. Lee, "Comparative Analysis of Four Types of High-Impedance Surfaces for Low Profile Antenna Applications," *2005 IEEE Antennas and Propagation Society International Symposium*, pp. 758–761, Jul. 2005.

[SUR09] W. Shin, M. Uzunkol, and G. Rebeiz, "Ultra Low Power 60 GHz ASK SiGe Receiver with 3-6 GBPS Capabilities," *2009 IEEE Compound Semiconductor Integrated Circuit Symposium (CISC 2009)*, pp. 1–4, Oct. 2009.

[Surrey] "Introducing the World's Premier 5G Innovation Centre," Surrey, UK, University of Surrey. http://www.surrey.ac.uk/ccsr/business/5GIC/.

[SV07] E. Sail and M. Vesterbacka, "Thermometer-to-Binary Decoders for Flash Analog-to-Digital Converters," *2007 European Conference on Circuit Theory and Design (ECCTD 2007)*, pp. 240–243, Aug. 2007.

[SV87] A. A. M. Saleh and R. A. Valenzuela, "A Statistical Model for Indoor Multipath Propagation," *IEEE Journal on Selected Areas in Communications*, vol. 5, no. 2, pp. 128–137, Feb. 1987.

[SVC09] S. Shahramian, S. Voinigescu, and A. Carusone, "A 35-GS/s, 4-Bit Flash ADC With Active Data and Clock Distribution Trees," *IEEE Journal of Solid-State Circuits*, vol. 44, no. 6, pp. 1709–1720, Jun. 2009.

[SW92] P. F. M. Smulders and A. G. Wagemans, "Wideband Indoor Radio Propagation Measurements at 58 GHz," *Electronics Letters*, vol. 28, 1992.

[SW94a] N. Seshadri and J. Winters, "Two Signaling Schemes for Improving the Error Performance of Frequency-Division-Duplex (FDD) Transmission Systems using Transmitted Antenna Diversity," *International Journal on Wireless Information Networks*, vol. 1, no. 1, pp. 49–60, Jan. 1994.

[SW94b] P. F. M. Smulders and A. G. Wagemans, "Biconical Horn Antennas for Near Uniform Coverage in Indoor Areas at Mm-Wave Frequencies," *IEEE Journal of Vehicular Technology*, vol. 43, no. 4, pp. 897–901, 1994.

[SW07] D. C. Schultz and B. Walke, "Fixed Relays for Cost Efficient 4G Network Deployments: An Evaluation," *2007 IEEE International Symposium on Personal, Indoor and Mobile Radio Communications (PIMRC 2007)*, Athens, Greece, Sep. 2007.

[SWA+13] M. Samimi, K. Wang, Y. Azar, G. N. Wong, R. Mayzus, H. Zhao, J. K. Schulz, S. Sun, J. F. Gutierrez, and T. S. Rappaport, "28 GHz Angle of Arrival and Angle of Departure Analysis for Outdoor Cellular Communications Using Steerable Beam Antennas in New York City," *2013 IEEE Vehicular Technology Conference (VTC 2013-Spring)*, pp. 1–6, Jun. 2013.

[SWF00] D. Seo, A. Weil, and M. Feng, "A 14 bit, 1 GS/s Digital-to-Analog Converter with Improved Dynamic Performances," in *2000 IEEE International Symposium on Circuits and Systems (ISCAS 2000 Geneva)*, vol. 5, pp. 541–544, 2000.

[SYON09] H. Singh, S.-K. Yong, J. Oh, and C. Ngo, "Principles of IEEE 802.15.3c: Multi-Gigabit Millimeter-Wave Wireless PAN," *2009 International Conference on Computer Communications and Networks (ICCCN 2009)*, pp. 1–6, Aug. 2009.

[SZB+99] D. Sievenpiper, L. Zhang, R. F. J. Broas, N. G. Alexopolous, and E. Yablonovitch, "High-Impedance Electromagnetics Surfaces with a Forbidden Frequency Band," *IEEE Transactions on Microwave Theory and Techniques*, vol. 47, no. 11, pp. 2059–2074, Nov. 1999.

[SZC+08] M. Sun, Y. P. Zhang, K. M. Chua, L. L. Wai, D. Liu, and B. P. Gaucher, "Integration of Yagi Antenna in LTCC Package for Differential 60 GHz Radio," *IEEE Transactions on Antennas and Propagation*, vol. 56, no. 8, pp. 2780–2783, Aug. 2008.

[SZG+09] M. Sun, Y. P. Zhang, Y. X. Guo, K. M. Chua, and L. L. Wai, "Integration of Grid Array Antenna in Chip Package for Highly Integrated 60 GHz Radios," *IEEE Letters on Antennas and Wireless Propagation*, vol. 8, pp. 1364–1366, Dec. 2009.

[SZM+09] S. Singh, F. Ziliotto, U. Madhow, E. Belding, and M. Rodwell, "Blockage and Directivity in 60 GHz Wireless Personal Area Networks: From Cross-Layer Model to Multihop MAC Design," *IEEE Journal on Selected Areas in Communications*, vol. 27, no. 8, pp. 1400–1413, Oct. 2009.

[SZW08] R. Schoenen, W. Zirwas, and B. H. Walke, "Raising Coverage and Capacity Using Fixed Relays in a Realistic Scenario," *2008 European Wireless Conference (EW 2008)*, pp. 1–6, Jun. 2008.

[TAMR06] E. Torkildson, B. Ananthasubramaniam, U. Madhow, and M. Rodwell, "Millimeter-Wave MIMO: Wireless Links at Optical Speeds," *44th Allerton Conference on Communication, Control and Computing*, 2006.

[Tan81] R. Tanner, "A Recursive Approach to Low Complexity Codes," *IEEE Transactions on Information Theory*, vol. 27, no. 5, pp. 533–547, Sep. 1981.

[Tan02] A. S. Tanenbaum, *Computer Networks*, 4th ed. Upper Saddle River, NJ, Prentice Hall, 2002.

[Tan06] S. Y. Tan, "Regulatory Update in Europe for Gigabit Application in the 60 GHz Range," *IEEE 802.15.3c Meeting Contributions*, Jul. 2006. https://mentor.ieee.org/802.15/file/06/15-06-0308-00-003c-regulatory-update-in-europe-gigabit-application-in-60ghz-range.ppt.

[TAY+09] A. Tomkins, R. Aroca, T. Yamamoto, S. Nicolson, Y. Doi, and S. Voinigescu, "A Zero-IF 60 GHz 65 nm CMOS Transceiver With Direct BPSK Modulation Demonstrating up to 6 Gb/s Data Rates Over a 2 m Wireless Link," *IEEE Journal of Solid-State Circuits*, vol. 44, no. 8, pp. 2085–2099, 2009.

[TCM+11] M. Tabesh, J. Chen, C. Marcu, L.-K. Kong, E. Alon, and A. M. Niknejad, "A 65nm CMOS 4-Element Sub-34mW/Element 60 GHz Phased Array Transceiver," *2011 IEEE International Solid-State Circuits Conference (ISSCC)*, pp. 166–167, Feb. 2011.

[TDS+09] T. Taris, Y. Deval, R. Severino, C. Ameziane, D. Belot, and J.-B. Begueret, "Millimeter-Waves Building Block Design Methodology in BiCMOS Technology," *2009 IEEE International Conference on Electronics, Circuits, and Systems (ICECS 2009)*, pp. 968–971, Dec. 2009.

[Tel99] I. E. Telatar, "Capacity of Multi-Antenna Gaussian Channels," *European Transactions on Telecommunications*, vol. 10, no. 6, pp. 585–595, 1999.

[TGTN+07] P. Triverio, S. Grivet-Talocia, M. Nakhla, F. Canavero, and R. Achar, "Stability, Causality, and Passivity in Electrical Interconnect Models," *IEEE Transactions on Advanced Packaging*, vol. 30, no. 4, pp. 795–808, Nov. 2007.

[TH07] J.-H. Tsai and T.-W. Huang, "35 - 65 GHz CMOS Broadband Modulator and Demodulator With Sub-Harmonic Pumping for MMW Wireless Gigabit Applications," *IEEE Transactions on Microwave Theory and Techniques*, vol. 55, no. 10, pp. 2075–2085, Oct. 2007.

[TH13] J. Thornton, K. C. Huang, *Modern Lens Antennas for Communications Engineering.* IEEE Press, Wiley Interscience, 2013.

[The10] The Wireless Gigabit Alliance, "WiGig White Paper: Defining the Future of Multi-Gigabit Wireless Communications," Jul. 2010. http://wilocity.com/resources/WiGigWhitepaper_FINAL5.pdf.

[TI13] "AN-1558 Clocking High-Speed A/D Converters," *Texas Instruments Application Report SNAA036B*, Jan. 2007, revised May 2013.

[TK08] W.-H. Tu and T.-H. Kang, "A 1.2V 30mW 8b 800MS/s Time-Interleaved ADC in 65nm CMOS," *2008 IEEE Symposium on VLSI Circuits*, pp. 72–73, Jun. 2008.

[TKJ+12] C. Thakkar, L. Kong, K. Jung, A. Frappe, and E. Alon, "A 10 Gb/s 45 mW Adaptive 60 GHz Baseband in 65 nm CMOS," *IEEE Journal of Solid-State Circuits*, vol. 47, no. 4, pp. 952–968, Apr. 2012.

[TM88] A. R. Tharek and J. P. McGeehan, "Propagation and Bit Error Rate Measurements within Buildings in Millimeter Wave Band about 60 GHz," *1988 IEEE Electrotechnics Conference (EUROCON 1988)*, pp. 318–321, 1988.

[TMRB02] M. Takai, J. Martin, A. Ren, and R. Bagrodia, "Directional Virtual Carrier Sensing for Directional Antennas in Mobile Ad Hoc Networks," *ACM MobiHoc*, pp. 183–193, 2002.

[TOIS09] K. Takahagi, M. Ohno, M. Ikebe, and E. Sano, "Ultra-Wideband Silicon On-Chip Antenna with Artificial Dielectric Layer," *2009 Intelligent Signal Processing and Communication Systems (ISPACS 2009)*, pp. 81–84, Jan. 2009.

[Tor02] J. A. Torres, "Method for Using Non-Squared QAM Constellations," May 2002. http://grouper.ieee.org/groups/802/16/tga/contrib/C80216a-02_66.pdf

[Toy06] I. Toyoda, "Reference Antenna Model with Sidelobe Level for TG3c Evaluation," *IEEE 802.15.06-0474-00-003c*, Oct. 2006.

[TP11] Y. Tsang and A. Poon, "Detecting Human Blockage and Device Movement in mmWave Communication System," *2011 IEEE Global Telecommunications Conference (GLOBECOM 2011)*, pp. 5718–5722, 2011.

[TPA11] Y. Tsang, A. Poon, and S. Addepalli, "Coding the Beams: Improving Beamforming Training in mmWave Communication System," *2011 IEEE Global Telecommunications Conference (GLOBECOM 2011)*, pp. 4386–4390, 2011.

[TS05] J. Tan and G. Stuber, "Frequency-Domain Equalization for Continuous Phase Modulation," *IEEE Transactions on Wireless Communications*, vol. 4, no. 5, pp. 2479–2490, Sep. 2005.

[TSC98] V. Tarokh, N. Seshadri, and A. Calderbank, "Space-Time Codes for High Data Rate Wireless Communication: Performance Criterion and Code Construction," *IEEE Transactions on Information Theory*, vol. 44, no. 2, pp. 744–765, 1998.

[TSMR09] E. Torkildson, C. Sheldon, U. Madhow, and M. Rodwell, "Millimeter-Wave Spatial Multiplexing in an Indoor Environment," *2009 IEEE GLOBECOM Workshops*, pp. 1–6, 2009.

[Tso01] G. V. Tsoulos, Ed., *Adaptive Antennas for Wireless Communications*. New York, Wiley-IEEE Press, 2001.

[TSRK04] W. H. Tranter, K. S. Shanmugan, T. S. Rappaport, and K. L. Kosbar, *Principles of Communication Systems Simulation*. Upper Saddle River, NJ, Prentice Hall, 2004.

[TWMA10] H. Tuinhout, N. Wils, M. Meijer, and P. Andricciola, "Methodology to Evaluate Long Channel Matching Deterioration and Effects of Transistor Segmentation on MOSFET Matching," *2010 IEEE International Conference on Microelectronic Test Structures (ICMTS)*, pp. 176–181, Mar. 2010.

[TWS+07] K. To, P. Welch, D. Scheitlin, B. Brown, D. Hammock, M. Tutt, D. Morgan, S. Braithwaite, J. John, J. Kirchgessner, and W. Huang, "60 GHz LNA and 15GHz VCO Design for Use in Broadband Millimeter-Wave WPAN System," *2007 IEEE Bipolar/BiCMOS Circuits and Technology Meeting (BCTM 2007)*, pp. 210–213, Sep. 2007.

[UK688] Great Britain, Department of Trade and Industry, Radiocommunications Division, "The Use of the Radio Frequency Spectrum above 30 GHz: A Consultative Document," Great Britain Department of Trade and Industry, Radiocommunications Division, Dec. 1988.

[Ung74] G. Ungerboeck, "Adaptive Maximum-Likelihood Receiver for Carrier-Modulated Data-Transmission Systems," *IEEE Transactions on Communications*, vol. 22, no. 5, pp. 624–636, May 1974.

[US02] K. Uyttenhove and M. S. J. Steyaert, "Speed-Power-Accuracy Tradeoff in High-Speed CMOS ADCs," *IEEE Transactions on Circuits and Systems II: Analog and Digital Signal Processing*, vol. 49, no. 4, pp. 280–287, Apr. 2002.

[VGNL+10] A. Valdes-Garcia, S. T. Nicolson, J.-W. Lai, A. Natarajan, P.-Y. Chen, S. K. Reynolds, J.-H. C. Zhan, D. G. Kam, D. Liu, and B. Floyd, "A Fully Integrated 16-Element Phased-Array Transmitter in SiGe BiCMOS for 60 GHz Communications," *IEEE Journal of Solid-State Circuits*, vol. 45, no. 12, pp. 2757–2773, 2010.

[VKKH08] M. Varonen, M. Karkkainen, M. Kantanen, and K. A. I. Halonen, "Millimeter-Wave Integrated Circuits in 65-nm CMOS," *IEEE Journal of Solid-State Circuits*, vol. 43, no. 9, pp. 1991–2002, Sep. 2008.

[VKR+05] M. Varonen, M. Karkkainen, J. Riska, P. Kangaslahti, and K. Halonen, "Resistive HEMT Mixers for 60 GHz Broad-Band Telecommunication," *IEEE*

Transactions on Microwave Theory and Techniques, vol. 53, no. 4, pp. 1322–1330, Apr. 2005.

[VKTS09] S. Veeramachanen, A. Kumar, V. Tummala, and M. Srinivas, "Design of a Low Power, Variable-Resolution Flash ADC," *2009 International Conference on VLSI Design*, pp. 117–122, Jan. 2009.

[VLC96] P. Voois, I. Lee, and J. Cioffi, "The Effect of Decision Delay in Finite-Length Decision Feedback Equalization," *IEEE Transactions on Information Theory*, vol. 42, no. 2, pp. 618–621, Mar. 1996.

[VLR13] N. Valliappan, A. Lozano, and R. W. Heath Jr. "Antenna Subset Modulation for Secure Millimeter-Wave Wireless Communication," *IEEE Transactions on Communications*, vol. 61, no. 8, pp. 3231–3245, Aug. 2013.

[VM05] H. Viswanathan and S. Mukherjee, "Performance of Cellular Networks with Relays and Centralized Scheduling," *IEEE Transactions on Wireless Communications*, vol. 4, no. 5, pp. 2318–2328, Sep. 2005.

[VWZP89] A. Viterbi, J. Wolf, E. Zehavi, and R. Padovani, "A Pragmatic Approach to Trellis-Coded Modulation," *IEEE Communications Magazine*, vol. 27, no. 7, pp. 11–19, Jul. 1989.

[Wad91] B. C. Wadell, *Transmission Line Design Handbook*, 1st ed. Norwood, MA, Artech House, 1991.

[Wal99] R. Walden, "Analog-to-Digital Converter Survey and Analysis," *IEEE Journal on Selected Areas in Communications*, vol. 17, no. 4, pp. 539–550, 1999.

[WAN97] M. R. Williamson, G. E. Athanasiadou, and A. R. Nix, "Investigating the Effects of Antenna Directivity on Wireless Indoor Communication at 60 GHz," *1997 IEEE International Symposium on Personal, Indoor and Mobile Radio Communications (PIMRC 1997)*, vol. 2, pp. 635–639, Sep. 1997.

[Wan01] H. Wang, "A 50 GHz VCO in 0.25μm CMOS," *2001 IEEE International Solid-State Circuits Conference (ISSCC)*, pp. 372–373, 2001.

[WBF+09] J. Wang, T. Baykas, R. Funada, C.-S. Sum, A. Rahman, Z. Lan, H. Harada, and S. Kato, "A SNR Mapping Scheme for ZF/MMSE Based SC-FDE Structured WPANs," *2009 IEEE Vehicular Technology Conference (VTC 2009-Spring)*, pp. 1–5, Apr. 2009.

[Wei09] S. Weinstein, "The History of Orthogonal Frequency-Division Multiplexing [History of Communications]," *IEEE Communications Magazine*, vol. 47, no. 11, pp. 26–35, Nov. 2009.

[Wel09] J. Wells, "Faster Than Fiber: The Future of Multi-G/s Wireless," *IEEE Microwave Magazine*, vol. 10, no. 3, pp. 104–112, May 2009.

[Wi14] "Wireless Networking and Communications Group (WNCG)," 2014, http://www.wncg.org.

[Wid65] B. Widrow, "Adaptive Filters I," Stanford Electronics Lab, Technical Report, Dec. 1965.

[Winner2] IST-WINNER D1.1.2 P. Kyosti, et al., "WINNER II Channel Models," ver1.1, Sep. 2007.

[Wir10] WirelessHD LLC, "WirelessHD Specification Version 1.1 Overview," May 2010. http://www.wirelesshd.org/pdfs/WirelessHD-Specification-Overview-v1.1May2010.pdf

[Wit91] A. Wittneben, "Basestation Modulation Diversity for Digital Simulcast," *1991 IEEE Vehicular Technology Conference (VTC 1991)*, pp. 848–853, St. Louis, MO, May 1991.

[WLP+10] J. Wang, Z. Lan, C.-W. Pyo, T. Baykas, C.-S. Sum, M. Azizur Rahman, J. Gao, R. Funada, F. Kojima, H. Harada, and S. Kato, "A Pro-Active Beamforming Protocol for Multi-Gbps Millimeter-Wave WPAN Systems," *2010 IEEE Wireless Communications and Networking Conference (WCNC)*, pp. 1–5, 2010.

[WLwP+09] J. Wang, Z. Lan, C. Pyo, T. Baykas, C.-S. Sum, M. A. Rahman, J. Gao, R. Funada, F. Kojima, H. Harada, and S. Kato, "Beam Codebook based Beamforming Protocol for Multi-Gbps Millimeter-Wave WPAN Systems," *IEEE Journal on Selected Areas in Communications*, vol. 27, no. 8, pp. 1390–1399, Oct. 2009.

[WMG04] Z. Wang, X. Ma, and G. Giannakis, "OFDM or Single-Carrier Block Transmissions?" *IEEE Transactions on Communications*, vol. 52, no. 3, pp. 380–394, 2004.

[WMGG67] B. Widrow, P. Mantey, L. Griffiths, and B. Goode, "Adaptive Antenna Systems," *Proceedings of the IEEE*, vol. 55, no. 12, pp. 2143–2159, Dec. 1967.

[WP03] Z. Wang and K. Parhi, "High Performance, High Throughput Turbo/SOVA Decoder Design," *IEEE Transactions on Communications*, vol. 51, no. 4, pp. 570–579, Apr. 2003.

[WPS08] X. Wu, P. Palmers, and M. Steyaert, "A 130 nm CMOS 6-bit Full Nyquist 3 GS/s DAC," *IEEE Journal of Solid-State Circuits*, vol. 43, no. 11, pp. 2396–2403, Nov. 2008.

[WR05] H. Wang and T. S. Rappaport, "A Parametric Formulation of the UTD Diffraction Coefficient for Real-Time Propagation Prediction Modeling," *IEEE Antennas and Wireless Propagation Letters*, vol. 4, no. 1, pp. 253–257, 2005.

[WRBD91] S. Wales, D. Rickard, M. Beach, and R. Davies, "Measurement and Modelling of Short Range Broadband Millimetric Mobile Communication Channels," *IEE Colloquium on Radiocommunications in the Range 30-60 GHz*, pp. 12/1–12/6, Jan. 1991.

[WSC+10] S.-C. Wang, P. Su, K.-M. Chen, K.-H. Liao, B.-Y. Chen, S.-Y. Huang, C.-C. Hung, and G.-W. Huang, "Comprehensive Noise Characterization and Modeling for 65-nm MOSFETs for Millimeter-Wave Applications," *IEEE Transactions on Microwave Theory and Techniques*, vol. 58, no. 4, pp. 740–746, Apr. 2010.

[WSE08] B. Wicks, E. Skafidas, and R. Evans, "A 60 GHz Fully-Integrated Doherty Power Amplifier Based on 0.13-μm CMOS Process," *2008 IEEE Radio Frequency Integrated Circuits Symposium (RFIC)*, pp. 69–72, Jun. 2008.

[WSJ09] C. Wagner, A. Stelzer, and H. Jager, "A Phased-Array Transmitter Based on 77 GHz Cascadable Transceivers," *2009 IEEE MTT-S International Microwave Symposium (IMS)*, pp. 73–77, Jun. 2009.

[WT99] J. Wikner and N. Tan, "Modeling of CMOS Digital-to-Analog Converters for Telecommunication," *IEEE Transactions on Circuits and Systems II: Analog and Digital Signal Processing*, vol. 46, no. 5, pp. 489–499, May 1999.

[WVH+09] T. Wirth, V. Venkatkumar, T. Haustein, E. Schulz, and R. Halfmann, "LTE-Advanced Relaying for Outdoor Range Extension," *2009 IEEE Vehicular Technology Conference (VTC 2009-Fall)*, pp. 1–4, Sep. 2009.

[WZL10] Y. Wu, Y. Zhao, and H. Li, "Constellation Design for Odd-Bit Quadrature Amplitude Modulation," *2010 IEEE Wireless Communications and Networking Conference Workshops (WCNCW)*, pp. 1–4, Apr. 2010.

[Xia11] P. Xiaoming, "China WPAN mmWave Liaison w/ 802.11, IEEE 802.11-11/1034r0," *IEEE 802.11ad Standard Contribution*, 2011.

[XK08] H. Xu and K. Kenneth, "High-Thick-Gate-Oxide MOS Varactors With Subdesign-Rule Channel Lengths for Millimeter-Wave Applications," *IEEE Electron Device Letters*, vol. 29, no. 4, pp. 363–365, Apr. 2008.

[XKR00] H. Xu, V. Kukshya, and T. S. Rappaport, "Spatial and Temporal Characterization of 60 GHz Channels," *2000 IEEE Vehicular Technology Conference (VTC 2000-Fall)*, Sep. 2000.

[XKR02] H. Xu, V. Kukshya, and T. S. Rappaport, "Spatial and Temporal Characteristics of 60 GHz Indoor Channels," *IEEE Journal on Selected Areas in Communications*, vol. 20, no. 3, pp. 620–630, Apr. 2002.

[XRBS00] H. Xu, T. S. Rappaport, R. J. Boyle, and J. H. Schaffner, "Measurements and Models for 38 GHz Point-to-Multipoint Radiowave Propagation," *IEEE Journal on Selected Areas of Communication*, vol. 18, no. 3, pp. 310–321, Mar. 2000.

[XYON08] P. Xia, S.-K. Yong, J. Oh, and C. Ngo, "A Practical SDMA Protocol for 60 GHz Millimeter Wave Communications," *2008 Asilomar Conference on Signals, Systems and Computers*, pp. 2019–2023, Oct. 2008.

[YA87] H. Yang and N. Alexopoulos, "Gain Enchancement Methods for Printed Circuit Antennas Through Multiple Superstrates," *IEEE Transactions on Antennas and Propagation*, vol. 35, no. 7, pp. 860–863, Jul. 1987.

[YC07] S.-K. Yong and C.-C. Chong, "An Overview of Multigigabit Wireless through Millimeter Wave Technology: Potentials and Technical Challenges," *EURASIP Journal on Wireless Communications and Networking*, vol. 2007, no. 1, p. 078907, 2007. http://jwcn.eurasipjournals.com/content/2007/1/078907.

[YCP+09] D. Yeh, A. Chowdhury, R. Pelard, S. Pinel, S. Sarkar, P. Sen, B. Perumana, D. Dawn, E. Juntunen, M. Leung, H.-C. Chien, Y.-T. Hsueh, Z. Jia, J. Laskar, and G.-K. Chang, "Millimeter-Wave Multi-Gigabit IC Technologies for Super-Broadband Wireless over Fiber Systems," *2009 Conference on Optical Fiber Communication (OFC 2009)*, pp. 1–3, Mar. 2009.

[YCWL08] C.-Y. Yu, W.-Z. Chen, C.-Y. Wu, and T.-Y. Lu, "A 60 GHz, 14% Tuning Range, Multi-Band VCO with a Single Variable Inductor," *2008 IEEE Asian Solid-State Circuits Conference (A-SSCC 2008)*, pp. 129–132, Nov. 2008.

[YGT+07] T. Yao, M. Q. Gordon, K. K. W. Tang, K. H. K. Yau, M.-T. Yang, P. Schvan, and S. P. Voinigescu, "Algorithmic Design of CMOS LNAs and PAs for 60 GHz Radio," *IEEE Journal of Solid-State Circuits*, vol. 42, no. 5, pp. 1044–1057, May 2007.

[YGY+06] T. Yao, M. Gordon, K. Yau, M.-T. Yang, and S. P. Voinigescu, "60 GHz PA and LNA in 90-nm RF-CMOS," *2006 IEEE Radio Frequency Integrated Circuits Symposium (RFIC 2006)*, p. 4, Jun. 2006.

[YH92] T.-S. Yum and K.-W. Hung, "Design Algorithms for Multihop Packet Radio Networks with Multiple Directional Antennas Stations," *IEEE Transactions on Communications*, vol. 40, no. 11, pp. 1716–1724, Nov. 1992.

[YHS05] H. Yang, M. H. A. J. Herben, and P. F. M. Smulders, "Impact of Antenna Pattern and Reflective Environment on 60 GHz Indoor Channel Characteristics," *Antenas and Wireless Propagation Letters*, vol. 4, pp. 300–303, 2005.

[YHXM09] Y. Yang, H. Hu, J. Xu, and G. Mao, "Relay Technologies for WiMAX and LTE-advanced Mobile Systems," *IEEE Communications Magazine*, vol. 47, no. 10, pp. 100–105, Oct. 2009.

[YIMT02] S. Yano, S. Ide, T. Mitsuhashi, and H. Thwaites, "A Study of Visual Fatigue and Visual Comfort for 3D HDTV/HDTV Images," *Displays*, vol. 23, no. 4, pp. 191–201, 2002.

[YKG09] E. Yilmaz, R. Knopp, and D. Gesbert, "On the Gains of Fixed Relays in Cellular Networks with Intercell Interference," *IEEE Signal Processing Advances in Wireless Communications*, pp. 603–607, Jun. 2009.

[YLLK10] C. Yoon, H. Lee, W. Y. Lee, and J. Kang, "Joint Frame/Frequency Synchronization and Channel Estimation for Single-Carrier FDE 60 GHz WPAN System," *2010 IEEE Consumer Communications and Networking Conference (CCNC 2010)*, pp. 1–5, Jan. 2010.

[YMM+94] A. Youssef, J. P. Mon, O. Meynard, H. Nkwawo, S. Meyer, and J. C. Leost, "Indoor Wireless Data Systems Channel at 60 GHz Modeling by a Ray-Tracing Method," *1994 IEEE Vehicular Technology Conference (VTC 1994)*, pp. 914–918, 1994.

[Yng91] S. Yngvesson, *Microwave Semiconductor Devices*, 1st ed. New York, Kluwer Academic Publishers, 1991.

[Yon07] S. Yong, "TG3c Channel Modeling Sub-Committee Final Report," *IEEE 802.15-07-0584-00-003c,* Jan. 2007.

[Yos07] S. Yoshitomi, "Challenges to Accuracy for the Design of Deep-Submicron RF-CMOS Circuits," *2007 Asia and South Pacific Design Automation Conference (ASP-DAC 2007)*, pp. 438–441, Jan. 2007.

[YP08] L. Yang and M. Park, "Applications and Challenges of Multi-band Gigabit Mesh Networks," *2008 Second International Conference on Sensor Technologies and Applications (SENSORCOMM 2008)*, pp. 813–818, Aug. 2008.

[YS02] Y. Yoo and M. Song, "Design of a 1.8V 10bit 300MSPS CMOS Digital-to-Analog Converter with a Novel Deglitching Circuit and Inverse Thermometer Decoder," *2002 Asia-Pacific Conference on Circuits and Systems (APCCAS 2002)*, vol. 2, pp. 311–314, 2002.

[YSH07] H. Yang, P. Smulders, and M. Herben, "Channel Characteristics and Transmission Performance for Various Channel Configurations at 60 GHz," *EURASIP Journal on Wireless Communications and Networking*, vol. 2007, no. 1, p. 019613, 2007. http://jwcn.eurasipjournals.com/content/2007/1/019613.

[YTKY+06] T. Yao, L. T.-Kebir, O. Yuryevich, M. Gordon, and S. P. Voinigescu, "65 GHz Doppler Sensor with On-Chip Antenna in $0.18 \mu m$ SiGe BiCMOS," *2006 IEEE MTT-S International Microwave Symposium (IMS)*, pp. 1493–1496, Jun. 2006.

[YXVG11] S.-K. Young, P. Xia, and A. Valdes-Garcia, *60 GHz Technology for Gbps WLAN and WPAN: From Theory to Practice*. West Sussex, UK, Wiley Press, 2011, ch. 2 and 8.

[Zaj13] A. Zajic, *Mobile-to-Mobile Wireless Channels*. Norwood, MA, Artech House, 2013.

[ZBN05] T. Zwick, T. J. Beukema, and H. Nam, "Wideband Channel Sounder with Measurements and Model for the 60 GHz Indoor Radio Channel," *IEEE Journal of Vehicular Communications*, vol. 54, no. 4, pp. 1266–1277, 2005.

[ZBP+04] T. Zwick, C. Baks, U. R. Pfeiffer, D. Liu, and B. P. Gaucher, "Probe Based MMW Antenna Measurements Setup," *2004 IEEE Antennas and Propagation Society International Symposium*, vol. 1, pp. 20–25, Jun. 2004.

[ZCB+06] T. Zwick, A. Chandrasekhar, C. W. Baks, U. R. Pfeiffer, S. Brebels, and B. P. Gaucher, "Determination of the Complex Permittivity of Packaging Materials at Millimeter-Wave Frequencies," *IEEE Transactions on Microwave Theory and Techniques*, vol. 54, no. 3, pp. 1001–1010, Mar. 2006.

[ZGV+03] X. Zhao, S. Geng, L. Vuokko, J. Kivinen, and P. Vainikainen, "Polarization Behaviors at 2, 5, and 60 GHz for Indoor Mobile Communications," *Wireless Personal Communications*, vol. 27, pp. 99–115, 2003.

[Zha09] Y. P. Zhang, "Enrichment of Package Antenna Approach With Dual Feeds, Guard Ring, and Fences of Vias," *IEEE Transactions on Advanced Packaging*, vol. 32, pp. 612–618, Aug. 2009.

[ZL06] Q. Zhao and J. Li, "Rain Attenuation in Millimeter Wave Ranges," *2006 IEEE International Symposium on Antennas, Propagation & EM Theory*, pp. 1–4, Guilin, China, Oct. 2006.

[ZL09] Y. P. Zhang and D. Liu, "Antenna-on-Chip and Antenna-in-Package Solutions to Highly-Integrated Millimeter-Wave Devices for Wireless Communications," *IEEE Transactions on Antennas and Propagation*, vol. 57, no. 10, pp. 2830–2841, Oct. 2009.

[ZLG06] T. Zwick, D. Liu, and B. P. Gaucher, "Broadband Planar Superstrate Antenna for Integrated Millimeterwave Transceivers," *IEEE Transactions on Antennas and Propagation*, vol. 54, no. 10, pp. 2790–2796, Oct. 2006.

[ZMK05] X. Zhang, A. Molisch, and S.-Y. Kung, "Variable-Phase-Shift-Based RF-Baseband Codesign for MIMO Antenna Selection," *IEEE Transactions on Signal Processing*, vol. 53, no. 11, pp. 4091–4103, Nov. 2005.

[ZMS+13] H. Zhao, R. Mayzus, S. Sun, M. Samimi, J. K. Schulz, Y. Azar, K. Wang, G. N. Wong, F. Gutierrez Jr., and T. S. Rappaport, "28 GHz Millimeter Wave Cellular Communication Measurements for Reflection and Penetration Loss in and around Buildings in New York City," *2013 IEEE International Conference on Communications (ICC)*, Jun. 2013.

[ZO12] L. Zhou and Y. Ohashi, "Efficient Codebook Based MIMO Beamforming for Millimeter-Wave WLANs," *2012 IEEE International Symposium on Personal, Indoor and Mobile Radio Communications (PIMRC 2012)*, pp. 1885–1889, 2012.

[ZPDG07] T. Zwick, U. Pfeiffer, L. Duixian, and B. Gaucher, "Broadband Planar Millimeter-Wave Dipole with Flip-Chip Interconnect," *2007 IEEE Antennas and Propagation Society International Symposium*, pp. 5047–5050, Jun. 2007.

[ZS09a] Y. P. Zhang and M. Sun, "An Overview of Recent Antenna Array Designs for Highly Integrated 60 GHz Radios," *2009 European Conference on Antennas and Propagation (EuCAP 2009)*, pp. 3783–3786, Mar. 2009.

[ZS09b] Y. Zhang and M. Sun, "An Overview of Recent Antenna Array Designs for Highly-Integrated 60 GHz Radios," *2009 European Conference on Antennas and Propagation (EuCAP 2009)*, pp. 3783–3786, Mar. 2009.

[ZSCWL08] Y. P. Zhang, M. Sun, K. P. Chua, L. L. Wai, D. X. Liu, "Integration of Slot Antenna in LTCC Package for 60 GHz Radios," *IET Electronics Letters*, vol. 5, no. 44, Feb. 2008.

[ZSG05] Y. P. Zhang, M. Sun, and L. H. Guo, "On-Chip Antenna for 60 GHz Radios in Silicon Technology," *IEEE Transactions on Electron Devices*, vol. 52, no. 7, pp. 1664–1668, Jul. 2005.

[ZSS07] F. Zhang, E. Skafidas, and W. Shieh, "A 60 GHz Double-Balanced Gilbert Cell Down-Conversion Mixer on 130-nm CMOS," *2007 IEEE Radio Frequency Integrated Circuits Symposium (RFIC 2007)*, Jun. 2007.

[ZSS08] F. Zhang, E. Skafidas, and W. Shieh, "60 GHz Double-Balanced Up-Conversion Mixer on 130 nm CMOS Technology," *Electronics Letters*, vol. 44, no. 10, pp. 633–634, May 2008.

[ZWS07] S. Zhan, R. J. Weber, and J. Song, "Effects of Frequency Selective Surface (FSS) on Enhancing the Radiation of Metal-Surface Mounted Dipole Antenna," *2007 IEEE MTT-S International Microwave Symposium (IMS)*, pp. 1659–1662, Jun. 2007.

[ZXLS07] X. Zheng, Y. Xie, J. Li, and P. Stoica, "MIMO Transmit Beamforming under Uniform Elemental Power Constraint," *IEEE Transactions on Signal Processing*, vol. 55, no. 11, pp. 5395–5406, 2007.

[ZYY+10] Q. Zou, K.-S. Yeo, J. Yan, B. Kumar, and K. Ma, "A Fully Symmetrical 60 GHz Transceiver Architecture for IEEE 802.15.3c Application," *2010 IEEE International Conference on Solid-State and Integrated Circuit Technology (ICSICT)*, pp. 713–715, Nov. 2010.

[ZZH07] H. Zhong, T. Zhong, and E. F. Haratsch, "Quasi-Cyclic LDPC Codes for the Magnetic Recording Channel: Code Design and VLSI Implementation," *IEEE Transactions on Magnetics*, vol. 43, no. 3, pp. 1118–1123, Mar. 2007.

List of Abbreviations

3-D three-dimensional
3GPP Third Generation Partnership Project
3WDP three-wave with diffuse power distribution
4G fourth-generation
5G fifth-generation
5GPPP Fifth Generation Public Private Partnership
A-BFT Association Beamforming Training
AC alternate current
ACK acknowledgement
ADC analog-to-digital converter
AES advanced encryption standard
AF amplify-and-forward
AGC automatic gain control
AIC Akaike information criterion
AM amplitude modulation
ANSI American National Standards Institute
ANT antenna training field
AOA angle of arrival
AOD angle of departure
AP access point
ARQs automatic repeat requests
AS asynchronous
ASIC application specific integrated circuit
ASK amplitude shift keying
ATI announcement transmission interval
ATS antenna sequence training
A/V audio-visual
AWGN additive white Gaussian noise
BCC binary convolutional inner code
BCH Bose, Ray-Chaudhuri, and Hocquenghem [codes]
BEOL Back End of Line
BER bit error rate
BFSK binary frequency shift keying
BICM bit interleaved coded modulation
BJT bipolar junction transistor
BPSK binary phase shift keying
BRP beam refinement protocol
BTI beacon transmission interval
BW bandwidth
CAD computer aided design
CAP contention access period
CBAP contention based access period
CCMP cipher block chaining message authentication protocol
CDF cumulative distribution
CES channel estimation sequence
CEPT Conference of Postal and Telecommunications Administrations
CF Consumption Factor
CFO carrier frequency offset
CIR channel impulse response
CMOS complementary metal oxide semiconductor
CMP chemical mechanical polishing
CMS common mode signaling
CPM continuous phase modulation
CPW co-planar waveguide
CPHY Control PHY
CQI channel quality indicator
CRC cyclic redundancy check
CRS common reference signal
CSMA/CA carrier sense multiple access with collision avoidance
CSF channel state feedback
CC convolutional code
CS compressive sensing
CSI channel state information
CSIT channel state information at the transmitter

CSIR channel state information at the receiver
CTAP channel time allocation period
CTS clear to send
DAC digital-to-analog converter
DAMI dual alternate mark inversion
dB decibel
DBPSK differential binary shift keying
DC direct current
DCA distributed contention access
DCF distributed coordination function
DCM dual carrier modulation
DEV device
DF decode-and-forward
DFE decision feedback equalizer
DFT discrete Fourier transform
DIBL drain induced barrier lowering
DMG directional multi-gigabit
DNL differential non-linearity
DOA direction of arrival
DOD direction of departure
DQPSK differential quadrature phase shift keying
DRP distributed reservation protocol
DS device-synchronized
DSP digital signal processor
DTI data transfer interval
DTP dynamic tone pairing
EEP equal error protection
EIRP effective isotropic radiated power
ENOB effective number of bits
ESPRIT Estimation of Signal Parameters via Rotational Invariance Techniques
ETRI Electronics and Telecommunications Research Institute
ETSI European Telecommunications Standards Institute
EVD eigenvalue decomposition
EVM error vector magnitude
FCC Federal Communications Commission
FDD frequency division duplex
FDE frequency domain equalization
FDTD finite difference time domain
FEC forward error correction
FET field-effect transistor
FEM finite element method
FFT fast Fourier transform
FIR finite impulse response

FM frequency modulation
FPGA field-programmable gate array
FSK frequency shift keying
FSS frequency selective screen
FTP file transfer protocol
GaAs gallium arsenide
Gbps gigabit per second
GCMP Galois message authentication code protocol
GF Galois field
GHz gigahertz (not spelled out)
GMSK Gaussian minimum shift keying
GOP group of pictures
GPI group pair index
GSM Global Systems for Mobile Communications
Gs/s gigasamples per second
HDMI high-definition multimedia interface
HDTV high-definition television
HIS high-impedance surface
HIPERLAN high-performance radio local area network
HPBW half power beamwidth
HRP high-rate PHY
HSI high-speed interface
HTTP hypertext transfer protocol
IC integrated circuit
ICI inter-carrier interference
ID identifier
IDFT inverse DFT
IEEE Institute of Electrical and Electronics Engineers
IF intermediate frequency
IFFT inverse fast Fourier transform
i.i.d. independent and identically distributed
IIR infinite impulse response
ILFD injection-locked frequency divider
INL integral non-linearity
IP Internet protocol
IQ in-phase/quadrature
ISI intersymbol interference
ISM industrial, scientific, and medical (bands)
ISO International Organization for Standardization
ITRS International Roadmap for Semiconductors

List of Abbreviations

ITU International Telecommunication Union
IWPC International Wireless Industry Consortium
KAIST Korea Advanced Institute of Science and Technology
kHz kilohertz
LAN local area network
LAS lobe angle spread
LCP liquid crystal polymer
LDPC low-density parity check
LLC logical link control
LLR log-likelihood ratio
LMDS local multipoint distribution service
LMS least mean squares
LNA low noise amplifier
LO local oscillator
LOS line of sight
LOVA list output Viterbi decoder algorithm
LP PHY low power PHY
LQI link quality index
LRP low rate PHY
LSB least significant bit
LTCC low-temperature co-fired ceramic
LTE Long Term Evolution
M-ASK M-amplitude shift keying
M-PSK M-phase shift keying
M-QAM M-quadrature amplitude modulation
MAC medium access control (protocol)
MAN metropolitan area network
MAP maximum a posteriori
MCS modulation and coding scheme
MCTA management operation
MHz megahertz
MIME multipurpose internet mail extensions
MIMO multiple input multiple output
ML maximum likelihood
MISO multiple input single output
MLSE maximum likelihood sequence estimator
MMSE minimum mean squared error
mmWave millimeter wave wireless communication
MoM method of moments
MOSCAP metal oxide semiconductor capacitor
MOSFET metal oxide semiconductor field-effect transistors
MPC multipath component
MPR mandatory PHY rate
MUSIC multiple signal classification
NetBIOS network basic input output system
NF noise factor or noise figure
NLOS non-line of sight
NMOS N-type metal oxide semiconductor
NS8-QAM non-square 8-QAM
NTIA National Telecommunications and Information Administration
NYU WIRELESS New York University wireless research center
OfCom Office of Communications
OFDM orthogonal frequency division multiplexing
OFDMA orthogonal frequency division multiple access
OOB out-of-band
OOK on-off keying
OQPSK offset quadrature phase shift keying
OSI Open Systems Interconnection
PA power amplifier
PAE power added efficiency
PAF power efficiency factor
PAM pulse amplitude modulation
PAN personal area network
PAPR peak-to-average-power ratio
PAS power angle spectrum
PBSS personal base station mode
PCB printed circuit board
PCES pilot channel estimation sequence
PCF point coordination function
PCP personal basic service set control point
PCS personal communication systems
PDF probability density function
PDK process design kit
PEF power-efficiency factor (H)
PET pattern estimation and tracking
PHY physical layer
PLE path loss exponent
PLL phase-locked loop
PMI precoding matrix indicator
PMOS P-type metal oxide semiconductor
PN pseudorandom

PNC piconet coordinator
PPP Poisson point process
PS piconet-synchronized
PSD power spectral density
QAM quadrature amplitude modulation
QoS quality of service
QPSK quadrature phase shift keying
RCS radar cross section
RF radio frequency
RI rank indicator
RLS recursive least squares
RMS root mean square
RPC remote procedure call
RS Reed-Solomon (code)
RS reference signal
RSSI received signal strength indicator
RTS request to send
SA successive approximation
SAS symmetric antenna set
SCBT single carrier block transmission
SC-FDE single carrier modulation with frequency domain equalization
SCP session control protocol
SC-PHY single carrier PHY
SCTP stream control transmission protocol
SD steepest descent
SFD static frequency divider
SFDR spurious-free dynamic range
SiGe silicon germanium
SINR signal-to-interference-plus-noise ratio
SIP Session Initiated Protocol
SIRCIM Simulation of Indoor Radio Channel Impulse Response Models
SISO single-input single-output
SIMO single-input multiple-output
SJNR signal-to-jitter-noise radio
SNDR signal to noise and distortion ratio
SNR signal-to-noise ratio
SoC system-on-chip
SOI silicon on insulator
SOVA soft output Viterbi algorithm
SP scheduled service period
SQNR signal-to-quantization-noise ratio
SRF self-resonant frequency

SSCM statistical spatial channel model
STA station
STF short training field
STP static tone pairing
S-V Saleh-Valenzuela (model)
SWR standing wave ratio
TB transport block
TCM trellis coded modulation
TCP transmission control protocol
TDD time division duplex
TDMA time division multiple access
TEM transverse electromagnetic
THz terahertz
TIA Telecommunications Industry Association
TIQ threshold inverter quantization
TRL through line reflect
TWDP two wave with diffuse power distribution
U-NII Unlicensed National Information Infrastructure
UDP user datagram protocol
UEP-QPSK unequal error protection QPSK
UHF ultra high frequency
USB universal serial bus
UWB ultrawideband
V2I vehicle-to-infrastructure
V2V vehicle-to-vehicle
VCO voltage control oscillator
VGA variable-gain amplifier
VHF very high frequency
VNA vector network analyzer
VSWR voltage standing wave ratio
WAN wide area network
WHDI wireless home digital interface
WiFi wireless fidelity, usually in the context of the WiFi Alliance
WiGig Wireless Gigabit Alliance
WLAN wireless local area network
WNCG Wireless Networking and Communications Group
WPAN wireless personal area network
WRC World Radiocommunication Conference
ZF zero forcing

Index

A

ABCD-parameters, 263–267
    transmission lines, 293
Absorption effect, 7, 106–107
Access points (APs)
    aircraft, 26
    backhaul, 108
    DMG, 564
    in-room architecture, 489
    repeaters/relays, 487–488
    STA-AP, 184
    WiFi, 500
    WLANs, 11–12
Accuracy
    ADCs, 405
    ray tracing, 124
ACK messages
    frame control, 515
    IEEE 802.11ad, 562–564
    lost, 569
    sector sweep, 568
Active impedances in on-chip antennas, 208
Active mixers, 349
Adaptive antenna arrays, 235–237
    beam steering, 237–242
    beamforming, 242–249, 481
    case studies, 251–252
    gain, 103–105
Adaptive equalization, 446
Adaptive training, 461
ADCs. *See* Analog-to-digital converters
    (ADCs)
Additive white Gaussian noise (AWGN), 41–43
    consumption factor, 376–377
    linear equalization, 52–53
    OFDM and SC-FDE, 59, 448–449
    phase noise, 443
    symbol detection, 41–43
    trellis coded modulation, 69
Adjustable gain antennas, 110
Admittance
    low noise amplifiers, 339–341
    Y-parameters, 264

Advanced Design System (ADS) design tool,
    268
Aerospace applications, 26–27
AF (amplify-and-forward) repeaters, 488–489,
    492
AGC (Automatic Gain Control)
    common mode signaling, 523
    radios, 316
    transceivers, 437–438
Aggregation in DMG, 575
Alamouti code, 85
Aliasing in ADC sampling, 388–389
Alternation relay transmission mode, 569
AM (Amplitude modulation), 35
AM-AM and AM-PM distortion profiles,
    439–440
Amplifiers
    low noise, 324–326, 334–342
    op-amps, 396–397
    power. *See* Power amplifiers (PAs)
    track-and-hold, 397–403
    variable gain, 239, 412
Amplify-and-forward (AF) repeaters, 488–489,
    492
Amplitude modulation (AM), 35
Amplitude shift keying (ASK), 45, 462
Amplitude thresholding, 437
Analog baseband phase shifting, 237–240
Analog components, 323
    consumption factor theory, 370–382
    low noise amplifiers, 334–342
    metrics, 317–323
    mixers, 342–349
    phase-locked loops, 364–370
    power amplifiers, 323–334
    power-efficiency factor, 374–375
    VCOs. *See* Voltage-controlled oscillators
        (VCOs)
Analog equalization, 50
Analog-to-digital converters (ADCs), 383–384
    baseband representation, 39
    case studies, 420–421
    comparators, 394–397

Analog-to-digital converters (*continued*)
    design challenges and goals, 403–406
    device mismatch, 393–394
    encoders, 407–409
    flash and folded-flash, 413–420
    non-linearity, 406–407
    sampling, 384–392
    successive approximation, 410–411
    time-interleaved, 411–413
    transceivers, 436–438
        trends and architectures, 409
        ultra-low resolution, 464–466
Angle-dependent reflection, 112
Angle of arrival (AOA) information
    indoor channel models, 173–174
    multipath, 135–138, 161–164
    statistics, 137
Angle of departure (AOD)
    indoor channel models, 173–174
    multipath models, 161
Angular segment combining, 132–133
Angular spreads, 135–138, 174
Announcement Transmission Interval (ATI), 565–566
ANT (antenna training field), 576
Antenna sequence training (ATS), 558
Antennas and arrays, 187
    adaptive arrays. *See* Adaptive antenna arrays
    array factor, 196
    beam patterns, 473
    beamforming. *See* Beamforming
    dipole, 104, 212–215
    directional. *See* Directional antennas
    effective area, 102
    fundamentals, 191–198
    gain. *See* Gain
    height, impact of, 145–146
    in-package, 188–191, 209–211, 255–257
    indoor channel models, 166–169
    integrated lens, 231–235
    large-scale path loss models, 159–160
    MIMO beamforming, 86
    multipath models, 163–164
    on-chip. *See* On-chip antennas
    patch, 218–222
    path loss data, 156–160
    pointing angle, 145–148
    polarization. *See* Polarization
    steerable, 105
    topologies, 212–225
    Yagi, 256–257
    Yagi-Uda, 222–225
AOA (angle of arrival)
    indoor channel models, 173–174
    multipath, 135–138, 161–164
    statistics, 137

AOD (angle of departure)
    indoor channel models, 173–174
    multipath models, 161
Aperture
    area, 102
    defined, 472
    delay, 403
    jitter, 403
Appended Packet in DMG, 575
Application Layer, 92–93
Applications
    emerging. *See* Emerging applications
    WirelessHD, 550–551
APs. *See* Access points (APs)
ARQs (automatic repeat requests), 495
Array factor, 196
Arrays. *See* Antennas and arrays
Artificial magnetic conductors, 230
AS (Asynchronous) mode, 519
ASK (amplitude shift keying), 45, 462
Association Beamforming Training (A-BFT), 565, 569
Association in MAC, 519–520
Asynchronous (AS) mode, 519
Asynchronous CTAP, 516
ATI (Announcement Transmission Interval), 565–566
Atmospheric effects, 7, 106–107
ATS (antenna sequence training), 558
Attenuation
    CMOS chips, 204–205
    radio waves. *See* Radio wave propagation
Audio/visual mode (AV PHY), 521, 531–537
    UEP, 541–542
    WirelessHD, 552–553
Audio/visual (A/V) packetizer, 553
Australia, 60 GHz spectrum regulation in, 511
Auto-correlation in channel estimation, 459
Automatic Gain Control (AGC)
    common mode signaling, 523
    radios, 316
    transceivers, 437–438
Automatic repeat requests (ARQs), 495
AV PHY (audio/visual mode). *See* Audio/visual mode (AV PHY)
AWGN. *See* Additive white Gaussian noise (AWGN)
Azimuth angles
    indoor channel models, 173–174
    outdoor channel models, 145–148, 158–160

B

B-frames, 494, 497
Back End of Line (BEOL) process, 201–202, 206, 270, 277
Back-off region, 319, 333

Index

Backhaul, 13–14
    beam adaptation protocols, 484–485
    large-scale path loss models, 157–161
    weather effects, 108–109
Backside lenses, 234
Backward isolation, 267
Balanced mixers, 349
Balanced transmission lines, 298
Baluns, 298
Band diversity, 498
Bandlimited signals, 35
Bandwidth
    ADCs, 404, 411
    antennas, 193, 211
    coherence, 142
    OFDM and SC-FDE, 455
    passband, 22
Baseband representation, 34–39
BCCs (binary convolutional inner codes), 531, 533–535, 541–543
BCH (Bose, Ray-Chaudhuri, and Hocquenghem) codes, 64
Beacon Transmission Interval (BTI), 564, 567
Beam adaptation protocols, 481
    backhaul, 484–485
    channel estimation, 484–487
    IEEE 802.11ad, 483
    IEEE 802.15.3c, 482–483
Beam coding, 483
Beam combining, 103, 132–135
Beam feedback, 546
Beam-level searches, 482
Beam refinement protocol (BRP), 568–569
Beam steering, 18–19
    antenna arrays, 196, 237–242
    blockages, 478
    ECMA-387, 555
    IEEE 802.1d, 562
    indoor channel models, 166
    MAC, 483
    methods, 481
    on-chip antennas, 189–190
    outdoor channel models, 148
Beam-to-hi-res-beam exchange, 546–547
Beam tracking, 547, 575
Beam training, 546
Beamforming, 473
    antenna arrays, 236, 242–249
    cellular networks, 502–503
    directional, 16, 501–502
    hybrid architectures, 16
    MAC, 567–569
    MIMO systems, 85–87
    OFDM and SC-FDE, 454–455
    PHY, 444–445, 543–548
    protocols, 543–545
BEOL. *See* Back End of Line (BEOL)

BER (bit error rate)
    OFDM and SC-FDE, 449–458
    radios, 317
Berlekamp-Massey algorithm, 64–65
Best straight-line INL, 407
Beyond 4G, 6, 25
BFSK (binary frequency shift keying), 462
Binary convolutional inner codes (BCCs), 531, 533–535, 541–543
Binary frequency shift keying (BFSK), 462
Binary phase shift keying (BPSK), 43–45
Bipolar junction transistors (BJTs), 202
Bistatic cross section, 117–118
Bit error rate (BER)
    OFDM and SC-FDE, 449–458
    probability, 69
    radios, 317
Bit flipping, 66
Bit labeling, 43–48
Block codes in error control coding, 63–64
Block fading, 37
Block transmissions, 527–528
Blockage in cellular networks, 501
Blu-Ray Discs, 494
Body ties, 283
Bond wire packaging inductances, 304
Bonded channels in ECMA-387, 557
Boresight alignment, 473
Bose, Jagadish Chandra, 4
Bose, Ray-Chaudhuri, and Hocquenghem (BCH) codes, 64
BPSK (binary phase shift keying), 43–45
Break-point distance in LOS, 149
Broadband access in cellular, 15
Broadside slot orientation, 230–231
BRP (beam refinement protocol), 568–569
BSIM model, 285–287
BTI (Beacon Transmission Interval), 564, 567
Bubble errors, 397, 404
Burst errors, 64

C

Cadence Virtuoso Analog Design Environment design tool, 268
CAP (contention access period), 514
Capacitances
    BEOL structures, 270
    diode-varactors, 356
    dipole antennas, 212–213
    flash ADCs, 418–419
    gate-source and gate-drain, 284
    interconnects, 308
    low noise amplifiers, 341
    Miller, 310–311
    parasitic, 281–283, 308
    track-and-hold amplifiers, 401
    transmission lines, 296

Capacity studies in ultra-low ADC resolution, 465
Carrier feedthrough in homodyne receivers, 89
Carrier frequency in baseband representation, 34
Carrier frequency offset (CFO), 73–74, 76
Carrier sense multiple access with collision avoidance (CSMA/CA), 477
    DMG, 564
    IEEE 802.11ad, 562
    MAC, 516
Cascaded devices, 320–321
Cascode design
    low noise amplifiers, 341
    transistors, 308–309
CBAP (Contention-Based Access Period), 566
CCMP (cipher block chaining message authentication protocol), 572
CDF (cumulative distribution function), 140–141
CE (channel estimation) training, 73, 573
Cellular communications, 6
    applications, 13–15
    backhaul costs, 108
    future, 25–26
    performance, 500–504
    repeaters/relays, 493
Center frequency in baseband representation, 34
Centralized clustering, 567
CEPT (European Conference of Postal and Telecommunication Administrations), 510
CES field, 550
CF (consumption factor) theory, 370–374
    definition, 376–381
    description, 259–260
    power-efficiency factor example, 374–375
CFO (carrier frequency offset), 73–74, 76
Channel estimation, 73, 79–81
    beam adaptation, 484–487
    high-throughput PHYs, 459–461
    OOK, 550
    training, 73, 573
Channel Impulse Response (CIR), 171–173
Channel noise, 285
Channel order in channel estimation, 81
Channel surface approach, 288
Channel time allocation periods (CTAPs), 514
Channels
    ECMA-387, 556–557
    human blockage, 478–480
    indoor. *See* Indoor environments
    multipath models, 160–164
    OOB, 557
    outdoor, 152–160
    PHY, 521–522
    spatial reuse, 480–481

Characteristic impedance of transmission lines, 292–293
Charge-based modeling, 287
Chemical Mechanical Polishing (CMP)
    CMOS chips, 202
    transmission lines, 296–297
Chien algorithm, 64–65
Child piconets, 517–518
China, 60 GHz spectrum regulation in, 511
Chip-to-chip communication, 20
Cipher block chaining message authentication protocol (CCMP)
CIR (Channel Impulse Response), 171–173
Circuit fabrication, 187
Circuit layout, 267–273
Circular polarization
    antennas, 138–139
    indoor channel models, 175
Classes
    ECMA-387 devices, 555–556
    power amplifiers, 329–333
Clear glass reflection, 112
Clear to send (CTS) messages, 477, 562
Clocks
    ADCs, 409, 411–412
    frame synchronization, 73
    jitter, 403
    skew, 401
Close-in free-space reference distance path loss model, 152–160
Closed-loop functions in VCOs, 361
Closed-loop gain
    transistors, 313
    VCOs, 359, 363
Clustering in PCP/AP, 566–567
CMOS (complementary metal oxide semiconductor) technology
    on-chip antennas, 200–209
    production, 267–273
    transistors, 260
CMP (Chemical Mechanical Polishing)
    CMOS chips, 202
    transmission lines, 296–297
CMS (common mode signaling) frames, 523
Co-channel interference, 165
Co-planar designs for transmission lines, 293
Co-planar waveguides (CPW) topologies
    probes, 270
    transmission lines, 293–296
Code book approaches
    beam steering, 240–241
    beamforming protocols, 482, 484–485
Code swapping, 495
Codewords in DACs, 422
Coding
    DMG, 574
    error control. *See* Error control coding
    high-throughput PHYs, 445–447

Coexistence in MAC, 518
Coherence bandwidth, 142
Collision detection and avoidance, 476–478
    DMG, 564
    IEEE 802.11ad, 562
    MAC, 516
Colored additive white Gaussian noise model, 443
Colpitts oscillators, 364
Column spaces, 249
Combining
    beams, 103, 132–135
    vectors, 85, 87
Common definitions, 31
Common-drain transistors, 308–309
Common-gate transistors, 308–309
Common mode operation in PHY, 522–524
Common mode signaling (CMS) frames, 523
Common-source amplifiers, 311, 328
Common-source transistors, 308–309
Comparators in ADCs, 394–397, 409
Complementary metal oxide semiconductor (CMOS) technology. *See* CMOS
    on-chip antennas, 200–209
    production, 267–273
    transistors, 260
Complex baseband representation, 34–39
Complex envelopes, 35
Compress-and-forward strategy, 488–489
Compressed video, 493–495
Compression requirements, 11
Compressive channel estimation, 485–486
Concatenated coding techniques, 67
Concrete reflection, 112
Conduction angles in power amplifiers, 330, 332
Conduction losses, 208–209
Conference room environments, 183–184
Conjugate matching in transistors, 309–310
Constellations
    AV PHY, 531
    HSI PHY, 528
    NS8-QAM, 49
    OFDM and SC-FDE, 447–458
    SC-PHY, 527–528
    skewing, 543
    symbols from, 40
    unit-form, 40
Consumption factor (CF) theory, 370–374
    definition, 376–381
    description, 259–260
    power-efficiency factor example, 374–375
Content delivery in future, 23
Contention access period (CAP), 514
Contention-Based Access Period (CBAP), 566
Continuous phase modulation (CPM), 463–464

Control PHY (CPHY), 572, 577
Conversion
    ADCs. *See* Analog-to-digital converters (ADCs)
    DACs. *See* Digital-to-analog converters (DACs)
    heterodyne receivers, 90–91
    mixer gain, 342, 345–346
    radios, 316
Convolution in equalization, 57
Convolutional codes, 67–69
Cooling systems in data centers, 20
Cosine components in baseband representation, 36
Coupling transformer windings, 306
Covariance
    array beamforming, 246–248
    linear equalization, 52–53
Coverage extension, relaying for, 487–493
CPHY (Control PHY), 572, 577
CPM (continuous phase modulation), 463–464
CPW (co-planar waveguides) topologies
    probes, 270
    transmission lines, 293–296
Crane model, 109
CRC (cyclic redundancy check) codes, 64, 495
Cross correlation, 285
Cross-polarized signals, 107
Cross section modeling, 117–119
Cross talk, 277
CSMA/CA (carrier sense multiple access with collision avoidance), 477
    DMG, 564
    IEEE 802.11ad, 562
    MAC, 516
CTAPs (channel time allocation periods), 514
CTS (clear to send) messages, 477, 562
Cumulative distribution function (CDF), 140–141
Current compensation circuits, 402
Current DACs, 422–429
Current return paths, 341
Currents, scattering parameters for, 264–265
Cursers, 171
Cyclic prefix, 57
Cyclic redundancy check (CRC) codes, 64, 495

D

D-flip-flops in phase-locked loops, 366–367, 370
DACs. *See* Digital-to-analog converters (DACs)
DAMI (dual alternate mark inversion) signaling, 549–550
Data centers, 19–21
Data Link Layer in OSI model, 92–94
DATA messages, 562–563

Data rates
    AV PHY, 531
    DMG, 574
    HSI PHY, 529
    Type A PHY in ECMA-387, 559
    Type B PHY in ECMA-387, 561
    Type C PHY in ECMA-387, 561
    UEP, 541–542
Data Transfer Interval (DTI), 565–566
DBPSK (differential binary shift keying), 577
DC offset in homodyne receivers, 88–89
DCA (distributed contention access) protocol, 557
DCF (distributed coordination function), 564
DCM (dual carrier modulation), 579
Deafness problem, 474, 476
Decentralized clustering, 566
Decision feedback equalizers (DFEs), 53–55
    high-throughput PHYs, 446
    time domain equalization, 50
Decode-and-forward (DF) repeaters, 488–493
Decoders
    flash ADCs, 416–417
    high-throughput PHYs, 445
    LOVA, 69
    Reed-Solomon, 495–496
    soft decision, 64
De-embedding structures in test-chip layouts, 289
Deep fades, 140
Definitions, 31
Deflection routing, 490–491
Delay, 18
    ADCs, 403
    beam steering, 239
    DACs, 428–429
    encoders, 408
    equalization in time domain, 50
    RMS delay spread. See Root-mean-square (RMS) delay spread
Demodulate-and-forward strategy, 488–489
Depolarization, path, 106
Desired transmission line impedance, 295
Desktop environments, 182
Detection, symbol, 41–43
Device classes in ECMA-387, 555–556
Device discovery, 475–476
Device mismatch in ADCs and DACs
    ADCs, 411–412, 418–419
    DACs, 427
    description, 393–394
Device-Synchronized (DS) mode, 519
DF (decode-and-forward) repeaters, 488–493
DFEs. See decision feedback equalizers (DFEs)
DFT. See Discrete Fourier transform (DFT)

DIBL (Drain Induced Barrier Lowering), 284
dielectric constant, 189
Differential binary shift keying (DBPSK), 577
Differential feeds in dipole antennas, 213
Differential non-linearity (DNL), 406–407
Differential QPSK (DQPSK), 44–46
Differential transmission lines, 298
Diffraction, 111, 176
Digital baseband phase-shifting, 239
Digital beamforming, 454–455
Digital modulation, 39
    amplitude shift keying, 45
    binary phase shift keying, 43–45
    phase shift keying, 47
    quadrature amplitude modulation, 48–49
    quadrature phase shift keying, 45–47
    symbols, 40–43
Digital-to-analog converters (DACs), 383–384, 421–422
    baseband representation, 39
    case studies, 429–430
    current steering, 422–429
    device mismatch, 393–394
    sampling, 384–392
Digitization in ADC sampling, 388
Diodes, 260
Dipole antennas, 104, 212–215
Dirac delta function, 330
Direct conversion receivers, 88
Direct conversion transmitters, 90
Directional antennas
    indoor channel models, 167, 175
    PHY, 472–474
    pointing angle, 145–148
    ray-tracing models, 169
    received power, 106
Directional multi-gigabit (DMG) PHY, 564–566
    common aspects, 572–576
    low-power SC-PHY, 581–582
    OFDM PHY, 578–581
    SC-PHY, 577–578
Directions of arrival (DOAs), 138
    3GPP-style models, 152
    array beamforming, 242–245, 248–251
    wave vectors, 198
Directions of departure (DODs), 152
Directivity of antennas
    gain relationship, 192, 196
    indoor channel models, 167
    integrated lens, 234
Disassociation in MAC, 519–520
Discovery complexity in high-throughput PHYs, 446
Discrete Fourier transform (DFT)
    beam adaptation, 483

Index

frequency domain equalization, 57–59
    spatial multiplexing, 83
Dispersion relationship, 198
Distortion
    power amplifiers, 439–440
    sample-and-hold amplifiers, 400
Distributed contention access (DCA) protocol, 557
Distributed coordination function (DCF), 564
Distributed reservation protocol (DRP), 557
Diversity
    band, 492–493, 498
    spatial, 84–85
Divide-by-two injection-locked frequency dividers, 370
DMG. *See* Directional multi-gigabit (DMG) PHY
DNL (differential non-linearity), 406–407
DOAs (directions of arrival), 138
    3GPP-style models, 152
    array beamforming, 242–245, 248–251
    wave vectors, 198
DODs (directions of departure), 152
Doherty design in power amplifiers, 332–333
Doped silicon, 260–262
Doping concentration, 405
Doppler effects, 130–132
    human blockage, 480
    indoor channel models, 175
    temporal fading, 142, 144
    vehicular applications, 24
Double-balanced mixers, 349
Double pi-models, 302
Double-segmented approach, 430
Downconversion
    baseband representation, 35–38
    frequency synchronization, 73–74, 76
    heterodyne receivers, 89–91
    mixers, 342–346, 349
DQPSK (differential QPSK), 44–46
Drain in CMOS transistors, 260
Drain Induced Barrier Lowering (DIBL), 284
DRP (distributed reservation protocol), 557
Drywall reflection, 112
DS (Device-Synchronized) mode, 519
DTI (Data Transfer Interval), 565–566
DTP (dynamic tone pairing), 576, 579–581
Dual alternate mark inversion (DAMI) signaling, 549–550
Dual carrier modulation (DCM), 579
Dual ground planes in transmission lines, 297
Dummy preamplifiers in flash ADCs, 415
Durgin's TWDP distribution, 131, 170
Duty cycle in sample-and-hold circuitry, 401
Dynamic channels in MAC, 518
Dynamic errors in DACs, 424–425

Dynamic ranges
    ADC sampling, 388–389, 391–392
    encoders, 405
Dynamic tone pairing (DTP), 576, 579–581

E

ECMA/ECMA-387
    channelization, 556–557
    device classes, 555–556
    encoders, 68
    second edition, 561
    spreading factor, 70
    standard, 555–561
    trellis coded modulation, 70
    Type A PHY, 559–560
    Type B PHY, 560–561
    Type C PHY, 561
EEP (equal error protection), 495
Effective area, 102
Effective isotropic radiated power (EIRP), 102
Effective number of bits (ENOB) in ADCs, 403–404
Effective permeability of inductors, 300
Efficiency
    antennas, 191–192
    consumption factor theory, 371–374
    energy, 519
    in-package antennas, 210
    on-chip antennas, 206–209
Eigenvalue decomposition (EVD), 249–250
EIRP (effective isotropic radiated power), 102
EKV model, 277, 284–285, 287–288
Electrically large antenna arrays, 189
Electro-bandgap surfaces, 227–228
Electromagnetic field polarization, 167
Electromagnetic simulators, 204, 267–268
Electromagnetic (EM) waves, 99
Element spacing in arrays, 196–197
Elevation angle, 171–172
Ellipsoidal lenses, 232–233
EM (electromagnetic) waves, 99
Emerging applications, 19
    aerospace, 26–27
    cellular and personal mobile communications, 25–26
    data centers, 19–21
    home and office, 23–24
    information showers, 22–23
    replacing wired interconnects on chips, 21–22
    vehicular, 24–25
Encoders
    ADC, 407–409, 416
    ECMA, 68
End-point INL, 407
Energy efficiency in MAC, 519

Energy loss in on-chip antennas, 207–208
ENOB (effective number of bits) in ADCs, 403–404
Ensemble design tool, 268
Equal error protection (EEP), 495
Equalization algorithms in high-throughput PHYs, 445–446
Equalization in frequency domain, 56–57
    OFDM modulation, 60–62
    single carrier frequency domain, 57–60
Equalization in high-throughput PHYs, 445–447
Equalization in time domain, 49–51
    decision feedback equalizers, 53–55
    linear equalization, 51–53
    maximum likelihood sequence estimator, 55–56
Error concealment, 494–495
Error control coding, 62–63
    AV PHY, 531
    block codes, 63–64
    code rate, 495
    convolutional codes, 67–69
    DMG, 574
    HSI PHY, 528–529
    low density parity check codes, 65–67
    Reed-Solomon codes, 64–65
    SC-PHY, 524, 529
    symbol detection, 43
    time domain spreading, 70–71
    trellis coded modulation, 69–70
    Type A PHY in ECMA-387, 559
    Type B PHY in ECMA-387, 561
    Type C PHY in ECMA-387, 561
    UEP, 541–542
    unequal error protection, 71
Error vector magnitude (EVM), 241
Errors
    DACs, 423–425
    flash ADCs, 418–419
ESPRIT (Estimation of Signal Parameters via Rotational Invariance Techniques), 249–251
Estimation and synchronization, 72–73
    channel estimation, 79–81
    frame synchronization, 78–79
    frequency offset synchronization, 75–77
    structure, 73–75
ETSI (European Telecommunications Standards Institute) standard, 510
Euclidean distance metric, 42
Europe, 60 GHz spectrum regulation in, 510
European Conference of Postal and Telecommunication Administrations (CEPT), 510
European Telecommunications Standards Institute (ETSI) standard, 510
EVD (Eigenvalue decomposition), 249–250

EVM (error vector magnitude), 241
Exclusive or gates (XOR) in phase-locked loops, 366
Exposed node problem, 474, 477

F

Factor graphs, 66
Fading, 140
    baseband representation, 37
    indoor channel models, 176
    multiwave, 169–170
Far-field region, 192, 195
Fast Fourier transform (FFT)
    frequency domain equalization, 57
    high-throughput PHYs, 446
Fat-tree encoders, 408, 416–417
FCC (Federal Communications Commission), 509–510
FDD (Frequency Division Duplexing), 444
FDE (frequency domain equalization), 56–60
    frame synchronization, 75
    high-throughput PHYs, 446–447
    vs. OFDM, 447–458
    time domain, 50
FDTD (finite difference time domain), 268–269
FEC. *See* Forward error correction
Federal Communications Commission (FCC), 509–510
Feedback
    beamforming, 86, 546
    comparators, 396–397
    phase-locked loops, 365
    spatial diversity, 85
    transistors, 311–313
FEKO design tool, 268
FEM (finite element method), 191, 268–269
FFT (Fast Fourier transform)
    frequency domain equalization, 57
    high-throughput PHYs, 446
Fidelity design tool, 268
Field-effect transistors (FETs), 261–262
    mismatch, 393–394
    operation, 273–279
Fifth-generation (5G) cellular networks, 14, 16
Figures of merit
    ADCs, 403
    antennas, 211
    consumption factor theory, 373
    low noise amplifiers, 334, 338
    noise performance, 285
    phase-locked loops, 364
    power amplifiers, 323
    transistors, 277
Filters
    baseband representation, 36, 39, 42
    symbol detection, 42

Index 665

Fingers of MOSFETs, 279–281
Finite difference time domain (FDTD), 268–269
Finite element method (FEM), 191, 268–269
Finite impulse response equalizers, 51
5G, 6, 25
5th Generation Public Private Partnership (5GPPP) organization, 152
Fixed antennas, 188
Flash ADCs, 413–420
Flicker noise
    homodyne receivers, 89
    low noise amplifiers, 337
Flip-flops in phase-locked loops, 366–367, 369–370
Floating intercept model, 153, 156, 158
Folding-flash ADCs, 417–418
Folding factor, 418
Foliage effects, 119–122
Forney approach in MLSE, 55–56
Forward error correction (FEC). *See* Error control coding
4-ASK, 45
Fourier transform in digital modulation, 40
Fragmentation Control field, 515
Frame Check Sequence field, 515
Frame Control field, 514–515
Frame format in ECMA-387, 558–559
Frame Length field in DMG, 575
Frames
    DMG, 573, 575
    MAC, 514–515
    synchronization, 72–74, 78–79
Frank-Zado-Chu sequences
    channel estimation, 81
    in frame synchronization, 74–75, 79
Frank-Zafoff (FZ) sequence, 559
Free space path loss, 7
    close-in free-space reference distance model, 152–160
    directional antennas, 472
    equation, 101–102
    frequency relationship, 103–105
    log-distance models, 106
    ray tracing, 124
Frequency
    60 GHz spectrum allocations, 511
    antenna gain, 192
    Doppler effects, 130–132
    multiplexing, 454
    in path loss, 103–105, 175
    power amplifiers, 328
Frequency dividers in phase-locked loops, 366
Frequency Division Duplexing (FDD), 444
Frequency domain equalization (FDE), 56–60
    frame synchronization, 75
    high-throughput PHYs, 446–447
    vs. OFDM, 447–458
    time domain, 50
Frequency flat channels, 130
Frequency offset synchronization, 75–77
Frequency selectivity
    channels, 50, 130
    fading, 140–141, 175
    OFDM and SC-FDE, 452–453
    screens in on-chip antenna gain, 227–228
Frequency shift keying (FSK), 462
Frequency spreading, PHYs, 444
Frequency synthesizers, 364–370
Frequency tuning in transistors, 313–315
Fresnel zones, 121–122
Friis free space propagation theory, 102–104, 184
Front haul, 14
FSK (frequency shift keying), 462
f_T (transit frequency) in transistors, 10, 278, 284
Full duplex relaying, 488, 569
Full-MAC model, 498
Fully balanced mixers, 349
Fused silica in in-package antennas, 210–211
FZ (Frank-Zafoff) sequence, 559

G

GaAs (Gallium Arsenide) substrates, 209
Gain
    ADCs, 411–412
    AGC. *See* Automatic Gain Control (AGC)
    antenna arrays, 103–105, 196
    antennas, 472–473
    channel estimation, 81
    expressing, 102
    frequency of resonance, 192
    in-package antennas, 210–211
    mixers, 342, 345–346
    models for, 279
    on-chip antennas, 225–235, 259
    power amplifiers, 323, 328–329
    transistors, 313
    VCOs, 350, 353–355, 359, 363
Gain-bandwidth product, 211
Gallium Arsenide (GaAs) substrates, 209
Galois fields in block codes, 63
Galois message authentication code protocol (GCMP), 572
Gate in transistors, 260
    capacitance, 284
    length, 260–261
    thermal noise, 285
Gate resistance in layout effects, 281–282

Gating circuits, 260
Gaussian distribution in multipath channel models, 174
Gaussian Minimum Shift Keying (GMSK), 44
Gaussian noise. *See* Additive white Gaussian noise (AWGN)
GCMP (Galois message authentication code protocol), 572
Geometry configurations in beamforming, 544
Glass reflection, 112–113
Glitches in DACs, 424
Global spectrum allocations, 6
GMSK (Gaussian Minimum Shift Keying), 44
Golay sequences
    channel estimation, 81, 459–460
    DMG, 573
    frame synchronization, 74–75
    OOK, 550
    SC-PHY, 527, 578
GOPs (groups of pictures), 494
GPI (Group pair index), 580
Grating lobes, 191, 196
Gray coding, 408
Ground planes in patch antennas, 218–221
Ground return paths
    low noise amplifiers, 341
    transmission lines, 293
Ground shields in antenna arrays, 251
Grounding transmission lines, 296–297
Group pair index (GPI), 580
Group velocity, 198
Groups of pictures (GOPs), 494
Guard rings in CMOS chips, 202
Guard tones in HSI PHY, 530, 532
Guided waves in CMOS chips, 205–206

H

H.264 standard, 494, 497
Hail events, 109–110
Half duplex relaying, 488, 569
Half-wave dipole antennas, 104, 213
Hallways, 106
Handover model, 498
Handover rules in WirelessHD, 553–554
Hardware architectures, 88–91
Hardware Layer in OSI model, 94–95
Harmonic distortion in S/H amplifiers, 400
HDMI (high-definition multimedia interface), 11–12, 493–494, 556
Header Check Sequence in DMG, 576
Heating requirements in data centers, 20
Height of transmitter antennas, 145–146
Helmholtz equation for electric fields, 197
Hemispherical lenses, 231–235
Heterodyne receivers, 89–91
HFSS design tool, 268, 290
Hidden node problem, 19, 474, 476–477

High-definition multimedia interface (HDMI), 11–12, 493–494, 556
High efficiency PHYs, 461–464
High-frequency effects in transmission lines, 289
High-impedance surfaces (HIS), 227–230
High-k dielectric constants, 277
High-rate PHY (HRP), 531–535, 541–542
High-speed interface (HSI) PHY, 521, 528–530, 540–541
High-throughput PHYs, 444–445
    modulation, coding, and equalization, 445–447
    OFDM vs. and SC-FDE, 447–458
    synchronization and channel estimation, 459–461
Higher layer design considerations, 471–472
    beam adaptation protocols, 481–487
    cellular networks, 500–504
    multiband considerations, 497–500
    multimedia transmission, 493–497
    networking devices. *See* Networking devices
    relaying, 487–493
HIS (high-impedance surfaces), 227–230
Hold capacitor size in T/H amplifiers, 401
Home of the future, 23–24
Homodyne receivers, 88–89
Horizontal polarization, 139
HRP (high-rate PHY), 531–535, 541–542
HSI PHY (high-speed interface mode), 521, 528–530, 540–541
Human blockage events, 119–122
    channel reliability, 478–480
    effects, 176
Huygens' principle
    magnetic conductors, 227–228
    microstrip patch antennas, 218–219
Hybrid approach in channel estimation, 486
Hybrid architectures for beamforming, 16
Hybrid precoding in MIMO, 87–88
Hyperhemispherical lenses, 232–234

I

I-frames, 494, 497
IC (Industry Canada), 510
ICI (inter-carrier interference), 62, 444
IDFT (inverse DFT), 57
IE3D design tool, 268
IEEE 802.11ac standard, 13
IEEE 802.11ad standard, 13, 177, 183–184, 562, 572–582
    background, 562–564
    beam adaptation protocols, 483
    indoor environments, 171–177
    MAC features, 564–572

Index

IEEE 802.15.3c standard, 177–182, 512
    beam adaptation protocols, 482–483
    indoor environments, 171–177
    MAC, 512–520
    PHY. *See* Physical layer (PHY)
IF frequencies
    array beam steering, 238, 240
    heterodyne receivers, 89–90
    mixers, 342–346, 349
IIR (infinite impulse response) equalization, 52–53
ILFDs (injection-locked frequency dividers), 366–367, 370
Images for heterodyne receivers, 89
Impedance
    conjugate matching, 310
    DAC current sources, 426
    dipole antennas, 212–213
    input impedance matching of LNAs, 339–340
    low noise amplifiers, 339, 341
    on-chip antennas, 208
    transmission lines, 292–297
Impinging wave fronts in array beamforming, 243
In-package antennas, 188, 209–211
    fundamentals, 189–191
    probe station characterizations, 255–257
In-phase components in baseband representation, 35
In-phase/quadrature (IQ) mismatch, 89
Indoor environments, 166–169
    delay spread, 129
    human effects, 120
    IEEE 802.15.3c and IEEE 802.11ad models, 171–184
    line-of-sight, 106
    penetration losses, 113–116
    ray-tracing models, 169
    Rayleigh, Rician, and multiwave fading models, 169–170
    reflection, 111–112
Inductors and inductances
    interconnects, 308
    low noise amplifiers, 339
    parasitic, 304
    transformers, 305–307
    transmission lines, 289–290, 298–304
Industrial Scientific, and Medical (ISM) bands, 3
Industry Canada (IC), 510
Infinite impulse response (IIR) equalization, 51–53
Information Exchange, 516
Information showers, 22–23
Injection-locked frequency dividers (ILFDs), 366–367, 370
Injection locking in antenna arrays, 251

INL (integral non-linearity), 406–407
Input bandwidth of ADCs, 404
Input capacitance
    flash ADCs, 418–419
    track-and-hold amplifiers, 401
Input impedance of low noise amplifiers, 341
Input-referred 1-dB compression points, 319–320
Input referred noise voltage in low noise amplifiers, 339
Input referred offsets in flash ADCs, 414
Input symbols in Reed-Solomon codes, 64
Insertion loss parameters, 266–267
Integral non-linearity (INL), 406–407
Integrated lens antennas, 231–235
Integrity Code field in MAC, 515
Intelligent repeaters, 487–488
Inter-arrival times, 171
Inter-carrier interference (ICI), 62, 444
Inter-chip wireless interconnects, 212
Intercept points, non-linear, 317–322
Interconnects, 21–22, 212, 308
    delay, 22
Interference in MAC, 518
Interleaving in convolutional codes, 68
Intermodulation, 319, 323
Internal reference voltages in S/H amplifiers, 397
International recommendations for 60 GHz spectrum, 509
International Roadmap for Semiconductors (ITRS), 22
International Telecommunication Union (ITU), 5, 509
International Wireless Industry Consortium (IWPC), 152
Interpolating factor, 415
Interpolating flash ADCs, 415, 420
Intersatellite communication links, 26–27
Intersymbol interference (ISI), 49, 51
Intravehicle communications, 24
Inverse DFT (IDFT), 57
Inversion layers, 262
Inversion regions in MOSFETs, 274–275
Inverters in comparators, 396
IQ (in-phase/quadrature) mismatch, 89
ISI (intersymbol interference), 49, 51
ISM (Industrial Scientific, and Medical) bands, 3
Isolation
    defined, 267
    mixers, 342, 347, 349
    track-and-hold amplifiers, 402
Isotropic radiators, 102
IST Broadway project, 498–499
Isynchronous CTAP, 516
ITRS (International Roadmap for Semiconductors), 22

ITU (International Telecommunication Union), 5, 509
IWPC (International Wireless Industry Consortium), 152

J

Japan, 60 GHz spectrum regulation in, 510
Jitter in ADCs, 391–392, 403, 411
JK-flip-flops in phase-locked loops, 369

K

K-factor, 171–173
KAIST (Korea Advanced Institute of Science and Technology), 17
Kalman filters in linear equalization, 53
Kiosk environments, 183
Knife-edge diffraction model, 176
Korea, 60 GHz spectrum regulation in, 511
Korea Advanced Institute of Science and Technology (KAIST), 17
Kronecker product in frame synchronization, 75

L

Laplacian distribution in multipath channel models, 174
Large-scale path loss models, 152–160
Large-scale propagation channel effects, 101–105
    atmospheric, 106–107
    diffraction, 111
    log-distance path loss models, 105–106
    ray tracing and site-specific propagation prediction, 122–126
    reflection and penetration, 111–117
    scattering and radar cross section modeling, 117–119
    surrounding objects, humans, and foliage, 119–122
    weather, 108–110
Large-scale regime, 277
LAS (lobe angle spread), 164
Last RSSI in DMG, 576
Latch up, 202
Layering model in WirelessHD, 552–553
Layout, circuit, 267–273
    effects, 281–282
LC VCOs, 350–361
LCP (Liquid Crystal Polymer), 210–211
LDPC. *See* Low-density parity check (LDPC)
Least mean squares (LMS)
    decision feedback equalizers, 54
    linear equalization, 53
Least significant bits (LSBs)
    ADC sampling, 390–391
    UEP, 494–496, 538, 541

Least-squares solution
    channel estimation, 79–80
    linear equalization, 52
Lebedew, Pyotr N., 4
Levison-Durbin recursion, 53
Library environments, 182
Limited feedback communications, 85–86
Line-of-sight (LOS)
    28 GHz path loss, 156–157
    backhaul systems, 14
    beam-combining, 133–135
    break-point distance, 149
    cellular networks, 16, 149, 501–504
    decode-and-forward repeaters, 492–493
    description, 132
    diffraction, 111
    directional antennas, 473
    indoor environments, 106, 171–175, 184
    intersatellite communications, 26
    large-scale path loss models, 156–159
    MIMO channels, 467–470
    polarization mismatch, 139
    Rician models, 169
    spatial characterization, 132–135
    surrounding objects, 119–121
    transmitter to receiver separation, 149–151
Linear block codes in LDPC codes, 66
Linear equalization
    high-throughput PHYs, 446
    time domain, 51–53
Linear polarization, 139
Linearity
    ADC non-linearity, 406–407
    cascaded devices, 320
    mixers, 342, 345, 347
    radios, 316
Link adaptation in MAC, 571
Link budget analysis, 314–317
Link disruption, 569
Link quality indices (LQIs), 546
Link switching, 569
Liquid Crystal Polymer (LCP), 210–211
List output Viterbi decoder algorithm (LOVA), 69
LLC (Logical Link Control) in OSI model, 92–94
LLR (log-likelihood ratio), 43
LMDS (Local Multipoint Distribution Service), 3, 120, 122
LMS (least mean squares)
    decision feedback equalizers, 54
    linear equalization, 53
LNAs. *See* Low noise amplifiers (LNAs)
Load-pull analysis, 334
Lobe angle spread (LAS), 164

Lobes in multipath models, 162–165
Local Multipoint Distribution Service
  (LMDS), 3, 120, 122
Local oscillator (LO), 9
  array beam steering, 237–240
  distortion, 441
  heterodyne receivers, 89–90
  leakage, 89
  mixers, 342, 345–349
  phase-locked loops, 364
  transceivers, 441, 444
  VCOs, 350
local oscillator pull, 90
Locking range in phase-locked loops, 367–368
Log-distance path loss models, 105–106
Log-likelihood ratio (LLR), 43
Log-normal shadowing, 174
Logical Link Control (LLC) in OSI model, 92–94
Look-ahead logic scheme, 410
Loops
  inductors, 300–301
  PLLs, 364–370, 441
  transformers, 304–305
LOS. See Line-of-sight (LOS)
Loss tangents
  antennas, 193
  BEOL structures, 270–271
  in-package antennas, 211
  transmission lines, 272–273
Losses. See Radio wave propagation
LOVA (list output Viterbi decoder algorithm), 69
Low complexity PHYs, 461–464
Low-density parity check (LDPC), 65–67, 521
  control PHY, 577
  HSI PHY, 528–529
  OFDM PHY, 578, 581
  SC-PHY, 524, 577
  UEP, 540–541
Low-k dielectric constants, 277
Low noise amplifiers (LNAs), 334–342
  key characteristics, 324–326
Low pass filters, 36, 39
Low-power (LP) PHY, 572
Low-power (LP) SC-PHY, 581–582
Low-rate PHY (LRP), 531–536
Low-temperature cofired ceramic (LTCC)
  materials, 200
LQIs (link quality indices), 546
LSBs (least significant bits)
  ADC sampling, 390–391
  UEP, 494–496, 538, 541
LTCC (Low-temperature cofired ceramic)
  materials, 200
Lumped components model for Q factor, 291

M

M-ary quadrature amplitude modulation
  (M-QAM), 45
M-ASK (M-amplitude shift keying), 45
M-Phase Shift Keying (M-PSK), 47
MAC (medium access control) protocol, 564
  association and disassociation, 519–520
  beam adaptation, 487
  beam steering, 19
  beamforming, 567–569
  DMG, 564–566
  ECMA-387, 556–559
  energy efficiency, 519
  frame structure, 514–515
  IEEE 802.11ad, 564–572
  IEEE 802.15.3c, 512–514
  Information Exchange, 516
  interference and coexistence, 518
  link adaptation, 571
  multi-band operation, 492, 571
  OSI model, 92–94
  PCP/AP clustering, 566–567
  piconets, 515–517
  relaying, 569–571
  security, 517, 572
  UEP, 538
MAFIA (Microwave Studio) design tool, 268
Magnetic conductors in on-chip antenna gain, 227–228
Management operation (MCTA) frames, 514–515
Mandatory PHY rate (MPR), 524
MAP (maximum a posteriori) decoding
  algorithm, 69
Mapping symbols, 40
Marconi, Guglielmo, 3
Markov property, 68
Matched filters in symbol detection, 42
Material penetration losses, 113–116
Maximum a posteriori (MAP) decoding
  algorithm, 69
Maximum antenna gain in power amplifiers, 328
Maximum distance separable in Reed-Solomon
  codes, 64
Maximum effective area, 102
Maximum excess delay time, 129
Maximum frequency of operation, 278
Maximum likelihood estimate, 79–80
Maximum likelihood (ML) receivers, 42–43
Maximum likelihood sequence estimator
  (MLSE), 55–56, 68
Maximum transmit power in power amplifiers, 328
Maxwell's time-harmonic equations, 197
MCS. See Modulation and coding scheme
  (MCS)

MCTA (management operation) frames, 514–515
Medium access control. *See* MAC (medium access control) protocol
Medium inversion, 262
Memory storage in future, 23
Message passing in LDPC, 66
Metal oxide semiconductor capacitors (MOSCAPs), 354–356
Metal oxide semiconductor field-effect transistors (MOSFETs), 261–262
    mismatch, 393–394
    operation, 273–279
Metal oxide semiconductor (MOS) materials, 261–262
Metastability windows in comparators, 395
Method of moments (MoM), 191, 268–269
Metrics for analog devices
    noise figure and noise factor, 322–323
    non-linear intercept points, 317–322
Micropatch design tool, 268
Microstrip patch antennas, 218–222
Microstrip transmission lines, 294
Microwave Studio (MAFIA) design tool, 268
Miller capacitance, 310–311, 341
MIMO. *See* Multiple Input Multiple Output (MIMO) communication
Miniaturization of antennas, 187
Minimum distance decoding in symbol detection, 42
Minimum excess delay spread, 129
Minimum feature size in in-package antennas, 211
Minimum mean squared error (MMSE) equalizers
    decision feedback, 54
    linear, 52–53
    single carrier frequency domain equalization, 59
Mismatch, ADCs, 405
Mismatch
    ADCs, 393–394, 411–412, 418–419
    DACs, 393–394, 427
    IQ, 89
Mixed-signal equalization, 18
Mixers, 342–349
ML (maximum likelihood) receivers, 42–43
MLSE (maximum likelihood sequence estimator), 55–56, 68
MMSE (minimum mean squared error) equalizers
    decision feedback, 54
    linear, 52–53
    single carrier frequency domain equalization, 59
Moderate inversion, 262
Modified Rapp model, 439–440

Modulation
    AM, 35
    AV PHY, 531
    CPM, 463–464
    DCM, 579
    digital. *See* Digital modulation
    DMG, 574
    high-throughput PHYs, 445–447
    HSI PHY, 529
    OFDM, 60–62, 75, 555
    physical layer, 435
    radios, 314–315
    TCM, 69–70
    Type A PHY in ECMA-387, 559
    Type B PHY in ECMA-387, 561
    Type C PHY in ECMA-387, 561
    UEP, 539–541
Modulation and coding scheme (MCS)
    DMG, 575
    frequency selective fading, 175
    HSI PHY, 529
    OFDM, 559–560
    PHY, 524, 559–561
    UEP, 495, 538–542
Modulation index, 463
Moment method in simulation, 191, 268–269
Momentum design tool, 268
Monostatic cross section, 117–118
Moore's Law, 8, 277–278
MOS (metal oxide semiconductor) materials, 261–262
MOSCAPs (metal oxide semiconductor capacitors), 354–356
MOSFETs (metal oxide semiconductor field-effect transistors), 261–262
    aspect ratio, 275, 280, 329
    biased, 275
    mismatch, 393–394
    operation, 273–279
Most significant bits (MSBs) in UEP, 494–496, 538–541
MPCs (multipath components), 171, 173
MPR (mandatory PHY rate), 524
Multi-band diversity, 492–493
Multi-band operation, 571
Multi-beam solution, 497
Multi-beam transmission, 479
Multi-gigabit transmission
    ADCs. *See* Analog-to-digital converters (ADCs)
    DACs. *See* Digital-to-analog converters (DACs)
    DMG. *See* Directional multi-gigabit (DMG) PHY
Multi-hop communication, 479, 489
Multi-step conversion in heterodyne receivers, 90–91

Multi-user beamforming, 503
Multiband considerations, 497–500
Multifinger transistors, 394
Multimedia transmission, 493–497
Multipath
    angle of arrival, 135–138, 161–164
    baseband representation, 37–38
    channel estimation, 487
    delay spread, 129
    outdoor channels, 160–164
    spatial characterization, 132–135
    time delays, 152
    time dispersion, 129
    urban environments, 8
Multipath components (MPCs), 171, 173
Multiple Input Multiple Output (MIMO) communication, 13–14, 81–82
    antennas, 188
    beamforming, 85–87
    cellular networks, 503–504
    hybrid precoding, 87–88
    path loss, 103, 105
    spatial diversity, 84–85
    spatial multiplexing, 82–84, 466–469
Multiple phased elements in on-chip antenna gain, 230
Multiple Signal Classification (MUSIC), 249–251
Multiplexer-based thermometer-to-binary encoders, 408
Multiplexing
    frequency, 454
    OFDM. *See* Orthogonal frequency division multiplexing (OFDM)
    spatial, 82–84, 466–469, 503
Multiwave fading model, 169–170
MUSIC (Multiple Signal Classification), 249–251
Mutual inductance, 303, 305–306
Mux-based decoders, 416

N

N-type metal oxide semiconductor (NMOS), 273
N-type silicon, 260–261
Narrowband communication, 34
Narrowband CW signals, 144
National Telecommunications and Information Administration (NATA) study, 120
Neighbor discovery, 475–476
Neighbor piconets, 517–518
Nested piconets, 517–518
Network Layer in OSI model, 93
Networking devices, 472
    channel utilization and spatial reuse, 480–481
    collision detection and avoidance, 476–478
    device discovery, 475–476
    directional antennas at PHY, 472–474
    human blockage, 478–480
NLOS. *See* Non-line of sight (NLOS)
NMOS (N-type metal oxide semiconductor), 273–275
Noise
    1/f, 356–357
    DACs, 424–425
    gate, 285
    Gaussian. *See* Additive white Gaussian noise (AWGN)
    homodyne receivers, 89
    linear equalization, 52–53
    low noise amplifiers, 334–339
    OFDM and SC-FDE, 451
    radios, 317
    symbol detection, 41–43
    thermal, 356–357
    transceivers, 441–444
    VCOs, 356–361
Noise covariance matrix in linear equalization, 53
Noise figure and noise factor (NF), 322–323, 338
Noise-free ground planes, 297
Noise rejection circuits in VCOs, 360
Noise subspaces, 249
Non-line of sight (NLOS), 8
    angle of arrival, 136
    beam-combining, 132–135
    cellular networks, 15–16, 501–504
    decode-and-forward repeaters, 492–493
    diffraction, 111
    directional antennas, 473
    frequency selectivity, 449–452, 455
    HSI PHY, 529
    indoor environments, 106, 171, 175–184
    large-scale path loss models, 156–161
    outdoor environments, 119, 140–141, 149
    Rician models, 170
    spatial multiplexing, 467
    surrounding objects, 120–122
    transmitter to receiver separation, 149
    vegetation effects, 120
Non-linear equalization in high-throughput PHYs, 446
Non-linear intercept points, 317–322
Non-linearity in power amplifiers, 439–441
Non-quasi static effects, 280, 285
Non-relaying mode, 497
Non-square 8-QAM (NS8-QAM) constellations, 49
Non-UEP operation mode, 542
Normal relay transmission mode, 569

Normalization factor in phase shift keying, 47
North America, 60 GHz spectrum regulation in, 509–510
Notations, 31
NS8-QAM (non-square 8-QAM) constellations, 49
Null cyclic prefix single carrier modulation (null CP SC), 454–458
Nyquist frequency and sampling
    ADCs, 388
    baseband representation, 36, 39
    DACs, 422
    equalization in time domain, 50
    heterodyne receivers, 90
    maximum likelihood sequence estimator, 56
    sample-and-hold amplifiers, 397, 399
NYU WIRELESS research center, 16

O

O_2 absorption effect, 7
OfCom (Office of Communications), 7
OFDM. *See* Orthogonal frequency division multiplexing (OFDM)
Office environments
    evaluation parameters, 181
    future, 23–24
Office of Communications (OfCom), 7
Offset QPSK (OQPSK), 46
Offsets in flash ADCs, 418–419
Ohmic contacts, 283
Omnidirectional antennas
    beam patterns, 473
    large-scale path loss models, 159–160
    multipath models, 163–164
    path loss data, 156–160
On-chip antennas, 187
    case studies, 253–255
    CMOS technology, 200–209
    connections, 21
    environment, 198–200
    fundamentals, 189–191
    gain, 225–235, 259
    probe station characterizations, 255–257
    spiral inductors, 303
On-Off Keying (OOK), 45, 463, 549–550
Ones-counter decoders, 416
OOB (out-of-band) control channels, 557
OOBE (out of band emission), 165
OOK (On-Off Keying), 45, 463, 549–550
Open loop circuits in S/H, 400
Open-loop gain
    transistors, 313
    VCOs, 350, 353–355, 359, 363
Open-loop transfer function, 361
Open space scenarios, 106
Open Systems Interconnection (OSI) model, 91–95

Operating consumption factor, 376
Operating frequency
    baseband representation, 34
    in gain, 102
Operational amplifiers (op-amps) in comparators, 396–397
Optical interconnects in data centers, 20
OQPSK (Offset QPSK), 46
Orthogonal frequency division multiplexing (OFDM), 572
    AV PHY, 531–535
    channel estimation, 80–81
    DMG, 578–581
    ECMA-387, 555
    frame synchronization, 75
    frequency domain equalization, 57, 60–62
    frequency offset synchronization, 77
    high-throughput PHYs, 446–447
    HSI PHY, 528–530
    radios, 317
    vs. SC-FDE, 447–458
    SC-PHY, 524
    spatial multiplexing, 83
    time domain equalization, 50
    UEP, 540
Oscillators
    LO. *See* Local oscillator (LO)
    sidebands, 444
    VCOs. *See* Voltage-controlled oscillators (VCOs)
OSI (Open Systems Interconnection) model, 91–95
Out-of-band (OOB) control channels, 557
Out of band emission (OOBE), 165
Outdoor environments, 139–149
    3GPP-style. *See* 3GPP-style outdoor propagation models
    cellular networks, 501
    temporal and spatial multipath models, 160–164
Output 1 dB compression points, 319
Output impedance in DAC current sources, 426
Output power in power amplifiers, 323
Output symbols in Reed-Solomon codes, 64
Overdrive voltage for MOSFETs, 273
Oxygen absorption, 106–107

P

P-frames, 494, 497
P-type metal oxide semiconductor (PMOS), 273
P-type silicon, 260–261
Packaged antennas, 188, 209–211
    fundamentals, 189–191
    probe station characterizations, 255–257
Packaging parasitic inductances, 304

Packet detection
    frame synchronization, 79
    mmWave PHYs, 460–461
Packet Type field for DMG, 575
Pad capacitance in BEOL structures, 270
Pad rings in CMOS chips, 202
PAE (power added efficiency) of power
        amplifiers, 323, 328–329
PANs (personal area networks), 190
PAPRs (peak-to-average-power ratios), 333
Parallel ADCs, 407–408, 411–412
Parasitics, 281–283
    bond wire packaging, 304
    interconnects, 308
    low noise amplifiers, 341–342
Parent piconets, 517–518
Parity bits in block codes, 64
Partition losses overview, 126, 128
PAS (Power Angle Spectrum) plots, 166
Passband bandwidth, 22, 193
Passband signals, 34–35
Passive mixers, 349
Passives. See Transmission lines and passives
Patch antennas, 218–222
Path blockage, 474
Path depolarization, 106
Path loss, 7
    cellular networks, 15, 502
    overview. See Radio wave propagation
Path loss exponent (PLE), 105, 133–134, 162, 175
Pattern Estimation and Tracking (PET)
        protocol, 547–548
Patterns for beam adaptation protocols, 482–483
Payload field in MAC, 515
PBSS (personal base station mode), 499
PCAAD design tool, 268
PCESs (pilot channel estimation sequences), 528–529
PCP/AP clustering, 566–567
PCPs (personal basic service set control
        points), 564
PDF (probability density function), 169–170, 391
Peak signal-to-noise ratio (PSNR), 496
Peak-to-average-power ratio (PAPR), 333
PeakView design tool, 268
Peer-to-peer channels, 145
Pelgrom mismatch models, 418–419
Penetration losses, 111–117, 169
Perfect electrical conductors, 290
Performance
    cellular networks, 500–504
    on-chip antennas, 252–253
Permeability of inductors, 300
Permittivity
    antennas, 193
    BEOL structures, 270–271

CMOS chips, 204
in-package antennas, 210–211
materials, 113, 115, 117
on-chip antenna gain, 227
patch antennas, 219
simulation, 269
transmission lines, 272–273
Personal area networks (PANs), 190
Personal base station mode (PBSS), 499
Personal basic service set control points
        (PCPs), 564
Personal mobile communications in future, 25–26
Personal security in WirelessHD, 554
PET (Pattern Estimation and Tracking)
        protocol, 547–548
Phantom cells, 476, 499–500
Phase
    ADCs, 411
    antenna arrays, 194–195, 236–237, 473–474, 522
    in baseband representation, 35
    beam steering, 237–242
    beamforming, 243–246
    VCOs, 353
Phase detectors in phase-locked loops, 366
Phase error in channel estimation, 81
Phase fronts in antenna arrays, 194–195
Phase lock in antenna arrays, 251
Phase-locked loops (PLLs), 364–370, 441
Phase margin of amplifiers, 312
Phase noise
    OFDM and SC-FDE, 451
    PLL reference signal, 365
    transceivers, 441–444
    VCOs, 356–361
Phase shift keying in digital modulation, 47
Phased arrays, 473–474, 522
Phased elements in on-chip antenna gain, 230
Physical layer (PHY), 520–521
    AV, 531–537
    beamforming, 543–548
    challenges, 17–19
    channelization and spectral mask, 521–522
    common mode operation, 522–524
    directional antennas, 472–474
    DMG. See Directional multi-gigabit
        (DMG) PHY
    ECMA-387. See ECMA-387 standard
    in frame synchronization, 74
    high-throughput. See High-throughput
        PHYs
    HSI, 521, 528–530, 540–541
    low complexity, high efficiency, 461–464
    On-Off keying, 549–550
    OSI model, 91–92, 94
    practical transceivers, 436–444
    SC-PHY, 524–528

Physical layer (*continued*)
  spatial multiplexing, 466–469
  ultra-low ADC resolution, 464–466
  unequal error protection, 537–543
Pi-model for inductors, 302–303
PiCasso design tool, 268
Picocellular network applications, 16
Piconet controllers (PNCs), 491
Piconet ID field in MAC, 515
Piconet-Synchronized (PS) mode, 519
Piconets, 512–514
  nested, 517–518, 520
  starting, 516
Pilot carriers in frame synchronization, 75
Pilot channel estimation sequences (PCESs), 528–529
Pilot signals in channel estimation, 73
Pilot words in SC-PHY data, 527–528
Pinched off MOSFETs, 275
Pipeline ADC, 409–410
Pixel partitioning, 495–496
Planar antennas, 212
PLE (path loss exponent), 105, 133–134, 162, 175
PLLs (phase-locked loops), 364–370, 441
PMOS (P-type metal oxide semiconductor), 273–275
PNC handover, 518, 520
PNCs (Piconet controllers), 491
Point coordination function in OSI model, 94
Pointing angles with antennas, 145–148
Poisson point process (PPP), 500–501
Polarization, 113, 138–139
  indoor channel models, 167, 172–173, 175
  on-chip antenna mismatch, 252
Polarization vectors, 197–198
Poles, transistor, 311–313
Portability across technology nodes, 279
Positive feedback techniques in comparators, 397
Post-curser rays, 171–172
Power
  ADCs, 404–405, 407–409
  data centers, 20
  EKV model, 288
Power added efficiency (PAE) of power amplifiers, 323, 328–329
Power amplifiers (PAs), 323–334
  key characteristics, 324–326
  non-linearity, 439–441
  radios, 316
Power Angle Spectrum (PAS) plots, 166
Power consumption, ADCs, 404–405
Power control in MAC, 518
Power-efficiency factor
  consumption factor theory, 373–374
  numerical example, 374–375

Power flux density, 102
Power spectral density (PSD), 442–443
Poynting vector, 204
PPP (Poisson point process), 500–501
Practical transceivers, 436
  phase noise, 441–444
  power amplifier non-linearity, 439–441
  signal clipping and quantization, 436–439
Pre-curser arrival rate, 171–172
Preamplifiers in flash ADCs, 415
Precipitation attenuation, 106–107
Precoding in MIMO, 87–88
Preferential radiation in CMOS chips, 204–205
Presentation Layer in OSI model, 93
Primary loops in transformers, 305–307
Primary ray-tracing technique, 125
Probability density function (PDF), 169–170, 391
Probability of error in symbol detection, 42
Probe-based technique for on-chip antennas, 253–254
Probe stations, 188, 252–257
Process nodes, 269
Processing complexity in mmWave PHYs, 446
Propagation constants in transmission lines, 292
Propagation losses. *See* Radio wave propagation
PS (Piconet-Synchronized) mode, 519
PSD (power spectral density), 442–443
Pseudo-complex baseband equivalent channels, 38
PSNR (peak signal-to-noise ratio), 496
Pulse-shaping filters, 40
Puncturing convolutional codes, 68

Q

Quadrature amplitude modulation (QAM), 48–49
  baseband representation, 40
Quadrature components in baseband representation, 35
Quadrature phase shift keying (QPSK), 45–47
  DMG, 579–581
  in OFDM and SC-FDE, 448–455, 457
Quality factor (Q)
  antennas, 193
  circuits, 288–289
  inductors, 300–302
  LC tank circuits, 360
  passive components, 291
  transformer windings, 306–307
  VCOs, 356
Quality-of-service for video, 497
Quantization
  ADC sampling, 390–391
  transceivers, 436–439

Quasi-omni patterns in beam adaptation, 482
Quasi-omni transmission mode, 476
Quasi-transverse electromagnetic transmission lines, 296

R

Rack-to-rack communication, 20
Radar cross section (RCS) model, 117–119
Radiation efficiency
    description, 191–192
    on-chip antennas, 206–209
Radiation patterns for patch antennas, 221
Radio wave propagation, 99–100
    angle spread and multipath angle of arrival, 135–138
    antenna polarization, 138–139
    indoor environments. *See* Indoor environments
    large-scale. *See* Large-scale propagation channel effects
    outdoor environments. *See* Outdoor environments
    small-scale. *See* Small-scale channel effects
Radios
    non-linear intercept points, 317–322
    sensitivity and link budget analysis, 314–317
Rain attenuation, 106–110
Random access channels, 478
Random shape theory, 501
Range extension, 135
Rapp model, 439–440
Ray tracers, 122–123
Ray tracing models, 122–126
    indoor channels, 169
    outdoor channels, 149
    scattering, 117
Ray tubes, 122–123
Rayleigh model, 169–170
Rayleigh range parameter, 468
RCS (radar cross section) model, 117–119
Read-only memory decoders in flash ADCs, 416
Received power in free space, 102
Received signals in baseband representation, 37
Reception spheres, 122–123
Reconstructed signals in array beamforming, 243
Recursive least squares (RLS) algorithm, 54
Redundancy through repetition, 70
Reed-Solomon (RS) codes, 64–65
Reed-Solomon decoders for error concealment, 495–496
Reference signals in PLL, 365

Reference voltages
    flash ADCs, 416
    sample-and-hold amplifiers, 397
Reflections, 111–117
Regions of inversion in MOSFETs, 274–275
Relative bandwidth, 193
Relative permittivity, 189
Relaying
    coverage extension, 487–493
    MAC, 569–571
Reliability of power amplifiers, 333
Repeaters, 487–493
Request to send (RTS) messages, 477, 562
Research programs in universities, 16–17
Residential environments, 180
Resistances, parasitic, 281–283
Resistive averaging in flash ADCs, 414–415, 419
Resistive ladders in flash ADCs, 415–416
Resistive networks, 415
Resistor divider circuits, 397
Resistors in low noise amplifiers, 334
Resolution in ADCs
    flash ADCs, 420
    ultra-low, 464–466
Resonances in dipole antennas, 212–213
Return loss in scattering parameters, 266
Return paths
    low noise amplifiers, 341
    transmission lines, 293
Reverse isolation, 267
Reverse short channel effect, 284
RF beamforming in OFDM and SC-FDE, 455
RF phase-shifting in array beam steering, 237–238
Rician distribution, 120
Rician model, 169–170
RLS (recursive least squares) algorithm, 54
RMS delay spread. *See* Root-mean-square (RMS) delay spread
RMS (root mean square) values
    ADC sampling, 389
    quantization error, 391
ROM decoders in flash ADCs, 416
Root-mean-square (RMS) delay spread
    description, 129
    directional antennas, 176
    human blockage, 120
    multipath, 124
    outdoor environments, 144–148
    time domain equalization, 50
    vehicle-to-vehicle models, 165
Root mean square (RMS) values
    ADC sampling, 389
    quantization error, 391
Roughness, surface, 118–119
RS (Reed-Solomon) codes, 64–65

RS FEC, 524
RTS (request to send) messages, 477, 562
Rule of eight, 260

S

S-APs (synchronization APs), 566
S/H (sample-and-hold) amplifiers, 397–403
S-parameters, 263–267, 314
S-PCPs (synchronization PCPs), 566
S/T (Spectrum Management and
 Telecommunications Sector), 510
SA (successive approximation) ADCs, 410–411
Saleh-Valenzuela (S-V) model, 169–171, 177
Sample-and-hold (S/H) amplifiers, 397–403
Sample-and-hold circuits, 384
Sampling
 ADCs and DACs, 384–392
 comparators, 395
SAS (symmetric antenna set), 543
Satellite communications, 108
Saturation region in MOSFETs, 275
SC-FDE (single carrier frequency domain
 equalization), 57–60
 in frame synchronization, 75
 vs. OFDM, 447–458
 time domain equalization, 50
SC MCS, 560–561
SC-PHY (single carrier mode), 521, 524–528
 DMG, 572, 577–578, 581–582
 UEP, 538–541
Scaled transistors, ADCs, 404–405
Scaling rules, 279
Scatter plots for path loss models, 158–159
Scattering effects, 15
 parameters, 263–267
 radar cross section modeling, 117–119
SCBT (single-carrier block transmission), 57
Scheduled Service Period (SP), 566
Scheduled time slots in MAC, 516
Schmitt trigger comparators, 396
Scrambler Initialization field in DMG, 575
SD (steepest descent) type algorithms, 243
Secondary loops in transformers, 305–307
Sector feedback in beamforming, 545–546
Sector patterns in beam adaptation protocols,
 482–483
Sector sweeps in beamforming, 567–568
Sector-to-beam exchange in beamforming, 546
Sector training in beamforming, 544–545
Secure Frame Counter field in MAC, 515
Secure Session ID field in MAC, 515
Security
 MAC, 517, 572
 WirelessHD, 554
Segmentation
 DACs, 428–429
 MOSFET mismatch, 395

Selective channels for time domain
 equalization, 50
Self-resonant frequency (SRF)
 inductors, 300–303
 passive devices, 290
Sensitivity analysis for radios, 314–317
Separation, transmitter to receiver, 149–151,
 153–155, 158–162, 168
Sequences in high-throughput PHYs, 459–461
Session Layer in OSI model, 93
Settling errors in DACs, 424–425
Severed ground shields in antenna arrays, 251
SFD field in OOK, 549–550
SFDR (spurious-free dynamic range)
 DACs, 422–423, 426
 differential non-linearity, 406
SFDs (static frequency dividers), 366–368
Shadows
 losses, 125–126
 in models, 174–175
Shannon limits
 consumption factor, 377–378
 LDPC codes, 65
Shelf-to-shelf communication, 20
Shields
 antenna arrays, 251
 inductors, 301
 transmission lines, 294
Short-channel effects, 283–284
Short training field (STF) in DMG, 573
Shot noise in low noise amplifiers, 336
Sidebands, oscillator, 444
Sievenpiper surfaces, 229
Signal clipping, 436–439
Signal path components in consumption factor
 theory, 371
Signal propagation in baseband representation,
 37
Signal subspaces, 249
Signal-to-interference-plus-noise ratios (SINRs)
 cellular networks, 501–504
 repeaters/relays, 492
Signal-to-jitter-noise ratio (SJNR), 392
Signal to noise and distortion ratio (SNDR)
 ADCs, 403
 DACs, 422, 426
Signal-to-noise ratio (SNR)
 antenna arrays, 235
 consumption factor, 376–377
 differential non-linearity, 406
 ESPRIT, 250
 noise performance, 322
 OFDM and SC-FDE, 449–458
 radios, 314, 316–317
 symbol detection, 42
Signal-to-quantization-noise-ratio (SQNR), 391
Silicon, 260–262

Index

Silicon on Insulator (SOI) processes, 17, 269
Simulations
    cellular networks, 502–503
    circuit production, 267–273
    passive components, 290
    ray tracing models, 122–123
SINAD (signal power to noise and distortion power)
    ADCs, 403
    DACs, 422, 426
Sine components in baseband representation, 36
Single amplify-and-forward relay mode, 497
Single-balanced mixers, 349
Single-carrier block transmission (SCBT), 57
Single carrier frequency domain equalization (SC-FDE), 57–60
    in frame synchronization, 75
    vs. OFDM, 447–458
    time domain equalization, 50
Single carrier mode (SC-PHY), 521, 524–528
    DMG, 572, 577–578, 581–582
    UEP, 538–541
Single decode-and-forward relay mode, 497
Single-ended mixers, 349
Single-ended transmission lines, 298
Single IF receivers, 89
Single input single output (SISO) CIR model, 173
SINRs (signal-to-interference-plus-noise ratios)
    cellular networks, 501–504
    repeaters/relays, 492
SISO (single input single output) CIR model, 173
Site-specific propagation prediction, 122–126
60 GHz spectrum regulation, 508–512
Skew
    ADCs, 403
    constellation, 543
Slant-path satellite communication links, 108
Sliding correlator systems, 139–140
Slot antennas, 215–219
Small-scale channel effects, 101, 126–129
    delay spread characteristics, 129–130
    Doppler effects, 130–132
    spatial characterization of multipath and beam combining, 132–135
Small-scale regime, 276
Smooth surfaces, 118–119
SNDR (signal to noise and distortion ratio)
    ADCs, 403
    DACs, 422, 426
SNR. *See* Signal-to-noise ratio (SNR)
SoC (system-on-chip) radios, 9
Soft decision decoding, 64
Soft information in symbol detection, 43
Soft output Viterbi algorithm (SOVA), 69

SOI (Silicon on Insulator) processes, 17, 269
Source/Destination fields in MAC, 515
Source follower transistors, 308–309
Source in CMOS transistors, 260
Source packetizers in WirelessHD, 552–553
SOVA (soft output Viterbi algorithm), 69
SP (Scheduled Service Period), 566
Sparse beamforming approach, 486
Sparsity in channel estimation, 81
Spatial characteristics of multipath and beam combining, 132–135
Spatial diversity
    AV PHY, 535
    MIMO, 84–85
Spatial-ESPRIT, 250
Spatial fading, 143
Spatial filtering
    antenna arrays, 235–236
    directional antennas, 175
Spatial Fourier transform, 192
Spatial lobes, 162
Spatial multipath models, 160–164
Spatial multiplexing
    cellular networks, 503
    MIMO, 82–84, 466–469
Spatial reuse
    overuse, 480–481
    repeaters/relays for, 490–491
Spatial statistics in vehicle-to-vehicle models, 166
Spectral masks, 521–522, 572–573
Spectrum allocations, 3–6, 15
Spectrum Management and Telecommunications Sector (S/T), 510
Speed of ADCs, 405
SPICE circuit simulators, 267
Spiral inductors, 299, 301, 303
Spread, delay. *See* Root-mean-square (RMS) delay spread
Spread QSPK (SQSPK), 579–581
Spreading factor
    PHYs, 444
    time domain spreading, 70
Spurious-free dynamic range (SFDR)
    DACs, 422–423, 426
    differential non-linearity, 406
SQNR (signal-to-quantization-noise-ratio), 391
SQSPK (spread QSPK), 579–581
SRF (self-resonant frequency)
    inductors, 300–303
    passive devices, 290
SSCMs (statistical spatial channel models), 160, 164–165
STA (station)
    beamforming, 568–569
    DMG, 564–565

STA-to-relay-to-STA links, 569
STA-to-STA links, 569
Staggered QPSK, 46
Standardization, 507–508
    60 GHz spectrum regulations, 508–512
    ECMA-387, 555–561
    IEEE 802.11ad. *See* IEEE 802.11ad standard
    IEEE 802.15.3c. *See* IEEE 802.15.3c standard
    WiGi, 581
    WirelessHD, 550–554
Standing wave ratio (SWR), 263
Starting Piconets in MAC, 516
Static frequency dividers (SFDs), 366–368
Static tone pairing (STP), 579–580
Station (STA)
    beamforming, 568–569
    DMG, 564–565
Station-to-Access Point (STA-AP), 184
Station-to-Station (STA-STA), 184
Statistical spatial channel models (SSCMs), 160, 164–165
Steepest descent (SD) type algorithms, 243
Steerable antennas, 105
steering vectors, 248
STF (short training field) in DMG, 573
Storage mechanisms in on-chip antennas, 208
STP (static tone pairing), 579–580
Stream Index field in MAC, 515
Strong inversion, 262
Structure in communication, 73–75
Sub-terahertz frequencies, 6
Subharmonic VCOs, 361–363
Substrate ties, 283
Substrates in on-chip antenna gain, 225–226
Subsymbols in OFDM, 61
Successive approximation (SA) ADCs, 410–411
Superframes in MAC, 514
Superstrates, 225–226
Surface roughness, 118–119
Surface waves in CMOS chips, 205–208
Surrounding object effects, 119–122
Switched beam systems, 481
Switched-capacitor circuits, 400
Switched capacitor comparators, 396–397
Switching noise in DACs, 425
SWR (standing wave ratio), 263
Symbol detection, 41–43
Symbol energy, 40
Symbols, 31
    block codes, 63
    digital modulation, 40–43
    linear equalization, 51
    mapping, 40
    Reed-Solomon codes, 64
Symmetric antenna set (SAS), 543

SYNC field in OOK, 549
Synchronization, 72–73
    frame, 78–79
    frequency offset, 75–77
    high-throughput PHYs, 459–461
    structure, 73–75
Synchronization APs (S-APs), 566
Synchronization PCPs (S-PCPs), 566
System architecture, 91–95
System-on-chip (SoC) radios, 9

T

T/H (track-and-hold) amplifiers, 397–403
T-model of transmission lines, 295
TCM (trellis coded modulation), 69–70
TDD (Time Division Duplexing), 25, 444
Teflon for in-package antennas, 210–211
Telecommunications Industry Association (TIA), 152
TEM (transverse electromagnetic) transmission lines, 296
Temporal and spatial variations, 18
Temporal clusters, 163–164
Temporal-ESPRIT, 250
Temporal fading, 142, 144
Temporal multipath models, 160–164
Temporal statistics in vehicle-to-vehicle models, 166
Terahertz frequencies, 8
Thermal noise
    gate, 285
    low noise amplifiers, 334–336
    OFDM and SC-FDE, 451
Thermometer code
    flash ADCs, 413
    multiplexer-based encoders, 408
3rd Generation Partnership Project (3GPP) organization, 152
Three-antenna technique for on-chip antennas, 254–255
3GPP-style outdoor propagation models, 149–152
    large-scale path loss models, 152–160
    temporal and spatial multipath models, 160–164
    vehicle-to-vehicle models, 165, 168
Threshold Inverter Quantization (TIQ) method, 420
Threshold voltage in MOSFETs, 273
TIA (Telecommunications Industry Association), 152
Time delay elements in array beam steering, 239
Time delay spread characteristics, 129–130
Time Division Duplexing (TDD), 25, 444
Time domain equalization, 49–51
    decision feedback equalizers, 53–55

linear equalization, 51–55
maximum likelihood sequence estimator, 55–56
Time-interleaved ADC, 411–413
Time-of-arrival in indoor channel models, 173
Time spreading
    PHYs, 444
    time domain spreading, 70–71
Timing offsets
    flash ADCs, 418–419
    frame synchronization, 78
Tinted glass reflection, 112–113
TIQ (Threshold Inverter Quantization) method, 420
Tone Pairing Type in DMG, 575
Topologies for antennas, 212–225
Total power consumption, 378
Total signal power, 371–372
Track-and-hold (T/H) amplifiers, 397–403
Tracking antenna beamforming, 547
Training
    antenna beamforming, 544–545
    channel estimation, 73, 573
    frame synchronization, 74–75
    high-throughput PHYs, 461
Training Length in DMG, 575
Transcapacitance, 287
Transceivers, 436
    phase noise, 441–444
    power amplifier non-linearity, 439–441
    signal clipping and quantization, 436–439
Transconductance in MOSFETs, 275
Transconductance model, 284
Transfer functions
    folding amplifiers, 418
    mixers, 345
    VCOs, 353, 361
Transformers, 304–308
Transistors
    ADCs, 404–405
    basics, 273–279
    bias point of, 330
    BSIM model, 285–287
    concepts, 260–262
    configurations, 308–309
    conjugate matching, 309–310
    EKV model, 287–288
    frequency tuning, 313–315
    Miller capacitance, 310–311
    mismatch, 394
    models, 279–286
    MOSFETs, 261–262, 273–279, 393–394
    poles and feedback, 311–313
    transistors, 10
    transit frequency, 10, 278, 284
Transmission lines and passives, 272–273, 288–298
    differential vs. single-ended, 298

inductors, 298–304
interconnects, 308
Transmit power in power amplifiers, 328
Transmitter antenna heights, 145–146
Transmitter to receiver separation, 149–151, 153–155, 158–162, 168
Transport Layer in OSI model, 93
Transverse electromagnetic (TEM) transmission lines, 296
Trellis coded modulation (TCM), 69–70
Trellis codes, 63
TWDP (Two Wave with Diffuse Power), 131, 170
2-ASK, 45
Two-tone tests, 319
Two Wave with Diffuse Power (TWDP), 131, 170
Type 1 UEP, 538
Type 2 UEP, 538
Type 3 UEP, 538–539
Type A ECMA-387 devices, 555–556, 559–560
Type B ECMA-387 devices, 555–556, 560–561
Type C ECMA-387 devices, 555–556, 561

U

U-NII (Unlicensed National Information Infrastructure) bands, 5
UEP. *See* Unequal error protection
UEP-QPSK (unequal error protection QPSK), 46–47
Ultra-low ADC resolution, 464–466
Ultra-wide band (UWB), 6
Umbrella architectures, 499–500
Uncompressed video, 493–495
Underutilization problem, 474
Unequal error protection (UEP), 71, 537–543
    ECMA-387 devices, 555, 560
    multimedia transmission, 494–497
Unequal error protection QPSK (UEP-QPSK), 46–47
Ungerboek form in MLSE, 55–56
Unit-form constellations, 40
University research programs, 16–17
Unlicensed National Information Infrastructure (U-NII) bands, 5
Unlicensed spectrum, 7
Upconversion
    baseband representation, 35–36
    heterodyne receivers, 90
    mixers, 342
Urban environment multipath, 8
UWB (ultra-wide band), 6

V

V2I (vehicle-to-infrastructure) links, 24
V2V (vehicle-to-vehicle) communication, 24
Valence electrons, 260–261

Varactors
    array beam steering, 239
    phase-locked loops, 369
    VCOs, 354, 356, 360
Variable branch coding rates, 542–543
Variable-gain amplifiers (VGAs)
    ADCs, 412
    array beam steering, 239
Variance in device parameters, 393
VCOs. See Voltage-controlled oscillators (VCOs)
Vector Network Analyzer (VNA) measurement systems, 139
Vegetation effects, 119–122
Vehicle-to-infrastructure (V2I) links, 24
Vehicle-to-vehicle (V2V) communication, 24
Vehicle-to-vehicle models, 165, 168
Vehicular applications, 24–25
Vertical polarization, 139
VGAs (variable gain amplifiers)
    ADCs, 412
    array beam steering, 239
Vias, 269
Video, 493–497
Viterbi algorithm and decoders
    convolutional codes, 68
    high-throughput PHYs, 445
    maximum likelihood sequence estimator, 55
VNA (Vector Network Analyzer) measurement systems, 139
Voltage-controlled oscillators (VCOs), 350
    array beam steering, 237, 240
    basic, 350–361
    Colpitts oscillators, 364
    phase-locked loops, 366–367, 369
    return loss, 266
    subharmonic, 361–363
    transceivers, 441–442
Voltage gain in power amplifiers, 329
Voltage offsets in flash ADCs, 418–419
Voltage standing wave ratio (VSWR), 263
Voltages in scattering parameters, 264–265

W

Wall material delay spread, 129
Wallace tree design, 416
Water vapor, 107
Wave vectors, 197–198
Waveform synthesis in digital modulation, 40
Wavelengths, 99
Weak inversion, 262
Weather effects, 108–110
Well ties, 283
WHDI (wireless home digital interface) standard, 12

Whisper radios, 7, 101
White Gaussian noise model. See Additive white Gaussian noise (AWGN)
WiGig standard, 581
Wireless communication background, 33–34
    complex baseband representation, 34–39
    digital modulation. See Digital modulation
    error control coding. See Error control coding
    estimation and synchronization, 72–81
    frequency domain equalization, 56–62
    hardware architectures, 89–91
    MIMO. See Multiple Input Multiple Output (MIMO) communication
    time domain equalization, 49–56
Wireless fiber, 108
Wireless home digital interface (WHDI) standard, 12
Wireless Networking and Communications Group (WNCG), 17
Wireless personal area networks (WPANs), 10–11
WirelessHD, 550
    application focus, 550–551
    next generation, 554
    technical specification, 551–554
WLANs, 11–12, 14
WNCG (Wireless Networking and Communications Group) research center, 17
World Radiocommunication Conference (WRC), 4
WPANs (wireless personal area networks), 10–11

X

XOR (exclusive or gates) in phase-locked loops, 366

Y

Y-Parameters, 263–267
Yagi antennas, 256–257
Yagi-Uda antennas, 222–225

Z

Z-parameters, 263–267
    transformers, 307
    transistors, 313
Zero-forcing
    decision feedback equalizers, 54
    linear equalization, 52
    networks, 503
Zero forcing (ZF)-precoders, 503
Zero-padding in SC-FDE, 59
Zigzag dipole antennas, 214